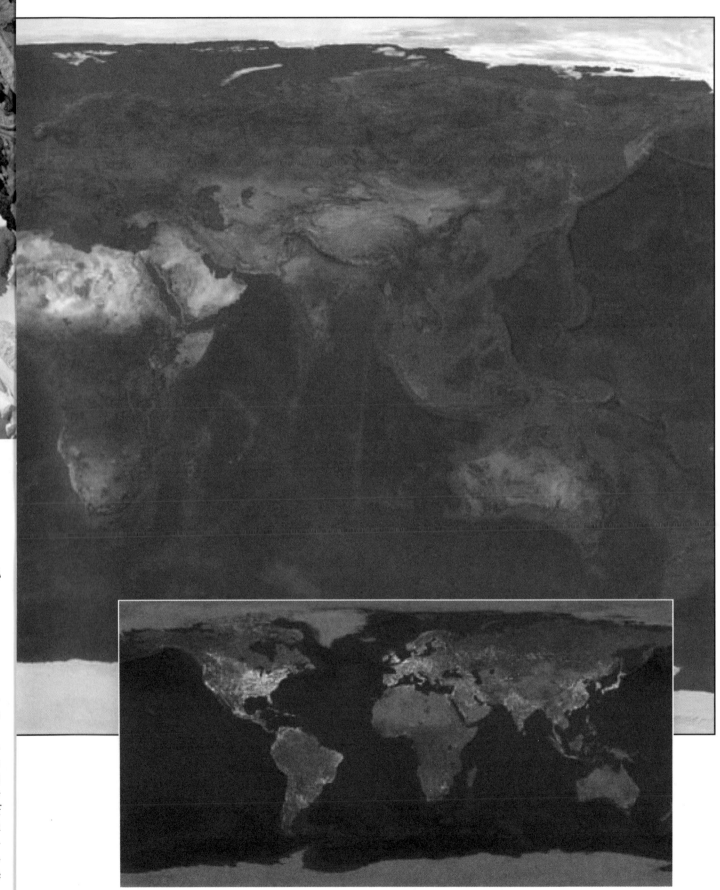

Mount Everest at New Heights

Astronaut Thomas D. Jones, Ph.D., made this photo of Mount Everest during Space Shuttle flight STS-80, November 1996. This startling view caught the mountain's triangular East Face bathed in morning light (east is to the lower left, north is to the lower right, south at the upper left; see locator map direction arrow). The North Face is in the shadow. Several glaciers are visible flowing outward from the mountains, darkened by rock and debris.

Mount Everest is at a higher elevation than previously thought! A revised elevation measurement using a global positioning system (GPS) instrument came in November 1999 at the opening reception of the 87th annual meeting of the American Alpine Club in Washington, DC. The National Geographic Society hosted the press conference. (See News Report 1.1 for more on GPS and Everest.)

In May 1999, mountaineers Pete Athans and Bill Crouse reached the summit with five Sherpas. The climbers operated a Trimble GPS unit on Everest's summit and found the precise height of the world's tallest mountain.

A photograph in News Report 1.1 shows a GPS unit placed by climber Wally Berg 18 m (60 ft) below the summit in 1998—Earth's highest benchmark, a permanent

metal marker was installed in the rock. This is a continuation of a measurement effort begun in 1995 (see News Report 12.1 for more on this geographic challenge).

The new measurement is 8,850 m (29,035 ft)—see Figure 12.2 for height comparisons. Everest's new elevation is close to the previous official measure of 8,848 m (29,028 ft) set in 1954 by the Survey of India. In addition, the new measurements determined that the horizontal position of Everest is moving steadily northeastward, about 3 to 6 mm a year (up to 0.25 in. a year). The mountain range is plowing further into Asia due to plate tectonics—the continuing collision of Indian and Asian landmasses (see Figure 11.16).

Geosystems

Combination of satellite *Terra* MODIS sensor image and *GOES* satellite image produces a true-color view of North and South America from 35,000 km (22,000 mi) in space. [Courtesy of MODIS Science Team, Goddard Space Flight Center, NASA and NOAA.]

Geosystems

An Introduction to Physical Geography

Fifth Edition

Robert W. Christopherson

Prentice
Hall

Pearson Education, Inc.
Upper Saddle River, New Jersey 07458

Library of Congress Cataloging-in-Publication Data:

Christopherson, Robert W.
  Geosystems: an introduction to physical geography / Robert W. Christopherson.—5th ed.
    p. cm.
  Includes bibliographical references
  ISBN 0-13-066824-9 (alk. paper)
    1. Physical geography. I. Title.

GB54.5 .C48 2003
910'.02—dc21   2002025779

Geosciences Executive Editor: *Daniel E. Kaveney*
Editor in Chief, Life and Geosciences: *Sheri L. Snavely*
Vice President/Director of Production and Manufacturing: *David W. Riccardi*
Executive Managing Editor: *Kathleen Schiaparelli*
Assistant Managing Editor: *Beth Sweeten*
Production Management and Composition: *Elm Street Publishing Services, Inc.*
Production Assistant: *Nancy Bauer*
Production Assistant to the Author: *Bobbé Christopherson*
Assistant Manufacturing Manager: *Michael Bell*
Manufacturing Manager: *Trudy Pisciotti*
Associate Editor: *Amanda Griffith*
Senior Marketing Manager: *Christine Henry*
Media Editor: *Chris Rapp*
Assistant Managing Editor, Science Media: *Nicole Bush*
Editorial Assistant: *Margaret Ziegler*
Art Director: *Maureen Eide*
Interior Designer: *Joseph Sengotta*
Cover Designer: *Joseph Sengotta*
AV Editor: *Adam Velthaus*
Art Studio: *Precision Graphics*
Director of Design: *Carole Anson*
Director of Creative Services: *Paul Belfanti*
Photo Researcher: *Picture Research and Editing, South Salem, NY*
Photo Editor: *Beth Boyd*
Front Cover Image: *True-color satellite view of the Pacific Northwest (see back cover).* [Terra *satellite image, August 11, 2001, courtesy of MODIS Land Response Team, GSFC/NASA; see* **http://terra.nasa.gov/**]
Back Cover Image: *Full Earth photo by Apollo 17 astronauts, December 1972.* [*NASA.*]

ISBN 0-13-066824-9

Pearson Education LTD., *London*
Pearson Education Australia PTY, Limited, *Sydney*
Pearson Education Singapore, Pte. Ltd
Pearson Education North Asia Ltd, *Hong Kong*
Pearson Education Canada, Ltd., *Toronto*
Pearson Educación de Mexico, S.A. de C.V.
Pearson Education—Japan, *Tokyo*
Pearson Education Malaysia, Pte. Ltd

*To all the students and teachers of Earth, our home planet,
and a sustainable future.*

*And to both our moms
for their inspiring lives and for all they have given us.*

*Earth is not ours, it is a treasure we hold in trust for
our children and their children.*

—an ancient African Proverb

Brief Contents

Contents

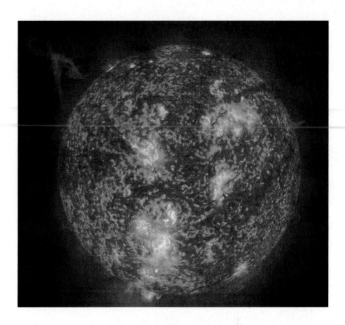

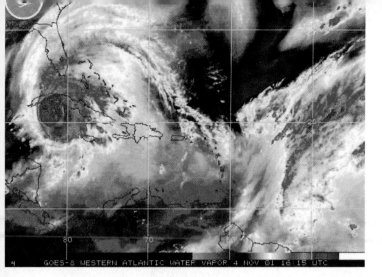

10 Global Climate Systems   275

9 Water Resources   245

PART 3
The Earth–Atmosphere Interface 320

11 The Dynamic Planet 323

12 Tectonics, Earthquakes, and Volcanism 357

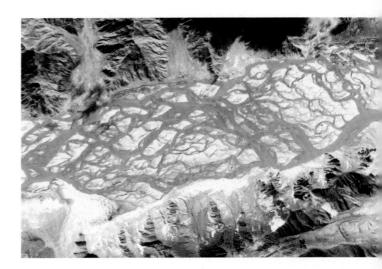

Preface

Welcome to physical geography and the fifth edition of *Geosystems*! We begin this new century with a sense that the world community is responding to global concerns as to the status and conditions of Earth's physical, biological, and chemical systems. The globalization of the world economies seems paralleled by a global scientific inquiry into the state of the environment. Marking this awareness is the second Earth Summit held in 2002 in Johannesburg, South Africa, with an agenda including climate change, freshwater, and the five Rio Conventions, among other topics.

Armed with the spatial analysis tools of geographic science, physical geographers are well equipped to participate in a planetary understanding of environmental conditions. U.N. Secretary General Kofi Annan, recipient of the 2001 Nobel Peace Prize, spoke to the Association of American Geographers annual meeting in 2001, stating,

> As you know only too well the signs of severe environmental distress are all around us. . . . The idea of interdependence is old hat to geographers, but for most people it is a new garment they are only now trying on for size . . . I look forward to working with you in that all-important journey.

This edition of *Geosystems*, with its contemporary global scope, builds on the success achieved by the first four editions in the United States, Canada, and elsewhere. Students and teachers alike continue to express appreciation for the systems organization, readability, scientific accuracy, up-to-date coverage and relevancy, clarity of the summary and review sections, the functional beauty of the photographs, art, cartography, and the many integrated figures in the text that combine media.

Geosystems Communicates the Science of Physical Geography

The goal of physical geography is to explain the spatial dimension of Earth's dynamic systems—its energy, air, water, weather, climate, tectonics, landforms, rocks, soils, plants, ecosystems, and biomes. Understanding human-Earth relations is part of the challenge of physical geography—to create a holistic (or complete) view of the planet and its inhabitants.

Geosystems analyzes the worldwide impact of environmental events, synthesizing many physical factors into a complete picture of Earth system operations. A good example is the eruption of Mount Pinatubo in the Philippines. The global implications of this major event (one of the largest eruptions in the 20th century) are woven through seven chapters of the book (see Figure 1.6 for a summary). Our update on global climate change and its re-

lated potential effects is part of the fabric in six chapters. These content threads weave together the variety of interesting and diverse topics crucial to a thorough understanding of physical geography.

This edition of *Geosystems* features more than 500 photographs from across the globe and 105 remote-sensing images from a wide variety of orbital platforms. Twenty-nine of these images are from the *Terra* satellite and its five sensor packages. To assist with spatial analysis and location, 121 maps are utilized, and more than 300 illustrations explain concepts.

Systems Organization Makes *Geosystems* Flow

Each section of this book is organized around the flow of energy, materials, and information. *Geosystems* presents subjects in the same sequence in which they occur in nature. In this way you and your teacher logically progress through topics which unfold according to the flow of individual systems, or in accord with time and the flow of events. See Figure 1.7 in the text for an illustration of this systems organization.

For flexibility, *Geosystems* is divided into four parts, each containing chapters that link content in logical groupings. The diagram on the next page from Figure 1.8 illustrates our part structure. A quick check of the Table of Contents and this illustration shows you the order of chapters within these four parts.

The text culminates with Chapter 21, "Earth and the Human Denominator," a unique capstone chapter that summarizes physical geography as an important discipline to help us understand Earth's present status and possible future. Think of the world's population and the totality of our impact as the *human denominator*. Just as the denominator in a fraction tells how many parts a whole is divided into, so the growing human population and the increasing demand for resources and rising planetary impact suggest how much the whole Earth system must adjust. This chapter is sure to stimulate further thought and discussion, dealing as it does with the most profound issue of our time, Earth's stewardship.

Geosystems is a Text That Teaches

Teaching and learning begin with the front and back cover images of the Pacific Northwest, two detail photos, and locator map. *Geosystems* is written to assist you in the learning process. Three heading levels are used throughout the

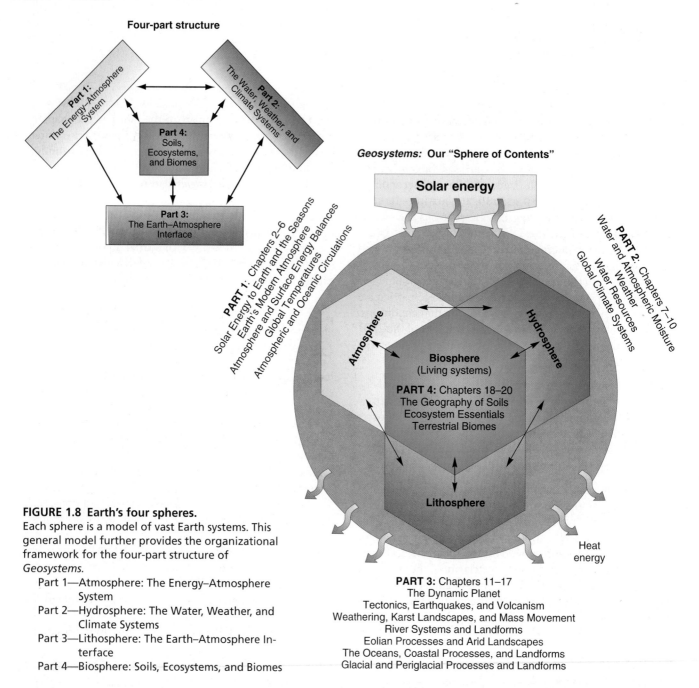

Four-part structure

FIGURE 1.8 Earth's four spheres.
Each sphere is a model of vast Earth systems. This general model further provides the organizational framework for the four-part structure of *Geosystems*.

    Part 1—Atmosphere: The Energy–Atmosphere System

    Part 2—Hydrosphere: The Water, Weather, and Climate Systems

    Part 3—Lithosphere: The Earth–Atmosphere Interface

    Part 4—Biosphere: Soils, Ecosystems, and Biomes

text and precise topic sentences begin each paragraph to help you outline and review material. **Boldface** words are defined where they first appear in the text. These terms and concepts are collected in the Glossary alphabetically, with a chapter-number reference. *Italics* are used in the text to emphasize other words and phrases of importance. Every figure has a title that summarizes the caption. Also, in the introduction to each chapter, a new feature called "In this chapter:" gives you an overview.

An important continuing feature is a list of Key Learning Concepts that opens each chapter, stating what you should be able to do upon completing the chapter. These objectives are keyed to the main headings in the chapter. At the end of each chapter is a unique Summary and Review

section that corresponds to the Key Learning Concepts. Grouped under each learning concept is a narrative review that redefines the boldfaced terms, a key terms list with page numbers, and specific review questions for that concept. You can conveniently review each concept, test your understanding with review questions, and check key terms in the glossary, then return to the chapter and the next learning concept. In this way, the chapter content is woven together using specific concepts.

A Critical Thinking section ends each chapter, challenging you to take the next step with information from the chapter. The key learning concepts help you determine what you want to learn, the text helps you develop information and more questions, the summary and review helps you assess what you have learned and what more you might want to know about the subject, and the critical thinking provokes action and application.

New *Career Link* essays feature geographers and other scientists in a variety of professional fields practicing their spatial analysis craft. You will read about an astronaut with over 1200 hours in orbit, a weather forecaster at the Forecast Systems Lab, an environmental scientist, a hydrologist with the National Weather Service, a snow avalanche specialist, and an expert on global scale ecosystems, among others.

Coverage of Canadian physical geography in the text and figures throughout the book continues. Canadian data on a variety of subjects are portrayed on 30 different maps in combination with the United States—physical geography does not stop at the United States–Canadian border!

 Twenty "Focus Study" essays, some completely revised and several new to this edition, provide additional explanation of key topics. A few examples from this diverse collection include: the stratospheric ozone predicament, solar energy collection and wind power, the newly (2001) calibrated wind-chill chart, forecasting the near-record 1995–2000 hurricane seasons, the 1997–1998 El Niño phenomenon, geothermal energy development, status of the High Plains Aquifer using new maps, floodplain strategies, an environmental approach to shoreline planning, the Mount St. Helens eruption, the 2001 status of the Colorado River, and the continuing global loss of biodiversity.

 Forty-three "News Reports" relate topics of special interest. For example: GPS, careers in GIS, a 34-kilometer sky dive to study the atmosphere, jet streams and airline flight times, how one culture harvests fog, the UV Index, coordination of global climate change research with many URLs presented, the disappearing Nile Delta, water issues in the Middle East, artificial scouring of the Grand Canyon to restore beaches and habitats, how sea turtles read Earth's magnetic field, alien and exotic plant and animal invasions, and threats to the Arctic National Wildlife Refuge.

We now live on a planet served by the Internet and its World Wide Web, a resource that weaves threads of information from around the globe into a vast fabric. The fact that we have Internet access into almost all the compartments aboard Spaceship Earth is clearly evident in *Geosystems*. Many entry points link directly from the words in a chapter to an Internet source allowing you to be up-to-the minute in understanding the facts. You will find more than 200 URLs (Internet addresses) in the body of the text (printed in blue color and boldface). Given the fluid nature of the Internet, URLs were rechecked at press time for accuracy. If some URLs changed since publication, you can most likely find the new location using elements of the old address. Textbooks, especially in dynamic fields like geography must be tapped into these streams of scientific dis-coveries and environmental events. This Internet link begins with a new Table 1.1 presenting the URLs for major geography organizations.

The Geosystems Learning/ Teaching Package

The fifth edition provides a complete physical geography program for you and your teacher.

For You the Student:

- *Student Study Guide*, Fifth Edition (ISBN: 0-13-034822-8), by Robert Christopherson. The study guide includes additional learning objectives, a complete chapter outline, critical thinking exercises, problems and short essay work using actual figures from the text, and a self-test with answer key in the back.
- *Geosystems WWW Site*: This site gives you the opportunity to further explore topics presented in the book using the Internet. The site contains numerous review exercises (from which you get immediate feedback), exercises to expand students' understanding of physical geography and resources for further exploration. This web site provides an excellent opportunity from which to start using the Internet for the study of geography. Please visit the site at **www.prenhall.com/christopherson**.
- *Science on the Internet*: A Student's Guide (ISBN: 0-13-021308-X) by Andrew T. Stull is a guide to the Internet and World Wide Web specific to geography.

 Virtual Fieldtrip 3.0 CD, (ISBN: 0-13-009750-0) a CD-ROM by Jeremy Dunning, Larry Onesti, and Robert Christopherson. This revised CD contains support material for every chapter in the text and more than a dozen virtual field trips. If your instructor chooses, the *Virtual Field Trip 3.0 CD-ROM* can be packaged free of charge with *Geosystems*, *5e*. Please see your local Prentice Hall representative for details.

For You the Teacher:

Geosystems is designed to give you flexibility in presenting your course. The text is comprehensive in that it is true to each scientific discipline from which it draws subject matter. This diversity is a strength of physical geography, yet makes it difficult to cover the entire book in a school term. You should feel free to customize use of the text based on your specialty or emphasis. The four-part structure of chapters, systems organization within each chapter, focus study and news report features, all will assist you in sampling some chapters while covering others to greater depth. The following materials are available to assist you—have a great class!

- *Instructor's Resource Manual*, Fifth Edition (ISBN: 0-13-034808-2), by Robert Christopherson. The Instructor's Resource Manual, intended as a resource for both new and experienced teachers, includes lecture outlines and key terms, additional source materials, teaching tips, complete annotation of chapter review questions, and a list of overhead transparencies.

- *Digital Image Gallery for Interactive Teaching (D.I.G.I.T.)*, Fifth Edition (ISBN: 0-13-034814-7), contains most of the figures and some of the photographs from the text. Images are high-resolution, low compression in high quality digital form. The software makes customizing your multimedia presentations easy. You can organize figures in any order you want, add labels, lines, and your own artwork to them using an overlay tool, integrate materials from other sources, and edit and annotate lecture notes. New to this edition is a complete *PowerPoint* presentation for the book, ready for classroom presentation, prepared by Charlie Thomsen.

- *Test Item File* (ISBN: 0-13-034809-0), the Test Item File contains many test questions drawn from the book, available in printed format.

- *Test Manager Geosystems Test Bank* (ISBN: 0-13-034800-2), by Robert Christopherson and Charlie Thomsen. This collaboration has produced the most extensive and fully revised test item file available in physical geography. This test bank employs **TestGen-EQ** software. **TestGen-EQ** is a computerized test generator that lets you view and edit test bank questions, transfer questions to tests, and print customized formats. Included is the **QuizMaster-EQ** program that lets you administer tests on a computer network, record student scores, and print diagnostic reports. Mac and IBM/DOS computer formats are served.

- *Overhead Transparencies* (ISBN: 0-13-034811-2) includes more than 350 illustrations from the text on 300 transparencies, all enlarged for excellent classroom visibility. And, *Slide Set* (ISBN: 0-13-034812-0) includes the complete set of figures from the transparency set selected from the text.

- *Applied Physical Geography—Geosystems in the Laboratory*, Fifth Edition (ISBN: 0-13-034823-6), by Robert Christopherson and Gail Hobbs of Pierce College. Reviewer comments and the feedback from users were very positive for the third edition. The new fifth edition is the result of a careful revision. Twenty lab exercises, divided into logical sections, allow flexibility in presentation. Each exercise comes with a list of learning concepts. Our manual is the only one that comes with its own complete glossary and stereolenses and stereomaps for viewing photo stereopairs in the manual. A complete *Solutions and Answers Manual* is available to teachers (ISBN: 0-13-034815-5).

 Prentice Hall–New York Times Themes of the Times supplements, especially *The Changing Earth*, and also *Geography*, reprint significant recent articles on related topics. These are available at no charge from your local Prentice Hall representative, in quantities for all your students.

Acknowledgments

As in all past editions, I recognize my family, for they have endured our work load, yet never wavered in supporting *Geosystems'* goals—both our moms, my sister Lynne, brothers Randy and Marty, and our children Keri, Matt, Reneé, and Steve. And now the next generation: Chavon, Bryce, Payton, Brock, Trevor, and our newest Blake. When I look into our grandchildren's faces it tells me why we need to work toward a sustainable future; one for the children.

I give special gratitude to all the students, and my colleagues, over 29 years at American River College for defining the importance of Earth's future, for their questions, and their enthusiasm. To all students and teachers, and to both our moms, this text is dedicated.

My thanks go to the many authors and scientists who published research, articles and books that enriched my work. To all the colleagues who served as reviewers on one or more editions, who participated in our focus groups, or who offered helpful suggestions at our national and regional geography meetings. And, although unnamed here, to all the correspondence received from students and teachers from across the globe who shared with me over the Internet, e-mail, FAX, and phone—a continuing appreciated dialogue. I am grateful to all of them for their generosity of ideas and sacrifice of time. Here is a master list of all our reviewers.

Ted J. Alsop, *Utah State University*
Ward Barrett, *University of Minnesota*
David Berner, *Normandale Community College*
Peter D. Blanken, *University of Colorado, Boulder*
David R. Butler, *Southwest Texas State University*
Ian A. Campbell, *University of Alberta–Edmonton*
Fred Chambers, *University of Colorado, Boulder*
Muncel Chang, *Butte College and California State University–Chico*
Andrew Comrie, *University of Arizona*
C. Mark Cowell, *Indiana State University*
Richard A. Crooker, *Kutztown University*
Armando M. da Silva, *Towson State University*
Dirk H. de Boer, *University of Saskatchewan*
Mario P. Delisio, *Boise State University*
Joseph R. Desloges, *University of Toronto*
Lee R. Dexter, *Northern Arizona University*
Don W. Duckson, Jr., *Frostburg State University*
Christopher H. Exline, *University of Nevada–Reno*
Michael M. Folsom, *Eastern Washington University*

Mark Francek, *Central Michigan University*
Glen Fredlund, *University of Wisconsin–Milwaukee*
David E. Greenland, *University of North Carolina–Chapel Hill*
Duane Griffin, *Bucknell University*
John W. Hall, *Louisiana State University–Shreveport*
Vern Harnapp, *University of Akron*
Gail Hobbs, *Pierce College*
David A. Howarth, *University of Louisville*
Patricia G. Humbertson, *Youngstown State University*
David W. Icenogle, *Auburn University*
Philip L. Jackson, *Oregon State University*
J. Peter Johnson, Jr., *Carleton University*
Guy King, *California State University–Chico*
Ronald G. Knapp, *SUNY–The College at New Paltz*
Peter W. Knightes, *Central Texas College*
Thomas Krabacher, *California State University–Sacramento*
Richard Kurzhals, *Grand Rapids Junior College*
Steve Ladochy, *California State University, Los Angeles*
Robert D. Larson, *Southwest Texas State University*
Joyce Lundberg, *Carleton University*
W. Andrew Marcus, *Montana State University*
Elliot G. McIntire, *California State University, Northridge*
Norman Meek, *California State University, San Bernardino*
Sherry Morea-Oaks, *Boulder, CO*
Lawrence C. Nkemdirim, *University of Calgary*
John E. Oliver, *Indiana State University*
Bradley M. Opdyke, *Michigan State University*
Patrick Pease, *East Carolina University*
James Penn, *Southeastern Louisiana University*
Greg Pope, *Montclair State University*
Robin J. Rapai, *University of North Dakota*
Philip D. Renner, *American River College*
William C. Rense, *Shippensburg University*
Dar Roberts, *University of California–Santa Barbara*
Wolf Roder, *University of Cincinnati*
Robert Rohli, *Louisiana State University*
Bill Russell, *L.A. Pierce College*
Dorothy Sack, *Ohio University*
Glenn R. Sebastian, *University of South Alabama*
Daniel A. Selwa, *U.S.C. Coastal Carolina College*
Thomas W. Small, *Frostburg State University*
Daniel J. Smith, *University of Victoria*
Stephen J. Stadler, *Oklahoma State University*
Susanna T.Y. Tong, *University of Cincinnati*
Suzanne Traub-Metlay, *Front Range Community College*
David Weide, *University of Nevada–Las Vegas*
Brenton M. Yarnal, *Pennsylvania State University*
Stephen R. Yool, *University of Arizona*

I extend my continuing gratitude to the editorial, production, and sales staff of Prentice Hall. Thanks to ESM President Paul Corey for his inspiring questions and challenges from the beginning. Thanks to Dan Kaveney, Geosciences Executive Editor, who is a dedicated, innovative, and energetic geographer and friend, for playing such a positive role in the *Geosystems* texts and in geographic education in general. Compliments to Amanda Griffith who oversees my ancillaries and Chris Rapp for expertise on the CD projects and Ginger Birkeland for work on the *Geosystems* Web site. Thanks to Margaret Ziegler's organizational skills in the Geosciences office. And to all the staff for allowing us to participate in the entire publishing process.

My thanks to Ingrid Mount and Sue Langguth at Elm Street Publishing Services for such expertise and care in converting all my materials into this textbook. Thanks to the art and design team at Prentice Hall for this powerful cover and beautiful text design and for letting us in on many decisions. The many sales representatives that spend months in the field communicating the *Geosystems* approach are a tremendous asset to the book, thanks and safe travels to them.

My continuing partnership with a special collaborator, photographer, and production assistant, Bobbé Christopherson, is at the heart of the success of *Geosystems*. My wife has worked tirelessly on all the *Geosystems'* projects. She catalogues photographs, prepares our extensive figure and photo logs, processes satellite imagery and crops photos, copy edits, obtains permissions, and assists me in proofing art and final pages. Bobbé's love for the smallest of living things is evident in the 139 beautiful and insightful photographs she has in this edition. And, she is my best friend and colleague.

Physical geography teaches us a holistic view of the intricate supporting web that is Earth's environment and our place in it. Dramatic changes that demand our understanding are occurring in many human-Earth relations, as we alter physical, chemical, and biological systems. All things considered, this is a critical time for you to be enrolled in a physical geography course! The best to you in your studies—and ***carpe diem!***

Robert W. Christopherson
P. O. Box 128
Lincoln, California 95648
E-mail: bobobbe@aol.com
Web site: **http://www.prenhall.com/christopherson**

The Mid-Atlantic region in true-color imagery by the MODIS sensor aboard *Terra* (October 11, 2000). The panorama sweeps from South Carolina to New York City and west across the Appalachian Mountains. Note the fall colors in the mountains to the north and west, the coastal barrier islands, and the outflow of sediment. [*Terra* image courtesy of MODIS Land Science Team, NASA/GSFC.]

1 Essentials of Geography

Key Learning Concepts

After reading the chapter, you should be able to:

- *Define* geography and physical geography in particular.
- *Describe* systems analysis, open and closed systems, feedback information, and system operations and *relate* those concepts to Earth systems.
- *Explain* Earth's reference grid: latitude, longitude, latitudinal geographic zones, and time.
- *Define* cartography and mapping basics: map scale and map projections.
- *Describe* remote sensing and *explain* geographic information system (GIS) methodology as a tool used in geographic analysis.

Welcome to physical geography and *Geosystems*, fifth edition! This is a scientifically exciting and challenging time to be enrolled in a physical geography course. Physical geography deals with our environment and the powerful forces that influence our lives and the many ways we are altering Earth's systems. Most natural ecosystems now bear the imprint of civilization.

This book is an assessment of the operation of Earth's physical systems. Earth's physical and natural diversity continues to fascinate us: lush rain forests straddle the equator, stark deserts bake in the subtropics, cool moist climates dominate northwestern coastlines, and perpetual cold hugs the polar regions. In many ways human societies take for granted the services Earth systems do for us. This natural "work" of Earth's physical, chemical, and biological systems has an estimated worth of some $35 trillion annually. Ironically, the investment costs of sustaining these life-support operations are comparatively low.

Physical geography provides essential information needed to understand and sustain our planetary journey in this new century. United Nations Secretary General and recipient of the 2001 Nobel Peace Prize, Kofi Annan, speaking to the Association of American Geographers annual meeting on March 1, 2001, offered this assessment,

> As you know only too well the signs of severe environmental distress are all around us. Unsustainable practices are woven deeply into the fabric of modern life. Land degradation threatens food security. Forest destruction threatens biodiversity. Water pollution threatens public health, and fierce competition for freshwater may well become a source of conflict and wars in the future. . . . the overwhelming majority of scientific experts have concluded that climate change is occurring, that humans are contributing, and that we cannot wait any longer to take action . . . environmental problems build up over time, and take an equally long time to remedy.

Scientists and governments scramble to understand the worldwide impact of global changes in weather, climate, water resources, and the landscape. We are experiencing unprecedented environmental conditions: increasing air and ocean temperatures, stratospheric ozone losses, increased rates of sea-level rise, heightened thunderstorm intensity, large losses of glacial ice worldwide, increasing plant and animal species extinctions, weather-related damage with massive floods in one region contrasted with severe droughts elsewhere, and related wildfires. As an example, weather-related damage in 1998 topped $90 billion, nine times greater than the previous annual average.

Why do all of these conditions occur? How are these events different from past experience? Why does the environment and rates of global change vary from equator to midlatitudes, between deserts and polar regions? How does solar energy influence the distribution of trees, soils, climates, and lifestyles? How does energy produce the patterns of wind, weather, and ocean currents? How do natural systems affect human populations, and, in turn, what impact are humans having on natural systems? In this book, we explore those questions, and more, through geography's unique perspective. Once again, welcome to an exploration of physical geography!

We live in an extraordinary era of **Earth systems science**. This science contributes to our emerging view of Earth as a complete entity—an interacting set of physical, chemical, and biological systems that produce a whole Earth. Physical geography is at the heart of Earth systems science as we answer the *spatial* questions concerning Earth's physical systems and their interaction with living things.

Physical geographers analyze interactions and changes that are occurring in natural systems to better quantify Earth systems over diverse *temporal* (time) and *spatial* (space or area) scales, and to address the question of how these changes might affect life on Earth. In an editorial in the journal *Science*, John Lawton stated,

> One of the great scientific challenges of the twenty-first century is to forecast the future of planet Earth . . . we find ourselves, literally, in uncharted territory, performing an uncontrolled experiment with planet Earth. . . . Wrestling to understand these challenges is the young, and still emerging, discipline of Earth systems science.*

*John Lawton, "Earth systems science," *Science*, 292 (June 15, 2001): 1965.

In this chapter: Our study of *Geosystems*—Earth systems—begins in this chapter with a look at the science of physical geography and the geographic tools we use. Physical geography is key to studying entire Earth systems because of its integrative approach.

Physical geographers analyze systems to study the environment. Therefore, we discuss systems and the feedback mechanisms that influence system operations. We then consider location, a key theme of geographic inquiry—the latitude, longitude, and time coordinates that inscribe Earth's surface, and the new technologies in use to measure them. The study of longitude and a universal time system provide us with interesting insights into geography. Next, we examine maps as critical tools that geographers use to portray physical and cultural information. This chapter concludes with an overview of the technology that is adding exciting new dimensions to geography: remote sensing from space and computer-based geographic information systems (GIS).

The Science of Geography

Geography (from *geo*, "Earth," and *graphein*, "to write") is the science that studies the relationships among natural systems, geographic areas, society, cultural activities, and the interdependence of all of these *over space*. The term **spatial** refers to the nature and character of physical space, its measurement, and the distribution of things within it. For example, think of your own route to the classroom or library today and how you used your knowledge of street patterns, traffic trouble spots, parking spaces, or bike rack locations to minimize walking distance. All these are spatial considerations.

Humans are spatial actors, both affecting and being affected by Earth. We profoundly influence vast areas because of our mobility and access to energy and technology. In turn, Earth's systems influence our activities in a most obvious way—these systems give us life.

To guide geographic education, the "National Geography Standards" for learning were prepared by the Association of American Geographers (AAG) and the National Council for Geographic Education (NCGE) in response to *Goals 2000*, the Educate America Act of 1994. In these standards, geographic science is divided into essential elements, critical skills, and geographic standards (see *Geography for Life*, prepared by the Geography Education Project for the AAG, NCGE, American Geographical Society, and National Geographic Society, 1994). For a listing of some important geography organizations and their URLs, see Table 1.1.

We simplify the standards of geographic science using five important spatial themes: **location**, **region**, **human–Earth relationships**, **movement**, and **place**, which are illustrated and defined in Figure 1.1. *Geosystems* draws on each theme.

Geographic Analysis

Within these five geographic themes, geography is governed by a *method* rather than a specific body of knowledge, and the method is **spatial analysis**. Using this method, geography synthesizes (brings together) knowledge from many fields, integrating information to form a whole Earth concept. Geographers view phenomena as occurring in

Table 1.1	
A Few Geography Organizations	**URL Addresses**
American Geographical Society	http://www.geography.unr.edu/GR/ags/ags.html
Association of American Geographers*	http://www.aag.org/
National Council for Geographic Education	http://www.ncge.org/
National Geographic Society	http://www.nationalgeographic.com/
Canadian Association of Geographers	http://venus.uwindsor.ca/cag/cagindex.html
Royal Canadian Geographical Society	http://www.rcgs.org/
Institute of Australian Geographers	http://www.ssn.flinders.edu.au/geog/iag/
Australian Geography Teachers Association	http://www.agta.asn.au/
European Geography Association	http://egea.geog.uu.nl/
Royal Geographical Society, Institute of British Geographers	http://www.dti.gov.uk/ost/rgs.htm
Complete global listing of geography organizations	http://www.geoggeol.fau.edu/prof_org/GeographyPO.html

*Includes nine regional divisions: East Lakes, Great Plains/Rocky Mountain, Middle Atlantic, Middle States, New England—St. Lawrence Valley, Pacific Coast Regional, Southeastern, Southwestern, West Lakes.

Location

Absolute and relative location on Earth. Location answers the question *Where?*—the specific planetary address of a location. This road sign is posted along the I-5 freeway in Oregon.

Region

Portions of Earth's surface that have uniform characteristics; how they form and change; how they are related to other regions. The Northwest region of the United States and southern Canada features dramatic coasts and the Channel Islands.

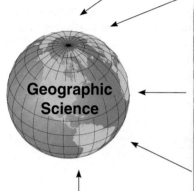

Geographic Science

Human–Earth Relationships

Humans and the environment: resource exploitation, hazard perception, and environmental modification—the oldest theme of geographic inquiry. The village on the Amazon River is constructed from materials derived from the surrounding area.

Place

Tangible and intangible living and nonliving characteristics that make each place unique. No two places on Earth are exactly alike. New York City's Central Park is unique in its answer to the place of nature in the human city.

Movement

Communication, movement, circulation, and diffusion across Earth's surface. Global interdependence links all regions and places—both physical and human systems. Winds and ocean currents form circulations of energy and water. The ferocious winds of Hurricane Mitch devastated tracts of land in Honduras.

FIGURE 1.1 Five themes of geographic science.
Definitions of five fundamental themes in geographic science with examples of each—location, region, human–Earth relationships, movement, and place. Drawing from your own experience, can you think of several examples of each theme? [Regions and place photos by Bobbé Christopherson; Human–Earth relationship photo by Gael Summer-Hebden; Movement by Gregory Bull/AP/Wide World Photos. Location photo by author.]

spaces and areas having distinctive characteristics. The language of geography reflects this spatial view: *space, territory, zone, pattern, distribution, place, location, region, sphere, province,* and *distance.* Geographers analyze the differences and similarities among places and locations.

Process, a set of actions or mechanisms that operate in some special order, governed by physical, chemical, and biological laws, is central to geographic analysis. As examples in *Geosystems,* numerous processes are involved in Earth's vast water–atmosphere–weather system, or in continental crust movements and earthquake occurrences, or in ecosystem functions. Geographers use spatial analysis to examine how Earth's processes interact over space or area.

Therefore, **physical geography** is the *spatial analysis of all the physical elements and processes that make up the environment: energy, air, water, weather, climate, landforms, soils, animals, plants, microorganisms, and Earth itself.* We add to this the oldest theme in the geographic tradition, that of human activity. As a science, geographers employ the **scientific method,** a methodology important to physical

geography. Focus Study 1.1 explains this essential process of science.

The Geographic Continuum

Geography is eclectic, integrating a wide range of subject matter from diverse fields; virtually any subject can be examined geographically. Figure 1.2 shows a continuous distribution—a continuum—along which the content of geography is arranged. Disciplines in the physical and life sciences are at one end, and those in the human and cultural sciences are at the other. As the figure shows, various specialties within geography draw from these subject areas.

The continuum in Figure 1.2 reflects a basic duality, or split, within geography—*physical geography* versus *human and cultural geography.* This duality is paralleled in society by the tendency of those of us who live in more developed countries (MDCs) to distance ourselves from our life-sustaining environment, to think of ourselves as exempt from physical Earth processes—like actors not paying attention to their stage, props, and lighting. In contrast,

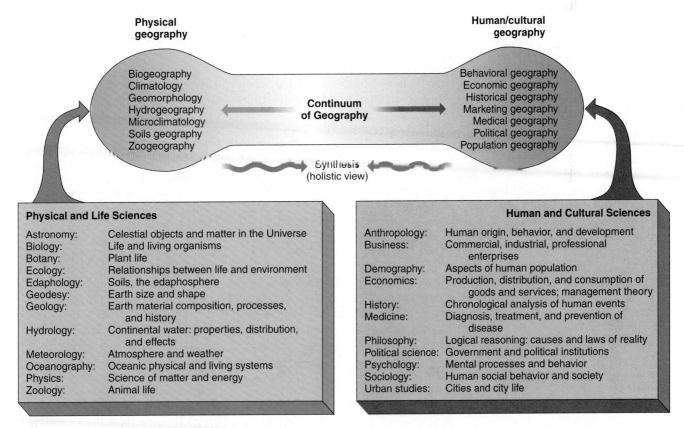

FIGURE 1.2 The content of geography.
Distribution of geographic content along a continuum (a continuous distribution). Geography derives subject matter from many different sciences. The focus of this book is physical geography, but we also integrate some human and cultural components. Synthesis of Earth topics and human topics is suggested by movement toward the middle of the continuum—a *holistic,* or balanced, view. Examine the subjects listed in the two boxes for any course you completed. Do you find any subjects you have taken? Think for a moment and recall any spatial aspects you remember studying in that course.

The Scientific Method

The term *scientific method* may have an aura of complexity that it should not. The scientific method is simply the application of common sense in an organized and objective manner. A scientist observes, makes a general statement to summarize the observations, formulates a hypothesis, conducts experiments to test the hypothesis, and develops a theory and governing scientific laws. Sir Isaac Newton (1642–1727) developed this method of discovering the patterns of nature, although the term *scientific method* was applied later.

Scientists are curious about nature and appreciate the challenge of problem solving. Society depends on the discoveries of science to decipher mysteries, find solutions, and foster progress. *Complexity* dominates nature, making several outcomes possible as a system operates. Science serves an important function to reduce such uncertainty. Yet, the more knowledge we have, the more the uncertainty and awareness of other possible scenarios (outcomes and events) increases. This in turn demands more precise and aggressive science. A danger in society occurs when scientific uncertainty fuels an anti-science viewpoint. As we realize scientific principles of complexity and chaos in natural and human-made systems, the need for critical thinking and the scientific method deepens in all aspects of life.

Follow the scientific method illustration in Figure 1 as you read. The scientific method begins with our perception of the real world and a determination of what we know, what we want to know, and the many unanswered questions that exist. Scientists who study the physical environment turn to nature for clues that they can observe and measure. They discern what data are needed and begin to collect those data. Then, these observations and data are analyzed to identify coherent patterns that may be present. This search for patterns requires *inductive reasoning*, or the process of drawing generalizations from specific facts. This step is important in modern Earth systems sciences, in which the goal is to understand *a whole functioning Earth*, rather than isolated, small compartments of information. Such understanding allows the scientist to construct models that simulate general operations of Earth systems.

If patterns are discovered, the researcher may formulate a *hypothesis*—a formal generalization of a principle. Examples include the planetesimal hypothesis, nuclear-winter hypothesis, and moisture-benefits-from-hurricanes hypothesis. Further observations are related to the general principles established by the hypothesis. Further data gathered may support or refute the hypothesis, or predictions made according to it may prove accurate or inaccurate. All these findings provide feedback to adjust data collection and model building and to refine the hypothesis statement. Verification of the hypothesis after exhaustive testing may lead to its elevation to the status of a *theory*.

A theory is constructed on the basis of several extensively tested hypotheses. Theories represent truly broad general principles—unifying concepts that tie together the laws that govern nature (for example, the theory of relativity, theory of evolution, atomic theory, Big Bang theory, stratospheric ozone depletion theory, or plate tectonics theory). A theory is a powerful device with which to understand both the order and chaos (disorder) in nature. Using a theory allows predictions to be made about things not yet known, the effects of which can be tested and verified or disproved through tangible evidence. The value of a theory is the continued observation, testing, understanding, and pursuit of knowledge that the theory stimulates. A general theory reinforces our perception of the real world, acting as positive feedback.

Pure science does not make value judgments. Instead, pure science provides people and their institutions with objective information on which to base their own value judgments. Social and political judgments about the applications of science are increasingly critical as Earth's natural systems respond to the impact of modern civilization. Again from Lubchenco's 1997 AAAS Presidential Address:

> Science alone does not hold the power to achieve the goal of greater sustainability, but scientific knowledge and wisdom are needed to help inform decisions that will enable society to move toward that end.

The growing awareness that human activity is producing global change places increasing pressure on scientists to participate in decision making. Numerous editorials in scientific journals have called for such *applied science* involvement. An example, discussed in Chapter 3, is provided by F. Sherwood Rowland and Mario Molina, who first proposed a hypothesis that certain human-made chemicals caused damaging reactions in the stratosphere. They hypothesized in 1974—along with Paul Crutzen, who also contributed to these discoveries—that chlorine-containing products such as chlorofluorocarbons (CFCs), commonly used as aerosol propellants, refrigerants, foaming agents, and cleaning solvents, were depleting our protective stratospheric ozone (O_3) layer.

Subsequently, surface, atmosphere, and satellite measurements confirmed the photochemical reactions and provided data to map the real losses that were occurring. International treaties and agreements to ban the chemical culprits followed. In 1995 the Royal Swedish Academy of Sciences awarded these three scientists the Nobel Prize for chemistry for their pioneering work. Such successful applied science is strengthening society's resolve and provoking treaties to ban problem chemicals and practices and introduce sustainable substitutes.

Focus Study 1.1 *(continued)*

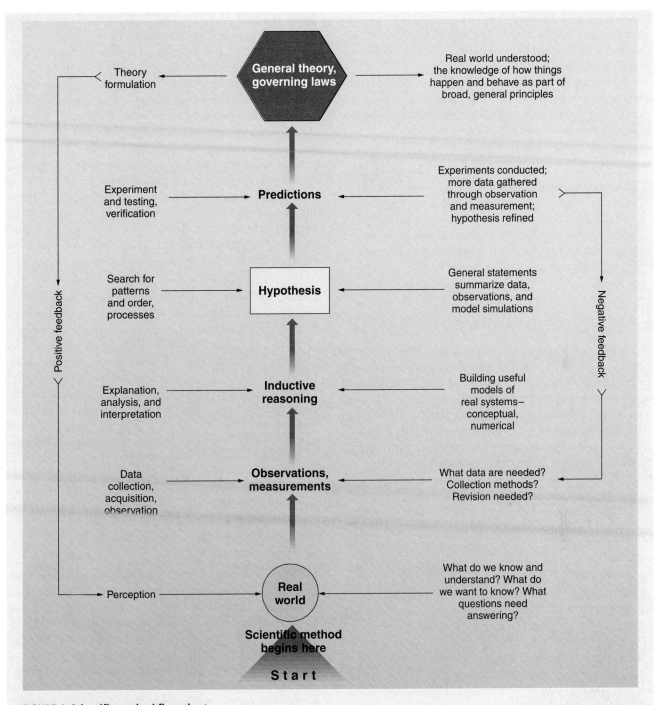

FIGURE 1 Scientific method flow chart.
The scientific method process: from perceptions, to observations, reasoning, hypothesis, predictions, and possibly to general theory and natural laws.

many people in less-developed countries (LDCs) live closer to nature and are acutely aware of its importance in their daily lives.

Regardless of our philosophies toward Earth, we all depend on Earth's systems to provide oxygen, water, nutrients, energy, and materials to support life. The growing complexity of the human–Earth relation requires that we shift our study of geographic processes, and perhaps our

philosophies, toward the center of the continuum in Figure 1.2 to attain a more *holistic*, or balanced, perspective—such is the thrust of Earth systems science.

In the millennium ahead we face unique patterns of spatial change, for we are taxing Earth's systems in new ways. More people are alive today than ever before in Earth's history and their lifestyles are increasing in complexity and resource demands. We surpassed the population milestone of 6 billion in 1999 with virtually all population growth occurring in LDCs—present forecasts are for 7.8 billion people by 2025 and 9 billion by 2050 (a 47 percent increase over 2001). Again quoting from UN Secretary General Kofi Annan, "Environmental concerns are the national security issues of the future."

Some past civilizations adapted to crises, whereas others failed. Perhaps this ability to adapt is the key. If so, understanding our relationship to Earth's physical geography is important to human survival.

> . . . during the last few decades, humans have emerged as a new force of nature. We are modifying physical, chemical, and biological systems in new ways, at faster rates, and over larger *spatial* scales than ever recorded on Earth. Humans have unwittingly embarked upon a grand experiment with our planet. The outcome of this experiment is unknown, but has profound implications for all of life on Earth.*

Earth Systems Concepts

The word *system* pervades our lives daily: "Check the car's cooling system"; "How does the grading system work?"; "There is a weather system approaching." Systems of many kinds surround us. Not surprisingly, *systems analysis* has moved to the forefront as a method for understanding operational behavior in many disciplines. The technique began with studies of energy and temperature (thermodynamics) systems in the nineteenth century and was further developed in engineering during World War II. Today, geographers use systems methodology as an analytical tool. In this book's 4 parts and 21 chapters, content is organized along logical flow paths consistent with systems thinking.

Systems Theory

Simply stated, a **system** is any ordered, interrelated set of things and their attributes, linked by flows of energy and matter, as distinct from the surrounding environment outside the system. The elements within a system may be arranged in a series or interwoven with one another. A system comprises any number of subsystems. Within Earth's systems, both matter and energy are stored and retrieved, and energy is transformed from one type to another. (Remember: *matter* is mass that assumes a physical shape and occupies space; *energy* is a capacity to change the motion of, or to do work on, matter.)

Open Systems   Systems in nature are generally not self-contained: Inputs of energy and matter flow into the system, and outputs of energy and matter flow from the system. Such a system is called an **open system**. Within a system, the parts function in an interrelated manner, acting

---

* Jane Lubchenco, Presidential Address, American Association for the Advancement of Science, February 15, 1997.

together in a way that gives each system its character. Earth is an open system *in terms of energy*, for solar energy enters freely and heat energy leaves freely back into space. Most natural systems are open in terms of energy. Figure 1.3 schematically illustrates an open system and presents the inputs and outputs of an automobile as an example.

Most Earth systems are dynamic (energetic, in motion) because of the tremendous infusion of radiant energy from thermonuclear reactions deep within the Sun. This energy penetrates the outermost edge of Earth's atmosphere and cascades through the terrestrial systems, being transformed along the way into various forms of energy, such as *kinetic energy* (of motion), *potential energy* (of position), or other expressions as chemical or mechanical energy. Eventually, Earth radiates this energy back to the cold vacuum of space as heat energy. Researchers are examining Earth's energy system to distinguish natural operations from those changes being forced by human activities.

Closed Systems   A system that is shut off from the surrounding environment so that it is self-contained is a **closed system**. Although such closed systems are rarely found in nature, Earth is essentially a closed system *in terms of physical matter and resources*—air, water, and material resources. The only exceptions are the slow escape of light-weight gases (such as hydrogen) from the atmosphere into space and the input of frequent but tiny meteors and cosmic and meteoric dust. Since the initial formation of our planet, no significant quantities of new resources have entered the system. This is it! Earth's physical materials are finite (limited). No matter how numerous and daring the technological reorganizations of matter become, our physical base is, for all practical purposes, fixed. The fact that Earth is a closed material system makes recycling efforts inevitable if we want a sustainable economy.

System Example   Figure 1.4 illustrates a simple open-flow system, using plant photosynthesis and respiration as an example. In *photosynthesis* (Figure 1.4a), plants use an energy input (certain wavelengths of sunlight) and material inputs of water, nutrients, and carbon dioxide. The photosynthetic process converts these inputs to stored chemical energy in the form of plant sugars (carbohydrates). The process also releases an output from the plant system: the oxygen we breathe.

Reversing the process, plants derive energy for their operations from respiration. In *respiration*, the plant consumes inputs of chemical energy (carbohydrates) and oxygen and releases outputs of carbon dioxide, water, and heat energy into the environment (Figure 1.4b). Thus, a plant acts as an open system, in which both energy and materials freely flow into and out of the plant. (Photosynthesis and respiration processes are discussed further in Chapter 19.)

System Feedback   As a system operates, it generates outputs that influence its own operations. These outputs

FIGURE 1.3 An open system.
In an open system, inputs of energy and matter undergo conversions and are stored as the system operates. Outputs include energy and matter and heat energy (waste) that flow from the system. See how the various inputs and outputs are related to the operation of a car: Matter and energy are converted, stored, and produced in an automobile—an open system. Expand your viewpoint to the entire system of auto production, from raw materials, to assembly, to sales, to car accidents, to junkyards. Can you identify other open systems that you encounter in your daily life?

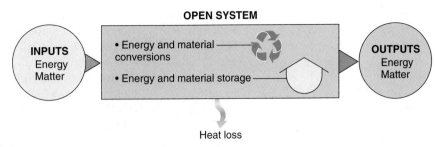

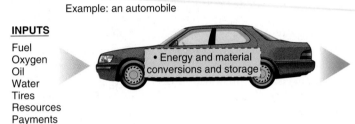

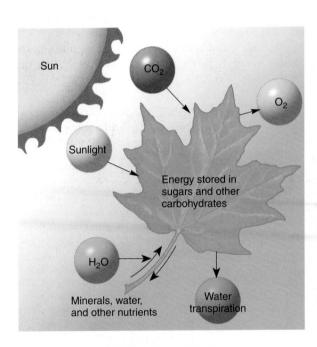
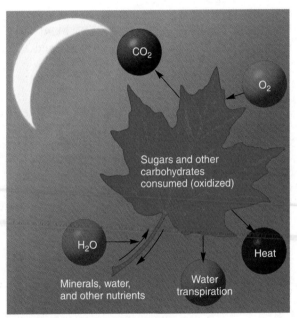

(a) Plant photosynthesis

(b) Plant respiration

FIGURE 1.4 A leaf is a natural open system.
A plant leaf provides an example of a natural open system. (a) In the process of photosynthesis, plants consume light, carbon dioxide (CO_2), nutrients, and water (H_2O), and produce outputs of oxygen (O_2) and carbohydrates (sugars) as stored chemical energy. (b) Plant respiration, illustrated here at night, approximately reverses this process and produces outputs of carbon dioxide (CO_2) and water (H_2O), using oxygen and consuming (oxidizing) carbohydrates to produce energy for cell operations.

function as "information" that is returned to various points in the system via pathways called **feedback loops.** Feedback information can control (or at least guide) further system operations. In the plant's photosynthetic system (see Figure 1.4), any increase or decrease in daylength (sunlight availability), carbon dioxide, or water will produce feedback that elicits (causes) specific responses in the plant. For example, decreasing the water input will slow the growth

process; increasing daylength will increase the growth process within limits.

If the feedback *information* discourages response in the system, it is called **negative feedback**—like bad reviews affecting ticket sales for a film or play. *Further production in the system decreases the growth of the system.* Such negative feedback causes self-regulation in a natural system, stabilizing and maintaining the system. Think of a weight-loss diet—the human body is an open system. When you stand on the scales, the good or bad news reported about your weight both act as negative feedback to you. This, in turn, provides you with guidance (negative feedback) as to how inputs (food Calories) need to be adjusted (such as not buying and eating that burrito grande in the college cafeteria), or how exercise and increased metabolism is needed to burn excess Calories.

If feedback *information* encourages increased response in the system, it is called **positive feedback**—like good reviews affecting ticket sales for a film. *Further production in the system stimulates the growth of the system.* In finance, a compound interest-bearing account provides an example: The larger the account becomes, the more interest it earns, thus the larger the account becomes, and so on. Unchecked positive feedback in a system can create a runaway ("snowballing") condition. In natural systems, such unchecked growth will reach a critical limit, leading to instability, disruption, or death of organisms.

Think of the numerous devastating wildfires that occurred in Indonesia, Mexico, and Florida in 1998, and the western United States between 1999 and 2001, as examples of positive feedback. As the fires burned, they dried wet shrubs and green wood around the fire, thus providing more fuel for combustion. The greater the fire, the greater becomes the availability of fuel, and thus more fire is possible—a positive feedback for the fire. Control of the input of flammable fuel and oxygen is the key to extinguishing such fires. Knowing this process allows us to design strategies for controlling fuel availability, regulating excessive landscaping in vulnerable urban areas, and practicing "control burns."

Can you describe possible negative and positive feedback in the automobile illustration in Figure 1.3? For instance, how would the inputs and outputs be affected if tailpipe exhaust emission standards are weakened or strengthened? What if the automobile is operated at high elevation? Or if the car is a sports utility vehicle (SUV) that weighs 2300 kg (5070 lb)? Or if the price of gasoline increases or decreases? Take a moment to assess the affect of such changes on automobile system operations.

System Equilibrium Most systems maintain structure and character over time. An energy and material system that remains balanced over time, in which conditions are constant or recur, is in a *steady-state condition.* When the

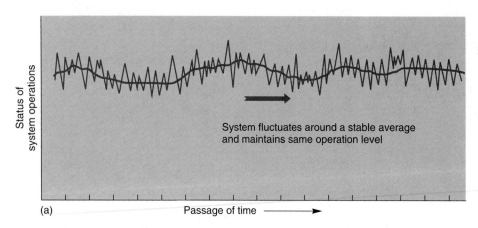

System fluctuates around a stable average
and maintains same operation level

(a)   Passage of time ⟶

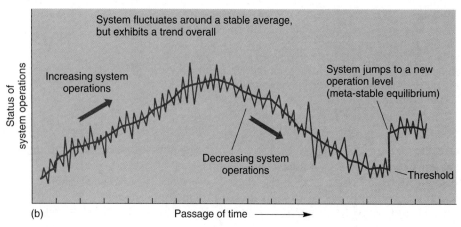

System fluctuates around a stable average,
but exhibits a trend overall

Increasing system
operations

Decreasing system
operations

System jumps to a new
operation level
(meta-stable equilibrium)

Threshold

(b)   Passage of time ⟶

FIGURE 1.5 System equilibria: steady-state and dynamic.
Some systems exhibit a steady-state equilibrium over time; system operations fluctuate around a stable average (a). Other systems are in a condition of dynamic equilibrium, with an increasing or decreasing operational trend (b). Rather than changing gradually, some systems may reach a threshold at which system operations lurch (change abruptly) to a new set of relations. The 1998 landslides along the Pacific Coast south of San Francisco destroyed homes and property as a threshold was reached and the bluffs collapsed. [Inset photo by Sam Sargent/ Liaison Agency Inc.]

Coastal systems pass a threshold.

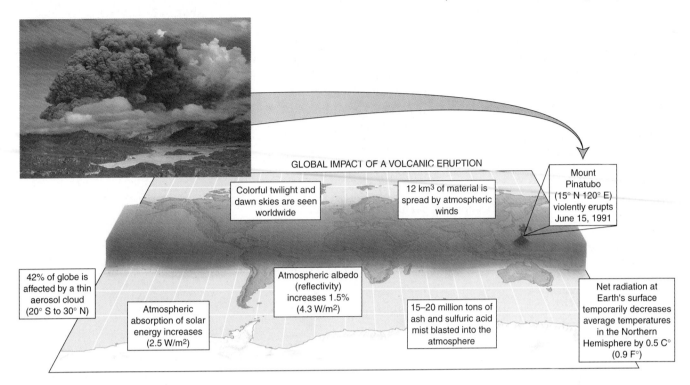

FIGURE 1.6 The eruption of Mount Pinatubo.
The 1991 Mount Pinatubo eruption, one of the largest volcanic eruptions in the twentieth century, widely affected the Earth–atmosphere system. Geographers and other scientists use the latest technology to study how such eruptions affect the atmosphere's dynamic equilibrium. For a summary of the impacts from this eruption you can refer to Chapter 12. [Inset photo by Van Cappellen/REA/SABA.]

rates of inputs and outputs in the system are equal and the amounts of energy and matter in storage within the system are constant (or more realistically, as they fluctuate around a stable average, such as a person's body weight), the system is in **steady-state equilibrium**.

However, a steady-state system fluctuating around an average value may demonstrate a changing trend over time, a condition described as **dynamic equilibrium**. These changing trends of either increasing or decreasing system operations may appear gradual. Examples of dynamic equilibrium include long-term climatic changes and the present pattern of increasing temperatures in the atmosphere and ocean, or the erosion and loss of many beaches along the West Coast in the late 1990s, or the ongoing loss of barrier islands along the East and Gulf coastlines. The present rate of species extinction exhibits a downward trend in numbers of living species. Figure 1.5 illustrates these two states.

Note that given the nature of systems to maintain their operations, they tend to resist abrupt change. However, a system may reach a *threshold* at which it can no longer maintain its character, so it lurches to a new operational level. This abrupt change places the system in a *metastable equilibrium*. An example of such a condition is a landscape, such as a hillside or coastal bluff, that adjusts after a sudden landslide. A new equilibrium is eventually achieved among slope, materials, and energy over time.

This threshold concept raises concern in the scientific community, especially if some natural systems reach such threshold limits. The relatively sudden collapse of ice shelves surrounding a portion of Antarctica serves as an example of systems reaching a threshold and changing to a new status, that of disintegration. The bleaching (death) of living coral reefs worldwide accelerated dramatically during 1997 to 2001. Warming conditions and some pollution in the ocean led to such a threshold and coral system collapse. A sudden change to a new equilibrium arrangement may not be as desirable or supportive to us as present conditions (see inset photo in Figure 1.5).

Mount Pinatubo—Global System Impact A particularly dramatic example of interactions between volcanic eruptions and Earth systems illustrates the strength of spatial analysis in physical geography and the systems organization of this textbook. Mount Pinatubo in the Philippines erupted violently in 1991, injecting 15–20 million tons of ash and sulfuric acid mist into the upper atmosphere (Figure 1.6). This was the second greatest eruption during the twentieth century; Mount Katmai in Alaska (1912) is the only one greater. The eruption materials from Mount Pinatubo affected Earth systems in several ways noted on the map.

As you progress through this book, you will see the story of Mount Pinatubo and its implications woven through seven chapters: Chapter 1 (systems theory), Chapter 4 (effects on energy budgets in the atmosphere), Chapter 6 (satellite images of the spread of debris by atmospheric winds), Chapter 10 (temporary effect on global atmospheric temperatures),

Systems in *Geosystems*

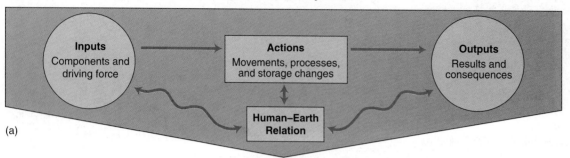

(a)

Examples of systems organization in text:

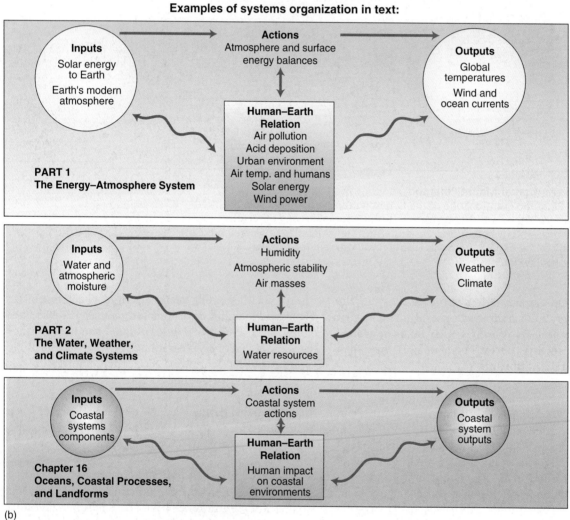

(b)

FIGURE 1.7 The systems in *Geosystems*.
Chapters, sections, and topics are organized around simple flow systems, or around time and the flow of events, at various scales (a). This systems structure as it is applied to Parts 1 and 2 (b). As a chapter example, turn to Chapter 16, which examines coastal processes. Note the outline headings used: Coastal System Components, Coastal System Actions, Coastal System Outputs, and Human Impact on Coastal Environments. Thus, Chapter 16 is organized along the systems flow in (a).

Chapters 11 and 12 (volcanic processes), and Chapter 17 (past climatic effects of volcanoes).

Systems in *Geosystems*   To further illustrate the conceptual systems organization used in this textbook you will notice chapters and portions of chapters presented in sim-

ple sequences organized around the flow of energy, materials, and information (Figure 1.7). This book presents subjects in the same sequence in which they occur in nature. In this way a logical progression is followed through topics that unfold according to the flow within individual systems, or in accord with time and the flow of events.

The sequence (Figure 1.7a)—*input* (components and driving force), *actions* (movements, processes, and storage changes), *outputs* (results and consequences), and *human impacts/impact on humans* (measure of relevance)—is seen in Part 1 as an example. The Sun begins Chapter 2, the energy flows across space to the top of the atmosphere, through the atmosphere to the surface and surface energy budgets (Chapters 3 and 4). Then we look at the outputs of temperature (Chapter 5) and winds and ocean currents (Chapter 6). Note the same logical systems flow in the other three parts and in each chapter of this text—Part 2 and Chapter 16 also are shown as examples.

Models of Systems   A **model** is a simplified, idealized representation of part of the real world. Models are designed with varying degrees of generalization. You may have built or drawn a model of something in your life that was a simplification of the real thing. Physical geographers construct simple system models to demonstrate complex associations in the environment. The simplicity of a model makes a system easier to comprehend and to simulate in experiments (for example, the leaf model in Figure 1.4). A good example is a model of the *hydrologic system*, which models Earth's entire water system, its related energy flows, and atmosphere, surface, and subsurface environments through which water moves.

Adjusting the variables in a model produces differing conditions and allows predictions of possible system operations to be made. A general circulation model (GCM) of the atmosphere (see Figure 10.31), operated by the Goddard Institute, accurately predicted the effect of Mount Pinatubo's ash on the atmosphere, the lowering and subsequent recovery of global air temperatures. However, predictions are only as good as the assumptions and accuracy built into the model. It is best to view a model for what it is—a simplification to help us understand a complex process.

We discuss many *system models* in this text, including the hydrologic system, water balance, surface energy budgets, earthquakes and faulting as outputs of Earth systems, river drainage basins, glacier mass budgets, soil profiles, and various ecosystems. Computer-based models are in use to study most natural systems—from the upper atmosphere, to climate change, to Earth's interior. Each of Earth's four "spheres" represents such a model.

Earth's Four "Spheres"

Earth's surface is a vast area of 500 million square kilometers (193 million square miles) where four immense open systems interact. Figure 1.8 shows a simple model of three **abiotic** (nonliving) systems overlapping to form the realm of the **biotic** (living) system. The abiotic spheres are the *atmosphere*, *hydrosphere*, and *lithosphere*. The biotic sphere is called the *biosphere*. Because these four systems are not independent units in nature, their boundaries must be understood as transition zones rather than sharp barriers. As noted in the figure, these four spheres form the part structure in which chapters are grouped in this text.

Content in each part and among chapters overlap and interrelate as we build our discussion of Earth's abiotic and biotic systems.

Atmosphere (Part 1, Chapters 2–6)   The **atmosphere** is a thin, gaseous veil surrounding Earth, held to the planet by the force of gravity. Formed by gases arising from within Earth's crust and interior, and the exhalations of all life over time, the lower atmosphere is unique in the Solar System. It is a combination of nitrogen, oxygen, argon, carbon dioxide, water vapor, and trace gases.

Hydrosphere (Part 2, Chapters 7–10)   Earth's waters exist in the atmosphere, on the surface, and in the crust near the surface. Collectively, these areas form the **hydrosphere**. Water of the hydrosphere exists in all three states: liquid, solid (the frozen cryosphere), and gaseous (water vapor). Water occurs in two general chemical conditions, fresh and saline (salty). It exhibits important heat-storage properties. And water is an extraordinary solvent. Water is the medium of life. Among the planets in the Solar System, only Earth possesses surface water in such quantity, adding to Earth's uniqueness among the planets.

Lithosphere (Part 3, Chapters 11–17)   Earth's crust and a portion of the upper mantle directly below the crust form the **lithosphere**. The crust is quite brittle compared with the layers deep beneath the surface, which move slowly in response to an uneven distribution of heat energy and pressure. An important component of the lithosphere is *soil*, which generally covers Earth's land surfaces; the soil layer sometimes is referred to as the *edaphosphere*. (In a broad sense, the term *lithosphere* sometimes refers to the entire solid planet.)

Biosphere (Part 4, Chapters 18–20)   The intricate, interconnected web that links all organisms with their physical environment is the **biosphere**. Sometimes called the **ecosphere**, the biosphere is the area in which physical and chemical factors form the context of life. The biosphere exists in the overlap among the abiotic spheres, extending from the seafloor and even the upper layers of the crustal rock to about 8 km (5 mi) into the atmosphere. Life is sustainable within these natural limits. In turn, life processes have powerfully shaped the other three spheres through various interactive processes. The biosphere evolves, reorganizes itself at times, faces some extinctions, and manages to flourish. Earth's biosphere is the only one known in the Solar System; thus, life as we know it is unique to Earth.

A Spherical Planet

We have all heard that some people believed Earth to be flat. Yet Earth's sphericity is not as modern a concept as many think. For instance, more than two millennia ago, the Greek mathematician and philosopher Pythagoras (ca. 580–500 B.C.) determined through observation that

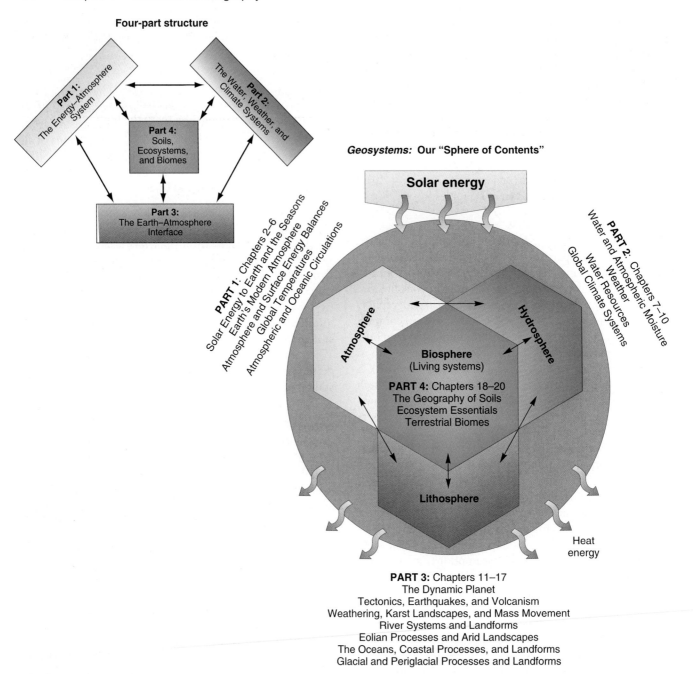

FIGURE 1.8 Earth's four spheres.
Each sphere is a model of vast Earth systems. This general model further provides the organizational framework for the four-part structure of *Geosystems*.

    Part 1—Atmosphere: The Energy–Atmosphere System
    Part 2—Hydrosphere: The Water, Weather, and Climate Systems
    Part 3—Lithosphere: The Earth–Atmosphere Interface
    Part 4—Biosphere: Soils, Ecosystems, and Biomes

Earth is spherical. We do not know what observations led Pythagoras to this conclusion. Can you guess at what he saw to deduce Earth's roundness?

    He might have noticed ships sailing beyond the horizon and apparently sinking below the water's surface, only to arrive back at port with dry decks. Perhaps he noticed Earth's curved shadow cast on the lunar surface during an eclipse of the Moon. He might have deduced that the Sun and Moon are not just the flat disks they appear to be in the sky but are spherical and that Earth must be a sphere as well.

    Earth's sphericity was generally accepted by the educated populace as early as the first century A.D. Christopher Columbus, for example, knew he was sailing around a sphere in 1492; that is one reason why he thought he had reached the East Indies.

Earth as a Geoid   Until 1687, the spherical-perfection model was a basic assumption of **geodesy**, the science that determines Earth's shape and size by surveys and mathematical calculations. But in that year, Sir Isaac Newton postulated that the round Earth, along with the other planets, could not be perfectly spherical. Newton reasoned that the more-rapid rotational speed at the equator—the equator being farthest from the central axis of the planet and therefore moving faster—would produce an equatorial bulge in response to a greater centrifugal force, which, in effect, pulls Earth's surface outward. He was convinced that Earth is slightly misshapen into what he termed an *oblate spheroid*, or more correctly an *oblate ellipsoid* (*oblate* means "flattened"), with the oblateness occurring at the poles.

Today, Earth's equatorial bulge and its polar oblateness are universally accepted and confirmed by satellite observations. Our modern era of Earth measurement is one of tremendous precision and is called the "geoidal epoch" because Earth is a **geoid**, meaning literally that "the shape of Earth is Earth-shaped." Imagine Earth's geoid as a sea-level surface that is extended uniformly worldwide, beneath the continents. Both heights on land and depths in the oceans are measured from this hypothetical surface. Think of the geoid surface as a balance between the gravitational attraction of Earth's mass and the centrifugal pull caused by Earth's rotation.

Figure 1.9 gives Earth's polar and equatorial circumferences and diameters. Earth's polar circumference was first measured more than 2200 years ago by the Greek geographer, astronomer, and librarian Eratosthenes (ca. 276–195 B.C.). The ingenious reasoning by which he arrived at his calculation follows.

Measuring Earth in 247 B.C.

Foremost among early geographers, Eratosthenes served as the librarian of Alexandria in Egypt during the third century B.C. He was in a position of scientific leadership, for Alexandria's library was the finest in the ancient world. Among his achievements was calculation of Earth's polar circumference to a high level of accuracy, quite a feat for 247 B.C. Here's how he did it. As you read, follow along on Figure 1.10.

Travelers told Eratosthenes that on June 21 they had seen the Sun's rays shine directly to the bottom of a well at Syene, the location of present-day Aswan, Egypt. This meant that the Sun had to be directly overhead. North of Syene in Alexandria, Eratosthenes knew from his own observations that the Sun's rays never were directly overhead, even at noon on June 21. This day is the longest day of the year and the day on which the Sun is at its northernmost position in the sky. Unlike objects in Syene, he knew objects in Alexandria always cast a noontime shadow. Using the considerable geometric knowledge of the era, Eratosthenes conducted an experiment.

In Alexandria at noon on June 21, he measured the angle of a shadow cast by an obelisk (a perpendicular column). Knowing the height of the obelisk and measuring the length of the Sun's shadow from its base, he solved the triangle for the angle of the Sun's rays, which he determined to be 7.2° off from directly overhead. However, at Syene on the same day, the angle of the Sun's rays was 0° from a perpendicular—that is, the Sun was directly overhead.

Geometric principles told Eratosthenes that the distance on the ground between Alexandria and Syene formed an arc of Earth's circumference equal to the angle of the Sun's rays at Alexandria. Since 7.2° is roughly 1/50 of the 360° (360° ÷ 7.2° = 50) in Earth's total circumference, the distance between Alexandria and Syene must represent approximately 1/50 of Earth's total circumference.

Next, Eratosthenes determined the surface distance between the two cities as 5000 stadia. A *stadium*, a Greek unit of measure, equals approximately 185 m (607 ft). He then multiplied 5000 stadia by 50 to determine that Earth's polar circumference is about 250,000 stadia. Eratosthenes'

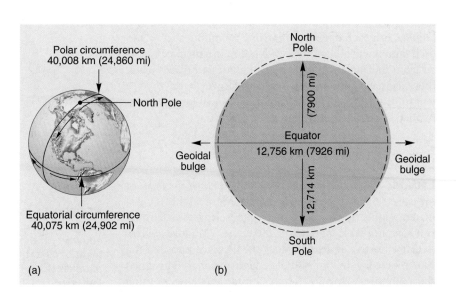

FIGURE 1.9 Earth's dimensions.
Earth's circumference (a) and diameter (b)—equatorial and polar—are shown. The dashed line is a perfect circle for reference to Earth's geoid.

Polar circumference
40,008 km (24,860 mi)

North Pole

Equatorial circumference
40,075 km (24,902 mi)

(a)

North Pole

(7900 mi)

Equator
12,756 km (7926 mi)

Geoidal bulge

12,714 km

Geoidal bulge

South Pole

(b)

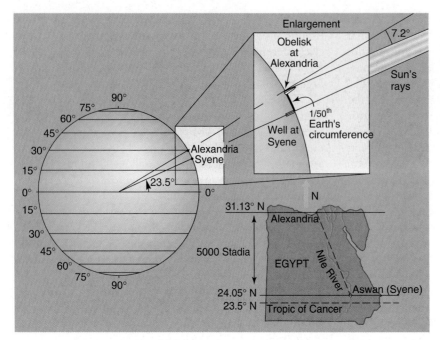

FIGURE 1.10 Eratosthenes' calculation. Eratosthenes' work teaches the value of observing carefully and integrating all observations with previous learning. Calculating Earth's circumference required application of his knowledge of Earth–Sun relationships, geometry, and geography to his keen observations. His estimate was remarkably close to modern measurements. (Several values can be used for the distance of a Greek stadium—the 185 m used here represents an average value.)

calculations convert to roughly 46,250 km (28,738 mi), which is remarkably close to the correct value of 40,008 km (24,860 mi) for Earth's polar circumference. Not bad for 247 B.C.!

Location and Time on Earth

An essential for geographic science is a coordinated grid system to determine location on Earth, a system of coordinates agreed to by all peoples. The terms *latitude* and *longitude* were in use on maps as early as the first century A.D., with the concepts themselves dating back to Eratosthenes and others.

The geographer, astronomer, and mathematician Ptolemy (ca. A.D. 90–168) contributed greatly to modern maps, and many of his terms and configurations are still used today. Ptolemy divided the circle into 360 degrees (360°), with each degree comprising 60 minutes (60′) and each minute including 60 seconds (60″), in a manner adapted from the ancient Babylonians. He located places using these degrees, minutes, and seconds. However, the precise length of a degree of latitude and a degree of longitude remained unresolved for the next 17 centuries.

Latitude

Latitude *is an angular distance north or south of the equator*, measured from the center of Earth (Figure 1.11a). On a map or globe, the lines designating these angles of latitude run east and west, parallel to the equator (Figure 1.11b). Because Earth's equator divides the distance between the North Pole and the South Pole exactly in half, it is assigned the value of 0° latitude. Thus, latitude increases in value

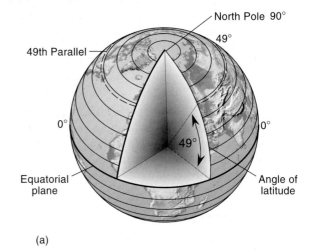

(a)

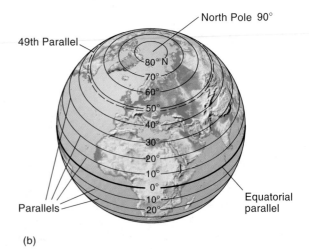

(b)

FIGURE 1.11 Parallels of latitude.
(a) Latitude is measured in degrees north or south of the equator, which is 0°. Each pole is at 90°. Note the measurement of 49° latitude. (b) These angles of latitude determine parallels along Earth's surface. Do you know your present latitude?

from the equator northward to the North Pole, at 90° north latitude, and southward to the South Pole, at 90° south latitude.

A line connecting all points along the same latitudinal angle is called a **parallel**. In the figure, an angle of 49° north latitude is measured, and, by connecting all points at this latitude, we have the 49th parallel. Thus, *latitude* is the name of the angle (49° north latitude), *parallel* names the line (49th parallel), and both indicate distance north of the equator. The 49th parallel is a significant one in the Western Hemisphere, for it forms the boundary between Canada and the United States from Minnesota to the Pacific Ocean.

Latitude is readily determined by reference to *fixed celestial objects* such as the Sun or the stars, a method dating to ancient times. During daylight hours, the angle of the Sun above the horizon indicates the observer's latitude, after adjustment is made for the seasonal tilt of Earth and for the time of day. Because Polaris (the North Star) is almost directly overhead at the North Pole, persons anywhere in the Northern Hemisphere can determine their latitude at night simply by sighting Polaris and measuring its angle above the local horizon (Figure 1.12). The angle of elevation of Polaris above the horizon equals the latitude of the observation point.

In the Southern Hemisphere, Polaris cannot be seen, because it is below the horizon. Instead, latitude measurement south of the equator is accomplished by sighting on a constellation that points to a celestial location above the South Pole. This indicator constellation is the Southern Cross (Crux Australis).

Latitudinal Geographic Zones   Natural environments differ dramatically from the equator to the poles, in both their processes and their appearance. These differences result from the amount of solar energy received, which varies by latitude and season of the year. As a convenience, geographers identify *latitudinal geographic zones* as regions with fairly consistent qualities. Figure 1.13 portrays these zones, their locations, and their names: *equatorial* and *tropical*, *subtropical*, *midlatitude*, *subarctic* or *subantarctic*, and *arctic* or *antarctic*. "Lower latitudes" are those nearer the equator, whereas "higher latitudes" are those nearer the poles. These generalized latitudinal zones are useful for reference and comparison, but don't think of them as having rigid boundaries.

The *Tropic of Cancer* (23.5° north parallel) and the *Tropic of Capricorn* (23.5° south parallel) are the most extreme northern and southern parallels that experience perpendicular (directly overhead) rays of the Sun at local noon. When the Sun arrives overhead at these tropics, it marks the first day of summer in each hemisphere. (The tropics are discussed further in Chapter 2.) The *Arctic Circle* (66.5° north parallel) and the *Antarctic Circle* (66.5° south parallel) are the parallels farthest from the poles that still experience 24 uninterrupted hours of night during local winter or of day during local summer.

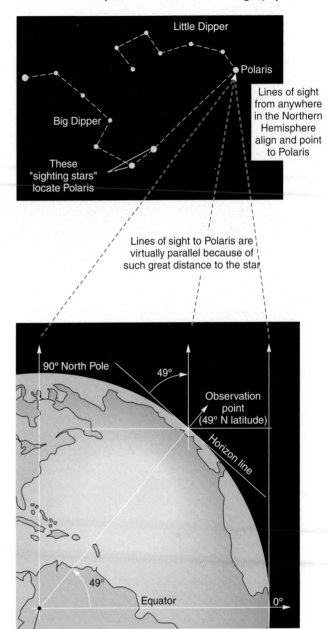

FIGURE 1.12 Determining latitude by using Polaris (the North Star).
To locate Polaris from anywhere in the Northern Hemisphere, you can use the "sighting stars" in the Big Dipper constellation. Next, measure the angular distance between Polaris and the local horizon. This angular distance of Polaris above the horizon is the same as your latitude. In this case Polaris appears 49° above the horizon, so you are standing at 49° north latitude. (Note that Polaris is at such a great distance from Earth that lines of sight from anywhere in the Northern Hemisphere can be considered parallel.) On the next clear night, take a protractor and stick or ruler and sight on Polaris to determine your latitude manually.

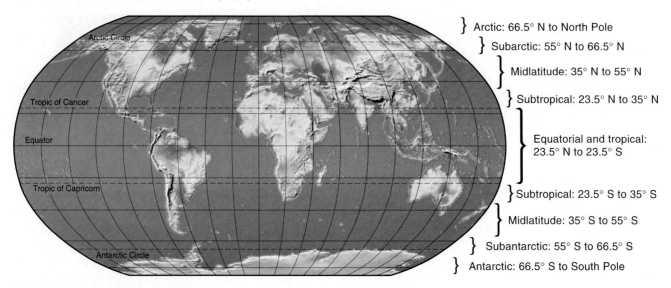

FIGURE 1.13 Latitudinal geographic zones.
Geographic zones are generalizations that characterize various regions by latitude. Think of these as transitional into one another over broad areas.

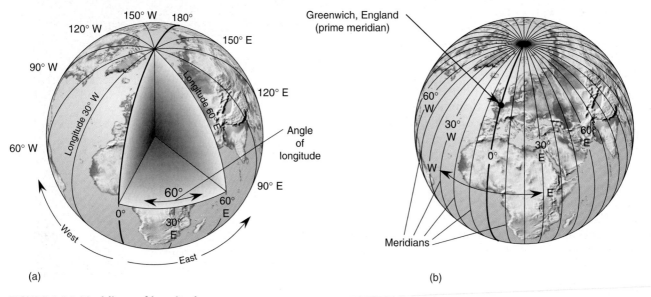

(a)

(b)

FIGURE 1.14 Meridians of longitude.
(a) Longitude is measured in degrees east or west of a 0° starting line, the prime meridian. Note the measurement of 60° E longitude. (b) Angles of longitude measured from this prime meridian determine other meridians. The prime meridian is drawn from the North Pole through the Royal Observatory in Greenwich, England, to the South Pole. North America is west of Greenwich; therefore, it is in the Western Hemisphere. Do you know your present longitude?

Longitude

Longitude *is an angular distance east or west of a point on Earth's surface*, measured from the center of Earth (Figure 1.14a). On a map or globe, the lines designating these angles of longitude run north and south (Figure 1.14b). A line connecting all points along the same longitude is a **meridian**. In the figure a longitudinal angle of 60° E is measured. These meridians run at right angles (90°) to all parallels, including the equator. Thus, *longitude* is the name of the angle, *meridian* names the line, and both indicate dis-

tance east or west of an arbitrary **prime meridian**—a meridian designated as zero degrees (Figure 1.14b). Earth's prime meridian passes through the old Royal Observatory at Greenwich, England, as set by treaty—the *Greenwich prime meridian.*

Determination of Latitude and Longitude Table 1.2 compares the length of latitude and longitude degrees. Because meridians of longitude converge toward the poles, the actual distance on the ground spanned by a degree of

Table 1.2	Physical Distances Represented by Degrees of Latitude and Longitude			
Latitudinal Location	Latitude Degree Length		Longitude Degree Length	
	km	(mi)	km	(mi)
90° (poles)	111.70	(69.41)	0	(0)
60°	111.42	(69.23)	55.80	(34.67)
50°	111.23	(69.12)	71.70	(44.55)
40°	111.04	(69.00)	85.40	(53.07)
30°	110.86	(68.89)	96.49	(59.96)
0° (equator)	110.58	(68.71)	111.32	(69.17)

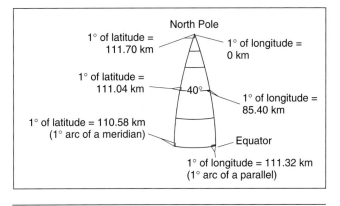

longitude is greatest at the equator (where meridians separate to their widest distance apart) and diminishes to zero at the poles (where meridians converge). Note the consistent distance represented by a degree of latitude from equator to poles, yet the decreasing value a degree of longitude covers as meridians converge toward each pole.

We have noted that latitude is easily determined by sighting the Sun or the North Star or by using the Southern Cross as a pointer. In contrast, a method of accurately determining longitude, especially at sea, remained a major difficulty in navigation until the late 1700s. The key to measuring the longitude of a place lies in accurately knowing time. The relation between time and longitude and an exciting chapter in human discovery is the topic of Focus Study 1.2.

Today, using a hand-held instrument that reads radio signals from satellites, you can accurately calibrate latitude, longitude, and elevation. This is the **Global Positioning System (GPS)** technology. News Report 1.1 discusses the dramatic applications of the GPS.

Great Circles and Small Circles

Great circles and small circles are important concepts that help summarize latitude and longitude (Figure 1.15). A **great circle** is any circle of Earth's circumference whose

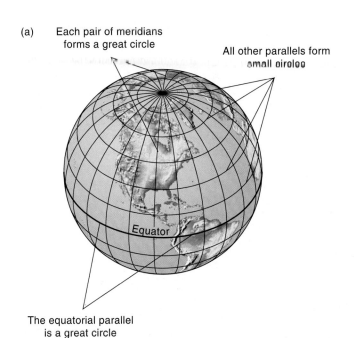

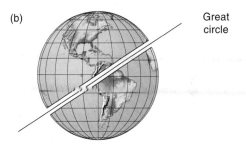

A plane intersecting the globe along a great circle divides the globe into equal halves and passes through its center

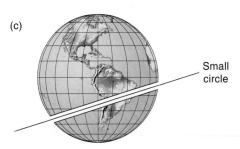

A plane that intersects the globe along a small circle splits the globe into unequal sections—this plane does not pass through the center of the globe

FIGURE 1.15 Great circles and small circles.
(a) Examples of great circles and small circles on Earth. (b) Any plane that divides Earth into equal halves will intersect the globe along a great circle; this great circle is a full circumference of the globe and is the shortest distance between any two surface points. (c) Any plane that splits the globe into unequal portions will intersect the globe along a small circle.

The Timely Search for Longitude

Unlike latitude, longitude cannot be determined readily from fixed celestial bodies. The problem is Earth's rotation, which constantly changes the apparent position of the Sun and stars. Determining longitude is particularly critical at sea, where no landmarks are visible.

In his historical novel *Shogun*, author James Clavell expressed the frustration of the longitude problem through his pilot, Blackthorn: "Find how to fix longitude and you're the richest man in the world. . . . The Queen, God bless her, 'll give you ten thousand pound and dukedom for answer to the riddle. . . . Out of sight of land you're always lost, lad."*

In the early 1600s, Galileo explained that longitude could be measured by using two clocks. Any point on Earth takes 24 hours to travel around the full 360° of one rotation (one day). If you divide 360° by 24 hours, you find that any point on Earth travels through 15° of longitude every hour. Thus, if there were a way to measure time accurately at sea, a comparison of two clocks could give a value for longitude. One clock would indicate the time back at home port (Figure 1). The other clock would be reset at local noon each day, as determined by the highest Sun position in the sky (solar zenith). The time difference then would indicate the longitudinal difference traveled: 1 hour for each 15° of longitude. The principle was sound; all that was needed were accurate clocks. Unfortunately, the pendulum clock invented by Christian Huygens in 1656 did not work on the rolling deck of a ship at sea!

In 1707 the British lost four ships and 2000 men in a sea tragedy that was blamed specifically on the longitude problem. In response, Parliament passed an act in 1714—"Publik Reward . . . to Discover the Longitude at Sea"—and authorized a prize worth

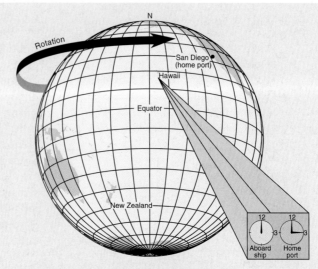

FIGURE 1 Clock time determines longitude.
Using two clocks on a ship to determine longitude. For example, if the shipboard clock reads local noon and the clock set for home port reads 3:00 P.M., ship time is 3 hours earlier than home time. Therefore, calculating 3 hours at 15° per hour puts the ship at 45° west longitude from home port.

more than $2 million in today's dollars to the first successful inventor of an accurate seafaring clock. The Board of Longitude was established to judge any devices submitted.

John Harrison, a self-taught country clockmaker, began work on the problem in 1728 and finally produced his brilliant marine chronometer, known as Number 4, in 1760. The clock was tested on a voyage to Jamaica in 1761. When taken ashore and compared to land-based longitude, Harrison's ingenious Number 4 was only 5 seconds slow, an error that translates to only 1.25′ or 2.3 km (1.4 mi), well within Parliament's standard. After many delays, Harrison finally received most of the prize money in his last years of life.

With his marine clocks, John Harrison tested the waters of space–time. He succeeded, against all odds, in using the fourth—temporal—dimension to link points

on the three-dimensional globe. He wrested the world's whereabouts from the stars, and locked the secret in a pocket watch.†

From that time it was possible to determine longitude accurately on land and sea, as long as everyone agreed upon a meridian to use as a reference for time comparisons—the Royal Observatory in Greenwich, England. In this modern era of atomic clocks and satellites in mathematically precise orbits, we have far greater accuracy available for the determination of longitude on Earth's surface and a basis for precise navigation.

*From J. Clavell, *Shogun*, Copyright © 1975 by James Clavell (New York: Delacorte Press, a division of Dell Publishing Group, Inc.), p. 10.
†From Dava Sobel, *Longitude, The Story of a Lone Genius Who Solved the Greatest Scientific Problem of His Time* (New York: Walker and Co., 1995), p. 175.

News Report 1.1

GPS: A Personal Locator

The Global Positioning System (GPS) comprises 24 orbiting satellites, in six orbital planes, that transmit navigational signals for Earth-bound use (backup GPS satellites are in orbital storage as replacements). Originally devised in the 1970s by the Department of Defense for military purposes, the present system is commercially available worldwide.

A small receiver, some about the size of a pocket radio, receives signals from four or more satellites at the same time, calculates latitude and longitude within 10-m accuracy (33 ft) and elevation within 15 m (49 ft), and displays the results. With the shutdown in 2000 of the Pentagon Selective Availability, commercial resolution is the same as for military applications and its Precise Positioning Service (PPS). *Differential GPS (DGPS)* increases accuracy by comparing readings with another base station (reference receiver) for differential correction (Figure 1).

GPS is useful for diverse applications, such as ocean navigation, land surveying, tracking small changes in Earth's crust, managing the movement of fleets of trucks, mining and resource mapping, tracking wildlife migration and behavior, and environmental planning. Relative to earthquakes in southern California, JPL operates the GPS Observation Office that monitors a network of 250 seismic stations. GPS also is useful to the backpacker and sportsperson. Commercial airlines use GPS to improve accuracy of routes flown and thus increase fuel efficiency. Scientists used GPS to accurately determine the height of Mount Everest in the Himalayan Mountains—now 8850 m compared to the former 8848 m (29,035 ft, 29,028 ft) (Figure 2). In contrast, GPS measurements of Mount Kilimanjaro lowered its summit from 5895 m to a lower 5892 m (19,340 ft, 19,330 ft).

FIGURE 1 GPS in action.
A GPS unit in operation for surveying and location analysis in an environmental study. [Photo courtesy of Trimble Navigation Ltd., Sunnyvale, California.]

FIGURE 2 GPS used to measure Everest's summit height.
Installation of a Trimble 4800 GPS unit at Earth's highest bench mark, only 18 m (60 ft) below the 8850 m (29,035 ft) summit of Mount Everest. This is a GPS-based revised height measurement announced in 1999. Scientists are using the data collected from a regional GPS network to accurately analyze the height of the world's tallest mountain and the rate that the mountain range is moving due to tectonic forces. [Photo and GPS installation by Wally Berg, May 20, 1998.]

Farmers use GPS to determine crop yields on specific parts of their farms. A detailed plot map is made to guide the farmer to where more fertilizer, proper seed distribution, irrigation applications, or other work is needed. A computer and GPS unit on board the farm equipment guides the work. This is the science of *variable-rate technology*,

(continued)

News Report 1.1 *(continued)*

made possible by GPS. Hawkeye Community College in Waterloo, Iowa, offers a certificate program in Precision Agriculture. (See http://www.ag.hawkeye.cc.ia.us/PrecisAg.htm.)

The importance of GPS to geography is obvious because this precise technology reduces the need to maintain ground control points for location,

mapping, and spatial analysis. Instead, geographers working in the field can determine their position accurately as they work. Boundaries and data points in a study area are easily determined and entered into a data base, reducing the need for traditional surveys. For this and myriad other applications, GPS sales are exceeding $10 billion a

year. As additional frequencies are added in 2003 and 2006, accuracy will increase significantly. Also, the European Union plans to launch its own GPS system of 20 satellites beginning in 2005. (For a GPS overview, see http://www.colorado.edu/geography/gcraft/notes/gps/gps_f.html.)

center coincides with the center of Earth. An infinite number of great circles can be drawn on Earth. Every meridian is one-half of a great circle that passes through the poles. On flat maps, airline and shipping routes appear to arch their way across oceans and landmasses. These are *great circle routes*, the shortest distance between two points on Earth (discussion with Figure 1.22 is just ahead).

Only one parallel is a great circle—the equatorial parallel. All other parallels diminish in length toward the poles and, along with any other non-great circles that one might draw, constitute **small circles**. These circles have centers that do not coincide with Earth's center.

Figure 1.16 combines latitude and parallels with longitude and meridians to illustrate Earth's complete coordinate grid system. Note the dot that marks our measurement of 49° N by 60° E, a location in western Kazakstan. Next time you see a world globe, follow the parallel and meridian that converge on your location.

Prime Meridian and Standard Time

Coordination of international trade, airline schedules, business and agricultural activities, and daily living depends on a worldwide time system. Today we take for granted standard time zones and an agreed-upon prime meridian, but such a standard is a relatively recent development.

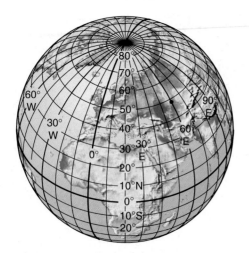

FIGURE 1.16 Earth's coordinate grid system.
Latitude and parallels, longitude and meridians, allow all places on Earth to be precisely located. The dot is at 49° N latitude by 60° E longitude.

Setting time was not a great problem in small European countries, most of which are less than 15° wide. But in North America, which spans more than 90° of longitude (the equivalent of six 15° time zones), the problem was serious. In 1870 railroad travelers going from Maine to San Francisco made 22 adjustments to their watches to stay consistent with local time!

In Canada, Sir Sanford Fleming led the fight for standard time and for an international agreement on a prime meridian. His struggle led the United States and Canada to adopt a standard time in 1883. Today, only three adjustments are needed in the continental United States—from Eastern Standard Time to Central, Mountain, and Pacific—and four changes across Canada.

Twenty-seven countries attended the 1884 International Meridian Conference in Washington, DC. Before that year, most nations used their own national capital as a prime meridian for their land maps, whereas more than 70% of the world's merchant ships were using Greenwich as a prime meridian on marine charts. After lengthy debate at the conference, most participating nations chose the highly respected Royal Observatory at Greenwich, London, England, as the place for the prime meridian of 0° longitude for all maps. Thus, a world standard was set— **Greenwich Mean Time (GMT)**—and a consistent Universal Time was established. (See http://www.gmt2000.co.uk/meridian/place/plco0a1.htm.)

The basis of time is that Earth revolves 360° every 24 hours, or 15° per hour (360° ÷ 24 = 15°). Thus, a time zone of 1 hour is established for each 15° increment of longitude, or 7.5° on either side of a *central meridian*. Assuming it is 9:00 P.M. in Greenwich, then it is 4:00 P.M. in Baltimore (+5 hrs), 3:00 P.M. in Oklahoma City (+6 hrs), 2:00 P.M. in Salt Lake City (+7 hrs), 1:00 P.M. in Seattle and Los Angeles (+8 hrs), noon in Anchorage (+9 hrs), and 11:00 A.M. in Honolulu (+10 hrs). To the east, it is midnight in Ar Riyāḍ, Saudi Arabia (−3 hrs). (The designation A.M. is for *ante meridiem*, "before noon," whereas P.M. is for *post meridiem*, "after noon." A 24-hour clock avoids the use of these designations.)

As you can see from the modern international time zones in Figure 1.17, national boundaries and political considerations distort time zone boundaries. For example, China spans four time zones, but its government decided to keep the entire country operating at the same time.

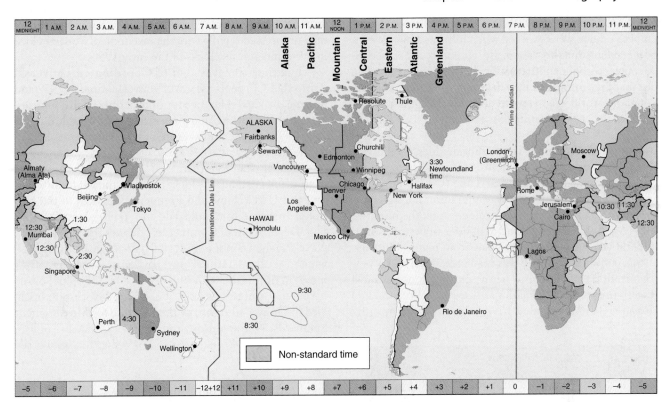

FIGURE 1.17 Modern international standard time zones.
Numbers along the bottom of the map indicate how many hours each zone is earlier (plus sign) or later (minus sign) than Coordinated Universal Time (UTC) at the prime meridian. The United States has five time zones; Canada is divided into six. If it is 7 P.M. in Greenwich, determine the present time in Moscow, London, Halifax, Chicago, Winnipeg, Denver, Los Angeles, Fairbanks, Honolulu, Tokyo, and Singapore. The island country of Kiribati moved the International Date Line to its eastern margin (150° west longitude) to be the first to experience each new day. These distortions of the IDL only apply to the countries and their territorial waters and not to international waters between them and the 180th meridian. [Adapted from Standard Time Zone Chart of the World, Defense Mapping Agency, Bethesda, Maryland.]

Thus, in some parts of China clocks are several hours off from what the Sun is doing. In the United States parts of Florida and west Texas are in the same time zone.

International Date Line   An important corollary of the prime meridian is the 180° meridian on the opposite side of the planet. This meridian is called the **International Date Line** and marks the place where each day officially begins (at 12:01 A.M.). From this "line" the new day sweeps westward. This *westward* movement of time is created by Earth's turning *eastward* on its axis.

At the International Date Line, the west side of the line is always one day ahead of the east side. No matter what time of day it is when the line is crossed, the calendar changes a day (Figure 1.18). Note in the illustration how the IDL deviates from the 180° meridian; this deviation is due to local administrative and political preferences, as described in the caption to Figure 1.19.

Locating the date line in the sparsely populated Pacific Ocean minimizes most local confusion. However, early explorers before the date-line concept were "lost." For example, Magellan's crew returned from the first circumnavigation of Earth in 1522, confident from their ship's log

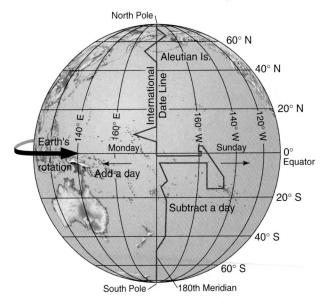

FIGURE 1.18 International Date Line.
International Date Line location, approximately along the 180th meridian (see the IDL location on Figure 1.17). Note that it is officially one day later west of the IDL.

that the day of their arrival home was Wednesday, September 7. They were shocked when informed by insistent local residents that it was actually Thursday, September 8! Of course, without an International Date Line they had no idea that they must advance a day somewhere when sailing around the world in a westward direction. Imagine the confusion as the crew accounted for each day in their log!

Coordinated Universal Time   For decades, Greenwich Mean Time from the Royal Observatory's astronomical clocks was the world's Universal Time (UT) standard for accuracy. GMT was broadcast using radio time signals as early as 1910. The French government took the initiative in 1912 and called a gathering of nations to better coordinate the various radio time signals from many countries. At this conference, GMT was made standard, and a new organization established to be the custodian of the most "exact" time—the International Bureau of Weights and Measures (BIPM) outside Paris (see **http://www.bipm.fr/enus/welcome.html**). Progress in accurately measuring time progressed rapidly with the invention of a quartz clock in 1939 and atomic clocks in the early 1950s.

The time signal system of **Coordinated Universal Time (UTC*)** replaced GMT universal time in 1972 and became the legal reference for official time in all countries. Although the prime meridian still runs through Greenwich, UTC is based on average time calculations collected by the BIPM near Paris and broadcast worldwide.

Regular vibrations (natural frequency) of cesium atoms in six primary standard clocks measure the length of a second and UTC. Time and Frequency Services of the National Institute for Standards and Technology (NIST), U.S. Department of Commerce, operates the newest clock *NIST-F1*, shown in Figure 1.19. *NIST-F1* replaced *NIST-7*, which is still operational and participates in UTC. (For more on time, call 303-499-7111 or 808-335-4363; or see **http://nist.time.gov/** for UTC.) Three clocks are operated in Ottawa, Ontario, by the Institute for Measurement Standards, National Research Council Canada (English, 613-745-1578; French, 613-745-9426; for more see **http://www.nrc.ca/**).

Daylight Saving Time   In many countries, time is set ahead 1 hour in the spring and set back 1 hour in the fall—a practice known as **daylight saving time**. The idea to extend daylight for early evening activities (at the expense of daylight in the morning) was first proposed by Benjamin Franklin. It was not adopted until World War I and again in World War II, when Great Britain, Australia, Germany, Canada, and the United States used the practice to save energy (one less hour of artificial lighting needed).

In 1986 the United States and Canada increased daylight saving time. Time now "springs forward" 1 hour on the first Sunday in April and "falls back" an hour on the last Sunday in October, except in a few places that do not use daylight saving time (Hawai'i, Arizona, portions of Indiana, and Saskatchewan). In Europe, the last Sundays in March and October are used to begin and end what is called "summer time." (See **http://deil.lang.uiuc.edu/web.pages/holidays/DST.html**.)

FIGURE 1.19 The United States primary standard clock, *NIST-F1*.
Located in Boulder, Colorado, *NIST-F1* calculates time by tracking the vertical flow of laser-cooled cesium atoms through a detector. New in 2000, *NIST-F1* is an important part of the determination of UTC. Accuracy is down to one second in 20 million years, more than three times more accurate than the previous *NIST-7*. See **http://www.boulder.nist.gov/timefreq/cesium/fountain.htm**. [Photo by Bobbé Christopherson.]

Maps, Scales, and Projections

The earliest known graphic map presentations date to 2300 B.C., when the Babylonians used clay tablets to record information about the region of the Tigris and Euphrates Rivers (the area of modern-day Iraq). Today, the making of maps and charts is a specialized science as well as an art, blending aspects of geography, engineering, mathematics, graphics, computer science, and artistic specialties. It is similar in ways to architecture, in which aesthetics and utility combine to produce a useful product.

*UTC is in use because agreement was not reached on whether to use the English word order, CUT, or the French order, TUC. UTC was the compromise and is recommended for all timekeeping applications; use of the term GMT is discouraged.

A **map** is a generalized view of an area, usually some portion of Earth's surface, as seen from above and greatly reduced in size. The part of geography that embodies mapmaking is called **cartography**. Maps are critical tools with which geographers depict spatial information and analyze spatial relationships.

We all use maps at some time to visualize our location and our relationship to other places, or maybe to plan a trip, or to coordinate commercial and economic activities. Have you found yourself looking at a map, planning real and imagined adventures to far-distant places? Maps are wonderful tools! Understanding a few basics about maps is essential to our study of physical geography.

The Scale of Maps

Architects, toy designers, and mapmakers have something in common: they all create scale models. They reduce real things and places to the more convenient scale of a drawing, a model car, train, or plane, a diagram, or a map. An architect renders a blueprint of a structure to guide the building contractors, selecting a scale so that one centimeter (or inch) on the drawing represents so many meters (or feet) on the proposed building. Often, the drawing is 1/50 to 1/100 of real size.

The cartographer does the same thing in preparing a map. The ratio of the image on a map to the real world is called **scale**; it relates a unit on the map to a similar unit on the ground. A 1:1 scale means that a centimeter on the map represents a centimeter on the ground (although this is certainly an impractical map scale, for the map is as large as the area mapped!). A more appropriate scale for a local map is 1:24,000, in which 1 unit on the map represents 24,000 identical units on the ground.

Map scales are presented in several ways: as a written scale, a representative fraction, or a graphic scale (Figure 1.20). A *written scale* simply states the ratio—for example, "one centimeter to one kilometer" or "one inch to one mile." A *representative fraction* (*RF*, or *fractional scale*) is expressed with either a colon or a slash, as in 1:125,000 or 1/125,000. No actual units of measurement are mentioned because any

Table 1.3 Sample Representative Fractions and Written Scales for Small-, Medium-, and Large-Scale Maps

System	Scale Size	Representative Fraction	Written Scale
English	Small	1:3,168,000	1 in. = 50 mi
		1:2,500,000	1 in. = 40 mi
		1:1,000,000	1 in. = 16 mi
		1:500,000	1 in. = 8 mi
		1:250,000	1 in. = 4 mi
	Medium	1:125,000	1 in. = 2 mi
		1:63,360	1 in. = 1 mi
		(or 1:62,500)	
		1:31,680	1 in. = 0.5 mi
		1:30,000	1 in. = 2500 ft
	Large	1:24,000	1 in. = 2000 ft

System	Representative Fraction	Written Scale
Metric	1:1,000,000	1 cm = 10.0 km
	1:50,000	1 cm = 0.50 km
	1:25,000	1 cm = 0.25 km
	1:20,000	1 cm = 0.20 km

unit is applicable as long as both parts of the fraction are in the same unit: 1 cm to 125,000 cm, 1 in. to 125,000 in., or even 1 arm length to 125,000 arm lengths, and so on.

A *graphic scale*, or bar scale, is a bar graph with units to allow measurement of distances on the map. An important advantage of a graphic scale is that, if the map is enlarged or reduced, the graphic scale enlarges or reduces along with the map. In contrast, written and fractional scales become incorrect with enlargement or reduction. As an example, you can shrink a map from 1:24,000 to 1:63,360, but the scale will still say "1 in. to 2000 ft," instead of the new correct scale of 1 in. to 5280 ft (1 mi).

Scales are *small*, *medium*, and *large*, depending on the ratio described. In relative terms, a scale of 1:24,000 is a large scale, whereas a scale of 1:50,000,000 is a small scale. The greater the denominator in a fractional scale (or the number on the right in a ratio expression), the smaller the scale and the more abstract the map is in relation to what is being mapped. Examples of selected representative fractions and written scales are listed in Table 1.3 for small-, medium-, and large-scale maps. In Chapter 12, a small-scale image of a portion of the Mid-Atlantic states is at a 1:4,000,000 scale. Compare to the smaller-scale image inside the front cover which is at a scale of 1:80,00,000. Note the increased detail at the larger scale (see Figure 12.18).

If there is a globe or map available in your library or classroom, check to see the scale at which it was drawn. See if you can find examples of written, representative, and

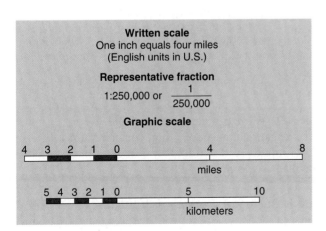

FIGURE 1.20 Map scale.
Three common expressions of map scale—written scale, representative fraction, and graphic scale.

graphic scales on wall maps, highway maps, and in atlases. In general, do you think a world globe is a small- or a large-scale map of Earth's surface?

Map Projections

A globe is not always a helpful map representation of Earth. When you go on a trip, you need more detailed information than a globe can provide. Consequently, to provide local detail, cartographers prepare large-scale flat maps, which are two-dimensional representations (scale models) of our three-dimensional Earth. Unfortunately, such conversion from three dimensions to two causes distortion.

A globe is the only true representation of *distance, direction, area, shape,* and *proximity.* A flat map distorts those properties. Therefore, in preparing a flat map, the cartographer must decide which characteristic to preserve, which to distort, and how much distortion is acceptable.

To understand this problem, consider these important properties of a *globe*:

- Parallels always are parallel to each other, always are evenly spaced along meridians, and always decrease in length toward the poles.
- Meridians converge at both poles and are evenly spaced along any individual parallel.
- The distance between meridians decreases toward poles, with the spacing between meridians at the 60th parallel equal to one-half the equatorial spacing.
- Parallels and meridians always cross each other at right angles.

The problem is that all these qualities cannot be reproduced on a flat surface. Simply taking a globe apart and laying it flat on a table illustrates the challenge faced by cartographers (Figure 1.21). You can see the empty spaces that open up between the sections, or gores, of the globe. This reduction of the spherical Earth to a flat surface is called a **map projection**. Thus, no flat map projection of Earth can ever have all the features of a globe. Flat maps always possess some degree of distortion—much less for large-scale maps representing a few kilometers; much more for small-scale maps covering individual countries, continents, or the entire world.

Properties of Projections There are many projections, four of which are shown in Figure 1.22. *The best projection is always determined by its intended use.* The major decisions in selecting a map projection involve the properties of **equal area** (equivalence) and **true shape** (conformality). If a cartographer selects equal area as the desired trait—for example, for a map showing the distribution of world climates—then true shape must be sacrificed by *stretching* and *shearing,* which allows parallels and meridians to cross at other than right angles. On an equal-area map, a coin covers the same amount of surface area no matter where you place it on the map. If, on the other hand, a cartogra-

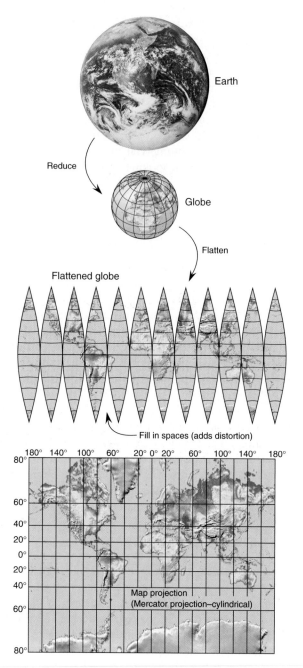

FIGURE 1.21 From globe to flat map.
Conversion of the globe to a flat map projection requires decisions about which properties to preserve and the amount of distortion that is acceptable. [NASA astronaut photo.]

pher selects the property of true shape, as for a map used for navigational purposes, then equal area must be sacrificed, and the scale will actually change from one region of the map to another.

The Nature and Classes of Projections Despite the fact that modern cartographic technology uses mathematical constructions and computer-assisted graphics, the word *projection* is still used. The term comes from times past,

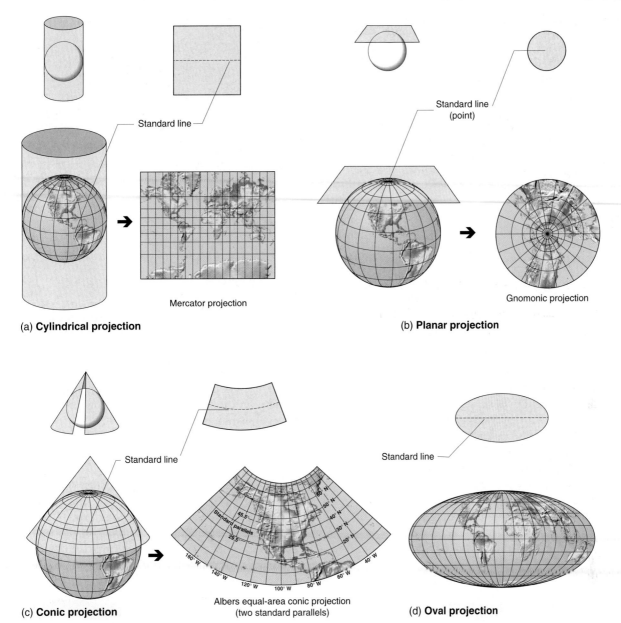

(a) **Cylindrical projection**

Standard line

Mercator projection

(b) **Planar projection**

Standard line
(point)

Gnomonic projection

(c) **Conic projection**

Standard line

Albers equal-area conic projection
(two standard parallels)

(d) **Oval projection**

Standard line

FIGURE 1.22 Classes of map projections.
Four general classes and perspectives of map projections—cylindrical, planar, conic, and oval projections.

when geographers actually projected the shadow of a wire-skeleton globe onto a geometric surface. The wires represented parallels, meridians, and outlines of the continents. A light source then cast a shadow pattern of latitude and longitude lines from the globe onto various geometric surfaces, such as a *cylinder*, *plane*, or *cone*.

Figure 1.22 illustrates the derivation of the general classes of map projections and the perspectives from which they are generated. The classes shown include the cylindrical, planar (or azimuthal), and conic. Another class of projections, which cannot be derived from this physical-perspective approach, is the nonperspective oval shape. Still other projections are derived from purely mathematical calculations.

With projections, the contact line or contact point between the wire globe and the projection surface—called a *standard line* or *standard point*—is *the only place where all globe properties are preserved*. Thus, a *standard parallel* or *standard meridian* is a standard line true to scale along its entire length without any distortion. Areas away from this critical tangent line or point become increasingly distorted. Consequently, this area of optimum spatial properties should be centered on the region of interest so that greatest accuracy is preserved there.

The commonly used **Mercator projection** (from Gerardus Mercator, A.D. 1569) is a cylindrical projection (Figure 1.22a). The Mercator is a true-shape projection, with meridians appearing as equally spaced straight lines

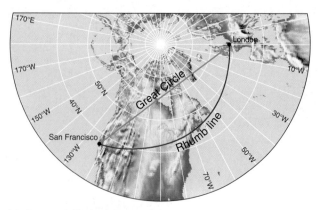

(a) Gnomonic Projection

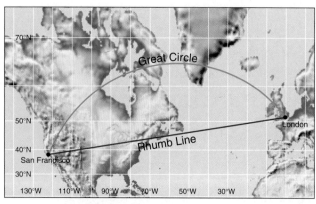

(b) Mercator Projection (conformal, true shape)

FIGURE 1.23 Determining great circle routes.
A gnomonic projection (a) is used to determine the shortest distance—great circle route—between San Francisco and London, because on this projection the arc of a great circle is a straight line. This great circle route is then plotted on a Mercator projection (b), which has true compass direction. Note that straight lines of constant direction (or bearing) on a Mercator projection—rhumb lines—are not the shortest route in terms of distance.

and parallels appearing as straight lines that are spaced closer together near the equator. The poles are infinitely stretched, with the 84th north parallel and 84th south parallel fixed at the same length as that of the equator. Note in Figures 1.21 and 1.22a that the Mercator projection is cut around the 80th parallel in each hemisphere because of the severe distortion at higher latitudes.

Unfortunately, Mercator classroom maps present false notions of the size (area) of midlatitude and poleward landmasses. A dramatic example on the Mercator projection is Greenland, which looks bigger than all of South America. In reality, Greenland is only one-eighth the size of South America and is actually 20% smaller than Argentina alone!

The advantage of the Mercator projection is that lines of constant direction, called **rhumb lines**, are straight and thus facilitate plotting directions between two points (see Figure 1.23). Thus, the Mercator projection is useful in navigation and is the standard for nautical charts prepared by the National Ocean Service since 1910 (formerly U.S. Coast and Geodetic Survey).

The *gnomonic*, or *planar projection* in Figure 1.22b is generated by projecting a light source at the center of a globe onto a plane that is tangent to (touching) the globe's surface. The resulting severe distortion prevents showing a full hemisphere on one projection. However, a valuable feature is derived: All great circle routes, which are the shortest distance between two points on Earth's surface, are projected as straight lines (Figure 1.23a). The great circle routes plotted on a gnomonic projection then can be transferred to a true-direction projection, such as the Mercator, for determination of precise compass headings (Figure 1.23b).

For more information on maps used in this text and standard map symbols, turn to Appendix A, "Maps in This Text and Topographic Maps." Topographic maps are essential tools of landscape analysis. Geographers, other scientists, travelers, and anyone visiting the outdoors may use topographic maps. U.S. Geological Survey (USGS) topographic maps appear in several chapters of this text because they are useful in depicting the tremendously varied features of the physical landscape. Perhaps you have used a "topo" map in planning a hike.

Remote Sensing and GIS

Geographers probe, analyze, and map our home planet through remote sensing and geographic information systems (GIS). These technologies enhance our understanding of Earth. Geographers use remote-sensing data to study humid and arid lands, natural and economic vegetation, snow and ice, Earth energy budgets, seasonal variation of atmospheric and oceanic circulation, sea-level measurements, atmospheric chemistry, geologic features and events, changes in the timing of seasons, and the human activities that produce global change.

Remote Sensing

In this era of observations from orbit outside the atmosphere and from aircraft within it, scientists obtain a wide array of remotely sensed data (Figure 1.24). Remote sensing is nothing new to humans; we do it with our eyes all the time. When we scan the environment with our eyes, we are sensing the shape, size, and color of objects from a distance, registering energy from the visible-wavelength portion of the electromagnetic spectrum. Similarly, when a camera views the wavelengths for which its film or sensor is designed (visible light or infrared), it remotely senses energy that is reflected or emitted from a scene.

Our eyes and cameras are familiar means of obtaining **remote-sensing** information about a distant subject without having physical contact. Aerial photographs have been used for years to improve the accuracy of surface

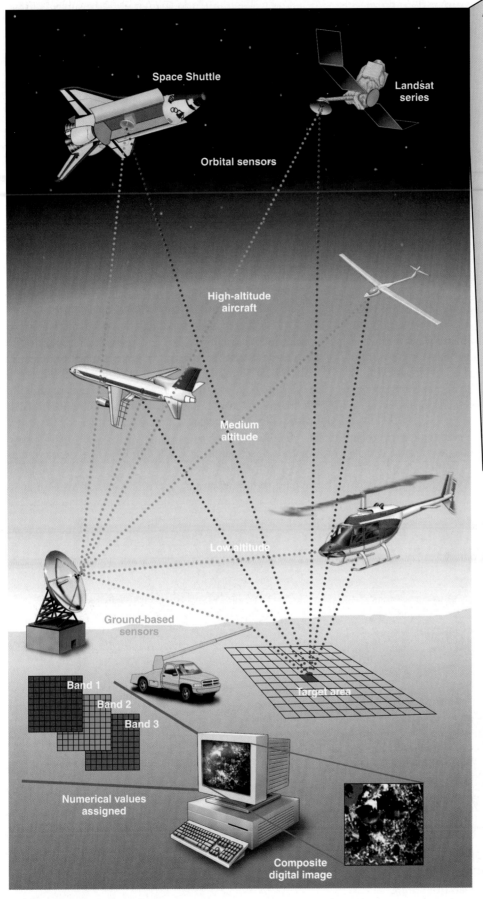

FIGURE 1.24 Remote-sensing technologies.
Remote-sensing technology is used to measure and monitor Earth's systems from orbiting spacecraft, aircraft in the atmosphere, and ground-based sensors. Various wavelengths (bands) are collected from sensors. Computers process the data to produce digital images for analysis. Many of the physical systems discussed in this text are studied using this technology. A sample of remote sensing platforms is in the margin. The Space Shuttle is shown in its inverted orbital flight mode. (Illustration is not to scale.)

maps faster and more cheaply than can be done by on-site surveys. Deriving accurate measurements from photographs is the realm of **photogrammetry**, an important application of remote sensing.

Remote sensors on satellites, the International Space Station, and other craft sense a broader range of wavelengths than can our eyes. They can be designed to "see" wavelengths shorter than visible light (ultraviolet) and wavelengths longer than visible light (infrared and microwave radar).

Satellites do not take conventional-film photographs. Rather, they record *images* that are transmitted to Earth-based receivers in a manner similar to television satellite transmissions, or a digital camera. A scene is scanned and broken down into pixels (*pic*ture *el*ements) each identified by coordinates named *lines* (horizontal rows) and *samples* (vertical columns). For example, a grid of 6000 lines and 7000 samples forms 42,000,000 pixels, providing great detail. The large amount of data needed to produce a single image requires computer processing and data storage at ground stations.

Digital data are processed in many ways to enhance their utility: simulated natural color, "false" color to highlight a particular feature, enhanced contrast, signal filtering, and different levels of sampling and resolution. Active and passive are two types of remote-sensing systems.

Active Remote Sensing   Active systems direct a beam of energy at a surface and analyze the energy reflected back. An example is *radar* (*ra*dio *d*etection *a*nd *r*anging). A radar transmitter emits short bursts of energy that have relatively long wavelengths (0.3 to 10 m) toward the subject terrain, penetrating clouds and darkness. Energy reflected back, known as *backscatter*, is received by a radar receiver and analyzed. An example is the computer image of wind and sea-surface patterns over the Pacific in Figure 6.6a, developed from 150,000 radar-derived measurements made on a single day by the *Seasat* satellite.

In addition, NASA sent imaging radar systems into orbit on several Space Shuttles. The subjects of study included oceanography, landforms and geology, and biogeography. Shuttle missions in 1994 by *Endeavour* and *Atlantis* marked dramatic contributions to Earth observations using radar and other sensors to study stratospheric ozone, weather, volcanic activity, earthquakes, and water resources, among many subjects.

One Space Shuttle mission in September 1994 was appropriately loaded with radar sensors to study volcanoes. Only 8 hours after launch, the Kliuchevskoi Volcano on the Kamchatka Peninsula of Russia erupted unexpectedly. Previously this volcano had erupted in 1737 and 1945. The shuttle radar was able to see through ash and smoke and expose lava flows and the volcanic eruption in dramatic images (Figure 1.25). Astronaut Mission Specialist Dr. Thomas Jones operated the radar and camera to make the image and photo in the figure. He is profiled in a Career Link at the end of this chapter. Chapter 12 discusses volcanic processes.

The European Space Agency (ESA) now operates two Earth resource satellites (*ERS 1* and *2*). They work in tandem, producing a spectacular 10-cm (3.9-in.) resolution, imaging the same area at different times. Pairs of images

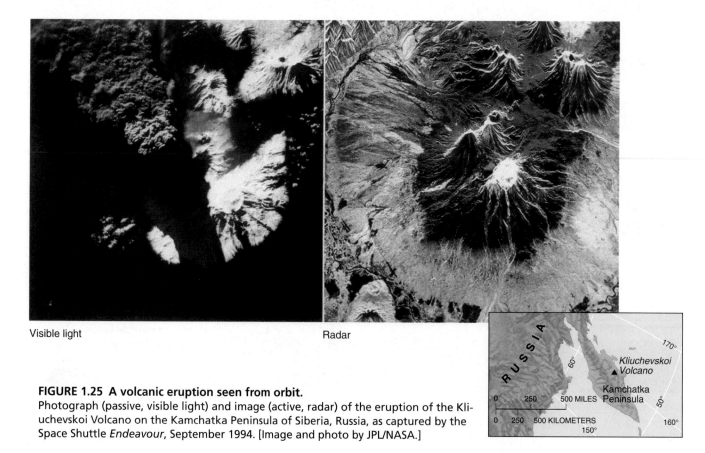

Visible light   Radar

FIGURE 1.25 A volcanic eruption seen from orbit.
Photograph (passive, visible light) and image (active, radar) of the eruption of the Kliuchevskoi Volcano on the Kamchatka Peninsula of Siberia, Russia, as captured by the Space Shuttle *Endeavour*, September 1994. [Image and photo by JPL/NASA.]

produce a digital three-dimensional data set. Along with the ESA satellites, the Canadian satellite, *Radarsat*, and the Japanese, *JERS-1*, image in radar wavelengths.

Passive Remote Sensing   Passive remote-sensing systems record energy radiated from a surface, particularly visible light and infrared. Our own eyes are passive remote sensors, as was the *Apollo 17* astronaut camera that made the picture of Earth on the back cover of this book from a distance of 37,000 km (23,000 mi).

Passive remote sensors on five *Landsat* satellites, launched by the United States, provide a variety of data as shown in images of the Appalachian Mountains in Chapter 12, river deltas in Chapter 14, and the Malaspina and Kuskulana glaciers in Alaska in Chapter 17. Three *Landsats* remain operational (4, 5, and the newest 7) although *Landsat 4* no longer gathers images and is used for orbital tests. *Landsat 5* remains threatened with shut-off due to budget cuts. See http://geo.arc.nasa.gov/sge/landsat/landsat.html for the *Landsat* HomePage and other links.

The National Oceanic and Atmospheric Administration's (NOAA, see http://noaa.gov/) polar-orbiting satellites carrying the *advanced very high resolution radiometer* (*AVHRR*) sensors on *NOAA-14*, *NOAA-15*, and *NOAA-16* are now operational. These sensors are sensitive to visible and infrared wavelengths. The incredible images of Hurricane Andrew (Chapter 8), among others in this text, were produced by an AVHRR system. In Chapter 19, an AVHRR image portrays clear-cutting of trees and production of biomass in the Pacific Northwest. These examples of resource analysis were impossible to perform at such a scale just a few years ago. In addition, these satellites measure ozone concentrations and temperatures in the stratosphere.

Key to NASA's Earth Observing System (EOS) is satellite *Terra*, which began beaming back data and images in 2000 (see http://terra.nasa.gov/), followed by another satellite in the series called *Aqua*. Five instrument packages observe Earth systems in detail, exploring the atmosphere, landscapes, oceans, environmental change, and climate, among other abilities. For example, the Clouds and the Earth's Radiant Energy System (CERES) instruments aboard *Terra* monitor the Earth's energy balance, giving new insights into climate change (see Chapter 4). These monitors offer the most accurate global radiation and energy measurements ever available. Another instrument set, the Moderate-resolution Imaging Spectroradiometer (MODIS), sees all of Earth's surface every 1–2 days in 36 spectral bands, thereby expanding on AVHRR capabilities (Figure 1.26). *Geosystems'* cover is a *Terra* MODIS image.

FIGURE 1.26 Satellite *Terra* imagery.
A portion of northern Arizona, the Grand Canyon, Lake Mead and Hoover Dam (left), and Lake Powell and Glen Canyon Dam (north) as captured by the MODIS sensor aboard *Terra*, June 2001. [Satellite *Terra* image, a portion of VE ID 9315, courtesy of MODIS Land Rapid Response Team, Jacques Descloitres, NASA.]

FIRST GOES-12
VISIBLE IMAGE
AUGUST 17, 2001 18:00 UTC (2:00 PM EDT)

FIGURE 1.27 *GOES-12* **first image.**
New environmental satellite in 2001, the first image of *GOES-12* demonstrates excellent image quality from its 37,500 km (23,300 mi) orbital post. This satellite, along with *GOES-11*, is stored in orbit to replace either of the existing *GOES* satellites as needed. [Image courtesy of NOAA.]

Of the 105 remote-sensing images presented throughout this text, 29 are from the *Terra* satellite.

One of two commercial systems includes the three French satellites (numbered 1, 2, and 4) called *SPOT* (Système Probatoire d'Observation de la Terre; see http://spot4.cnes.fr/waiting.htm), that resolve objects on Earth down to 10 to 20 m (33 to 66 ft), depending on which sensors are used. The other commercial system is Space Imaging, Inc., that offers 1–4 meter resolution from its *IKONOS* satellite in a Sun-synchronous orbit at 680 km (420 mi) altitude.

In addition, previously unavailable intelligence ("spy") satellite images are now becoming available. These images from a CIA program code-named *Corona* were actually photographs taken by satellite cameras with film returned from orbit. Comparative analysis of Earth's surface over time is a potentially valuable use of these photographs that go back to 1960. These images are released through the National Reconnaissance Office, Department of Defense (see http://www.nro.odci.gov/index5.html). This declassified satellite imagery is available through the USGS Eros Data Center, Sioux Falls, South Dakota.

The Geostationary Operational Environmental Satellites, known as *GOES*, became operational in late 1994, providing frequent infrared and visible images—the ones you see on television weather reports. Geostationary satellites stay in semipermanent positions because they keep pace with Earth's rotational speed at their altitude of 35,400 km (22,000 mi).

GOES-10, on-line in 1998, operates above 135° W longitude to monitor the West Coast and the eastern Pacific

Ocean. *GOES-8* is positioned above 75° W longitude to monitor central and eastern North America and the western Atlantic. *GOES-11* and *GOES-12*, launched in 2000 and 2001 respectively, are being stored in orbit to replace the older satellites in the *GOES* series, as needed (Figure 1.27). See the Geostationary Satellite Server at http://www.goes.noaa.gov/. The *GOES* Project Science appears at http://rsd.gsfc.nasa.gov/goes/, or see http://www.ghcc.msfc.nasa.gov/GOES/. The image of Earth on the half-title page of this book includes a cloud snapshot from *GOES* added on *Terra* MODIS images over a 16-day period in 2000.

Other satellites used for weather include Japan's *GMS-5* weather satellite and China's *Feng Yun-2* covering the Far East, and *METEOSAT-7* for Europe and Africa, operated by the European Space Agency. (See the Remote Sensing Virtual Library at http://www.vtt.fi/aut/rs/virtual/ for remote sensing links; click on "Satellite Data" for specific coverage.)

Geographic Information Systems (GIS)

Remote sensing is an important tool for acquiring large volumes of spatial data. The next step is storing, processing, and retrieving those data in useful ways. The value of remote sensing rests on the ability to provide data to powerful information-handling systems. Computers allow the integration of geographic information from direct surveys (on-the-ground mapping) and remote sensing in complex ways never before possible.

A **geographic information system (GIS)** is a computer-based, data-processing tool for gathering, ma-

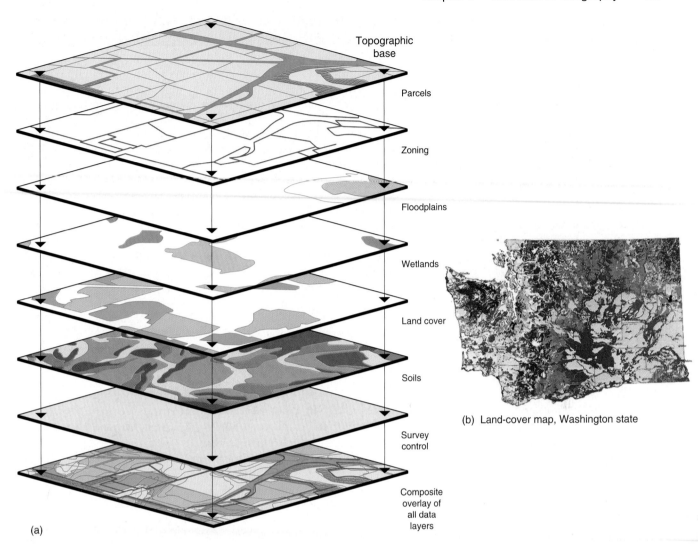

Topographic base

Parcels

Zoning

Floodplains

Wetlands

Land cover

Soils

Survey control

Composite overlay of all data layers

(a)

(b) Land-cover map, Washington state

FIGURE 1.28 A geographic information system (GIS) model.
(a) Layered spatial data in a geographic information system (GIS) format. (b) This comprehensive land-cover map is an important component of a statewide GIS analysis for Washington state. The goal is to evaluate the protection of species and biodiversity. [(a) after USGS. (b) GIS map courtesy of Dr. Kelly M. Cassidy, *Gap Analysis of Washington State*, v. 5, Map 3. Seattle: Washington Cooperative Fish and Wildlife Research Unit, University of Washington, 1997.]

nipulating, and analyzing geographic information. Through a GIS, Earth and human phenomena are analyzed over time. An example is shown in Figure 1.28. GIS is a rapidly expanding career field in many sectors of the economy. Regardless of your academic major, the ability to analyze data spatially is important. Be sure to check some of the URLs listed in News Report 1.2 for information on professional career directions in this exciting field.

The beginning component for any GIS is a coordinate system such as latitude–longitude, which establishes reference points against which to position data. The coordinate system is digitized, along with all areas, points, and lines. Remotely sensed imagery and data are then added on the coordinate system.

A GIS is capable of analyzing patterns and relationships within a single data plane, such as the floodplain or soil layer in Figure 1.28. The GIS also can generate an overlay analysis where two or more data planes interact. When the layers are combined, the resulting synthesis—a *composite overlay*—is a valuable product, ready for use in analyzing complex problems. A research study may follow specific points or areas through the complex of overlay planes. The utility of a GIS compared with that of a fixed map is the ability to manipulate the variables for analysis—to constantly change the map!

Before the advent of computers, an environmental-impact analysis required someone to gather data and painstakingly hand-produce overlays of information to determine positive and negative impacts of a project or event. Today, this layered information is handled by a computer-driven GIS, which assesses the complex interconnections among different components. In this way, subtle changes in one element of a landscape may be identified as having a powerful impact elsewhere.

News Report 1.2

Careers in GIS

Geographic information system (GIS) methodology offers great career opportunities in industry, government, business, marketing, teaching, sales, military, and other fields. Right now, geographers trained in GIS analyze ozone depletion, deforestation, soil erosion, and acid deposition. They map ecosystems and monitor the declining diversity of plant and animal species. Geographers plan, design, and survey urban developments, follow the trends of global warming, study the impact of human population, and analyze air and water pollution.

We are in the midst of a growing GIS revolution. Potent new careers are emerging in almost every academic area as GIS programs are implemented at many universities, colleges, and community colleges. GIS applications are used in environmental analysis, weather forecasting, natural hazard assessment, business and marketing, industry location analysis, criminal justice, and natural resources exploration, among many examples. (For a list of current trends in GIS, simply enter this topic in your search engine; for an alphabetized GIS resources list, check out http://www.geo.ed.ac.uk/home/giswww.html.)

GIS degree programs are available at many colleges and universities. GIS curriculum and certificate programs are now available at many community colleges. A consortium of three universities forms the National Center for Geographic Information and Analysis (NCGIA) for GIS education, research, outreach, and model generation. These GIS centers are Department of Geography, University of California Santa Barbara, Santa Barbara, CA 93106; Department of Surveying and Engineering, University of Maine, Orono, ME 04669; and State University of New York–Buffalo, Buffalo, NY 14260. (See http:// www.ncgia.ucsb.edu/ or http://www.ncgia.maine.edu/ or http://www.geog.buffalo.edu/ncgia.)

GIS is particularly helpful in analyzing natural hazards and society. An example is the European earthquake catalog that records more than 20,000 earthquakes, dating back to 500 B.C. Having this data base installed in a GIS permits detailed spatial analysis of these events along with country boundaries, human settlements, rivers, lakes, and seas, nuclear power plant locations, hazardous materials storage sites, and other economic considerations.

Scientists at NASA's Goddard Space Flight Center completed a comprehensive GIS of Brazil in an effort to better understand land-use patterns—specifically, loss of the rain forest. The flood-prone country of Bangladesh is undergoing analysis using a GIS. GIS models are used to analyze regions as to their vulnerability to wildfire. The GIS discloses those high-risk areas that can guide officials in the placement of fire-fighting equipment and crews. (For GIS resources and information see http://www.geo. ed.ac.uk/home/giswww.html and http://www.usgs.gov/research/gis/title.html.)

One of the most extensive and longest-operating systems is the Canada Geographic Information System (CGIS). Environmental data about natural features, resources, and land use were taken from maps, aerial photographs, and orbital sources, reduced to map segments, and entered into the CGIS. The development of this system has progressed with the ongoing Canada land-inventory project. As an example, see the use of GIS in Canada's national park system at http://parkscanada.pch.gc.ca/natress/inf_pa1/gis/gis_e.htm.

Summary and Review—Essentials of Geography

To assist you, here is a review of the Key Learning Concepts listed on this chapter's title page, in handy summary form. Each concept review concludes with a list of the key terms from the chapter, their page numbers, and review questions. Such summary and review sections follow each chapter in the book.

● *Define* **geography and physical geography in particular.**

Geography brings together disciplines from the physical and life sciences with the cultural and human sciences to attain a holistic view of Earth—physical geography is an essential aspect of the emerging **Earth systems sciences**. **Geography** is a science of method, a special way of analyzing phenomena over space; **spatial** refers to the nature and character of physical space. Geography integrates a wide range of subject matter. Geographic education recognizes five major themes: **location**, **region**, **human–Earth relationships**, **movement**, and **place** (including environmental concerns). Geography's method is **spatial analysis**, used to study the interdependence among geographic areas, natural systems, society, and cultural activities over space. **Process**—that is, analyzing a set of actions or mechanisms that operate in some special order—is central to geographic synthesis.

Physical geography applies spatial analysis to all the physical elements and processes that make up the environment: energy, air, water, weather, climate, landforms, soils, animals, plants, microorganisms, and Earth itself. Understanding the complex relations among these elements is important to human survival because Earth's physical systems and human society are so intertwined. The development of hypotheses and theories about the Universe, Earth, and life involves the **scientific method**.

Earth systems science (p. 2)
geography (p. 3)

spatial (p. 3)
location (p. 3)
region (p. 3)
human–Earth relationships (p. 3)
movement (p. 3)
place (p. 3)
spatial analysis (p. 3)
process (p. 5)
physical geography (p. 5)
scientific method (p. 5)

1. What is unique about the science of geography? On the basis of information in this chapter, define physical geography and review the geographic approach.

2. In general terms, how might a physical geographer analyze water pollution in the Great Lakes?

3. Assess your geographic literacy by examining atlases and maps. What types of maps have you used—political? physical? topographic? Do you know what projections they employed? Do you know the names and locations of the four oceans, seven continents, and most individual countries? Can you identify the new countries that have emerged since 1990?

4. Suggest a representative example for each of the five geographic themes and use that theme in a sentence.

5. Have you made decisions today that involve geographic concepts discussed within the five themes presented? Explain briefly.

● *Describe* **systems analysis, open and closed systems, feedback information, and system operations and** *relate* **those concepts to Earth systems.**

A **system** is any ordered, related set of things and their attributes, as distinct from their surrounding environment. Systems analysis is an important organizational and analytical tool used by geographers. Earth is an **open system** in terms of energy, receiving energy from the Sun, but it is essentially a **closed system** in terms of matter and physical resources.

As a system operates, "information" is returned to various points in the system via pathways called **feedback loops**. If the feedback information discourages response in the system, it is called **negative feedback**. (Further production in the system decreases the growth of the system.) If feedback information encourages response in the system, it is called **positive feedback**. (Further production in the system stimulates the growth of the system.) When the rates of inputs and outputs in the system are equal and the amounts of energy and matter in storage within the system are constant (or as they fluctuate around a stable average), the system is in **steady-state equilibrium**. A system that demonstrates a steady increase or decrease in system operations—a trend over time—is in **dynamic equilibrium**. Geographers often construct simplified **models** of natural systems to better understand them.

Four immense open systems powerfully interact at Earth's surface: three nonliving **abiotic** systems (**atmosphere, hydrosphere**, and **lithosphere**) and a living **biotic** system (**biosphere**, or **ecosphere**).

system (p. 8)
open system (p. 8)

closed system (p. 8)
feedback loops (p. 9)
negative feedback (p. 9)
positive feedback (p. 10)
steady-state equilibrium (p. 11)
dynamic equilibrium (p. 11)
model (p. 13)
abiotic (p. 13)
biotic (p. 13)
atmosphere (p. 13)
hydrosphere (p. 13)
lithosphere (p. 13)
biosphere (p. 13)
ecosphere (p. 13)

6. Define systems theory as an organizational strategy. What are open systems, closed systems, and negative feedback? When is a system in a steady-state equilibrium condition? What type of system (open or closed) is a human body? A lake? A wheat plant?

7. Describe Earth as a system in terms of both energy and matter—use simple diagrams to illustrate your description.

8. What are the three abiotic spheres (nonliving) that make up Earth's environment? Relate these to the biotic (living) sphere: the biosphere.

● *Explain* **Earth's reference grid: latitude, longitude, and latitudinal geographic zones and time.**

The science that studies Earth's shape and size is **geodesy**. Earth bulges slightly through the equator and is oblate (flattened) at the poles, producing a misshapen spheroid called a **geoid**. Absolute location on Earth is described with a specific reference grid of **parallels** of **latitude** (measuring distances north and south of the equator) and **meridians** of **longitude** (measuring distances east and west of a prime meridian). A historic breakthrough in navigation and timekeeping occurred with the establishment of an international **prime meridian** (0° through Greenwich, England) and the invention of precise chronometers that enabled accurate measurement of longitude. Latitude, longitude, and elevation are accurately calibrated using a hand-held **global positioning system** (**GPS**) instrument that reads radio signals from satellites. A **great circle** is any circle of Earth's circumference whose center coincides with the center of Earth. Great circle routes are the shortest distance between two points on Earth. **Small circles** are those whose centers do not coincide with Earth's center.

The prime meridian provided the basis for **Greenwich Mean Time** (**GMT**), the world's first universal time system. A corollary of the prime meridian is the 180° meridian, the **International Date Line**, which marks the place where each day officially begins. Today, **Coordinated Universal Time** (**UTC**) is the worldwide standard and the basis for international time zones. **Daylight saving time** is a seasonal change of clocks by one hour in summer months.

geodesy (p. 15)
geoid (p. 15)
latitude (p. 16)
parallel (p. 17)
longitude (p. 18)
meridian (p. 18)

9. Draw a simple sketch describing Earth's shape and size.

10. How did Eratosthenes use Sun angles to figure out that the 5000-stadia distance between Alexandria and Syene was 1/50 of Earth's circumference? Once he knew this fraction of Earth's circumference, how did he calculate the distance of Earth's circumference?

11. What are the latitude and longitude coordinates (in degrees, minutes, and seconds) of your present location? Where can you find this information?

12. Define latitude and parallel and define longitude and meridian using a simple sketch with labels.

13. Define a great circle, great circle routes, and a small circle. In terms of these concepts, describe the equator, other parallels, and meridians.

14. Identify the various latitudinal geographic zones that roughly subdivide Earth's surface. In which zone do you live?

15. What does timekeeping have to do with longitude? Explain this relationship. How is Coordinated Universal Time (UTC) determined on Earth?

16. What and where is the prime meridian? How was the location originally selected? Describe the meridian that is opposite the prime meridian on Earth's surface.

17. What is GPS and how does it assist you in finding location and elevation on Earth? Give a couple of examples where it was utilized to correct heights for some famous mountains.

● **_Define_ cartography and mapping basics: map scale and map projections.**

A **map** is a generalized view of an area, usually some portion of Earth's surface, as seen from above, and greatly reduced in size. The science and art of mapmaking is called **cartography**. Maps are used by geographers for the spatial portrayal of Earth's physical systems. **Scale** is the ratio of the image on a map to the real world; it relates a unit on the map to an identical unit on the ground. Cartographers create **map projections** for specific purposes, selecting the best compromise of projection for each application. Compromise is always necessary because Earth's round, three-dimensional surface cannot be exactly duplicated on a flat, two-dimensional map. **Equal area** (equivalence), **true shape** (conformality), true direction, and true distance are all considerations in selecting a projection. **Rhumb lines** are lines of constant direction and appear as straight lines on the **Mercator projection**.

map (p. 25)
cartography (p. 25)

scale (p. 25)
map projections (p. 26)
equal area (p. 26)
true shape (p. 26)
Mercator projection (p. 27)
rhumb lines (p. 28)

18. Define cartography. Explain why it is an integrative discipline.

19. What is map scale? In what three ways may it be expressed on a map?

20. State whether each of the following ratios is a large scale, medium scale, or small scale: 1:3,168,000, 1:24,000, 1:250,000.

21. Describe the differences between the characteristics of a globe and those that result when a flat map is prepared.

22. What type of map projection is used in Figure 1.13? In Figure 1.17? (See Appendix A.)

● **_Describe_ remote sensing and _explain_ geographic information system (GIS) methodology as a tool used in geographic analysis.**

The operation of Earth's systems is disclosed through orbital and aerial **remote sensing**. Satellites do not take photographs but record images that are transmitted to Earth-based receivers. Satellite images are recorded in digital form for later processing, enhancement, and generation. Aerial photographs have been used for years to improve the accuracy of surface maps. This is the realm of **photogrammetry**, an important application of remote sensing.

The mountain of data already collected has led to the development of **geographic information system (GIS)** technology. Computers process geographic information from direct surveys and remote sensing in complex ways never before possible. GIS methodology is an important step in better understanding Earth's systems and is a vital career opportunity for geographers.

The science of physical geography is in a unique position to synthesize the spatial, environmental, and human aspects of our increasingly complex relationship with our home planet—Earth.

remote sensing (p. 28)
photogrammetry (p. 30)
geographic information system (GIS) (p. 32)

23. What is remote sensing? What are you viewing when you observe a weather satellite image on TV or in the newspaper? Explain.

24. Describe _Terra_, _Landsat_, _GOES_, and _NOAA_ satellites, and explain them using several examples.

25. If you were in charge of planning for development of a large tract of land, how would GIS methodologies assist you? How might planning and zoning be affected if a portion of the tract in the GIS is a floodplain or prime agricultural land?

NetWork

The *Geosystems* Home Page provides on-line resources for this chapter on the World Wide Web. To begin: Once at the Home Page, click on the cover of this textbook, scroll the Table of Contents menu, select this chapter, and click "Begin." You will find self-tests that are graded, review exercises, spe-

cific updates for items in the chapter, and in "Destinations" many links to interesting related pathways on the Internet. *Geosystems* Home Page is found at http://www.prenhall.com/ christopherson.

Critical Thinking

A. Select a location (for example, your campus, home, workplace, a public place, or a city) and determine the following: latitude, longitude, and elevation. Describe the resources you used to gather this geographic information. Have you ever used a GPS unit to determine these aspects of your location?

B. Let's say there is a world globe in the library or geography department that is 61 cm (24 in.) in diameter. We know that Earth has an equatorial diameter of 12,756 km (7926 mi), so the scale of the globe is the ratio of 61 cm to 12,756 km. We divide Earth's actual diameter by the globe's diameter (12,756 km ÷ 61 cm) and determine that 1 cm of the globe's diameter equals about ___ cm of

Earth's actual diameter. Thus, the representative fraction for the globe is expressed in centimeters as 1/___. (Hint: 1 km = 1000 m, 1 m = 100 cm; therefore Earth's diameter 12,756 km represents 1,275,600,000 cm.)

C. The various geographic information technologies discussed in your text (GIS, GPS, remote sensing) promise to revolutionize many aspects of modern life. How are they used now, and what sorts of changes might we expect in the near future? Use the Net Search and Destinations sections to find examples, and write a brief description of what you find. Be sure to include any URLs for the sites you visit.

Career Link 1.1

Thomas D. Jones, Ph.D., Astronaut, Earth Observer, and Geographer

I first met Dr. Thomas Jones at an annual meeting of the National Council for Geographic Education (NCGE) in Indianapolis. Earth was the feature as the audience orbited the planet through NASA photos and imagery piloted by his enthusiastic geographic analysis. At one point the photo and radar image of the Kliuchevskoi volcanic eruption, on the Kamchatka Peninsula, appeared on the screen. The same photo and image are in Figure 1.25 in *Geosystems*. Afterward, I asked astronaut Tom Jones to sign the figure in a copy of the book—a real thrill.

Here was the man that made the photo and operated the Spaceborne Imaging Radar (SIR-C/X-Band Synthetic Aperture Radar) in the Space Radar Laboratory during flight STS-68. His enthusiasm for Earth observation and his spatial analysis of natural and human phenomena—from cities, to rock formations, to hydrology, and

weather—makes him a real friend of geographic education. I interviewed Tom in June 2001, at the Johnson Space Center (JSC), in Building 9,

where the simulators for both the Shuttle and components of the International Space Station (ISS) are housed for training [Figure 1].

FIGURE 1 Astronaut Thomas Jones.
Astronaut Mission Specialist Thomas Jones stands in front of a Space Shuttle simulator at the Johnson Space Center, Houston, Texas. Dr. Jones logged 1272 hours in orbit aboard four Shuttle flights. [Photo by Bobbé Christopherson.]

(continued)

Career Link 1.1 (*continued*)

Tom was born and raised in Maryland, living there through high school graduation, becoming a National Merit Scholar. I asked him if he had an early interest in geography. Tom said, "Yes, elementary school geography was fascinating. Even as early as first grade, I can remember looking at maps with my older second-grade friend. We were fascinated by landforms, straits of water and isthmus formations, and shapes of continents, as we turned through each page of an old atlas. In Boy Scouts, I loved compass work and map reading, and all the outdoor activities." Tom said, "These early interests feed right into piloting where you have a map right on your knee and must observe geography."

He received his B.S. degree as a Distinguished Graduate from the U.S. Air Force Academy and served for 6 years on active duty as pilot and commander on B-52D strategic bombers. Tom completed more than 2000 hours of flying time, achieving the rank of Captain. He stated, "Spending thousands of hours in aircraft is a good way to learn how to look carefully at landscapes from above. I have always thought it valuable in my life to have so many hours viewing Earth from above, first in planes, then from orbit."

Tom Jones continued his education at the University of Arizona, earning a Ph.D. in planetary science. He said, "I used remote sensing to study the water and mineral content of asteroids—those chunks of rock between Mars and Jupiter that are the source of many meteors that Earth encounters. We did telescopic surveys of dark asteroids, searching for water. I tried to link these spectroscopic fingerprints to meteorites (meteors that hit Earth) that we had in the lab. Any water and minerals found on asteroids are possible future resources for space travel."

Tom joined NASA in 1990 and became an astronaut in 1991. He flew on Space Shuttle *Endeavour* as a Mission Specialist on STS-59 (April 1994) and STS-68 (October 1994). This later flight is when he captured the volcanic eruption image and photo. Tom next went into space aboard *Columbia* in STS-80 (1996), where he operated the robotic arm to launch a satellite and, incidentally, made the photo of Mount Everest at dawn on this book's title page. On his fourth flight, aboard *Atlantis* in STS-98 (February 2001), he worked installing the Destiny Laboratory Module for the International Space Station (ISS).

I asked him what it felt like as he fled gravity? Tom answered, "After the main engine cuts off some 8.5 minutes out, you go to 0 Gs and weightlessness. With the disappearance of acceleration, you are instantly in free-fall conditions. We are strapped in so tight we don't notice much. So the first thing I did on my initial flight was unzip my glove and let it float in front of me in the cabin. 'Yep! I am really here in orbit,' I remember thinking."

I inquired about the *Endeavour* flight in October 1994. Tom said, "We were going to study volcanoes with the SIR-C/X-SAR, including the volcanic complex on Kamchatka. So it was on our charts and flight plan. We never expected that the biggest volcano in Asia would erupt right after launch and present us with such a wonderful opportunity. We saw it on our first orbit! This huge smudge on the horizon looked like the strangest thunderstorm. Then we realized that Kliuchevskoi had blown.

"Engineers on Earth quickly reprogrammed the radar between orbits and we got right over it on the second pass. We tended the radar and used our cameras to photograph it. On subsequent days a storm covered the area, and when it cleared you could see the lava and hot mud streaking the fresh snow—quite a sight." Tom added, "Earth gave us a gift. We were supposed to fly 6 weeks earlier and would have missed this chance to see an active eruption."

When working outside the Shuttle above the protective layers of the atmosphere, an astronaut must wear protection. The spacesuit must regulate the temperature differences experienced, from nearly 120°C (156°F) where the Sun strikes the spacesuit, to −250°C (−156°F) when in shadows. Oxygen and water must be provided and carbon dioxide buildup managed.

Here on Earth's surface the atmosphere does all this for us. Imagine designing a spacesuit that does everything portrayed in Figure 3.2 in this book.

Tom Jones completed three extravehicular activities (EVAs) on his last flight, totaling 19 hours (Figure 2a). He and his partner Bob Curbeam installed the U.S. Destiny Laboratory Module on the ISS. Relative to his spacesuit (extravehicular mobility unit, or EMU), Tom said, "The limiting factor in a spacesuit is the carbon dioxide scrubber because you can't replenish that during a space walk. You can replenish your water supply, oxygen supply, and electrical battery charge by simply plugging in your umbilical for a short time. But the CO_2-scrubber needs to be replaced; and that requires going back in after 8 hours."

I asked him about any feeling of vulnerability during his EVA while he was an Earth satellite drifting in orbit at more than 28,100 kmph (17,500 mph)? Tom answered, "Intellectually you are aware of that, and you can stop and think about your independence from things as you look over the brilliant Earth, but most of the time you are focused on the work and the spacesuit is almost invisible to you. It works so well that you get quite comfortable, you forget that you are next to a vacuum travelling ten times faster than a bullet, or in such a harsh environment.

He continued, "Despite the thermal protection of the EMU spacesuit, when the Sun rises, which it does every 90 minutes in orbit, I could feel the warmth when the light hit me. The thermal tubing in the suit that regulates temperature goes throughout the suit—ankles, to wrists, to neckline. If it does get warm in the suit, you can adjust your thermostat. With sunset, I felt that warmth drop away and when my feet were on a work platform, I felt the coldness of the metal plate and the heat energy conduct out through the feet of the suit, despite the bulky socks I wore. If you feel your feet chilling, you adjust the suit temperature up."

I asked about Earth observation time. "Everyone gets training for Earth

(continued)

Career Link 1.1 *(continued)*

(a)

(b)

FIGURE 2 Space walk February 2001.
(a) Astronaut Thomas Jones completed three space walks, totaling 19 hours, working on the International Space Station *Destiny* module installation. He is waving to crew members inside *Atlantis*. His partner in the space walks was Robert L. Curbeam. (b) Insignia for the *Atlantis* mission, note the reflection of Earth in the large window in the Destiny module. See http://www.spaceflight.nasa. gov/gallery/images/shuttle/sts-98/ndxpage1.html. [Space Shuttle photograph STS98-E-5195 and mission Insignia courtesy of NASA.]

observing, including classroom work in physical and human geography and geology. For my last mission we attended lectures and got experience doing actual camera practice. We have NASA's *Space Shuttle Physiographic Atlas* (1:10,000,000 scale), from the Earth Observation Project, organized along flight paths (west to east). Clear, bold labels used in the atlas help us orient the camera to specific photographic targets. Decal black circles are placed on charts so you know what needs to be photographed.

"The Earth Sciences Team chooses three dozen or more candidates for intense observation, depending on research needs. We have electronic maps on our laptops, so we can click on a site and it will tell us the time when the target is within range, and it will suggest camera and film to use. Houston sends a daily Earth Observation Bulletin that lists the times for certain scenes, mission elapsed time to the target, correlating with map notations."

I asked, "After 1272 hours in Earth orbit, more than 52 days total, what are some of your thoughts about Earth?" Tom answered, "Earth never fails to amaze me and captivate me with the beauty it presents, the ever-changing aspects of light, the changing vision, for you are always seeing Earth in a new way. Every time I looked out the window I saw a different angle or lighting that changes something I might have seen many times before. I always found something new to be amazed at! Looking at Earth is relaxing and enjoyable; it refreshes you instantly."

Tom added, "My days in orbit are a privilege, to see Earth from that distance is a life-long memory and one that will never leave me. Yet, it comes with a sense of regret because on this last mission I know there was so much I could have seen, but we had an important mission to complete and there wasn't time to linger by the windows. This is the mixed blessing of working in space."

As to the future, Tom wants to focus on further research related to his dissertation topic and nonfiction writing for the general public on space travel and possible missions to asteroids. He said, "The Moon is a close-by testing place for equipment, so we will no doubt be returning. We need to build support for future space exploration."

In closing, I asked Tom about the Insignia patch for STS-98 (Figure 2b). He answered, "The crew thought it was important to show the Earth observing window in the Destiny Laboratory Module, with Earth reflected in it. Uncovering the protective coating to begin operations of this crystal-clear viewing portal was a real thrill for Bob and me on our EVA." This says it all about Tom and the enrichment he gives to geographic education—he has opened a portal for us to better see Earth, our Home Planet.

The Energy–Atmosphere System

Sunset over the ocean as seen through smoke from Western wildfires in 2001. Several sunspot groups are visible on the Sun. [Photo by author.]

Our planet and our lives are powered by radiant energy from the star that is closest to Earth—the Sun. For more than 4.6 billion years, solar energy has traveled across interplanetary space to Earth, where a small portion of the solar output is intercepted. Because of Earth's curvature, the energy at the top of the atmosphere is unevenly distributed, creating imbalances from the equator to each pole—the equatorial region experiences energy surpluses; the polar regions experience energy deficits. Also, the annual pulse of seasonal change varies the distribution of energy during the year.

Earth's atmosphere acts as an efficient filter, absorbing most harmful radiation, charged particles, and space debris so that they do not reach Earth's surface. In the lower atmosphere the unevenness of daily energy receipt empowers atmospheric and surface energy balances, giving rise to global patterns of temperature and circulation of wind and ocean currents. Each of us depends on many systems that are set into motion by energy from the Sun. These systems are the subject of Part 1, Chapters 2 through 6.

SOLAR ENERGY

Atmosphere

Hydrosphere

Biosphere

Lithosphere

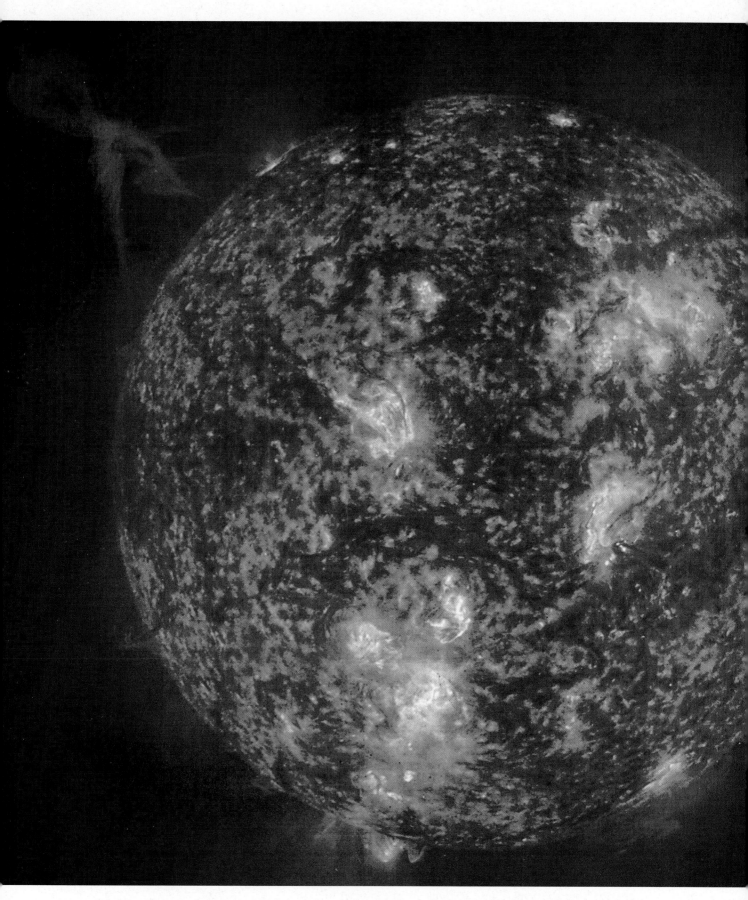

A dramatic Sun captured by instruments aboard the *SOHO* satellite, February 12, 2001. A twirling prominence rises into the Sun's corona, and other prominences are also visible. [Image courtesy of *SOHO/EIT* (Solar and Heliospheric Observatory/ Extreme Ultraviolet Imaging Telescope) Consortium. *SOHO* is an international project of cooperation between the European Space Agency and NASA. See http://sohowww.nascom.nasa.gov/.]

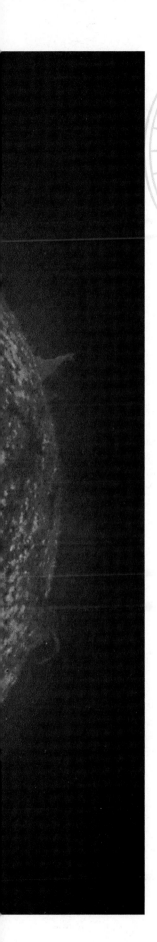

2

Solar Energy to Earth and the Seasons

Key Learning Concepts

After reading the chapter, you should be able to:

- *Distinguish* among galaxies, stars, and planets and *locate* Earth.
- *Overview* the origin, formation, and development of Earth and *construct* Earth's annual orbit about the Sun.
- *Describe* the Sun's operation and *explain* the characteristics of the solar wind and the electromagnetic spectrum of radiant energy.
- *Portray* the intercepted solar energy and its uneven distribution at the top of the atmosphere.
- *Define* solar altitude, solar declination, and daylength and *describe* the annual variability of each—Earth's seasonality.

The Universe is populated with billions of galaxies. One of these is our own **Milky Way Galaxy**, consisting of billions of stars. Among these stars is an average yellow star we call the Sun, although the dramatic *SOHO* satellite image that opens this chapter seems anything but average! Our Sun radiates energy in all directions and upon its family of orbiting planets. Of special interest to us is the solar energy that falls on the third planet from the Sun.

In this chapter: Incoming solar energy that arrives at the top of Earth's atmosphere establishes the pattern of energy input that drives Earth's physical systems and that daily influences our lives. This solar energy input to the atmosphere, plus Earth's tilt and rotation, produce daily, seasonal, and annual patterns of changing daylength and Sun angle. The Sun is the ultimate energy source for most life processes in our biosphere.

The Solar System, Sun, and Earth

Our Solar System is located on a remote, trailing edge of the Milky Way Galaxy, a flattened, disk-shaped mass estimated to contain nearly 400 billion stars (Figure 2.1a, b). Our Solar System is embedded more than halfway out from the galactic center, in one of the Milky Way's spiral arms, called the Orion Arm. A supermassive black hole some two million solar masses in size, named Sagittarius A*, sits in the center (pronounced *Sagittarius A Star*). From our Earth-bound perspective in the Milky Way, the Galaxy appears to stretch across the night sky like a narrow band of hazy light. On a clear night, the unaided eye can see only a few thousand of these billions of stars.

Solar System Formation and Structure

According to prevailing theory, our Solar System condensed from a large, slowly rotating and collapsing cloud of dust and gas called a *nebula*. **Gravity**, the mutual attracting force exerted by the mass of an object upon all other objects, was the key force in this condensing solar nebula. As the nebular cloud organized and flattened into a disk shape, the early *protosun* grew in mass at the center, drawing more matter to it. Small accretion (accumulation) eddies swirled at varying distances from the center of the solar nebula; these were the *protoplanets*.

The early protoplanets, called *planetesimals*, orbited at approximately the same distances from the Sun as the planets are today. The beginnings of the Sun and its Solar System are estimated to have occurred more than 4.6 billion years ago. The explanation of how suns condense from nebular clouds with planetesimals forming in orbits around their central masses is called the **planetesimal hypothesis**, or *dust-cloud hypothesis*. Astronomers study this formation process in other parts of the Galaxy, where planets are observed orbiting distant stars; the *Hubble Space Telescope* is a major scientific tool in these discoveries. (See http://www.seds.org/hst/ and http://www.ncc.com/misc/hubble_sites.html. A commercial site offers views of the Solar System at http://www.solarviews.com/.)

Dimensions and Distances   The **speed of light** is 300,000 kmps (kilometers per second), or 186,000 mps (miles per second)—in other words about 9.5 trillion kilometers, or nearly 6 trillion miles, per year. (In more precise numbers, light speed is 299,792 kmps, or 186,282 mps.) This tremendous distance that light travels in a year is known as a *light-year*, and it is used as a unit of measurement for the vast Universe.

The known Universe that is observable from Earth stretches approximately 12 billion light-years in all directions. The Milky Way Galaxy is about 100,000 light-years from edge to edge (Figure 2.1b). For spatial comparison, our entire Solar System of nine planets is approximately 11 hours in diameter, as measured by light speed (Figure 2.1c). The Moon is an average distance of 384,400 km (238,866 mi) from Earth, or about 1.28 seconds from Earth in terms of light speed—for the *Apollo* astronauts this was a three-day journey. (See a Solar System simulator at http://space.jpl.nasa.gov/.)

Earth's Orbit   Earth's orbit around the Sun is presently *elliptical*—a closed, oval path (Figure 2.1d). Earth's average distance from the Sun is approximately 150 million kilometers (93 million miles), which means that light reaches Earth from the Sun in an average of 8 minutes and 20 seconds. Earth is at **perihelion** (its closest position to the Sun) during the Northern Hemisphere winter (January 3 at 147,255,000 km, or 91,500,000 mi). It is at **aphelion** (its farthest position from the Sun) during the Northern Hemisphere summer (July 4 at 152,083,000 km, or 94,500,000 mi). This seasonal difference in distance from the Sun causes a slight variation in the solar energy intercepted by Earth, but is not an immediate reason for seasonal change.

A plane touching all points of Earth's orbit is termed the **plane of the ecliptic**. Earth's tilted axis remains fixed relative to this plane as Earth revolves around the Sun. The plane of the ecliptic is important to our discussion of Earth's seasons. The structure of Earth's orbit is not a constant but instead exhibits change over long periods. As shown in Chapter 17 in Figure 17.31, Earth's distance from the Sun varies more than 17.7 million kilometers (11 million miles) during a 100,000-year cycle, placing it closer or farther at different periods in the cycle. This variation is thought to be one of several factors that create Earth's cyclical pattern of glaciations (colder) and interglacial (warmer) periods.

Solar Energy: From Sun to Earth

Our Sun is unique to us but is a commonplace star in the Galaxy. It is only average in temperature, size, and color when compared with other stars, yet it is the ultimate energy source for most life processes in our biosphere.

The Sun captured about 99.9% of the matter from the original nebula. The remaining 0.1% of the matter formed all the planets, their satellites, asteroids, comets, and debris. Consequently, the dominant object in our region of space is the Sun. In the entire Solar System, it is the only object having the enormous mass needed to create the internal temperature and pressure to sustain a nuclear reaction and produce radiant energy.

The solar mass produces tremendous pressure and high temperatures deep in its dense interior. Under these conditions, the Sun's abundant hydrogen atoms, the lightest of all the natural elements, are forced together, and pairs of hydrogen nuclei are joined in a process called **fusion**. In the fusion reaction, hydrogen nuclei form helium, the

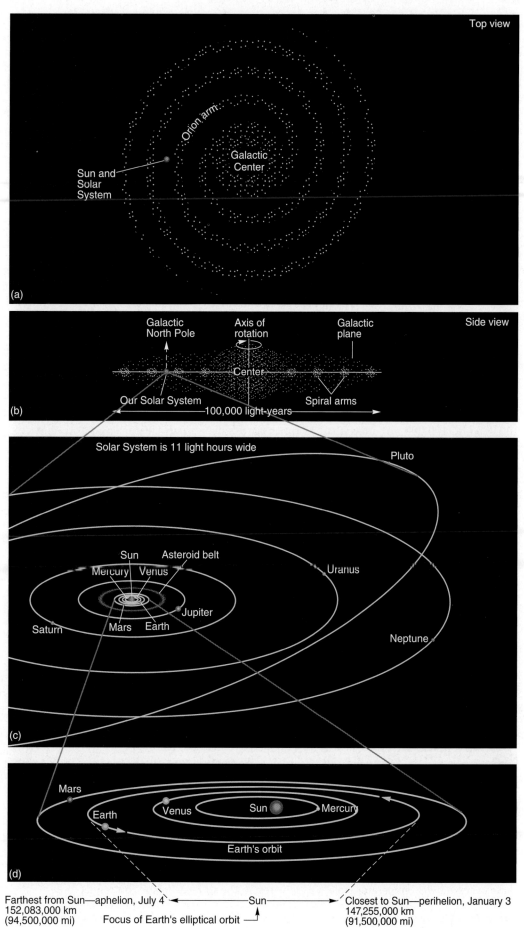

FIGURE 2.1 Milky Way Galaxy, Solar System, and Earth's orbit.
The Milky Way Galaxy viewed from above (a) and cross-section side view (b). Our Solar System of nine planets and asteroids is some 30,000 light-years from the center of the Galaxy. All of the planets except Pluto have orbits closely aligned to the plane of the ecliptic (c). The four inner terrestrial planets and the structure of Earth's elliptical orbit, illustrating perihelion (closest) and aphelion (farthest) positions during the year, are given in (d). Have you ever observed the Milky Way Galaxy in the night sky?

second-lightest element in nature, and enormous quantities of energy are liberated—literally, disappearing solar mass becomes energy.

A sunny day can seem so peaceful, certainly belying the violence taking place on the Sun. The Sun's principal outputs consist of the solar wind and radiant energy in portions of the electromagnetic spectrum. Let us trace each of these emissions across space to Earth.

Solar Activity and Solar Wind

The Sun constantly emits clouds of electrically charged particles (principally hydrogen nuclei and free electrons) that surge outward in all directions from the Sun's surface. This stream of energetic material travels much more slowly than light—only about 50 million kilometers (31 million miles) a day—taking approximately 3 days to reach Earth. The term **solar wind** was first applied to this phenomenon in 1958. Solar wind extends from the Sun to a distance beyond Pluto's orbit. The *Voyager* and *Pioneer* spacecraft launched in the 1970s are now far beyond our Solar System and have yet to escape the solar wind.

The Sun's most conspicuous features are large **sunspots**, caused by magnetic storms on the Sun. Individual sunspots may range in diameter from 10,000 to 50,000 km (6200 to 31,000 mi), with some growing as large as 160,000 km (100,000 mi), more than 12 times Earth's diameter (Figure 2.2). These surface disturbances produce flares and prominences. In addition, outbursts of charged material referred to as *coronal mass ejections* contribute to the flow of material to space in the solar wind.

A regular cycle exists for sunspot occurrences, averaging 11 years from maximum to maximum; however, the cycle may vary from 7 to 17 years. In recent cycles, a solar minimum occurred in 1976 and a solar maximum took place during 1979, with more than 100 sunspots visible at maximum. Another minimum was reached in 1986, and an extremely active solar maximum followed in 1990, with more than 200 sunspots visible at some time during the year. In fact, the 1990–1991 maximum was the most intense ever observed—11 years after the previous maximum. A sunspot minimum in 1997 and an intense maximum in 2001 maintain the average. (For more on the sunspot cycle, see **http://www.ssl.msfc.nasa.gov/ssL/pad/solar/sunspots.htm**.) The order and patterns we see in nature are the result of chaotic interactions of many systems, which are discussed in News Report 2.1.

Solar Wind Effects   The charged particles of the solar wind first interact with Earth's magnetic field as they approach Earth. The **magnetosphere** is a magnetic field surrounding Earth, generated by dynamo-like motions within our planet. The magnetosphere deflects the solar wind toward both of Earth's poles so that only a small portion of it enters the atmosphere.

Because the solar wind does not reach Earth's surface, research on this phenomenon must be conducted in space. In 1969 the *Apollo XI* astronauts exposed a piece of foil on the lunar surface as a solar wind experiment (Figure 2.3). When examined back on Earth, the exposed foil exhibited particle impacts that confirmed the presence and character of the solar wind.

Interaction of the solar wind and the upper layers of Earth's atmosphere produces the remarkable **auroras** that occur toward both poles. These lighting effects are the *aurora borealis* (northern lights) and *aurora australis* (southern lights) in the upper atmosphere, 80–500 km (50–300 mi) above Earth's surface (Figure 2.4). During the 2001 solar maximum, auroras were visible as far south as Jamaica,

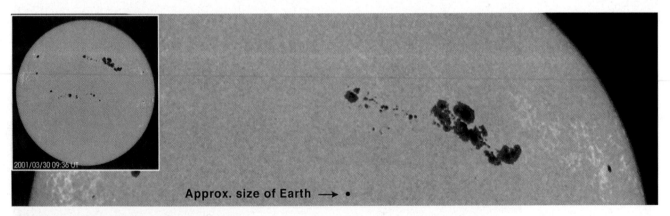

2001/03/30 09:36 UT

Approx. size of Earth → •

FIGURE 2.2 Image of the Sun and sunspots.
The Sun and large sunspot group in a recent active cycle, imaged March 30, 2001, by the MDI (Michelson Doppler Imager) instrument aboard satellite *SOHO*. This group was the source of numerous flares and coronal mass ejections, including the largest flare in 25 years on April 2. The area within the sunspot group is more than 13 times the entire surface area of Earth (13 times 500 million km^2)—Earth is shown for scale. Sunspots appear as visible dark patches because of their lower temperature relative to the rest of the surface. [Image courtesy of *SOHO/MDI* Consortium. *SOHO* (Solar and Heliospheric Observatory) is a cooperative effort between the European Space Agency and NASA. MDI is from Stanford-Lockheed/Martin Institute for Space Research.]

The Nature of Order is Chaos

In 1960, Edward Lorenz, an MIT scientist, rocked the scientific world with the statement that a butterfly flapping its wings in Brazil might produce a tornado in Texas. He used this strange example to suggest that the interaction of orderly and deterministic systems may produce chaotic and unpredictable results.

For example, ice has a rigid internal structure, forced by bonding between water molecules. This structure dictates that all ice crystals are six-sided, yet no two ice crystals, despite this similarity, are identical. Beneath this chaos of design exists an order dictated by physical principles (see Figure 7.5).

A major shift in our real-world view occurred with the advent of *chaos theory*, a revolution in science that considers the nonlinear and unpredictable behavior of operational systems. This theory suggests that the scientific method must consider the coexistence of disorder and order, randomness and pattern, and symmetry and chaos in natural systems—the science of complexity and complex systems.

Consider the weather: Mathematical models and numerical equations describe the behavior of water vapor, temperature, and pressure patterns. Yet weather systems are sensitive to very small fluctuations in any of those ingredients. Therefore, it is difficult to exactly predict how a weather system will develop, what track it will follow, or how severe it might be (see Chapter 8). Two similar chaotic weather systems might produce a similar result, although it is not possible to say exactly what the output will be. This understanding of the role of chaos is helping scientists to improve forecasting of weather phenomena. In a dynamic weather system, chaos is the rule, just as it is in chemical and biological systems.

Chaos theory is useful in studying all of Earth's physical environments. For example, a river flows in branched channels over a floodplain in a pattern to conserve energy. The channels constantly shift in a randomness that is irregular and difficult to predict. See the *Terra* satellite photograph of the many mouths of the Ganges River in Figure 14.24 for an example of such *fractal branching*—irregular, curving channels that may or may not repeat their pattern.

This may seem strange, but the nature of order we observe is the result of chaos and the almost infinite interaction among physical elements! Chaos theory is a new dimension of the scientific method and physical geography.

Texas, and California. Also, the solar wind disrupts certain radio broadcasts and some satellite transmissions, causes overloads on Earth-based electrical systems, and may affect weather patterns.

Our understanding of the solar wind is increasing dramatically as data are collected by a variety of satellites: the *SOHO* (*Solar and Heliospheric Observatory*), *FAST* (*Fast Auroral Snapshot*), *WIND*, *Ulysses*, the *Dynamics Explorer*, and the earlier *Voyager-2* and *Pioneers-10* and *11* launched in the 1970s. All satellite data are available on the Internet. (For auroral activity, see http://www.sel.noaa.gov/pmap/ and for forecasts, see http://www.gi.alaska.edu.)

Weather Effects   Another effect of the solar wind in the atmosphere is its possible influence on weather and climate cycles. Why do wetter periods in some midlatitude areas tend to coincide with every other solar maximum? Why do droughts often occur near the time of every other solar minimum? For example, sunspot cycles during more than

FIGURE 2.3 Astronaut and solar wind experiment.
Without a protective atmosphere, the lunar surface receives charged particles of the solar wind and all of the Sun's electromagnetic radiation. A sheet of foil is deployed by an *Apollo XI* astronaut in 1969 in the solar wind experiment. Earth-bound scientists analyzed the foil that revealed the composition of the solar wind. Why wouldn't this experiment work if deployed on Earth's surface? [NASA photo.]

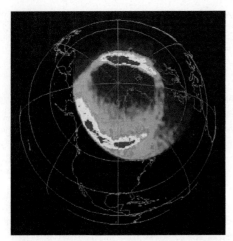

(a)

(b)

FIGURE 2.4 Auroras from an orbital perspective and from the ground in Alaska.
(a) *Polar* satellite false-color image of auroral halo over Earth's North Pole region from April 1996—the UVI sensor is able to capture the aurora on the day and night sides of Earth. (b) Surface view of an aurora borealis in the night sky over central Alaska in April 2000. [(a) Image from the Ultraviolet Imager (UVI) aboard *Polar* satellite courtesy of NASA. (b) Photo by Jan Curtis, all rights reserved.]

48　Part One　The Energy–Atmosphere System

250 years from 1740 to 1998 coincided with periods of wetness and drought, as estimated by an analysis of tree growth rings for that period throughout the western United States and elsewhere, as well as instrumental weather records. These variations in weather tend to lag 2 or 3 years behind the solar maximum or minimum.

Regardless of the *cause* for the cyclical patterns of drought and wetness that do occur, a remarkable failure in current planning worldwide is the lack of *attention* given to them. Preparing for such patterns could reduce property loss and casualties. Cyclical drought could be offset

through widespread water conservation and more efficient water use. Wet spells might require strengthening of levees along river channels, floodplain zoning to restrict development, and better reservoir management to reduce flooding. As knowledge of the solar wind–weather relation improves, it will demand the attention of policy makers and the public.

Electromagnetic Spectrum of Radiant Energy

The key essential solar input to life is electromagnetic energy of various wavelengths. Solar radiation occupies a portion of the **electromagnetic spectrum** of radiant energy. This radiant energy travels at the speed of light to Earth. The total spectrum of this radiant energy is made up of different wavelengths. Figure 2.5 shows that a **wavelength** is the distance between corresponding points on any two successive waves. The number of waves passing a fixed point in one second is the *frequency*.

The Sun emits radiant energy composed of 8% ultraviolet, X-ray, and gamma-ray wavelengths; 47% visible light wavelengths; and 45% infrared wavelengths. A portion of the electromagnetic spectrum is illustrated in Figure 2.6, with wavelengths increasing from the top of the illustration to the bottom. Note the wavelengths at which various phenomena and human applications of energy occur.

An important physical law states that all objects radiate energy in wavelengths related to their individual surface temperatures: the hotter the object, the shorter the wavelengths emitted. This law holds true for the Sun and Earth. Figure 2.7 shows that the hot Sun radiates shorter wavelength energy, concentrated around 0.4–$0.5 \mu m$ (micrometer).

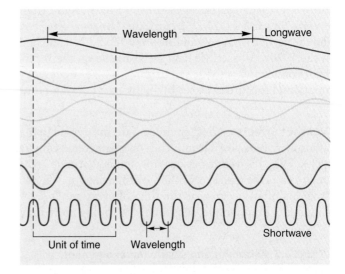

FIGURE 2.5 Wavelength and frequency.
Wavelength and frequency are two ways of describing the same phenomenon—electromagnetic wave motion. More short wavelengths pass a given point during a unit of time, so they are higher in frequency, whereas fewer long wavelengths pass a point in a unit of time, so they are lower in frequency.

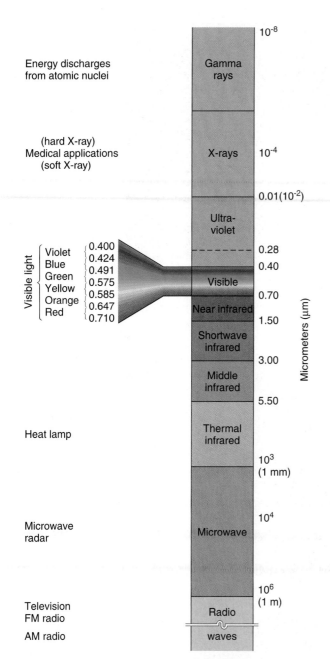

FIGURE 2.6 A portion of the electromagnetic spectrum of radiant energy.
The spectrum is oriented with shorter wavelengths toward the top and longer wavelengths toward the bottom.

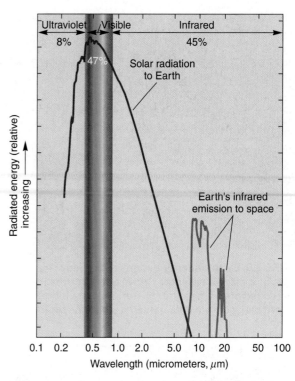

FIGURE 2.7 Solar and terrestrial energy distribution by wavelength.
The solar output peaks in shorter wavelengths of visible light in relation to its higher surface temperature, whereas Earth's emissions are concentrated in the infrared portion of the spectrum in relation to its lower surface temperature.

The Sun's surface temperature is about 6000°C (11,000°F), and its emission curve shown in the figure is similar to that predicted for an idealized 6000°C surface, or *blackbody radiator.* An ideal blackbody emits as much radiant energy as it absorbs—the hotter the blackbody, the more radiation it emits at all wavelengths, with shorter wavelengths dominant at higher temperatures. The Sun emits a much greater amount of energy per unit area of its surface than does a similar area of Earth's environment.

Earth is a cooler radiating body, so longer wavelengths are emitted. In comparison to a shorter-wavelength emitting hot body, lower temperatures at Earth's surface produce radiation mostly in the infrared portion of the spectrum. Figure 2.7 shows that the radiation emitted by Earth occurs in longer wavelengths, centered around $10\,\mu m$ and entirely within the infrared portion of the spectrum. Figure 2.8 illustrates the flows of energy into and out of Earth systems.

To summarize, the solar spectrum is *shortwave radiation* that peaks in the short visible wavelengths, whereas Earth's radiated energy is *longwave radiation* concentrated in infrared wavelengths. In Chapter 4, we see that Earth, clouds, sky, ground, and things that are terrestrial are cool-body radiators in contrast to the Sun.

Intercepted Energy at the Top of the Atmosphere

The region at the top of the atmosphere, approximately 480 km (300 mi) above Earth's surface, is termed the **thermopause**. It is the outer boundary of Earth's energy system and provides a useful point at which to assess the arriving solar radiation before it is diminished by scattering and absorption in passage through the atmosphere.

Earth's distance from the Sun results in its interception of only one two-billionth of the Sun's total energy output. Nevertheless, this tiny fraction of the Sun's overall output is an enormous amount of energy input to Earth's systems. Solar radiation that reaches a horizontal plane at Earth is called **insolation** (*in*tercepted *sol*ar radi*ation*), a term specifically applied to radiation arriving at Earth's atmosphere and surface. Insolation at the top of the atmosphere is expressed as the *solar constant.*

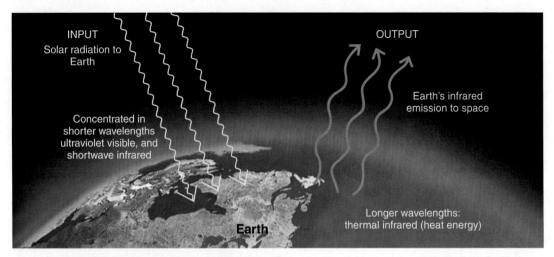

FIGURE 2.8 Earth's energy budget simplified.
Inputs of shorter wavelengths arrive at Earth from the Sun. Outputs of longer wavelengths of infrared radiate to space from Earth. The data plotted in Figure 2.10 and the map in Figure 2.11 are derived from data gathered along the top of the atmosphere.

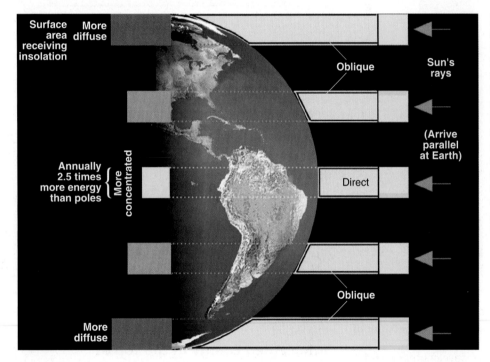

FIGURE 2.9 Insolation receipts and Earth's curved surface.
Solar insolation angles determine the concentration of energy receipts by latitude. Lower latitudes receive more concentrated energy from a more direct solar beam. Higher latitudes receive slanting (oblique) rays and more diffuse energy. Note the area covered by identical columns of solar energy arriving at Earth's surface at higher latitudes (more diffuse, larger area covered) and at lower latitudes (more concentrated, smaller are covered).

Solar Constant Knowing the amount of insolation intercepted by Earth is important to climatologists and other scientists. The **solar constant** is the average insolation received at the thermopause when Earth is at its average distance from the Sun. That value is 1372 W/m^2 (watts per square meter).* The constancy of the solar constant over time is important, for small variations of even 0.5% or 1.0% could prove dramatic for Earth's energy system.

As we follow insolation through the atmosphere to Earth's surface (Chapters 3 and 4), we see that the value of the solar constant is reduced by half or more through reflection, scattering, and absorption of shortwave radiation.

Uneven Distribution of Insolation Earth's curved surface presents a continually varying angle to the incoming parallel rays of insolation (Figure 2.9). Differences in the angle of solar rays at each latitude result in an uneven distribution of insolation and heating. The place receiving maximum insolation is the point where insolation rays are

*A watt is equal to 1 joule (a unit of energy) per second and is the standard unit of power in the SI–metric system. (See Appendix C of this text for more information on measurement conversions.) In nonmetric calorie heat units, the solar constant is expressed as approximately 2 calories per square centimeter per minute, or 2 langleys per minute (a langley is 1 cal/cm^2). A calorie is the amount of energy required to raise the temperature of 1 g of water (at 15°C) 1 degree Celsius and is equal to 4.184 joules.

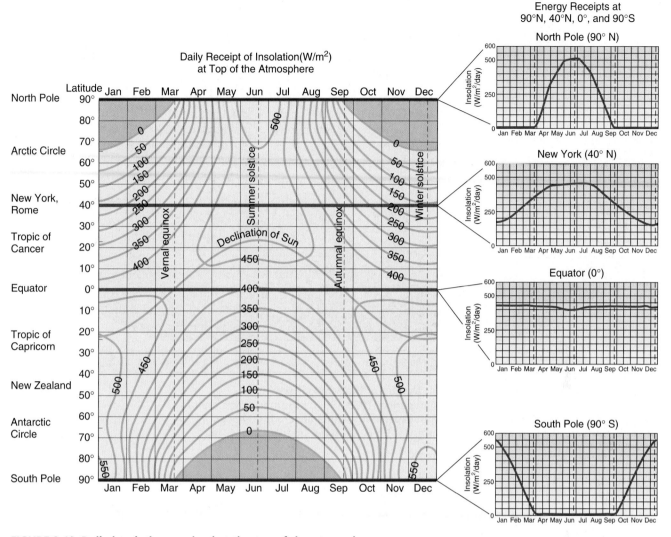

FIGURE 2.10 Daily insolation received at the top of the atmosphere.
The total daily insolation received at the top of the atmosphere is charted in watts per square meter per day by latitude and month (1 W/m²/day = 2.064 cal/cm²/day). A profile of annual energy receipts is graphed to the right for the North Pole, along 40° north latitude, along the equator, and for the South Pole. [Reproduced by permission of the Smithsonian Institution Press from *Smithsonian Miscellaneous Collections: Smithsonian Meteorological Tables*, vol. 114, 6th edition. Robert List, ed. (Washington, DC: Smithsonian Institution, 1984), p. 419, Table 134.]

perpendicular to the surface (radiating from directly over-head), called the **subsolar point.** All other places receive insolation at less than a 90° angle and thus experience more diffuse energy receipts. Solar beam angles become more pronounced at higher latitudes. As a result, during a year's time, the thermopause above the equatorial region receives 2.5 times more insolation than the thermopause above the poles. Lower-angle solar rays toward the poles must pass through a greater thickness of atmosphere, resulting in further losses of energy due to scattering, absorption, and reflection.

Figure 2.10 illustrates the daily variations throughout the year of energy at the top of the atmosphere for various latitudes in watts per square meter (W/m²). The chart shows a decrease in insolation from the equatorial

regions northward and southward toward the poles. However, in June, the North Pole receives more than 500 W/m² per day, which is more than is ever received at 40° N latitude or at the equator. Such high values result from the duration of exposure at the poles in summer: 24 hours a day, compared with only 15 hours of daylight at 40° N latitude and 12 hours at the equator. However, at the poles the summertime Sun at noon is low in the sky, so a daylength twice that of the equator yields only about 100 W/m² difference.

In December, the pattern reverses. Note that the top of the atmosphere at the South Pole receives even more insolation than the North Pole does in June (more than 550 W/m²). This is a function of Earth's closer location to the Sun at perihelion (January 3 on Figure 2.1d).

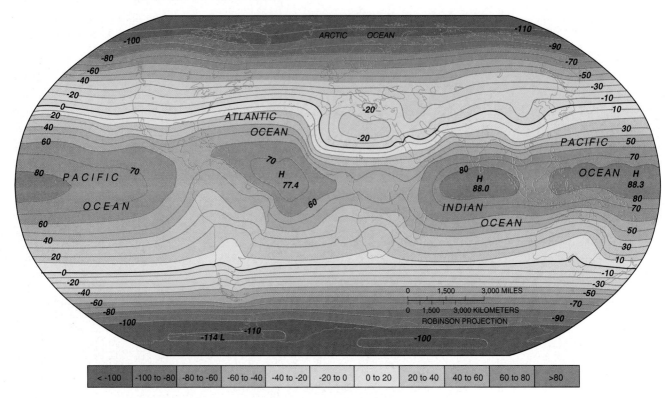

FIGURE 2.11 Daily net radiation patterns at the top of the atmosphere.
Averaged daily net radiation flows for a 9-year period (1979–1987), measured at the top of the atmosphere by the Earth Radiation Budget experiment (ERBE) aboard the *Nimbus-7* satellite. Units are W/m². [Data for map courtesy of Dr. H. Lee Kyle, Goddard Space Flight Center, NASA.]

Along the equator, two maximum periods of approximately 430 W/m² occur at the spring and fall equinoxes, when the subsolar point is at the equator. Find your latitude on the graph and follow across the months to determine the seasonal variation of insolation where you live. The four graphs to the right show plots of the energy received at the North Pole, along 40° N, along the equator, and at the South Pole during the year to give you energy profiles to compare.

Global Net Radiation *Earth Radiation Budget (ERBE)* instruments aboard several satellites measured shortwave and longwave flows of energy at the top of the atmosphere. ERBE sensors collected the data used to develop the map in Figure 2.11. This map shows *net radiation*, or the balance between incoming shortwave and outgoing longwave radiation—energy inputs minus energy outputs. See News Report 2.2 for more on this experiment.

First, note the latitudinal energy imbalance in net radiation on the map—positive values in lower latitudes (energy sources for Earth systems) and negative values toward the poles (energy sinks). In middle and high latitudes, approximately poleward of 36° north and south latitudes, net radiation is negative. The reason for this in these higher latitudes is that Earth's climate system loses more energy to space than it gains from the Sun, as measured at the top of the atmosphere. In the lower atmosphere, these polar energy deficits are offset by flows of energy from tropical energy surpluses (as we will see in Chapters 4 and 6). The largest net radiation values, averaging 80 W/m², are above the tropical oceans along a narrow equatorial zone. Net radiation minimums are lowest over Antarctica.

Of interest is the −20 W/m² area over the Sahara region of North Africa. Here, typically clear skies—which permit great longwave radiation losses from Earth's surface—and light-colored reflective surfaces work together to reduce net radiation values at the thermopause. In other regions, clouds and atmospheric pollution in the lower atmosphere also affect net radiation patterns at the top of the atmosphere by reflecting more shortwave energy to space.

The atmosphere and ocean form a giant heat engine, driven by differences in energy from place to place and causing major circulations within the lower atmosphere and in the ocean. These circulations include global winds, ocean currents, and weather systems—subjects to follow in Chapters 6 and 8. As you go about your daily activities, let these dynamic natural systems remind you of the constant flow of solar energy through the environment.

Having examined the flow of solar energy to Earth and the top of the atmosphere, let us now look at how seasonal changes affect the distribution of insolation as Earth orbits the Sun during the year.

News Report 2.2

Monitoring Earth Radiation Budget

Earth's weather and climate are a direct result of the balance between sunlight received from space and energy reflected and radiated to space. Patterns of insolation absorption and energy reradiation by Earth systems produce an energy budget. Since 1978, the Earth–atmosphere energy budget was monitored by the Earth Radiation Budget (ERB) package on board the *Nimbus-7* satellite, and later by the *ERB, NOAA-9,* and *NOAA-10* satellites. More than a decade of data were collected by these satellites. ERB sensors mapped the complex exchanges of energy among atmosphere, ocean, and land.

Measurements included monthly energy budget and variations, seasonal shifts in energy between equator and poles, and daily energy budgets at the regional scale. The map in Figure 2.11 is a direct result of ERBE, showing regions that absorb more energy than radiated (a heat-energy source, rose color on map in Figure 2.11) and that radiate more energy than received (a heat-energy sink, purple color).

A new tool in understanding Earth's radiation budget is the CERES (Clouds and Earth Radiant Energy System) sensor aboard satellite *Terra*. CERES monitors shortwave (light) and longwave (heat energy), and thus, Earth's energy balance of incoming and outgoing radiation. A pair of CERES images from March 2000 are in Chapter 4.

In addition, ERB measurements determined that solar irradiance, or the luminous brightness of solar radiation, is directly correlated to the sunspot cycle discussed earlier. During the solar maximums of 1979 and 1991, the solar constant exceeded 1374 W/m^2; the 1986 minimum produced a constant of 1371 W/m^2 (remember the solar constant averages 1372 W/m^2). The ERBE findings help scientists better understand the systems described in Part 1 in this textbook (see http://asd-www.larc.nasa.gov/erbe/ASDerbe.html).

The Seasons

Earth's periodic rhythms of warmth and cold, dawn and daylight, twilight and night, have fascinated humans for centuries. In fact, many ancient societies demonstrated an intense awareness of seasonal change and formally commemorated these natural energy rhythms with festivals, monuments, and calendars (Figure 2.12). Such ancient seasonal monuments and calendar markings occur worldwide, including thousands of sites in North America.

Seasonality

Seasonality refers to both the seasonal variation of the Sun's position above the horizon and changing daylengths during the year. Seasonal variations are a response to changes in the Sun's **altitude**, or the angle between the horizon and the Sun. At sunrise or sunset, the Sun is at the horizon, so its altitude is 0°. During the day, if the Sun reaches halfway between the horizon and directly overhead, it is at 45° altitude. If the Sun reaches the point directly overhead, it is at 90° altitude. The Sun is directly overhead (90° altitude, or *zenith*) only at the *subsolar point*, where insolation

FIGURE 2.12 Ancient calendar in stone.
Stonehenge, Salisbury Plain in Wiltshire, England. Here, rocks weighing 25 metric tons (28 tons) were hauled about 500 km (300 mi) and placed in patterns that evidently mark seasonal changes, predicted by this 3500-year-old "calendar" monument. [Photo by Robert Llewellyn.]

is at a maximum. At all other surface points, the Sun is at a lower angle, producing more-diffuse insolation.

The Sun's **declination** is the latitude of the subsolar point. Declination annually migrates through 47° of latitude, moving between the *Tropic of Cancer* at 23.5° N and the *Tropic of Capricorn* at 23.5° S latitude. Other than Hawai'i, which is between 19° N and 22° N, the subsolar point does not reach the continental United States or Canada; all other states and provinces are too far north.

In addition to changing Sun altitude and declination, seasonality means changing **daylength**, or duration of exposure. Daylength varies during the year, depending on latitude. The equator always receives equal hours of day and night: If you live in Ecuador, Kenya, or Singapore, every day and night is 12 hours long, year-round. People living along 40° N latitude (Philadelphia, Denver, Madrid, Beijing), or 40° S latitude (Buenos Aires, Capetown, Melbourne), experience about 6 hours' difference in daylight between winter and summer. Those at 50° N or S latitude (Winnipeg, Paris, Falkland Islands) experience almost 8 hours of annual daylength variation.

At the North and South Poles, the range of daylength is extreme and extends from a 6-month period of no insolation (ranging from twilight to darkness to dawn) to a 6-month period of continuous 24-hour insolation (daylight)—literally the poles experience one long day and one long night each year! This is evident in Figure 2.15, if you note the illumination of the North Pole in June (to the left) and South Pole in December (to the right). Given this observation, turn to the back cover of this textbook and see whether you can determine the month during which the Apollo astronaut made the Earth photo. (The answer is on the copyright page.)

Reasons for Seasons

Seasons result from variations in the Sun's *altitude* above the horizon, the Sun's *declination* (latitude of the subsolar point), and *daylength* during the year. These in turn are created by several physical factors that operate in concert: Earth's *revolution* in orbit around the Sun, its daily *rotation* on its axis, its *tilted axis*, the unchanging *orientation of its axis*, and its *sphericity* (Table 2.1). Of course, the essential ingredient is having a single source of radiant energy—the Sun. We now look at each of these factors individually. As

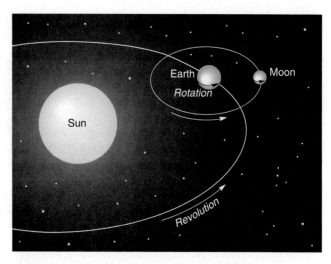

FIGURE 2.13 Earth's revolution and rotation.
Earth's *revolution* about the Sun and *rotation* on its axis, as viewed from above Earth's orbit. Note the Moon's rotation on its axis and revolution about Earth are counterclockwise as well.

we do, please note the distinction between revolution—Earth's travel around the Sun—and rotation—Earth's spinning on its axis (Figure 2.13).

Revolution Earth's orbital **revolution** about the Sun is shown in Figure 2.1d. Earth's speed in orbit averages 107,280 kmph (66,660 mph). This speed, together with Earth's distance from the Sun, determines the time required for one revolution around the Sun and, therefore, the length of the year and duration of the seasons. Earth completes its annual revolution in 365.2422 days. This number is based on a *tropical year*, measured from equinox to equinox, or the elapsed time between two crossings of the equator by the Sun.

The Earth-to-Sun distance (aphelion to perihelion) might seem a seasonal factor, but it is not significant, even though it varies about 3% (4.8 million kilometers or 3 million miles) during the year, amounting to a 50 W/m² difference between local polar summers. Remember that the distance averages 150 million kilometers (93 million miles).

Rotation Earth's **rotation**, or turning on its axis, is a complex motion that averages 24 hours in duration. Rota-

Table 2.1 Five Reasons for Seasons

Factor	Description
Revolution	Orbit around the Sun; requires 365.24 days to complete at 107,280 kmph (66,660 mph)
Rotation	Earth turning on its axis; takes approximately 24 hours to complete at 1675 kmph (1041 mph) at the equator
Tilt	Axis is aligned at a 23.5° angle from a perpendicular to the plane of the ecliptic (the plane of Earth's orbit)
Axial parallelism	Remains in a fixed alignment, with Polaris directly overhead at the North Pole throughout the year
Sphericity	Appears as an oblate spheroid to the Sun's parallel rays; the geoid

tion determines daylength, creates the apparent deflection of winds and ocean currents, and produces the twice-daily rise and fall of the ocean tides in relation to the gravitational pull of the Sun and the Moon.

Earth rotates about its **axis**, an imaginary line extending through the planet from the geographic North Pole to the South Pole. When viewed from above the North Pole, Earth rotates counterclockwise around this axis. Viewed from above the equator, Earth rotates west to east, or eastward. This eastward rotation creates the Sun's apparent westward daily journey from sunrise in the east to sunset in the west. Of course, the Sun actually remains in a fixed position in the center of the Solar System. (Note in Figure 2.13 that the Moon both revolves around Earth and rotates on its axis in a counterclockwise direction.)

Although every point on Earth takes the same 24 hours to complete one rotation, the linear velocity of rotation at any point on Earth's surface varies dramatically with latitude. The equator is 40,075 km (24,902 mi) long; therefore, rotational velocity at the equator must be approximately 1675 kmph (1041 mph) to cover that distance in one day. At 60° latitude, a parallel is only half the length of the equator, or 20,038 km (12,451 mi) long, so the rotational velocity there is 838 kmph (521 mph). At the poles, the velocity is 0. (This variation in rotational velocity establishes the effect of the Coriolis force, discussed in Chapter 6.) Table 2.2 lists the speed of rotation for several selected latitudes.

Earth's rotation produces the continually changing daily pattern of day and night. Half of Earth is in sunlight and half is in darkness at any moment. The traveling boundary that divides daylight and darkness is called the **circle of illumination** (as illustrated in Figure 2.15). Because this day–night dividing circle of illumination is a great circle that intersects the equator, which is another great circle, daylength at the equator is always evenly divided—

12 hours of day and 12 hours of night. (Any two great circles on a sphere bisect one another.)

A true day varies slightly from 24 hours, but by international agreement a day is defined as exactly 24 hours, or 86,400 seconds. This average, called *mean solar time*, eliminates predictable variations in rotation and revolution that cause the solar day to change slightly in length throughout the year. The complexity of Earth's rotation is now exactly measured by satellites that are in precise mathematical orbits: GPS (global positioning system, see News Report 1.1, Chapter 1), SLR (satellite-laser ranging), and VLBI (very-long baseline interferometry). All contribute to our knowledge of Earth's rotation. (Monthly and annual reports are issued by the International Earth Rotation Service at http://www.iers.org/.)

Tilt of Earth's Axis   To understand Earth's **axial tilt**, imagine a plane (a flat surface) that intersects Earth's elliptical orbit about the Sun, with half of the Sun and Earth above the plane and half below. This flat surface is termed the *plane of the ecliptic*. Now, imagine a perpendicular (at a 90° angle) line passing through the plane. Earth's axis and equatorial plane are tilted 23.5° from this perpendicular to the plane of the ecliptic. Another way of looking at it is that Earth's axis forms a 66.5° angle from the plane itself (Figure 2.14).

Hypothetically, if Earth were tilted on its side, with its axis parallel to the plane of the ecliptic, we would experience a maximum variation in seasons worldwide. On the other hand, if Earth's axis were perpendicular to the plane of its orbit—that is, with no tilt—we would experience no seasonal changes, just a perpetual spring or fall season, and all latitudes would experience 12-hour days and nights.

Axial Parallelism   Throughout our annual journey around the Sun, Earth's axis *maintains the same alignment*

Table 2.2	Speed of Rotation at Selected Latitudes		
Latitude	**Speed kmph**	**(mph)**	**Cities at Approximate Latitudes**
90°	0	(0)	North Pole
60°	838	(521)	Seward, Alaska; Oslo, Norway; Saint Petersburg, Russia
50°	1078	(670)	Chibougamau, Québec; Kyyîv (Kiev), Ukraine
40°	1284	(798)	Valdivia, Chile; Columbus, Ohio; Beijing, China
30°	1452	(902)	Pôrto Alegre, Brazil; New Orleans, Louisiana
0°	1675	(1041)	Quito, Ecuador; Pontianak, Indonesia

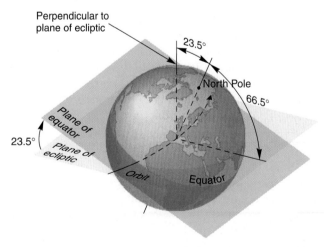

FIGURE 2.14 The plane of Earth's orbit—the ecliptic—and Earth's axial tilt.
Note that the plane of the equator is inclined at 23.5° to the plane of the ecliptic.

relative to the plane of the ecliptic and to Polaris and other stars. You can see this consistent alignment in Figure 2.15. If we compared the axis in different months, it would always appear parallel to itself, a condition known as **axial parallelism**.

Sphericity Earth's *sphericity*, discussed in Chapter 1, is also part of seasonality, for it produces the uneven receipt of insolation from pole to pole shown in Figures 2.9 and 2.10. All five reasons for seasons are summarized in Table 2.1: revolution, rotation, tilt, axial parallelism, and sphericity. Now, considering all these factors operating together, let us examine the annual march of the seasons.

Annual March of the Seasons

During the annual march of the seasons on Earth, daylength is the most obvious way of sensing changes in season at latitudes away from the equator. Daylength is the interval between **sunrise**, the moment when the disk of the Sun first appears above the horizon in the east, and **sunset**, that moment when it totally disappears below the horizon in the west. Table 2.3 lists the average times of sunrise and sunset and the daylength for various latitudes and seasons in the Northern Hemisphere. (For the Southern Hemisphere merely switch the solstice column headings and switch the equinox column headings.)

The extremes of daylength occur in December and June. The times around December 21 and June 21 are termed *solstices*. Strictly speaking, however, the solstices are specific

points in time at which the Sun's declination is at its position farthest north (Tropic of Cancer at 23.5° N) or south (Tropic of Capricorn at 23.5° S). "Tropic" is from *tropicus*, meaning a turn or change, so a tropic latitude is where the Sun's declination appears to stand still briefly (Sun stance, or *sol stice*); then it "turns" and heads toward the other tropic.

Table 2.4 presents the key seasonal anniversary dates during which the specific time of the equinoxes or solstices occur, their names, and the subsolar point location (declination). During the year, places on Earth outside of the equatorial region experience a continuous but gradual shift in daylength, a few minutes each day, and the Sun's altitude increases or decreases a small amount. You may have noticed that these daily variations become more pronounced in spring and autumn, when the Sun's declination changes at a faster rate.

Figure 2.15 demonstrates the annual march of the seasons and illustrates Earth's relationship to the Sun during the year. Let us begin with December. On December 21 or 22, at the moment of the **winter solstice** ("winter Sun stance"), or **December solstice**, the circle of illumination excludes the North Pole region from sunlight but includes the South Pole region. The subsolar point is at 23.5° S latitude, the parallel called the Tropic of Capricorn. The Northern Hemisphere is tilted away from these more direct rays of sunlight—our northern winter—thereby creating a lower angle for the incoming solar rays and thus a more diffuse pattern of insolation.

From 66.5° N latitude to 90° N (the North Pole), the Sun remains below the horizon the entire day. This

Table 2.3 Daylength Times (Sunrise and Sunset) at Selected Latitudes (Northern Hemisphere)

Latitude	Winter Solstice (December Solstice) December 21–22			Vernal Equinox (March Equinox) March 20–21			Summer Solstice (June Solstice) June 20–21			Autumnal Equinox (September Equinox) September 22–23		
	A.M.	P.M.	Daylength	A.M.	P.M.	Daylength	A.M.	P.M.	Daylength	A.M.	P.M.	Daylength
0°	6:00	6:00	12:00	6:00	6:00	12:00	6:00	6:00	12:00	6:00	6:00	12:00
30°	6:58	5:02	10:04	6:00	6:00	12:00	5:02	6:58	13:56	6:00	6:00	12:00
40°	7:30	4:30	9:00	6:00	6:00	12:00	4:30	7:30	15:00	6:00	6:00	12:00
50°	8:05	3:55	7:50	6:00	6:00	12:00	3:55	8:05	16:10	6:00	6:00	12:00
60°	9:15	2:45	5:30	6:00	6:00	12:00	2:45	9:15	18:30	6:00	6:00	12:00
90°	No sunlight			Rising Sun			Continuous sunlight			Setting Sun		

Note: All times are standard and do not consider the local option of daylight saving time.

Table 2.4 Annual March of the Seasons

Approximate Date	Northern Hemisphere Name	Location of the Subsolar Point
December 21–22	Winter solstice (December solstice)	23° S latitude (Tropic of Capricorn)
March 20–21	Vernal equinox (March equinox)	0° (equator)
June 20–21	Summer solstice (June solstice)	23.5° N latitude (Tropic of Cancer)
September 22–23	Autumnal equinox (September equinox)	0° (equator)

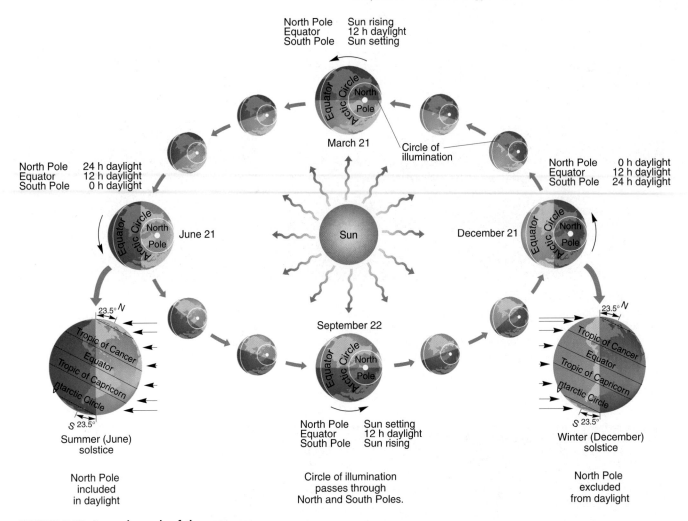

FIGURE 2.15 Annual march of the seasons.
Annual march of the seasons as Earth revolves about the Sun. Shading indicates the changing
position of the circle of illumination. Note the hours of daylight for the equator and the poles.
To follow the text, begin on the right side at December 21 and move counterclockwise.

latitude (66.5° N) marks the *Arctic Circle*, the southern-most parallel (in the Northern Hemisphere) that experiences a 24-hour period of darkness. During the following 3 months, daylength and solar angles gradually increase in the Northern Hemisphere as Earth completes one-fourth of its orbit.

The moment of the **vernal equinox**, or **March equinox**, occurs on March 20 or 21. At that time, the circle of illumination passes through both poles so that all locations on Earth experience a 12-hour day and a 12-hour night. People living around 40° N latitude (New York, Denver) have gained 3 hours of daylight since the December solstice. At the North Pole, the Sun peeks above the horizon for the first time since the previous September; at the South Pole the Sun is setting—a dramatic moment for the people working there.

From March, the seasons move on to June 20 or 21 and the moment of the **summer solstice**, or **June solstice**. The subsolar point now has shifted from the equator to 23.5° N latitude, the Tropic of Cancer. Because

the circle of illumination now includes the North Polar region, everything north of the Arctic Circle receives 24 hours of daylight—the "midnight Sun." Figure 2.16 is a multiple-image photo of the midnight Sun as seen north of the Arctic Circle. In contrast, the region from the Antarctic Circle to the South Pole (66.5°–90° S latitude) is in darkness the entire 24 hours. Those working in Antarctica call this *Midwinter's Day.*

September 22 or 23 is the moment in time of the **autumnal equinox**, or **September equinox**, when Earth's orientation is such that the circle of illumination again passes through both poles so that all parts of the globe experience a 12-hour day and a 12-hour night. The subsolar point has returned to the equator, with days growing shorter to the north and longer to the south. Researchers stationed at the South Pole see the disk of the Sun just rising, ending their 6 months of darkness. In the Northern Hemisphere, autumn arrives, a time of many colorful changes in the landscape, whereas in the Southern Hemisphere it is spring.

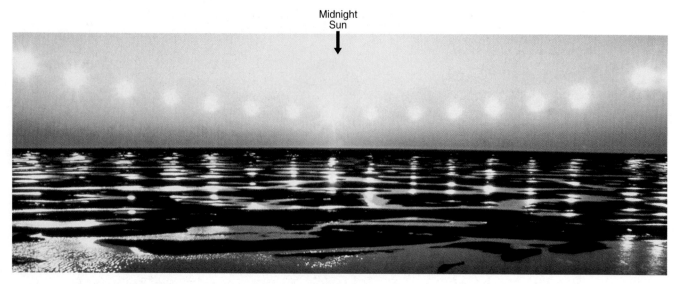

FIGURE 2.16 The midnight Sun.
The midnight Sun north of the Arctic Circle captured in a series of 18 exposures on the same piece of film. The camera is facing due north. Midnight is the exposure showing the Sun closest to the horizon. The photographer removed the lens cap at regular intervals to make the multiple exposures. [Photo by Gary Braasch/Tony Stone Images.]

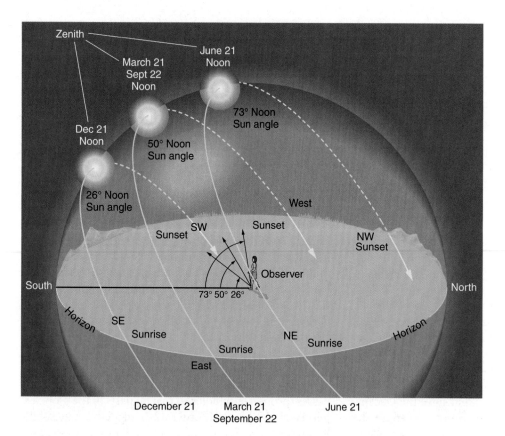

FIGURE 2.17 Seasonal observations—sunrise, noon, and sunset through the year.
Seasonal observations are at 40° N latitude for the December solstice, March equinox, June solstice, and September equinox. The Sun's altitude increases from 26° in December to 73° above the horizon in June—a difference of 47°. Note the changing position of sunrise and sunset along the horizon during the year. For a useful sunrise and sunset calculator for any location, go to http://www.srrb.noaa.gov/highlights/sunrise/sunrise.html.

(a)   January

(b)   April

(c)   July

(d)   November

FIGURE 2.18 The four seasons.
Seasonality produces dramatic change in the leaves of an ornamental pear tree (*Pyrus calleryana*) in January, April, July, and November. [Photos by author.]

Dawn and Twilight   *Dawn* is the period of diffused light that occurs before sunrise. The corresponding evening period after sunset is *twilight*. During both periods, light is scattered by molecules of atmospheric gases and reflected by dust and moisture illuminating the atmosphere. The duration of both is a function of latitude, because the angle of the Sun's path above the horizon determines the thickness of the atmosphere through which the Sun's rays must pass. This effect may be enhanced by the presence of pollution aerosols and suspended particles from volcanic eruptions or forest and grassland fires.

At the equator, where the Sun's rays are almost directly above the horizon throughout the year, dawn and twilight are limited to 30–45 minutes each. These times increase to 1–2 hours each at 40° latitude, and at 60° latitude they each range upward from 2.5 hours, with little true night in summer. The poles experience about 7 weeks of dawn and 7 weeks of twilight, leaving only 2.5 months of darkness during the 6 months when the Sun is completely below the horizon.

Seasonal Observations   In the midlatitudes of the Northern Hemisphere, the position of sunrise on the horizon migrates from day to day, from the southeast in December to the northeast in June. Over the same period, the point of sunset migrates from the southwest to the northwest. The Sun's altitude at local noon at 40° N latitude increases from a 26° angle above the horizon at the winter (December) solstice to a 73° angle above the horizon at the summer (June) solstice—a range of 47° (Figure 2.17).

Seasonal change is quite noticeable across the landscape away from the equator, as shown by the four photos in Figure 2.18. Figure 2.19 presents two composite images of vegetation cover in winter and late summer as recorded by the AVHRR sensors aboard polar-orbiting satellites. Recently, the timing of the seasons is changing as global climates shift in middle and high latitudes. Spring and leafing out is occurring as much as two weeks earlier than expected from average conditions. Think back over the past year. What seasonal changes have you observed in vegetation, temperatures, and weather?

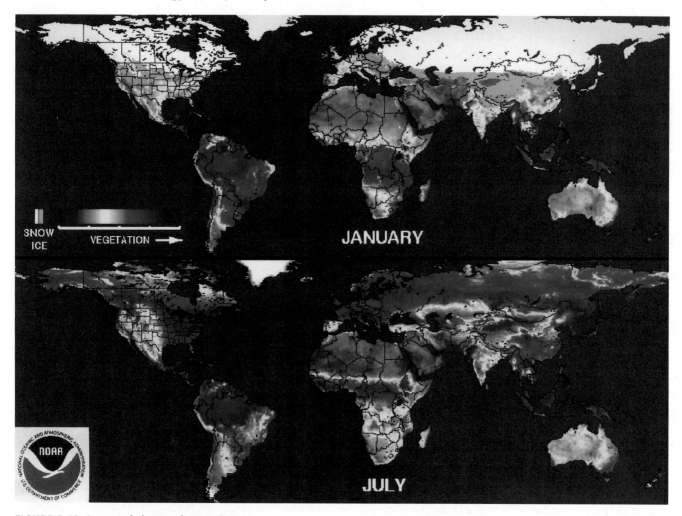

FIGURE 2.19 Seasonal change from orbit.
Seasonal change is monitored and measured by sensors aboard polar-orbiting satellites. Compare and contrast similar regions for January and July: the American Midwest, China, Argentina, and Europe. Orbital remote sensing is tracking an earlier spring season world-wide. [Images courtesy of Garik Gutman, NOAA, National Environmental Satellite, Data, and Information Service.]

Summary and Review—Solar Energy to Earth and the Seasons

● *Distinguish* among galaxies, stars, and planets and *locate* Earth.

Our Solar System—Sun and nine planets—is located on a remote, trailing edge of the **Milky Way Galaxy**, a flattened, disk-shaped mass estimated to contain up to 400 billion stars. **Gravity**, the mutual attracting force exerted by the mass of an object upon all other objects, is an organizing force in the Universe. The process of suns (stars) condensing from nebular clouds with planetesimals (protoplanets) forming in orbits around their central masses is the **planetesimal hypothesis**.

Milky Way Galaxy (p. 43)
gravity (p. 44)
planetesimal hypothesis (p. 44)

1. Describe the Sun's status among stars in the Milky Way Galaxy. Describe the Sun's location, size, and relationship to its planets.
2. If you have seen the Milky Way at night, briefly describe it. Use specifics from the text in your description.
3. Briefly describe Earth's origin as part of the Solar System.
4. Compare the locations of the nine planets of the Solar System.

● *Overview* the origin, formation, and development of Earth and *construct* Earth's annual orbit about the Sun.

The Solar System, planets, and Earth began to condense from a nebular cloud of dust, gas, debris, and icy comets approximately 4.6 billion years ago. Distances in space are so vast that the **speed of light** (300,000 kmps, or 186,000 mps, which is about 9.5 trillion kilometers, or nearly 6 trillion miles, per year) is used to express distance.

In its orbit, Earth is at **perihelion** (its closest position to the Sun) during our Northern Hemisphere winter (January 3 at 147,255,000 km, or 91,500,000 mi). It is at **aphelion** (its farthest position from the Sun) during our Northern Hemisphere summer (July 4 at 152,083,000 km, or 94,500,000 mi). Earth's average distance from the Sun is approximately 8 minutes and 20 seconds in terms of light speed. In the Solar System, an imaginary plane touching all points of Earth's orbit is termed the **plane of the ecliptic**.

> speed of light (p. 44)
> perihelion (p. 44)
> aphelion (p. 44)
> plane of the ecliptic (p. 44)

5. How far is Earth from the Sun in terms of light speed? In terms of kilometers and miles?

6. Briefly describe the relationship among these concepts: Universe, Milky Way Galaxy, Solar System, Sun, and Planet Earth.

7. Diagram in a simple sketch Earth's orbit about the Sun. How much does it vary during the course of a year?

● *Describe* the Sun's operation and *explain* the characteristics of the solar wind and the electromagnetic spectrum of radiant energy.

The **fusion** process—hydrogen atoms forced together under tremendous temperature and pressure in the Sun's interior—generates incredible quantities of energy. The Sun's most conspicuous features are large **sunspots**, caused by magnetic disturbances. Solar energy in the form of charged particles of **solar wind** travels out in all directions from disturbances on the Sun. Solar wind is deflected by Earth's **magnetosphere**, producing various effects in the upper atmosphere, including spectacular **auroras**, the northern and southern lights, which surge across the skies at higher latitudes. Another effect of the solar wind in the atmosphere is its possible influence on weather.

The **electromagnetic spectrum** of radiant energy travels outward in all directions from the Sun. The total spectrum of this radiant energy is made up of different **wavelengths**—the distance between corresponding points on any two successive waves of radiant energy. Eventually some of this radiant energy reaches Earth's surface.

> fusion (p. 44)
> solar wind (p. 46)
> sunspots (p. 46)
> magnetosphere (p. 46)
> auroras (p. 46)
> electromagnetic spectrum (p. 48)
> wavelength (p. 48)

8. How does the Sun produce such tremendous quantities of energy?

9. What is the sunspot cycle? At what stage was the cycle in the year 2001?

10. Describe Earth's magnetosphere and its effects on the solar wind and the electromagnetic spectrum.

11. Summarize the presently known effects of the solar wind relative to Earth's environment.

12. Describe the various segments of the electromagnetic spectrum, from shortest to longest wavelength. What are the main wavelengths produced by the Sun? Which are principally radiated by Earth to space?

● *Portray* the intercepted solar energy and its uneven distribution at the top of the atmosphere.

Electromagnetic radiation from the Sun passes through Earth's magnetic field to the top of the atmosphere—the **thermopause**, at approximately 500 km (300 mi) altitude. Solar radiation that reaches a horizontal plane at Earth is called **insolation**, a term specifically applied to radiation arriving at Earth's surface and atmosphere. Insolation at the top of the atmosphere is expressed as the **solar constant**: the average insolation received at the thermopause when Earth is at its average distance from the Sun. The solar constant is measured as 1372 W/m^2 (2.0 cal/cm^2/min; 2 langleys/min). The place receiving maximum insolation is the **subsolar point**, where solar rays are perpendicular to the surface (radiating from directly overhead). All other locations away from the subsolar point receive slanting rays and more diffuse energy.

> thermopause (p. 49)
> insolation (p. 49)
> solar constant (p. 50)
> subsolar point (p. 51)

13. What is the solar constant? Why is it important to know?

14. Select 40° or 50° north latitude on Figure 2.10 and plot the amount of energy in watts per square meter (W/m^2) per day on a graph for each month throughout the year. Compare this with the amount at the North Pole and at the equator.

15. If Earth were flat and oriented perpendicularly to incoming solar radiation (insolation), what would be the latitudinal distribution of solar energy at the top of the atmosphere?

● *Define* solar altitude, solar declination, and daylength and *describe* the annual variability of each—Earth's seasonality.

The angle between the Sun and the horizon is the Sun's **altitude**. The Sun's **declination** is the latitude of the subsolar point. Declination annually migrates through 47° of latitude, moving between the *Tropic of Cancer* at 23.5° N (June) and the *Tropic of Capricorn* at 23.5° S latitude (December). Seasonality means an annual change in the Sun's altitude and changing **daylength**, or duration of exposure.

Earth's distinct seasons are produced by interactions of **revolution** (annual orbit about the Sun), **rotation** (turning on the axis), **axial tilt** (23.5° from a perpendicular to the plane of the ecliptic), **axial parallelism** (the parallel alignment of the axis throughout the year), and sphericity. Earth rotates about its **axis**, an imaginary line extending through the planet from the geographic North Pole to the South Pole. As it rotates, the traveling boundary that divides daylight and darkness is called the **circle of illumination**. Earth's **axial tilt** is 23.5° from a perpendicular to the plane of the ecliptic and it remains oriented to the stars in the same direction throughout the year, or **axial parallelism**. Daylength is the interval between **sunrise**, the moment when the disk of the Sun first appears above the horizon in the east, and **sunset**, that moment when it totally disappears below the horizon in the west.

On December 21 or 22, at the moment of the **winter solstice** ("winter Sun stance"), or **December solstice**, the circle of illumination excludes the North Pole but includes the South Pole. The subsolar point is at 23.5° S latitude, the parallel called the Tropic of Capricorn. The moment of the **vernal equinox**, or **March equinox**, occurs on March 20 or 21. At that time, the circle of illumination passes through both poles so that all locations on Earth experience a 12-hour day and a 12-hour night.

June 20 or 21 is the moment of the **summer solstice**, or **June solstice**. The subsolar point now has shifted from the equator to 23.5° N latitude, the Tropic of Cancer. Because the circle of illumination now includes the North Polar region, everything north of the Arctic Circle receives 24 hours of daylight—the "midnight Sun." September 22 or 23 is the time of the **autumnal equinox**, or **September equinox**, when Earth's orientation is such that the circle of illumination again passes through both poles so that all parts of the globe experience a 12-hour day and a 12-hour night.

altitude (p. 53)
declination (p. 54)
daylength (p. 54)
revolution (p. 54)
rotation (p. 54)
axis (p. 55)
circle of illumination (p. 55)
axial tilt (p. 55)
axial parallelism (p. 56)
sunrise (p. 56)
sunset (p. 56)
winter solstice, December solstice (p. 56)
vernal equinox, March equinox (p. 57)
summer solstice, June solstice (p. 57)
autumnal equinox, September equinox (p. 57)

16. How is the Stonehenge monument related to seasons?

17. The concept of seasonality refers to what specific phenomena? How do these two aspects of seasonality change during a year at 0° latitude? At 40°? At 90°?

18. Differentiate between the Sun's altitude and its declination at Earth's surface.

19. For the latitude at which you live, how does daylength vary during the year? How does the Sun's altitude vary? Does your local newspaper publish a weather calendar containing such information?

20. List the five physical factors that operate together to produce seasons.

21. Describe Earth's revolution and rotation, and differentiate between them.

22. Define Earth's present tilt relative to its orbit about the Sun.

23. Describe seasonal conditions at each of the four key seasonal anniversary dates during the year. What are the solstices and equinoxes, and what is the Sun's declination at these times?

NetWork

The *Geosystems* Home Page provides on-line resources for this chapter on the World Wide Web. To begin: Once on the Home Page, click on this textbook, scroll the Table of Contents menu, select this chapter, and click "Begin." You will find self-tests that are graded, review exercises, specific updates for items in the chapter, and many links to interesting related pathways on the Internet under "Destinations." *Geosystems* is at **http://www.prenhall.com/christopherson**.

Critical Thinking

A. Using the concepts in Figure 2.17, use a protractor and stick or ruler to measure the angle of the Sun's altitude at noon (or 1 P.M., if in daylight saving time). Do not look at the Sun; rather, with your back to the Sun, align the stick so that it casts no shadow as you measure its rays against the protractor. Place this measurement in your notebook and affix a Post-It® in the textbook to remind you to repeat the measurement near the end of the semester. Compare and analyze seasonal change using your different measurements of the Sun's altitude.

B. Also, in reference to the concepts in Figure 2.17, post a reminder in your notebook to check the position of sunrise and sunset at least twice during the semester. If you have a magnetic compass, note the degrees from north for sunrise and sunset (*azimuth* is read from north in a clockwise direction—0° and 360° being the same point, north). Find some place where you can see the horizon. If new to such observations, it will be a surprise as to the degree of location change over the span of months in a school term.

C. The variability of Earth's axial tilt, orbit about the Sun, and a wobble to the axis is described in Figure 17.31. Please refer to this figure and compare these changing conditions to the information in Table 2.1 and the related figures in this chapter. Speculate on the effects of these changes on the annual march of the seasons.

A powerful cumulonimbus storm cloud system over the Great Salt Lake, Utah. The atmosphere is the setting for dramatic displays of energy and water. [Photo by Bobbé Christopherson.]

3

Earth's Modern Atmosphere

Key Learning Concepts

After reading the chapter, you should be able to:

- *Construct* a general model of the atmosphere based on the criteria composition, temperature, and function, and *diagram* this model in a simple sketch.

- *List* the stable components of the modern atmosphere and their relative percentage contributions by volume, and *describe* each.

- *Describe* conditions within the stratosphere; specifically, *review* the function and status of the ozonosphere (ozone layer).

- *Distinguish* between natural and anthropogenic variable gases and materials in the lower atmosphere.

- *Describe* the sources and effects of carbon monoxide, nitrogen dioxide, and sulfur dioxide, and *construct* a simple equation that illustrates photochemical reactions that produce ozone, peroxyacetyl nitrates, nitric acid, and sulfuric acid.

Earth's atmosphere is a unique reservoir of gases, the product of nearly 5 billion years of development. It sustains us and protects us from hostile radiation and particles from the Sun and beyond—the atmosphere serves as an efficient filter. When astronauts work in space, they must wear a bulky spacesuit that does everything to sustain and protect them that the atmosphere does for us all the time.

In this chapter: We examine the modern atmosphere through its composition, temperature, and function. Our consideration of the atmosphere also includes the spatial aspects of human-induced inputs, such as air pollution. We all participate in the atmosphere with each breath we take, the energy we consume, the traveling we do, and the products we buy. Human activities caused the stratospheric ozone predicament and the blight of acid deposition on the landscape. These topics are essential, for we are influencing the atmosphere of the future.

Atmospheric Composition, Temperature, and Function

The *modern* atmosphere probably is the fourth general atmosphere in Earth's history. Therefore, this modern atmosphere is a gaseous mixture of ancient origin, the sum of all the exhalations and inhalations of life on Earth throughout time. The principal substance of this atmosphere is air, the medium of life as well as a major industrial and chemical raw material. *Air* is a simple mixture of gases that is naturally odorless, colorless, tasteless, and formless, blended so thoroughly that it behaves as if it were a single gas.

In his book *The Lives of a Cell*, the late physician and self-styled "biology watcher" Lewis Thomas compared the atmosphere of Earth to an enormous cell membrane. The membrane around a cell regulates the interactions between the cell's delicate inner workings and the potentially disruptive outer environment. Each cell membrane is very selective as to what it will allow to pass through. The modern atmosphere acts as Earth's protective membrane, as Thomas described so vividly (Figure 3.1).

As a practical matter, we consider the top of our atmosphere to be around 480 km (300 mi) above Earth's surface, the same altitude we use in Chapter 2 for measuring the solar constant and insolation receipt. Beyond that altitude, the atmosphere is rarefied (nearly a vacuum) and is called the **exosphere**, which means "outer sphere." It contains scarce lightweight hydrogen and helium atoms, weakly bound by gravity as far as 32,000 km (20,000 mi) from Earth.

Atmospheric Profile

Earth's modern atmosphere is in a series of imperfectly shaped concentric "shells" or "spheres" that grade into one another, all bound to the planet by gravity. As critical as the atmosphere is to us, it represents only the thinnest envelope, amounting to less than one-millionth of Earth's total mass. We study the atmosphere by viewing it in layers that have distinctive properties and purposes. Figure 3.2 charts essential aspects of the atmosphere in a vertical cross section, or side view, and is key to the following discussion. We simplify this complexity by using three atmospheric criteria: *composition*, *temperature*, and *function*—noted along the left side of Figure 3.2a with the relations among the criteria detailed in Table 3.1.

Earth's atmosphere exerts its weight, pressing downward under the pull of gravity. Air molecules create air pressure through their motion, size, and number. Pressure is exerted on all surfaces in contact with the air. The weight

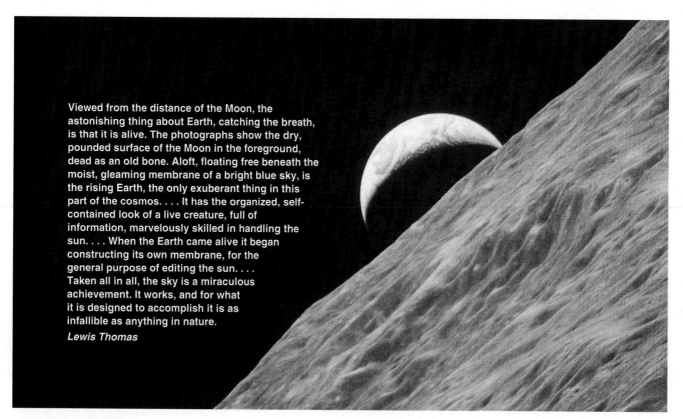

Viewed from the distance of the Moon, the astonishing thing about Earth, catching the breath, is that it is alive. The photographs show the dry, pounded surface of the Moon in the foreground, dead as an old bone. Aloft, floating free beneath the moist, gleaming membrane of a bright blue sky, is the rising Earth, the only exuberant thing in this part of the cosmos. . . . It has the organized, self-contained look of a live creature, full of information, marvelously skilled in handling the sun. . . . When the Earth came alive it began constructing its own membrane, for the general purpose of editing the sun. . . . Taken all in all, the sky is a miraculous achievement. It works, and for what it is designed to accomplish it is as infallible as anything in nature.
Lewis Thomas

FIGURE 3.1 Earthrise.
Earthrise over the stark and lifeless lunar surface. [Photo by NASA; quotation from "The World's Biggest Membrane" from *The Lives of a Cell* by Lewis Thomas. Copyright © 1973 by the Massachusetts Medical Society. Originally published in the *New England Journal of Medicine*. Reprinted by permission of Viking Penguin, a division of Penguin Books USA, Inc.]

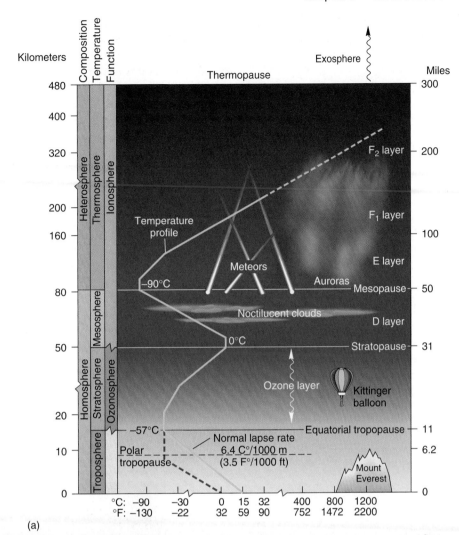

(b)

FIGURE 3.2 Profile of the modern atmosphere.
(a) An integrated chart of our modern atmosphere. The chart spans the atmosphere from
Earth's surface to the top of the atmosphere at 480 km (300 mi). The columns along the left
side show the division of the atmosphere by composition, temperature, and function. The
plot of temperature and the scale along the bottom permit you to tell the temperature at
any altitude. (The small balloon shows the height achieved by Kittinger discussed in News Report 3.1.) (b) Space Shuttle astronauts captured a dramatic sunset through various atmospheric layers across the "edge" of our planet—called *Earth's limb*. A silhouetted
cumulonimbus thunderhead cloud is seen topping out at the tropopause. [Space Shuttle
photo from NASA.]

(force over a unit area) of the atmosphere, or **air pressure**, pushes in on all of us. Fortunately, that same pressure also exists inside us, pushing outward; otherwise we would be crushed by the mass of air around us.

The atmosphere exerts an average force of approximately 1 kg/cm^2 (14.7 lb/in.^{2}) at sea level. Gravity compresses air, making it denser near Earth's surface; it thins rapidly with increasing altitude (Figure 3.3a). Consequently,

Table 3.1 **Atmospheric Criterion**		
Composition	**Temperature**	**Function**
Heterosphere	Thermosphere	Ionosphere
Homosphere	Mesosphere	Ozonosphere
	Stratosphere	
	Troposphere	

over half the total mass of the atmosphere is compressed below 5500 m (18,000 ft), 75% is compressed below 10,700 m (35,100 ft), and 90% is below 16,000 m (52,500 ft). All but 0.1% of the atmosphere exists below an altitude of 50 km (31 mi), as shown in the pressure profile in Figure 3.3b (percentage column is farthest to the right).

Few people are aware that in routine air travel they are sitting above 80% of the total atmospheric volume! To better understand this pressure profile, imagine sky diving from a high-altitude balloon 33 km (20 mi) above Earth. What would you experience? How fast would you fall? What sounds would you hear? See News Report 3.1 to read about someone who did this very thing.

At sea level, the atmosphere exerts a pressure of 1013.2 mb (*millibar*, *mb*, force per square meter of surface area) or 29.92 in. of mercury, as measured by a barometer. In Canada and other countries normal air pressure is expressed as 101.32 kPa (*kilopascal*; 1 kPa = 10 mb). More information on air pressure, the instruments that measure it, and the role air pressure plays in generating winds is in Chapter 6.

Atmospheric Composition Criterion

Using chemical *composition* as a criterion, the atmosphere divides into two broad regions, the *heterosphere* (80 to 480 km altitude) and the *homosphere* (Earth's surface to 80 km altitude), as shown along the left side of Figure 3.2a. As you read, note that we follow the same path that incoming solar radiation travels through the atmosphere to Earth's surface.

Heterosphere   The **heterosphere** is the outer atmosphere in terms of composition. It begins at about 80 km (50 mi) altitude and extends outward to the transition to the exosphere and interplanetary space (see Figure 3.2). The International Space Station and most Space Shuttle missions orbit in the upper heterosphere.

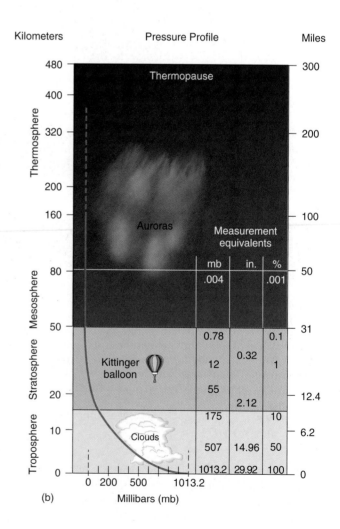

FIGURE 3.3   Density decreases with altitude.
(a) The atmosphere, denser near Earth's surface, rapidly decreases in density with altitude. The difference in density is easily measured because air exerts its weight as pressure. Have you experienced pressure changes that you could feel on your eardrums? How high above sea level were you at the time?
(b) Pressure profile of the atmosphere. Note the rapid decrease in atmospheric pressure with altitude and that about 90% of the atmospheric mass resides in the troposphere.

News Report 3.1

Falling Through the Atmosphere—The Highest Sky Dive

Imagine a small, unpressurized compartment large enough for only one person, floating at 31.3 km (19.5 mi) altitude, carried to such height by a helium-filled balloon. Even though it is daytime, the sky is almost black for this is above all but 1.0% of the atmospheric mass. The air pressure at 31,300 m (102,800 ft) is barely measurable—this altitude is the beginning of space in experimental aircraft testing. The year is 1961. (See the position of his balloon on Figures 3.2 and 3.3b.)

Captain Joseph Kittinger, Jr., an Air Force officer, stood looking out of the opening in his capsule. He paused for the view of Earth's curved horizon, the New Mexico landscape over 31 km below, and the dark heavens above, and then leaped into the stratospheric void (Figure 1). He placed himself at tremendous personal risk for an experimental reentry into the atmosphere. What do you think he would initially hear and feel? At what speed would he fall? What would the temperature be? How long would he fall before opening his parachute?

Initially frightening to him was that he heard nothing, no rushing sound, for there was not enough air to produce any sound. The fabric of his pressure suit did not flutter, for there was not enough air to create friction against the fibers. He had no sensation of movement until he glanced back at the balloon and capsule. They retreated rapidly from his motionless perception. This lack of a sense of speed as he fell was unnerving.

His speed was remarkable. Near the ground, objects fall at "terminal velocity," reaching about 200 kmph (125 mph) in a few seconds. However, in the rarefied stratosphere the lack of air resistance permits incredible speeds. Captain Kittinger quickly accelerated to 988 kmph (614 mph), slightly less than the speed of sound at sea level.

FIGURE 1   Stratospheric leap into history.
Moments into Captain Joseph Kittinger's historic exploration of the atmosphere, captured by a remotely triggered camera. The lanyard cord is attached to his parachute pack and snaps to set a timer that opens a small stabilization chute 16 seconds into the free fall. The clouds are more than 26,000 m (85,000 ft) below him. He carries an instrument pack on his seat, his main chute, and pure oxygen for his breathing mask.
[Volkmar Wentzel/NGS Image Collection, used by permission from the National Geographic Society, *National Geographic Magazine*, Dec. 1960, p. 855. All rights reserved.]

His free fall took him through the stratosphere and its ozone layer. As he encountered denser layers of atmospheric gases, frictional drag slowed his body. As he fell past 15,240 m (50,000 ft) he slowed to 400 kmph (250 mph). He dropped into the lower atmosphere, finally falling below airplane flying altitudes. His free fall lasted 4 minutes and 25 seconds, slowing to terminal velocity and the opening of his main chute at 5500 m (18,000 ft). He safely drifted to Earth's surface. This remarkable 13-minute 35-second voyage through 99% of the atmospheric mass remains a record to this day. Kittinger placed himself in harm's way in this experiment to better understand the atmosphere in the early beginnings of human space exploration.

As the prefix *hetero-* implies, this region is not uniform—its gases are *not evenly mixed*. This distribution is quite different from the nicely blended gases we breathe near Earth's surface, in the homosphere. Gases in the heterosphere occur in distinct layers sorted by gravity according to their atomic weight, with the lightest elements (hydrogen and helium) at the margins of outer space and the heavier elements (oxygen and nitrogen) dominant in the lower heterosphere. Less than 0.001% of the atmosphere's mass is in this rarified heterosphere.

Homosphere Below the heterosphere is the other compositional shell of the atmosphere, the **homosphere**. This region extends from an altitude of 80 km (50 mi) to Earth's surface. Even though the atmosphere rapidly changes density in the homosphere—increasing toward Earth's surface—the blend (proportion) of gases is nearly uniform throughout the homosphere. The only exceptions are the concentration of ozone (O_3) in the "ozone layer," from 19 to 50 km (12 to 31 mi), and the variations in water vapor, pollutants, and some trace chemicals in the lowest portion of the atmosphere.

The stable mixture of gases comprising air in the homosphere evolved slowly. The present proportion, which includes oxygen, evolved approximately 500 million years ago. Table 3.2 lists by volume the stable ingredients that constitute dry, clean air in the homosphere.

The homosphere is a vast reservoir of relatively inert *nitrogen*, originating principally from volcanic sources. Nitrogen is a key element of life, yet we exhale all the nitrogen we inhale. The explanation for this contradiction is that nitrogen integrates into our bodies not from the air we breathe but through compounds in food. In the soil, nitrogen is bound to these compounds by nitrogen-fixing

FIGURE 3.4 An air-mining operation.
Air is a major industrial and chemical raw material that is extracted from the atmosphere by air-mining companies. Nitrogen, oxygen, and argon are extracted using a cryogenic process (very low temperatures). [Photo by author.]

bacteria, and it returns to the atmosphere by denitrifying bacteria that remove nitrogen from organic materials. A complete discussion of the nitrogen cycle is in Chapter 19.

Oxygen, a by-product of photosynthesis, also is essential for life processes. Slight spatial variations occur in the percentage of oxygen in the atmosphere because of variations in photosynthetic rates with latitude, seasonal changes, and the lag time as atmospheric circulation slowly mixes the air. Although it forms about one-fifth of the atmosphere, oxygen forms compounds that compose about half of Earth's crust. Oxygen readily reacts with many elements to form these materials. Both nitrogen and oxygen reserves in the atmosphere are so extensive that, at present, they far exceed human capabilities to disrupt or deplete them.

The gas *argon*, constituting about 1% of the homosphere, is completely inert (an unreactive "noble" gas) and therefore is unusable in life processes. Argon is a residue from the radioactive decay of an isotope (form) of potassium called potassium-40 (symbolized ^{40}K). Slow accumulation over millions of years accounts for all the argon present in the modern atmosphere. Because industry has found uses for inert argon (in light bulbs, welding, and some lasers), it is extracted or "mined" from the atmosphere, in addition to nitrogen and oxygen, for commercial and industrial uses (Figure 3.4).

Carbon dioxide is a natural by-product of life processes. It is essentially a stable atmospheric component, qualifying it for inclusion in Table 3.2. Although its present percentage in the atmosphere is small at more than 0.037%, it is important in maintaining global temperatures. Its per-

Table 3.2	Stable Components of the Modern Homosphere	
Gas (Symbol)	Percentage by Volume	Parts per Million (ppm)
Nitrogen (N_2)	78.084	780,840
Oxygen (O_2)	20.946	209,460
Argon (Ar)	0.934	9,340
Carbon dioxide (CO_2)*	0.037	369.7
Neon (Ne)	0.001818	18
Helium (He)	0.000525	5
Methane (CH_4)	0.00014	1.4
Krypton (Kr)	0.00010	1.0
Ozone (O_3)	Variable	
Nitrous oxide (N_2O)	Trace	
Hydrogen (H)	Trace	
Xenon (Xe)	Trace	

*2000 average measured at Mauna Loa, Hawai'i (see: http://cdiac.esd.ornl.gov/ftp/maunaloa-co2/maunaloa.co2).

centage has increased over the past 200 years as a result of human activities. Chapter 10 discusses the implications of this increase in global warming and climate change.

Atmospheric Temperature Criterion

Shifting to temperature as a criterion, the atmosphere has four distinct temperature zones—the *thermosphere, mesosphere, stratosphere,* and *troposphere* (labeled in Figure 3.2).

Thermosphere   We define the **thermosphere** ("heat sphere") as roughly corresponding to the heterosphere (80 km out to 480 km, or 50–300 mi). The upper limit of the thermosphere is called the **thermopause** (the suffix *-pause* means "to change"). During periods of a less active Sun (fewer sunspots and coronal bursts), the thermopause may lower in altitude from the average 480 km (300 mi) to only 250 km altitude (155 mi). An active Sun will cause the outer atmosphere to swell to an altitude of 550 km (340 mi), where it can create frictional drag on satellites in low orbit.

The temperature profile in Figure 3.2 (yellow curve) shows that temperatures rise sharply in the thermosphere, to 1200°C (2200°F) and higher. Despite such high temperatures, the thermosphere is not "hot" in the way you might expect. Temperature and heat are different concepts. The intense solar radiation in this portion of the atmosphere excites individual molecules (principally nitrogen and oxygen) to high levels of vibration. This **kinetic energy**, the energy of motion, is the vibrational energy that we measure as *temperature*.

However, the actual heat involved is very small. The reason is that the density of molecules is so low. There is little actual **heat** produced, or the flow of kinetic energy from one body to another because of a temperature difference between them. Heating in the atmosphere near Earth's surface is different because the greater number of molecules in the denser atmosphere transmits their kinetic energy as **sensible heat**, meaning that we can measure it. (Density, temperature, and heat capacity determine the sensible heat of a substance.) There is more on this topic in Chapters 4 and 5.

Mesosphere   The **mesosphere** is the area from 50 to 80 km (30 to 50 mi) above Earth and is the highest in altitude of the three temperature regions within the homosphere. As Figure 3.2 shows, the mesosphere's outer boundary, the *mesopause,* is the coldest portion of the atmosphere, averaging −90°C (−130°F), although that temperature may vary considerably (25–30 C°, or 45–54 F°). Note in Figure 3.3b the extremely low pressures (low density of molecules) in the mesosphere.

The mesosphere sometimes receives cosmic or meteoric dust, acting as nuclei around which fine ice crystals form. At high latitudes, an observer may see these bands of crystals glow in rare and unusual night clouds called **noctilucent clouds**. For reasons not clearly understood

these unique clouds are on the increase. See, among several sites, http://lasp.colorado.edu/noctilucent_clouds/.

Stratosphere   The **stratosphere** extends from 18 to 50 km (11 to 31 mi) from Earth's surface. Temperatures increase with altitude throughout the stratosphere, from −57°C (−70°F) at 18 km (tropopause), warming to 0°C (32°F) at 50 km at the stratosphere's outer boundary, the *stratopause.*

Troposphere   The **troposphere** is the final layer encountered by incoming solar radiation as it surges through the atmosphere to the surface. It is the home of the biosphere, the atmospheric layer that supports life, and the region of principal weather activity.

Approximately 90% of the total mass of the atmosphere and the bulk of all water vapor, clouds, air pollution, and life forms are within the troposphere. The **tropopause**, its upper limit, is defined by an average temperature of −57°C (−70°F), but its exact elevation varies with the season, latitude, and surface temperatures and pressures. Near the equator, because of intense heating from the surface, the tropopause occurs at 18 km (11 mi); in the middle latitudes, it occurs at 12 km (8 mi); and at the North and South Poles it is only 8 km (5 mi) or less above Earth's surface.

Figure 3.5 illustrates the normal temperature profile within the troposphere during daytime. As the graph shows, temperatures decrease rapidly with increasing altitude at an average of 6.4 C° per kilometer (3.5 F° per 1000 ft), a rate known as the **normal lapse rate**. This temperature plot also appears in Figure 3.2.

The normal lapse rate is an average. The actual lapse rate at any particular time and place, which may deviate considerably because of local weather conditions, is called the **environmental lapse rate**. This variation in temperature gradient in the lower troposphere is central to our discussion of weather processes (Chapter 7).

In the stratosphere, the marked warming with increasing altitude causes the tropopause to act like a lid, essentially preventing whatever is in the cooler (denser) air below from mixing into the warmer (less dense) stratosphere. However, the tropopause may be disrupted above the midlatitudes wherever jet streams produce vertical turbulence and an interchange between the troposphere and the stratosphere (Chapter 6). Also, hurricanes occasionally inject moisture above this inverted temperature layer at the tropopause, and powerful volcanic eruptions may loft ash and sulfuric acid mists into the stratosphere, as did the Mount Pinatubo eruptions in 1991.

Atmospheric Function Criterion

Looking at our final atmospheric criterion of *function,* the atmosphere has two specific zones that remove most of the harmful wavelengths of incoming solar radiation and charged particles: the *ionosphere* and the *ozonosphere*

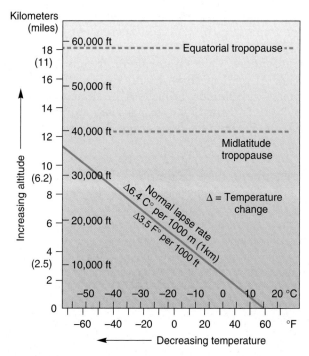

FIGURE 3.5 The temperature profile of the troposphere. During daytime, temperature decreases with increased altitude at a rate known as the *normal lapse rate*. Scientists use a concept called the *standard atmosphere* as an accepted description of air temperature and pressure changes with altitude. The values in this profile are part of the standard atmosphere. Note the approximate locations of the equatorial and midlatitude tropopauses.

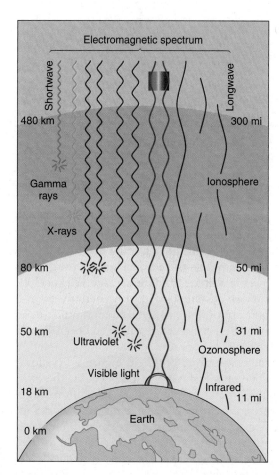

FIGURE 3.6 The atmosphere protects Earth's surface. As solar energy passes through the atmosphere the shortest wavelengths are absorbed. Only a fraction of the ultraviolet radiation, and most of the visible light and infrared, reaches Earth's surface. When above these protective layers astronauts must wear a spacesuit that duplicates this filtering process for them to survive.

(*ozone layer*). Figure 3.6 depicts in a general way the absorption of radiation by the various functional layers of the atmosphere.

Ionosphere   The outer functional layer, the **ionosphere**, extends throughout the thermosphere and into the mesosphere below (Figure 3.2). The ionosphere absorbs cosmic rays, gamma rays, X-rays, and shorter wavelengths of ultraviolet radiation, changing atoms to positively charged ions and giving the ionosphere its name. The glowing auroral lights discussed in Chapter 2 occur principally within the ionosphere.

Figure 3.2 shows the average daytime altitudes of four regions within the ionosphere, known as the D, E, F_1, and F_2 layers. These are important to broadcast communications for they reflect certain radio wavelengths, including AM radio and other shortwave broadcasts, especially at night. However, during the day the ionosphere actively absorbs arriving radio signals, preventing distant reception. Normally unaffected are FM or television broadcast wavelengths, which pass through to space, thus necessitating use of communication satellites.

Ozonosphere   That portion of the stratosphere that contains an increased level of ozone is the **ozonosphere**, or

ozone layer. Ozone is a highly reactive oxygen molecule made up of three oxygen atoms (O_3) instead of the usual two atoms (O_2) that make up oxygen gas. Ozone absorbs wavelengths of ultraviolet light (0.1–0.3 μm) and subsequently reradiates this energy at longer wavelengths, as infrared radiation. This process converts most harmful ultraviolet radiation, effectively "filtering" it and safeguarding life at Earth's surface.

The ozone layer is presumed to have been relatively stable over the past several hundred million years (allowing for daily and seasonal fluctuations). Today, however, it is in a state of continuous change. Focus Study 3.1 presents an analysis of the crisis in this critical portion of our atmosphere. Fortunately, international treaties to prevent further losses appear to be working. Nobel Prizes were awarded to the pioneering scientists who led the way to these environmental and political successes (see News Report 3.2).

Stratospheric Ozone Losses: A Worldwide Health Hazard

Consider these news reports and conditions:

- The depletion of stratospheric ozone above Antarctica reached record losses in 2000, and comparable levels in 2001, covering an area three times larger than the United States.
- An international scientific consensus confirmed previous assessments of the anthropogenic (human-caused) disruption of the ozone layer—chlorine atoms and chlorine monoxide molecules in the stratosphere are of human origin. (See *Scientific Assessment of Ozone Depletion*, by NASA, NOAA, United Nations Environment Programme, and World Meteorological Organization.)
- During the months of Antarctic spring (September to November), ozone concentrations approach total depletion in the stratosphere over an area twice the size of the Antarctic continent. Each year, the "ozone hole" (actually a thinning) widens and deepens. The protective ozone layer is being depleted over southern South America, southern Africa, Australia, and New Zealand.
- At Earth's opposite pole, a similar ozone depletion over the Arctic annually exceeds 30%, and it dropped to a record 45% loss during spring 1997. Canadian and U.S. governments report an "ultraviolet index" to help the public protect themselves.
- Environment Canada operates a scientific observatory to monitor ozone losses over Canada. This high-Arctic facility is at a remote weather station near Eureka on Ellesmere Island, Nunavut, Canada, about 1000 km (620 mi) from the North Pole (see http://www.ec.gc.ca/ozone/indexe.htm).
- Overall ozone losses in the midlatitudes are continuing at 6% to 8% per decade. In North America, related skin cancers are increasing, totaling more than 1 million cases a year, of which some 41,000 are malignant melanomas averaging 10,000 deaths. Skin cancer is increasing at 4.0% per year. Most affected are

light-skinned persons who live at higher elevations and those who work principally outdoors (see http://www.cancer.org/).

- Increased ultraviolet is affecting atmospheric chemistry, biological systems, oceanic phytoplankton (small photosynthetic organisms that form the basis of the ocean's primary food production) and fisheries, crop yields, and human skin, eye tissues, and immunity.

More ultraviolet radiation than ever before is breaking through Earth's protective ozone layer. What is happening in the stratosphere everywhere and in the polar regions specifically? Why is this happening, and how are people and their governments responding? What effects does this have on you personally?

Monitoring Earth's Fragile Safety Screen

A sample from the ozone layer's densest part (at 29 km, or 18 mi altitude) contains only 1 part ozone per 4 million parts of air—compressed to surface pressure, it would be only 3 mm thick. Yet, this rarefied layer was in steady-state equilibrium for several hundred million years, absorbing intense ultraviolet radiation and permitting life to proceed safely on Earth.

The ozone layer has been monitored since the 1920s. Ground stations

EP/TOMS Total Ozone for Sep 25, 2001

Dobson Units
Dark Gray < 100, red > 500 DU

FIGURE 1 The Antarctic ozone hole. TOMS image for September 21, 2001, uses a revised color scale to represent ozone concentrations in Dobson units, with purples for amounts less than 180, black below 125. (One Dobson unit equals 2.69×10^{16} molecules of O_3/cm^2.) Measurements have dropped below 100 Dobson units. The ozone "hole" has grown larger since 1979, covering a record 25 million km² (9 million mi²) or more than double the surface area of Antarctica—an area larger than Canada, the United States, and Mexico combined. [NASA.]

with instrumented balloons, aircraft, orbiting satellites, and a 30-station ozone-monitoring network (mostly in North America) observe stratospheric ozone. The total ozone mapping spectrometer (TOMS) began operations in 1978 aboard *Nimbus-7*, later on the *Upper Atmosphere Research Satellite* (*UARS*), *Adeos*, and *Meteor-3*, and, at the time of this writing, aboard the *Earth Probe* satellite. The September 2001 image is in Figure 1. (See http://jwocky.gsfc.nasa.gov/.)

Ozone Losses Explained

What is causing the decline in stratospheric ozone? In 1974, two atmospheric chemists, F. Sherwood Rowland and Mario Molina, hypothesized that some synthetic chemicals were releasing chlorine atoms that decompose ozone. These **chlorofluorocarbons**, or **CFCs**, are synthetic molecules of chlorine, fluorine, and carbon. (See Rowland and Molina's report: "Stratospheric sink for chlorofluoromethanes: Chlorine atom catalyzed destruction of ozone," *Nature* 249 (1974): 810.)

CFCs are stable (inert) under conditions at Earth's surface and they possess remarkable heat properties. Both qualities made them valuable as propellants in aerosol sprays and as refrigerants. Also, some 45% of CFCs were solvents in the electronics industry and

(continued)

Focus Study 3.1 (continued)

as foaming agents. Being inert, CFC molecules do not dissolve in water and do not break down in biological processes. (In contrast, chlorine compounds derived from volcanic eruptions and the ocean are water-soluble and rarely reach the stratosphere.)

Researchers Rowland and Molina hypothesized that stable CFC molecules slowly migrate into the stratosphere, where intense ultraviolet radiation splits them, freeing chlorine (Cl) atoms. This process produces a complex set of reactions that breaks up ozone molecules (O_3) and leaves oxygen gas molecules (O_2) in their place. The effect is severe, for a single chlorine atom decomposes more than 100,000 ozone molecules. The long residence time of chlorine atoms in the ozone layer (40 to 100 years) is likely to produce long-term consequences through this century from the chlorine already in place. More than 22 million metric tons (24 million tons) of CFCs were sold worldwide and subsequently released into the atmosphere by 1998. Of many identified reactions, here are simplified equations that show chlorine monoxide-produced (ClO) and chlorine-produced ozone losses. Note that chlorine monoxide (ClO) reacts with ozone (O_3) to produce a free chlorine atom (Cl) and two oxygen gas molecules (O_2). The free chlorine atom then reacts with an ozone molecule (O_3) to produce more chlorine monoxide (ClO) and oxygen gas (O_2).

Ozone depletion by ClO and by Cl:

$$ClO + O_3 \rightarrow C + 2O_2$$
$$Cl + O_3 \rightarrow ClO + O_2$$

The chlorine monoxide is then available to go through the cycle all over again. The *net reaction*, with the chlorine catalyst removed, is a loss of ozone (O_3) and its conversion to oxygen gas molecules. The problem then is that O_2 (oxygen gas), unlike O_3, is transparent to ultraviolet radiation. (See http://www.epa.gov/ozone/index.html.)

Political Realities—An International Response and the Future

Between 1976 and 1979, Canada, Sweden, Norway, and the state of Oregon banned CFC propellants in aerosols, and a U.S. federal ban began in 1978. However, more than half of U.S. production was exempted, including CFCs used as air-conditioning refrigerants and to make polyurethane foam. CFC sales initially dropped in the late 1970s, but they rose again under a 1981 presidential order that permitted the export and sales of banned products. Sales increased and hit a new peak in 1987 at 1.2 million metric tons (1.32 million tons).

Chemical manufacturers once claimed that no hard evidence existed to prove the ozone-depletion model, and they successfully delayed remedial action for 15 years. Today, with extensive scientific evidence and verification of losses, even the CFC manufacturers admit that the problem is serious. A few remaining critics are outside of the scientific community.

The *Montreal Protocol on Substances That Deplete the Ozone Layer*, as amended three times in 1990, 1992, and 1997, aims to reduce and eliminate CFC damage to the ozone layer. CFC sales are declining as many countries reduce demand and as worldwide industry phases in alternative chemicals. By the end of 2000, production of problem CFCs was a fraction of the record-sales year of 1987. Action on an international scale produced these positive results. All production of harmful CFCs will cease by 2010. If the Protocol is fully implemented, it is estimated that the stratosphere will return to nominal conditions by the end of this century. (See the United Nations Ozone Secretariat at http://www.unep.org/ozone/index-en.shtml and the Montreal Protocol at http://www.ec.gc.ca/ozone/tocmontr.htm.)

The Protocol also bans other chemicals, such as bromine, that are damaging to the atmosphere, using different target dates. Bromine sources include halon-charged fire extinguishers and methyl bromide pesticides. These bromine compounds, like CFCs, are inert in the lower atmosphere and reach the stratosphere by slow transport. They are about 50 times more effective than chlorine as catalysts at

causing reactions that destroy ozone, although they have only a residence time of two years in the atmosphere.

Mario Molina stated, "It was frustrating for many years, but it really paid off with the Protocol, which was a marvelous example of what the international community can do working together. We can see from atmospheric measurements that it is already working" (*Nature* 389 (September 18, 1997): 219).

Ozone Losses over the Poles

How do Northern Hemisphere CFCs become concentrated over the South Pole? Evidently, chlorine freed in the Northern Hemisphere midlatitudes concentrates over Antarctica through the work of atmospheric winds. Persistent cold temperatures over the South Pole and the presence of thin, icy clouds in the stratosphere promote development of the ozone hole. Over the North Pole, conditions are more changeable, so the hole is smaller, although growing each year.

Polar stratospheric clouds (PSCs) are thin clouds that are important catalysts in the release of chlorine for ozone-depleting reactions. During the long, cold winter months, a tight circulation pattern forms over the Antarctic continent—the polar vortex. Chlorine that is freed from droplets in PSCs triggers the breakdown of the otherwise inert molecules and frees the chlorine for catalytic reactions. Figure 2a graphs the negative correlation between ClO and O_3 over the Antarctic continent within the polar vortex. You can see that concentrations of ClO increase toward the pole as O_3 levels decline (Figure 2b). The ozone hole usually fills by early December with stratospheric ozone from lower latitudes, thus thinning ozone in middle and high latitudes. Figure 2c graphs the concentrations of ozone before formation of the hole and at the peak of depletion in 1997.

UV Index Helps Save Your Skin

The television weather report and the newspaper weather page regularly include the *UV Index* in forecasts. This

(continued)

Focus Study 3.1 *(continued)*

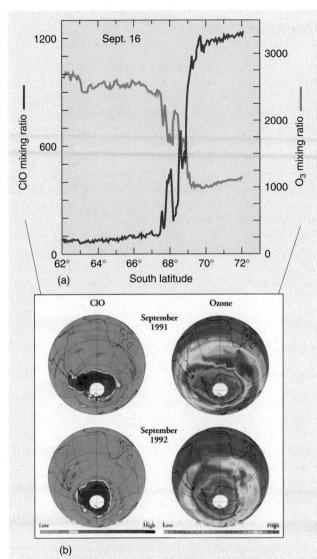

(a)

(b)

(c)

FIGURE 2 Chemical evidence of ozone damage by humans.
(a) The negative correlation between ClO and O_3 over the Antarctic continent poleward of 68° S latitude. The data were collected from flights during September 1987 at stratospheric altitudes. (b) Chlorine monoxide (ClO) and ozone (O_3) concentrations above 20 km (12.5 mi) measured during September 1991 and 1992—the beginning of the Antarctic spring—confirm causal relationship. (c) Ozone concentrations measured before and during the ozone hole development in 1997. [(a) Data from NASA; (b) Jet Propulsion Laboratory and Goddard Space Flight Center; (c) data from NOAA.]

Table 1 UV Index (EPA)		
Exposure Category/ Index Value	Minutes to Burn for "Never Tans" (Most Susceptible)	Minutes to Burn for "Rarely Burns" (Least Susceptible)
Minimal 0–2	30 minutes	> 120 minutes
Low 4	15 minutes	75 minutes
Moderate 6	10 minutes	50 minutes
High 8	7.5 minutes	35 minutes
Very high 10	6 minutes	30 minutes
15	< 4 minutes	20 minutes

item is reported widely by the National Weather Service (NWS) and the Environmental Protection Agency (EPA). Table 1 presents a sampling of the ultraviolet index numbers and their application to two skin types.

As stratospheric ozone levels continue to thin, surface exposure to cancer-causing radiation climbs. The public now is alerted to take extra precautions in the form of sunscreens, hats, and sunglasses. Remember, damage is cumulative and it may be decades before you experience the ill effects triggered by this summer's sunburn. (For more information, contact the American Cancer Society, 800-227-2345, http://www.cancer.org/; or write the American Academy of Dermatology, P.O. Box 681069, Schaumburg, IL 60168-1069.) The scientific community hopes that, with the international actions taken and the hazard to life reduced, science will have scored a significant victory.

News Report

News Report 3.2

1995 Nobel Chemistry Prize for Ozone Depletion Researchers

For the first time, Nobel Prizes were awarded to atmospheric chemists in 1995. The three recipients discovered that human-produced chemicals were destroying the stratospheric ozone layer. In making the award, the Royal Swedish Academy of Sciences said,

By explaining the chemical mechanism that affects the thickness of the ozone layer, the three researchers have contributed to our salvation from a global environmental problem that could have catastrophic consequences. It has been possible to make far-reaching decisions on prohibiting the release of the gases that destroy ozone.

Paul Crutzen is a native of Amsterdam and a researcher at the Max Plank Institute for Chemistry in Mainz, Germany. In 1970 he described the catalytic effects of nitric oxides and nitrogen dioxide in reducing stratospheric ozone. And in 1987 he discovered the role of polar stratospheric clouds and how they enhance the action of chlorine. F. Sherwood Rowland, University of California at Irvine, and Mario Molina, Massachusetts Institute of Technology, told the world that chlorofluorocarbons (CFCs) were migrating into the stratosphere and disrupting Earth's protective shield in a self-sustaining chain reaction.

Part of the prize is for the persistence of these scientists who endured

years of criticism from the affected industries. All three scientists pressed for international accords, which began with the Montreal Protocols in 1987. Subsequent strengthening of the treaty, as more serious evidence accumulated, led to total ban of the most dangerous gases by 1996. The Protocol is widely regarded as successful in slowing the accumulation of harmful compounds in the stratosphere.

The world without this treaty might prove challenging. Imagine possible stratospheric ozone losses of 50% in the midlatitudes, producing 1.5 million more cases of malignant melanoma (a skin cancer). We owe much to doctors Rowland, Molina, and Crutzen.

Variable Atmospheric Components

The troposphere contains natural and human-caused variable gases, particles, and other chemicals. The spatial aspects of these variables are important applied topics in physical geography.

Air pollution is not a new problem. Romans complained more than 2000 years ago about the foul air of their cities. Filling Roman air was the stench of open sewers, smoke from fires, and fumes from ceramic-making kilns and smelters (furnaces) that converted ores into metals. In human experience, cities are always the place where the environment's natural ability to process and recycle waste is most taxed. Air pollution is closely linked to our production and consumption of energy and resources.

Solutions require regional, national, and international strategies, because the pollution sources often are distant from the observed impact—moving across political boundaries and even crossing oceans. Regulations to curb human-caused air pollution have met with great success, although much remains to be done. Before we discuss these topics, let's examine some natural pollution sources.

Natural Sources

Natural air pollution sources produce a greater quantity of pollutants—nitrogen oxides, carbon monoxide, hydrocarbons from plants and trees, and carbon dioxide—than do human-made sources. Table 3.3 lists some of these natural sources and the substances they contribute to the air.

However, any attempt to diminish the impact of human-made air pollution through a comparison with natural sources is irrelevant, for we evolved with and adapted to the natural ingredients in the air. We did not evolve in relation to the comparatively recent concentrations of *anthropogenic* (human-caused) contaminants in our metropolitan regions.

A dramatic natural source of pollution was the 1991 eruption of Mount Pinatubo in the Philippines (15° N 120° E), perhaps the century's largest eruption. This event injected between 15 and 20 million tons of sulfur dioxide (SO_2) into the stratosphere. The spread of these emissions is shown in a sequence of satellite images in Figure 6.1, Chapter 6.

The devastating wildfires in Florida, Indonesia, and Mexico in 1998, California and Nevada in 1999, Africa in 2000, and the American West in 2001 are examples of nat-

Table 3.3 Sources of Natural Variable Gases and Materials

Source	Contribution
Volcanoes	Sulfur oxides, particulates
Forest fires	Carbon monoxide and dioxide, nitrogen oxides, particulates
Plants	Hydrocarbons, pollens
Decaying plants	Methane, hydrogen sulfides
Soil	Dust and viruses
Ocean	Salt spray and particulates

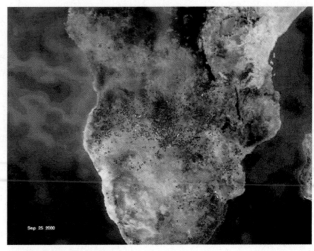

(a) Florida, 1998

(b) Southern Africa, 2000

FIGURE 3.7 Florida and Africa wildfires fill the atmosphere with smoke.
(a) Wildfires in rain-starved portions of east-central Florida are imaged from a satellite in June 1998. More than 70,000 people were evacuated at the height of the fires, which charred 121,000 hectares (300,000 acres). Damage was in the hundreds of millions of dollars. (b) Fires in drought-plagued portions of Southern Africa are caught in a satellite image from September 2000. [(a) *NOAA-8* image courtesy of NOAA/NESDIS and (b) *NOAA-14* image courtesy of NOAA/GSFC.]

ural air pollution. Soot, ash, and gases darkened skies and damaged health in affected regions. Wildfire smoke contains particulate matter (dust, smoke, soot, ash), nitrogen oxides, carbon monoxide, and volatile organic compounds. Wind spreads the pollution from the fires to nearby cities, closing airports and forcing evacuations to avoid the health-related dangers (Figure 3.7).

Natural Factors That Affect Air Pollution

The problems resulting from both natural and human-made atmospheric contaminants are made worse by several important natural factors. Among these are wind, local and regional landscape characteristics, and temperature inversions in the troposphere.

Winds   Winds gather and move pollutants from one area to another, sometimes reducing the concentration of pollution in one location while increasing it in another. Wind can produce dramatic episodes of dust movement. (Dust is defined as particles less than *62* μm, or 0.0025 in.) Traveling on prevailing winds, dust from Africa contributes to the soils of South America, and Texas dust ends up in Europe. Such movement is confirmed by chemical analysis, frequently employed by scientists to track dust to its source area (Figure 3.8).

Such air movements make the atmosphere's condition an international issue. Indeed, prevailing winds transport air pollution from the United States to Canada, causing much complaint and negotiation between the two governments.

Pollution in North America is tracked to Europe, adding to European air pollution problems. In Europe, the cross-boundary drift of pollution is a major issue because of the proximity of countries. This issue has led in part to Europe's unification and the European Union (EU).

A remarkable attribute of industrialization in the Northern Hemisphere is the haze (concentration of microscopic particles and air pollution that diminishes air clarity) across the unpopulated Arctic region, especially from November to April. This winter haze is worse toward Alaska's North Slope and extends up to 8 km (5 mi) in altitude. Simply, winds of atmospheric circulation transport pollution to sites far distant from points of origin. There is no comparable haze over the Antarctic continent.

Local and Regional Landscapes   Local and regional landscapes are another important factor in air pollution. Surrounding mountains and hills can form barriers to air movement or can direct pollutants from one area to another. Some of the worst incidents have resulted when local landscapes have trapped and concentrated air pollution.

Places such as Iceland and Hawai'i have their own natural pollution with which to deal. During periods of sustained volcanic activity at Kilauea, some 2000 metric tons (2200 tons) of sulfur dioxide are produced a day. Concentrations are sometimes high enough to merit broadcast warnings about health concerns, losses to agriculture, and other economic impacts from volcanic smog and acid rain. Hawaiians coined the word *vog* to describe their *vol*canic smog.

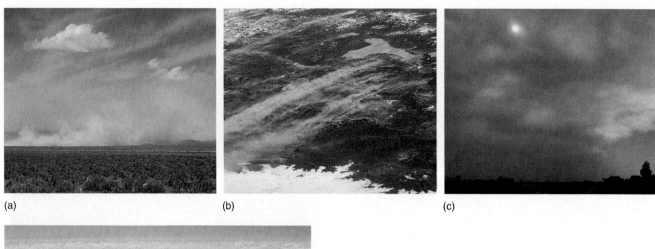

(a)   (b)   (c)

(d)

FIGURE 3.8 Natural variable dust in the atmosphere.
(a) A dust storm in central Nevada. (b) An orbital view of wind-blown silt and alkali dust rising from high interior-drainage basins in the Andes Mountains of Chile and Argentina and blowing far over the Atlantic Ocean. (c) A wall of dust from many kilometers distant blocks the sky as it moves over Melbourne, Australia. (d) Alkali dust, a serious air pollutant, rises from the exposed shorelines of Mono Lake, California.
[(a) Photo by author; (b) Space Shuttle photo from NASA; (c) photo by Bill Bachman/Photo Researchers, Inc.; (d) photo by Bobbé Christopherson.]

Temperature Inversion   Vertical temperature and atmospheric density distribution in the troposphere also can worsen pollution conditions. A **temperature inversion** occurs when the normal temperature decrease with altitude (normal lapse rate) begins to *increase* at some altitude. This can happen at any point from ground level to several thousand meters. Figure 3.9 compares a normal temperature profile with that of a temperature inversion. The normal profile (Figure 3.9a) permits warmer (less dense) air at the surface to rise, ventilating the valley and moderating surface pollution. But the warm air inversion (Figure 3.9b) prevents the rise of cooler (denser) air beneath, halting the vertical mixing of pollutants with other atmospheric gases. Thus, instead of being carried away, pollutants are trapped under the *inversion layer*.

Inversions most often result from certain weather conditions, such as when the air near the ground is radiatively cooled on clear nights, or from topographic situations that produce cold-air drainage into valleys. In addition, the air above snow-covered surfaces or beneath subsiding air in a high-pressure system may cause a temperature inversion. (The concepts of high- and low-pressure systems are in Chapter 6.)

Anthropogenic Pollution

Anthropogenic, or human-caused, air pollution remains most prevalent in urbanized regions. Approximately 2% of annual deaths in the United States are attributable to air pollution—some 50,000 people. Comparable risks are identified in Canada, Europe, Mexico, Asia, and elsewhere.

The human population is moving to cities and is therefore coming in increasing contact with air pollution. By the year 2010, approximately 3.3 billion people (48% of world population) will live in metropolitan regions, some one-third with unhealthful levels of air pollution. This represents a potentially massive public health issue in this century.

Table 3.4 lists the names, chemical symbols, principal sources, and impacts of variable anthropogenic components in the air. The first seven pollutants in the table result from combustion of fossil fuels in transportation (specifically automobiles) and at stationary sources such as power plants and factories. Overall, automobiles contribute more than 60% of United States and 50% of Canadian human-caused air pollution. Figure 3.10 identifies major human-caused pollutants and their proportional sources in the United

FIGURE 3.9 Normal and inverted temperature profiles. A comparison of a normal temperature profile in the atmosphere (a) with a temperature inversion in the lower atmosphere (b). Note how the warmer air layer prevents mixing of the denser (cooler) air below the inversion, thereby trapping pollution. An inversion layer is visible in the morning hours over Utah Valley (c). [Photo by author.]

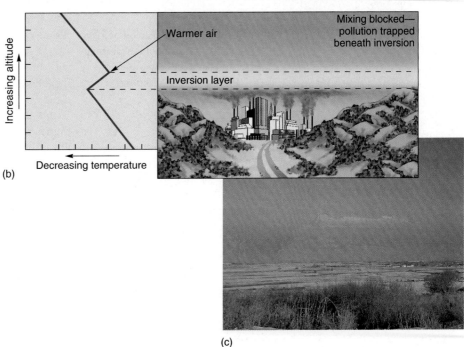

States in 1999 (the latest report). The proportion of sources for these pollutants may be considered representative for most developed countries (see http://www.epa.gov/).

In the United States in 1999, transportation produced 77% of the carbon monoxide, 47% of volatile organic compounds (VOC), 56% of the nitrogen oxides, and 25% of particulates. Environment Canada, an agency similar to the U.S. Environmental Protection Agency, reported in 1995 (latest available) that with Canada's smaller transportation fleet, Canadian automobile emissions contribute 40% of the carbon monoxide, 21% of the VOCs, and 60% of the nitrogen oxides in Canada (see http://www.ec.gc.ca/). The United States and Canada are negotiating an Air Quality Agreement, a process formally begun in 1991 and coordinated by an International Joint Commission. Acid deposition reduction is the main focus of the effort (see http://www.ijc.org/).

The apparent manageability of this transportation-pollution problem is interesting. In one California study, only 7% of the vehicles contributed half of the carbon monoxide and only 10% contributed half of the VOC pollution. These "gross polluting" vehicles are not old cars—

a common misconception—but include new cars! In random highway checks, 41% of vehicles had pollution equipment that was deliberately tampered with and 25% had defective or missing emission controls. Reducing air pollution from the transportation sector does not pose many mysteries.

Stationary sources, such as electric power plants and industrial plants that use fossil fuels, contribute the most sulfur oxides and particulates. For this reason, concentrations are focused in the Northern Hemisphere and the industrial, developed countries.

The last three gases shown in Table 3.4 are discussed elsewhere in this text: water vapor is examined with water and weather (Chapters 7 and 8); carbon dioxide and methane are discussed with greenhouse gases and climate (Chapters 4, 5, and 10).

Carbon Monoxide Pollution Carbon monoxide (CO) is a combination of one atom of carbon and one of oxygen. Carbon monoxide is produced by incomplete combustion (burning with limited oxygen) of fuels or other

Table 3.4 Anthropogenic Gases and Materials in the Lower Atmosphere

Name	Symbol	Source	Description and Effects of Criteria Pollutants
Carbon monoxide	CO	Incomplete combustion of fuels	Odorless, colorless, tasteless gas Toxicity: affinity for hemoglobin Displaces O_2 in bloodstream 50 to 100 ppm causes headaches, vision and judgement losses
Nitrogen oxides	NO_x (NO, NO_2)	High temperature/pressure combustion	Reddish-brown choking gas Inflames respiratory system, destroys lung tissue Damages plants 3 to 5 ppm is dangerous
Volatile organic compounds	VOC	Incomplete combustion of fossil fuels such as gasoline; cleaning and paint solvents	Prime agents of ozone formation
Ozone	O_3	Photochemical reactions	Highly reactive, unstable gas Oxidizes surfaces, dries rubber and elastic Damages plants at 0.01 to 0.09 ppm Agricultural losses at 0.1 ppm 0.3 to 1.0 ppm irritates eyes, nose, throat
Peroxyacetyl nitrates	PAN	Photochemical reactions	Produced by NO + VOC photochemistry No human health effects Major damage to plants, forests, crops
Sulfur oxides	SO_x (SO_2, SO_3)	Combustion of sulfur-containing fuels	Colorless; irritating smell 0.1 to 1 ppm impairs breathing, taste threshold Human asthma, bronchitis, emphysema Leads to acid deposition
Particulate matter	PM	Dust, dirt, soot, salt, metals, organics; fugitive dust from agriculture, construction, roads, and wind erosion	Complex mixture of solid and aerosol particles Dust, smoke, and haze affect visibility Various health effects: bronchitis, pulmonary function PM_{10} negative health effects established by researchers
Carbon dioxide	CO_2	Complete combustion, mainly from fossil fuel consumption	Principal greenhouse gas Atmospheric concentration increasing 64% of greenhouse warming effect
Methane	CH_4	Organic processes	Secondary greenhouse gas Atmospheric concentration increasing 19% of greenhouse warming effect
Water vapor	H_2O vapor	Combustion processes, steam	See Chapter 7 for more on the role of water vapor in the atmosphere.

carbon-containing substances. A log decaying in the woods produces carbon monoxide, as does a forest fire or other organic decomposition. Natural sources produce up to 90% of existing carbon monoxide, whereas anthropogenic sources, principally transportation, produce the other 10%. A dangerous point source of carbon monoxide for individuals is from primary and secondary tobacco smoke.

In the tropics of south-central Africa and the Amazon region of South America, an interesting source of carbon monoxide is the widespread burning of biomass (trees, grasses, brush) from August to October. This carbon monoxide spreads throughout the Southern Hemisphere.

Figure 3.11 compares carbon monoxide emissions from both human-caused and wildfire biomass burning for March and September 2000 in South America. Generally, the carbon monoxide from human sources is concentrated in urban areas, where it directly affects human health (see Table 3.4).

Photochemical Smog Pollution Photochemical smog was not generally experienced in the past but developed with the advent of the automobile. Today it is the major component of anthropogenic air pollution (Figure 3.12). **Photochemical smog** results from the interaction of sunlight and the combustion products in automobile exhaust

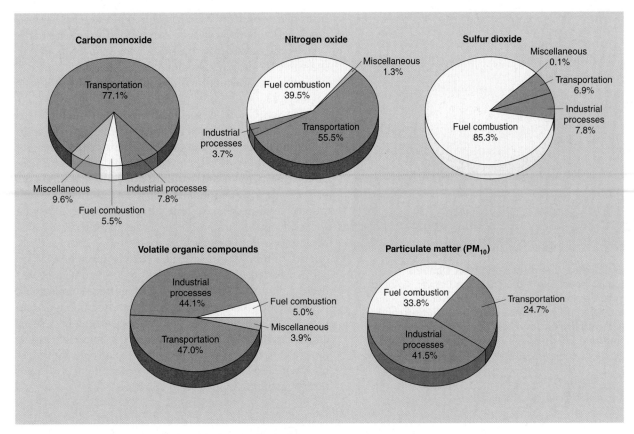

FIGURE 3.10 Human-caused air pollution and sources.
Major types of human-caused air pollution and their sources in the United States in 1999.
These proportions are typical of developed, industrialized countries. [Adapted from Office of
Air Quality, *National Air Quality and Emission Trends Report, 1999*, U.S. EPA (March 2001),
Figures 2-4, 2-19, 2-33, 2-40, and 2-67, EPA 454/R-01-004.]

**FIGURE 3.11 Biomass burning—a source of
global carbon monoxide (CO).**
Images from March and September, 2000, show
the onset of biomass burning. In March, the
slightly higher CO levels over the equatorial latitudes are from African fires, transported by trade
winds across the Atlantic. By September a large
CO plume rises from Brazil, primarily from the
burning of rain forest in the Amazon Basin. The
soot and ash is lifted high in the atmosphere by
strong cloud convection. Measurement is at the
700 millibar pressure level, approximately 3650 m
(12,000 ft). [Images from MOPITT (Measurement
of Pollution in the Troposphere) sensor aboard
the *Terra* satellite courtesy of David Edwards,
John Gille, and the MOPITT Science Team, UCAR.]

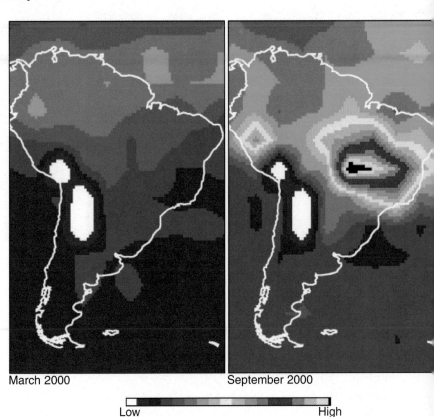

March 2000   September 2000

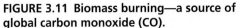

Low   High

FIGURE 3.12 Photochemical smog near Denver.
Photochemical reactions in the skies over Denver, Colorado, produce the blanket of smog that moves over mountains and forest. What do you notice about air pollution conditions where you live or go to college? [Photo by Bobbé Christopherson.]

(nitrogen oxides and VOCs). Although the term *smog*—a combination of the words *smoke* and *fog*—is a misnomer, it is generally used to describe this phenomenon. Smog is responsible for the hazy sky and reduced sunlight in many of our cities, as shown occurring in the contrasting photos in Figure 3.13.

Mexico City is notorious for poor air quality as its 22 million inhabitants and 3.5 million vehicles work, commute, and live in the world's second largest metropolitan region. In 1999 World Resources Institute named it as the unhealthiest city in the world for children; this was based on 1995 data. Conditions are worsened by frequent subtropical high-pressure systems (descending, stable air) that act as effective air traps over the Valley of Mexico, in which the city lies. The contrast between a rare, clear day and frequent polluted days in Mexico City is dramatic (Figure 3.13). Mexico enacted new laws in 1990 to reduce unhealthy photochemical smog conditions: more public rapid transportation, controls on automobiles, limitations on factory operations, and more pollution-absorbing park space and trees. Based on some success and an increase in clear days, a new 10-year plan began in 2000 directed at reducing ozone and particulates. The government has spent nearly $2 billion on pollution controls.

(a)

(b)

FIGURE 3.13 Contrasts in the air over Mexico City.
A day of clear air (a) in contrast to a more typical day of photochemical smog pollution (b) in the skies over Mexico City. [Photos by Larry Reider/Sipa Press.]

FIGURE 3.14 Photochemical reactions.

The interaction of automobile exhaust (NO$_2$, VOCs, CO) and ultraviolet radiation in sunlight causes photochemical reactions. The high temperatures in modern automobile engines produce reactions that form nitrogen dioxide (NO$_2$). This nitrogen dioxide, derived from automobiles and to a lesser extent from power plants, is highly reactive with ultraviolet light. The reaction liberates atomic oxygen (O) and a nitric oxide (NO) molecule from the NO$_2$. The free oxygen atom combines with an oxygen molecule (O$_2$) to form the oxidant ozone (O$_3$). In addition, the nitric oxide (NO) molecule reacts with VOCs to produce a family of chemicals called *peroxyacetyl nitrates* (*PAN*). To the left, note the formation of nitric acid and acid deposition.

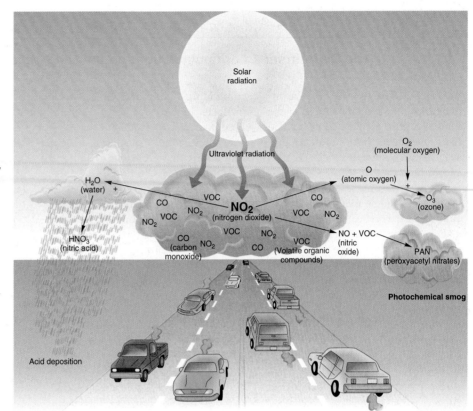

The connection between automobile exhaust and smog was not determined until 1953 in Los Angeles, long after society had established its dependence upon individualized transportation. Despite this discovery, widespread mass transit declined, the railroads dwindled, and the polluting individual automobile remains America's preferred transportation. Strangely, the U.S. fleet of 2002 model cars and trucks worsened in gas mileage from 2001—decreasing from 21.0 mpg to 20.4 mpg. This poor showing is principally due to gas-guzzling sport-utility vehicles and pickups that account for more than half of new sales. Efficiency standards for this class of vehicles have not changed since 1975; lower than for cars.

Figure 3.14 summarizes how car exhaust is converted into major air pollutants—ozone, **peroxyacetyl nitrates** (**PAN**), and nitric acid. PAN produces no known health effect in humans, but it is particularly damaging to plants, including both agricultural crops and forests. Damage in California is estimated to exceed $1 billion a year and several billion dollars nationwide in the farming and forestry sectors.

Worldwide, the problem with **nitrogen dioxide** production is its concentration in metropolitan regions. North American urban areas may have from 10 to 100 times higher nitrogen dioxide concentrations than nonurban areas. Nitrogen dioxide interacts with water vapor to form nitric acid (HNO$_3$), a contributor to acid deposition by precipitation, the subject of Focus Study 3.2.

Ozone is the primary ingredient in photochemical smog. (This is the same gas that is beneficial to us in the stratosphere in absorbing ultraviolet radiation.) The reactivity of ozone causes health concerns, for it damages biological tissues. For several reasons children are at greatest risk from ozone pollution—one in four children in U.S. cities is at risk of developing health problems from ozone pollution. This ratio is significant; it means that more than 12 million children are vulnerable in those cities with the worst polluted air (Los Angeles, New York City, Atlanta, Houston, and Detroit).

The **volatile organic compounds** (**VOCs**), including hydrocarbons from gasoline, surface coatings, and combustion at electric utilities, are important factors in ozone formation. States such as California base their standards for control of ozone pollution on VOC emission controls—a scientifically accurate emphasis.

Industrial Smog and Sulfur Oxides   Over the past 300 years, except in some developing countries, coal slowly replaced wood as the basic fuel used by society. The Industrial Revolution required high-grade energy to run machines. The changes involved conversion from *animate* energy (energy from animal sources, such as animal-powered farm equipment) to *inanimate* energy (energy from nonliving sources, such as coal, steam, and water). The air pollution associated with coal-burning industries

Focus Study 3.2

Acid Deposition: A Continuing Blight on the Landscape

Acid deposition is a major environmental problem in some areas of the United States, Canada, Europe, and Asia. Such deposition is most familiar as "acid rain," but it also occurs as "acid snow" and in dry form as dust or aerosols. (Aerosols are tiny liquid droplets or solid particles.) In addition, winds can carry the acid-producing chemicals many kilometers from their sources before they settle on the landscape, where they enter streams and lakes as runoff and groundwater flows.

Acid deposition is causally linked to serious problems: declining fish populations and fish kills in the northeastern United States, southeastern Canada, Sweden, and Norway; widespread forest damage in these same places and Germany; widespread changes in soil chemistry; and damage

to buildings, sculptures, and historic artifacts. In New Hampshire's Hubbard Brook Experimental Forest, a study covering 1960 to the present found half the nutrient calcium and magnesium base cations (see Chapter 18) were leached from the soil. Excess acids are the cause of the decline.

Despite scientific agreement about the problem, which the U.S. General Accounting Office calls a "combination of meteorological, chemical, and biological phenomena," corrective action was delayed by its complexity and politics.

The acidity of precipitation is measured on the pH scale, which expresses the relative abundance of free hydrogen ions (H^+) in a solution. Free hydrogen ions in a solution are what make an acid corrosive, for they easily

combine with other ions. The pH scale is logarithmic: each whole number represents a 10-fold change. A pH of 7.0 is neutral (neither acidic nor basic). Values less than 7.0 are increasingly *acidic*, and values greater than 7.0 are increasingly *basic*, or *alkaline*. (A pH scale for soil acidity and alkalinity is portrayed graphically in Chapter 18.)

Natural precipitation dissolves carbon dioxide from the atmosphere to form carbonic acid. This process releases hydrogen ions and produces an average pH reading for precipitation of 5.65. The normal range for precipitation is 5.3–6.0. Thus, normal precipitation is always slightly acidic.

Some anthropogenic gases are converted to acids in the atmosphere and then are removed by wet and dry deposition processes. Specifically, nitrogen and sulfur oxides released in the combustion of fossil fuels can produce nitric acid (HNO_3) and sulfuric acid (H_2SO_4) in the atmosphere.

Acid Precipitation Damage

Precipitation as acidic as pH 2.0 has fallen in the eastern United States, Scandinavia, and Europe. By comparison, vinegar and lemon juice register slightly less than 3.0. Aquatic plant and animal life perishes when lakes drop below pH 4.8.

More than 50,000 lakes and some 100,000 km (62,000 mi) of streams in the United States and Canada are at a

(continued)

(a)

(b)

FIGURE 1 The blight of acid deposition. The harm done to forests and crops by acid deposition is well established, especially in Europe, here in the Czech Republic (a) and in the forests of the Appalachian Mountains in the United States, here in the forests on Mount Mitchell (b). [(a) Photo by Simon Fraser/Science Photo Library/Photo Researchers, Inc.; (b) Will and Deni McIntyre/Photo Researchers, Inc.]

Focus Study 3.2 *(continued)*

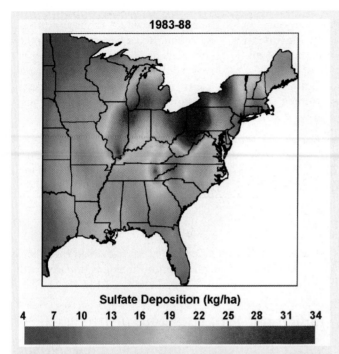

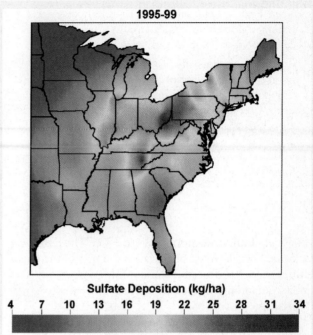

FIGURE 2 Improvement in sulfate wet deposition rate.
Spatial portrayal of annual sulfate (principally SO₄) wet deposition on the landscape, 1983–88 and 1995–99 in kilograms per hectare. Clean Air Act regulations have lowered levels of emissions that add acid to the environment. According to a new study more improvement is necessary to help ecosystems recover. [Maps courtesy of J.W. Lynch and J.A. Grimm, U.S. Forest Service, Northeast Forest Experimental Station, Northern Global Change Research Program.]

pH level below normal (i.e., below 5.3), with several hundred lakes incapable of supporting any aquatic life—15% of the lakes in New England and 41% in the Adirondack Mountains. Acid deposition causes the release of aluminum and magnesium from clay minerals in the soil, and both of these are harmful to fish and plant communities.

Also, relatively harmless mercury deposits in lake-bottom sediments convert in acidified lake waters into highly toxic *methylmercury*, which is deadly to aquatic life. Local health advisories in two provinces and 22 U.S. states are regularly issued to warn those who fish of the methylmercury problem. Mercury atoms rapidly bond with carbon and move through biological systems as an *organometallic compound*.

Damage to forests results from the rearrangement of soil nutrients, the death of soil microorganisms, and an aluminum-induced calcium deficiency that is currently under investigation. The most advanced impact is seen in

forests in Europe, especially in eastern Europe, principally because of its long history of burning coal and the density of industrial activity. In Germany and Poland up to 50% of the forests are dead or damaged; in Switzerland 30% are afflicted (Figure 1a).

In the United States, regional-scale decline in forest cover is significant, especially red spruce and sugar maples. In some maples, aluminum is collecting around rootlets; in spruce, acid fogs and rains leach calcium from needles directly. Affected trees are susceptible to winter cold, insects, and droughts. In New England, some stands of spruce are as much as 75% affected, as evidenced through analysis of tree-growth rings, which become narrower in adverse growing years. An indicator of forest damage is the reduction by almost half of the annual production of U.S. and Canadian maple sugar. Trees at higher elevations in the Appalachians are injured by acid-laden cloud cover (Figure 1b).

Government estimates of damage in the United States, Canada, and Europe exceed $50 billion annually. Because wind and weather patterns are international, efforts at reducing acidic deposition also must be international in scope. The decline of sulfur dioxide by more than 40% between 1973 and the present is a result of the U.S. Clean Air Act. If goals are met, by 2010 sulfur dioxide emissions should be less than half this earlier level. Researchers found a correlation between these reductions and a reduction in the geographic area affected by wet deposition of sulfur. Figure 2 maps the reduction in sulfate deposition between the 1983–1988 and 1995–1999 periods. However, this progress is only a beginning. According to a study in *BioScience*, power plants and other sources must cut emissions 80% beyond the Clean Air Act mandate.

(continued)

Focus Study 3.2 *(continued)*

Ten leading acid deposition researchers reported in an extensive study in *BioScience*, March 2001,

Model calculations suggest that the greater the reduction in atmospheric sulfur deposition ... , the greater the magnitude and rate of chemical recovery. Less aggressive proposals for controls of sulfur emissions will result in slower chemical and biological re-

covery and in delays in regaining the services of a fully functional ecosystem. ... North America and Europe are in the midst of a large-scale experiment. Sulfuric and nitric acids have acidified soils, lakes, and streams, thereby stressing or killing terrestrial and aquatic biota.*

At best, acid deposition is an issue of global spatial significance for which

science is providing strong incentives for action. Reductions in troublesome emissions are closely tied to energy conservation and therefore directly related to production of greenhouse gases and global warming concerns—thus, linking these environmental issues.

*Driscoll, C. T., *et al.*, "Acidic deposition in the northeastern United States: Sources and Inputs, Ecosystems Effects, and Management Strategies," *BioScience* 51 (March 2001): 195.

is known as **industrial smog** (Figure 3.15). The term *smog* was coined by a London physician at the turn of the twentieth century to describe the combination of fog and smoke containing sulfur gases (sulfur is an impurity in fossil fuels).

Once in the atmosphere, **sulfur dioxide (SO_2)** reacts with oxygen (O) to form sulfur trioxide (SO_3), which is highly reactive and, in the presence of water or water vapor, forms **sulfate aerosols**, tiny particles about 0.1 to 1 μm in diameter. Sulfuric acid (H_2SO_4) can form even in moderately polluted air at normal temperatures. Sulfur dioxide-laden air is dangerous to health, corrodes metals, and deteriorates stone building materials at accelerated rates. Focus Study 3.2 discusses this vital atmospheric issue.

In the United States, coal-burning electric utilities and steel manufacturing are the main sources of sulfur dioxide, principally in the East and Midwest. And, because of the prevailing movement of air masses, they are the main

sources of sulfur dioxide in adjacent Canadian regions. As much as 70% of Canadian sulfur dioxide is initiated within the United States.

Particulates   **Particulate matter (PM)** is a diverse mixture of fine particles, both solid and aerosol, that impact human health. Haze, smoke, and dust are visible reminders of particulate material in the air we breathe. PM_{10}, particulate smaller than 10 microns (10 μm or less) in diameter, was designated a matter for concern in 1987. $PM_{2.5}$ is currently being debated as an appropriate standard for human health. Studies in Provo and Orem, Utah (1989), Philadelphia (1992), and other cities, and by the American Cancer Society (1995), established links between PM pollution and health. Nationally, a study of major cities disclosed a 26% greater risk of premature death due to respirable particulate pollution as compared with nonpolluted air—further driving up medical costs and related expenses.

In Utah County, Utah, researchers correlated PM_{10} concentrations with increased rates of hospitalization for bronchitis, asthma, pneumonia, and pleurisy (especially in children), and greatly increased medical costs. Similar studies of children affected by related illnesses in seven other cities revealed sickness rates twice as high for the city having the dirtiest air as for the city having the cleanest air. Pollution sources dispute these findings and the concern generated by such studies.

Certainly, society cannot simply halt two centuries of industrialization to slow air pollution; the resulting economic chaos would be devastating. But neither can it permit pollution production to continue unabated, for catastrophic environmental and human health damage, and continued climate changes, inevitably will result. We are now contributing significantly to the creation of the **anthropogenic atmosphere**, a tentative label for Earth's next atmosphere. The urban air we breathe today may be just a preview. What is the air quality like where you live, work, and go to college? Where could you go to find out its status?

FIGURE 3.15 Typical industrial smog.
Pollution generated by industry differs from that produced by transportation. Industrial pollution has high concentrations of sulfur oxides, particulates, and carbon dioxide. [Photo by author.]

FIGURE 3.16 Trends in air pollutants.
Air pollution trends between 1970 and 1999: sulfur dioxide, −39% (due to scrubbers on smokestacks and emission controls); particulate matter, −75%; nitrogen oxide increased, +17%; volatile organic compounds, −42%; carbon monoxide, −25% (the last three involve exhaust emission controls); and a huge lead reduction of −98%, or 217,000 tons (as a result of unleaded gas and reduced industrial emissions). [Adapted from Office of Air Quality, *National Air Quality and Emission Trends Report, 1999*, U.S. EPA (March 2001), from Data Tables, Appendix A, EPA 454/R-01-004.]

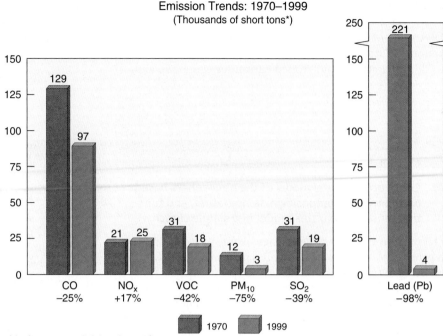

Emission Trends: 1970–1999
(Thousands of short tons*)

*1 short ton x 0.91 = 1 metric ton

Benefits of the Clean Air Act

Concentration of many air pollutants declined over the past several decades because of Clean Air Act legislation (1970, 1977, 1990), resulting in the saving of trillions of dollars in avoided health, economic, and environmental losses. Despite this reality, air pollution controls are subject to a continuing political debate and threats to weaken standards. According to a report prepared by the EPA, the *National Air Pollutant Emission Trends, 1999*, the 1999 emissions of five primary pollutants totaled 162 million metric tons (178 million tons), compared to 224 million metric tons (246 million tons) in 1970, the first year of the Clean Air Act (CAA). Figure 3.16 illustrates the trends in CO, NO_x, VOCs, SO_2, PM_{10}, and lead (Pb). Only nitrogen oxides increased between 1970 and 1999. In Canada, between 1980 and 1997, sulfur dioxide emissions decreased 44%.

Significant lead (Pb) reductions have global beneficial impact. Prior to the CAA lead was added to gasoline, emitted in the exhaust, traveled great distances, and settled in living tissues, especially in children.

To be justified and worthwhile, abatement (mitigation and prevention) costs must not exceed the financial benefits derived from reducing pollution damage. Compliance with the CAA affected patterns of industrial production, employment, and capital investment. Although these expenditures must be viewed as investments that generated benefits and opportunities, the dislocation and job loss in some regions was severe: reductions in high-sulfur coal mining and cutbacks in polluting industries such as steel, for example.

In 1990, Congress requested the Environmental Protection Agency (EPA) to answer the question: How do the overall health, welfare, ecological, and economic benefits of CAA programs compare with the costs of these programs? In response, the EPA performed an exhaustive cost–benefit analysis and published a report in 1997. *The Benefits of the Clean Air Act, 1970 to 1990* (Office of Policy, Planning, and Evaluation, U.S. EPA) calculated the following:

- The *total direct cost* to implement the Clean Air Act for all federal, state, and local rules from 1970 to 1990 was *$523 billion* (in 1990-value dollars). This cost was borne by businesses, consumers, and government entities.
- The estimate of *direct monetized benefits* from the Clean Air Act from 1970 to 1990 falls in a range from $5.6 to $49.4 trillion with a central mean of *$22.2 trillion.* (The uncertainty of the assessment is indicated by the range of benefit estimates.)
- Therefore, the estimated *net financial benefit* of the Clean Air Act is *$21.7 trillion!* "The finding is overwhelming. The benefits far exceed the costs of the CAA in the first 20 years," said Richard Morgenstern, associate administrator for policy planning and evaluation at the EPA.

The benefits to society, directly and indirectly, have been widespread across the entire population, including improved health and environment, less lead to harm children, lowered cancer rates, less acid deposition, and an estimated 206,000 fewer deaths related to air pollution in 1990 alone. These benefits took place during a period in which the U.S. population grew by 22% and the economy expanded by 70%. The benefits continued between 1990 and 2000 as air quality continued to improve.

As you reflect on this chapter and our modern atmosphere, the treaties to protect stratospheric ozone, and the EPA study of benefits from the CAA, you should feel encouraged. Scientists did the research, society knew what to do, took action, and reaped enormous economic and health benefits. An important role for physical geographers is to explain these global impacts through spatial analysis and to guide an informed citizenry toward a better understanding.

Summary and Review—Earth's Modern Atmosphere

● *Construct* a general model of the atmosphere based on the criteria composition, temperature, and function, and *diagram* this model in a simple sketch.

Our modern atmosphere is a gaseous mixture so evenly mixed it behaves as if it were a single gas. It is naturally odorless, colorless, tasteless, and formless. The principal substance of this atmosphere is air—the medium of life.

Above 480 km (300 mi) altitude, the atmosphere is rarefied (nearly a vacuum) and is called the **exosphere**, which means "outer sphere." The weight (force over a unit area) of the atmosphere, exerted on all surfaces, is termed **air pressure**. It decreases rapidly with altitude.

By *composition*, we divide the atmosphere into the **heterosphere**, extending from 480 km (300 mi) to 80 km (50 mi), and the **homosphere**, extending from 80 km to Earth's surface. Within the heterosphere and using *temperature* as a criterion, we identify the **thermosphere**. Its upper limit, called the **thermopause**, is at approximately 480 km altitude. **Kinetic energy**, the energy of motion, is the vibrational energy that we measure and call temperature. The actual heat produced in the thermosphere is very small. The density of the molecules is so low that little actual **heat**, the flow of kinetic energy from one body to another because of a temperature difference between them, is produced. Nearer Earth's surface the greater number of molecules in the denser atmosphere transmits their kinetic energy as **sensible heat**, meaning that we can feel it.

The homosphere includes the **mesosphere**, **stratosphere**, and **troposphere**, as defined by temperature criteria. Within the mesosphere, cosmic or meteoric dust acts as nuclei around which fine ice crystals form to produce rare and unusual night clouds called **noctilucent clouds**.

The normal temperature profile within the troposphere during the daytime decreases rapidly with increasing altitude at an average of 6.4 C° per kilometer (3.5 F° per 1000 ft), a rate known as the **normal lapse rate**. The top of the troposphere is wherever a temperature of −57°C (−70°F) is recorded, a transition known as the **tropopause**. The actual lapse rate at any particular time and place may deviate considerably because of local weather conditions and is called the **environmental lapse rate**.

We distinguish a region in the heterosphere by its *function*. The **ionosphere** absorbs cosmic rays, gamma rays, X-rays, and shorter wavelengths of ultraviolet radiation and converts them into kinetic energy. A functional region within the stratosphere is the **ozonosphere**, or **ozone layer**, which absorbs life-threatening ultraviolet radiation, subsequently raising the temperature of the stratosphere.

exosphere (p. 66)
air pressure (p. 67)
heterosphere (p. 68)
homosphere (p. 70)
thermosphere (p. 71)
thermopause (p. 71)
kinetic energy (p. 71)
heat (p. 71)
sensible heat (p. 71)
mesosphere (p. 71)
noctilucent clouds (p. 71)
stratosphere (p. 71)
troposphere (p. 71)
tropopause (p. 71)
normal lapse rate (p. 71)
environmental lapse rate (p. 71)
ionosphere (p. 72)
ozonosphere, ozone layer (p. 72)

1. What is air? Where did the components in Earth's present atmosphere originate?
2. In view of the analogy by Lewis Thomas, characterize the various functions the atmosphere performs that protect the surface environment.
3. What three distinct criteria are employed in dividing the atmosphere for study?
4. Describe the overall temperature profile of the atmosphere, and list the four layers defined by temperature.
5. Describe the two divisions of the atmosphere on the basis of composition.
6. What are the two primary functional layers of the atmosphere and what does each do?

● *List* the stable components of the modern atmosphere and their relative percentage contributions by volume, and *describe* each.

Even though the atmosphere's density decreases with increasing altitude in the homosphere, the blend (proportion) of gases is nearly uniform. This stable mixture of gases has evolved slowly.

The homosphere is a vast reservoir of relatively inert *nitrogen*, originating principally from volcanic sources and from bacterial action in the soil; *oxygen*, a by-product of photosynthesis; *argon*, constituting about 1% of the homosphere and completely inert; and *carbon dioxide*, a natural by-product of life processes and fuel combustion.

7. Name the four most prevalent stable gases in the homosphere. Where did each originate? Is the prevalence of any of these changing at this time?

● *Describe* **conditions within the stratosphere; specifically,** *review* **the function and status of the ozonosphere (ozone layer).**

The overall reduction of the stratospheric ozonosphere, or ozone layer, during the past several decades represents a hazard for society and many natural systems and is caused by chemicals introduced into the atmosphere by humans. Since World War II quantities of human-made **chlorofluorocarbons** (**CFCs**) and bromine-containing compounds have made their way into the stratosphere. The increased ultraviolet light at those altitudes breaks down these stable chemical compounds, thus freeing chlorine and bromine atoms. These atoms act as catalysts in reactions that destroy ozone molecules.

chlorofluorocarbons (CFCs) (p. 73)

8. Why is stratospheric ozone (O_3) so important? Describe the effects created by increases in ultraviolet light reaching the surface.
9. Summarize the ozone predicament and present trends and any treaties that intend to protect the ozone layer.
10. Evaluate Crutzen, Rowland, and Molina's use of the scientific method in investigating stratospheric ozone depletion.

● *Distinguish* **between natural and anthropogenic variable gases and materials in the lower atmosphere.**

Within the troposphere, both natural and human-caused variable gases, particles, and other chemicals are part of the atmosphere. We coevolved with natural "pollution" and thus are adapted to it. But we are not adapted to cope with our own anthropogenic pollution. It constitutes a major health threat, particularly where people are concentrated in cities.

Vertical temperature and atmospheric density distribution in the troposphere can worsen pollution conditions. A **temperature inversion** occurs when the normal temperature decrease with altitude (normal lapse rate) reverses. In other words, temperature begins to increase at some altitude.

temperature inversion (p. 78)

11. Why are anthropogenic gases more significant to human health than are those produced from natural sources?
12. In what ways does a temperature inversion worsen an air pollution episode? Why?

● *Describe* **the sources and effects of carbon monoxide, nitrogen dioxide, and sulfur dioxide, and** *construct* **a simple equation that illustrates photochemical reactions that produce ozone, peroxyacetyl nitrates, nitric acid, and sulfuric acid.**

Odorless, colorless, and tasteless, **carbon monoxide** (**CO**) is produced by incomplete combustion (burning with limited oxygen) of fuels or other carbon-containing substances. Transportation is the major human-caused source for carbon monoxide. The toxicity of carbon monoxide is due to its affinity for blood hemoglobin, which is the oxygen-carrying pigment in red blood cells. In the presence of carbon monoxide, the oxygen is displaced and the blood becomes deoxygenated (Table 3.4).

Photochemical smog results from the interaction of sunlight and the products of automobile exhaust, the single largest contributor of pollution that produces smog. Car exhaust, containing *nitrogen dioxide* and *volatile organic compounds* (*VOCs*), in the presence of ultraviolet light in sunlight converts into major air pollutants—*ozone, peroxyacetyl nitrates* (*PAN*), and *nitric acid*. The principal photochemical by-products include *ozone* (O_3), which causes negative health effects, oxidizes surfaces, and kills or damages plants; and **peroxyacetyl nitrates** (**PAN**), which produce no known health effects in humans but are particularly damaging to plants, including both agricultural crops and forests. **Nitrogen dioxide** (NO_2) damages and inflames human respiratory systems, destroys lung tissue, and damages plants. Nitric oxides participate in reactions that form nitric acid (HNO_3) in the atmosphere, forming both wet and dry acidic deposition. The **volatile organic compounds** (**VOCs**), including hydrocarbons from gasoline, surface coatings, and electric utility combustion, are important factors in ozone formation.

The distribution of human-produced **industrial smog** over North America, Europe, and Asia is related to transportation and electrical production. Such characteristic pollution contains **sulfur dioxide**. Sulfur dioxide in the atmosphere reacts to produce **sulfate aerosols**, which produce sulfuric acid (H_2SO_4) deposition, which can be dangerous to health, and affect Earth's energy budget by scattering and reflecting solar energy. **Particulate matter** (**PM**) consists of dirt, dust, soot, and ash from industrial and natural sources.

Energy conservation and efficiency and reducing emissions are essential strategies for abating air pollution. Earth's next atmosphere most accurately may be described as the **anthropogenic atmosphere** (human-influenced atmosphere).

carbon monoxide (CO) (p. 79)
photochemical smog (p. 80)
peroxyacetyl nitrates (PAN) (p. 83)
nitrogen dioxide (p. 83)
volatile organic compounds (VOCs) (p. 83)
industrial smog (p. 86)
sulfur dioxide (p. 86)
sulfate aerosols (p. 86)
particulate matter (PM) (p. 86)
anthropogenic atmosphere (p. 86)

13. What is the difference between industrial smog and photochemical smog?

14. Describe the relationship between automobiles and the production of ozone and PAN in city air. What are the principal negative impacts of these gases?

15. How are sulfur impurities in fossil fuels related to the formation of acid in the atmosphere and acid deposition on the land?

16. In summary, what are the results from the first 20 years under Clean Air Act regulations?

NetWork

The *Geosystems* Home Page provides on-line resources for this chapter on the World Wide Web. To begin: Once on the Home Page, click on this textbook, scroll the Table of Contents menu, select this chapter, and click "Begin." You will find self-tests that are graded, review exercises, specific updates for items in the chapter, and many links to interesting related pathways on the Internet under "Destinations." *Geosystems* is at **http://www.prenhall.com/christopherson.**

Critical Thinking

A. A scientific study, *The Benefits of the Clean Air Act, 1970 to 1990* (Office of Policy, Planning, and Evaluation, U.S. EPA, October 1997), determined that the Clean Air Act provided the American people with benefits. Health, welfare, ecological, and economic benefits rated 42 to 1 over the costs of CAA implementation (estimated at $21.7 trillion in benefit compared to $523 billion in costs). Do you think this is significant? Should this be part of the debate about future weakening or strengthening of the Clean Air Act? Do you think such benefits might result from other regulations such as the Clean Water Act, or groundwater protection measures?

B. Relative to item A: In your opinion, why is the public generally unaware of these details? What are the difficulties in instructing the public? Why is there such media attention given to antiscience and nonscientific opinions? Take a moment and brainstorm recommendations for action, education, and public awareness on these issues.

C. To determine the total ozone column at your present location, go to the TOMS Home Page at http://toms.gsfc.nasa.gov/teacher/ozone_overhead.html, "What was the total ozone column at your house?" Select a point on the map or enter your latitude and longitude, and the date you want to check. The ozone column refers principally to stratospheric ozone and not to the photochemical pollutant in the lower troposphere. Note the instrument and satellite platform used. (Note also the limitations listed on the extent of data availability.) For several different dates, when do the lowest values occur? The highest values? Briefly explain what your results mean. How do you interpret the values found?

Surface energy budgets in a metropolitan area differ from the surrounding countryside, creating an urban heat island. Park space offers some relief, such as in Central Park in New York City, comprising 840 acres (340 ha), stretching some 4 km north to south by 0.8 km wide (2.5 by 0.5 mi). The park consists of myriad forest, lakes, brambles, trails, and lawn, here seen in March before the greening of spring. [Photo by Bobbé Christopherson.]

4

Atmosphere and Surface Energy Balances

Key Learning Concepts

After reading the chapter, you should be able to:

- *Identify* the pathways of solar energy through the troposphere to Earth's surface: transmission, scattering, diffuse radiation, refraction, albedo (reflectivity), conduction, convection, and advection.

- *Describe* what happens to insolation when clouds are in the atmosphere and *analyze* the effect of clouds and air pollution on solar radiation received at ground level.

- *Review* the energy pathways in the Earth–atmosphere system, the greenhouse effect, and the patterns of global net radiation.

- *Plot* the daily radiation curves for Earth's surface and *label* the key aspects of incoming radiation, air temperature, and the daily temperature lag.

- *Portray* typical urban heat island conditions and *contrast* the microclimatology of urban areas with that of surrounding rural environments.

Earth's biosphere pulses with flows of solar energy that sustain our lives and empower natural systems. Changing seasons, Earth's variety of climates, and daily weather fluctuations, remind us of this constant flow of energy cascading through the atmosphere. Earth's shifting seasonal rhythms are covered in Chapter 2. Energy and moisture exchanges between Earth's surface and the atmosphere are essential elements of weather and climate, and these are discussed in later chapters.

In this chapter: This chapter follows solar energy through the troposphere to Earth's surface. The *input* of insolation is countered by the *outputs* of reflected light and emitted infrared energy from the atmosphere and surface environment. Together, this input and output determines the net energy available to perform work. We examine surface energy budgets and analyze how net radiation is spent. The chapter concludes with a look at the unique energy and moisture environment in our cities. The view across a hot parking lot in summer, of cars and pavement, is all too familiar—the air shimmers as heat energy is radiated skyward. The climate of our urban areas differs measurably from that of surrounding rural areas. Focus Study 4.1 discusses solar energy, a renewable energy resource of great potential.

Energy Essentials

When you look at a photograph of Earth taken from space, you can clearly see the surface receipts of incoming insolation (see Earth on the back cover of this book). Land and water surfaces, clouds, and atmospheric gases and dust intercept solar energy. The flows of energy are manifest in swirling weather patterns, powerful oceanic currents, and the varied distribution of vegetation. Specific energy patterns differ for deserts, oceans, mountaintops, plains, rain forests, and ice-covered landscapes. In addition, the presence or absence of clouds may make a 75% difference in the amount of energy that reaches the surface, because clouds reflect incoming energy.

Energy Pathways and Principles

Earth's atmosphere and surface are heated by solar energy, which is unevenly distributed by latitude and which fluctuates seasonally. Figure 4.1 is a simplified flow diagram of shortwave and longwave radiation in the Earth–atmosphere system, discussed in the pages that follow. You will find it helpful to refer to this figure, and the more detailed energy balance illustration in Figure 4.13, as you read through the following section. We first look at some important pathways and principles for insolation as it passes through the atmosphere to Earth's surface.

Transmission **Transmission** refers to the passage of shortwave and longwave energy through either the atmosphere or water. Our budget of atmospheric energy comprises shortwave radiation *inputs* (ultraviolet light, visible light, and near-infrared wavelengths) and longwave radiation *outputs* (thermal infrared) that pass through the atmosphere by transmission. Atmospheric gases and dust physically interact with insolation through processes of scattering, a redirection of energy through refraction and reflection.

Scattering Insolation encounters an increasing density of atmospheric gases as it travels toward the surface. The gas molecules redirect radiation, changing the direction of the light's movement *without altering its wavelengths*. This phenomenon is known as **scattering** and represents 7% of Earth's reflectivity, or albedo (see Figure 4.13). Dust particles, pollutants, ice, cloud droplets, and water vapor produce further scattering.

Have you wondered why Earth's sky is blue? And why sunsets and sunrises are often red? These simple questions have an interesting explanation, based upon a principle known as Rayleigh scattering (named for English physicist Lord Rayleigh, who stated the principle in 1881). This principle relates wavelength to the size of molecules or particles that cause the scattering. The general rule is: *The shorter the wavelength, the greater the scattering, and the longer the wavelength, the less the scattering.* Small gas molecules in the air scatter shorter wavelengths of light. Thus, the shorter wavelengths of visible light—the blues and violets—scatter the most and dominate the lower atmosphere. And, because there are more blue than violet wavelengths

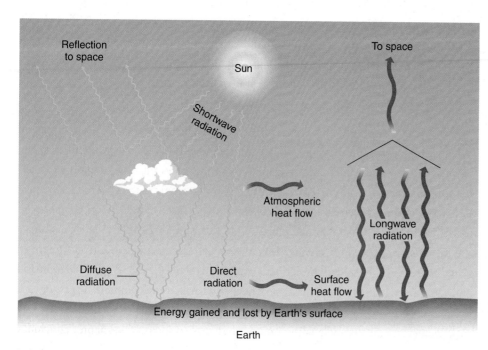

FIGURE 4.1 Energy gained and lost by Earth's surface and atmosphere.
Simplified view of the Earth–atmosphere energy system. Circuits include incoming shortwave insolation, reflected shortwave radiation, and outgoing longwave radiation.

in sunlight, a blue sky prevails. A sky filled with smog and haze appears almost white because the larger particles associated with air pollution act to scatter all wavelengths of the visible light.

The angle of the Sun's rays determines the thickness of atmosphere they must pass through to reach the surface. Therefore, direct rays (from overhead) experience less scattering and absorption than do low, oblique-angle rays that must travel farther through the atmosphere. Insolation from the low-altitude Sun undergoes more scattering of shorter wavelengths, leaving only the residual oranges and reds to reach the observer at sunset or sunrise.

Some incoming insolation is diffused by clouds and atmosphere and is transmitted to Earth as **diffuse radiation**, the downward component of scattered light (labeled in Figure 4.13). This light is multidirectional and thus casts shadowless light on the ground.

Refraction   As insolation enters the atmosphere, it passes from one medium to another, from virtually empty space into atmospheric gases, or from air into water. This transition subjects the insolation to a change of speed, which also shifts its direction, a bending action called **refraction**. In the same way, a crystal or prism refracts light passing through it, bending different wavelengths to different angles, separating the light into its component colors to display the spectrum. A rainbow is created when visible light passes through myriad raindrops and is refracted and reflected toward the observer at a precise angle (see Figure 4.2).

Another example of refraction is a mirage, an image that appears near the horizon where light waves are refracted by

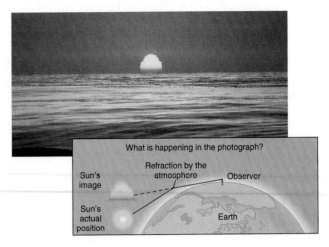

FIGURE 4.3 Sun refraction.
The distorted appearance of the Sun nearing sunset over the ocean is produced by refraction of the Sun's image in the atmosphere. Have you ever noticed this effect? [Photo by author.]

layers of air of different temperatures (and consequently of different densities) on a hot day. The atmospheric distortion of the setting Sun in Figure 4.3 is also a product of refraction—light from the Sun low in the sky must penetrate more air than when the Sun is high; it is refracted through air layers of different densities on its way to the observer.

An interesting function of refraction is that it adds approximately 8 minutes of daylight that we would lack if Earth had no atmosphere. Sunlight refracts in its passage from space through the atmosphere, and so, at sunrise, we see the Sun's image about 4 minutes before the Sun actually peeks over the horizon. Similarly, the Sun actually sets at sunset but refracts its image from over the horizon for about 4 minutes afterward, so we see sunset later than it truly happens. To this day, science cannot predict the exact time of visible sunrise or sunset within these 4 minutes because the degree of refraction continually varies with atmospheric temperature, moisture, and pollutants.

Insolation Input   *Insolation* is the single energy input driving the Earth–atmosphere system. The world map in Figure 4.4 shows the distribution of average annual solar energy received at Earth's surface. It includes all the radiation that arrives at Earth's surface, both direct and diffuse (scattered by the atmosphere).

Several patterns are notable on the map. Insolation decreases poleward from about 25° latitude in both the Northern and Southern Hemispheres. Consistent daylength and high Sun altitude produce average annual values of 180–220 W/m² throughout the equatorial and tropical latitudes. In general, greater insolation of 240–280 W/m² occurs in low-latitude deserts worldwide because of frequently cloudless skies. Note this energy pattern in the cloudless subtropical deserts in both hemispheres (for example, the Sonoran, Saharan, Arabian, Gobi, Atacama, Namib, Kalahari, and Australian deserts).

FIGURE 4.2 A rainbow.
Raindrops refract and reflect light to produce a primary rainbow. Note that the color order in the rainbow is distributed with the shortest wavelengths on the inside of the bow and the longest wavelengths on the outside of the bow. [Photo by author.]

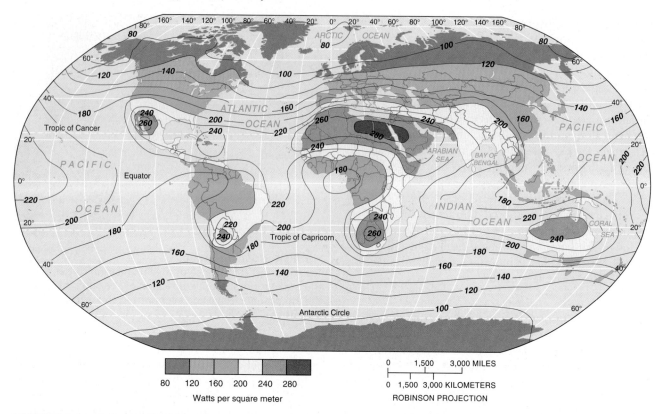

FIGURE 4.4 Insolation at Earth's surface.
Average annual solar radiation received on a horizontal surface at ground level in watts per square meter (100 W/m² = 75 kcal/cm²/year). [After M. I. Budyko, *The Heat Balance of the Earth's Surface* (Washington, DC: U.S. Department of Commerce, 1958), p. 99.]

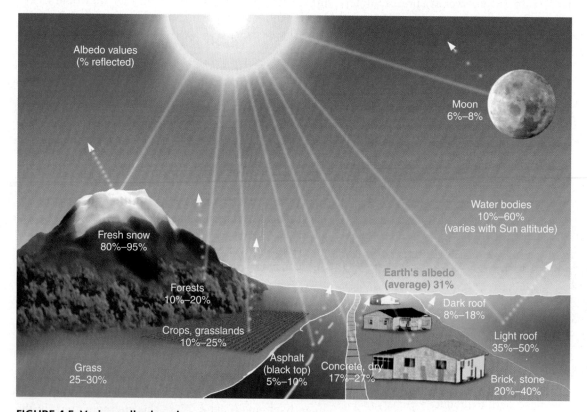

FIGURE 4.5 Various albedo values.
Different surfaces have different albedo values. In general, light surfaces are more reflective than dark surfaces and thus have higher albedo values. [Data from M. I. Budyko, *The Heat Balance of the Earth's Surface* (Washington, DC: U.S. Department of Commerce, 1958), p. 36.]

FIGURE 4.6 Sunlight reflected off the ocean.
An astronaut's view of reflected sunlight off the Mozambique Channel, between the East coast of Africa and the island country of Madagascar. Ocean surface albedo values increase with lower Sun angles and calmer seas. [Space Shuttle photo from NASA.]

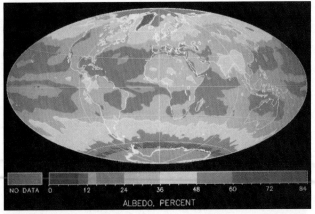

(a) July

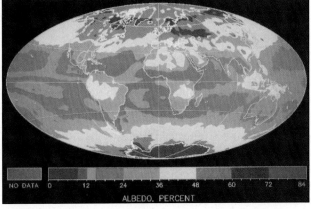

(b) January

FIGURE 4.7 Total albedos for July and January.
Total albedos for July 1985 (a) and January 1986 (b) as measured by the Earth Radiation Budget experiments (ERB) aboard satellites *Nimbus-7, NOAA-9,* and *ERBS* (*ERB Satellite*). Each color represents a 12% interval in albedo values. (The map base is a modified elliptical equal-area projection.) [Courtesy of Radiation Services Branch, Langley Research Center, NASA.]

Albedo and Reflection A portion of arriving energy bounces directly back into space without being absorbed or performing any work. This returned energy is called **reflection.** **Albedo** is the reflective quality (intrinsic brightness) of a surface. It is an important control over the amount of insolation that is available for absorption by a surface. We state albedo as the percentage of insolation that is reflected (zero percent is total absorption; 100% is total reflectance). Examine the different surfaces and their albedo values in Figure 4.5.

In the visible wavelengths, darker colors have lower albedos, and lighter colors have higher albedos. On water surfaces, the angle of the solar rays also affects albedo values; lower angles produce a greater reflection than do higher angles (Figure 4.6). In addition, smooth surfaces increase albedo, whereas rougher surfaces reduce it. Specific locations experience highly variable albedo values during the year in response to changes in cloud and ground cover. Earth Radiation Budget (ERB) sensors aboard the *Nimbus-7* satellite measured average albedos of 19%–38% for all surfaces between the tropics (23.5° N to 23.5° S) to as high as 80% in the polar regions.

Earth and its atmosphere reflect 31% of all insolation when averaged over a year. Looking ahead to Figure 4.13, you can see that Earth's average albedo is a combination of 21% reflected by clouds, 3% reflected by the ground (combined land and ocean surfaces), and 7% reflected and scattered by the atmosphere. By comparison, a full Moon, which is bright enough to read by under clear skies, has only a 6%–8% albedo value. Thus, with *earthshine* being four times brighter than moonlight (four times the albedo), and with Earth four times greater in diameter than the Moon, it is no surprise that astronauts report how startling our planet looks from space.

Figure 4.7 portrays total albedos for July 1985 and January 1986 as measured by the Earth Radiation Budget experiments aboard several satellites. These patterns are typical of most years. As compared with July albedos, January albedos are higher poleward of 40° N, because of the snow and ice that cover the ground. Tropical forests are characteristically low in albedo (15%), whereas generally cloudless deserts have high albedos (35%). The southward-shifting cloud cover over equatorial Africa is quite apparent on the January map. What other seasonal changes do you see on these ERB albedo maps?

Clouds, Aerosols, and the Atmosphere's Albedo An unpredictable factor in the tropospheric energy budget, and therefore in refining climatic models, is the role of clouds. Clouds reflect insolation and thus cool Earth's surface. An *increase in albedo* caused by clouds is described by the term **cloud-albedo forcing.** Yet clouds act as insulation, trapping longwave radiation from Earth and raising minimum temperatures. An *increase in greenhouse warming*

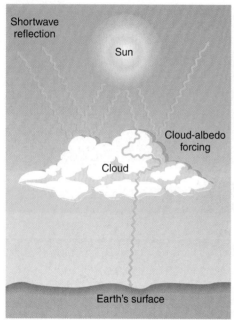

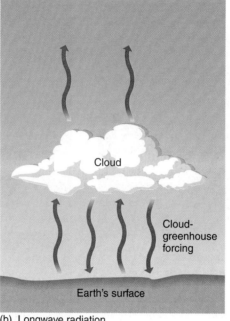

(a) Shortwave radiation

(b) Longwave radiation

FIGURE 4.8 The effects of clouds on shortwave and longwave radiation.
Shortwave radiation is reflected and scattered by clouds; a high percentage is returned to space (a). Longwave radiation emitted by Earth is absorbed and reradiated by clouds; some infrared energy is radiated to space and some back toward the surface (b).

caused by clouds is described as **cloud-greenhouse forcing**. Figure 4.8 illustrates the general effects of clouds on shortwave radiation and longwave radiation. More on clouds is presented later in this chapter and in a detailed section in Chapter 7.

Other mechanisms affect atmospheric albedo and therefore atmospheric and surface energy budgets. Industrialization is producing a haze of pollution that is increasing the reflectivity of the atmosphere, including sulfate aerosols, soot and fly-ash, and black carbon. Emissions of sulfur dioxide and the subsequent chemical reactions in the atmosphere form *sulfate aerosols*. These aerosols act as an insolation-reflecting haze in clear-sky conditions. The result is both an atmospheric warming through absorption by the pollutants and a surface cooling through reduction in insolation reaching ground and water surfaces.

Figure 4.9 is from the CERES (Clouds and the Earth Radiant Energy System) sensors aboard satellite *Terra* and shows these effects on the atmospheric energy budget. The increased aerosols, containing black carbon, cause higher albedos (reflection off the atmosphere), an increase in atmospheric warming (absorption of energy by aerosols), and an increase in surface cooling (less insolation reaching Earth's surface). These aerosols reduce surface insolation by 10% and increase energy absorption in the atmosphere by 50%. Such pollution effects in southern Asia will affect the dynamics of the Asian monsoon through alteration of the region's Earth–atmosphere energy budget (see monsoons in Chapter 6). A weakening monsoonal flow will negatively impact regional water resources and agriculture. This research is part of the international, multi-agency Indian Ocean Experiment (INODOEX, see http://www-indoex.ucsd.edu/).

The eruption of Mount Pinatubo in the Philippines, beginning explosively during June 1991, illustrates how Earth's internal processes affect the atmosphere. Approximately 15–20 megatons of sulfur dioxide were injected into the stratosphere; winds rapidly spread this aerosol (tiny droplets) worldwide (see the images in Figure 6.1). As a result, atmospheric albedo increased worldwide and produced a temporary average cooling of 0.5 C° (0.9 F°).

Absorption   **Absorption** is the assimilation of radiation by molecules of matter and its conversion from one form of energy to another. Insolation (both direct and diffuse) that is not part of the 31% reflected from Earth's surface and atmosphere is absorbed. It is converted into either infrared radiation or chemical energy (by plants in photosynthesis). The temperature of the absorbing surface is raised in the process, and that warmer surface radiates more total energy at shorter wavelengths—thus, *the hotter the surface, the shorter the wavelengths that are emitted*. In addition to absorption by land and water surfaces (about 45% of incoming), absorption also occurs in atmospheric gases, dust, clouds, and stratospheric ozone (about 24% of incoming insolation). Figure 4.13 summarizes the pathways of insolation and the flow of heat in the atmosphere and at the surface.

Conduction, Convection, and Advection   Several means transfer heat energy in a system. **Conduction** is the molecule-to-molecule transfer of heat energy as it diffuses through a substance. As molecules warm, their vibration increases, causing collisions that produce motion in neighboring molecules, thus transferring heat from warmer to cooler materials.

Different materials (gases, liquids, and solids) conduct sensible heat directionally from areas of higher temperature to those of lower temperature. This heat flow transfers en-

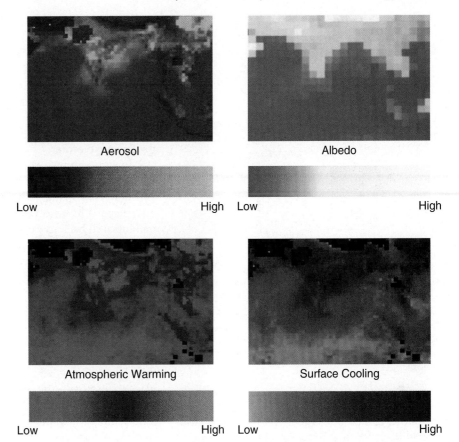

FIGURE 4.9 Aerosols impact Earth–atmosphere energy budgets.
Aerosols including black carbon both reflect and absorb incoming insolation as identified in these four images by the CERES sensors aboard satellite *Terra*, taken between January and March 2001 for southern Asia and the Indian Ocean. Note the color scale for each image rating from low to high aerosol levels, albedo values, amount of atmospheric warming, and resultant surface cooling. The increased aerosol levels from human activities produce higher albedos and atmospheric absorption of energy, resulting in a lowering of surface temperatures. [CERES images aboard *Terra* courtesy of Goddard Space Flight Center, NASA.]

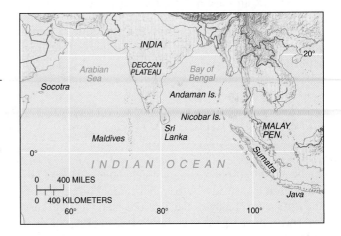

ergy through matter at varying rates, depending on the conductivity of the material—Earth's land surface is a better conductor than air; moist air is a slightly better conductor than dry air.

Gases and liquids also transfer energy by movements called **convection**, when the physical mixing involves a strong vertical motion. When a lateral (horizontal) motion dominates, the term **advection** applies. In the atmosphere or in bodies of water, warmer (less dense) masses tend to rise and cooler (denser) masses tend to sink, establishing patterns of convection. Sensible heat transports physically through the medium in this way.

You commonly experience such energy flows in the kitchen: Energy is conducted through the handle of a pan, or boiling water bubbles in the saucepan in convective motions (Figure 4.10). Also in the kitchen you may use a convection oven that uses a fan to circulate heated air to uniformly cook food.

In physical geography, we find many examples of each physical transfer mechanism.

- *Conduction* (surface energy budgets, temperature differences between land and water bodies, the heating of surfaces and overlying air, soil temperatures);
- *Convection* (atmospheric and oceanic circulation, air mass movements and weather systems, internal motions deep within Planet Earth that produce a magnetic field, and movements in the crust); and

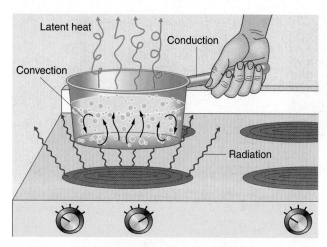

FIGURE 4.10 Heat energy transfer processes.
A pan of water on the stove illustrates heat transfer. Infrared energy *radiates* from the burner to the saucepan and the air. Energy *conducts* through the molecules of the pan and the handle. The water physically mixes, carrying heat energy by *convection*. The energy in the water and handle is measurable as *sensible heat*. The vapor leaving the surface of the water contains the *latent heat* absorbed in the change of water to a vapor.

- *Advection* (horizontal movement of winds from land to sea and back, fog that forms and moves to another area, air mass movements from source regions).

Now with these pathways and principles in mind, we put it all together in the energy budgets of the lower atmosphere.

Energy Balance in the Troposphere

The Earth–atmosphere energy system budget naturally balances itself in a steady-state equilibrium. The atmosphere and surface eventually radiate infrared energy back to space, and this energy, together with reflected energy, equals the initial solar input—think of cash flows into and out of a checking account and the desired balance when deposits and withdrawals are equal.

Greenhouse gases in the atmosphere effectively delay losses to space and act to warm the lower atmosphere. We examine this effect, and then develop an overall budget for the troposphere.

The Greenhouse Effect and Atmospheric Warming

Previously, we characterized Earth as a cool-body radiator, emitting energy in infrared wavelengths from its surface and atmosphere toward space. (In contrast, the Sun is a hot-body radiator, emitting shorter wavelengths from its

surface.) However, some of this infrared radiation is absorbed by carbon dioxide, water vapor, methane, nitrous oxide, chlorofluorocarbons (CFCs), and other gases in the lower atmosphere and then reradiated back toward Earth. This absorption and reradiation delays energy loss to space and is an important factor in warming the troposphere. The rough similarity between this process and the way a greenhouse operates gives the process its name—the **greenhouse effect**.

In a greenhouse, the glass is transparent to shortwave insolation, allowing light to pass through to the soil, plants, and materials inside, where absorption and conduction take place. The absorbed energy is then radiated as thermal infrared back toward the glass, but the glass physically traps both the longer infrared wavelengths and the warmed air inside the greenhouse. Thus, the glass acts as a one-way filter, allowing the light energy in but not allowing the heat energy out, except through conduction. The same process also is quite evident in a car parked in direct sunlight.

Opening the greenhouse roof vent, or the car windows, allows the air inside to mix with the outside environment, thereby removing heat by moving air physically from one place to another—the process of convection. It is surprising how hot the interior of a car gets, even on a day with mild temperatures outside. Many people place an opaque sunscreen across the windshield to prevent shortwave energy from entering the car to begin the greenhouse process. Have you taken steps to reduce the insolation input into your car's interior or into windows at home?

In the atmosphere, the greenhouse analogy does not fully apply because infrared radiation is not trapped as in a greenhouse. Rather, its passage to space is *delayed* as the infrared radiation is absorbed by certain gases, clouds, and dust in the atmosphere and reradiates to Earth's surface. According to scientific consensus, today's increasing carbon dioxide concentration is forcing more infrared radiation absorption in the lower atmosphere, thus producing a warming trend and changes in the Earth–atmosphere energy system.

Clouds and Earth's "Greenhouse"

Clouds affect the heating of the lower atmosphere in several ways, depending on cloud type. Not only is the percentage of cloud cover important, but the cloud type, height, and thickness (water content and density) also has an effect. High-altitude, ice-crystal clouds reflect insolation with albedos of about 50%, whereas thick, lower cloud cover reflects about 90% of incoming insolation.

To understand the actual effects on the atmosphere's energy budget, however, we must consider both transmission of shortwave and longwave radiation and cloud type. Figure 4.11a portrays the *cloud-greenhouse forcing* caused by high clouds (warming, because their greenhouse effects exceed their albedo effects); and Figure 4.11b portrays the *cloud-albedo forcing* produced by lower, thicker

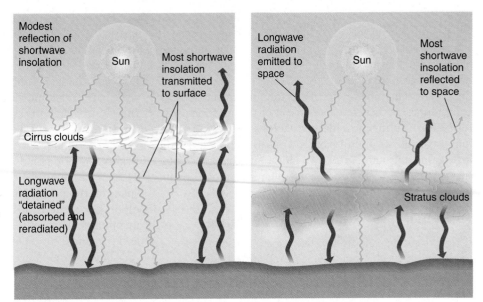

FIGURE 4.11 Energy effects of two cloud types.
Cloud effects vary depending on cloud type. (a) High, ice-crystal clouds (called *cirrus*) transmit most of the insolation. However, they absorb and delay losses of outgoing longwave infrared, producing a greater greenhouse forcing and a net warming of Earth. (b) Low, thick clouds (stratus) reflect most of the incoming insolation and radiate longwave infrared to space, producing a greater albedo forcing and a net cooling of Earth.

(a) High clouds: net greenhouse forcing and atmospheric warming

(b) Low clouds: net albedo forcing and atmospheric cooling

clouds (cooling, because albedo effects exceed greenhouse effects). Understanding the nature of global cloud cover is crucial in refining computer models that forecast global climate change.

To better understand the role of clouds, NASA is operating the CERES program, an acronym for Clouds and Earth's Radiant Energy System. CERES sensors aboard satellites began operating in 1997 (*TRMM*) and the newest in 2000 (*Terra*). In addition to understanding natural clouds, the role of jet contrails (soot, oxidized sulfur, nitrogen oxides, mixed in hot and humid exhaust gases) is also important for they cause the formation of high ice-

crystal cirrus clouds. CERES data are an important part of the study of aerosols and the monsoons in Figure 4.9. (See http://asd-www.larc.nasa.gov/ceres/ASDceres.html.)

Collection of data by CERES sensors cover both reflected and emitted radiation. Figure 4.12 shows the first global monthly images from satellite *Terra* for March 2000 in W/m² flux (meaning energy "flow"). In Figure 4.12a lighter regions indicate where more sunlight is reflected into space than is absorbed—for example, by light land surfaces such as deserts, or by cloud cover such as over tropical lands. Green and blue areas illustrate where less light was reflected.

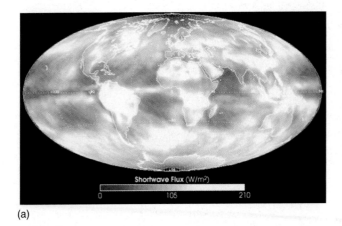

(a)

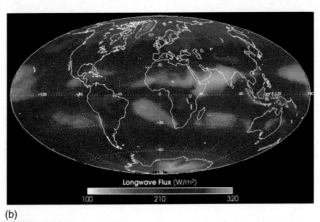

(b)

FIGURE 4.12 Shortwave and longwave images show Earth's radiation budget components.
The CERES sensors aboard *Terra* made these portraits in March 2000, capturing (a) outgoing shortwave energy flux reflected from clouds, land, and water—Earth's albedo; and, (b) longwave energy flux emitted by surfaces back to space. The scale beneath each image displays value gradations in watts per square meter. [Images courtesy of CERES Instrument Team, Langley Research Center, NASA.]

In Figure 4.12b orange and red pixels indicate regions where more heat was absorbed and emitted to space, whereas less heat energy is escaping in the blue and purple regions. Those blue regions of lower longwave emissions over tropical lands are due to tall, thick clouds along the equatorial convergence (Amazon, equatorial Africa, and Indonesia); these clouds also caused higher shortwave reflection. Subtropical desert regions exhibit greater longwave radiation emissions owing to the presence of little cloud cover and greater radiative energy losses from surfaces that have absorbed a lot of energy, as we saw in Figure 4.4.

Earth–Atmosphere Radiation Balance

If Earth's surface and its atmosphere are considered separately, neither exhibits a balanced radiation budget. The average annual energy distribution is positive (an energy surplus) for Earth's surface and negative (an energy deficit) for the atmosphere as it radiates energy to space. Considered together, these two equal each other, making it possible for us to construct an overall energy balance. However, regionally and seasonally Earth absorbs more energy in the tropics and less in the polar regions, establishing the imbalance that drives global circulation patterns.

Figure 4.13 summarizes the Earth–atmosphere radiation balance. It brings together all the elements discussed to this point in the chapter by following 100% of arriving insolation through the troposphere. The shortwave portion of the budget is on the left in the illustration; the longwave part of the budget is on the right. Of 100% of solar energy arriving, Earth's average albedo is 31%. Figure 4.12a shows this shortwave radiation reflected from the Earth–atmosphere system during March 2000. Absorption by atmospheric clouds, dust, and gases involves another 21% and accounts for the atmospheric heat input. Stratospheric ozone absorption and radiation accounts for another 3% of the atmospheric budget. About 45% of the incoming insolation transmits through to Earth's surface as direct and diffuse radiation.

The natural energy balance occurs through energy transfers from the surface that are both *nonradiative* (physical motion) and *radiative*. Nonradiative transfers include convection, conduction, and the latent heat of evaporation (energy that is absorbed and dissipated by water as it evaporates and condenses). Radiative transfer is by infrared radiation between the surface, the atmosphere, and space, as illustrated to the right in Figure 4.13's depiction of the greenhouse effect.

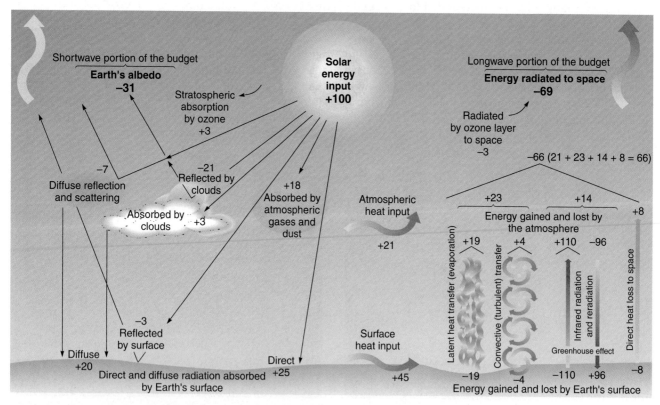

FIGURE 4.13 Detail of the Earth–atmosphere energy balance.
Solar energy cascades through the lower atmosphere (left-hand portion of the illustration), where it is absorbed, reflected, and scattered. Clouds, atmosphere, and the surface reflect 31% of this insolation back to space. Atmospheric gases and dust and Earth's surface absorb energy and radiate infrared radiation. Earth and atmosphere exchange energy through latent heat transfer in water vapor, convective transfer (moving air), and infrared radiation (right-hand portion of the illustration). Over time, Earth reradiates, on average, 69% of incoming energy to space. When added to Earth's average albedo (31%, reflected energy), this equals the total energy input from the Sun.

FIGURE 4.14 Energy budget by latitude.
Earth's energy surpluses and deficits by latitude
produce poleward transport of energy and mass in
each hemisphere—atmospheric circulation and
ocean currents.

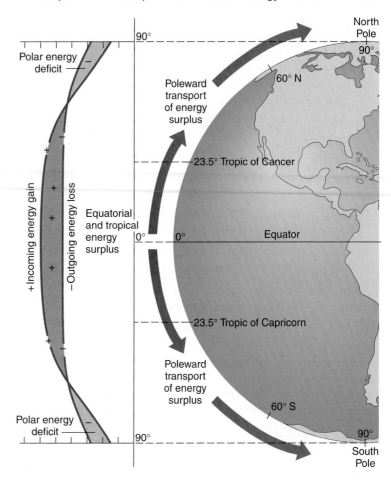

In summarizing the thermal infrared part of the budget, Earth eventually reradiates the remaining 69% into space: 21% (atmosphere heating) + 45% (surface heating) + 3% (ozone emission) = 69%. This longwave radiation from the Earth–atmosphere system (net outgoing longwave) is shown in Figure 4.12b.

Figure 4.14 summarizes the radiation balance for all shortwave and longwave energy by latitude:

- Between the tropics, the angle of incoming insolation is high and daylength is consistent, with little seasonal variation, so more energy is gained than lost—*energy surpluses dominate.*
- In the polar regions, the Sun is low in the sky, surfaces are light (ice and snow) and reflective, and for up to 6 months during the year no insolation is received, so more energy is lost than gained—*energy deficits prevail.*
- At around 36° latitude, a balance exists between energy gains and losses for the Earth–atmosphere system.

The imbalance of net radiation from tropical surpluses to the polar deficits drives a vast global circulation of both energy and mass. The meridional (north–south) transfer agents are winds, ocean currents, dynamic weather systems, and related phenomena. Dramatic examples of such energy and mass transfers are tropical cyclones—hurricanes

and typhoons. Forming in the tropics, these powerful storms mature and migrate to higher latitudes, carrying with them energy, water, and water vapor.

Having established the Earth–atmosphere radiation balance, we next focus on energy characteristics at Earth's surface. Essentially, we want to examine energy budgets along the ground level as portrayed in Figure 4.13.

Energy Balance at Earth's Surface

Solar energy is the principal heat source at Earth's surface. The direct and diffuse radiation and infrared radiation arriving at the ground surface are shown in Figure 4.13. These radiation patterns at Earth's surface are of great interest to geographers and should be of interest to everyone because these are the surface environments where we live.

Daily Radiation Patterns

The fluctuating daily pattern of incoming shortwave energy absorbed and the resultant air temperature is shown in Figure 4.15. This graph represents idealized conditions for bare soil on a cloudless day in the middle latitudes. Incoming energy arrives during daylight, beginning at sunrise, peaking at noon, and ending at sunset.

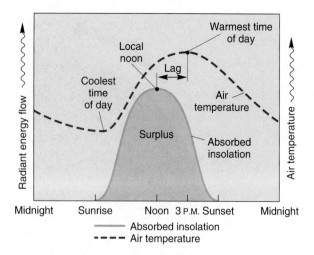

FIGURE 4.15 Daily radiation curves.
Sample radiation plot for a typical day shows the changes in insolation (orange line) and air temperature (dashed line). Comparing the curves demonstrates a lag between local noon (the insolation peak for the day) and the warmest time of day.

The shape and height of this insolation curve vary with season and latitude. The highest trend for such a curve occurs at the time of the summer solstice (around June 21 in the Northern Hemisphere and December 21 in the Southern Hemisphere). The air temperature plot also responds to seasons and variations in insolation input. Within a 24-hour day, air temperature generally peaks between 3:00 and 4:00 P.M. and dips to its lowest point right at or slightly after sunrise.

The relationship between the insolation curve and the air temperature curve on the graph is interesting—they do not align; there is a *lag*. The warmest time of day occurs not at the moment of maximum insolation but at the moment when a *maximum of insolation is absorbed* and reradiated to the atmosphere from the ground. As long as the incoming energy exceeds the outgoing energy, air temperature continues to increase, not peaking until the incoming energy begins to diminish in the afternoon as the Sun's altitude decreases. In contrast, if you have ever gone camping in the mountains, you no doubt experienced the coldest time of day with a wake-up chill at sunrise!

The annual pattern of insolation and air temperature exhibits a similar lag. For the Northern Hemisphere, January is usually the coldest month, occurring after the December solstice and the shortest days. Similarly, the warmest months of July and August occur after the June solstice and the longest days.

Simplified Surface Energy Balance

Earth's surface is supplied with energy that varies daily and seasonally. Energy and moisture are continually exchanged at the surface, creating worldwide "boundary layer climates" of great variety. Physical conditions at or near Earth's surface are studied in **microclimatology**—

the science of this lowest portion of the atmosphere. The following discussion is more meaningful if you visualize an actual surface—perhaps a park, a front yard, or a place on campus.

The surface receives visible light and infrared radiation, and it reflects light and radiates infrared according to the following simple scheme:

$$+SW\downarrow \quad -SW\uparrow \quad +LW\downarrow -LW\uparrow \quad = \quad NET\ R$$
(Insolation)(Reflection)(Infrared) (Infrared)   (NetRadiation)

We use SW for shortwave, LW for longwave for simplicity. You may come across other symbols in the microclimatology literature, such as Q^* for NET R, K for shortwave, and L for longwave.

Figure 4.16 shows the components of a surface energy balance. The soil column shown continues to a depth at which energy exchange with surrounding materials or with the surface becomes negligible, usually less than a meter. Sensible heat transfer in the soil is through conduction, predominantly downward during the day or in summer, and toward the surface at night or in winter. Energy from the atmosphere that is moving toward the surface is regarded as positive (a gain), and energy that is moving away from the surface, through sensible and latent heat transfers, is considered negative (a loss) to the surface account.

Adding and subtracting the energy flow at the surface completes the calculation of **net radiation** (**NET R**), or the balance of all radiation at Earth's surface. As the components of this simple equation vary with daylength through the seasons, cloudiness, and latitude, NET R

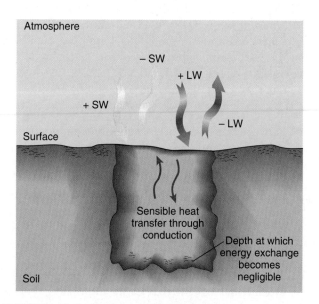

FIGURE 4.16 Surface energy budget.
Idealized input and output energy budget components for a surface and a soil column. Sensible heat transfer in the soil is through conduction, predominantly downward during the day or in summer, and toward the surface at night or in winter. (SW = shortwave, LW = longwave.)

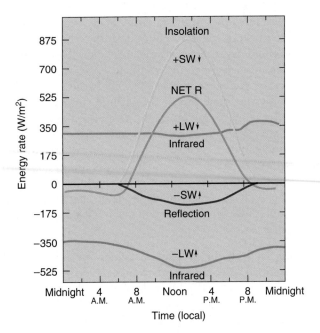

FIGURE 4.17 A day's radiation budget.
Radiation budget components and resulting net radiation
(NET R) on a typical summer day for a midlatitude location
(Matador in southern Saskatchewan, about 51° N, on July 30,
1971). [Adapted by permission from T. R. Oke, *Boundary
Layer Climates* (New York: Methuen & Co., 1978), p. 21.]

varies. Figure 4.17 illustrates the components of a surface
energy balance for a typical summer day at a midlatitude location showing the daily change in the components of net

radiation. The items plotted are direct and diffuse insolation (+SW↓), reflected energy (surface albedo value,
–SW↑), and infrared radiation arriving at (+LW↓) and
leaving from (–LW↑) the surface.

Surface albedo values (–SW↑) dictate the amount
of insolation reflected, and therefore not absorbed, at the
surface. As an example, imagine a snow-covered landscape as a surface energy system. Snow has high reflectivity, sending sunlight back to space and reducing the
amount of insolation absorbed. Consequently, surfaces
and air temperatures are lower, and thus the snow cover
does not melt. This is a system with positive feedback.
(Remember from Chapter 1 that *positive feedback* increases response in the system: more cooling–more snow–
increasing albedo–more cooling–then colder, drier air–then
less snow–and so on.)

At night, the net radiation value becomes negative
because the SW component ceases at sunset and the surface continues to lose infrared energy to the atmosphere.
The surface rarely reaches a zero net radiation value—a
perfect balance—at any one moment, but over time,
Earth's surface naturally balances incoming and outgoing
energies.

Net Radiation The net radiation (NET R of all wavelengths) available at Earth's surface is the final outcome of
the entire radiation-balance process discussed in this chapter. Figure 4.18 displays the global mean annual net radiation at ground level. The abrupt change in radiation
balance from ocean to land surfaces is evident on the map.

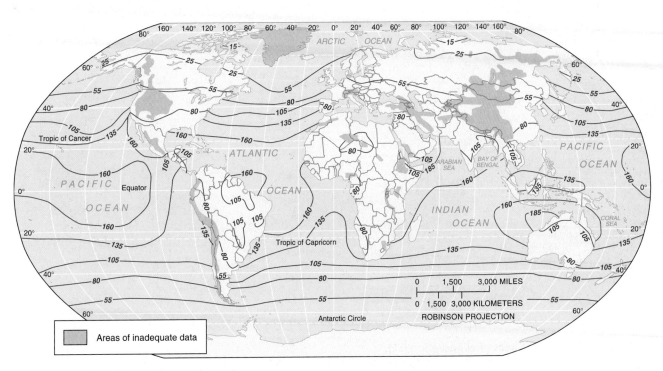

FIGURE 4.18 Global net radiation (NET R).
Distribution of global mean annual net radiation (NET R) at surface level in watts per square
meter (100 W/m² = 75 kcal/cm²/year). [After M. I. Budyko, *The Heat Balance of the Earth's
Surface* (Washington, DC: U.S. Department of Commerce, 1958), p. 106.]

Note that all values are positive; negative values probably occur only over ice-covered surfaces poleward of 70° latitude in both hemispheres. The highest net radiation occurs north of the equator in the Arabian Sea at 185 W/m² per year. Aside from the obvious interruption caused by landmasses, the pattern of values appears generally zonal, or parallel, decreasing away from the equator.

Net radiation is expended from a nonvegetated surface through three pathways:

- *LE*, or *latent heat of evaporation*, is the energy that is stored in water vapor as water evaporates. Large quantities of this *latent heat* are absorbed into water vapor, during water's change of state. This heat energy is thereby removed from the surface. Conversely, this heat energy releases to the environment when water vapor changes state back to a liquid (discussed in Chapter 7). Latent heat is the dominant expenditure of Earth's entire NET R, especially over water surfaces.

- *H*, or *sensible heat*, is the back-and-forth transfer between air and surface in turbulent eddies, through convection and conduction within materials. This activity depends on surface and boundary-layer temperatures and on the intensity of convective motion in the atmosphere. About one-fifth of Earth's entire NET R is mechanically radiated as sensible heat from the surface, especially over land.

- *G*, or *ground heating and cooling*, is the energy that flows into and out of the ground surface (land or water) by conduction. During a year, the overall *G* value is zero because the stored energy from spring and summer is equaled by losses in fall and winter. Another factor in ground heating is energy absorbed at the surface to melt snow or ice. In snow- or ice-covered landscapes, most available energy is in sensible and latent heat used in the melting and warming process.

On land, the highest annual values for latent heat of evaporation (LE) occur in the tropics and decrease toward the poles (Figure 4.19). Over the oceans, the highest LE values are over subtropical latitudes, where hot, dry air comes into contact with warm-ocean water.

The values for sensible heat (H) are distributed differently, being highest in the subtropics (Figure 4.20). Here, vast regions of subtropical deserts feature nearly waterless surfaces, cloudless skies, and almost vegetation-free land-

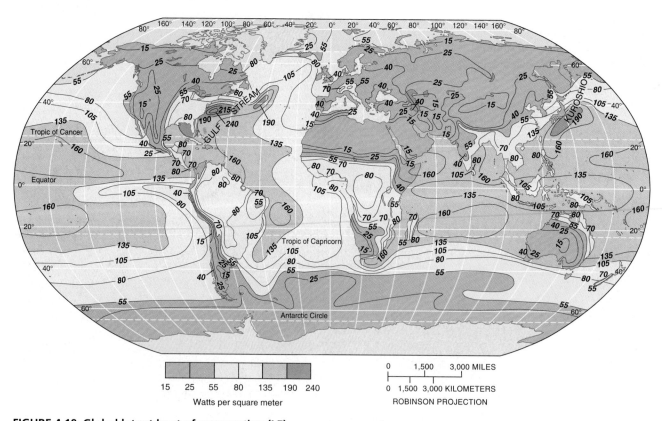

FIGURE 4.19 Global latent heat of evaporation (LE).
Distribution of annual energy expenditure as the latent heat of evaporation (LE) at surface level in watts per square meter (100 W/m² = 75 kcal/cm²/year). Note the high values associated with high sea-surface temperatures in the area of the Gulf Stream and Kuroshio currents. [Adapted by permission from M. I. Budyko, *The Earth's Climate Past and Future* (New York: Academic Press, 1982), p. 56.]

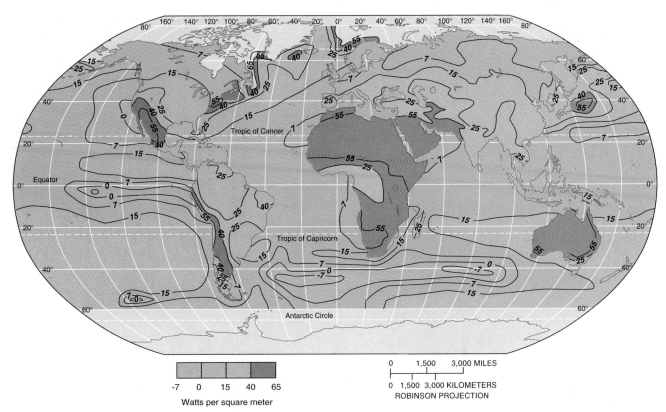

FIGURE 4.20 Global sensible heat (H).
Distribution of annual energy expenditure as sensible heat (H) at surface level in watts per square meter (100 W/m² = 75 kcal/cm²/year). [Adapted by permission from M. I. Budyko, *The Earth's Climate Past and Future* (New York: Academic Press, 1982), p. 59.]

scapes. The bulk of NET R is expended as sensible heat in these dry regions. Moist and vegetated surfaces expend less in H and more in LE, as you can see by comparing the maps in Figures 4.19 and 4.20.

Understanding net radiation is essential to solar energy technologies that concentrate shortwave energy for use. Solar energy offers great potential worldwide and is presently the fastest-growing form of energy conversion by humans, although still decades away from its possibilities. Focus Study 4.1 briefly reviews direct application of surface energy budgets.

Sample Stations   A couple of real locations brings what you just read into meaningful perspective. Variation in the expenditure of NET R among sensible heat (H, energy we can feel), latent heat (LE, energy for evaporation), and ground heating and cooling (G) produces the variety of environments we experience in nature. Let us examine the daily energy balance at two locations, El Mirage in California and Pitt Meadows in British Columbia.

El Mirage, at 35° N, is a hot desert location characterized by bare, dry soil with sparse vegetation (Figure 4.21a, b). Our sample is a clear summer day, with a light wind in the late afternoon. The NET R-value is lower than might

be expected, considering the Sun's position close to zenith (June solstice) and the absence of clouds. But the income of energy at this site is countered by surfaces of higher albedo than forest or cropland and by hot soil surfaces that radiate infrared back to the atmosphere throughout the afternoon.

El Mirage has little or no energy expenditure for evaporation (LE). With little water and sparse vegetation, most of the available radiant energy dissipates through turbulent transfer of sensible heat (H), warming air and soil to high temperatures. Over a 24-hour period, H is 90% of NET R; the remaining 10% are for ground heating (G). The G component is greatest in the morning, when winds are light and turbulent transfers are lowest. In the afternoon, heated air rises off the hot ground, and convective heat expenditures are accelerated as winds increase.

Compare El Mirage (Figure 4.21a) and Pitt Meadows (Figure 4.21c). Pitt Meadows is midlatitude (49° N), vegetated, and moist, and its energy expenditures differ greatly from those at El Mirage. The Pitt Meadows landscape is able to retain much more of its energy because of lower albedo values (less reflection), the presence of more water and plants, and lower surface temperatures than those of El Mirage.

The graph plots the energy balance data for Pitt Meadows for a cloudless summer day. Higher LE values result

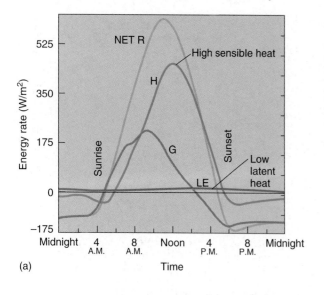

(a)

(b)

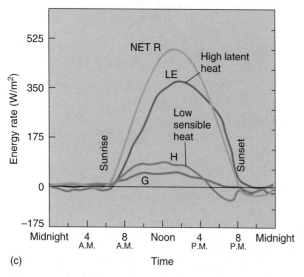

(c)

(d)

FIGURE 4.21 Radiation budget comparison for two stations.
Graph shows daily net radiation expenditure for El Mirage, California, east of Los Angeles, at about 35° N (a). Photo of the typical desert landscape near the site (b). Graph of the daily net radiation expenditure for Pitt Meadows in southern British Columbia, at about 49° N (c). Photo of irrigated blueberry orchards characteristic of the agricultural activity in this moist environment of moderate temperatures (d). (H = turbulent sensible heat transfer; LE = latent heat of evaporation; G = ground heating and cooling.) [(a) Adapted by permission from William D. Sellers, *Physical Climatology,* fig. 33, copyright © 1965 by The University of Chicago. All rights reserved.(c) Adapted by permission from T. R. Oke, *Boundary Layer Climates* (New York: Methuen & Co., 1978), p. 23. Photos by author.]

from the moist environment of rye grass and irrigated mixed-orchard ground cover for the sample area (Figure 4.21d), contributing to the more moderate sensible heat (H) levels during the day.

The Urban Environment

For most of you reading this book, an urban landscape produces the temperatures you feel each day. Urban microclimates generally differ from those of nearby nonurban

areas. In fact, the surface energy characteristics of urban areas are similar to desert locations. Because almost 50% of the world's population will live in cities by the year 2010, urban microclimatology and other specific environmental effects related to cities are important topics for physical geographers.

The physical characteristics of urbanized regions produce an **urban heat island** that has on average both maximum and minimum temperatures higher than nearby rural settings. Table 4.1 lists five urban characteristics and the

Table 4.1 Urban Physical Characteristics and Conditions

Urban characteristics	Results and conditions
Urban surfaces typically are metal, glass, asphalt, concrete, or stone, and their energy characteristics respond differently from natural surfaces	Albedos of urban surfaces are lower, leading to higher net radiation values Urban surfaces expend more energy as sensible heat than do nonurban areas (70% of the net radiation to H) Surfaces conduct up to three times more energy than wet, sandy soil and thus are warmer During the day and evening, temperatures above urban surfaces are higher than those above natural areas
Irregular geometric shapes in a city affect radiation patterns and winds	Incoming insolation is caught in mazelike reflection and radiation "canyons" Delayed energy is conducted into surface materials, thus increasing temperatures Buildings interrupt wind flows, diminishing heat loss through advective (horizontal) movement Maximum heat island effects occur on calm, clear days and nights
Human activity alters the heat characteristics of cities	In summer, urban electricity production and use of fossil fuels release energy equivalent to 25%–50% of insolation In winter, urban-generated sensible heat averages 250% greater than arriving insolation, reducing winter heating requirements
Many urban surfaces are sealed (built on and paved), so water cannot reach the soil	Central business district surfaces average 50% sealed, suburbs average 20% sealed, producing more water runoff Urban areas respond as a desert landscape: A storm may cause a flash flood over the hard, sparsely vegetated surfaces, to be followed by dry conditions a few hours later
Air pollution, including gases and aerosols, greater in urban areas than in comparable natural settings; increased convection and precipitation possible	Pollution increases the atmosphere's reflectivity above a city, reducing insolation and absorbing infrared radiation, reradiating infrared downward Increased particulates in pollution are condensation nuclei for water vapor, increasing cloud formation and precipitation Urban-stimulated increases in precipitation may occur downwind from cities

Focus Study 4.1

Solar Energy Collection and Concentration

Consider the following:

- Old photographs of residences in southern California and Florida show solar (flat-plate) water heaters on many rooftops. Early twentieth-century newspaper ads and merchandise catalogues featured the Climax solar water heater (1905) and the Day and Night water heater (1909). Applied solar energy principles and technologies are well established! However, low-priced natural gas and oil displaced many of these early applications.

- The insolation receipt in just 35 minutes at the surface of the United States exceeds the amount of energy derived from the burning of fossil fuels (coal, oil, natural gas) in a year.

- An average building in the United States receives 6 to 10 times more energy from the Sun hitting its exterior than is required to heat the inside.

- A 500-watt photovoltaic system that converts sunlight directly into electricity (including batteries) is cheaper to install at a rural site than a 2 km (1.2 mi) power line bringing in electricity. Such a system supplies more than enough electricity for lights, television, computer, a water pump, and some appliances. Photovoltaic prices dropped from $75 per watt in 1975 to $3.50 per watt in 2000.

- Installed photovoltaic capacity jumped 43% in 2000 (87 new megawatts in one year), worldwide capacity is now estimated at 288 megawatts. This production tripled 1996 levels.

Not only does insolation warm Earth's surface, it also provides an *(continued)*

Focus Study 4.1 (*continued*)

inexhaustible supply of energy far into the future for humanity. Sunlight is direct, pervasive (spreading widely), and renewable and has been collected for centuries through various technologies. Yet it is underutilized.

Rural villages in developing countries could benefit greatly from the simplest, most cost-effective solar application—the *solar-panel cooker*. For example, people in Kenya walk many kilometers collecting fuelwood for cooking fires (Figure 1a). Each village and refugee camp is surrounded by impoverished land, stripped of wood. Using solar cookers, villagers are able to cook meals and sanitize their drinking water without scavenging for wood (Figure 1b). (See Solar Cookers International at http://solarcooking.org/.)

In less-developed countries, the money for electrification (a *centralized* technology) is not available despite the push from more developed countries and energy corporations for large capital-intensive power projects. In such countries, the pressing need is for *decentralized* energy sources, appropriate in scale to everyday needs, such as cooking, heating water, and pasteurization. Net per capita (per person) cost for solar cookers is far less than for centralized electrical production, regardless of fuel source.

Collecting and Concentrating Solar Energy

Any surface that receives light from the Sun is a *solar collector*. But the diffuse nature of solar energy received at the surface requires that it be collected, concentrated, transformed, and stored to be useful. Space heating is the simplest application. Windows that are carefully designed and placed allow sunlight to shine into a building, where it is absorbed and converted into sensible heat. Here we have an everyday application of the greenhouse effect.

A *passive solar system* captures heat energy and stores it in a "thermal mass," such as water-filled tanks, adobe, tile, or concrete. Three key features of passive solar design are (1) large areas of glass facing south toward the Sun (facing north in the Southern Hemisphere), (2) a thermal storage medium, and (3) shades or screens to prevent light entry in the warm summer months. At night, shades or other devices can cover glass areas to prevent energy loss from the structure. An *active solar system* involves heating water or air in a collector and then pumping it through a plumbing system to a tank where it can provide hot water for direct use or for space heating.

Solar energy systems can generate heat energy of an appropriate scale for approximately half the present domestic applications in the United States (space heating and water heating). In marginal climates, solar-assisted water and space heating is feasible as a back-up; even in New England and the Northern Plains states, solar collection systems prove effective. (See the National Solar Radiation Data Base at http://rredc.nrel.gov/solar/pubs/NSRDB/.)

Focusing (concentrating) mirrors, such as Fresnel lenses, or parabolic (curved surface) troughs and dishes can be used to attain very high temperatures to heat water or other heat-storing fluids. Kramer Junction, California, about 225 km (140 mi) northeast of Los Angeles, in the Mojave Desert near Barstow, has the world's largest operating solar electric-generating facility. A capacity of 354 MW (megawatts; 354 million watts) operates in nine plants. This is a moderate-sized power plant. Long troughs of computer-guided curved mirrors concentrate sunlight to create temperatures of 390°C (735°F)

(a)

(b)

FIGURE 1 The solar-cooking solution.
(a) Five women haul firewood many miles to the Dadaab refugee camp in northeastern Kenya. (b) Kenyan women in training to use their solar panel cookers, which do not require scavenging the countryside for scarce fuelwood. These simple cookers collect direct and diffuse insolation through transparent glass or plastic and trap infrared radiation in an enclosed box or cooking bag. (This is a small-scale, efficient application of the greenhouse effect.) Construction is easy, using cardboard components. Temperatures easily exceed 105°C (220°F) for baking, boiling, purifying water, and sterilizing instruments. [Photos by Solar Cookers International, Sacramento, California.]

(*continued*)

Focus Study 4.1 *(continued)*

(a)   (b)   (c)

FIGURE 2 Solar thermal and photovoltaic energy production.
(a) Kramer Junction solar thermal energy installation in southern California. (b) NREL Outdoor Test Facility in Golden, Colorado, where a variety of photovoltaic cell arrays successfully convert sunlight directly into electricity. (c) The Sacramento Municipal Utility District installed a PV array that doubles as parking lot cover. [Photos by (a) Kramer Junction Operating Company, Los Angeles; (b) and (c) Bobbé Christopherson.]

in vacuum-sealed tubes filled with synthetic oil. The heated oil heats water; the heated water produces steam that rotates turbines to generate cost-effective electricity. The facility converts 23% of the sunlight it receives into electricity during peak hours (Figure 2a) and operation and maintenance costs continue to decrease.

The National Renewable Energy Laboratory (NREL, http://www.nrel. gov and the National Center for Photovoltaics at http://www.nrel.gov/ ncpv/) was established in 1974 to coordinate solar energy research, development, and testing in partnership with private industry. NREL's headquarters building in Golden, Colorado, features the latest in passive and active solar design. At NREL's Outdoor Test Facility, successful tests continue on arrays of prototype solar cells. Some of these panels have been in operation for more than 10 years (Figure 2b).

Electricity Directly from Sunlight

Producing electricity by *photovoltaic cells* (PVs) is a technology that has been used in spacecraft since 1958. Familiar to us all are the solar cells in pocket calculators (more than 100 million units now in use). When light shines upon a

semiconductor material in these cells, it stimulates a flow of electrons (an electrical current) in the cell. PV cells are arranged in modules that can be assembled in large arrays.

The efficiency of these cells has improved to the level that they are generally cost-competitive, especially if government policies and subsidies were to be balanced evenly among energy sources. NREL developed a copper–indium–gallium solar cell that achieves an astonishing 18.8% conversion rate of sunlight to electricity. New cells that are 50% more efficient than crystalline-silicon cells are now available for satellites.

Rooftop photovoltaic electrical generation is now cheaper than power line construction to rural sites. As of 1998, some 250,000 homes in Mexico, Indonesia, South Africa, India, and elsewhere have PV roof systems. A solar power project in the Philippines will bring electricity to 150 remote villages when completed. Norway, at the same high latitudes as Alaska, has 60,000 units operating and is adding about 8000 new PV systems a year. Some 200 photovoltaic power systems are operating in the Navajo Nation of northeastern Arizona. Innovative uses abound as shown in Figure 2c. (See the

Department of Energy's "Photovoltaic Home Page," at http://www.eren. doe.gov/pv/.)

Production leader Japan expanded its production of photovoltaics by 75% in 2001 to 224 MW. U.S. production lagged behind at 75 MW mostly for export to Japan and Europe, and in 2000, Europe produced 61 MW led by Germany. Despite this international boom in photovoltaic electric power generation, the United States' administration proposed sweeping budget cuts in its photovoltaic program for the 2002 budget.

Obvious drawbacks of both solar-heating and solar-electric systems are periods of cloudiness and night, which inhibit operations. Research is under way to enhance energy-storage technologies, such as hydrogen fuel production (using energy to extract hydrogen from water for later use in producing more energy) and to improve battery technology.

The Promise of Solar Energy

Solar energy is a wise choice for the future. It is directly available to the consumer; it is based on a renewable energy source of an appropriate scale for end-use needs (it matches them

(continued)

Focus Study 4.1 (continued)

well); and most solar strategies are labor-intensive (rather than capital-intensive as in centralized power production). Solar is preferable to further development of our decreasing fossil fuel reserves, further increases in oil imports and tanker spills, investment in foreign military incursions, or to adding more troubled nuclear power.

Whether or not we follow the alternative path of solar energy is a matter of political control and not technological innovation. Much of the technology is ready for installation and is cost-effective when all direct and indirect costs are considered for all the alternatives. NREL asserts in its mission statement that the laboratory wants to "lead the nation toward a sustainable energy future by developing renewable energy technologies, improving energy efficiency, advancing related science, and engineering commercialization." Unfortunately, political decisions slow progress, cut budgets, and create uneven investment incentives weighted toward fossil fuels.

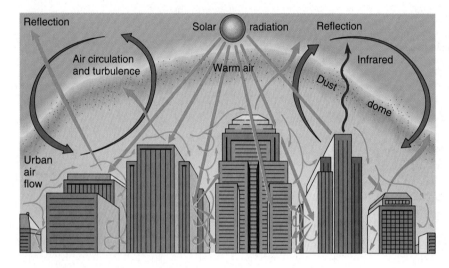

FIGURE 4.22 The urban environment. Insolation, wind movements, and dust dome in city environments.

resulting temperature and moisture effects produced. Figure 4.22 illustrates these traits. Every major city produces its own **dust dome** of airborne pollution, which can be blown from the city in elongated plumes; as noted in the table, such domes affect urban energy budgets.

Table 4.2 compares climatic factors of rural and urban environments. The worldwide trend toward greater urbanization is placing more and more people on urban heat islands. NASA launched its Urban Heat Island Pilot Project (UHIPP) in 1997, through its Global Hydrology and Climate Center, to better understand the role of cities in climate. If ways can be found to make cities cooler, this will reduce energy consumption and fossil fuel use, thus reducing greenhouse gas emissions.

Thermal infrared measurements were made in 1998 from NASA's research jet over Atlanta, Georgia; Sacramento, California; Baton Rouge, Louisiana; and Salt Lake City, Utah. Teachers and students on the ground assisted efforts by making temperature measurements at the same time as the flights. Instruments carried by balloons tracked vertical temperature, relative humidity, and air pressure profiles. (See http://www.ghcc.msfc.nasa.gov/ghcc_research.html and http://198.122.199.205/uhipp/urban_uhipp.html for an update.)

Figure 4.23 illustrates a generalized cross section of a typical urban heat island, showing increasing temperatures toward the downtown central business district. Note that temperatures drop over areas of trees and parks. Sensible heat is lessened because of latent heat of evaporation and plant effects (transpiration and shade). Urban forests are important factors in cooling cities. NASA measurements in an outdoor mall parking lot found temperatures of 48°C (118°F), but a small planter with trees in the same lot was significantly cooler at 32°C (90°F)—16 C° (29 F°) lower in temperature! In Central Park in New York City (see chapter-opening photo) daytime temperatures average 5–10 C° (9–18 F°) cooler than urban areas outside the park.

Figure 4.24 is a thermal infrared image of a portion of Sacramento, California (midday). The false colors of red and white are relatively hot areas (60°C, 140°F), blues and

Table 4.2   Average Differences in Climatic Elements Between Urban and Rural Environments

Element	Urban Compared with Rural Environs	Element	Urban Compared with Rural Environs
Contaminants		**Precipitation (cont.)**	
Condensation nuclei	10 times more	Snowfall, downwind (lee) of city	10% more
Particulates	10 times more		
Gaseous admixtures	5–25 times more	Thunderstorms	10%–15% more
Radiation		**Temperature**	
Total on horizontal surface	0%–20% less	Annual mean	0.5–3.0 C° (0.9–5.4 F°) more
Ultraviolet, winter	30% less	Winter minima (average)	1.0–2 C° (1.8–3.6 F°) more
Ultraviolet, summer	5% less	Summer maxima	1.0–3 C° (1.8–3.0 F°) more
Sunshine duration	5%–15% less	Heating degree days	10% less
Cloudiness		**Relative Humidity**	
Clouds	5%–10% more	Annual mean	6% less
Fog, winter	100% more	Winter	2% less
Fog, summer	30% more	Summer	8% less
Precipitation		**Wind Speed**	
Amounts	5%–15% more	Annual mean	20%–30% less
Days with < 5 mm (0.2 in.)	10% more	Extreme gusts	10%–20% less
Snowfall, inner city	5%–10% less	Calm	5%–20% more

Source: H. E. Landsberg, *The Urban Climate*, International Geophysics Series, vol. 28 (1981), p. 258. Reprinted by permission from Academic Press.

FIGURE 4.23 Typical urban heat island profile.
Generalized cross section of a typical urban heat island. The trend of the temperature gradient from rural to downtown is measurable. Temperatures steeply rise in urban settings, plateau over the suburban built-up area, and peak where temperature is highest at the urban core. Note the cooling over the park area and river.

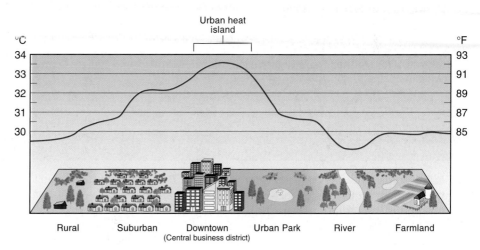

greens are relatively cool areas (29°–36°C, 85°–96°F); clearly buildings are the hottest objects in the scene. As you look across this landscape, what strategies do you think would reduce the urban heat island effect? Possibilities include lighter-colored surfaces for buildings, streets, and parking lots, more reflective roofs; and more trees, parks, and open space. (The images from all the test cities are available at **http://science.nasa.gov/newhome/headlines/essd01jul98%5F1.htm.**)

FIGURE 4.24 Heat island exposed.
Urban Heat Island Pilot Project (UHIPP) thermal infrared image of Sacramento, California, taken June 29, 1998, at 1 P.M. PDT from a NASA jet at 2000 m (6600 ft). The Sacramento River runs from north to south to the left of center of the frame, the American River is to the upper right, and the California state capitol building is the red dot in the green rectangle right of center from the river. False color tags white and red as relatively hot sites; greens and blues are relatively cool on this preliminary image. [Image courtesy of UHIPP, Marshall Space Flight Center, Global Hydrology and Climate Center, Huntsville, Alabama.]

Summary and Review—Atmosphere and Surface Energy Balances

● *Identify* **the pathways of solar energy through the troposphere to Earth's surface: transmission, scattering, diffuse radiation, refraction, albedo (reflectivity), conduction, convection, and advection.**

Earth's biosphere is powered by radiant energy from the Sun that cascades through complex circuits to the surface. Our budget of atmospheric energy comprises shortwave radiation *inputs* (ultraviolet light, visible light, and near-infrared wavelengths) and longwave radiation *outputs* (thermal infrared).

Transmission refers to the passage of shortwave and longwave energy through either the atmosphere or water. The gas molecules redirect radiation, changing the direction of the light's movement *without altering its wavelengths*. This phenomenon is known as **scattering** and represents 7% of Earth's reflectivity, or albedo. Dust particles, pollutants, ice, cloud droplets, and water vapor produce further scattering. Some incoming insolation is diffused by clouds and atmosphere and is transmitted to Earth as **diffuse radiation**, the downward component of scattered light. The speed of insolation entering the atmosphere changes as it passes from one medium to another; the change of speed causes a bending action called **refraction**.

A portion of arriving energy bounces directly back into space without being converted into heat or performing any work. This returned energy is called **reflection**. **Albedo** is the reflective quality (intrinsic brightness) of a surface. It is an important control over the amount of insolation that is available for absorption by a surface. We state albedo as the percentage of insolation that is reflected. Earth and its atmosphere reflect 31% of all insolation when averaged over a year.

An increase in albedo and reflection of shortwave radiation caused by clouds is described by the term **cloud-albedo forcing**. Also, clouds can act as insulation, thus trapping longwave radiation and raising minimum temperatures. An increase in greenhouse warming caused by clouds is described by the term **cloud-greenhouse forcing**.

Absorption is the assimilation of radiation by molecules of a substance and its conversion from one form to another—for example, visible light to infrared radiation. **Conduction** is the molecule-to-molecule transfer of energy as it diffuses through a substance.

Energy also is transferred in gases and liquids by **convection** (when the physical mixing involves a strong vertical motion) or **advection** (when the dominant motion is horizontal). In the atmosphere or bodies of water, warmer portions tend to rise (they are less dense) and cooler portions tend to sink (they are more dense), establishing patterns of convection.

transmission (p. 94)
scattering (p. 94)
diffuse radiation (p. 95)
refraction (p. 95)
reflection (p. 97)
albedo (p. 97)
cloud-albedo forcing (p. 97)
cloud-greenhouse forcing (p. 97)
absorption (p. 98)
conduction (p. 98)
convection (p. 99)
advection (p. 99)

1. Diagram a simple energy balance for the troposphere. Label each shortwave and longwave component and the directional aspects of related flows.
2. Define refraction. How is it related to daylength? To a rainbow? To the beautiful colors of a sunset?
3. List several types of surfaces and their albedo values. Explain the differences among these surfaces. What determines the reflectivity of a surface?
4. Using Figure 4.7, explain the seasonal differences in albedo values for each hemisphere. Be specific, using the albedo values given in Figure 4.5 where appropriate.
5. What would you expect the sky color to be at 50 km (30 mi) altitude? Why? Why is the lower atmosphere blue?
6. Define the concepts transmission, absorption, diffuse radiation, conduction, and convection.

● *Describe* what happens to insolation when clouds are in the atmosphere and *analyze* the effect of clouds and air pollution on solar radiation received at ground level.

Clouds reflect *insolation*, thus cooling Earth's surface—*cloud-albedo forcing*. Yet clouds also act as insulation, thus trapping longwave radiation and raising minimum temperatures—*cloud-greenhouse forcing*. Clouds affect the heating of the lower atmosphere, depending on cloud type, height, and thickness (water content and density). High-altitude, ice-crystal clouds reflect insolation with albedos of about 50%, producing a

net cloud-greenhouse forcing (warming); thick, lower cloud cover reflects about 90%, producing a net cloud-albedo forcing (cooling).

Emissions of sulfur dioxide and the subsequent chemical reactions in the atmosphere form *sulfate aerosols*, which act either as insolation-reflecting haze in clear-sky conditions or as a stimulus to condensation in clouds that increases reflectivity.

7. What role do clouds play in the Earth–atmosphere radiation balance? Is cloud type important? Compare high, thin cirrus clouds and lower, thick stratus clouds.
8. In what way does the presence of sulfate aerosols affect solar radiation received at ground level? How does it affect cloud formation?

● *Review* the energy pathways in the Earth–atmosphere system, the greenhouse effect, and the patterns of global net radiation.

The Earth–atmosphere energy system naturally balances itself in a steady-state equilibrium. It does so through energy transfers that are *nonradiative* (convection, conduction, and the latent heat of evaporation) and *radiative* (by infrared radiation between the surface, the atmosphere, and space).

Some infrared radiation is absorbed by carbon dioxide, water vapor, methane, CFCs (chlorofluorocarbons), and other gases in the lower atmosphere and is then reradiated to Earth, thus delaying energy loss to space. This process is the **greenhouse effect**. In the atmosphere, infrared radiation is not actually trapped, as it would be in a greenhouse, but its passage to space is delayed (heat energy is detained in the atmosphere) as it is absorbed and reradiated.

Between the tropics, high insolation angle and consistent daylength cause more energy to be gained than lost (there are energy surpluses). In the polar regions, an extremely low insolation angle, highly reflective surfaces, and up to 6 months of no insolation annually cause more energy to be lost (there are energy deficits). This imbalance of net radiation from tropical surpluses to the polar deficits drives a vast global circulation of both energy and mass.

Surface energy balances are used to summarize the energy expenditure for any location. Surface energy measurements are used as an analytical tool of **microclimatology**. Adding and subtracting the energy flow at the surface lets us calculate **net radiation** (**NET R**), or the balance of all radiation at Earth's surface—shortwave (SW) and longwave (LW).

greenhouse effect (p. 100)
microclimatology (p. 104)
net radiation (NET R) (p. 104)

9. What are the similarities and differences between an actual greenhouse and the gaseous atmospheric greenhouse? Why is Earth's greenhouse changing?
10. In terms of energy expenditures for latent heat of evaporation, describe the annual pattern as mapped in Figure 4.19.
11. Generalize the pattern of global net radiation. How might this pattern drive the atmospheric weather machine? (See Figures 4.14 and 4.18.)

12. In terms of surface energy balance, explain the term *net radiation* (NET R).

13. What are the expenditure pathways for surface net radiation? What kind of work is accomplished?

14. What is the role played by latent heat in surface energy budgets?

15. Compare the daily surface energy balances of El Mirage, California, and Pitt Meadows, British Columbia. Explain the differences.

● *Plot* the daily radiation curves for Earth's surface and *label* the key aspects of incoming radiation, air temperature, and the daily temperature lag.

The greatest insolation input occurs at the time of the summer solstice in each hemisphere. Air temperature responds to seasons and variations in insolation input. Within a 24-hour day, air temperature peaks between 3:00 and 4:00 P.M. and dips to its lowest point right at or slightly after sunrise.

Air temperature lags behind each day's peak insolation. The warmest time of day occurs not at the moment of maximum insolation but at that moment when a maximum of insolation is absorbed.

16. Why is there a temperature lag between the highest Sun altitude and the warmest time of day? Relate your answer to the insolation and temperature patterns during the day.

● *Portray* typical urban heat island conditions and *contrast* the microclimatology of urban areas with that of surrounding rural environments.

A growing percentage of Earth's people live in cities and experience their unique set of altered microclimatic effects: increased conduction, lower albedos, higher NET R values, increased water runoff, complex radiation and reflection patterns, anthropogenic heating, and the gases, dusts, and aerosols of urban pollution. Urban surfaces of metal, glass, asphalt, concrete, and stone conduct up to three times more energy than wet sandy soil and thus are warmed as described by the term **urban heat island**. Air pollution, including gases and aerosols, is greater in urban areas than in rural ones. Every major city produces its own **dust dome** of airborne pollution.

urban heat island (p. 108)
dust dome (p. 112)

17. What is the basis for the urban heat island concept? Describe the climatic effects attributable to urban as compared with nonurban environments. What did NASA determine from the UHIPP overflight of Sacramento (Figure 4.24)?

18. Which of the items in Table 4.2 have you yourself experienced? Explain.

19. Assess the potential for solar energy applications in our society. What are some negatives? What are some positives?

NetWork

The *Geosystems* Home Page provides on-line resources for this chapter on the World Wide Web. To begin: Once on the Home Page, click on this textbook, scroll the Table of Contents menu, select this chapter, and click "Begin." You will find self-tests that are graded, review exercises, specific updates for items in the chapter, and many links to interesting related pathways on the Internet under "Destinations." *Geosystems* is at http://www.prenhall.com/christopherson.

Critical Thinking

A. Given what you now know about reflection, albedo, absorption, and net radiation expenditures, assess your wardrobe (fabrics and colors), house or apartment (colors of walls, especially south and west facing; or, in the Southern Hemisphere, north and west facing), and roof (orientation relative to the Sun), automobile (color, use of sun shades), bicycle seat (color), and other aspects of your environment to determine a personal "Energy IQ." What grade do you give yourself? Be cool!

B. On the *Geosystems* Home Page, Chapter 4, "Destinations," and in this chapter of the text, there are several listings of URLs relating to solar energy applications. Take some time and explore the Internet for a personal assessment of these necessary technologies (solar thermal, solar-electric, photovoltaic cells, solar-box cookers, and the like). As we near the climatic limitations of the fossil fuel era, and the depletion of the resource itself, these available technologies will become part of the fabric of our lives. Briefly describe your search results. Given these findings, determine if there is availability of solar technology in your area.

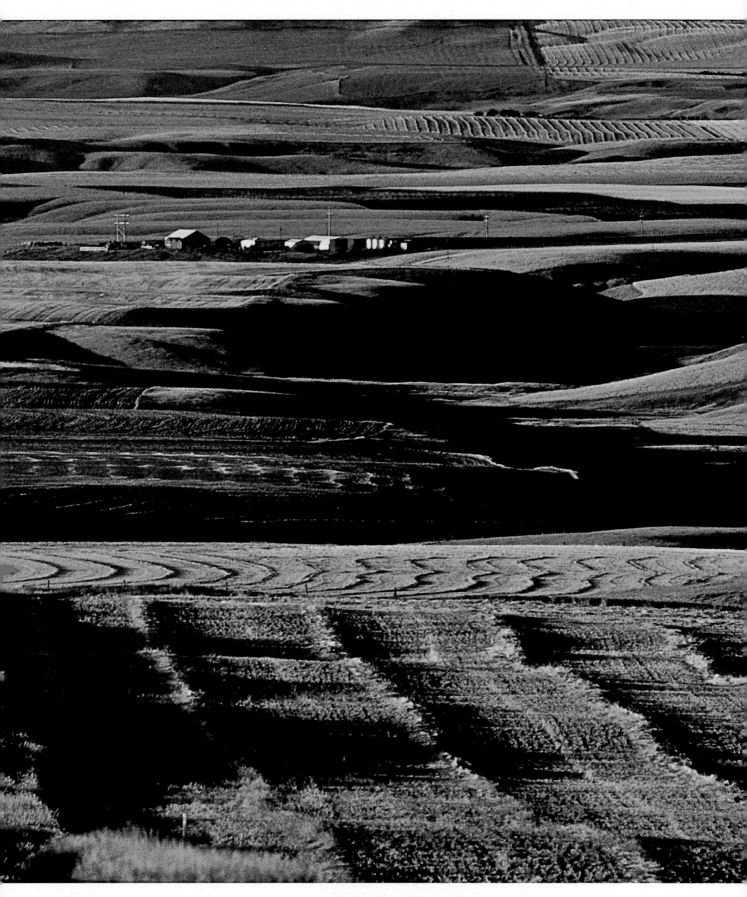

An autumn sunset bathes these Alberta, Canada, prairie farms. Warmer average temperatures are expanding the growing season in Canada and melting permafrost soil in the far north. Conditions of drought are a concern. [Photo by John Eastcott/Yva Momatiuk, DRK Photo.]

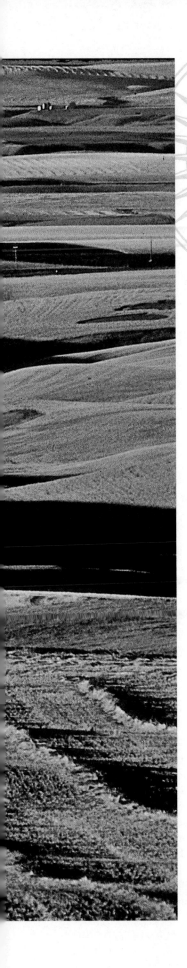

5

Global Temperatures

Key Learning Concepts

After reading the chapter, you should be able to:

- *Define* the concepts of temperature, kinetic energy, and sensible heat, and *distinguish* among Kelvin, Celsius, and Fahrenheit scales and how they are measured.

- *List* and *review* the principal controls and influences that produce global temperature patterns.

- *Review* the factors that produce different marine effects and continental effects as they influence temperatures and *utilize* several pairs of stations to illustrate these differences.

- *Interpret* the pattern of Earth's temperatures from their portrayal on January and July temperature maps and on a map of annual temperature ranges.

What is the temperature now—both indoors and outdoors—as you read these words? How is it measured, and what does the value mean? How is air temperature influencing your plans for the day? Air temperature plays a remarkable role in our lives, both at the micro level and at the macro level. Our bodies subjectively sense temperature and judge comfort and react to changing temperatures.

We read of heat waves and cold spells affecting people, crops, events, and energy consumption. For example, in 1995 devastating summer heat and high humidity killed more than 1000 people in the Midwest and East; the Southwest and Midwest suffered an intense heat wave in 1998 and again in 2001. Record warmth is reaching into the Arctic. Tree-ring analysis and other proxy measures indicate that present temperatures are warmer than Earth has experienced for the last 1000 years. This warming trend is the subject of much scientific, geographic, and political interest.

In this chapter: A variety of temperature regimes affect cultures, decision making, and resources consumed across the globe. Understanding some temperature concepts and measurements helps us begin our study of these Earth systems. We look at principal temperature controls of latitude, altitude, cloud cover, and land-water heating differences as they interact to produce Earth's temperature patterns. We end by examining the effect of temperature on the human body and the resultant spatial aspects of heat and cold on human experience.

Temperature Concepts and Measurement

Heat and temperature are not the same. *Heat* is a form of energy that flows from one system or object to another because the two are at different temperatures. **Temperature** is a measure of the average kinetic energy (motion) of individual molecules in matter. We feel the effect of temperature as the *sensible heat* transfer from warmer objects to cooler objects. For instance, when you jump into a cool lake you can sense the heat transfer from your skin to the water as kinetic energy leaves your body and flows to the water; a chill develops.

Temperature and heat are related because changes in temperature are caused by the absorption or emission (gain or loss) of heat energy. The term *heat energy* is frequently used to describe energy that is added to or removed from a system or substance.

Temperature Scales

The temperature at which all atomic and molecular motion in matter completely stops is called 0° absolute temperature (commonly, "absolute zero"). Its value on the different temperature-measuring scales is −273° Celsius (C), −459.4° Fahrenheit (F), and 0 Kelvin (K). Figure 5.1 compares these three scales. Formulas for converting between Celsius, SI (Système International), and English units are in Appendix C of this text.

The Fahrenheit scale places the melting point of ice at 32°F and the boiling point of water at 212°F. The scale is named for Daniel G. Fahrenheit, a German physicist (1686–1736). He used these odd values based on the coldest temperature he could achieve in his laboratory (which he called 0°F) and on the approximate temperature of the human body, thought to be about 100°F. His estimates placed the melting point of ice at 32°F, with 180 subdivisions in his scale to the boiling point of water at 212°F. (Note that there is only one melting point for ice, but there are many freezing points for water ranging from 32°F down to −40°F, depending on its purity and volume and certain conditions in the atmosphere.)

About a year after the adoption of the Fahrenheit scale, Swedish astronomer Anders Celsius (1701–1744) developed the Celsius scale (formerly called centigrade). He

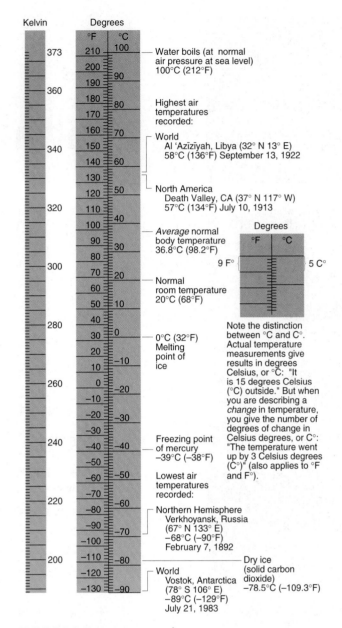

FIGURE 5.1 Temperature scales.
Scales for expressing temperature in Kelvins (K) and degrees Celsius (°C) and Fahrenheit (°F).

placed the melting point of ice at 0° and boiling temperature of water at sea level at 100°, dividing his scale into 100 degrees using a decimal system.

British physicist Lord Kelvin (born William Thomson, 1824–1907) proposed the Kelvin scale in 1848. Science uses this scale because temperature readings start at absolute zero and thus are proportional to the actual kinetic energy in a material. The Kelvin scale's melting point for ice is 273 K, and its boiling point of water is 373 K.

Most countries use the Celsius scale to express temperature. The United States remains the only major country still using the Fahrenheit scale. This textbook presents Celsius (with Fahrenheit equivalents in parentheses) throughout, to help bridge this transitional era in the United States. The continuing pressure from the scientific community and corporate interests, not to mention the rest of the world, makes adoption of Celsius and SI units inevitable for the United States. This author wishes it was sooner than later!

Measuring Temperature

A mercury thermometer or alcohol thermometer is a sealed glass tube that measures outdoor temperatures. (Fahrenheit invented the alcohol and mercury thermometers.) Cold climates demand alcohol thermometers because alcohol freezes at −112°C (−170°F), whereas mercury freezes at −39°C (−38.2°F). The principle of these thermometers is simple: When fluids are heated, they expand; upon cooling, they contract. A thermometer stores fluid in a small reservoir at one end and is marked with calibrations to measure the expansion or contraction of the fluid, which reflects the temperature of the thermometer's environment. *Thermistors* measure temperature by sensing the electrical resistance of a semi-conducting material, resistance changes at 4% per C° (a thermister is in the shelter in Figure 5.3).

Figure 5.2 shows a mercury minimum–maximum thermometer. It preserves readings of the day's highest and lowest temperatures until reset (by moving the markers with a magnet). Another type, the recording thermometer, creates an inked record on a turning drum; usually set for one full rotation every 24 hours or every 7 days.

Thermometers for standardized official readings are placed outdoors, in small shelters that are white (for high albedo) and louvered (for ventilation) to avoid overheating of the instruments (Figure 5.3). They are placed at least 1.2–1.8 m (4–6 ft) above a surface, usually on turf; in the

FIGURE 5.3 Instrument shelter.
This standard thermometer shelter is white (for high albedo) and louvered (for ventilation), replacing the traditional cotton-region shelter. [Photo by Bobbé Christopherson.]

United States 1.2 m (4 ft) is standard. Official temperature measurements are therefore in the shade to prevent the effect of direct insolation.

Temperature readings are taken daily, sometimes hourly, at more than 15,400 weather stations worldwide. One of the goals of the Global Climate Observing System (GCOS) is to establish a reference network of one station per 250,000 km² (95,800 mi²). Some stations with recording equipment also report the duration of temperatures, rates of rise or fall, and variation over time throughout the day and night. (See the World Meteorological Organization at http://www.wmo.ch/.)

The *daily mean temperature* is an average of daily minimum–maximum readings. The *monthly mean temperature* is the total of daily mean temperatures for the month divided by the number of days in the month. An *annual temperature range* expresses the difference between the lowest and highest monthly mean temperatures for a given year. If you install a thermometer for outdoor temperature reading, be sure to avoid direct sunlight on the instrument and place it in an area of good ventilation, and at least 1.2 m off the ground.

Principal Temperature Controls

The interaction of complex control mechanisms produces Earth's temperature patterns. These principal influences upon temperatures include latitude, altitude, cloud cover, and land–water heating differences.

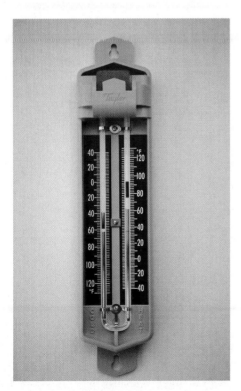

FIGURE 5.2 A minimum–maximum thermometer.
[Photo by Bobbé Christopherson.]

Latitude

Insolation is the single most important influence on temperature variations. Figure 2.9 shows how insolation intensity decreases as one moves away from the subsolar point—a point that migrates annually between the Tropic of Cancer and Tropic of Capricorn (between 23.5° N and 23.5° S). In addition, daylength and Sun angle change throughout the year, increasing seasonal effect with increasing latitude. The five cities graphed in Figure 5.4 demonstrate the effects of latitudinal position. From equator to poles, Earth ranges from continually warm, to seasonally variable, to continually cold.

Altitude

Within the troposphere, temperatures decrease with increasing altitude above Earth's surface. (Recall that the *normal lapse rate* of temperature change with altitude is 6.4 C°/1000 m, or 3.5 F°/1000 ft; see Figure 3.5). Thus, worldwide, mountainous areas experience lower temperatures than do regions nearer sea level, even at similar latitudes. The density of the atmosphere also diminishes with increasing altitude. In fact, the density of the atmosphere at an elevation of 5500 m (18,000 ft) is about half of that at sea level. As the atmosphere thins, its ability to absorb and radiate sensible heat is reduced.

The consequences are that, at high elevations, average air temperatures are lower, nighttime cooling is greater, and the temperature range between day and night is greater than at low elevations. The temperature difference between areas of sunlight and shadow is greater than at sea level.

You may have felt temperatures decrease noticeably in the shadows and shortly after sunset when you were in the mountains. Surfaces both gain energy rapidly and lose energy rapidly to the thinner atmosphere.

Also, at higher elevations, the insolation received is more intense because of the reduced mass of atmospheric gases. As a result of this intensity, the ultraviolet energy component makes sunburn a distinct hazard.

The snowline seen in mountain areas indicates where winter snowfall exceeds the amount of snow lost through summer melting and evaporation. The snowline's location is a function both of elevation and latitude, so glaciers can exist even at equatorial latitudes if the elevation is great enough. In equatorial mountains, the snowline occurs at approximately 5000 m (16,400 ft) because of the latitude (more insolation in the tropics produces temperatures that place the snowline at higher altitude). Permanent ice fields and glaciers exist on equatorial mountain summits in the Andes and East Africa. With increasing latitude, snowlines gradually lower in elevation from 2700 m (8850 ft) in the midlatitudes to lower than 900 m (2950 ft) in southern Greenland.

Two cities in Bolivia illustrate the interaction of the two temperature controls, latitude and altitude. Figure 5.5 displays temperature data for the cities of Concepción and La Paz, which are near the same latitude (about 16° S). Note the elevation, average annual temperature, and precipitation for each location noted on the figure.

The hot, humid climate of Concepción at its much lower elevation stands in marked contrast to the cool, dry climate of highland La Paz. People living around La Paz actually grow wheat, barley, and potatoes—crops character-

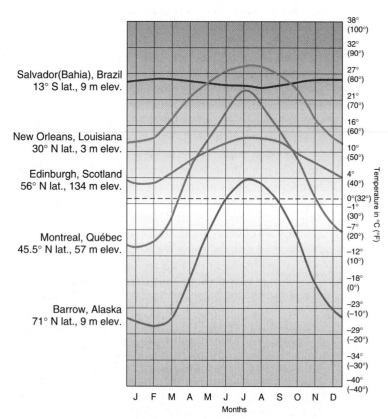

FIGURE 5.4 Latitude affects temperatures.
A comparison of five cities from near the equator to above the Arctic Circle demonstrates changing seasonality and increasing differences between average minimum and maximum temperatures.

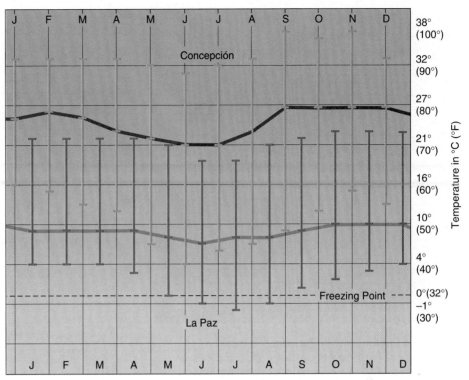

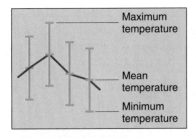

	Station	
	Concepción, Bolivia	**La Paz, Bolivia**
Latitude/longitude	16° 15' S 62° 03' W	16° 30' S 68° 10' W
Elevation	**490 m (1608 ft)**	**4103 m (13,461 ft)**
Avg. ann. temperature	24°C (75.2°F)	9°C (48.2°F)
Ann. temperature range	5 C° (9 F°)	3 C° (5.4 F°)
Ann. precipitation	121.2 cm (47.7 in.)	55.5 cm (21.9 in.)
Population	10,000	810,300 (Administrative division 1.6 million)

FIGURE 5.5 Effects of latitude and altitude.
Comparison of temperature patterns in two Bolivian cities: La Paz (4103 m, 13,461 ft) and
Concepción (490 m, 1608 ft).

istic of the cooler midlatitudes—despite the fact that La Paz is 4103 m (13,461 ft) above sea level (Figure 5.6). (For comparison, the summit of Pikes Peak in Colorado is at 4301 m or 14,111 ft, and Mount Rainier, Washington, is at 4392 m or 14,410 ft.)

The combination of elevation and low-latitude location guarantees La Paz nearly constant daylength and moderate temperatures, averaging about 9°C (48°F) every month. Such moderate temperature and moisture conditions lead to the formation of more fertile soils than those found in the warmer, wetter climate of Concepción.

Cloud Cover

Orbiting satellites reveal that approximately 50% of Earth is cloud covered at any given moment. Clouds moderate temperature, and their effect varies with cloud type, height, and density. Because their moisture reflects, absorbs, and liberates large amounts of energy, clouds reduce the insolation that reaches the surface. In general, they lower daily

maximum temperatures and raise nighttime minimum temperatures. Clouds also reduce latitudinal and seasonal temperature differences.

At night, clouds act as insulation and radiate longwave energy, preventing rapid energy loss. During the day, clouds reflect insolation as a result of their high albedo values. In the last chapter Figures 4.8 and 4.11 portrayed cloud-albedo forcing and cloud-greenhouse forcing as they relate to the presence of clouds and cloud types.

Clouds are the most variable factor influencing Earth's radiation budget, making them the subject of much investigation and simulation in computer models of atmospheric behavior. The International Satellite Cloud Climatology Project, part of the World Climate Research Programme, is presently in the midst of such research. The Earth Radiation Budget experiments (ERB), including the Clouds and the Earth Radiant Energy System (CERES) sensors aboard *TRMM* and *Terra* satellites, are assessing cloud effects on longwave, shortwave, and net radiation patterns as never before possible. (For more about clouds, see the ISCCP and

FIGURE 5.6 High-elevation farming.
People in these high-elevation villages grow potatoes and wheat in view of the permanent ice-covered peaks of the Andes Mountains. The combination of low latitude and high elevation creates consistent but moderate temperatures throughout the year, averaging about 9°C (48°F). [Photo by Mireille Vautier/Woodfin Camp & Associates.]

WCRP Home Pages at **http://isccp.giss.nasa.gov/** and **http://www.wmo.ch/web/wcrp/wcrp-home.html**.)

Land–Water Heating Differences

Another major control over temperature is the pronounced difference in the heating of land and water by insolation. Earth presents these two surfaces in an irregular arrange-

ment of continents and oceans. Each absorbs and stores energy differently than the other and therefore contributes to the global pattern of temperature. Moderate temperature patterns are associated with water bodies; extreme temperatures occur inland.

The physical nature of land (rock and soil) and water (oceans, seas, and lakes) is the reason for these **land–water heating differences**—land heats and cools faster than water. Figure 5.7 visually summarizes the following discussion of the five land–water temperature controls: evaporation, transparency, specific heat, movement and ocean currents, and sea-surface temperatures.

Evaporation   Evaporation consumes more of the energy arriving at the ocean's surface than is expended over a comparable area of land, simply because so much water is available. An estimated 84% of all evaporation on Earth is from the oceans. When water evaporates and thus changes to water vapor, *heat energy is absorbed in the process and is stored in the water vapor as latent heat.* We saw this process in Figure 4.19, the map of energy expended for the latent heat of evaporation. Latent heat is discussed fully in Chapter 7.

You can experience this evaporative heat loss (cooling) by wetting the back of your hand and then blowing on the moist skin. Sensible heat energy is drawn from your skin to supply some of the energy for evaporation, and you feel the cooling. Similarly, as surface water evaporates, it absorbs substantial energy from the immediate environment, resulting in a lowering of temperatures. Temperatures over land, with far less water, are not as moderated by evaporative cooling as are marine locations.

Transparency   The transmission of light obviously differs between soil and water: Solid ground is opaque, water is transparent. Consequently, light striking a soil surface does not penetrate but is absorbed, heating the ground sur-

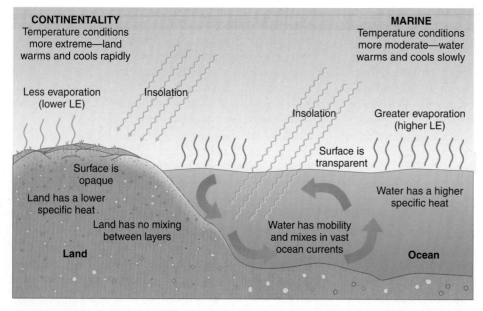

CONTINENTALITY
Temperature conditions more extreme—land warms and cools rapidly

Less evaporation (lower LE)

Insolation

Surface is opaque

Land has a lower specific heat

Land has no mixing between layers

Land

MARINE
Temperature conditions more moderate—water warms and cools slowly

Insolation

Greater evaporation (higher LE)

Surface is transparent

Water has a higher specific heat

Water has mobility and mixes in vast ocean currents

Ocean

FIGURE 5.7 Land–water heating differences.
The differential heating of land and water produces contrasting marine (more moderate) and continental (more extreme) temperature regimes. (LE, latent heat of evaporation, is the energy stored in water vapor.)

FIGURE 5.8 Land is opaque.
Profile of air and soil temperatures in Seabrook, New Jersey. Note that little temperature change occurs throughout the year at depths beyond a meter, whereas the daily extremes of temperature register along the surface, where insolation is absorbed. [Adapted from John R. Mather, *Climatology: Fundamentals and Applications* (New York: McGraw-Hill, 1974), p. 36. Adapted by permission.]

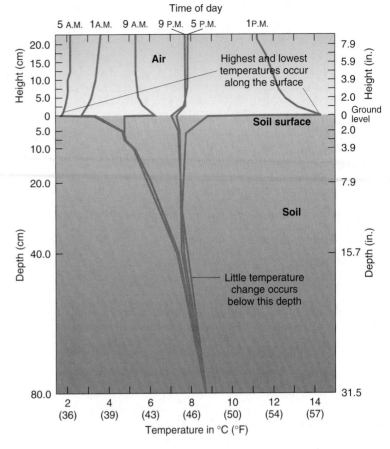

face. That energy is accumulated during times of exposure and is rapidly lost at night or in shadows.

Figure 5.8 shows the profile of diurnal (daily) temperatures for a column of soil and the atmosphere above it at a midlatitude location. You can see that maximum and minimum temperatures generally are experienced right at ground level. Below the surface, even at shallow depths, temperatures remain about the same throughout the day. This situation often exists at a beach, where surface sand may be painfully hot to your feet, but dig in your toes and you feel the sand a few centimeters below the surface is cooler, offering relief.

In contrast, when light reaches a body of water, it penetrates the surface because of water's **transparency**—water is clear and light transmits through it to an average depth of 60 m (200 ft) in the ocean (Figure 5.9). This illuminated zone is known as the *photic layer* and has been recorded in some ocean waters to depths of 300 m (1000 ft). This characteristic of water results in the distribution of available heat energy over a much greater depth and volume, forming a larger energy reservoir than that of the surface layers of the land.

FIGURE 5.9 Ocean is transparent.
The transparency of the ocean at Taveuni Reef near Fiji permits insolation to penetrate to average depths of 60 m (200 ft), greatly increasing the volume of water that absorbs energy. [Photo by Copr. F. Stuart Westmorland/Photo Researchers, Inc.]

Specific Heat   When comparing equal volumes of water and land, water requires far more energy to increase its temperature than does land. In other words, *water can hold more heat than can soil or rock*, and therefore water is said to have a higher **specific heat**, the heat capacity of a substance. On the average, the specific heat of water is about four times that of soil. A given volume of water represents a more substantial energy reservoir than an equal volume of land, so changing the temperature of the oceanic energy reservoir is a slower process than changing the temperature of land.

Likewise, for that oceanic heat reservoir to lose heat energy requires more time than would a similar volume of land. The temperature response of water bodies is "sluggish" in comparison with land surfaces. For this reason, day-to-day temperatures near a substantial body of water tend to be moderated.

Movement   Land is a rigid, solid material, whereas water is a fluid and is capable of movement. Differing temperatures and currents result in a mixing of cooler and warmer waters, and that mixing spreads the available energy over an even greater volume than if the water were still. Surface water and deeper waters mix, redistributing energy. Both ocean and land surfaces radiate longwave radiation at night, but land loses its energy more rapidly than does the greater mass of the moving oceanic energy reservoir.

Ocean Currents and Sea-Surface Temperatures   Warm water adds energy to overlying air through high evaporation rates and transfers of latent heat. Thus, in an air mass, the amount of water vapor that ocean temperature affects forms an interesting *negative feedback* mechanism. Across the globe, ocean water is rarely found warmer than 31°C (88°F). Higher ocean temperatures produce higher evaporation rates and more energy is lost from the ocean as latent heat. As water vapor content of the overlying air mass increases, the ability of the air to absorb longwave radiation also increases. Therefore, the air mass becomes warmer, enhancing the greenhouse effect. The warmer the air and the ocean become, the more evaporation will occur and the more water vapor will enter the air mass. More water vapor leads to cloud formation, which reflects insolation and produces lower temperatures. Lower temperatures of air and ocean reduce evaporation rates and the ability of the air mass to absorb water vapor.

As a specific example, the **Gulf Stream** (described in Chapter 6) moves northward off the east coast of North America, carrying warm water far into the North Atlantic (Figure 5.10). As a result, the southern third of Iceland experiences much milder temperatures than would be expected for a latitude of 65° N, just below the Arctic Circle (66.5°). In Reykjavík, on the southwestern coast of Iceland, monthly temperatures average above freezing during all months of the year. Similarly, the Gulf Stream moderates

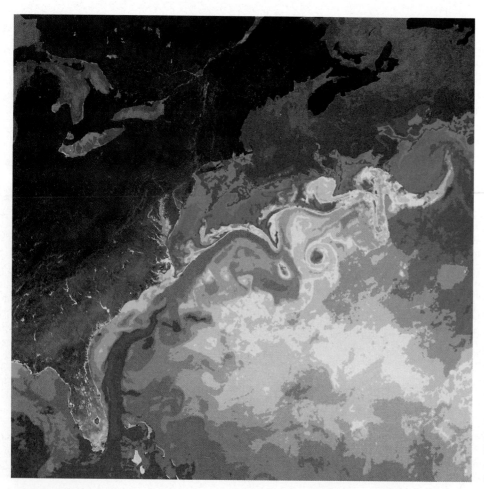

FIGURE 5.10 The Gulf Stream. Satellite image of the warm Gulf Stream as it flows northward along the North American eastern coast. It is the streamlike flow in red, orange, and yellow. Instruments sensitive to infrared wavelengths produced this remote-sensing image, which covers approximately 11.4 million square kilometers (4.4 million square miles). Temperature differences are distinguished by computer-enhanced coloration: reds/oranges = 25–29°C (76–84°F), yellows/greens = 17–24°C (63–75°F); blues = 10–16°C (50–61°F); and purples = 2–9°C (36–48°F). [Imagery by Rosenstiel School of Marine and Atmospheric Science, University of Miami.]

FIGURE 5.11 Sea-surface temperatures.
Average annual sea-surface temperatures for February (a) and July (b) 1999 from satellites in the NOAA/NASA Pathfinder AVHRR data set, aboard *NOAA-7, -9, -11,* and *-14.* The Western Pacific Warm Pool, the warmest area of all oceans, is well defined. These remotely sensed data are closely correlated with actual measurements of the ocean's surface temperature. [Satellite image data were obtained from the NASA Physical Oceanography Distributed Active Archive Center at the Jet Propulsion Laboratory, California Institute of Technology.]

(a) February

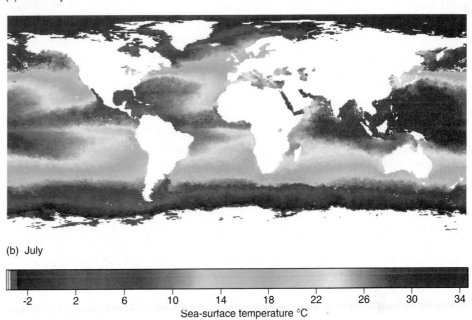

(b) July

temperatures in coastal Scandinavia and northwestern Europe. In the western Pacific Ocean, the Kuroshio, or Japan Current, similar to the Gulf Stream, functions much the same in its warming effect on Japan, the Aleutians, and the northwestern margin of North America.

In contrast, along midlatitude and subtropical west coasts, cool ocean currents influence air temperatures. When conditions in these regions are warm and moist, fog frequently forms in the chilled air over the cooler currents. Ocean currents are detailed further in Chapter 6.

Remote sensing from satellites is providing sea-surface temperature (SST) data that are well-correlated with actual sea-surface measurements. This correlation permits a thorough global assessment of SSTs in programs such as Tropical Ocean Global Atmosphere (TOGA) and Coupled Ocean-Atmosphere Response Experiment (COARE). (See http://lwf.ncdc.noaa.gov/oa/coare/index.html.)

The Physical Oceanography Distributed Active Archive Center (PO.DAAC, http://podaac.jpl.nasa.gov/), part of the Earth Observing System Data Information System (EOSDIS, http://eosdismain.gsfc.nasa.gov/eosinfo/welcome/), is responsible for storage and distribution of data relevant to the physical state of the ocean. Sea-level height, currents, and ocean temperatures are measured as part of the effort to understand oceans and climate interactions.

Figure 5.11 displays SST data for February and July 1999 from satellites in the NOAA/NASA Pathfinder AVHRR data set, aboard *NOAA-7, -9, -11,* and *-14.* The Western Pacific Warm Pool is the red and maroon area in the southwestern Pacific Ocean (north of New Guinea) with temperatures above 30°C (86°F). This region has the highest average ocean temperatures in the world. Note the seasonal change in ocean temperatures; for example compare the waters around Australia on the two images. Mean annual SSTs increased steadily from 1982 through 2000 to record levels for the past several hundred years.

Summary of Marine Effects vs. Continental Effects

As noted, Figure 5.7 summarizes the operation of the five land–water temperature controls presented: evaporation, transparency, specific heat, movement and ocean currents, and sea-surface temperatures. The term **marine effect**, or *maritime*, describes locations that exhibit the moderating influences of the ocean, usually along coastlines or on islands. **Continental effect**, a condition of *continentality*, refers to areas less affected by the sea and therefore having a greater range between maximum and minimum temperatures, daily and yearly.

The Canadian cities of Vancouver, British Columbia, and Winnipeg, Manitoba, exemplify marine and continental conditions (Figure 5.12). Both cities are at approximately 49° N latitude. However, Vancouver has a more moderate pattern of average maximum and minimum temperatures. Vancouver's annual range of 16.0 C° (28.8 F°) is far less than Winnipeg's 38.0 C° (68.4 F°) range. In fact, Winnipeg's continental temperature pattern is more extreme in every aspect than that of maritime Vancouver.

A similar comparison of San Francisco, California, and Wichita, Kansas, provides us another comparison of marine and continental conditions (Figure 5.13). Both cities are at approximately 37° 40′ N latitude. The cooling waters of the Pacific Ocean and San Francisco Bay surround San Francisco on three sides. Summer fog helps delay until September the warmest summer month in San Francisco. In 100 years of weather records, marine effects moderated temperatures so that only a few days a year have summer maximums that exceed 32.2°C (90°F). Winter minimums rarely drop below freezing. This influence is changing however, as ocean temperatures off the coast rise to record levels—20°C (68°F), some 7 C° (12 F°) above normal—before an El Niño began in 1997.

In contrast, Wichita is susceptible to freezes from late October to mid-April, with daily variations slightly increased by its elevation, experiencing −30°C (−22°F) as a record low. Wichita's temperature reaches 32.2°C (90°F) or higher at least 65 days each year, with 46°C (114°F) as a record high. In 1980, 1990, and again in 2000, temperatures exceeded 38°C (100°F) two to three weeks in a row. West of Wichita, winters increase in severity with increasing distance from the moderating influences of invading air masses from the Gulf of Mexico.

Earth's Temperature Patterns

Earth's temperature patterns result from the combined effect of the controlling factors in our discussion. Let us now look at temperatures portrayed on maps that show worldwide mean air temperatures for January (Figure 5.14) and July (Figure 5.16). To complete the analysis, Figure 5.17 presents the temperature range differences between the January and July maps, or the difference between averages of the coolest and warmest months.

We use maps for January and July instead of the solstice months of December and June because a lag occurs between insolation received and maximum or minimum temperatures experienced, as explained earlier in Chapter 4. The U.S. National Climate Data Center provided the data for this map preparation. Some ship reports go back to

1850 and land reports to 1890, although the bulk of the record is representative of conditions since 1950. An important consideration in using these maps is to remember that their small scale permits only generalizations about actual temperatures at specific locations.

The lines on temperature maps are known as *isotherms*. An **isotherm** is an isoline that connects points of equal temperature and portrays the temperature pattern, just as a contour line on a topographic map illustrates points of equal elevation. Geographers are concerned with spatial analysis of temperatures, and isotherms help with this analysis.

January Temperature Map

Figure 5.14 maps January's mean temperatures. In the Southern Hemisphere the higher Sun altitude causes longer days and summer weather conditions; in the Northern Hemisphere the lower Sun angle causes the short days of winter. Isotherms generally are zonal, trending east–west, parallel to the equator, and they appear to be interrupted by the presence of landmasses. Isotherms mark the general decrease in insolation and net radiation with distance from the equator.

The **thermal equator** (an isoline connecting all points of highest mean temperature, roughly 27°C, 80°F) trends southward into the interior of South America and Africa, indicating higher temperatures over landmasses. In the Northern Hemisphere, isotherms shift equatorward as cold air chills the continental interiors. The oceans, on the other hand, are more moderate, with warmer conditions extending farther north than over land at comparable latitudes.

As an example, follow along 50° north latitude (the 50th parallel) and compare isotherms: 2 to 4°C in the North Pacific and 4 to 10°C in the North Atlantic, as contrasted to −18°C in the interior of North America and −24 to −30°C in central Asia. Also, note the orientation of isotherms over areas where there are mountain ranges and how they illustrate the cooling effects of altitude. Check the South American Andes as an example of the effects of altitude.

Russia is the coldest area on the map, specifically northeastern Siberia. The intense cold results from winter conditions of consistent clear, dry, calm air, small insolation input, and an inland location far from moderating maritime effects. Verkhoyansk, Russia (located within the −48°C isotherm on the map), actually recorded a minimum temperature of −68°C (−90°F) and experiences a daily average of −50.5°C (−58.9°F) for January. Verkhoyansk (Figure 5.15) has 7 months of temperatures below freezing, including at least 4 months below −34°C (−30°F)! In contrast, this town has hit a maximum temperature of +37°C (+98°F) in July—an incredible 105 C° (189 F°) min–max range! People do live and work in Verkhoyansk, which has a population of 1400; the town has been occupied continuously since 1638 and is today a minor mining district.

FIGURE 5.12 Marine and continental cities—Canada.

Comparison of temperatures in coastal Vancouver, British Columbia, and continental Winnipeg, Manitoba. Note that the freezing levels on the two graphs are positioned differently to accommodate the contrasting data. [Vancouver waterfront photo by author; Winnipeg photo by John Eastcott/Yva Momatiak/The Image Works.]

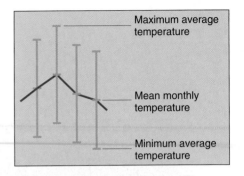

Maximum average temperature

Mean monthly temperature

Minimum average temperature

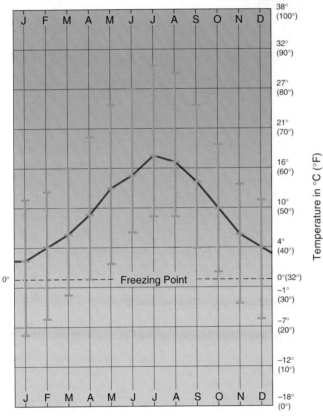

Temperature in °C (°F)

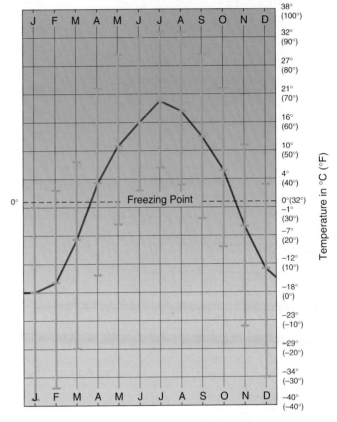

Temperature in °C (°F)

Station: Vancouver, British Columbia
Lat/long: 49° 11′ N 123° 10′ W
Avg. ann. temp.: 10°C (50°F)
Total ann. precip.:
   104.8 cm (41.3 in.)

Elevation: sea level
Population: 520,000
Ann. temp. range:
   16 C° (28.8 F°)

Station: Winnipeg, Manitoba
Lat/long: 49° 54′ N 97° 14′ W
Avg. ann. temp.: 2°C (35.6°F)
Total ann. precip.:
   51.7 cm (20.3 in.)

Elevation: 248 m (813.6 ft)
Population: 620,000
Ann. temp. range:
   38 C° (68.4 F°)

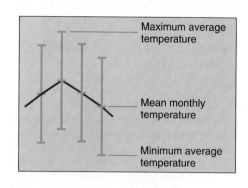

Maximum average temperature

Mean monthly temperature

Minimum average temperature

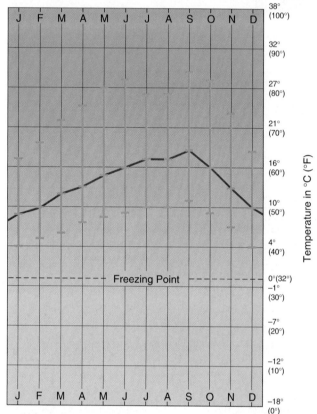

Station: San Francisco, California
Lat/long: 37° 37′ N 122° 23′ W
Avg. ann. temp.: 14°C (57.2°F)
Total ann. precip.:
   47.5 cm (18.7 in.)

Elevation: 5 m (16.4 ft)
Population: 777,000
Ann. temp. range:
   9 C° (16.2 F°)

Station: Wichita, Kansas
Lat/long: 37° 39′ N 97° 25′ W
Avg. ann. temp.: 13.7°C (56.6°F)
Total ann. precip.:
   72.2 cm (28.4 in.)

Elevation: 402.6 m (1321 ft)
Population: 327,000
Ann. temp. range:
   27 C° (48.6 F°)

FIGURE 5.13 Marine and continental cities—United States.
Comparison of temperatures in coastal San Francisco, California, and continental Wichita, Kansas. [San Francisco photo by author; Wichita photo by Garry D. McMichael/Photo Researchers, Inc.]

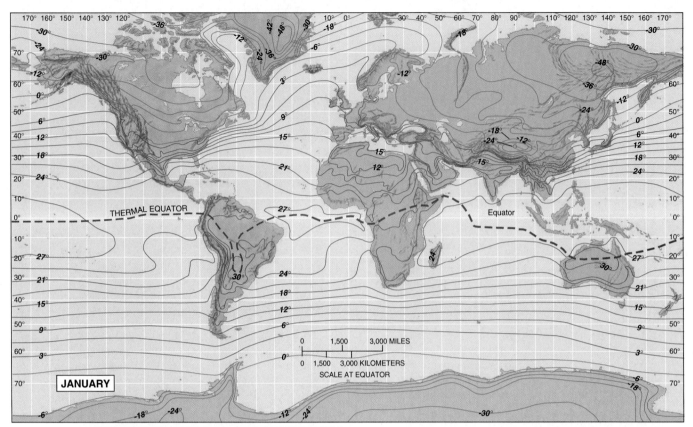

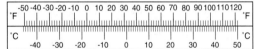

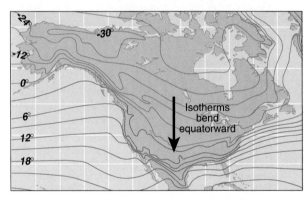

FIGURE 5.14 Global mean temperatures for January.
Temperatures are in Celsius (convertible to Fahrenheit by means of the scale) as taken from separate air temperature data bases for ocean and land. Note the inset map of North America and the equatorward-trending isotherms in the interior. (Compare with Figure 5.16.) [Adapted from National Climatic Data Center, *Monthly Climatic Data for the World*, 47 (January 1994). Prepared in cooperation with the World Meteorological Organization. Washington, DC: National Oceanic and Atmospheric Administration.]

Trondheim, Norway, is near the latitude of Verkhoyansk and at a similar elevation. But, Trondheim's coastal location moderates its annual temperature regime (see Figure 5.15). January minimum and maximum temperatures range between −17 and +8°C (+1.4 and +46°F), and the minimum–maximum range for July is from +5 to +27°C (+41 to +81°F). The most extreme minimum and maximum temperatures ever recorded in Trondheim are −30 and +35°C (−22 and +95°F)—quite a difference from the continentality extremes at Verkhoyansk.

July Temperature Map

Average July temperatures are presented in Figure 5.16. The longer days of summer and higher Sun altitude now are in the Northern Hemisphere. Winter dominates the Southern Hemisphere, although it is milder than winters north of the equator because continental landmasses are smaller and the dominant oceans and seas store and release more energy. The thermal equator shifts northward with the high summer Sun and reaches the Persian Gulf–Pakistan–Iran area. The Persian Gulf is the site of

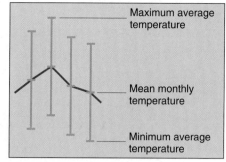

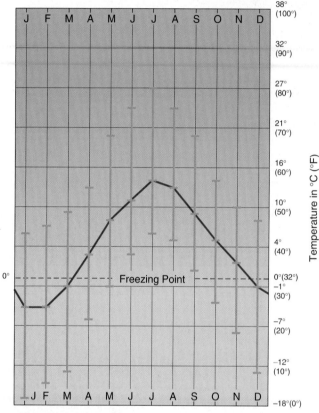

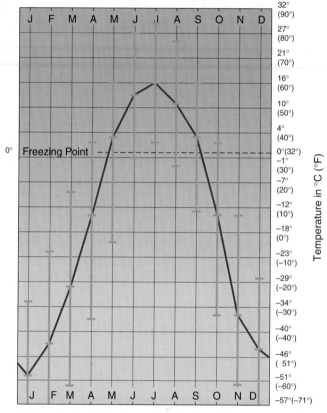

Station: Trondheim, Norway
Lat/long: 63° 25′ N 10° 27′ E
Avg. ann. temp.: 5°C (41°F)
Total ann. precip.:
  85.7 cm (33.7 in.)

Elevation: 115 m (377.3 ft)
Population: 139,000
Ann. temp. range:
  17 C° (30.6 F°)

Station: Verkhoyansk, Russia
Lat/long: 67° 35′ N 135° 23′ E
Avg. ann. temp.: −15°C (5°F)
Total ann. precip.:
  15.5 cm (6.1 in.)

Elevation: 137 m (449.5 ft)
Population: 1400
Ann. temp. range:
  63 C° (113.4 F°)

FIGURE 5.15 Marine and continental cities—Eurasia.
Comparison of temperatures in coastal Trondheim, Norway, and continental Siberian Russia.
Note that the freezing levels on the two graphs are positioned differently to accommodate the
contrasting data. [Trondheim photo by Norman Benton/Peter Arnold, Inc.; Verkhoyansk photo
by TASS/Sovfoto/Eastfoto.]

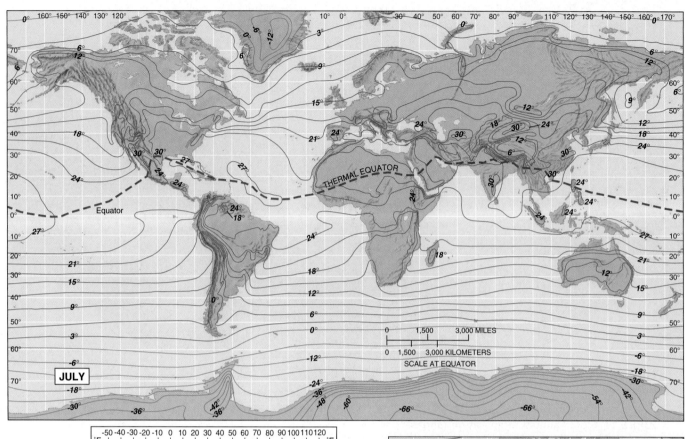

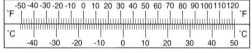

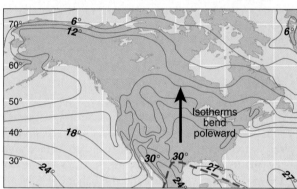

FIGURE 5.16 Global mean temperatures for July.
Temperatures are in Celsius (convertible to Fahrenheit by means of the scale) as taken from separate air temperature data bases for ocean and land. Note the inset map of North America and the poleward-trending isotherms in the interior. (Compare with Figure 5.14.) [Adapted from National Climatic Data Center, *Monthly Climatic Data for the World*, 47 (July 1994). Prepared in cooperation with the World Meteorological Organization. Washington, DC: National Oceanic and Atmospheric Administration.]

the highest recorded sea-surface temperature of 36°C (96°F), difficult to imagine for a large water body.

July is a time when nights in Antarctica are 24 hours long. This lack of insolation caused the lowest natural temperature reported on Earth, a frigid −89.2°C (−128.6°F), recorded on July 21, 1983, at the Russian research base at Vostok, Antarctica (78° 27′ S; elevation 3420 m, or 11,220 ft). Such a temperature is 11 C° (19.8 F°) colder than the freezing point of dry ice (solid carbon dioxide)! If the concentration of carbon dioxide were large enough, such a

cold temperature would theoretically freeze tiny carbon dioxide dry-ice particles out of the sky.

During July in the Northern Hemisphere, isotherms shift poleward over land, as higher temperatures dominate continental interiors. July temperatures in Verkhoyansk average more than 13°C (56°F), which represents a 63 C° (113 F°) seasonal range between winter and summer averages. The Verkhoyansk region of Siberia is probably Earth's most dramatic example of continental effects on temperature.

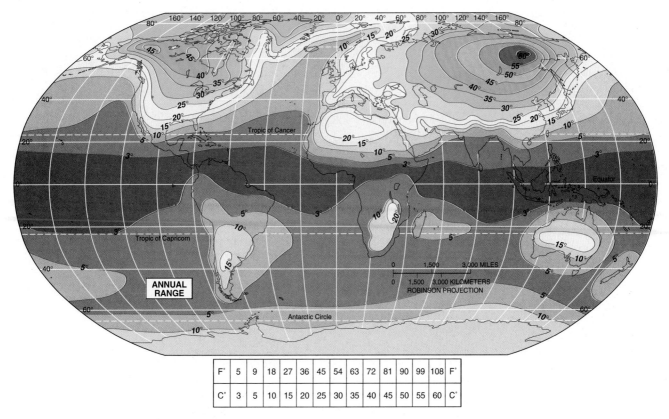

F°	5	9	18	27	36	45	54	63	72	81	90	99	108	F°
C°	3	5	10	15	20	25	30	35	40	45	50	55	60	C°

FIGURE 5.17 Global annual temperature ranges.
A generalized portrait of the annual range of global temperatures in Celsius degrees (conversions to Fahrenheit degrees shown on scale). The mapped data show the difference between January and July temperature means.

The hottest places on Earth occur in Northern Hemisphere deserts during July. The reasons are simple: clear skies, strong surface heating, virtually no surface water, and few plants. Prime examples are portions of the Sonoran Desert of North America and the Sahara of Africa. Africa recorded a shade temperature higher than 58°C (136°F), a record set on September 13, 1922, at Al 'Azīzīyah, Libya (32° 32′ N; 112 m, or 367 ft elevation).

The highest maximum and annual average temperatures in North America occurred in Death Valley, California, where the Greenland Ranch Station reached 57°C (134°F) in 1913. (The station is at 37° N and is −54.3 m, or −178 ft, below sea level). Such hot, arid lands are discussed further in Chapter 15.

Annual Temperature Range Map

The temperature range map helps identify areas that experience the greatest annual extremes (continental locations) and the most moderate temperature regimes (marine locations), as demonstrated in Figure 5.17. As you might expect, the largest temperature ranges occur in subpolar locations in North America and Asia, where average ranges of 64 C° (115 F°) are recorded. The Southern Hemisphere,

on the other hand, has little seasonal variation in mean temperatures, owing to the lack of large landmasses and vast expanses of water to moderate temperature extremes.

For example, in January (Figure 5.14), Australia is dominated by isotherms of 20–30°C (68–86°F), whereas in July (Figure 5.16), Australia is crossed by the 12°C (54°F) isotherm. Southern Hemisphere temperature patterns are generally maritime, and Northern Hemisphere patterns feature continentality. The Northern Hemisphere, with greater land area overall, registers a slightly higher average surface temperature than does the Southern Hemisphere.

Imagine living in some of these regions and the degree to which your personal wardrobe and other comfort adaptations would need to adjust. See Focus Study 5.1 for more on air temperature and the human body. Meanwhile the summer heat-wave deaths in 1995, and again in 1998, 2000, and 2001 in the United States, remain a powerful reminder of the role of temperatures in our lives.

There is a distinct possibility that in the future humans may experience greater temperature-related challenges owing to complex changes now under way in the lower atmosphere. See News Report 5.1 for an introduction to some of these global changes that are discussed more extensively in Chapter 10.

Focus Study 5.1

Air Temperature and the Human Body

We describe our perception of temperature with the term *apparent temperature*, or *sensible temperature*. This perception varies among individuals and cultures. Through complex mechanisms, our bodies maintain an average internal temperature within a degree of 36.8°C (98.2°F), slightly lower in the morning or in cold weather and slightly higher at emotional times or during exercise and work.*

The water vapor content of air, wind speed, and air temperature taken together affect each individual's sense of comfort. The most discomfort comes with high temperatures, high humidity, and low winds. More comfort comes with low humidity, moderate temperatures, and light winds.

Although modern heating and cooling systems, where available, reduce the impact of extreme temperatures indoors, the danger to human life from excessive heat or cold persists. A notable example was the deaths of nearly 1000 people in the midwestern and eastern United States during a summer heat wave in 1995.

When changes occur in the surrounding air, the human body reacts to maintain its core temperature and to protect the brain at all cost. Table 1 summarizes the human body's response to stress induced by exposure to *hypothermia* (low temperature) and *hyperthermia* (high temperature). Carefully review these two lists of responses for future reference.

Wind Chill

The wind-chill index is important to people who experience winters with freezing temperatures and was first

*The traditional value for "normal" body temperature, 37°C (98.6°F), was set in 1868 using old methods of measurement. According to Dr. Philip Mackowiak of the University of Maryland School of Medicine, a more accurate modern assessment places the normal value at 36.8°C (98.2°F), with a range of 2.7 C°(4.8 F°), for the human population overall (*Journal of the American Medical Association*, September 23–30, 1992).

Table 1   The Human Body's Response to Temperature Stress

At Low Temperatures (Heat-gaining mechanism)	At High Temperatures (Heat-dissipation mechanism)
Temperature Regulation Methods	
Constriction of surface blood vessels	Dilation of surface blood vessels
Concentration of blood	Dilution of blood
Flexing to reduce surface exposure	Extending to increase exposure
Increased muscle tone	Decreased muscle tone
Decrease in sweating	Sweating
Inclination to increase activity	Inclination to decrease activity
Shivering	
Increased cell metabolism	
Consequent Disturbances	
Increased urine volume	Decreased urine volume
Danger to inadequate blood supply to exposed parts; frostbite	Reduced blood supply to brain; dizziness, nausea, fainting
Discomfort leading to neuroses	Discomfort leading to neuroses
Increased appetite	Decreased appetite
	Mobilization of tissue fluid
	Thirst and dehydration
	Reduced chloride balance; heat cramps
Failure of Regulation	
Falling body temperature	Rising body temperature
Drowsiness	Impaired heat-regulating center
Cessation of heartbeat and respiration	Failure of nervous regulation; cessation of breathing

Source: Climatology—Arid Zone Research X. © UNESCO 1958. Reproduced by permission of UNESCO.

proposed by Antarctic explorer Paul Siple in 1939. The *wind-chill factor* indicates the enhanced rate at which body heat is lost to the air and at best represents an estimate of heat energy loss. As wind speeds increase, heat loss from the skin increases. A formula for calculating these relationships was developed in 1945 and the National Weather Service (NWS) began reporting wind-chill temperatures in 1970. This older index tended to overestimate heat loss from skin.

The NWS and the Meteorological Services of Canada (MSC, http://www.msc-smc.ec.gc.ca/) revised the

wind chill formula and standard assumptions for the 2001–2002 winter season. The new Wind Chill Temperature (WCT) Index is an effort to improve the accuracy of heat loss calculations. Computer modeling, clinical trials and testing, and advances in technology make this revision possible. Figure 1 is the new version that went into service November 2001.

For example, using the new chart, if the air temperature is −7°C (20°F) and the wind is blowing at 32 kmph (20 mph), skin temperatures will be at −16°C (4°F). The lower wind-chill
(continued)

Focus Study 5.1 *(continued)*

Actual Air Temperature in °C (°F)

Calm	4° (40°)	−1° (30°)	−7° (20°)	−12° (10°)	−18° (0°)	−23° (−10°)	−29° (−20°)	−34° (−30°)	−40° (−40°)
8 (5)	2° (36°)	−4° (25°)	−11° (13°)	−17° (1°)	−24° (−11°)	−30° (−22°)	−37° (−34°)	−43° (−46°)	−49° (−57°)
16 (10)	1° (34°)	−6° (21°)	−13° (9°)	−20° (−4°)	−27° (−16°)	−33° (−28°)	−41° (−41°)	−47° (−53°)	−54° (−66°)
24 (15)	0° (32°)	−7° (19°)	−14° (6°)	−22° (−7°)	−28° (−19°)	−36° (−32°)	−43° (−45°)	−50° (−58°)	−57° (−71°)
32 (20)	−1° (30°)	−8° (17°)	−16° (4°)	−23° (−9°)	−30° (−22°)	−37° (−35°)	−44° (−48°)	−52° (−61°)	−59° (−74°)
40 (25)	−2° (29°)	−9° (16°)	−16° (3°)	−24° (−11°)	−31° (−24°)	−38° (−37°)	−46° (−51°)	−53° (−64°)	−61° (−78°)
48 (30)	−2° (28°)	−9° (15°)	−17° (−1°)	−24° (−12°)	−32° (−26°)	−39° (−39°)	−47° (−53°)	−55° (−67°)	−62° (−80°)
56 (35)	−2° (28°)	−10° (14°)	−18° (0°)	−26° (−14°)	−33° (−27°)	−41° (−41°)	−48° (−55°)	−56° (−69°)	−63° (−82°)
64 (40)	−3° (27°)	−11° (13°)	−18° (−1°)	−26° (−15°)	−34° (−29°)	−42° (−43°)	−49° (−57°)	−57° (−71°)	−64° (−84°)
72 (45)	−3° (26°)	−11° (12°)	−19° (−2°)	−27° (−16°)	−34° (−30°)	−42° (−44°)	−50° (−58°)	−58° (−72°)	−66° (−86°)
80 (50)	−3° (26°)	−11° (12°)	−19° (−3°)	−27° (−17°)	−35° (−31°)	−43° (−45°)	−51° (−60°)	−59° (−74°)	−67° (−88°)

Wind speed, kmph (mph)

Frostbite times: ▢ 30 min. ▨ 10 min. ■ 5 min.

FIGURE 1 Wind Chill Temperature Index.
WCT Index factor for various temperatures and wind speeds. A new version of the wind-chill chart—effective November 2001. For quick calculations see: http://205.156.54.206/om/windchill/index.shtml#calculator. (In English units the new formula is: Wind chill (F°) = 35.74 + 0.6125T − 35.75($V^{0.16}$) + 0.4275T($V^{0.16}$); where T = air temperature in °F, V = wind speed in mph. In Canada, heat loss is expressed in watts per square meter.) [Adapted from the National Weather Service and Meteorological Services of Canada, version 11/01/01.]

values present a serious freezing hazard to exposed flesh. Imagine the wind chill experienced by a downhill ski racer going 130 kmph (80 mph)—frostbite is a definite possibility during a 2-minute run. The wind chill does omit consideration of sunlight intensity, a person's physical activity, and the use of protective clothing, such as a windbreaker, that prevents the wind access to your skin.

Heat Index

The *heat index* (*HI*) indicates the human body's reaction to the combination of air temperature and water vapor. The heat index, combining the effects of temperature and humidity, indicates how hot the air feels to an average person. Water vapor in air is expressed as relative humidity, a concept presented in Chapter 7. For now, it is sufficient to

say that the amount of water vapor in the air affects the evaporation rate of perspiration on the skin. The reason is simply available space: The more water vapor (higher humidity), the less space the air has to absorb water vapor evaporated from perspiration.

Figure 2 is an abbreviated version of the heat index that the NWS now includes in its daily weather summaries during summer months. The table beneath the graph describes the effects of heat-index categories on higher-risk groups. A combination of high temperature and high humidity can severely reduce the body's natural ability to regulate internal temperature (see http://weather.noaa.gov/weather/hwave.html).

Summer 1995 Heat-Index Deaths For nearly a week during July 1995,

Chicago's heat-index values went to category I in dwellings that lacked air conditioning. Chicago had never experienced a 48-hour period during which temperatures did not go below 31.5°C (89°F).

With high pressure (hot, stable air) dominating the Midwest to the Atlantic and moist air from the Gulf of Mexico, all the ingredients for this disaster were present. Afternoon temperatures were above 32°C (90°F) for a week in Chicago, 38°C (100°F) in South Dakota, and 40°C (104°F) in Toledo, Ohio, and New York City; the temperature hit 43°C (109°F) in Omaha, Nebraska. Cities in New England that had reached 37.7°C (100°F) only twice in their history exceeded that for 4 or 5 days during July 1995!

(continued)

Focus Study 5.1 *(continued)*

FIGURE 2 Heat index.
Heat index graph for various temperatures and relative humidity levels. [Courtesy of the National Weather Service.]

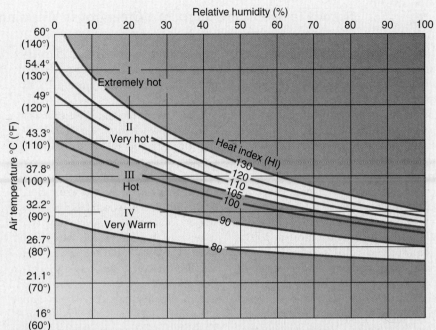

Level of concern	Category	Heat Index Apparent Temperature	General Effect of Heat Index on People in High-Risk Groups
Extreme danger	I	54°C (130°F) or higher	Heat/sunstroke highly likely with continued exposure
Danger	II	41° – 54°C (105° – 130°F)	Sunstroke, heat cramps, or heat exhaustion likely and heatstroke possible with prolonged exposure and/or physical activity
Extreme caution	III	32° – 41°C (90° – 105°F)	Sunstroke, heat cramps, and heat exhaustion possible with prolonged exposure and/or physical activity
Caution	IV	27° – 32°C (80° – 90°F)	Fatigue possible with prolonged exposure and/or physical activity

This combination of high temperatures and water vapor in the air produced stifling conditions for everyone, the sick and elderly in particular. On July 13 some apartments without air conditioning exceeded indoor heat-index temperatures of 54°C (130°F) when the official temperature at Midway Airport reached a record 41°C (106°F). Heat-index death toll totaled 700 people in Chicago. Nearly 1000 people died overall in the Midwest and East from these conditions.

The rush for medical help quickly swamped hospitals, which had to turn away hundreds of patients. The rate at which the heat claimed victims overwhelmed the coroner, so the dead were stored in many refrigerated trucks parked outside the morgue (Figure 3). We can only wonder how many lives would be saved if refrigeration for the living were available.

FIGURE 3 Chicago heat wave.
The Midwest and eastern United States were hit with a devastating heat wave during July 1995. City emergency, medical, and coroner services were overwhelmed by the tragedy. [Photo by Dave Weaver/Gamma Liaison Agency, Inc.]

News Report 5.1

Record Temperatures Suggest a Greenhouse Warming

Scientists agree that human activities are enhancing Earth's natural greenhouse. Certain radiatively active gases are absorbing longwave radiation and delaying losses of heat energy to space, thus throwing off the natural Earth–atmosphere energy equilibrium and producing a distinct warming trend.

The Intergovernmental Panel on Climate Change (IPCC), through major reports in 1990, 1992, 1995, and a Third Assessment Report in 2001, confirms that global warming is occurring and from their latest report that, ". . . there is new and stronger evidence that most of the warming observed over the past 50 years is attributable to human activities. . . . Both temperature and sea level are projected to continue to rise throughout the twenty-first cen-

tury for all scenarios studied." Uncertainty remains in forecasting the implications of this warming on Earth systems.

The period from 1970 to 2001 registered the warmest years in the history of instrumental measurements. Effects of this warming are the subject of many scientific studies: melting and retreat of mountain glaciers worldwide, disintegration of coastal ice shelves around Antarctica, sea level rising at a faster rate than previously observed, more intense thunderstorms and weather extremes, changes in vegetation on land and the distribution of marine organisms in the oceans, and heat-wave effects on grain production. The snowline in portions of the European Alps rose 100 m (330 ft) in elevation since 1980. Scientists determined

that 43% of the Arctic Ocean ice pack has disappeared since 1970.

An array of sophisticated satellites, remote-sensing capabilities, and powerful computers equip scientists to run global circulation models and decipher the climate trends. With the blame resting squarely on fossil fuel consumption and our modern technological society, the political fallout and debate is intense.

A discussion of global warming and climate change is in Chapter 10, bringing together topics throughout Parts 1 and 2 of the text. There we address what is known and unknown, how global models work, the radiatively active gases that are forcing the warming, the consequences to natural systems, and what international action is underway to slow the effects of such change.

Summary and Review—Global Temperatures

● ***Define*** **the concepts of temperature, kinetic energy, and sensible heat, and** ***distinguish*** **among Kelvin, Celsius, and Fahrenheit scales and how they are measured.**

Temperature is a measure of the average kinetic energy (motion) of individual molecules in matter. We feel the effect of temperature as the sensible heat transfer from warmer objects to cooler objects when these objects are touching. Temperature scales include:

- Kelvin scale: 100 units between ice's melting point (273 K) and water's boiling point (373 K).
- Celsius scale: 100 degrees between ice's melting point (0°C) and water's boiling point (100°C).
- Fahrenheit scale: 180 degrees between ice's melting point (32°F) and water's boiling point (212°F).

The Kelvin scale is used in scientific research because temperature readings start at absolute zero and thus are proportional to the actual kinetic energy in a material.

temperature (p. 120)

1. Distinguish between sensible heat and sensible temperature.
2. What does air temperature indicate about energy in the atmosphere?

3. Compare the three scales that express temperature. What is the basic assumption for each?
4. What is your source of daily temperature information? Describe the highest temperature you have experienced and the lowest temperature. From what we have discussed in this chapter, can you identify the factors that may have contributed to these temperatures?

● ***List*** **and** ***review*** **the principal controls and influences that produce global temperature patterns.**

Principal controls and influences upon temperature patterns include latitude (the distance north or south of the equator), altitude (location above sea level), cloud cover (reflect, absorb, and reradiate energy), and land–water heating differences (the nature of evaporation, transparency, specific heat, movement, and ocean currents and sea-surface temperatures).

5. Explain the effect of altitude on air temperature. Why is air at higher altitudes lower in temperature? Why does it feel cooler standing in shadows at higher altitude than at lower altitude?
6. What noticeable effect does air density have on the absorption and radiation of energy? What role does altitude play in that process?

7. How is it possible to grow moderate-climate-type crops such as wheat, barley, and potatoes at an elevation of 4103 m (13,460 ft) near La Paz, Bolivia, so near the equator?

8. Describe the effect of cloud cover with regard to Earth's temperature patterns. From the last chapter, review the cloud-albedo forcing and cloud-greenhouse forcing of different cloud types and relate the concepts with a simple sketch.

● *Review* the factors that produce different marine effects and continental effects as they influence temperatures and *utilize* several pairs of stations to illustrate these differences.

The physical nature of land (rock and soil) and water (oceans, seas, and lakes) is the reason for **land–water heating differences**, the fact that land heats and cools faster than water. Moderate temperature patterns are associated with water bodies, and extreme temperatures occur inland. The five controls that differ between land and water surfaces are evaporation, transparency, specific heat, movement and ocean currents, and sea-surface temperatures.

Light penetrates water because of its **transparency**. Water is clear and light transmits to an average depth of 60 m (200 ft) in the ocean. This penetration distributes available heat energy over a much greater volume than could occur on opaque land; thus, a larger energy reservoir is formed. When equal volumes of water and land are compared, water requires far more energy to increase its temperature than does land. In other words, water can hold more energy than can soil or rock, so water has a higher **specific heat**, the heat capacity of a substance, averaging about four times that of soil.

Ocean currents affect temperature. An example of the effect of ocean currents is the **Gulf Stream**, which moves northward off the east coast of North America, carrying warm water far into the North Atlantic. As a result, the southern third of Iceland experiences much milder temperatures than would be expected for a latitude of 65° N, just below the Arctic Circle (66.5°).

Marine effect, or maritime, describes locations that exhibit the moderating influences of the ocean, usually along coastlines or on islands. **Continental effect** refers to the condition of areas that are less affected by the sea and therefore have a greater range between maximum and minimum temperatures diurnally and yearly.

land–water heating differences (p. 124)
transparency (p. 125)
specific heat (p. 126)
Gulf Stream (p. 126)
marine effect (p. 128)
continental effect (p. 128)

9. List the physical aspects of land and water that produce their different responses to heating from absorption of insolation. What is the specific effect of transparency in a medium?

10. What is specific heat? Compare the specific heat of water and soil.

11. Describe the pattern of sea-surface temperatures (SSTs) as determined by satellite remote sensing. Where is the warmest ocean region on Earth?

12. What effect does sea-surface temperature have on air temperature? Describe the negative feedback mechanism created by higher sea-surface temperatures and evaporation rates.

13. Differentiate between marine and continental temperatures. Give geographic examples of each from the text: Canada, the United States, Norway, and Russia.

● *Interpret* the pattern of Earth's temperatures from their portrayal on January and July temperature maps and on a map of annual temperature ranges.

Maps for January and July instead of the solstice months of December and June are used for temperature comparison because of the natural lag that occurs between insolation received and maximum or minimum temperatures experienced. Each line on these temperature maps is an **isotherm**, an isoline that connects points of equal temperature. Isotherms portray temperature patterns.

Isotherms generally are zonal, trending east–west, parallel to the equator. They mark the general decrease in insolation and net radiation with distance from the equator. The **thermal equator** (isoline connecting all points of highest mean temperature) trends southward in January and shifts northward with the high summer Sun in July. In January it extends farther south into the interior of South America and Africa, indicating higher temperatures over landmasses.

In the Northern Hemisphere in January, isotherms shift equatorward as cold air chills the continental interiors. The coldest area on the map is in Russia, specifically northeastern Siberia. The intense cold experienced there results from winter conditions of consistent clear, dry, calm air, small insolation input, and an inland location far from any moderating maritime effects.

isotherm (p. 128)
thermal equator (p. 128)

14. What is the thermal equator? Describe its location in January and in July. Explain why it shifts position annually.

15. Observe trends in the pattern of isolines over North America and compare the January average temperature map with the July map. Why do the patterns shift locations?

16. Describe and explain the extreme temperature range experienced in north-central Siberia between January and July.

17. Where are the hottest places on Earth? Are they near the equator or elsewhere? Explain. Where is the coldest place on Earth?

18. From the maps in Figures 5.14, 5.16, and 5.17, determine the average temperature values and annual range of temperatures for your present location.

The *Geosystems* Home Page provides on-line resources for this chapter on the World Wide Web. To begin: Once on the Home Page, click on this textbook, scroll the Table of Contents menu, select this chapter, and click "Begin." You will find self-tests that are graded, review exercises, specific updates for items in the chapter, and many links to interesting related pathways on the Internet under "Destinations." *Geosystems* is at **http://www.prenhall.com/christopherson**.

Critical Thinking

A. With each temperature map (Figures 5.14, 5.16, and 5.17), begin by finding your own city or town and noting the temperatures indicated by the isotherms for January and July and the annual temperature range. Record the information from these maps in your notebook. As you work through the different maps throughout this text, note atmospheric pressure and winds, annual precipitation, climate type, landforms, soil orders, vegetation, and terrestrial biomes. By the end of the course you will have recorded a complete physical geography profile for your regional environment.

B. Have you ever experienced any of the different responses of the human body to low-temperature or high-temperature stress noted in Focus Study 5.1, Table 1? Where were you at the time of each experience? Make copies of the wind-chill and heat-index charts given in the Focus Study for later reference. On an appropriate day, check air temperature and wind speed and determine their combined effect on skin chilling. In contrast, on an appropriate day, check air temperature and relative humidity values and determine their combined effect on personal comfort. The *Geosystems* Home Page presents several interesting temperature links under "Destinations."

C. The Short Answer section of the *Geosystems* Home Page for Chapter 5 asks you to access a sea-surface temperature image map and analyze various aspects of the map. Click the "source" link and find out where this SST map originated. Compare this portrayal with the two SST image maps in Figure 5.11. Can you relate the seasonal change between the February and July maps with the one referenced in the Short Essay question?

The winds over Lake Michigan fill the sails of this four-masted sailing schooner, a class "B" tall ship, just offshore from Chicago. Serious questions arise about the feasibility of again harvesting the winds to assist in powering the world's merchant fleet and oil tankers. Estimates of fuel savings range from 15% to 50%. [Photo by Bobbé Christopherson.]

6

Atmospheric and Oceanic Circulations

Key Learning Concepts

After reading the chapter, you should be able to:

- *Define* the concept of air pressure and *describe* instruments used to measure air pressure.
- *Define* wind and *describe* how wind is measured, how wind direction is determined, and how winds are named.
- *Explain* the four driving forces within the atmosphere—gravity, pressure gradient force, Coriolis force, and friction force—and *describe* the primary high- and low-pressure areas and principal winds.
- *Describe* upper-air circulation and its support role for surface systems and *define* the jet streams.
- *Explain* several types of local winds: land–sea breezes, mountain–valley breezes, katabatic winds, and the regional monsoons.
- *Discern* the basic pattern of Earth's major surface and deep ocean currents.

Early in April 1815, on an island named Sumbawa in present-day Indonesia, the volcano Tambora erupted violently. It spewed an estimated 150 km³ (36 mi³) of material, 25 times the volume produced by the 1980 Mount St. Helens eruption in Washington State. Global atmospheric circulation carried material from Tambora worldwide, creating a stratospheric veil of dust and acid mist. Remarkable optical and meteorological effects troubled Earth's atmosphere years after the eruption—beautiful sunrises and sunsets and a temporary lowering of temperatures worldwide.

Scientists in 1815 lacked the remote-sensing capability of satellite technology, and they had no way of knowing the global impact of Tambora's eruption. Today,

143

(a)

Luzon Island, Philippines

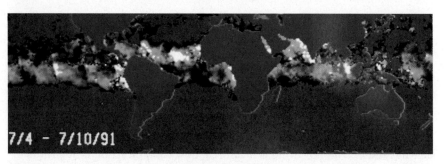

6/15 – 6/19/91

(b)

7/4 – 7/10/91

(c)

7/25 – 8/1/91

(d)

8/15 – 8/21/91

Effects cover
42% of globe in
just 60 days

(e)

FIGURE 6.1 Volcanic eruption effects spread worldwide by winds.
Satellite image shows Mount Pinatubo as it erupted (a). False-color images (b–e) show aerosols from Mount Pinatubo, smoke from fires, and dust storms, all swept about the globe by the general atmospheric circulation. The dramatic increase in aerosols over the oceans from mid-June 1991 (b) to mid-August (e) is due to the Mount Pinatubo eruption on June 15. The false color shows aerosol concentration, measured by the atmosphere's aerosol optical thickness (AOT): White is densest, dull yellow indicates medium values, and brown areas have the lowest aerosol concentration. In (b) note the dust moving westward from Africa, smoke from Kuwaiti oil well fires set during the Persian Gulf War, smoke from forest fires in Siberia, haze off the east coast of the United States, and the Pinatubo aerosol layer beginning to emerge north of Indonesia and expanding globally in (c) through (e). [(a) *AVHRR* satellite eruption image courtesy of U.S. Geological Survey, EROS Data Center. *AVHRR* satellite images of aerosols courtesy of Dr. Larry L. Stowe, National Environmental Satellite, Data, and Information Service, National Oceanic and Atmospheric Administration. Used by permission.]

technology permits a depth of analysis unknown in the past—satellites now track the atmospheric effects from dust storms, forest fires, industrial haze, warfare, and the dispersal of volcanic explosions, among many other things.

After 635 years of dormancy, Mount Pinatubo in the Philippines erupted in 1991. This event had tremendous atmospheric impact, lofting 15–20 million tons of ash, dust, and sulfur dioxide (SO_2) into the atmosphere. As the sulfur dioxide rose into the stratosphere, it quickly formed sulfuric acid (H_2SO_4) aerosols, which became concentrated at 16–25 km (10–15.5 mi) altitude. This debris increased atmospheric albedo about 1.5%, giving scientists an estimate of the aerosol volume generated by the eruption (see Figure 1.6).

The AVHRR instrument aboard *NOAA-11* monitored the reflected solar radiation from Mount Pinatubo's aerosols as global winds swept them around Earth. Figure 6.1 shows images made at three-week intervals that clearly track the spread of the debris worldwide (Figure 6.1b–d). Some 60 days after the eruption (the last satellite image in the sequence), the aerosol cloud spanned about 42% of the globe from 20° S to 30° N. Again, for almost 2 years colorful sunrises and sunsets and a small lowering of average temperatures followed. The eruption provided a unique insight into the dynamics of atmospheric circulation.

Global winds are certainly an important reason why the United States, the former Soviet Union, and Great Britain signed the 1963 Limited Test Ban Treaty. That treaty banned above-ground testing of nuclear weapons because atmospheric circulation spread radioactive contamination worldwide. Such agreements illustrate how the fluid movement of the atmosphere socializes humanity more than any other natural or cultural factor. Our atmosphere makes the world a spatially linked society—one person's or country's exhalation is another's inhalation.

In this chapter: we begin with a discussion of wind essentials, consisting of air pressure and its measurement and a description of wind. The driving forces that produce surface winds are pressure gradient, Coriolis, and friction. We examine the circulation of Earth's atmosphere and the patterns of global winds, including principal pressure systems and winds. We also consider Earth's wind-driven oceanic currents. The energy driving all this movement comes from one source: the Sun.

Wind Essentials

Earth's atmospheric circulation transfers both energy and mass on a grand scale. In the process, the imbalance between equatorial energy surpluses and polar energy deficits (Chapter 4) is partly resolved, Earth's weather patterns formed, and ocean currents produced. Air pollutants, whether natural or human-caused, are spread worldwide by atmospheric circulation, far from their point of origin.

Atmospheric circulation is generally categorized at three levels: *primary circulation* (general worldwide circulation), *secondary circulation* of migratory high-pressure and low-pressure systems, and *tertiary circulation* that includes local winds and temporal weather patterns. Winds that move principally north or south along meridians are known as *meridional flows*. Winds moving east or west along parallels of latitude are called *zonal flows*.

Air Pressure and Its Measurement

Important to an understanding of wind is the concept of air pressure, its measurement and expression. The molecules that constitute air create *air pressure* through their motion, size, and number—the factors that determine the temperature and density of the air. Air pressure, then, is a product of the temperature and density of a mass of air. Pressure is exerted on all surfaces in contact with the air.

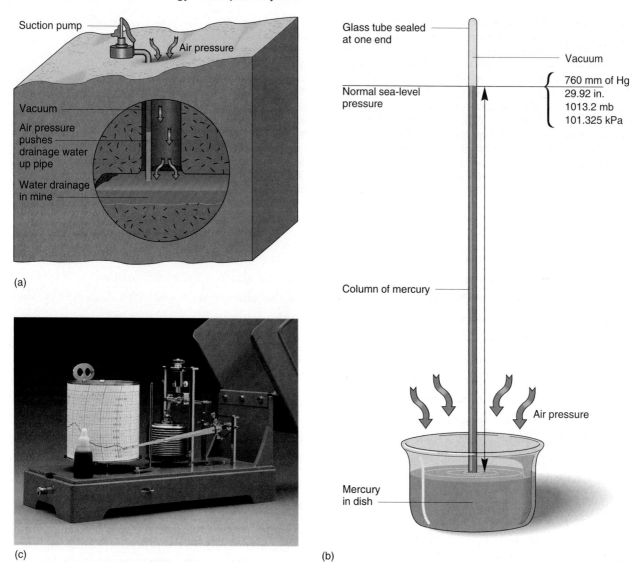

FIGURE 6.2 Developing the barometer.
Evangelista Torricelli developed the barometer to measure air pressure as a by-product of trying to solve a mine-drainage problem (a). Two types of instruments are used to measure
atmospheric pressure: (b) an idealized sketch of a mercury barometer, and (c) an aneroid
barometer. Have you used a barometer? If so, what type is it? Have you tried to reset it using a
local weather information source? [(c) Courtesy of Qualimetrics, Inc., Sacramento, California.]

In 1643, Galileo's pupil Evangelista Torricelli was
working on a mine-drainage problem. His work led him
to discover a method for measuring air pressure (Figure
6.2a). He knew that pumps in the mine were able to "pull"
water upward about 10 m (33 ft), but no higher, and he did
not know why. Careful observation led him to discover that
this limitation was caused not by weak pumps but by the atmosphere itself. Torricelli noted that the water level in the
vertical pipe fluctuated from day to day. He figured out
that air pressure varies with weather conditions.

To simulate the problem at the mine, Torricelli devised
an instrument at Galileo's suggestion, using a much denser
fluid than water—mercury (Hg)—in a glass laboratory tube
only 1 m high. Torricelli sealed the glass tube at one end,

filled it with mercury, and inverted it into a dish of mercury
(Figure 6.2b). He determined that the average height of
the column of mercury in the tube was 760 mm (29.92 in.)
and that it did vary day to day as the weather changed. He
concluded that the mass of surrounding air was exerting
pressure on the mercury in the dish that counterbalanced
the column of mercury.

Using similar instruments, scientists set a standard of
normal sea-level pressure at 1013.2 mb (millibar, mb, expresses force per square meter of surface area), or 29.92 in.
mercury. In Canada and other countries normal air pressure
is expressed as 101.32 kPa (kilopascal; 1 kPa = 10 mb).

Any instrument that measures air pressure is called
a *barometer* (from the Greek *baros*, meaning "weight").

Torricelli developed a **mercury barometer**. A more compact barometer design, which works without a meter-long tube of mercury, is the aneroid barometer, shown in Figure 6.2c. *Aneroid* means "using no liquid." The **aneroid barometer** principle is simple: Imagine a small chamber, partially emptied of air, which is sealed and connected to a mechanism attached to a needle on a dial. As air pressure increases, it presses on the chamber; as air pressure decreases, it relieves pressure from the chamber. The chamber responds to these changes in air pressure and moves the needle. An aircraft altimeter is a type of aneroid barometer. It accurately measures altitude because air pressure diminishes with elevation above sea level. For accuracy the altimeter must be adjusted for temperature changes.

Figure 6.3 illustrates comparative scales in millibars and inches used to express air pressure and its relative force. The normal range of Earth's atmospheric pressure from strong high pressure to deep low pressure is about 1050 to 980 mb (31.00 to 29.00 in.). The figure also indicates the extreme highest and lowest pressures ever recorded in the United States, Canada, and on Earth. Air pressure differences between places produce wind.

Wind: Description and Measurement

Simply stated, **wind** is generally the horizontal motion of air across Earth's surface. *It is produced by differences in air pressure from one location to another.* Wind's two principal properties are speed and direction, and instruments measure

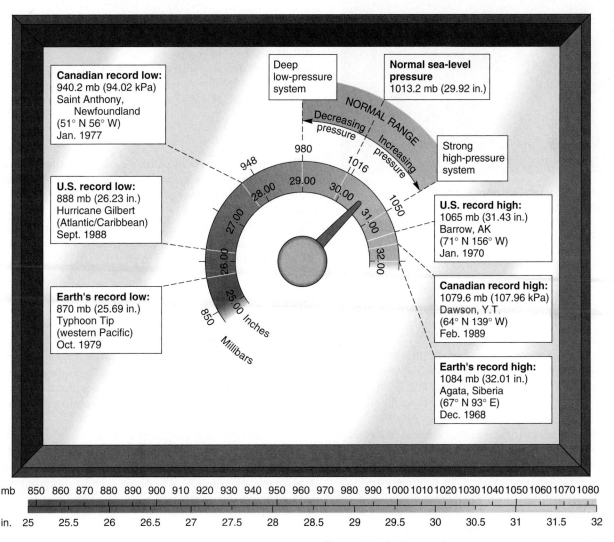

FIGURE 6.3 Air pressure readings and conversions.
Scales for expressing barometric air pressure in millibars and inches, with average air pressure values and recorded pressure extremes. Canadian values include kilopascal equivalents (10 mb = 1 kPa).

FIGURE 6.4 Wind vane and anemometer.
Instruments used to measure wind direction (wind vane, right) and wind speed (anemometer, left) at a weather station installation. [Photo by Belfort Instruments.]

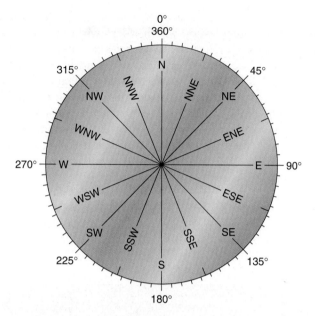

FIGURE 6.5 A wind compass.
Sixteen wind directions identified on a wind compass. Winds are named for the direction from which they originate. For example, a wind from the west is a westerly wind.

each. An **anemometer** measures wind speed in kilometers per hour (kmph), miles per hour (mph), meters per second (mps), or knots. (A *knot* is a nautical mile per hour, covering 1 minute of Earth's arc in an hour, equivalent to 1.85 kmph, or 1.15 mph.) A **wind vane** determines wind direction; the standard measurement is taken 10 m (33 ft) above the ground to reduce the effects of local topography on wind direction (Figure 6.4).

Winds are named for the direction *from which they originate*. For example, a wind from the west is a *westerly* wind (it blows eastward); a wind out of the south is a *southerly* wind (it blows northward). Figure 6.5 illustrates a simple wind compass, naming 16 principal wind directions used by meteorologists.

The traditional *Beaufort wind scale* is a descriptive scale useful in visually estimating wind speed. In 1806, Admiral Beaufort of the British Navy introduced his wind scale. In 1926, G. C. Simpson expanded Beaufort's scale to include wind speeds on land. The National Weather Service (old Weather Bureau) standardized the scale in 1955 (see http://www.crh.noaa.gov/lot/webpage/beaufort/. It is

still referenced on ocean charts, enabling estimation of wind speed without instruments, although most ships use sophisticated equipment to perform such measurements. Using no instruments, try using the Beaufort scale in Table 6.1 to estimate wind speed as you walk across campus today; moisten your finger to sense cooling and from which direction the wind is blowing.

Global Winds

The primary circulation of winds across Earth has fascinated travelers, sailors, and scientists for centuries, although only in the modern era is a true picture emerging of the pattern and causes of global winds. Breakthroughs in space-based observations and Earth-bound computer technology are refining models that simulate total atmospheric and oceanic circulation.

A remarkable portrait of surface winds across the Pacific Ocean was painstakingly assembled by scientists at the Jet Propulsion Laboratory and the University of California, Los Angeles (Figure 6.6). The *Seasat* satellite produced the image using radar to measure the motion and direction of ocean waves. Because wind drives waves on the ocean surface, wave patterns indicate winds.

The patterns in the figure are the result of specific forces at work in the atmosphere: *pressure gradient force*, *Coriolis force*, *friction force*, and *gravity*. These forces are our next topic. As we progress through this chapter, you may want to refer to this *Seasat* image to identify the winds, eddies, and vortexes it portrays.

Table 6.1   Beaufort Wind Scale

Wind Speed				Beaufort Wind Scale			
kmph	mph	knots	Beaufort Number	Wind Description	Observed Effects at Sea	Observed Effects on Land
<1	<1	<1	0	Calm	Glassy calm, like a mirror	Calm, no movement of leaves
1–5	1–3	1–3	1	Light air	Small ripples; wavelet scales; no foam on crests	Slight leaf movement; smoke drifts; wind vanes still
6–11	4–7	4–6	2	Light breeze	Small wavelets; glassy look to crests, which do not break	Leaves rustling; wind felt; wind vanes moving
12–19	8–12	7–10	3	Gentle breeze	Large wavelets; dispersed whitecaps as crests break	Leaves and twigs in motion; small flags and banners extended
20–29	13–18	11–16	4	Moderate breeze	Small, longer waves; numerous whitecaps	Small branches moving; raising dust, paper, litter, and dry leaves
30–38	19–24	17–21	5	Fresh breeze	Moderate, pronounced waves; many white-caps; some spray	Small trees and branches swaying; wavelets forming on inland waterways
39–49	25–31	22–27	6	Strong breeze	Large waves, white foam crests everywhere; some spray	Large branches swaying; overhead wires whistling; difficult to control an umbrella
50–61	32–38	28–33	7	Moderate (near) gale	Sea mounding up; foam and sea spray blown in streaks in the direction of the wind	Entire trees moving; difficult to walk into wind
62–74	39–46	34–40	8	Fresh gale (or gale)	Moderately high waves of greater length; breaking crests forming sea spray; well-marked foam streaks	Small branches breaking; difficult to walk; moving automobiles drifting and veering
75–87	47–54	41–47	9	Strong gale	High waves; wave crests tumbling and the sea beginning to roll; visibility reduced by blowing spray	Roof shingles blown away; slight damage to structures; broken branches littering the ground
88–101	55–63	48–55	10	Whole gale (or storm)	Very high waves and heavy, rolling seas; white appearance to foam-covered sea; overhanging waves; visibility reduced	Uprooted and broken trees; structural damage; considerable destruction; seldom occurring
102–116	64–73	56–63	11	Storm (or violent storm)	White foam covering a breaking sea of exceptionally high waves; small- and medium-sized ships lost from view in wave troughs; wave crests frothy	Widespread damage to structures and trees, a rare occurrence
>117	>74	>64	12–17	Hurricane	Driving foam and spray filling the air; white sea; visibility poor to nonexistent	Severe to catastrophic damage; devastation to affected society

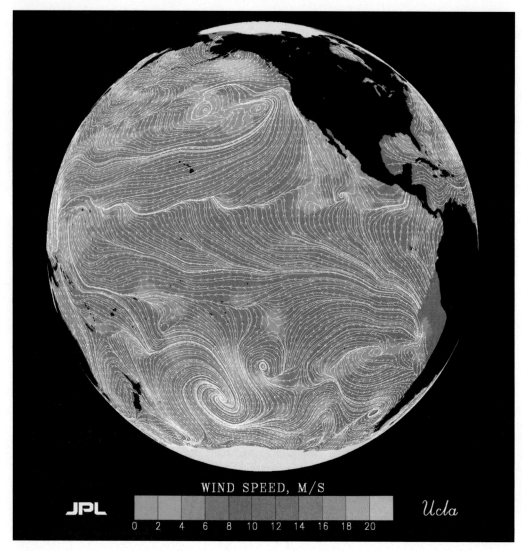

(a)

(b)

FIGURE 6.6 Wind portrait of the Pacific Ocean.
Surface wind measured by radar scatterometer aboard the *Seasat* satellite on a September day. Scientists analyzed 150,000 measurements to produce this image (a). Colors are correlated with wind speeds, and the white arrows denote wind direction. Try comparing the wind patterns with a visible-light image of the same region (b). Can you identify the pattern of trade winds, westerlies, high-pressure cells, and low-pressure cells from the cloud patterns on the image? [Wind portrait courtesy of Dr. Peter Woiceshyn, Jet Propulsion Laboratory, Pasadena, California. Satellite image inset from Laboratory of Planetary Studies, Cornell University. Used by permission.]

Driving Forces Within the Atmosphere

Four forces determine both speed and direction of winds:

- Earth's *gravitational force* on the atmosphere is virtually uniform. Gravity equally compresses the atmosphere worldwide, with the density decreasing as altitude increases. Without gravity, there would be no atmospheric pressure—or atmosphere, for that matter.

- **Pressure gradient force** drives air from areas of higher barometric pressure (more dense air) to areas of lower barometric pressure (less dense air), thereby causing winds. Without a pressure gradient force, there would be no wind.

- The **Coriolis force**, a deflective force, makes wind that travels in a straight path appear to be deflected in relation to Earth's rotating surface. The Coriolis force deflects wind to the right in the Northern Hemisphere and to the left in the Southern Hemisphere.

Without Coriolis force, winds would move along straight paths between high- and low-pressure areas.

- The **friction force** drags on the wind as it moves across surfaces; it decreases with height above the surface. Without friction, winds would simply move in paths parallel to isobars and at high rates of speed.

All four of these forces operate on moving air and ocean currents at Earth's surface and influence global wind circulation patterns. The following sections describe the actions of the pressure gradient, Coriolis, and friction forces. (The gravitational force operates uniformly worldwide.)

Pressure Gradient Force

High- and low-pressure areas exist in the atmosphere principally because Earth's surface is unequally heated. For example, cold, dense air at the poles exerts greater pressure than warm, less dense air along the equator. These pressure differences establish a *pressure gradient force*.

An **isobar** is an *isoline* (a line along which there is a constant value) plotted on a weather map to connect points of equal pressure. A pattern of isobars on a weather map provides a portrait of the pressure gradient between an area of higher pressure and one of lower pressure. The spacing between isobars indicates the intensity of the pressure difference, or *pressure gradient*.

Just as closer contour lines on a topographic map indicate a steeper slope on land, so do closer isobars denote a steepness in the pressure gradient. In Figure 6.7a, note the spacing of the isobars (green lines). A steep gradient causes faster air movement from a high-pressure area to a low-pressure area. Isobars spaced wider apart from one another mark a more gradual pressure gradient, one that creates a slower airflow. Along a horizontal surface, the pressure gradient force alone acts at right angles to the isobars, so wind blows across them at right angles. Note the location of steep ("strong winds") and gradual ("light winds") pressure gradients and their relationship to wind intensity on the map in Figure 6.7b.

Figure 6.8 illustrates the forces that direct the wind. Figure 6.8a shows the pressure gradient force acting alone. In a high-pressure area, as air descends, a field of subsiding, or sinking, air develops. Air diverges out of the high-pressure area at the surface, moving outward in all directions. On the other hand, in a low-pressure area, as air rises, it converges from all directions into the area of lower pressure at the surface. For instance, on a warm day, the temperature of the air increases and the air is less dense and more buoyant, so it rises. In contrast on a cold day, the temperature of the air is lower, the air is denser and less buoyant, so it descends. This behavior of a parcel of air is discussed further in Chapter 7.

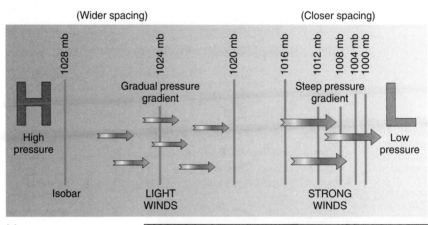

(a)

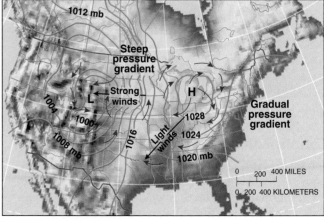

(b)

FIGURE 6.7 Pressure gradient determines wind speed. Pressure gradient (a). On a weather map (b), the closer spacing of isobars represents a steeper pressure gradient that produces stronger winds; wider spacing of isobars denotes a gradual pressure gradient that leads to lighter winds. Here we see surface winds spiraling clockwise out of a high-pressure system and spiraling counterclockwise into a low-pressure system.

Top view and side view of air movement in an idealized high-pressure area and low-pressure area on a nonrotating Earth.

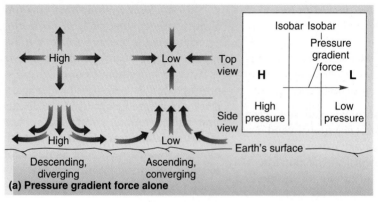

(a) Pressure gradient force alone

Earth's rotation adds the Coriolis force and a "twist" to air movements. High-pressure and low-pressure areas develop a rotary motion, and wind flowing between highs and lows flows parallel to isobars.

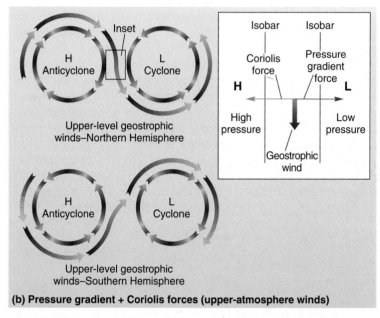

(b) Pressure gradient + Coriolis forces (upper-atmosphere winds)

Surface friction adds a countering force to Coriolis, producing winds that spiral out of a high-pressure area and into a low-pressure area. Surface winds cross isobars at an angle.

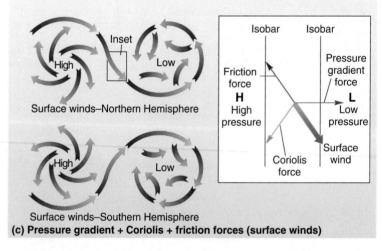

(c) Pressure gradient + Coriolis + friction forces (surface winds)

FIGURE 6.8 Three physical forces that produce winds.
Three physical forces integrate to produce wind patterns at the surface and aloft: (a) the pressure gradient force; (b) the Coriolis force counters the pressure gradient force, producing a geostrophic wind flow in the upper atmosphere; and (c) the friction force, which, combined with the other two forces, produces characteristic surface winds. The gravitational force is assumed. The three inset diagrams show the interaction of forces that form prevailing geostrophic and surface winds. In (b) and (c), note the reverse circulation pattern in the Southern Hemisphere.

Coriolis Force

You might expect surface winds to move in a straight line from areas of higher pressure to areas of lower pressure. On a nonrotating Earth, they would. But on our rotating planet, the Coriolis force deflects anything that flies or flows across Earth's surface—wind, an airplane, or ocean currents—from a straight path. This force is an effect of Earth's rotation.

Earth's rotational speed varies with latitude, increasing from 0 kmph at the poles to 1675 kmph (1041 mph) at the equator (see Table 2.2). The Coriolis force is zero along the equator, increases to half the maximum deflection at

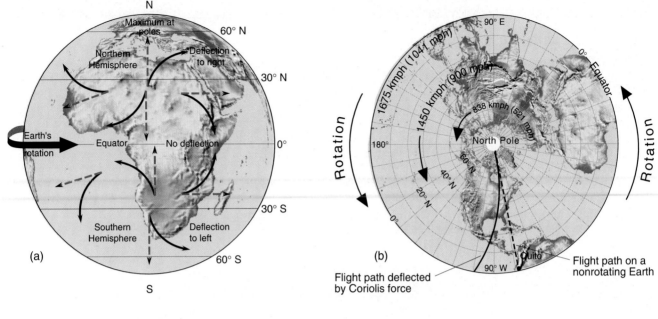

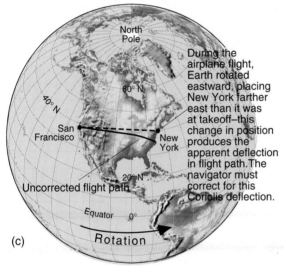

FIGURE 6.9 The Coriolis force—an apparent deflection.
Distribution of the Coriolis force on Earth: (a) apparent deflection to the right of a straight line in the Northern Hemisphere and apparent deflection to the left in the Southern Hemisphere; (b) Coriolis deflection of a flight path between the North Pole and Quito, Ecuador, which is on the equator; (c) deflection of a flight path between San Francisco and New York. Deflection from a straight path occurs regardless of the direction of movement.

30° N and 30° S latitudes, and reaches maximum deflection flowing away from the poles.

The deflection occurs regardless of the direction in which the object is moving. Because Earth rotates eastward, objects that move in an absolute straight line over a distance (such as winds and ocean currents) appear to curve to the right in the Northern Hemisphere and to the left in the Southern Hemisphere (Figure 6.9a). The effect of the Coriolis force increases as the speed of the moving object increases; thus, the faster the wind speed, the greater its apparent deflection. The Coriolis force does not normally affect small-scale motions that cover insignificant distance and time (see News Report 6.1).

The key to understanding Coriolis force is one's viewpoint. From the viewpoint of an airplane that is passing over Earth's surface, the surface can be seen to rotate slowly below. But, looking from the surface at the airplane,

the surface seems stationary, and the airplane appears to curve off course. The airplane does not actually deviate from a straight path, but it *appears* to do so because we are standing on Earth's rotating surface beneath the airplane. Because of this apparent deflection, the airplane must make constant corrections in flight path to maintain its "straight" heading.

As an example of the effect of this force, see Figure 6.9b. A pilot leaves the North Pole and flies due south toward Quito, Ecuador. If Earth were not rotating, the aircraft would simply travel along a meridian of longitude and arrive at Quito. But Earth is rotating eastward beneath the aircraft's flight path. If the pilot does not allow for this rotation, the plane will reach the equator over the ocean along an apparently curved path far to the west of the intended destination. Pilots must correct for this Coriolis deflection in their navigational calculations.

News Report 6.1

Coriolis, a Forceful Effect on Drains?

A common misconception about the Coriolis force is that it affects water draining out of a sink, tub, or toilet. Can the Coriolis force cause this twist? When a ship crosses the equator, does the direction of a draining spiral of water in a sink suddenly reverse?

Moving water or air must cover some distance across space and time before the Coriolis force noticeably

deflects it. Long-range artillery shells and guided missiles do exhibit small amounts of deflection that must be corrected for accuracy. But water movements down a drain are too small in spatial extent to be noticeably affected by this force.

Note that we call Coriolis a *force*. The label force is appropriate because, as the physicist Sir Isaac Newton

(1643–1727) stated, when something is accelerating over a space, a force is in operation (mass times acceleration). This *apparent force* (in classical mechanics, an inertial force) acts as an effect on moving objects. It is named for Gaspard Coriolis, a French mathematician and researcher of applied mechanics, who first described the phenomenon in 1831.

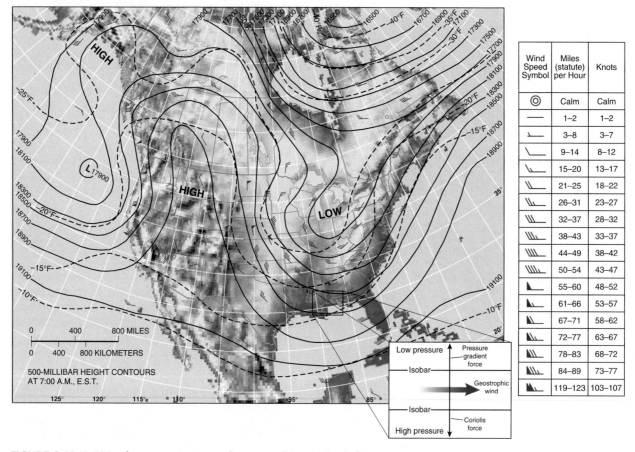

FIGURE 6.10 A 500-mb pressure map and geostrophic winds aloft.
Isobaric chart for an April day. Contours show elevation (in feet) at which 500-mb pressure occurs. The pattern of contours reveals geostrophic wind patterns in the troposphere at approximately 5500 m (18,000 ft) altitude. Note the "ridge" of high pressure over the Intermountain West (through the Rockies) and the "trough" of low pressure over the Great Lakes region. The inset diagram shows the interaction of forces that form prevailing geostrophic winds. [Data provided by National Weather Service, NOAA.]

This effect is in force regardless of the direction of the moving object. A flight from California to New York is shown in Figure 6.9c. The Coriolis deflection occurs because, as the airplane flew to New York, Earth continued to rotate eastward, so the destination moved farther to the east.

Unless the pilot corrected for Earth's rotational motion, the flight would end up somewhere in North Carolina.

How does the Coriolis force affect wind? As air rises from the surface through the lowest levels of the atmosphere, leaving the drag of surface friction behind, its speed increases. This increase in speed increases the Coriolis force, spiraling the winds to the right in the Northern Hemisphere or to the left in the Southern Hemisphere, generally producing upper-air westerly winds from the subtropics to the poles. In the upper troposphere, the Coriolis force just balances the pressure gradient force. Consequently, the winds between higher-pressure and lower-pressure areas aloft flow parallel to the isobars.

Figure 6.8b illustrates the combined effect of the pressure gradient force and the Coriolis force on air currents aloft. Together, they produce winds that do not flow directly from high to low, but *around* the pressure areas, remaining parallel to the isobars. Such winds are called **geostrophic winds** and are characteristic of upper tropospheric circulation. (The suffix -*strophic* means "to turn.") Geostrophic winds produce the characteristic pattern shown on the upper-air weather map in Figure 6.10. Note the inset illustration showing the effects of the pressure gradient and Coriolis forces that produce a geostrophic flow of air.

Friction Force

Figure 6.8c adds the effect of friction to the Coriolis and pressure gradient forces on wind movements; combining all three forces produces the wind patterns we see along Earth's surface. The effect of surface friction extends to a height of about 500 m (around 1600 ft) and varies with surface texture, wind speed, time of day and year, and atmospheric conditions. In general, rougher surfaces produce more friction.

Near the surface, friction disrupts the equilibrium established in geostrophic wind flows between the pressure gradient and Coriolis forces—note the inset illustration in Figure 6.8c. Because surface friction decreases wind speed, it reduces the effect of the Coriolis force and causes winds to move across isobars at an angle.

In Figure 6.8c, you can see that the Northern Hemisphere winds spiral out from a high-pressure area *clockwise* to form an **anticyclone** and spiral into a low-pressure area *counterclockwise* to form a **cyclone**. (In the Southern Hemisphere these circulation patterns are reversed, flowing out of high-pressure cells counterclockwise and into low-pressure cells clockwise.)

Atmospheric Patterns of Motion

With these forces and motions in mind, we are ready to build a general model of total atmospheric circulation. The warmer, less-dense air along the equator rises, creating low pressure at the surface, and the colder, denser air at the poles sinks, creating high pressure. If Earth did not rotate, the result would be a simple wind flow from the poles to the equator (a meridional flow), caused solely by pressure gradient.

However, Earth does rotate, creating a more complex flow system. On a rotating Earth the poles-to-equator flow is predominantly zonal (latitudinal), both at the surface and aloft. Winds are westerly (eastward-moving) in the middle and high latitudes and easterly (westward-moving) in the low latitudes toward the equator in both hemispheres. This system transfers thermal energy and air and water masses from equatorial energy surpluses to polar energy deficits, using waves, streams, and eddies on a planetary scale.

Primary High-Pressure and Low-Pressure Areas

The following discussion of Earth's pressure and wind patterns refers often to Figure 6.11, isobaric maps showing average surface barometric pressure in January and July. Indirectly, these maps indicate prevailing surface winds, which are suggested by the isobars.

The primary high- and low-pressure areas of Earth's general circulation appear on these maps as cells or uneven belts of similar pressure that stretch across the face of the planet, interrupted by landmasses. Between these areas flow the primary winds, which have been noted in adventure stories and hero myths throughout human experience.

Secondary highs and lows, from a few hundred to a few thousand kilometers in diameter and hundreds to thousands of meters high, are formed within these primary pressure areas. The secondary systems seasonally migrate to produce changing weather patterns in the regions over which they pass.

Four broad pressure areas cover the Northern Hemisphere and a similar set exists in the Southern Hemisphere. In each hemisphere, two of the pressure areas are stimulated by thermal (temperature) factors. These are the **equatorial low-pressure trough** (marked by the ITCZ line on the maps) and the weak **polar high-pressure cells** at the North and South Poles (not shown, as the maps are cut off at 80° N and 80° S). The other two pressure areas are formed by dynamic (mechanical) factors: the **subtropical high-pressure cells** (H) and **subpolar low-pressure cells** (L). Table 6.2 summarizes the characteristics of these pressure areas. We now examine each principal pressure region.

Table 6.2 **Four Hemispheric Pressure Areas**			
Name	**Cause**	**Location**	**Air Temperature/Moisture**
Polar high-pressure cells	Thermal	90° N 90° S	Cold/dry
Subpolar low-pressure cells	Dynamic	60° N 60° S	Cool/wet
Subtropical high-pressure cells	Dynamic	20° to 35° N and S	Hot/dry
Equatorial low-pressure trough	Thermal	10° N to 10° S	Warm/wet

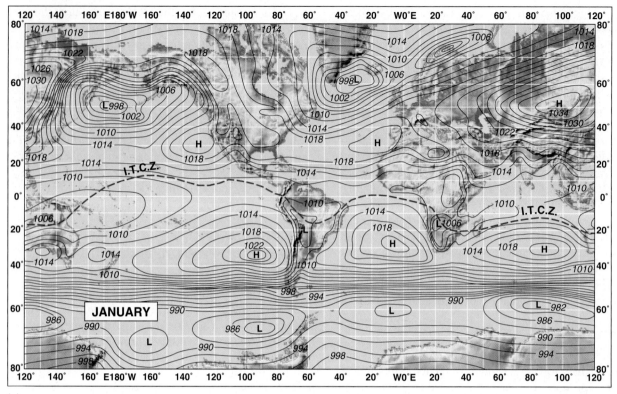

(a)

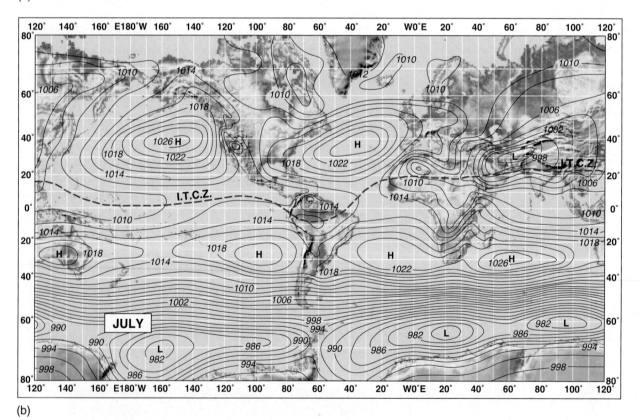

(b)

FIGURE 6.11 Global barometric pressures for January and July.
Long-term average surface barometric pressure (millibars) for (a) January and (b) July. The dashed line marks the general location of the intertropical convergence zone (ITCZ). Compare specific regions for January and July—for instance, the North Pacific, the North Atlantic, and the central Asian landmass. [Adapted from National Climatic Data Center, *Monthly Climatic Data for the World*, 46, no. 1, 1993 January and July issues. Prepared in cooperation with the World Meteorological Organization and the National Oceanic and Atmospheric Administration.]

FIGURE 6.12 Clouds portray equatorial and subtropical circulation patterns on a *Galileo* spacecraft image.
An interrupted band of clouds along the equator denotes the intertropical convergence zone (ITCZ), flanked to the north and south by several subtropical high-pressure systems and clear skies. Note the greater cloud development over land. This natural-color image was taken by the solid-state imaging instrument aboard the *Galileo* spacecraft during its December, 1990, flyby of Earth on its successful mission to the planet Jupiter in 1995. [The Solid State Imaging instrument (violet, green, and red filters) image courtesy of Dr. W. Reid Thompson, Laboratory of Planetary Studies, Cornell University. Used by permission.]

Equatorial Low-Pressure Trough—ITCZ: Clouds and Rain   Figure 6.12 is a satellite image showing the equatorial low-pressure trough. The broken band of clouds that straddles the equator across the middle of the image reveals this low-pressure system. The equatorial low-pressure trough is an elongated, undulating narrow band of low pressure (converging, ascending airflow) that nearly encircles the planet.

Constant high Sun altitude and consistent daylength (12 hours a day, year-round) make large amounts of energy available in this region throughout the year. The warming creates lighter, less-dense, ascending air, with surface winds converging along the entire extent of the low-pressure trough. This converging air is extremely moist and full of latent heat energy. As it rises, the air expands and cools, producing condensation; consequently, rainfall is heavy throughout this zone. Vertical cloud columns frequently reach the tropopause, in thunderous strength and intensity.

The combination of heating and convergence forces air aloft and forms the **intertropical convergence zone (ITCZ)**. The ITCZ is identified by bands of clouds associated with the convergence of winds along the equator and is noted on the January and July pressure maps (see Figure 6.11). During summer, a marked wet season accompanies the shifting ITCZ over various regions. The maps in Figure 6.11 show the ITCZ as a dashed line.

In January, the zone crosses northern Australia and dips southward in eastern Africa and South America. Note, using cloud patterns, this ITCZ location on the spacecraft image in Figure 6.12; for example, compare Australia in both figures. By July the zone shifts northward with the Sun, as far north as Pakistan and southern Asia. You easily can identify this ITCZ cloudiness across central Africa in the Earth photograph on this book's back cover, on the *GOES* images in Figure 1.27, and on this text's half-title page.

Figure 6.13 shows two views of Earth's general circulation. The winds converging on the equatorial low-pressure trough are known generally as the **trade winds**, or *trades*. *Northeast trade winds* blow in the Northern Hemisphere and *southeast trade winds* in the Southern Hemisphere. These are labeled in Figure 6.13a. The trade winds were named during the era of sailing ships carrying trade across the seas.

The trade winds pick up large quantities of moisture as they return through the Hadley circulation cell for another cycle of uplift and condensation (shown in cross section in Figure 6.13b). These *Hadley cells* in each hemisphere, named for the eighteenth-century English scientist who described the trade winds, denote the circuit completed by winds rising along the ITCZ, moving northward and southward into the subtropics, descending to the surface, and returning to the ITCZ as the trade winds. During the year in each hemisphere, this circulation pattern appears most vertically symmetrical near the equinoxes.

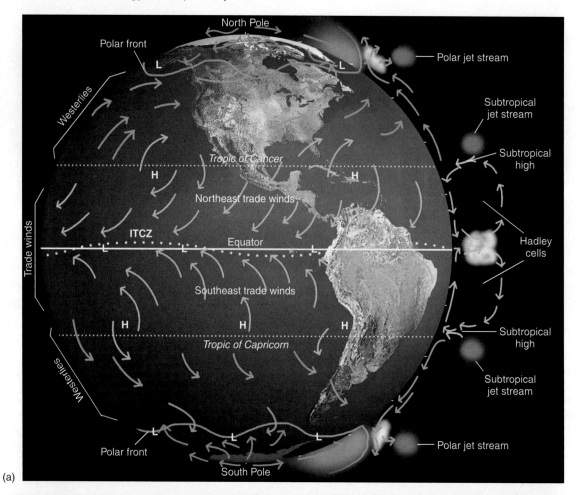

(a)

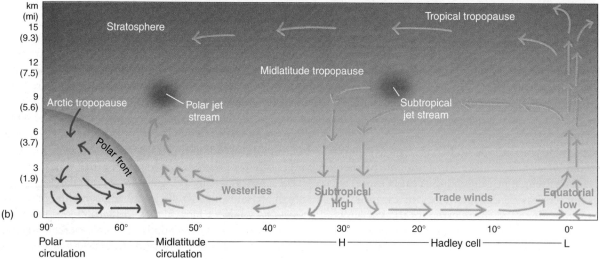

(b)

FIGURE 6.13 General atmospheric circulation model.
Two views of the general atmospheric circulation: (a) general circulation schematic;
(b) equator-to-pole cross section of the Northern Hemisphere. Both views show Hadley cells,
subtropical highs, polar front, the subpolar low-pressure cells, and approximate locations of
the subtropical and polar jet streams.

Within the ITCZ, winds are calm or mildly variable because of the even pressure gradient and the vertical ascent of air. These equatorial calms are the *doldrums*, a name formed from an older English word meaning "fool-ish," because of the difficulty sailing ships encountered when attempting to move through this zone. The rising air from the equatorial low-pressure area spirals upward into a geostrophic flow to the north and south. These upper-air winds turn eastward, flowing from west to east, beginning at about 20° N and 20° S, and form descending masses of air and high-pressure systems in the subtropical latitudes.

Subtropical High-Pressure Cells: Hot, Dry, Desert Air

Between 20° and 35° latitude in both hemispheres, a broad high-pressure zone of hot, dry air is evident across the globe (see Figures 6.11, 6.13). Clear, frequently cloudless skies over the Sahara and Arabian Desert and portions of the Indian Ocean dominate these regions. Can you identify these desert regions on the *Galileo* spacecraft image in Figure 6.12? How about on the photo of Earth on the text's back cover?

The dynamic cause of these subtropical anticyclones is too complex to detail here, but they generally form as air above the subtropics is mechanically pushed downward and heats by compression on its descent to the surface, as in Figure 6.13. Warmer air has a greater capacity to hold water vapor than does cooler air, making this descending warm air relatively dry (large water vapor holding capacity, low water vapor content). The air is also dry because heavy precipitation along the equatorial portion of the circulation removes moisture.

Surface air diverging from the subtropical high-pressure cells generates Earth's principal surface winds: the westerlies and the trade winds. The **westerlies** are the dominant surface winds from the subtropics to high latitudes. They diminish somewhat in summer and are stronger in winter.

As you examine the global pressure maps in Figure 6.11, you find several high-pressure areas. In the Northern Hemisphere, the Atlantic subtropical high-pressure cell is called the **Bermuda high** (in the western Atlantic) or the **Azores high** (when it migrates to the eastern Atlantic in winter). The area in the Atlantic under this subtropical high features clear, warm waters and large quantities of *Sargassum* (a seaweed), which gives the area its name—the Sargasso Sea.

The **Pacific high**, or *Hawaiian high*, dominates the Pacific in July, retreating southward in January. In the Southern Hemisphere, three large high-pressure centers dominate the Pacific, Atlantic, and Indian Oceans, especially in January, and tend to move along parallels of latitude in shifting zonal positions.

The entire high-pressure system migrates with the summer high Sun, fluctuating about 5–10° in latitude. The *eastern sides* (right-hand side) of these anticyclonic systems are drier and more stable (less convective activity) and feature cooler ocean currents than do the *western sides* (left-hand side). The drier eastern side of these systems and dry-summer conditions influence climate along subtropical and midlatitude west coasts (Figure 6.14). In fact, Earth's major deserts generally occur within the subtropical belt and extend to the west coast of each continent except Antarctica. In the figure, the desert regions of Africa come right to the shore in both hemispheres, with the cool, southward flowing *Canaries Current* offshore in the north,

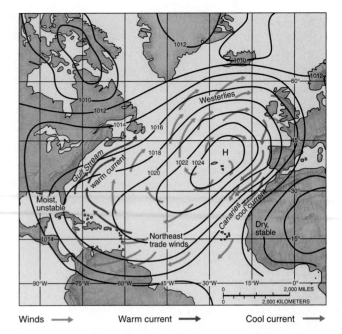

Winds ⟶   Warm current ⟶   Cool current ⟶

FIGURE 6.14 Subtropical high-pressure system in the Atlantic.
Characteristic circulation and climate conditions related to the Atlantic subtropical high-pressure anticyclone in the Northern Hemisphere. Note deserts extending to the shores of Africa with offshore cool currents, whereas the southeastern United States is moist and humid, with offshore warm currents.

and the cool, northward flowing *Benguela* current offshore in the south.

The western sides of subtropical high-pressure cells tend to be moist and unstable. These conditions cause warm, moist weather in Hawai'i, Japan, southeastern China, and the southeastern United States characterized in Figure 6.14.

Because the location of the subtropical belts are near 25° N and 25° S latitudes, these areas sometimes are known as the *calms of Cancer* and the *calms of Capricorn*. These zones of windless, hot, dry desert air, so deadly in the era of sailing ships, earned the name *horse latitudes*. The origin of this term is popularly attributed to becalmed and stranded sailing crews of past centuries, who destroyed the horses on board, not wanting to share food or water with the livestock. The term's true origin may never be known; the *Oxford English Dictionary* calls it "uncertain."

Subpolar Low-Pressure Cells: Cool and Moist Air

In January, two low-pressure cyclonic cells exist over the oceans around 60° N latitude, near their namesakes: the North Pacific **Aleutian low** and the North Atlantic **Icelandic low** (see Figure 6.11a). Both cells are dominant in winter and weaken or disappear in summer with the strengthening of high-pressure systems in the subtropics. The area of contrast between cold (from higher latitudes) and warm (from lower latitudes) air forms the **polar front**, an air mass battleground that encircles Earth and is focused in these low-pressure areas.

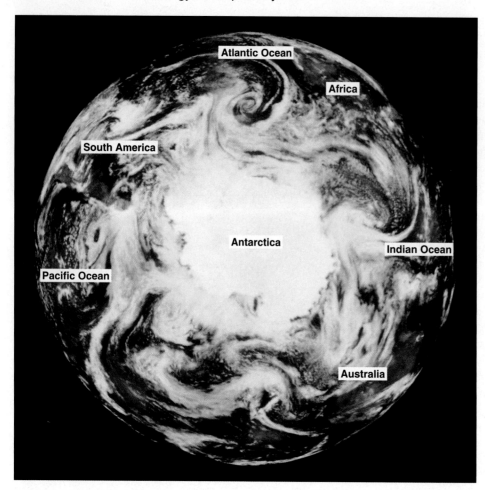

FIGURE 6.15 Clouds portray subpolar and polar circulation patterns.
Centered on Antarctica, this image shows a series of subpolar low-pressure cyclones in the Southern Hemisphere. Antarctica is fully illuminated by a mid-summer Sun as the continent approaches the December solstice. Imaging instrument aboard the *Galileo* spacecraft made this image during its December 1990 flyby of Earth. [The Solid State Imaging instrument (violet, green, and red filters) image courtesy of Dr. W. Reid Thompson, Laboratory of Planetary Studies, Cornell University. Used by permission.]

Figure 6.13 illustrates this confrontation between warm, moist air from the westerlies and cold, dry air from the polar and Arctic regions. The upward displacement of the warm air forces cooling and condensation in the lifted air. Low-pressure cyclonic storms migrate out of the Aleutian and Icelandic frontal areas and may produce precipitation in North America and Europe, respectively. Northwestern sections of North America and Europe generally are cool and moist as a result of the passage of these cyclonic systems onshore—consider the weather in British Columbia, Washington, Oregon, Ireland, and the United Kingdom.

In the Southern Hemisphere, a noncontinuous belt of subpolar low cyclonic pressure systems surrounds Antarctica. The spiraling cloud patterns produced by these cyclonic systems are visible on the spacecraft image in Figure 6.15. Severe cyclonic storms can cross Antarctica, producing strong winds and new snowfall. How many cyclonic systems can you identify on the image?

Polar High-Pressure Cells: Frigid, Dry Deserts   Polar high-pressure cells are weak. The polar atmospheric mass is small, receiving little energy to put it into motion. Variable winds, cold and dry, move away from the polar region in an anticyclonic direction. They descend and diverge

clockwise in the Northern Hemisphere (counterclockwise in the Southern Hemisphere) and form weak, variable winds named **polar easterlies**.

Of the two polar regions, the **Antarctic high** is stronger and more persistent, forming over the Antarctic landmass. Over the Arctic Ocean, a polar high-pressure cell is less pronounced. When it does form, it tends to locate over the colder northern continental areas in winter (Canadian and Siberian highs) rather than directly over the relatively warmer Arctic Ocean.

Upper Atmospheric Circulation

Circulation in the middle and upper troposphere is an important component of the atmosphere's general circulation. Just as sea level is a reference datum for evaluating air pressure at the surface, we use a pressure level such as 500 mb as a **constant isobaric surface** for a pressure-reference datum in the upper atmosphere.

On upper-air pressure maps, we plot the height above sea level at which an air pressure of 500 mb occurs. In contrast, on surface weather maps we plot different pressures at the fixed elevation of sea level—a *constant height surface*. Using the isobaric chart for an April day (see Figure 6.10), Figure 6.16a and b illustrate such an undulating isobaric surface, upon which all points have the same pressure.

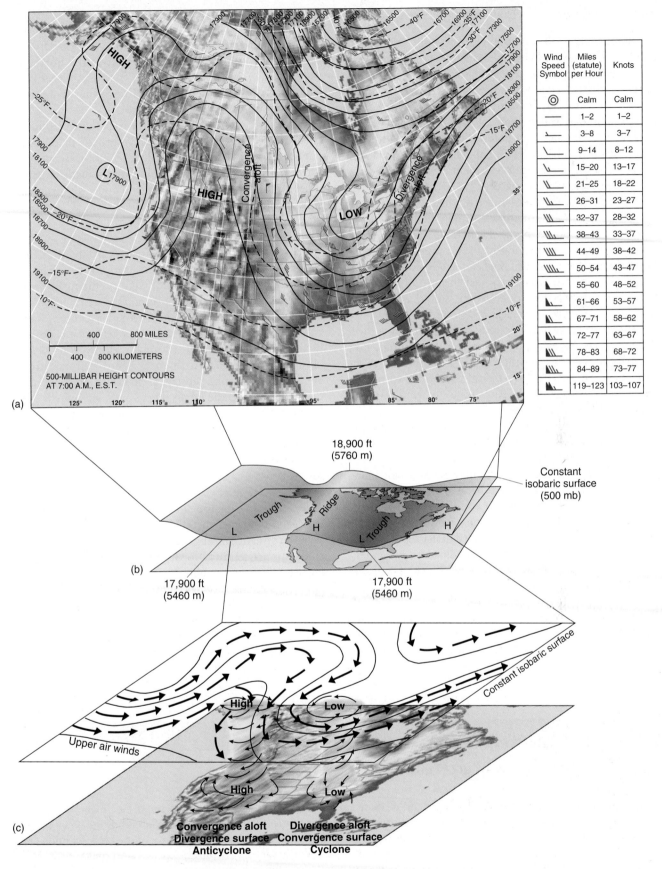

FIGURE 6.16 Analysis of a constant isobaric surface.
Isobaric chart for an April day. (a) Contours show elevation (in feet) at which 500-mb pressure occurs—a constant isobaric surface. The pattern of contours reveals a 500-mb isobaric surface and geostrophic wind patterns in the troposphere ranging from 16,500 to 19,100 ft elevation. (b) Note the "ridge" of high pressure over the Intermountain West, at 5760 m altitude, and the "trough" of low pressure over the Great Lakes region and off the Pacific Coast on the map, also at 5460 m altitude, and in the sketch beneath the chart. (c) Note areas of convergence aloft (corresponding to surface divergence) and divergence aloft (corresponding to surface convergence)—upper atmosphere conditions at the 500-mb level support surface cyclones and anticyclones. [Data for map in (a) from the National Weather Service, NOAA.]

161

We use this 500-mb level to analyze upper-air winds and possible support for surface weather conditions. Similar to surface maps, closer spacing of the isobars indicates faster winds; wider spacing indicates slower winds. On this isobaric pressure surface, altitude variations from the reference datum are called *ridges* for high pressure (with isobars on the map bending poleward) and *troughs* for low pressure (with isobars on the map bending equatorward). Looking at the figure, can you identify such ridges and troughs in the isobaric surface?

The pattern of ridges and troughs in the upper-air wind flow is important in sustaining surface cyclonic (low) and anticylonic (high) circulation. Frequently, the upper-air wind flow generates surface pressure systems. Near ridges in the isobaric surface, winds slow and converge (pile-up), whereas winds near the area of maximum wind speeds along the trough in the isobaric surface accelerate and diverge (spread out). Note the wind-speed indicators and labels in Figure 6.16 near the ridge (over Alberta, Saskatchewan, Montana, and Wyoming); now compare these with the wind-speed indicators around the trough (over Kentucky, West Virginia, the New England states, and the Maritimes). Also, note the wind relationships off the Pacific Coast.

Now look at Figure 6.16c. As wind moves along in geostrophic flow, it is constantly experiencing horizontal convergence (piling up) and divergence (spreading out). This divergence in the upper-air flow is important to cyclonic circulation at the surface because it creates an outflow of air aloft that stimulates an inflow of air into the low-pressure cyclone (such as what happens when you open a chimney damper to create an upward draft). Convergence aloft, on the other hand, drives descending airflows and divergent winds at the surface, moving out from high-pressure anticyclones.

Rossby Waves Within the westerly flow of geostrophic winds are great waving undulations called **Rossby waves**, named for meteorologist Carl G. Rossby who first described them mathematically in 1938. The polar front is the line of conflict between colder air to the north and warmer air to the south (Figure 6.17). Rossby waves bring tongues of cold air southward, with warmer tropical air moving northward.

The development of Rossby waves in the upper-air circulation is shown in the three-part figure. As these disturbances mature, distinct cyclonic circulation forms, with warmer air and colder air mixing along distinct fronts. These wave-and-eddy formations and upper-air divergence support cyclonic storm systems at the surface. Rossby waves develop along the flow axis of a jet stream.

Jet Streams The most prominent movement in these upper-level westerly wind flows is the **jet stream**, an irregular, concentrated band of wind occurring at several different locations that supports surface weather systems. (Figure 6.13a shows the location of four jet streams.) Rather flattened in vertical cross section, the jet streams normally are 160–480 km (100–300 mi) wide by 900–2150 m (3000–7000 ft) thick, with core speeds that can exceed 300 kmph (190 mph). Jet streams in each hemisphere tend to weaken during the hemisphere's summer and strengthen during winter as the streams shift closer to the equator. The pattern of ridges and troughs causes variation in jet stream speeds (convergence and divergence). These upper-level westerly wind systems also affect air transportation, see News Report 6.2.

The *polar jet stream* meanders between 30–70° N latitude, at the tropopause along the polar front, at altitudes between 7600 and 10,700 m (24,900 and 35,100 ft). The polar jet stream can migrate as far south as Texas, steering colder air masses into North America and influencing surface storm paths traveling eastward. In the summer, the polar jet stream exerts less influence on storms by staying far poleward. Figure 6.18 shows a map and a stylized view of a polar jet stream flow.

In subtropical latitudes, near the boundary between tropical and midlatitude air, another jet stream flows near

---

News Report 6.2

Jet Streams Affect Flight Times

As airplane travel increased during World War II, flight crews reported strong headwinds on routes to the west. In some cases, planes leaving San Francisco for the Pacific were turned back by opposing wind in the upper troposphere. These pilots had encountered the jet streams.

Next time you plan a flight, note that airline schedules reflect the presence of these upper-level westerly winds, for the airlines allot shorter flight times from west to east and longer flight times from east to west. For instance a recent round-trip between Sacramento, California, and Atlanta, Georgia, had a difference of 1 hour and 5 minutes between shorter eastbound and longer westbound flights—a difference caused by a particularly strong jet stream.

Also important to both military and civilian aircraft is the effect the jet streams have on fuel consumption and the existence of air turbulence. Seasonal adjustments are necessary because the jet streams in both hemispheres tend to weaken during each hemisphere's summer and strengthen during winter as the streams shift closer to the equator.

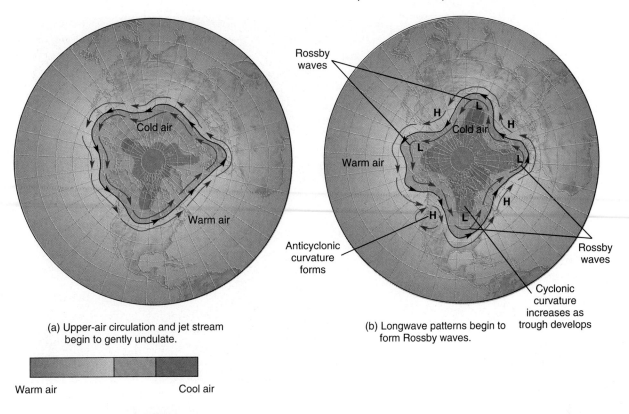

(a) Upper-air circulation and jet stream begin to gently undulate.

(b) Longwave patterns begin to form Rossby waves.

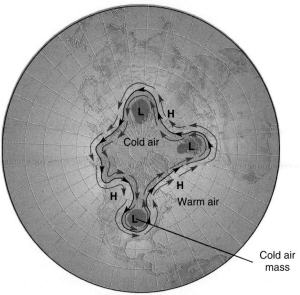

(c) Strong development of waves produces cells of cold and warm air—high-pressure ridges and low-pressure troughs.

FIGURE 6.17 Rossby upper-atmosphere waves.
Development of waves in the upper-air circulation first described by C. G. Rossby in 1938 and expanded by J. Namias in 1952. Note in (b) our Rossby waves are labeled "L" for low-pressure troughs.

the tropopause, the *subtropical jet stream* (see Figure 6.13). The subtropical jet stream meanders from 20–50° latitude and may occur over North America simultaneously with the polar jet stream—sometimes the two will actually merge for brief episodes. Now, let us move on to some important local winds.

Local Winds

Several winds form in response to local terrain. Passing weather systems through an area can, of course, overwhelm such local effects.

Land–sea breezes occur on most coastlines (Figure 6.19). Different heating characteristics of land and water

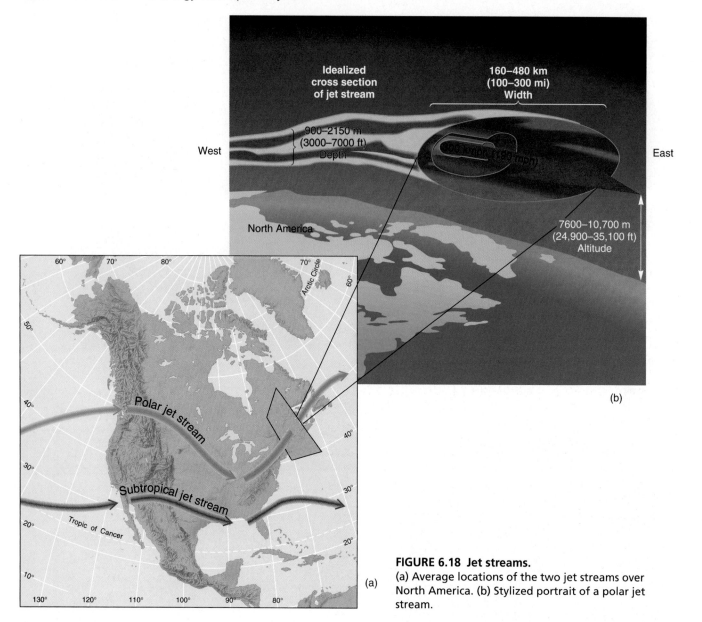

FIGURE 6.18 Jet streams.
(a) Average locations of the two jet streams over North America. (b) Stylized portrait of a polar jet stream.

surfaces create these breezes. During the day, land gains heat energy faster and becomes warmer than the water offshore. Because warm air is less dense, it rises and triggers an onshore flow of cooler marine air to replace the rising warm air—the flow is usually strongest in the afternoon. At night, inland areas cool (radiate heat energy) faster than offshore waters. As a result, the cooler air over the land subsides and flows offshore over the warmer water, where the air is lifted. This night pattern reverses the process that developed during the day.

Mountain–valley breezes result when mountain air cools rapidly at night, whereas valley air gains heat energy rapidly during the day (Figure 6.20). Thus, warm air rises upslope during the day, particularly in the afternoon; at night, cooler air subsides downslope into the valleys.

Katabatic winds, or gravity drainage winds, are of larger regional scale and usually stronger than moun-

tain–valley breezes, under certain conditions. An elevated plateau or highland is essential, where layers of air at the surface cool, become denser, and flow downslope. The ferocious winds that can blow off the ice sheets of Antarctica and Greenland are katabatic in nature.

Worldwide, a variety of terrains produce such winds and bear many local names. The *mistral* of the Rhône Valley in southern France can cause frost damage to vineyards as the cold north winds move over the region to the Gulf of Lions and the Mediterranean Sea. The frequently stronger *bora*, driven by the cold air of winter high-pressure systems inland, flows across the Adriatic Coast to the west and south. In Alaska such winds are called the *taku*.

Santa Ana winds are another local wind type, generated when high pressure builds over the Great Basin of the western United States. A strong, dry wind flows out across the desert to southern California coastal areas.

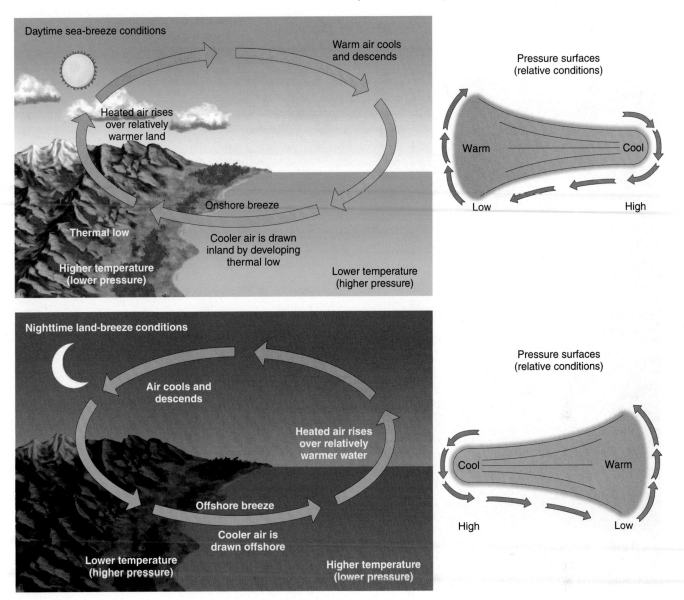

FIGURE 6.19 Land–sea breezes characteristic of day and night.

Compression heats the air as it flows from higher to lower elevations, and with increasing speed it moves through constricting valleys to the southwest. These winds irritate and provoke the population with dust, dryness, and heat.

Regionally, wind represents a significant source of renewable energy of great promise. Focus Study 6.1 briefly explores the potential for development of wind resources.

Monsoonal Winds

Some regional wind systems seasonally change direction. Intense, seasonally shifting wind systems occur in the tropics over Southeast Asia, Indonesia, India, northern Australia, and equatorial Africa. A milder version of such monsoonal-type flows affects the extreme southwestern United States. These winds involve an annual cycle of re-

turning precipitation with the summer Sun. The Arabic word for season, *mausim*, or **monsoon**, names them. (Specific monsoonal weather, associated climate types, and vegetation regions are discussed in Chapters 8, 10, and 20.) The location and size of the Asian landmass and its proximity to the Indian Ocean drive the monsoons of southern and eastern Asia (Figure 6.21). Also important to the generation of monsoonal flows are wind and pressure patterns in the upper-air circulation.

The extreme temperature range from summer to winter over the Asian landmass is due to its continentality (isolation from the modifying effects of the ocean). An intense high-pressure anticyclone dominates this continental landmass in winter (see Figures 6.11a and 6.21a), whereas the equatorial low-pressure trough (ITCZ) dominates the central area of the Indian Ocean. This pressure

Daytime valley-breeze conditions

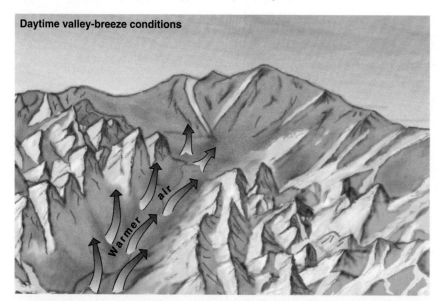

Nighttime mountain-breeze conditions

FIGURE 6.20 Pattern of mountain–valley breezes during day and night.

gradient produces cold, dry winds from the Asian interior over the Himalayas, downslope and across India. Average temperatures range between 15–20°C (60–68°F) at lower elevations. These winds desiccate (dry out) the landscape and then give way to hot weather from March through May. The photo in Chapter 1's News Report 1 was made May 20, 1998, 18 m (60 ft) below the summit of Mount Everest—the monsoons dictate climbing schedules.

During the June–September wet period, the subsolar point (direct overhead rays of sunlight) shifts northward to the Tropic of Cancer, near the mouths of the Indus and Ganges Rivers. The ITCZ shifts northward over southern Asia, and the Asian continental interior develops a thermal low pressure, associated with high average temperatures (remember the summer warmth in Verkhoyansk, Siberia, from Chapter 5). Meanwhile, subtropical high pressure dominates the Indian Ocean, with a surface temperature of 30°C (86°F). As a result of this reversed pres-

sure gradient, hot, dry subtropical air sweeps over the warm ocean, producing extremely high evaporation rates (Figure 6.21b).

By the time this air mass reaches India and the convergence zone, it is laden with moisture in thunderous, dark clouds. The warmth of the land lends additional lifting to the incoming air, as do the Himalayas, which force the air mass to higher altitudes. When the monsoonal rains arrive from June to September, they are welcome relief from the dust, heat, and parched land of Asia's springtime. Likewise, the annual monsoon is an integral part of Indian music, poetry, and life.

World-record rainfalls of the wet monsoon drench India. Cherrapunji, India, received both the second-highest average annual rainfall (1143 cm, or 450 in.) and the highest single-year rainfall (2647 cm, or 1042 in.).

The fact that the monsoons of southern Asia involve vast global pressure systems leads us to ask whether future

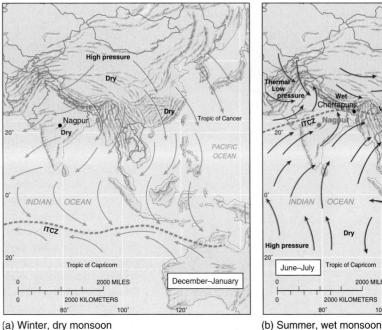

(a) Winter, dry monsoon

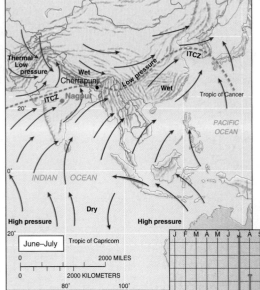

(b) Summer, wet monsoon

FIGURE 6.21 The Asian monsoons.
Asian monsoon pressure and wind patterns during (a) winter and (b) summer.
Note the shifting location of the ITCZ, the changing pressures over the Indian
Ocean, and the different conditions over the Asian landmass. The inset climo-
graph for Nagpur shows the severe contrast in seasonal precipitation.
[Adapted from Joseph E. Van Riper, *Man's Physical World*, p. 215. Copyright
1971 by McGraw-Hill. Adapted by permission.]

(c) Precipitation
at Nagpur,
India

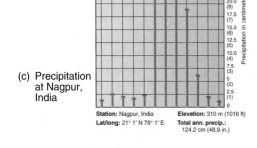

Station: Nagpur, India   Elevation: 310 m (1016 ft)
Lat/long: 21° 1' N 79° 1' E   Total ann. precip.:
124.2 cm (48.9 in.)

Focus Study 6.1

Wind Power: An Energy Resource for the Present and Future

The principles of wind power are ancient, but the technology is modern
and the benefits are worth pursuing.
In more-developed countries, energy
sources are dominated by the use of
nonrenewable fuels—coal, gas, oil,
nuclear—and centralized energy production. In less-developed countries,
however, many rely on renewable
energy—small hydroelectric plants,
wind-power systems, diminishing wood
supplies, solar energy—for cooking,
heating, and pumping water. These resources are considered renewable because they are not depleted in the span
of a human lifetime. Relative to wind
applications, rough estimates place the
global wind resource at 500% more
than present global energy use.

Wind Power Assessment

Wind-generated energy resources are
the fastest-growing energy technology,
in terms of new capacity, outside the
United States—there was a 1000% increase in capacity between 1990 and
2000, reaching 18,100 MW by the end
of 2000 (up from 7600 MW at the end
of 1997). Installations worldwide increased at almost a 30% per-year pace
through the late 1990s, as Germany,
Denmark, Spain, and India surge ahead
in annual additions to capacity (Figure
1). Germany now produces 2.5% of its
electricity from wind power. During
2000, Germany installed 40% of new
wind capacity (1670 MW out of global installations of 4200 MW), 17% in
Spain (713 MW), 14% in Denmark

(600 MW), followed by Italy (145
MW) and India (90 MW), among
other countries. The European Wind
Energy Association announced an installed capacity goal of 60,000 MW by
2010, more than three times the current worldwide total.

The United States installed only
0.01% of world installations in 2000
(53 MW) to bring its total to 2554
MW installed. The U.S. market tends
to swing widely in response to political decisions concerning tax credits
and incentives. This despite the fact
that the U.S. wind-energy potential
exceeds present electricity consumption by 300%. Scheduled for 2001 was
a record installation of 2000 MW,
(continued)

Focus Study 6.1 (continued)

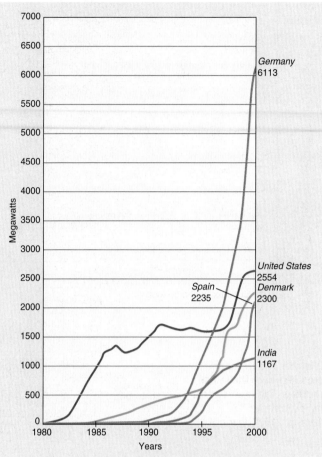

FIGURE 1 Installed wind-generating capacity.
In 1980 only 10 MW of wind-generated capacity was in operation in the world. By 1997 it rose to 7630 MW and by 2000 surged to 18,100 MW worldwide. Whereas new capacity is growing at record rates in Germany, Denmark, Spain, and India, note stagnation in the United States. [Sources: Lester Brown, et al., *Vital Signs 2001*, Worldwatch Institute, New York, NY, © 2001, as derived from the American Wind Energy Association, "Global Wind Energy Market Report," May 2001, and previous reports.]

although an extension of tax credits into 2002 failed. Texas, California, and the Pacific Northwest are home to four new wind farms, of 200 or more MW each.

To put these numbers in meaningful perspective, every 10,000 MW of wind-generation capacity reduces carbon dioxide emissions by 33 million metric tons if it replaces coal, or 21 million metric tons if it replaces mixed fossil fuels. As an example, if countries rally and create a proposed $400-billion-dollar industry by installing 500,000 MW of worldwide wind capacity by 2020, as much as a fourth

of carbon dioxide emissions would be avoided. For the United States, 225,000 wind turbines would reduce coal emissions of carbon dioxide by 59%. In the 1997 Kyoto Protocol the United States proposed reducing greenhouse gas emissions by just 7% below 1990 levels.

Economics Update
Natural gas prices in California reached 15 to 20 cents per kilowatt-hour (kWh) in the early 2000 price spike (electricity on the spot [short-term, daily] market hit 33 cents/kWh!), whereas wind-power contracts came in

at 3 cents/kWh in the same year. In Europe, smaller turbines produced at 4 cents per kWh. Two Stanford University engineers calculated that the contracted price of 3 cents/kWh in the United States is cheaper than coal (estimated at 5.5 to 8.3 cents per kWh), when considering all costs. Lower industry estimates for coal costs exclude consideration of coal-mine dust health impacts, acid deposition, smog, and global warming—significant when you consider that the black-lung disease program to date has cost $70 billion alone. (See M. Z. Jacobsen and G. M. Masters, "Exploiting Wind vs. Coal," *Science* 293 (August 24, 2001): 1438; also see American Wind Energy Association, "Global Wind Energy Market Report," May 2001.)

The Nature of Wind Energy
Power generation from wind is site-specific, because certain regions have adequate winds. Electricity is generated at wind farms, or groups of wind machines. Figure 2 shows such farms in California and Germany. Three basic settings favor wind resources: (1) along coastlines influenced by trade winds and westerly winds; (2) where mountain passes constrict airflow and interior valleys develop thermal low-pressure areas, thus drawing air across the landscape; or (3) where localized winds occur, such as katabatic or monsoonal flows. Cash-short developing countries are generally located in areas blessed by such steady winds. Where wind is reliable less than 25%–30% of the time, only small-scale uses are economically feasible.

The winds of North and South Dakota and Texas could meet all the U.S. electrical needs. In the California Coast Ranges, because of land-sea breezes between the Pacific and Central Valley, peak winds blow from April to October, matching peak electrical demands for residential, commercial, and agricultural water pumping.

One study done in Wyoming found that land presently selling for

(continued)

Focus Study 6.1 *(continued)*

$100 per hectare (2.47 acres) would yield about $25,000 worth of wind-generated electricity annually. The land for a wind farm is still available for multiple use such as livestock ranching. Farmers in Iowa and Minnesota receive $2000 in annual income from a leased turbine and $20,000 a year from an owned turbine that takes one-quarter acre to site. The corn harvested annually from that same site is worth about $100. Farmers are exploring this income source with assistance from the American Corn Growers Association (ACGA) and are forming wind-energy cooperatives. The limitation seems to be a shortage of transmission capacity from rural sites to cities. (See the ACGA report at http://www.acga.org/programs/2001WindCCS/.)

The Benefits of the Wind Resource

Economic reality should override further delaying actions, especially where peak winds are in concert with peak electrical demand for air cooling, space heating, or agricultural water pumping. Long-term and marginal costs favor wind-energy deployment. With *all* costs considered, wind energy is cost-competitive and actually cheaper than oil, coal, and nuclear power. The two Stanford engineers, in a study cited previously in this focus study, stated,

> Much of the recent energy debate in the United States has focused on increasing coal use. However, the cost of wind energy is now less than coal. Shifting from coal to wind would address health, environmental, and energy problems. (p. 1438)

(a)

(b)

FIGURE 2 Wind farms in California and Germany.
(a) Wind farm in the San Gorgonio Pass of southern California. (b) Wind farm near Harle, East Friesland, Germany. [Photos by (a) Bobbé Christopherson; (b) © Uwe Walz/CORBIS.]

Needless to say, whether or not governments and transnational energy corporations support full-scale planning and implementation of renewable resources, the energy realities of the near future will leave no alternative. By the middle of this century, wind-generated electricity will be routine, along with other renewable energy sources, conservation, and energy efficiency. Germany and Denmark are leading the way for now.

climate change might affect present monsoonal patterns. Researchers are conducting a field experiment in the Indian Ocean, the Indian Ocean Experiment (INDOEX) to examine this question (http://www-indoex.ucsd.edu/ProjDescription.html). Their model projections forecast higher rainfall if consideration is given to increases in carbon dioxide and its greenhouse warming alone. However, when consideration is given to increases in aerosols—principally sulfur compounds and black carbon—then a drop in precipitation of 7%–14% appears in the model. Air pollution reduces surface heating and therefore decreases the pressure differences at the heart of monsoonal flows. Considering that 70% of the annual precipitation for the entire region comes during the wet monsoon, such changes will force difficult societal adjustments to changing water resources. (See the satellite images in Figure 4.9 and review the text discussion.)

Oceanic Currents

The driving force for ocean currents is the frictional drag of the winds, thus linking the atmospheric and oceanic systems. Also important in shaping these currents is the interplay of the Coriolis force, density differences caused by temperature and salinity, the configuration of the continents and ocean floor, and astronomical forces (the tides).

Surface Currents

Figure 6.22 portrays the general patterns of major ocean currents. Because ocean currents flow over distance and through time, the Coriolis force deflects them. However, their pattern of deflection is not as tightly circular as that of the atmosphere. Compare this ocean-current map with the map showing Earth's pressure systems (see Figure 6.11) and you can see that ocean currents are driven by the circulation around subtropical high-pressure cells in both hemispheres. These circulation systems are known as *gyres* and generally appear to be offset toward the western side of each ocean basin. (Remember, in the Northern Hemisphere, winds and ocean currents move *clockwise* about high-pressure cells—note the currents in the North Pacific and North Atlantic on the map. In the Southern Hemisphere, circulation is *counterclockwise* around high-pressure cells, evident on the map.)

In sailing days the Spanish galleons (the *Manila Galleons*) would leave San Blas and Acapulco, Mexico (16.5° N), and sail southwest, catching the northeast trade winds across the Pacific to the Philippines and Manila (14° N). Goods would

be traded and new commodities obtained, and the ships would move northward to catch the westerlies in the mid-latitudes to be blown across the ocean to the shores of present-day Alaska or British Columbia down to Northern California. The galleons would sail south, fighting the rain, frequent fog, and the right-hand Coriolis deflection pushing them away from the coast. The journey would end back in Mexico with cargo from the Orient. Thus the history of the *Manilla Galleons* is a lesson in oceanic currents around the Pacific gyre. News Report 6.3 dramatically portrays this clockwise circulation around the Pacific Ocean.

Equatorial Currents   In Figure 6.22, you can see that trade winds drive the ocean surface waters westward in a concentrated channel along the equator. These currents, called *equatorial currents*, are kept near the equator by the Coriolis force, which diminishes to zero at the equator. As these surface currents approach the western margins of the oceans, the water actually piles up against the eastern shores of the continents. The average height of this pileup is 15 cm (6 in.). This phenomenon is the **western intensification**.

The piled-up ocean water then goes where it can, spilling northward and southward in strong currents, flowing in tight channels along the eastern shorelines. In the Northern Hemisphere, the *Gulf Stream* and the *Kuroshio* (current east of Japan) move forcefully northward as a result of western intensification. Their speed and depth are increased by the constriction of the area they occupy. The warm, deep, clear water of the ribbonlike Gulf Stream (Figure 5.10) usually is 50–80 km (30–50 mi) wide and 1.5–2.0 km (0.9–1.2 mi) deep, moving at 3–10 kmph

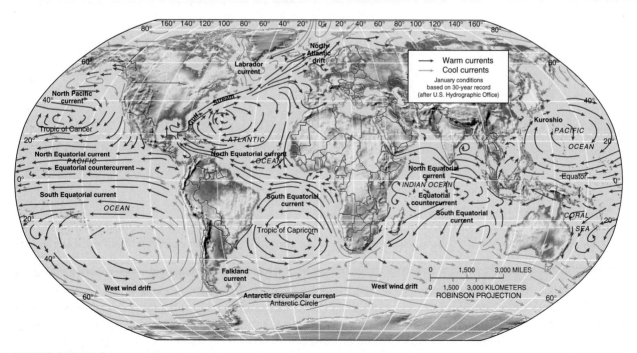

FIGURE 6.22 Major ocean currents.
[After the U. S. Navy Hydrographic Office.]

A Message in a Bottle and Rubber Duckies

To give you an idea of the dynamic circulation of the ocean, consider the following examples. A 9-year-old child at Dana Point, California (33.5° N), a small seaside community south of Los Angeles, placed a letter in a glass juice bottle in July 1992 and tossed it into the waves. Thoughts of distant lands and fabled characters filled the child's imagination as the bottle disappeared. The vast circulation around the Pacific high, clockwise-circulating gyre, now took command (Figure 1).

Three years passed before ocean currents carried the message in a bottle to the coral reefs and white sands of Mogmog, a small island in Micronesia (7° N). A 7-year-old child there had found a pen pal from afar and immediately sent a photo and card to the

sender of the message. Imagine the journey of that note from California—traveling through storms and calms, clear moonlit nights and typhoons, as it floated on ancient currents as the galleons once had.

In January 1994, a large container ship from Hong Kong loaded with toys and other goods was ravaged by a powerful storm. One of the containers on board split apart in the wind off the coast of Japan, dumping nearly 30,000 rubber ducks, turtles, and frogs into the North Pacific. Westerly winds and the North Pacific current swept this floating cargo across the ocean to the coast of Alaska, Canada, Oregon, and California. Other toys, still adrift, went through the Bering Sea and into the Arctic Ocean (this route is indicated by

the dashed line in Figure 1). These will eventually end up in the Atlantic Ocean as they drift around the Arctic Ocean frozen into pack ice.

The high-floating message bottle and rubber ducks offered a much better opportunity to study winds than did an earlier spill of 60,000 athletic shoes near Japan. The low-floating shoes tracked across the Pacific Ocean until they landed in the Pacific Northwest. By the way, the shoes that did not make landfall headed (or footed) back around the Pacific gyre into the tropics and westward, back toward Japan! Scientists are using incidents such as these to learn more about wind systems and ocean currents.

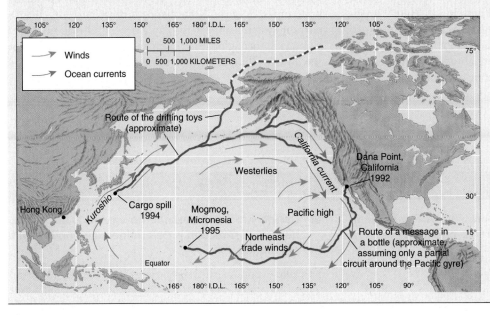

FIGURE 1 Pacific Ocean currents transport human artifacts. The approximate route of a message in a bottle from Dana Point, California, to Mogmog, Micronesia, assumes a partial circuit around the Pacific gyre. Given the 3-year travel time, we do not know if the message circumnavigated the Pacific Ocean more than once. The rubber duckies voyaged eastward across the Pacific and beyond.

(1.8–6.2 mph). In 24 hours, ocean water can move 70–240 km (40–150 mi) in the Gulf Stream.

Deep Currents

Where surface water is swept away from a coast, either by surface divergence (induced by the Coriolis force) or by offshore winds, an **upwelling current** occurs. This cool water generally is nutrient-rich and rises from great depths to replace the vacating water. Such cold upwelling currents exist off the Pacific coasts of North and South America and

the subtropical and midlatitude west coast of Africa. These areas are some of Earth's prime fishing regions.

In other regions where there is an accumulation of water—such as the western end of an equatorial current, or the Labrador Sea, or along the margins of Antarctica—the excess water gravitates downward in a **downwelling current**. These currents flow along the ocean floor and travel the full extent of the ocean basins, carrying heat energy and salinity.

To picture such a deep current, imagine a continuous channel of water beginning with cold water downwelling in

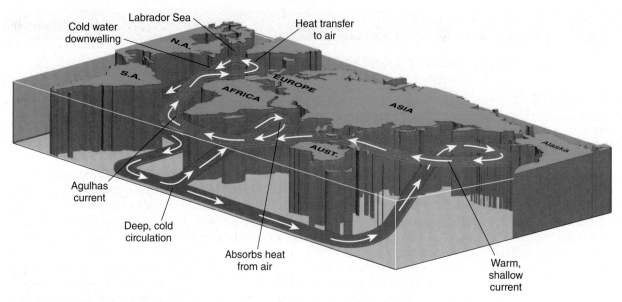

FIGURE 6.23 Deep-ocean circulation.
Scientists are deciphering centuries-long deep circulation in the oceans. This global circulation mimics a vast conveyor belt of water, drawing heat energy from some regions and transporting it for release in others.

the North Atlantic, flowing deep, and upwelling in the Indian Ocean and North Pacific (Figure 6.23). Here it warms and then is carried in surface currents back to the North Atlantic. A complete circuit of the current may require 1000 years from downwelling in the Labrador Sea off Greenland to its reemergence in the southern Indian Ocean and

return. Even deeper Antarctic bottom water flows northward in the Atlantic Basin beneath these currents. These current systems appear to play a profound role in global climate; in turn, global warming has the potential to disrupt the downwelling in the North Atlantic.

Summary and Review—Atmospheric and Oceanic Circulations

● *Define* the concept of air pressure and *describe* instruments used to measure air pressure.

The weight (created by motion, size, and number of molecules) of the atmosphere is *air pressure*, which exerts an average force of approximately 1 kg/cm^2 (14.7 lb/in.^{2}). A **mercury barometer** measures air pressure at the surface (mercury in a tube—closed at one end and open at the other, with the open end placed in a vessel of mercury—that changes level in response to pressure changes) or an **aneroid barometer** (a closed cell, partially evacuated of air, that detects changes in pressure).

mercury barometer (p. 147)
aneroid barometer (p. 147)

1. How does air exert pressure? Describe the basic instrument used to measure air pressure. Compare the operation of two different types of instruments discussed.

2. What is normal sea-level pressure in millimeters? Millibars? Inches? Kilopascals?

● *Define* wind and *describe* how wind is measured, how wind direction is determined, and how winds are named.

Volcanic eruptions such as those of Tambora in 1815 and Mount Pinatubo in 1991 dramatically demonstrate the power of global winds to disperse aerosols and pollution worldwide in a matter of weeks. Atmospheric circulation facilitates important transfers of energy and mass on Earth, thus maintaining Earth's natural balances. Earth's atmospheric and oceanic circulations represent a vast heat engine powered by the Sun.

Wind is the horizontal movement of air across Earth's surface. Its speed is measured with an **anemometer** (a device with cups that are pushed by the wind) and its direction with a **wind vane** (a flat blade or surface that is directed by the wind). A descriptive scale useful in visually estimating wind speed is the traditional *Beaufort wind scale*.

wind (p. 147)
anemometer (p. 148)
wind vane (p. 148)

3. What is a possible explanation for the beautiful sunrises and sunsets during the summer of 1992 in North America? Relate your answer to global circulation.

4. Explain this statement: "The atmosphere socializes humanity, making all the world a spatially linked society." Illustrate your answer with some examples.

5. Define wind. How is it measured? How is its direction determined?

6. Distinguish among primary, secondary, and tertiary general classifications of global atmospheric circulation.

7. What is the purpose of the Beaufort wind scale? Characterize winds given Beaufort numbers of 4, 8, and 12, giving effects over both water and land.

● *Explain* the four driving forces within the atmosphere—gravity, pressure gradient force, Coriolis force, and friction force—and *describe* the primary high- and low-pressure areas and principal winds.

Earth's gravitational force on the atmosphere operates uniformly worldwide. Winds are driven by the **pressure gradient force** (air moves from areas of high pressure to areas of low pressure), deflected by the **Coriolis force** (an apparent deflection in the path of winds or ocean currents caused by the rotation of Earth; deflecting objects to the right in the Northern Hemisphere and to the left in the Southern Hemisphere), and dragged by the **friction force** (Earth's varied surfaces exert a drag on wind movements in opposition to the pressure gradient). Maps portray air pressure patterns using the **isobar**—an isoline that connects points of equal pressure. A combination of the pressure gradient and Coriolis forces alone produces **geostrophic winds**, which move parallel to isobars, characteristic of winds above the surface frictional layer.

Winds descend and diverge, spiraling outward to form an **anticyclone** (clockwise in the Northern Hemisphere), and they converge and ascend, spiraling upward to form a **cyclone** (counterclockwise in the Northern Hemisphere). The pattern of high and low pressures on Earth in generalized belts in each hemisphere produces the distribution of specific wind systems. These primary pressure regions are the **equatorial low-pressure trough**, the weak **polar high-pressure cells** (at both the North and South Poles), and the **subtropical high-pressure cells** and **subpolar low-pressure cells**.

All along the equator winds converge into the equatorial low creating the **intertropical convergence zone (ITCZ)**. Air rises along the equator and descends in the subtropics in each hemisphere. The winds returning to the ITCZ from the northeast in the Northern Hemisphere and from the southeast in the Southern Hemisphere produce the **trade winds**.

Winds flowing out of the subtropics to higher latitudes produce the **westerlies** in either hemisphere. The subtropical high-pressure cells on Earth, generally between 20° and 35° in either hemisphere, are variously named the **Bermuda high**, **Azores high**, and **Pacific high**.

Along the polar front and the series of low-pressure cells, the **Aleutian low** and **Icelandic low** dominate the North Pacific and Atlantic, respectively. This region of contrast between colder air toward the poles and warmer air equatorward

is called the **polar front**. The weak and variable **polar easterlies** diverge from the polar high-pressure cells, particularly the **Antarctic high**.

> pressure gradient force (p. 150)
> Coriolis force (p. 150)
> friction force (p. 151)
> isobar (p. 151)
> geostrophic winds (p. 155)
> anticyclone (p. 155)
> cyclone (p. 155)
> equatorial low-pressure trough (p. 155)
> polar high-pressure cells (p. 155)
> subtropical high-pressure cells (p. 155)
> subpolar low-pressure cells (p. 155)
> intertropical convergence zone (ITCZ) (p. 157)
> trade winds (p. 157)
> westerlies (p. 159)
> Bermuda high (p. 159)
> Azores high (p. 159)
> Pacific high (p. 159)
> Aleutian low (p. 159)
> Icelandic low (p. 159)
> polar front (p. 159)
> polar easterlies (p. 160)
> Antarctic high (p. 160)

8. What does an isobaric map of surface air pressure portray? Contrast pressures over North America for January and July.

9. Describe the effect of the Coriolis force. Explain how it apparently deflects atmospheric and oceanic circulations.

10. What are geostrophic winds, and where are they encountered in the atmosphere?

11. Describe the horizontal and vertical air motions in a high-pressure anticyclone and in a low-pressure cyclone.

12. Construct a simple diagram of Earth's general circulation, including the four principal pressure belts or zones and the three principal wind systems.

13. How is the intertropical convergence zone (ITCZ) related to the equatorial low-pressure trough? How might it appear on a satellite image?

14. Characterize the belt of subtropical high pressure on Earth: Name the specific cells. Describe the generation of westerlies and trade winds. Discuss sailing conditions.

15. What is the relation among the Aleutian low, the Icelandic low, and migratory low-pressure cyclonic storms in North America? In Europe?

● *Describe* upper-air circulation and its support role for surface systems and *define* the jet streams.

Air pressure in the middle and upper troposphere is described using a **constant isobaric surface**, or surface along which the same pressure, such as 500 mb, is recorded regardless of altitude. The height of this surface above the ground is described in ridges and troughs that support the development of and help sustain surface pressure systems; lows are sustained by divergence aloft, and highs are sustained by convergence aloft.

Vast, flowing longwave undulations in these upper-air westerlies form wave motions called **Rossby waves**. Prominent streams of high-speed westerly winds in the upper-level troposphere are called the **jet streams**. Depending on their latitudinal position in either hemisphere, they are termed the *polar jet stream* or the *subtropical jet stream*.

constant isobaric surface (p. 160)
Rossby waves (p. 162)
jet streams (p. 162)

16. What is the relation between wind speed and the spacing of isobars?

17. How is the constant isobaric surface (ridges and troughs) related to surface pressure systems? To divergence aloft and surface lows? To convergence aloft and surface highs?

18. Relate the jet-stream phenomenon to general upper-air circulation. How is the presence of this circulation related to airline schedules from New York to San Francisco and the return trip to New York?

● *Explain* **several types of local winds: land–sea breezes, mountain–valley breezes, katabatic winds, and the regional monsoons.**

Different heating characteristics of land and water surfaces create **land–sea breezes. Mountain–valley breezes** are caused by temperature differences during the day and evening between valleys and mountain summits. **Katabatic winds**, or gravity drainage winds, are of larger regional scale and are usually stronger than mountain–valley breezes, under certain conditions. An elevated plateau or highland is essential, where layers of air at the surface cool, become denser, and flow downslope.

Intense, seasonally shifting wind systems occur in the tropics over Southeast Asia, Indonesia, India, northern Australia, equatorial Africa, and southern Arizona. These winds involve an annual cycle of returning precipitation with the summer Sun and are named after the Arabic word for season, *mausim*, or **monsoon**. The monsoons of southern and eastern Asia are driven by the location and size of the Asian landmass and its proximity to the Indian Ocean.

land–sea breezes (p. 163)
mountain–valley breezes (p. 164)
katabatic winds (p. 164)
monsoon (p. 165)

19. People living along coastlines generally experience variations in winds from day to night. Explain the factors that produce these changing wind patterns.

20. The arrangement of mountains and nearby valleys produces local wind patterns. Explain the day and night winds that might develop.

21. Describe the seasonal pressure patterns that produce the Asian monsoonal wind and precipitation patterns. Contrast January and July conditions.

● *Discern* **the basic pattern of Earth's major surface and deep ocean currents.**

Ocean currents are primarily caused by the frictional drag of wind and occur worldwide at varying intensities, temperatures, and speeds, both along the surface and at great depths in the oceanic basins. The circulation around subtropical high-pressure cells in both hemispheres is notable on the ocean circulation map—these *gyres* are usually offset toward the western side of each ocean basin.

The trade winds converge along the ITCZ and push enormous quantities of water in a process known as the **western intensification**. Where surface water is swept away from a coast, either by surface divergence (induced by the Coriolis force) or by offshore winds, an **upwelling current** occurs. This cool water generally is nutrient-rich and rises from great depths to replace the vacating water. In other portions of the sea where there is an accumulation of water, the excess water gravitates downward in a **downwelling current**. These currents generate important mixing currents that flow along the ocean floor and travel the full extent of the ocean basins, carrying heat energy and salinity.

western intensification (p. 170)
upwelling current (p. 171)
downwelling current (p. 171)

22. What is the relationship between global atmospheric circulation and ocean currents? Relate oceanic gyres to patterns of subtropical high pressure.

23. Define the western intensification. How is it related to the Gulf Stream and Kuroshio currents?

24. Where on Earth are upwelling currents experienced? What is the nature of these currents?

25. What is meant by deep-ocean circulation? At what rates do these currents flow? How might this circulation be related to the Gulf Stream in the western Atlantic Ocean?

NetWork

The *Geosystems* Home Page provides on-line resources for this chapter on the World Wide Web. To begin: Once on the Home Page, click on this textbook, scroll the Table of Contents menu, select this chapter, and click "Begin." You will find self-tests that are graded, review exercises, specific updates for items in the chapter, and many links to interesting related pathways on the Internet under "Destinations." *Geosystems* is at **http://www.prenhall.com/christopherson**.

Critical Thinking

A. Using Table 6.1 and Figure 6.5, on a day with wind, estimate wind speed and wind direction at least twice during the day and record them in your notebook. If possible, check these against the reports from some local source of weather information. What changes do you notice over several days? How did these changes relate to the weather you experienced?

B. Refer to the "Short Answer" section in Chapter 6 of the *Geosystems* Home Page. Questions 1 through 5 deal with satellite images of pressure systems, wind, and cloud patterns. Using information in this chapter, complete these five items. Can you identify any causes for these circulation patterns? Note the other nine short-answer items for further thought.

C. Go to **http://www.awea.org/** and the American Wind Energy Association. Sample the materials presented as you assess Focus Study 6.1 and the potential for wind-generated electricity. What are your thoughts concerning this resource—its potential, reasons for delays, and the competitive economics presented? What countries are leading the way? Propose a brief action plan for more rapid progress, or for more delays, depending on your point of view.

PART TWO

The Water, Weather, and Climate Systems

E arth is the water planet. Surface water in such quantity is unique in the solar system. Chapter 7 explains why water covers more than two-thirds of the globe and describes the remarkable qualities and properties it possesses. It also discusses the daily dynamics of the atmosphere—the powerful interaction of moisture and energy and the resulting stability and instability, and the variety of cloud forms—as a preamble to understanding weather. Chapter 8 examines weather and its causes. Topics include the interaction of air masses, understanding the daily weather map, and the violent phenomena of thunderstorms, tornadoes, and hurricanes and the recent trends for each of these.

Chapter 9 explains water circulation on and over Earth in the hydrologic cycle. We examine the water-budget concept, which is useful in understanding soil–moisture and water–resource relationships at all levels—global, regional, and local. In Chapter 10, we see the spatial implications over time of the energy–atmosphere and water–weather systems and the generation of Earth's climatic patterns. In this way Chapter 10 interconnects all the system elements from Chapters 2 through 9. Part 2 closes with a discussion of global climate change and future climate trends.

SOLAR ENERGY

Atmosphere

Hydrosphere

Biosphere

Lithosphere

Evaporation fog (sea smoke) at dawn on a cold morning at Donner Lake, California. Later that morning as temperatures rose, what do you think happened to the evaporation fog? [Photo by Bobbé Christopherson.]

7

Water and Atmospheric Moisture

Key Learning Concepts

After reading the chapter, you should be able to:

- *Describe* the origin of Earth's waters, *define* the quantity of water that exists today, and *list* the locations of Earth's freshwater supply.

- *Describe* the heat properties of water and *identify* the traits of its three phases: solid, liquid, and gas.

- *Define* humidity and the expressions of the relative humidity concept, and *explain* dew-point temperature and saturated conditions in the atmosphere.

- *Define* atmospheric stability and *relate* it to a parcel of air that is ascending or descending.

- *Illustrate* three atmospheric conditions—unstable, conditionally unstable, and stable—with a simple graph that relates the environmental lapse rate to the dry adiabatic rate (DAR) and moist adiabatic rate (MAR).

- *Identify* the requirements for cloud formation and *explain* the major cloud classes and types, including fog.

Walden is blue at one time and green at another, even from the same point of view. Lying between the earth and the heavens, it partakes of the color of both. . . . A lake is the landscape's most beautiful and expressive feature. It is earth's eye; looking into which the beholder measures the depth of his own nature. . . . Sky water. It needs no fence. Nations come and go without defiling it. It is a mirror which no stone can crack. . . . Nature continually repairs . . . a mirror in which all impurity presented to it sinks, swept and dusted by the sun's hazy brush. A field of water . . . is continually receiving new life and motion from above. It is intermediate in its nature between land and sky.*

Thus did Thoreau speak of the water so dear to him—Walden Pond in Massachusetts, where he lived along the shore.

Water is critical to our daily lives and is an extraordinary compound in nature. Several deep-space satellites detected ice beneath the poles of the Moon, erosional features and subsurface waters on Mars imaged by NASA's *Mars Global Surveyor* and *Odyssey* spacecrafts suggest evidence of water long thought lost at the Martian surface, and icy expanses can be seen on two of Jupiter's moons, Europa and Calisto—exciting new discoveries about water in our Solar System. Yet, in the Solar System water occurs in such significant quantities only on our planet. It covers 71% of Earth.

Pure water is colorless, odorless, and tasteless; yet, because it is a solvent (dissolves solids), pure water rarely occurs in nature. Water weighs 1 g/cm³ (gram per cubic centimeter), or 1 kg/L (kilogram per liter). In the English system water weighs 62.3 lb/ft³, or 8.337 lb/gal.

Water constitutes nearly 70% of our bodies by weight and is the major ingredient in plants, animals, and our food. A human can survive 50 to 60 days without food but only 2 or 3 days without water. The water we use must be adequate, both in quantity and quality, for its many tasks—everything from personal hygiene to vast national water projects. Water indeed occupies that place between land and sky, mediating energy and shaping both the lithosphere and atmosphere, as Thoreau revealed. Water is the medium of life.

In this chapter: We examine water on Earth and the dynamics of atmospheric moisture and stability—the essentials of weather. The key questions answered include: What were its origins? How much is there? And, where is it located? An irony exists with this most common of compounds that possesses the most uncommon physical characteristics. Water's unique heat properties and existence in all three states in nature are critical in powering Earth's weather systems. Condensation of water vapor and atmospheric conditions of stability and instability are key to cloud formation. We end with clouds, our beautiful indicators of the atmosphere's status.

*Reprinted by permission of Merrill, an imprint of Macmillan Publishing Company, from *Walden* by Henry David Thoreau, pp. 192, 202, 204. Copyright © 1969 by Merrill Publishing. Originally published 1854.

Water on Earth

Earth's hydrosphere contains about 1.36 billion cubic kilometers of water (specifically, 1,359,208,000 km³, or 326,074,000 mi³). According to scientific evidence, much of Earth's water originated from icy comets that formed part of the planetesimals that accreted (coalesced) to form the planet. The water within the planet reached the surface by outgassing. **Outgassing** is a continuing process by which water and water vapor emerge from layers deep within and below the crust, 25 km (15.5 mi) or more below Earth's surface (Figure 7.1).

In the early atmosphere, massive quantities of outgassed water vapor condensed and then fell to Earth in torrential rains. For water to remain on Earth's surface, land temperatures had to drop below the boiling point of 100°C (212°F), something that occurred about 3.8 billion years ago. The lowest places across the face of Earth then began to fill with water—first, ponds, then lakes and seas, and eventually ocean-sized bodies of water. Massive flows of water washed over the landscape, carrying both dissolved and solid materials to these early seas and oceans. Outgassing of water has continued ever since and is visible in volcanic eruptions, geysers, and seepage to the surface.

Worldwide Equilibrium

Today, water is the most common compound on the surface of Earth, having attained the present volume of 1.36 billion cubic kilometers approximately 2 billion years ago. This quantity has remained relatively constant, even though water is continuously being lost from the system. Water is lost when it dissociates into hydrogen and oxygen, and the hydrogen escapes Earth's gravity to space, or, when it breaks down and forms new compounds with other elements. Pristine water not previously at the surface, which emerges from within Earth's crust, replaces lost water in the system. The net result of these water inputs and outputs is that Earth's hydrosphere quantity is in a steady-state equilibrium.

Despite this overall net balance in water quantity, worldwide changes in sea level, called **eustasy** do occur. Eustatic changes relate to the water volume in the oceans and not changes in the overall quantity of planetary water. Some of these changes result when the amount of water stored in glaciers and ice sheets varies; these are called **glacio-eustatic** factors (see Chapter 17). In cooler times, as more water is bound up in glaciers (on mountains worldwide) and in ice sheets (Greenland and Antarctica), sea level lowers. In a warmer era, less water is stored as ice, so sea level rises.

Some 18,000 years ago, during the most recent ice-age pulse, sea level was more than 100 m (330 ft) lower than it is today; 40,000 years ago it was 150 m (about 500 ft) lower. Over the past 100 years, mean sea level has risen by 20–40 cm (8–16 in.) and is still rising worldwide as higher temperatures melt more ice.

FIGURE 7.1 Water outgassing from the crust.
Outgassing of water from Earth's crust in the geothermal
area near Wairakei on the North Island of New Zealand.
[Photo by Bill Bachman/Photo Researchers, Inc.]

Actual physical changes in landmasses, such as continental uplift or subsidence, cause apparent changes in sea
level relative to coastal environments. This change in land
elevation, called *isostasy*, is discussed in Chapter 11.

Distribution of Earth's Water Today

From a geographic point of view, ocean and land surfaces
are distributed unevenly. If you examine a globe, it is obvious that most of Earth's continental land is in the Northern Hemisphere, whereas water dominates the Southern
Hemisphere. In fact, from certain perspectives, Earth appears to have a distinct *oceanic hemisphere* and a distinct *land
hemisphere* (Figure 7.2).

An illustration of the present location of all of Earth's
liquid and frozen water—whether fresh or saline, surface
or underground—is in Figure 7.3. The oceans contain
97.22% of all water, about 1.321 billion cubic kilometers,
or 317 million cubic miles, of saltwater. A table in the figure lists and details the four oceans—Pacific, Atlantic, Indian, and Arctic. The extreme southern portions of the
Pacific, Atlantic, and Indian Oceans that surround the
Antarctic continent are sometimes collectively called the
"Southern Ocean."

Only 2.78% of all of Earth's water is freshwater (non-
saline and nonoceanic). The middle pie chart, along with
Table 7.1, details this freshwater portion—surface water
and subsurface water. Ice sheets and glaciers are the

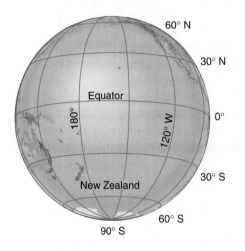

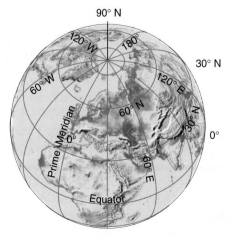

FIGURE 7.2 Land and water hemispheres.
Two perspectives that roughly divide Earth's surface into an
ocean hemisphere and a land hemisphere.

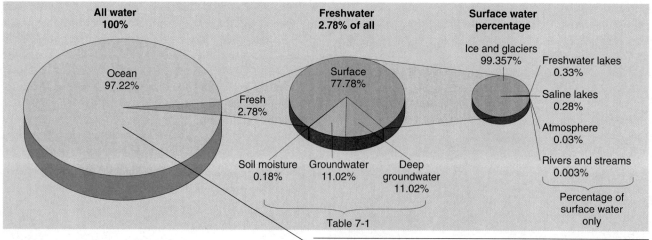

FIGURE 7.3 Ocean and freshwater distribution on Earth.
Pie diagrams show all water, freshwater only, and surface water percentages.

Table 7-1

Ocean	Earth's Ocean Area (%)	*Area (km² [mi²])	*Volume (km³ [mi³])	Mean Depth of Main Basin (m [ft])
Pacific	48	179,670 (69,370)	724,330 (173,700)	4280 (14,040)
Atlantic	28	106,450 (41,100)	355,280 (85,200)	3930 (12,890)
Indian	20	74,930 (28,930)	292,310 (70,100)	3960 (12,900)
Arctic	4	14,090 (5440)	17,100 (4100)	1205 (3950)

*Data in thousands (000): includes all marginal seas.

Table 7.1   Distribution of Freshwater on Earth

Location	Amount (km³ [mi³])		Percentage of Freshwater	Percentage of Total Water
Surface Water				
Ice sheets and glaciers	29,180,000	(7,000,000)	77.14	2.146
Freshwater lakes*	125,000	(30,000)	0.33	0.009
Saline lakes and inland seas	104,000	(25,000)	0.28	0.008
Atmosphere	13,000	(3,100)	0.03	0.001
Rivers and streams	1,250	(300)	0.003	0.0001
Total surface water	29,423,250	(7,058,400)	77.78	2.164
Subsurface Water				
Groundwater—surface to 762 m (2500 ft) depth	4,170,000	(1,000,000)	11.02	0.306
Groundwater—762 to 3962 m (2500 to 13,000 ft) depth	4,170,000	(1,000,000)	11.02	0.306
Soil moisture storage	67,000	(16,000)	0.18	0.005
Total subsurface water	8,407,000	(2,016,000)	22.22	0.617
Total Freshwater (rounded)	**37,800,000**	(9,070,000)	**100.00%**	2.78%

*Major Freshwater Lakes	Volume (km³ [mi³])		Surface Area (km² [mi²])		Depth (m [ft])	
Baykal (Russia)	22,000	(5280)	31,500	(12,160)	1620	(5315)
Tanganyika (Africa)	18,750	(4500)	39,900	(15,405)	1470	(4923)
Superior (U.S./Canada)	12,500	(3000)	83,290	(32,150)	397	(1301)
Michigan (U.S.)	4,920	(1180)	58,030	(22,400)	281	(922)
Huron (U.S./Canada)	3,545	(850)	60,620	(23,400)	229	(751)
Ontario (U.S./Canada)	1,640	(395)	19,570	(7,550)	237	(777)
Erie (U.S./Canada)	485	(115)	25,670	(9,910)	64	(210)

greatest single repository of surface freshwater; they contain 77.14% of all of Earth's freshwater. Adding subsurface groundwater to frozen surface water accounts for 99.36% of all freshwater.

The remaining freshwater, which resides in lakes, rivers, and streams so familiar to us, actually represents less than 1% of all water. All the world's freshwater lakes total only 125,000 km^3 (30,000 mi^3), with 80% of this volume in just 40 of the largest lakes and about 50% contained in just 7 lakes (listed with Table 7.1).

The greatest single volume of lakewater resides in 25-million-year-old Lake Baykal in Siberian Russia. It contains almost as much water as all five U.S. Great Lakes combined. Africa's Lake Tanganyika contains the next largest volume, followed by the five Great Lakes. Overall, 70% of lake water is in North America, Africa, and Asia, with about a fourth of lake water worldwide in innumerable small lakes. More than 3 million lakes exist in Alaska alone, and Canada has over 750 km^2 of lake surface.

Not connected to the ocean are saline lakes and salty inland seas. They usually exist in regions of interior river drainage (no outlet to the ocean), which allows salts to become concentrated. They contain 104,000 km^3 (25,000 mi^3) of water. Examples of such lakes include Utah's Great Salt Lake, California's Mono Lake, Southwest Asia's Caspian and Aral Seas, and the Dead Sea between Israel and Jordan.

Think of all the moisture in the atmosphere and Earth's thousands of flowing rivers and streams. Combined, they amount to only 14,250 km^3 (3400 mi^3), or only 0.033% of freshwater, or 0.0011% of all water! Yet, this small amount is very dynamic. A water molecule traveling through atmospheric and surface-water paths moves through the entire hydrologic cycle (ocean–atmosphere–precipitation–runoff) in less than two weeks. Contrast this to a water molecule in deep-ocean circulation, groundwater, or a glacier; moving slowly, taking thousands of years to migrate through the system. Chapter 9 explores the hydrologic system and water resources in more detail.

Unique Properties of Water

Earth's distance from the Sun places it within a most remarkable temperate zone when compared with the locations of the other planets. This temperate location allows all three states of water—ice, liquid, and vapor—to occur naturally on Earth and to change from one to another.

Even though water is the most common compound on Earth's surface, it exhibits most uncommon properties. Two atoms of hydrogen and one of oxygen, that readily bond, comprise each water molecule (suggested in Figure 7.4, upper left). Once hydrogen and oxygen atoms combine (in a covalent, or double, bond), they are difficult to separate, thereby producing a water molecule that remains stable in Earth's environment. This water molecule is a versatile solvent and possesses extraordinary heat characteristics.

The nature of the hydrogen–oxygen bond gives the hydrogen side of a water molecule a positive charge and

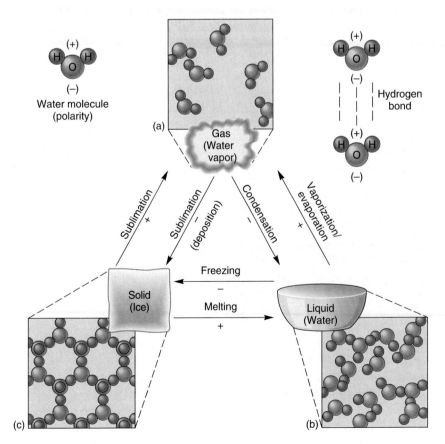

FIGURE 7.4 Three states of water and water's phase changes.
The three physical states of water: (a) gas, or water vapor, (b) water, and (c) ice. Note the molecular arrangement in each state and the terms that describe the changes from one phase to another. Also note how the polarity of water molecules bonds them to one another, loosely in the liquid state and firmly in the solid state. The plus and minus symbols denote whether heat energy is absorbed (+) or liberated (released) (−) during the phase change.

the oxygen side a negative charge (see Figure 7.4, upper right). As a result of this *polarity*, water molecules attract each other: The positive (hydrogen) side of a water molecule attracts the negative (oxygen) side of another. This bonding between water molecules is *hydrogen bonding*.

The polarity of water molecules also explains why water "acts wet" and sticks to things and dissolves so many substances. Because of this solvent ability, pure water is rare in nature for something is usually dissolved in it.

The effects of hydrogen bonding in water are observable in everyday life. Hydrogen bonding creates the *surface tension* that allows you to float a steel needle on the surface of water, even though steel is much denser than water. This *surface tension* allows you to slightly overfill a glass with water; the water is actually above the rim of the glass, held by a web of millions of hydrogen bonds.

Hydrogen bonding is the cause of *capillarity*, which you observe when you "dry" something with a paper towel. The towel draws water through its fibers because hydrogen bonds make each molecule pull on its neighbor. In chemistry laboratory classes, students observe the curved *meniscus*, or surface of the water, which forms in a cylinder or a test tube because hydrogen bonding allows the water to slightly "climb" the glass walls. *Capillary action* is an important component of soil–moisture processes, in Chapters 9 and 18. Without hydrogen bonding to hold molecules together in water and ice, water would be a gas at normal surface temperatures.

Heat Properties

For water to change from one state to another (solid, liquid, or vapor), *heat energy must be absorbed or liberated*. To cause a change of state, the amount of heat energy must be sufficient to affect the hydrogen bonds between molecules. This relation between water and heat energy is important to atmospheric processes. In fact, the heat exchanged between physical states of water provides more than 30% of the energy that powers the general circulation of the atmosphere.

Figure 7.4 presents the three states of water and the terms describing a change from one state to another, a **phase change**. Along the bottom of the illustration, *melting* and *freezing* describe the familiar phase change between solid and liquid. At the right, the terms *condensation* and *evaporation* (or *vaporization* at boiling temperature) apply to the change between liquid and vapor. At the left, the term **sublimation** refers to the direct change of ice to water vapor or water vapor to ice; although the latter is usually referred to as *deposition*, where water vapor attaches directly to an ice crystal. The deposition of water vapor to ice may form *frost* on surfaces.

Ice, the Solid Phase   As water cools, it behaves like most compounds and contracts in volume. However, it reaches its greatest density not as ice, but as water at 4°C (39°F). Below that temperature, water behaves differently from other compounds. It begins to expand as more hydrogen

---

News Report 7.1

Breaking Roads and Pipes and Sinking Ships

Road crews are busy in the summer in many parts of the country repairing winter damage to streets and freeways. A major contributor to this damage is the expansive phase change from water to ice. Rainwater seeps into roadway cracks and then expands as it freezes, thus breaking up the pavement (Figure 1). Perhaps you have noticed that

bridges suffer the greatest damage. The reason is that cold air can circulate beneath a bridge and produces more freeze–thaw cycles on the bridge than in the roadbed on rock and soil.

The expansion of freezing water exerts a tremendous force—enough to crack plumbing or an automobile radiator or engine block. Wrapping water pipes with insulation to avoid damage is a common winter task in many places. People living in very cold climates use antifreeze and engine

heaters to avoid damage to vehicles. Historically, this physical property of water was put to useful work in quarrying rock for building materials. Holes were drilled and filled with water before winter so that when cold weather arrived, the water would freeze and expand, cracking the rock into manageable shapes.

A major hazard to ships in higher latitudes is posed by floating ice. Since ice has 0.91 the density of water, an iceberg sits with approximately 10/11 of its mass below water level. The irregular edges of submarine ice can impact the side of a passing ship. Hitting an iceberg buckled plates and substandard rivets along the side of the RMS *Titanic* on its maiden voyage in 1912, causing its sinking and a shattering of society's faith in technology.

FIGURE 1 The power of freezing water. Damage to pavement is a result of the expansion that occurs as water temperature goes below 4°C (39°F). [Photo by Bobbé Christopherson.]

(a)

(b)

(c)

FIGURE 7.5 The uniqueness of ice forms.
(a) Computer-enhanced photo reveals ice-crystal patterns, which are dictated by the internal structure between water molecules. This open structure also explains the lower density of ice and why it floats in the denser water. Ice crystals demonstrate a unique interaction of chaos (all ice crystals are different) and the determinism of physical principles (all have a six-sided structure). (b) Icebergs (pinnacled and saddle-shaped forms) floating in Disko Bay, Greenland, calved into the ocean from the Jakobshavn glacier. (c) Frost—delicate ice crystals form through the deposit of water vapor directly as ice. [(a) Photo enhancement © Scott Camazine/Photo Researchers, Inc., after W. A. Bentley; (b) photo by George Hunter/Tony Stone Images; (c) photo by author.]

bonds form among the slower-moving molecules, creating the hexagonal (six-sided) structures shown in Figures 7.4c and 7.5a. This six-sided preference applies to ice crystals of all shapes: plates, columns, needles, and dendrites (branching or treelike forms). This expansion of water and ice that begins at 4°C continues to a temperature of −29°C (−20°F)—up to a 9% increase in volume is possible. This expansion is important in the weathering of rocks and in highway and pavement damage. News Report 7.1 discusses some effects of ice. (For more on ice crystals and snowflakes see **http://www.its.caltech.edu/~atomic/snowcrystals/physics/physics.htm** or see **http://www.lpsi.barc.usda.gov/emusnow/**.)

The expansion in volume that accompanies the freezing process results in a decrease in density (the same number of molecules occupy greater space). Specifically, ice has 0.91 times the density of water, so it floats. Without this change in density, much of Earth's freshwater would be bound in masses of ice on the ocean floor. Instead, we have floating icebergs, with approximately 1/11 (9%) of their mass exposed and 10/11 (91%) hidden beneath the ocean's surface (Figure 7.5b).

Chapter 13 discusses the freezing action of ice as an important physical weathering process. Chapter 17 discusses the freeze-and-thaw action of surface and subsurface water affecting approximately 30% of Earth's surface in periglacial landscapes, producing a variety of processes and landforms.

Water, the Liquid Phase   As a liquid, water assumes the shape of its container and is a noncompressible fluid. For ice to change to water, heat energy must increase the motion of the water molecules to break some of the hydrogen bonds (Figure 7.4b). Despite the fact that there is no change in sensible temperature between ice at 0°C (32°F) and water at 0°C, 80 calories of heat energy must be absorbed for the phase change of 1 g of ice to melt to 1 g of water (Figure 7.6, upper left). Heat energy involved in the phase change is **latent heat** and is hidden within the structure of water. It becomes liberated whenever the phase reverses and a gram of water freezes. The *latent heat of freezing* or the *latent heat of melting* involves 80 calories.

To raise the temperature of 1 g of water at 0°C (32°F) to boiling at 100°C (212°F), we must add 100 calories, gaining an increase of 1 C° (1.8 F°) for each calorie added.

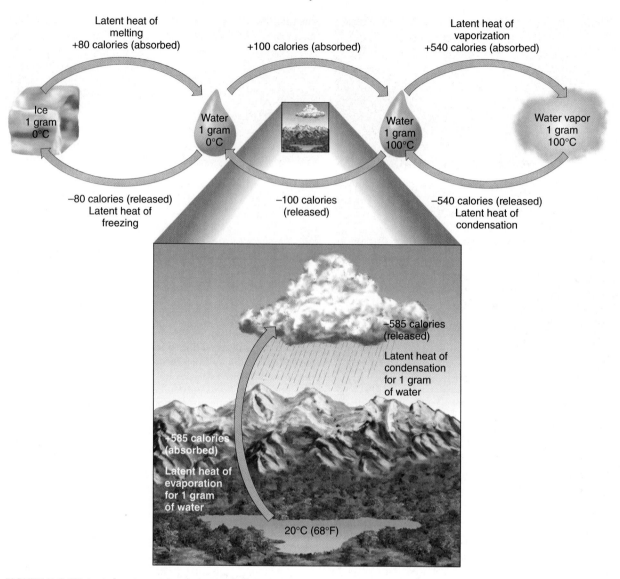

FIGURE 7.6 Water's heat-energy characteristics.
The phase changes of water absorb or release a lot of latent heat energy. To transform 1 g of ice at 0°C to 1 g of water vapor at 100°C requires 720 calories (80 + 100 + 540). The landscape illustrates phase changes between water (lake at 20°C) and water vapor under typical conditions in the environment.

Water Vapor, the Gas Phase   Water vapor is an invisible and compressible gas in which each molecule moves independently of the others (Figure 7.4a). The phase change from liquid to vapor at boiling temperature, under normal sea-level pressure, requires the addition of a much greater amount of heat energy than does the phase change from solid to liquid: each gram of water changed to water vapor requires 540 calories (Figure 7.6). Those calories are the **latent heat of vaporization**. When water vapor condenses to a liquid, each gram gives up its hidden 540 calories as the **latent heat of condensation**. Perhaps you have felt the liberation of the latent heat of condensation on your skin from steam, as when you drain steamed vegetables or pasta or fill a hot teakettle.

In summary, taking 1 g of ice at 0°C and changing its phase to water, then to water vapor at 100°C—from a solid, to a liquid, to a gas—*absorbs* 720 calories (80 cal + 100 cal + 540 cal). Reversing the process, or changing phase of 1 g of water vapor at 100°C to water, then to ice at 0°C, *liberates* 720 calories into the surrounding environment.

Heat Properties of Water in Nature

In a lake or stream or in soil water, at 20°C (68°F), every gram of water that breaks away from the surface through evaporation must absorb from the environment approximately 585 calories as the *latent heat of evaporation* (see the natural scene in Figure 7.6). This is slightly more energy than would be required if the water were at a higher temperature such as boiling (540 cal). You can feel this absorption of latent heat as evaporative cooling on your skin when it is wet. This latent heat exchange is the dominant cooling process in Earth's energy budget.

The process reverses when air cools and water vapor condenses back into the liquid state, forming moisture droplets and thus liberating 585 calories for every gram of water as the *latent heat of condensation*. When you realize that a small, puffy, fair-weather cumulus cloud holds 500–1000 tons of moisture droplets, think of the tremendous latent heat released when water vapor condensed to droplets! Government meteorologists estimated that the moisture in Hurricane Andrew (1992) weighed nearly 30 trillion metric tons at its maximum power and mass. With 585 calories released for every gram as the latent heat of condensation, you can see that a weather event such as a hurricane involves a staggering amount of energy.

The **latent heat of sublimation** absorbs 680 calories as a gram of ice transforms into vapor. Water vapor freezing directly to ice releases a comparable amount of energy.

Humidity

Humidity refers to water vapor in the air. The capacity of air to hold water vapor is primarily a function of temperature—the temperatures of both the air and the water vapor, which are usually the same. Warmer air has a greater capacity for holding water vapor than does cooler air.

We are all aware of humidity in the air, for its relationship to air temperature determines our sense of comfort. North Americans spend billions of dollars a year to adjust humidity, either with air conditioning (extracting water vapor and cooling) or with air humidifying (adding water vapor). We discussed the relation between humidity and temperature and the heat index in Chapter 5. To determine the energy available for powering weather, one has to know the water-vapor *content* of air and relate that to the air's *capacity* to hold water vapor at a given temperature.

Relative Humidity

After air temperature and barometric pressure, the most common piece of information in local weather broadcasts is relative humidity. **Relative humidity** is a ratio (expressed as a percentage) of the amount of water vapor that is *actually* in the air (content) compared to the maximum water vapor the air *could hold* at a given temperature (capacity). *It is water-vapor content compared with water-vapor capacity.*

If air is relatively dry compared to its capacity, the relative humidity percentage is low; if the air is relatively moist, the relative humidity percentage is higher; and if the air is saturated with all the moisture it can hold for its temperature, the percentage is 100% (Figure 7.7). The formula to calculate relative humidity and express it as a percentage is:

$$\text{Relative humidity} = \frac{\text{Actual water vapor } content \text{ of the air}}{\substack{\text{Maximum water vapor } capacity \\ \text{of the air at that temperature}}} \times 100$$

Relative humidity varies because of evaporation, condensation, or temperature changes. All three affect both the moisture content (the numerator) and the capacity (the denominator) of the air to hold water vapor. Relative humidity is an expression of an ongoing process between air and moist surfaces, for condensation and evaporation operate continuously—water molecules move back and forth between the air and bodies of water or ice.

To see how temperature affects relative humidity, think of warm air and cool air as sponges: Warm air is a large sponge that can hold a lot of water vapor; cool air is a small sponge that can hold little water vapor. A cool

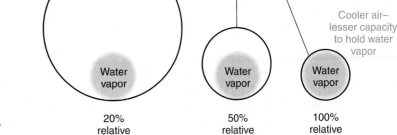

FIGURE 7.7 Water vapor content, capacity, and relative humidity.
The water vapor capacity of warm air is greater than the capacity of cold air, so relative humidity changes with temperature, even though in this example the actual water vapor content of the air stays the same during the day.

Warmer air– greater capacity to hold water vapor

Water-vapor capacity

Cooler air– lesser capacity to hold water vapor

Water vapor

Water vapor

Water vapor

20% relative humidity

50% relative humidity

100% relative humidity

5 P.M.

11 A.M.

5 A.M.

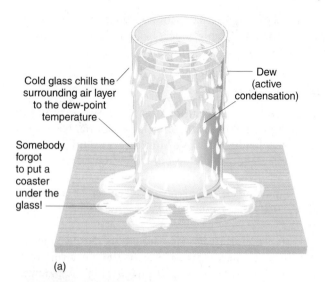

Cold glass chills the surrounding air layer to the dew-point temperature

Dew (active condensation)

Somebody forgot to put a coaster under the glass!

(a)

FIGURE 7.8 Dew-point temperature examples.
(a) The low temperature of the glass chills the surrounding air layer to the dew-point temperature and saturation. Thus, water vapor condenses out of the air and onto the glass as dew. (b) Cold air above the rain-soaked rocks is at the dew point and is saturated. Condensation of water shrouds the rock in a changing veil of clouds. (c) The cold ocean surface chills the moist air layer to the dew point and saturation. As water vapor condenses, a dense fog forms. In the evening when temperatures drop over coastal lands, fog forms there too, giving the appearance of moving inland. [Photos by author.]

(b)

(c)

body of air can be saturated—filled to capacity—by an amount of water vapor that will only partially fill the capacity of warm air.

Saturation   Air is **saturated**, or filled to capacity, when it contains all the water vapor that it can hold at a given temperature—that is 100% relative humidity. In saturated air, the net transfer of water molecules between a moist surface and air reaches an equilibrium. Saturation indicates that any further addition of water vapor (increase in content) or any decrease in temperature (reduction in capacity) results in active condensation (clouds, fog, or precipitation). Therefore, relative humidity indicates both the nearness of air to a saturated condition and when active condensation will begin.

The temperature at which a given mass of air becomes saturated is termed the **dew-point temperature**. In other words, *air is saturated when the dew-point temperature and the air temperature are the same.* A cold drink in a glass provides a common example of these conditions. The water droplets that form on the outside of the glass condense from the air

because the air layer next to the glass is chilled to below its dew-point temperature and thus becomes saturated (Figure 7.8). Figure 7.8 shows two additional examples of saturated air and active condensation above a rock surface and in a fog formation overlying a cool ocean surface.

Satellites now routinely sense water vapor content of the lower atmosphere. Water vapor absorbs infrared wavelengths, making it possible to distinguish areas of relatively high water-vapor content from areas of low water-vapor content, using infrared sensors. Figure 7.9 includes images of Hurricane Michelle and the Western Hemisphere showing water-vapor content recorded by sensors of the "water-vapor channel." This knowledge is important to forecasting because it shows the moisture available to weather systems and therefore the energy available (latent heat) and precipitation potential of those systems.

Daily and Seasonal Relative Humidity Patterns   An inverse relation occurs during a typical day between air temperature and relative humidity—as temperature rises, relative humidity falls (Figure 7.10a). Relative humidity is

FIGURE 7.9 Images of water vapor in the atmosphere.
Water-vapor content of the atmosphere as disclosed in a *GOES-8* infrared image. (a) On the color scale used for this image, the lighter-gray tones denote higher water-vapor content, and color denotes high-altitude cloud tops that are cooler. Over the Gulf of Mexico and Caribbean, Hurricane Michelle and its strong vertical development is clearly visible heading toward western Cuba and near Florida, November 4, 2001. (b) Water-vapor content over the full Western Hemisphere; note subpolar low-pressure circulation. [*GOES* images courtesy of NESDIS Satellite Services Division NOAA.]

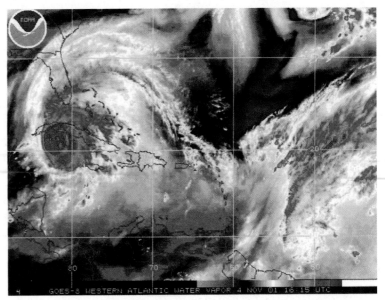

(a)

(b)

highest at dawn, when air temperature is lowest and the capacity of the air to hold water vapor is less. If you park outdoors, you know about the wetness of the dew that condenses on the car overnight. Relative humidity is lowest in the late afternoon, when higher air temperatures increase the capacity of the air to hold water vapor. The actual water-vapor content in the air may remain the same throughout the day, but because the temperature varies, relative humidity changes from morning to afternoon, as shown in Figure 7.7.

Weather records for Sacramento, California, show the seasonal variation in relative humidity by time of day, confirming the relationship of temperature and relative humidity (Figure 7.10b). January readings are higher than

July readings because air temperatures are lower overall in winter. Similar relative humidity records at most weather stations demonstrate the same relation among season, temperature, and relative humidity. In your own experience, you probably have noticed this pattern—the morning dew on windows, cars, and lawns evaporates by late morning as the capacity of the air to hold water vapor increases with air temperature.

Expressions of Relative Humidity

There are several ways to express humidity and relative humidity. Each has its own utility and application. Two measures involve vapor pressure and specific humidity.

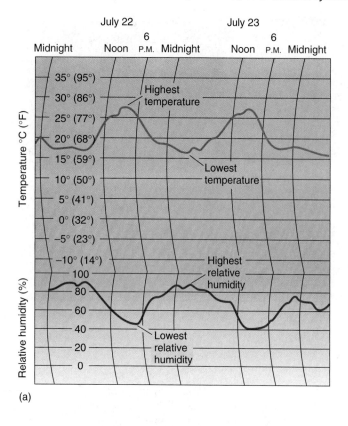

(a)

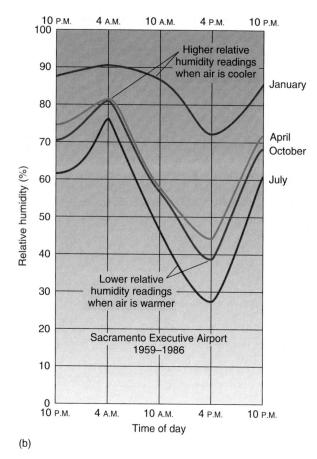

(b)

FIGURE 7.10 Daily and seasonal relative humidity patterns.
Relative humidity in Sacramento, California: (a) typical daily variations in temperature and
relative humidity; (b) seasonal variations in daily relative humidity.

Vapor Pressure As free water molecules evaporate from surfaces into the atmosphere, they become water vapor. Now part of the air, water vapor molecules become a part of the air pressure measurement for they exert a portion of the air pressure. The share of air pressure that is made up of water vapor molecules is **vapor pressure**. Millibars (mb) express vapor pressure, the same as air pressure.

Water-vapor molecules continue to evaporate from a moist surface, slowly diffusing into the air, until the increasing vapor pressure in the air causes some molecules to return to the surface. As explained earlier, *saturation* is reached when the movement of water molecules between surface and air is in equilibrium. The maximum capacity of the air at a given temperature is the *saturation vapor pressure* and indicates the maximum pressure that water vapor molecules can exert. Any temperature increase or decrease will change the saturation vapor pressure.

Figure 7.11 graphs the saturation vapor pressure at various air temperatures. The graph illustrates that, for every temperature increase of 10 C° (18 F°), the vapor pressure capacity of air nearly doubles. This relation explains why warm tropical air over the ocean can hold so much water vapor, thus providing great latent heat to power tropical

storms. It also explains why cold air is "dry" and why cold air toward the poles does not produce a lot of precipitation (it holds too little water vapor).

As the graph shows, air at 20°C (68°F) has a saturation vapor pressure of 24 mb; that is, the air is saturated if the water-vapor portion of the air pressure is at 24 mb. Thus, if the water-vapor content actually present is exerting a vapor pressure of only 12 mb in 20°C air, the relative humidity is 50% (12 mb ÷ 24 mb = 0.50 × 100 = 50%). The inset in Figure 7.11 compares saturation vapor pressure over water and over ice surfaces at subfreezing temperatures. You can see that saturation vapor pressure is greater above a water surface than over an ice surface—that is, it takes more water-vapor molecules to saturate air above water than it does above ice. This fact is important to condensation processes and rain-droplet formation, both of which appear later in this chapter.

Specific Humidity A useful humidity measure is one that remains constant as temperature and pressure change. **Specific humidity** is the mass of water vapor (in grams) per mass of air (in kilograms) at any specified temperature. Because it is measured in mass, specific humidity is not af-

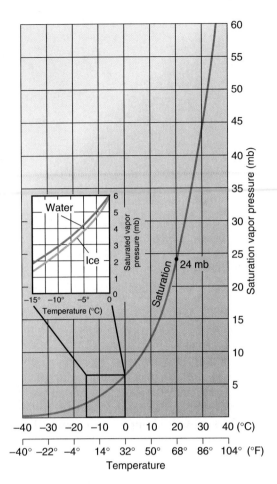

FIGURE 7.11 Saturation vapor pressure.
Saturation vapor pressure of air at various temperatures—
the maximum capacity of air to hold water vapor, expressed
in vapor pressure. Inset compares saturation vapor pressures
over water surfaces with those over surfaces at subfreezing
temperatures. Note the 24 mb label for the text discussion.

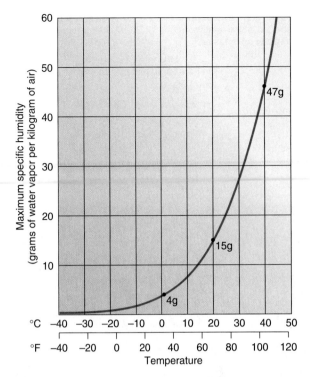

FIGURE 7.12 Maximum specific humidity.
Maximum specific humidity for a mass of air at various tem-
peratures—the maximum capacity of air to hold water vapor.
Note the 47 g, 15 g, and 4 g labels for the text discussion.

fected by changes in temperature or pressure, such as when
an air parcel rises to higher elevations. Specific humidity
stays constant despite volume changes.

The maximum mass of water vapor that a kilogram of air
can hold at any specified temperature is termed the *maximum
specific humidity*, plotted in Figure 7.12. The graph shows
that a kilogram of air could hold a maximum specific humidity of 47 g of water vapor at 40°C (104°F), 15 g at 20°C
(68°F), and about 4 g at 0°C (32°F). Therefore, if a kilogram
of air at 40°C has a specific humidity of 12 g, its relative humidity is 25.5% (12 g ÷ 47 g = 0.255 × 100 = 25.5%).
Specific humidity is useful in describing the moisture
content of large air masses that are interacting in a weather system, and it is necessary information for weather
forecasting.

Instruments for Measuring Humidity   Various instruments measure relative humidity. The **hair hygrometer**
uses the principle that human hair changes as much as 4%
in length between 0 and 100% relative humidity. The in-

strument connects a standardized bundle of human hair
through a mechanism to a gauge. As the hair absorbs or
loses water in the air, it changes length, indicating relative
humidity (Figure 7.13a).

Another instrument used to measure relative humidity is a **sling psychrometer**. Figure 7.13b shows this device, which has two thermometers mounted side by side on
a metal holder. One is the *dry-bulb thermometer*; it simply
records the ambient (surrounding) air temperature. The
other thermometer is the *wet-bulb thermometer*; it is set
lower in the holder and a moistened cloth wick covers its
bulb. The psychrometer is then spun ("slung") by its handle or placed where a fan forces air over the wet bulb.

The rate at which water evaporates from the wick depends on the relative saturation of the surrounding air. If
the air is dry, water evaporates quickly, *absorbing the latent
heat of evaporation from the wet-bulb thermometer and its wick*,
cooling the thermometer and causing its temperature to
drop (the *wet-bulb depression*). In conditions of high humidity, little water evaporates from the wick; in low humidity, more water evaporates. After being spun a minute
or two, the temperature on each bulb is compared on a relative humidity (psychrometric) chart, from which relative
humidity can be determined.

Now that you know something about atmospheric
moisture, dew point, and relative humidity, let us examine
the concept of stability in the atmosphere.

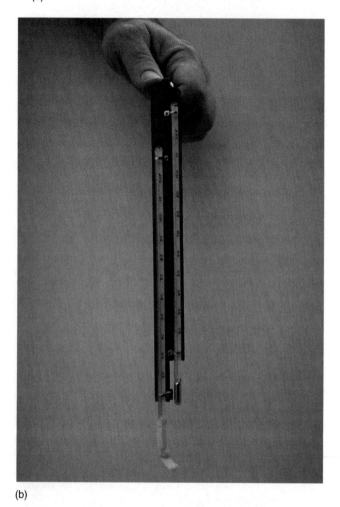

(a)

FIGURE 7.13 Instruments that measure relative humidity. (a) The principle of a hair hygrometer. (b) Sling psychrometer with wet and dry bulbs. [Photo by Bobbé Christopherson.]

(b)

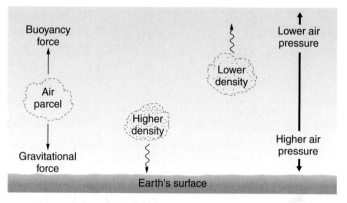

FIGURE 7.14 The forces acting on an air parcel. Buoyancy and gravitational forces work on an air parcel. Different densities produce rising or falling parcels in response to imbalance in these forces.

duces a lower density in a given volume of air; cold air produces a higher density. Two opposing forces work on a parcel of air: an upward *buoyant force* and a downward *gravitational force*. A parcel of lower density rises (is more buoyant); a rising parcel expands as external pressure decreases. A parcel of higher density descends (is less buoyant); a falling parcel compresses as external pressure increases. Figure 7.14 shows air parcels and illustrates these relationships. (In Chapter 8, we discuss air masses, which are larger parcels and regional in extent.)

An indication of weather conditions is the relative stability of air parcels and air masses in the atmosphere. **Stability** refers to the tendency of an air parcel, with its water-vapor cargo, either to remain in place or to change vertical position by ascending (rising) or descending (falling). An air parcel is *stable* if it resists displacement upward or, when disturbed, it tends to return to its starting place. An air parcel is *unstable* if it continues to rise until it reaches an altitude where the surrounding air has a density (air temperature) similar to its own.

To visualize this, imagine a hot-air balloon launch. The air-filled balloon sits on the ground with the same air temperature inside as in the surrounding environment, like a stable air parcel. As the burner ignites, the balloon fills with hot (less-dense) air and rises buoyantly (Figure 7.15), like an unstable parcel of air. The concepts of stable and unstable air allow us to examine the specific temperature characteristics that produce these conditions.

Adiabatic Processes

Determining the degree of stability or instability requires measuring two temperatures: the temperature inside an air parcel and the temperature in the air surrounding the parcel. The contrast of these two temperatures determines stability. Such temperature measurements are made daily with instrument packages called *radiosondes* carried aloft by helium-filled balloons at thousands of weather stations.

Atmospheric Stability

Meteorologists use the term *parcel* to describe a body of air that has specific temperature and humidity characteristics. Think of an air parcel as a volume of air, perhaps 300 m (1000 ft) in diameter, or more. Differences in temperature create changes in density within the parcel. Warm air pro-

FIGURE 7.15 Principles of air stability and balloon launches.
Hot-air balloons being launched in the Swiss Alps illustrate the principles of stability. As the temperature inside a balloon increases, the air in the balloon becomes less dense than the surrounding air and the buoyancy force causes the balloon to rise, acting like a warm air parcel. [Photo by Bap Vandystadt/Photo Researchers, Inc.]

The *normal lapse rate*, introduced in Chapter 3, is the average decrease in temperature with increasing altitude, a value of 6.4 C°/1000 m (3.5 F°/1000 ft). This rate of temperature change is for still, calm air, and it can vary greatly under different weather conditions. Consequently, the *environmental lapse rate* is the actual lapse rate at a particular place and time. It can vary by several degrees per thousand meters.

An ascending (rising) parcel of air cools by expansion, an expansion resulting from the reduced air pressure surrounding the parcel at higher altitudes (Figure 7.16a). A descending (falling) parcel heats by compression (Figure 7.16b).

Physical laws that govern the behavior of gases explain these temperature changes internal to a moving air parcel.

When pressure on a parcel decreases as it rises to higher altitudes, the parcel expands and its temperature and density decrease (the expansion process consumes sensible heat). Conversely, as an air parcel sinks toward the surface, air pressure increases, causing the parcel to compress. Compression increases the temperature and density in the parcel (compression produces sensible heat).

Temperature changes in both ascending and descending air occur *without any significant heat exchange between the surrounding environment and the vertically moving parcel of air.* The warming and cooling rates for a parcel of expanding or compressing air are termed **adiabatic**. (*Diabatic* means occurring with an exchange of heat; *adiabatic* means occurring *without a loss or gain* of heat energy to or from the environment.) There are two adiabatic rates, depending on moisture conditions in the vertically moving air parcel: a dry adiabatic rate (DAR) and a moist adiabatic rate (MAR).

Dry Adiabatic Rate   The **dry adiabatic rate (DAR)** is the rate at which "dry" air cools by expansion (if ascending) or heats by compression (if descending). "Dry" refers to air that is less than saturated (relative humidity is less than 100%). The average DAR is 10 C°/1000 m (5.5 F°/1000 ft).

The rising parcel at the left in Figure 7.17a illustrates the principle. To see how a specific example of dry air behaves, consider an unsaturated parcel of air at the surface with a temperature of 27°C (81°F). It rises, expands, and cools adiabatically at the DAR as it rises to 2500 m (approximately 8000 ft). What happens to the temperature of the parcel? Calculate the temperature change in the parcel, using the dry adiabatic rate:

(10 C°/1000 m) × 2500 m = 25 C° of total cooling
(5.5 F°/1000 ft) × 8000 ft = 44 F° of total cooling

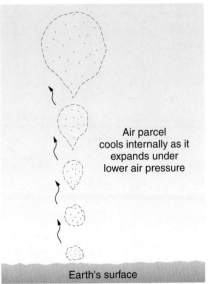

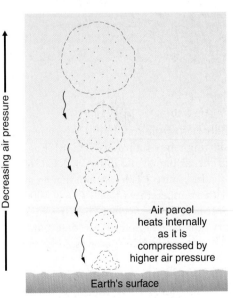

FIGURE 7.16 Vertically moving air experiences temperature changes.
(a) A rising air parcel cools by expansion. (b) A falling air parcel heats by compression.

Air parcel cools internally as it expands under lower air pressure

Decreasing air pressure

Air parcel heats internally as it is compressed by higher air pressure

Earth's surface

(a) Cooling by expansion

Earth's surface

(b) Heating by compression

(a) Air parcel *cools* adiabatically at the DAR

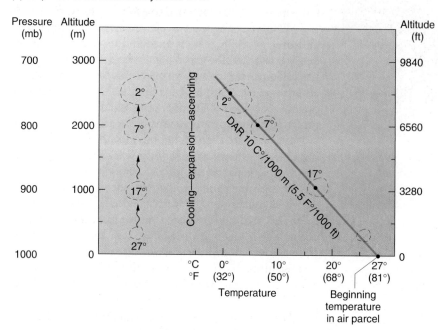

FIGURE 7.17 Adiabatic cooling and heating.
Vertically moving air parcels expand when they rise (because air pressure is less with increasing altitude) and are compressed when they descend. (a) An air parcel that is less than saturated cools adiabatically at the dry adiabatic rate (DAR). (b) A descending air parcel that is less than saturated heats adiabatically by compression at the DAR.

(b) Air parcel *heats* adiabatically at the DAR

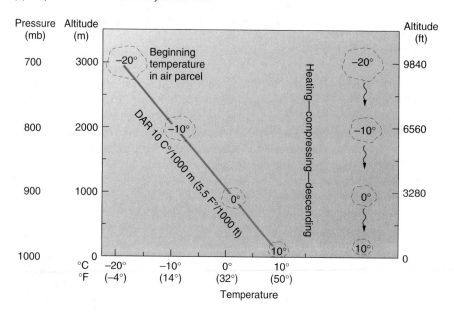

Subtracting the 25 C° (45 F°) of adiabatic cooling from the starting temperature of 27°C (81°F) gives the temperature in the air parcel at 2500 m of 2°C (36°F).

In Figure 7.17b, assume that an unsaturated air parcel with a temperature of −20°C is at 3000 m (−4°F at 9800 ft) descends to the surface, heating adiabatically. Using the dry adiabatic lapse rate, we determine the temperature of the air parcel when it arrives at the surface:

(10 C°/1000 m) × 3000 m = 30 C° of total warming
(5.5 F°/1000 ft) × 9800 ft = 54 F° of total warming

Adding the 30 C° of adiabatic warming from the starting temperature of −20°C gives the temperature in the air parcel at the surface of 10°C (50°F).

Moist Adiabatic Rate   The **moist adiabatic rate (MAR)** is the rate at which an ascending air parcel that is moist (saturated) cools by expansion or that a descending parcel warms by compression. The average MAR is 6 C°/1000 m (3.3 F°/1000 ft). This is roughly 4 C° (2 F°) less than the dry adiabatic rate. From this average, the MAR varies with moisture content and temperature and can range from 4 C° to 10 C° per 1000 m (2 F° to 5.5 F° per 1000 ft).

The cause of this variability, and the reason that the rate is lower than the DAR, is the latent heat of condensation. As water vapor condenses in the saturated air, sensible heat is liberated and the adiabatic rate of cooling lowers. The release of latent heat may vary with temperature and water vapor content. The MAR is much lower than the DAR in warm air, whereas the two rates are more similar in cold air.

Stable and Unstable Atmospheric Conditions

Now we bring this discussion together to determine atmospheric stability. The relation among the dry adiabatic rate (DAR), moist adiabatic rate (MAR), and the environmental (actual) lapse rate at a given time and place determines the stability of the atmosphere over an area. You see examples of possible stability relationships in Figure 7.18. Let's examine what produces this variation in atmospheric stability.

Temperature relationships in the atmosphere produce the three different conditions: unstable, conditionally unstable, and stable. For the sake of illustration, the three examples in Figure 7.19 begin with an air parcel at the surface at 25°C (77°F). In each example, compare the temperatures of the air parcel and the surrounding environment. Assume that there is a lifting mechanism present to get the parcel started (we examine lifting mechanisms in Chapter 8).

Given unstable conditions in Figure 7.19a, the air parcel continues to rise through the atmosphere because it is warmer (and, therefore, less dense and more buoyant) than

the surrounding environment. Note that the environmental lapse rate on this occasion is at 12 C°/1000 m (6.6 F°/1000 ft). That is, the air surrounding the air parcel is cooler by 12 C° for every 1000-m increase in altitude. By 1000 m (about 3300 ft), the lifting air parcel adiabatically cooled by expansion at the DAR from 25° to 15°C, while the surrounding air cooled from 25°C at the surface to 13°C. By comparing the temperature in the air parcel and the surrounding environment, you see that the temperature in the parcel is 2 C° (3.6 F°) warmer than the surrounding air at 1000 m. *Unstable* describes this condition because the less dense air parcel will continue to lift.

Eventually, as the air parcel continues lifting and cooling, it may achieve the dew-point temperature, saturation, and active condensation. This point of saturation forms the *lifting condensation level* that you see in the sky as the flat bottoms of clouds.

On the other hand, if the environmental lapse rate is at only 5 C°/1000 m (3 F°/1000 ft) on another day, stable conditions result, as shown in Figure 7.19c. An environmental lapse rate of 5 C°/1000 m is less than both the DAR and the MAR. This environmental lapse rate sets a condition in which the air parcel has a lower temperature (higher density, less buoyant) than in the surrounding environment. The relatively cooler air parcel settles back to its original position—it is *stable*. The denser air parcel resists lifting and the sky remains generally cloud-free. In regions experiencing air pollution, stable conditions in the atmosphere worsen the pollution by slowing exchanges in the surface air.

You may be wondering what stability condition exists if the environmental lapse rate is somewhere between the DAR and MAR and conditions, therefore, are neither unstable nor stable. In Figure 7.19b, the environmental lapse

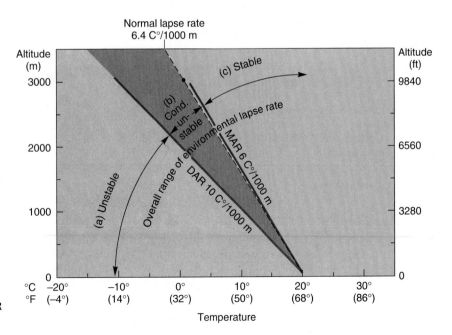

FIGURE 7.18 Temperature relationships and atmospheric stability.
The relationship between dry and moist adiabatic rates and environmental lapse rates produces three atmospheric conditions: (a) unstable (environmental lapse rate exceeds the DAR), (b) conditionally unstable (environmental lapse rate is between the DAR and MAR), and (c) stable (environmental lapse rate is less than DAR and MAR).

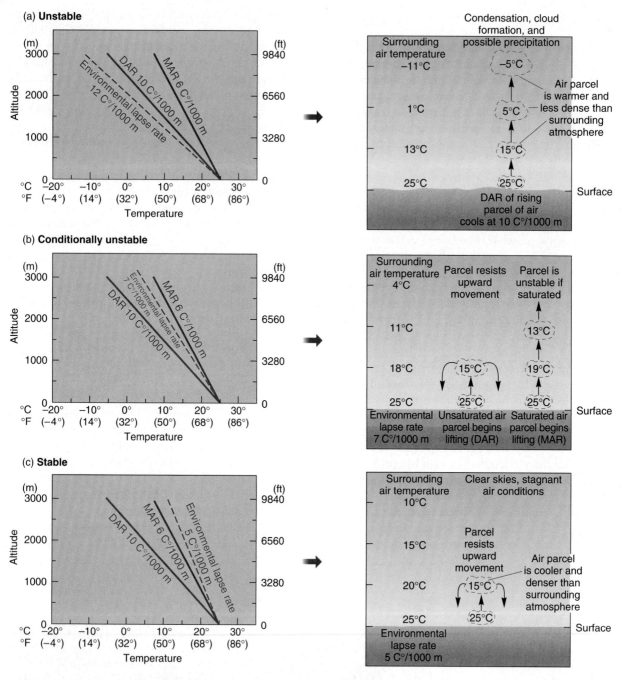

FIGURE 7.19 Stability—three examples.
Specific examples of (a) unstable, (b) conditionally unstable, and (c) stable conditions in the lower atmosphere. Note the response to these three conditions in the air parcel on the right side of each diagram.

rate is measured at 7 C°/1000 m. Under these conditions, the air parcel resists upward movement if it is less than saturated. But, if the air parcel becomes saturated and cools at the MAR, it acts unstable and continues to rise.

One example of such conditionally unstable air occurs when stable air lifts as it passes over a mountain range. As the air parcel lifts and cools to the dew point, the air be-

comes saturated and condensation begins. Now the MAR is in effect and the air parcel behaves in an unstable manner. The sky may be clear and without a cloud, yet huge clouds may develop over a nearby mountain range.

With these stability relationships in mind, let us look at the most visible expressions of stability and humidity in the atmosphere: clouds and fog.

Clouds and Fog

Clouds are more than whimsical, beautiful decorations in the sky; they are fundamental indicators of overall atmospheric conditions: stability, moisture content, and weather. They form as air becomes saturated with water. Clouds are the subject of much scientific inquiry, especially regarding their effect on net radiation patterns, as discussed in Chapters 4 and 5. With a little knowledge and practice, you can learn to "read" the atmosphere from its signature clouds.

A **cloud** is an aggregation of tiny moisture droplets and ice crystals that are suspended in air, great enough in volume and concentration to be visible. Fog is simply a cloud in contact with the ground. Cloud types are too numerous to fully describe here, so I offer the most common examples in a simple classification scheme.

Cloud Formation Processes

Clouds may contain raindrops, but not initially. At the outset, clouds are a great mass of moisture droplets, each invisible without magnification. A **moisture droplet** is approximately 20 μm (micrometers) in diameter (0.002 cm, or 0.0008 in.). It takes a million or more such droplets to form an average raindrop of 2000 μm in diameter (0.2 cm, or 0.078 in.), as shown in Figure 7.20.

Given unstable conditions an air parcel rises until it becomes saturated—that is, until the air cools to the dew-point temperature and relative humidity is 100%. (Under certain conditions, condensation may occur at slightly less or more than 100% relative humidity.) More lifting of the air parcel cools it further, producing condensation of water vapor into water. Water does not just condense among the air molecules. Condensation requires **cloud-condensation nuclei**, microscopic particles that always are present in the atmosphere.

Continental air masses average 10 billion cloud-condensation nuclei per cubic meter. Ordinary dust, soot, and ash from volcanoes and forest fires and particles from burned fuel, such as sulfate aerosols, typically provide these nuclei. The air over cities contains great concentrations of such nuclei. In maritime air masses, a high concentration of sea salts derived from ocean sprays, which average 1 billion nuclei per cubic meter, supply the needed nuclei. The lower atmosphere never lacks cloud-condensation nuclei.

Given the conditions of saturated air, availability of cloud-condensation nuclei, and the presence of cooling (lifting) mechanisms in the atmosphere, condensation occurs. Two principal processes account for the majority of the world's raindrops and snowflakes: the *collision-coalescence process* and the *Bergeron ice-crystal process*. Figure 7.21 summarizes these.

Cloud Types and Identification

As noted, a cloud is a collection of moisture droplets and ice crystals suspended in air in sufficient volume and concentration to be visible. In 1803, English biologist and amateur meteorologist Luke Howard, in his article "On the Modification of Clouds," established a classification system for clouds and coined Latin names for them that we still use. About Howard's accomplishment his biographer stated,

> Clouds were no longer exempt from human comprehension, and Howard, in contributing both a system of analysis and a full Latin nomenclature covering their families and genera, had contributed more than anyone to easing the path of understanding. . . . But the naming of clouds was a different kind of gesture for the hand of classification to have made. Here was the naming not of a solid, stable thing but of a series of self-canceling evanescences [disappearing entities]. Here was the naming of a fugitive presence that hastened to its onward dissolution. Here was the naming of clouds.*

Altitude and *shape* are key to cloud classification. Clouds occur in three basic forms—flat, puffy, and wispy—and in four primary altitude classes and ten basic cloud types. Horizontally developed clouds—flat and layered—are *stratiform* clouds. Vertically developed clouds—puffy and globular—are *cumuliform* clouds. Wispy clouds usually are quite high in altitude and are made of ice crystals; these are *cirroform*.

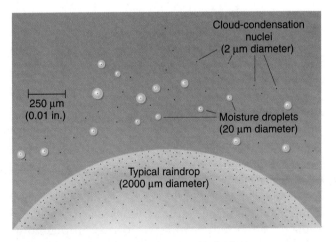

FIGURE 7.20 Moisture droplets and raindrops.
Cloud-condensation nuclei, moisture droplets, and a raindrop enlarged many times—compared at roughly the same scale.

*R. Hamblyn, *The Invention of Clouds, How An Amateur Meteorologist Forged the Language of the Skies* (New York: Farrar, Straus, and Giroux, 2001), pp. 165, 171.

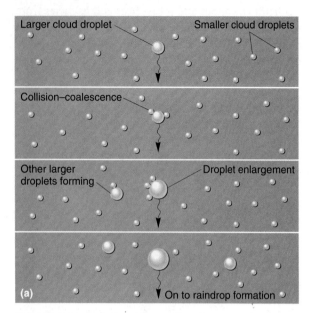

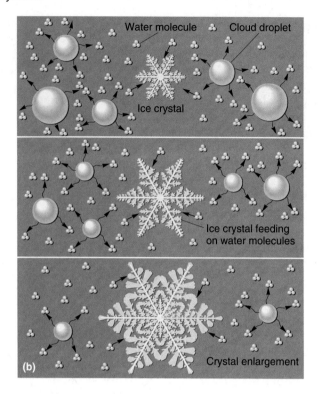

FIGURE 7.21 Raindrop formation.
Principal processes for raindrop and snowflake formation: the *collision–coalescence process* and the *ice-crystal process*. (a) The collision–coalescence process predominates in clouds that form at above-freezing temperatures, principally in the warm clouds of the tropics. Initially, simple condensation takes place on small nuclei, some of which are larger and produce larger water droplets. As those larger droplets respond to gravity and fall through a cloud, they combine with smaller droplets, gradually coalescing into a raindrop. (b) Supercooled water droplets (minute droplets of water that are below freezing and still in liquid form) will evaporate rapidly near ice crystals, which then absorb the vapor. The ice crystals feed on the supercooled cloud droplets, grow in size, and eventually fall as snow or rain. Precipitation in middle and high latitudes begins as ice and snow high in the clouds, then melts and gathers moisture as it falls through the warmer portions of the cloud. [Adapted from Frederick K. Lutgens and Edward J. Tarbuck, *The Atmosphere: An Introduction to Meteorology*, 3d. ed., copyright © 1986, p. 127. Reprinted by permission of Prentice Hall, Inc., Englewood Cliffs, NJ.]

These three basic forms occur in four altitudinal classes: low, middle, high, and those vertically developed through the troposphere. Table 7.2 presents the basic cloud classes and types. The symbols noted were of Luke Howard's invention. Figure 7.22 illustrates the general appearance of each type and includes representative photographs of most of them.

Low clouds, ranging from the surface up to 2000 m (6500 ft) in the middle latitudes, are simply called *stratus* or *cumulus* (Latin for "layer" and "heap," respectively). **Stratus** clouds appear dull, gray, and featureless. When they yield precipitation, they become **nimbostratus** (*nimbo-* denotes stormy or rainy), and their showers typically fall as drizzling rain (Figure 7.22e).

Cumulus clouds appear bright and puffy, like cotton balls. When they do not cover the sky, they float by in in-

finitely varied shapes. Vertically developed cumulus clouds are in a separate class in Table 7.2 because further vertical development can produce cumulus clouds that extend beyond low altitudes into middle and high altitudes (illustrated at the far right in Figure 7.22 and 7.22d).

Sometimes near the end of the day **stratocumulus** may fill the sky in patches, lumpy, grayish, low-level clouds. Near sunset, these spreading puffy stratiform remnants may catch and filter the Sun's rays, sometimes indicating clearing weather.

The prefix *alto-* (meaning "high") denotes middle-level clouds. They are made of water droplets and, when cold enough, can be mixed with ice crystals. **Altocumulus** clouds, in particular, represent a broad category that occurs in many different styles: patchy rows, wave patterns, a "mackerel sky," or lens-shaped (lenticular) clouds.

Table 7.2 Cloud Classes and Types

Class	Altitude/Composition at Midlatitudes	Type	Symbol	Description
Low clouds (C_L)	Up to 2000 m (6500 ft) Water	Stratus (St)		Uniform, featureless, gray, like high fog
		Stratocumulus (Sc)		Soft, gray, globular masses in lines, groups, or waves, heavy rolls, irregular overcast patterns
		Nimbostratus (Ns)		Gray, dark, low, with drizzling rain
Middle clouds (C_M)	2000–6000 m (6500–20,000 ft) Ice and water	Altostratus (As)		Thin to thick, no halos, Sun's outline just visible, gray day
		Altocumulus (Ac)		Patches of cotton balls, dappled, arranged in lines or groups, rippling waves, the lenticular clouds associated with mountains
High clouds (C_H)	6000–13,000 m (20,000–43,000 ft) Ice	Cirrus (Ci)		Mares'-tails, wispy, feathery, hairlike, delicate fibers, streaks, or plumes
		Cirrostratus (Cs)		Veil of fused sheets of ice crystals, milky, with Sun and Moon halos
		Cirrocumulus (Cc)		Dappled, "mackerel sky," small white flakes, tufts, in lines or groups, sometimes in ripples
Vertically developed clouds	Near surface to 13,000 m (43,000 ft) Water below, ice above	Cumulus (Cu)		Sharply outlined, puffy, billowy, flat-based, swelling tops, fair weather
		Cumulonimbus (Cb)		Dense, heavy, massive, dark thunderstorms, hard showers, explosive top, great vertical development, towering, cirrus-topped plume blown into anvil-shaped head

Ice crystals in thin concentrations compose clouds occurring above 6000 m (20,000 ft). These wispy filaments, usually white except when colored by sunrise or sunset, are **cirrus** clouds (Latin for "curl of hair"), sometimes dubbed mares'-tails. Cirrus clouds look as though an artist took a brush and added delicate feathery strokes high in the sky. Cirrus clouds indicate an oncoming storm, especially if they thicken and lower in elevation. The prefix *cirro-*, as in cirrostratus and cirrocumulus, indicates other high clouds that form a thin veil or puffy appearance, respectively.

A cumulus cloud can develop into a towering giant called **cumulonimbus** (again, *-nimbus* in Latin denotes rain storm or thundercloud; Figure 7.23). Such clouds are *thunderheads* because of their shape and associated lightning and thunder. Note the surface wind gusts, updrafts and downdrafts, heavy rain, and the presence of ice crystals at the top of the rising cloud column. High-altitude winds may then shear the top of the cloud into the characteristic anvil shape of the mature thunderhead.

Fog

By international definition, **fog** is a cloud layer on the ground, with visibility restricted to less than 1 km (3300 ft).

The presence of fog tells us that the air temperature and the dew-point temperature at ground level are nearly identical, indicating saturated conditions. An inversion layer generally caps a fog layer, with as much as 22 C° (40 F°) difference in air temperature between the cool ground under the fog and the warmer, sunny skies above.

Almost all fog is warm—that is, its moisture droplets are above freezing. Supercooled fog, which occurs when the moisture droplets are below freezing, is special because it can be dispersed by means of artificial seeding with ice crystals or other crystals that mimic ice, following the principles of the ice-crystal formation process described earlier. Let's briefly look at several types of fog.

Advection Fog **Advection fog** forms when air in one place *migrates* to another place where conditions are right for saturation. For example, when warm, moist air overlays cooler ocean currents, lake surfaces, or snow masses, the layer of migrating air directly above the surface becomes chilled to the dew point, and fog develops. Off all subtropical west coasts in the world, summer fog forms in the manner just described (Figure 7.24). Some coastal desert communities actually extract usable water from such fog formations, as described in News Report 7.2.

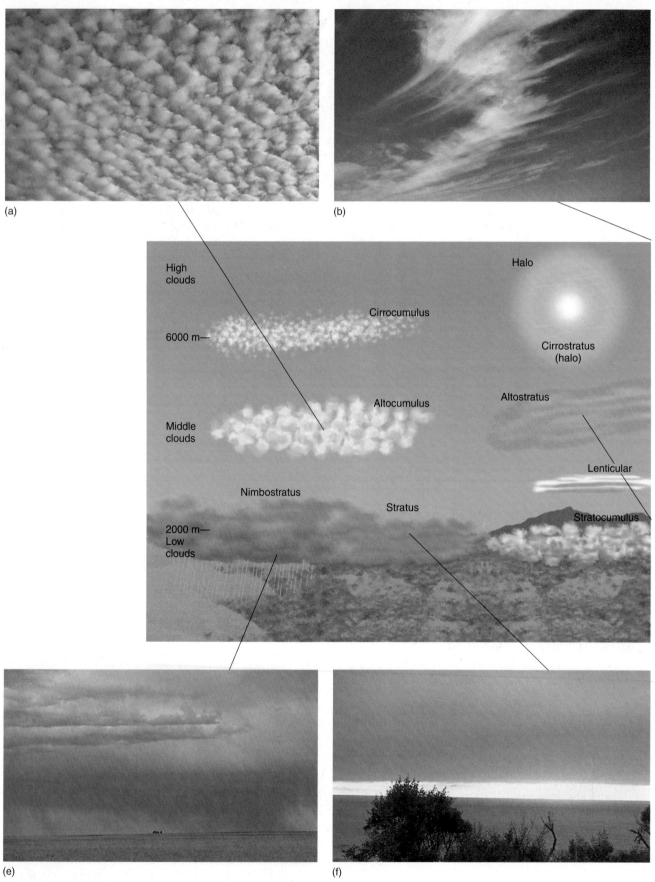

(a)

(b)

High
clouds

6000 m—

Cirrocumulus

Middle
clouds

Altocumulus

Halo

Cirrostratus
(halo)

Altostratus

Lenticular

Nimbostratus

Stratus

Stratocumulus

2000 m—
Low
clouds

(e)

(f)

FIGURE 7.22 Principal cloud types.
Principal cloud types, classified by form (cirroform, stratiform, and cumuliform) and altitude
(low, middle, high, and vertically developed across altitude): (a) altocumulus, (b) cirrus,
(c) cirrostratus, (d) cumulonimbus, (e) nimbostratus, (f) stratus, (g) altostratus, and
(h) cumulus. [Photos by author, except (b), (d), and (h) by Bobbé Christopherson.]

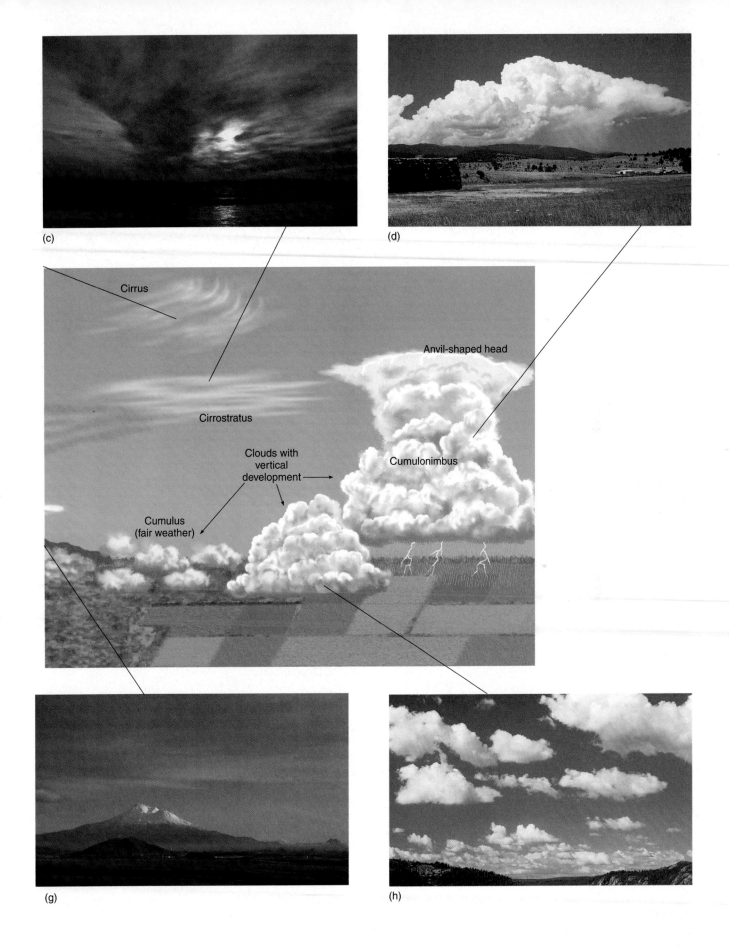

(c)

(d)

Cirrus

Cirrostratus

Anvil-shaped head

Clouds with
vertical
development

Cumulonimbus

Cumulus
(fair weather)

(g)

(h)

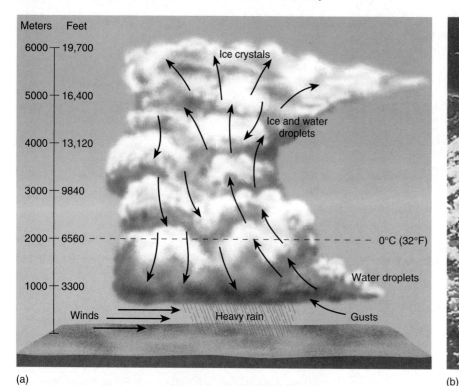

(a)

(b)

(c)

FIGURE 7.23 Cumulonimbus thunderhead.
(a) Structure and form of a cumulonimbus cloud. Violent up-drafts and downdrafts mark the circulation within the cloud. Blustery wind gusts occur along the ground. (b) Space shuttle astronauts capture a dramatic cumulonimbus thunderhead as it moves over Galveston Bay, Texas. (c) Few acts of nature can match the sheer power released by an intense thunderstorm. [(b) Space Shuttle photo from NASA; (c) photo by Bobbé Christopherson.]

Another type of advection fog forms when cold air lies over the warm water of a lake, ocean surface, or even a swimming pool. This wispy **evaporation fog**, or *steam fog*, may form as water molecules evaporate from the water surface into the cold overlying air, effectively humidifying the air, on to saturation and then condensation to form fog. When evaporation fog happens at sea, it is a shipping hazard called *sea smoke* (see the chapter-opening photo).

A type of advection fog forms when moist air flows to higher elevations along a hill or mountain. This upslope lifting leads to adiabatic cooling by expansion as the air rises. The resulting **upslope fog** forms a stratus cloud at the condensation level of saturation. Along the Appalachians and the eastern slopes of the Rockies such fog is common in winter and spring. Another advection fog associated with topography is **valley fog**. Because cool air is denser than warm air, it settles in low-lying areas, producing a fog in the chilled, saturated layer near the ground (Figure 7.25).

Radiation Fog **Radiation fog** forms when radiative cooling of a surface chills the air layer directly above that surface to the dew-point temperature, creating saturated conditions and fog. This fog occurs over moist ground especially on clear nights; it does not occur over water, because water does not cool appreciably overnight. Slight movements of air deliver even more moisture to the cooled area for more fog formation of greater depth (Figure 7.26).

FIGURE 7.24 Advection fog.
San Francisco's Golden Gate Bridge shrouded by an invading advection fog characteristic of summer conditions along a western coast. [Photo by author.]

FIGURE 7.25 Valley fog.
Cold air settles in the valleys of the Appalachian Mountains, chilling the air to the dew point and forming a valley fog. [Photo by author.]

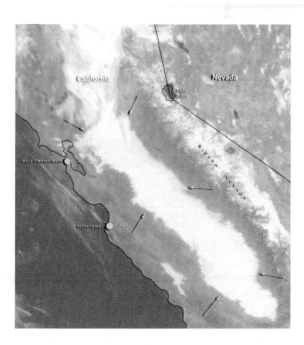

FIGURE 7.26 Radiation fog.
Satellite image of a radiation fog in Central California, June 11, 2001. This fog is locally known as a tule fog (pronounced "toolee") because of its association with the tule (bulrush) plants that line the low-elevation islands and marshes of the Sacramento River and San Joaquin River delta regions. [*GOES-10* image courtesy of National Oceanic and Atmospheric Administration.]

News Report 7.2

Harvesting Fog

Desert organisms have adapted remarkably to the presence of coastal fog along western coastlines in subtropical latitudes. For example, sand beetles in the Namib Desert in extreme southwestern Africa harvest water from the fog. They hold up their wings so condensation collects and runs down to their mouths. As the day's heat arrives, they burrow into the sand, only to emerge the next night in morning when the advection fog brings in more water for harvesting. For centuries, coastal villages in the deserts of Oman collected water drips deposited on trees by coastal fogs.

In the Atacama Desert of Chile and Peru, residents stretch large nets to intercept the fog; moisture condenses on the netting and drips into trays, flowing through pipes to a 100,000-liter (26,000-gallon) reservoir. Large sheets of plastic mesh along a ridge of the El Tofo mountains harvest water from advection fog (Figure 1). Chungungo, Chile, receives 10,000 liters (2,600 gallons) of water from 80 fog-harvesting collectors in a project developed by Canadian (International Development Research Center) and

FIGURE 1 Fog harvesting.
In the mountains inland from Chungungo, Chile, polypropylene mesh stretched between two posts capture advection fog for local drinking water supplies. [Photo by Robert S. Schemenauer.]

Chilean interests and made operational in 1993. At least 30 countries across the globe experience conditions suitable for this water resource technology. (See http://www.idrc.ca/nayudamma/fogcatc_72e.html.)

Every year the media carry stories of multi-car pileups on stretches of highway where tailgating vehicles continued to speed in foggy conditions. These totally avoidable crash scenes can involve dozens of cars and trucks. Fog is a hazard to drivers, pilots, sailors, pedestrians, and cyclists, and its conditions of formation are quite predictable. The spatial aspects of fog occurrence should be a planning element for any proposed airport, harbor facility, or highway. The prevalence of fog throughout the United States and Canada is shown in Figure 7.27.

Summary and Review—Water and Atmospheric Moisture

● **Describe** the origin of Earth's waters, **define** the quantity of water that exists today, and **list** the locations of Earth's freshwater supply.

The next time it rains where you live, pause and reflect on the journey each of those water molecules has made. Water molecules came from within Earth over a period of billions of years, in a process called **outgassing**. Thus began endless cycling of water through the hydrologic system of evaporation–condensation–precipitation. Water covers about 71% of

Earth. Approximately 97% of it is salty seawater, and the remaining 3% is freshwater—most of it frozen.

The present volume of water on Earth is estimated at 1.36 billion cubic kilometers (326 million cubic miles), an amount achieved roughly 2 billion years ago. This overall steady-state equilibrium might seem in conflict with the many changes in sea level that have occurred over Earth's history, but is not. Worldwide changes in sea level are called **eustasy** and are related to the change in volume of water in the oceans. Some of these changes are explained by the amount of water stored in

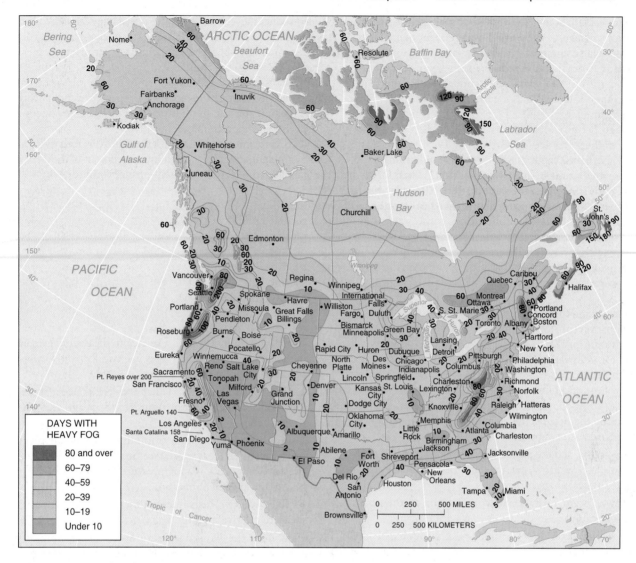

FIGURE 7.27 Fog incidence map.
Mean annual number of days with heavy fog in the United States and Canada. Officially, fog is declared if visibility is less than 1 km (3300 ft). The foggiest spot in the United States is the mouth of the Columbia River where it enters the Pacific Ocean at Cape Disappointment, Washington. One of the foggiest places in the world is Newfoundland's Avalon Peninsula, specifically Argentia and Belle Isle, which regularly exceed 200 days of fog each year. [Data courtesy of National Weather Service; Map Series 3, *Climatic Atlas of Canada*, Atmospheric Environment Service Canada, and *The Climates of Canada*, compiled by David Philips, Senior Climatologist, Environment Canada, 1990.]

glaciers and ice sheets, called **glacio-eustatic** factors. At present, sea level is rising because of increases in the temperature of the oceans and the record melting of glacial ice.

outgassing (p. 180)
eustasy (p. 180)
glacio-eustatic (p. 180)

1. Approximately where and when did Earth's water originate?

2. If the quantity of water on Earth has been quite constant in volume for at least 2 billion years, how can sea level have fluctuated? Explain.

3. Describe the locations of Earth's water, both oceanic and fresh. What is the largest repository of freshwater at this time? In what ways is this distribution of water significant to modern society?

4. Why might you describe Earth as the water planet? Explain.

● ***Describe* the heat properties of water and *identify* the traits of its three phases: solid, liquid, and gas.**

Water is the most common compound on the surface of Earth, and it possesses unusual solvent and heat characteristics. Part of Earth's uniqueness is that its water exists naturally in all three states—solid, liquid, and gas—owing to Earth's temperate position relative to the Sun. A change from one state to another is a **phase change**. The change from solid to vapor is **sublimation**; from liquid to solid, freezing; from solid to liquid, melting; from vapor to liquid, condensation; and from liquid to vapor, vaporization, or evaporation.

The heat energy required for water to change phase is **latent heat**, because, once absorbed, it is hidden within the structure of the water, ice, or water vapor. For 1 g of water to become 1 g of water vapor at boiling requires addition of 540 calories, or the **latent heat of vaporization**. When this 1 g of water vapor condenses, the same amount of heat energy is liberated, as 540 calories of the **latent heat of condensation**. The **latent heat of sublimation** is the energy exchanged in the phase change from ice to vapor and vapor to ice. Weather is powered by the tremendous amount of latent heat energy involved in the phase changes between the three states of water.

phase change (p. 184)
sublimation (p. 184)
latent heat (p. 185)
latent heat of vaporization (p. 186)
latent heat of condensation (p. 186)
latent heat of sublimation (p. 187)

5. Describe the three states of matter as they apply to ice, water, and water vapor.

6. What happens to the physical structure of water as it cools below 4°C (39°F)? What are some visible indications of these physical changes?

7. What is latent heat? How is it involved in the phase changes of water?

8. Take 1 g of water at 0°C and follow it through to 1 g of water vapor at 100°C, describing what happens along the way. What amounts of energy are involved in the changes that take place?

● *Define* humidity and the expressions of the relative humidity concept, and *explain* dew-point temperature and saturated conditions in the atmosphere.

The amount of water vapor in the atmosphere is **humidity**. The ability of air to hold water vapor is principally a function of the temperature of the air and of the water vapor (usually the same). **Relative humidity** is a percentage expression of the humidity content of the air compared with the capacity of the air to hold water vapor at a given temperature—content compared with capacity. Relatively dry air has a lower relative humidity value; relatively moist air a higher percentage. Air is said to be **saturated**, or filled to capacity, if it contains all the water vapor it can hold at a given temperature (100% relative humidity). The temperature at which air achieves saturation is the **dew-point temperature**.

Among the various ways to express humidity and relative humidity are vapor pressure and specific humidity. **Vapor pressure** is that portion of the atmospheric pressure that is produced by the presence of water vapor. A comparison of vapor pressure with the saturation vapor pressure at any moment produces a relative humidity percentage. **Specific humidity** is the mass of water vapor (in grams) per mass of air (in kilograms) at any specified temperature. Because it is measured as a mass, specific humidity does not change as temperature or pressure changes, making it a valuable measurement in weather forecasting. A comparison of specific humidity with the maximum specific humidity at any moment produces a relative humidity percentage.

Two instruments that are used to measure relative humidity, and indirectly the actual humidity content of the air, are the **hair hygrometer** and the **sling psychrometer**.

humidity (p. 187)
relative humidity (p. 187)
saturated (p. 188)
dew-point temperature (p. 188)
vapor pressure (p. 190)
specific humidity (p. 190)
hair hygrometer (p. 191)
sling psychrometer (p. 191)

9. What is humidity? How is it related to the energy present in the atmosphere? To our personal comfort and how we perceive apparent temperatures?

10. Define relative humidity. What does the concept represent? What is meant by the terms *saturation* and *dew-point temperature*?

11. Using different measures of humidity in the air given in the chapter, derive relative humidity values (vapor pressure/saturation vapor pressure; specific humidity/maximum specific humidity).

12. How do the two instruments described in this chapter measure relative humidity?

13. How does the daily distribution of relative humidity compare with the daily distribution of air temperature?

● *Define* atmospheric stability and *relate* it to a parcel of air that is ascending or descending.

Meteorologists use the term *parcel* to describe a body of air that has specific temperature and humidity characteristics. Think of a parcel of air as a volume of air, perhaps 300 m (1000 ft) in diameter, that is subjected to temperature differences that create changes in density within the parcel. Warm air has a lower density in a given volume of air; cold air has a higher density.

Stability refers to the tendency of an air parcel, with its water-vapor cargo, either to remain in place or to change vertical position by ascending (rising) or descending (falling). An air parcel is *stable* if it resists displacement upward or, when disturbed, it tends to return to its starting place. An air parcel is *unstable* if it continues to rise until it reaches an altitude where the surrounding air has a density (air temperature) similar to its own.

stability (p. 192)

14. Differentiate between stability and instability relative to a parcel of air rising vertically in the atmosphere.

15. What are the forces acting on a vertically moving parcel of air? How are they affected by the density of the air parcel?

● *Illustrate* **three atmospheric conditions—unstable, conditionally unstable, and stable—with a simple graph that relates the environmental lapse rate to the dry adiabatic rate (DAR) and moist adiabatic rate (MAR).**

An ascending (rising) parcel of air cools by expansion, responding to the reduced air pressure at higher altitudes. A descending (falling) parcel heats by compression. These temperature changes internal to a moving air parcel are explained by physical laws that govern the behavior of gases. Temperature changes in both ascending and descending air parcels occur without any significant heat exchange between the surrounding environment and the vertically moving parcel of air. The warming and cooling rates for a parcel of expanding or compressing air are termed **adiabatic**.

The **dry adiabatic rate (DAR)** is the rate at which "dry" air cools by expansion (if ascending) or heats by compression (if descending). The term *dry* is used when air is less than saturated (relative humidity less than 100%). The DAR is 10 C°/1000 m (5.5 F°/1000 ft). The **moist adiabatic rate (MAR)** is the average rate at which ascending air that is moist (saturated) cools by expansion, or descending air warms by compression. The average MAR is 6 C°/1000 m (3.3 F°/1000 ft). This is roughly 4 C° (2 F°) less than the dry rate. The MAR, however, varies with moisture content and temperature and can range from 4 to 10 C° per 1000 m (2 to 5.5 F° per 1000 ft).

A simple comparison of the dry adiabatic rate (DAR) and moist adiabatic rate (MAR) in a vertically moving parcel of air with that of the environmental lapse rate in the surrounding air determines the atmosphere's stability—whether it is unstable (lifting of air parcels continues), stable (air parcels resist vertical displacement), or conditionally unstable (air parcel behaves as though unstable if the MAR is in operation and stable otherwise).

adiabatic (p. 193)
dry adiabatic rate (DAR) (p. 193)
moist adiabatic rate (MAR) (p. 194)

16. How do the adiabatic rates of heating or cooling in a vertically displaced air parcel differ from the normal lapse rate and environmental lapse rate?

17. Why is there a difference between the dry adiabatic rate (DAR) and the moist adiabatic rate (MAR)?

18. What would atmospheric temperature and moisture conditions be on a day when the weather is unstable? When it is stable? Relate in your answer what you would experience if you were outside watching.

● *Identify* **the requirements for cloud formation and** *explain* **the major cloud classes and types, including fog.**

A **cloud** is an aggregation of tiny moisture droplets and ice crystals suspended in the air. Clouds are a constant reminder of the powerful heat-exchange system in the environment. **Moisture droplets** in a cloud form when saturated air and the presence of **cloud-condensation nuclei** lead to *condensation*. Raindrops are formed from moisture droplets through either the *collision–coalescence process* or the *Bergeron ice-crystal process*.

Low clouds, ranging from the surface up to 2000 m (6500 ft) in the middle latitudes, are **stratus** (flat clouds, in layers) or **cumulus** (puffy clouds, in heaps). When stratus clouds yield precipitation, they are **nimbostratus**. Sometimes near the end of the day, lumpy, grayish, low-level clouds called **stratocumulus** may fill the sky in patches. Middle-level clouds are denoted by the prefix *alto-*. **Altocumulus** clouds, in particular, represent a broad category that occurs in many different styles. Clouds at high altitude, principally composed of ice crystals, are called **cirrus**. A cumulus cloud can develop into a towering giant **cumulonimbus** cloud (-*nimbus* in Latin denotes rain storm or thundercloud). Such clouds are called *thunderheads* because of their shape and their associated lightning, thunder, surface wind gusts, updrafts and downdrafts, heavy rain, and hail.

Fog is a cloud that occurs at ground level. **Advection fog** forms when air in one place migrates to another place where conditions exist that can cause saturation—for example, when warm, moist air moves over cooler ocean currents. Another type of advection fog forms when cold air flows over the warm water of a lake, ocean surface, or swimming pool. This **evaporation fog**, or steam fog, may form as the water molecules evaporate from the water surface into the cold overlying air. **Upslope fog** is produced when moist air is forced to higher elevations along a hill or mountain. Another fog caused by topography is **valley fog**, formed because cool, denser air settles in low-lying areas, producing fog in the chilled, saturated layer near the ground. Radiative cooling of a surface that chills the air layer directly above the surface to the dew-point temperature, creating saturated conditions and a **radiation fog**.

cloud (p. 197)
moisture droplet (p. 197)
cloud-condensation nuclei (p. 197)
stratus (p. 198)
nimbostratus (p. 198)
cumulus (p. 198)
stratocumulus (p. 198)
altocumulus (p. 198)
cirrus (p. 199)
cumulonimbus (p. 199)
fog (p. 199)
advection fog (p. 199)
evaporation fog (p. 202)
upslope fog (p. 202)
valley fog (p. 202)
radiation fog (p. 202)

19. Specifically, what is a cloud? Describe the droplets that form a cloud.

20. Explain the condensation process: What are the requirements? What two principal processes are discussed in this chapter?

21. What are the basic forms of clouds? Using Table 7.2 describe how the basic cloud forms vary with altitude.

22. Explain how clouds might be used as indicators of the conditions of the atmosphere and of expected weather.

23. What type of cloud is fog? List and define the principal types of fog.

24. Describe the occurrence of fog in the United States and Canada. Where are the regions of highest incidence?

NetWork

The *Geosystems* Home Page provides on-line resources for this chapter on the World Wide Web. To begin: Once on the Home Page, click on this textbook, scroll the Table of Contents menu, select this chapter, and click "Begin." You will find self-tests that are graded, short essay and review exercises, specific updates for items in the chapter, and many links to interesting related pathways on the Internet under "Destinations." *Geosystems* is at **http://www.prenhall.com/ christopherson**.

Critical Thinking

A. Examine the iceberg photograph in Figure 7.5b. The photo was taken in Disko Bay, Greenland. Determine what caused the "shoreline" watermark above the ocean surface around the iceberg. Why do you think the iceberg appears to be riding higher in the water than several weeks earlier? In rough terms, how much ice (in percent) do you think is beneath the surface in comparison to the amount you see above sea level? Explain why this physical trait can be a hazard to shipping.

B. Using Figure 7.22, begin to observe clouds on a regular basis. See if you can relate the cloud type to particular weather conditions at the time of observation. You may want to keep a log in your notebook during this physical geography course. Seeing clouds in this way will help you to understand weather and make learning Chapter 8 easier.

Given our discussion of relative humidity, air temperature, dew-point temperature, and saturation, please refer to the "Short Answer" section of Chapter 7, items 1 and 2, on the *Geosystems* Home Page. View the dew-point temperature map in relation to the map of air temperatures. As you compare and contrast the maps, describe in general terms what relative humidity conditions you find in the Northwest and along the Gulf Coast in the Southeast.

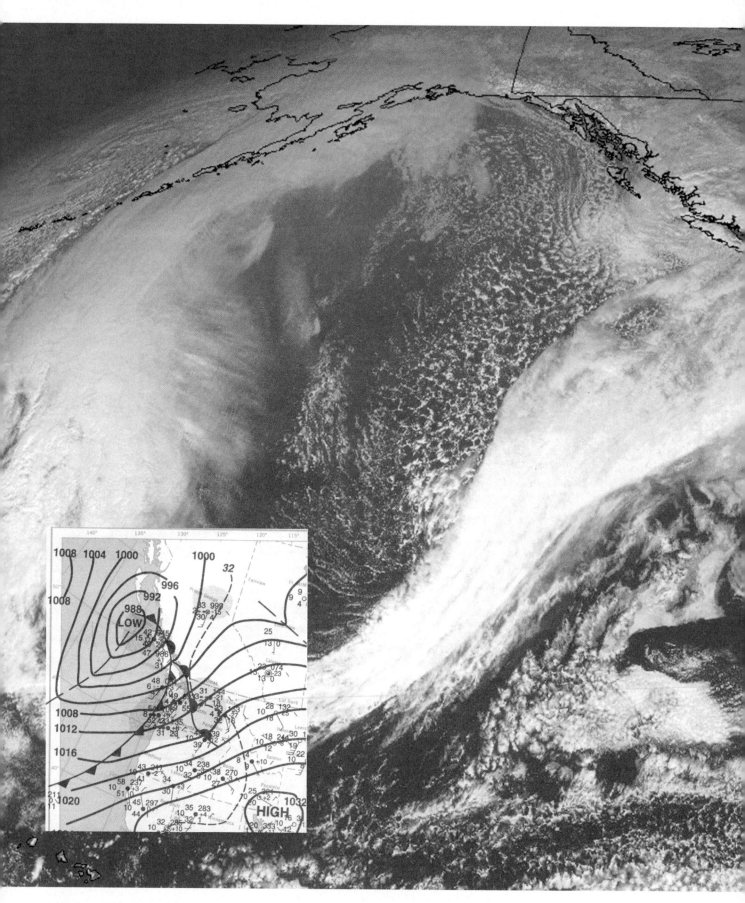

February 1, 2000: Parts of Washington and Oregon receive heavy snow in the mountains and rain elsewhere from a jet-stream-driven storm system. Wind gusts reach 110 kmph (70 mph) along the Oregon Coast as air rushes into the low-pressure center. The next system is far to the west over the Pacific Ocean. [*GOES-10* image courtesy of NESDIS/NOAA; weather map segment inset courtesy of "Daily Weather Maps," National Weather Service, NOAA.]

8
Weather

Key Learning Concepts

After reading the chapter, you should be able to:

- *Describe* air masses that affect North America and *relate* their qualities to source regions.
- *Identify* types of atmospheric lifting mechanisms and *describe* four principal examples.
- *Analyze* the pattern of orographic precipitation and *describe* the link between this pattern and global topography.
- *List* the measurable elements that contribute to weather and *describe* the life cycle of a midlatitude cyclonic storm system.
- *Analyze* various forms of violent weather and the characteristics of each.

Water has a leading role in the vast drama played out daily on Earth's stage. It affects the stability of air masses and their interactions and produces powerful and beautiful special effects in the lower atmosphere. Air masses come into conflict; they move and shift, dominating now one region then another, varying in strength and characteristics. Think of the weather as a play, North America the stage, and the air masses as actors of varying ability.

Weather is the short-term, day-to-day condition of the atmosphere, contrasted with *climate*, which is the long-term average (over decades) of weather conditions and extremes in a region. Weather is, at the same time, both a "snapshot" of atmospheric conditions and a technical status report of the Earth–atmosphere heat-energy budget. Important elements that contribute to the weather are temperature, air pressure, relative humidity, wind speed and direction, and insolation receipt related to daylength and Sun angle—in other words, the seasons.

We turn for the weather forecast to the National Weather Service in the United States (http://www. nws.noaa.gov) or Canadian Meteorological Centre, a branch of the Meteorological Service of Canada (MSC), (http://www.msc-smc.ec.gc.ca/cmc/index_e.html), to see current satellite images and to hear weather analysis. Internationally, the World Meteorological Organization coordinates weather information (see http://www.wmo.ch/). Many sources of weather information and related topics are found in the "Destinations" section for this chapter on the *Geosystems* Home Page.

Meteorology is the scientific study of the atmosphere. (*Meteor* means "heavenly" or "of the atmosphere.") Meteorologists study the atmosphere's physical characteristics and motions, related chemical, physical, and geologic processes, the complex linkages of atmospheric systems, and weather forecasting. Computers handle volumes of data for accurate forecasting of near-term weather and for studying trends in long-term weather, climatology, and climatic change.

New developments in supercomputing, an Earth-bound instrument network in the Automated Surface Observing System (ASOS) arrays, orbiting observation systems, and Doppler radar installations are rapidly advancing the science of the atmosphere. By the end of 2001, 155 WSR-88D (*W*eather *S*urveillance *R*adar) Doppler radar systems as part of the NEXRAD (Next Generation Weather Radar) program were operational through the National Weather Service (122), in conjunction with the Federal Aviation Administration (12) and the Department of Defense (21) (Figure 8.1). In Canada, the National Radar Project will have 30 CWSR-98 radars in service.

Doppler radar detects the direction of moisture droplets toward or away from the radar, indicating wind direction and speed. This information is critical to weather forecasting and severe storm warnings. An essential part of this modernization deployment is the Advanced Weather Interactive Processing System (AWIPS, software version 5.2 introduced in 2002) that will eventually include 148 stations.

Weather-related destruction has risen more than 500% over the past two decades, from an average of $2 billion to $10 billion annually—considering storms, floods, droughts, and wildfires. For instance, 1998 weather damage alone topped $90 billion worldwide, which exceeds the total for all the 1980s, even when adjusted for inflation! Floods, droughts, ice storms, tropical cyclones and hurricanes, tornadoes, coastal storm surges, and heat waves all contribute to these totals.

FIGURE 8.1 Weather installation.
Doppler radar installation at the Indianapolis International Airport operated by the National Weather Service. The radar antenna is sheltered within the dome structure. [Photo by Bobbé Christopherson.]

In this chapter: We follow huge air masses across North America, observe powerful lifting mechanisms in the atmosphere, revisit the concepts of stable and unstable conditions, examine migrating cyclonic systems with attendant cold and warm fronts, and conclude with a portrait of violent and dramatic weather. Water, with its ability to absorb and release vast quantities of heat energy, drives this daily drama in the atmosphere. The spatial implications of these weather phenomena and their relationship to human activities strongly link meteorology to physical geography and this chapter.

Air Masses

Each area of Earth's surface imparts its temperature and moisture characteristics to the air it touches. The effect of the surface on the air creates regional masses of air having specific conditions of temperature, humidity, and stability. These masses of air interact to produce weather patterns—in essence, they are the actors in our weather drama. Such a distinctive body of air is called an **air mass**, and it initially reflects the characteristics of its *source region*. For example, weather forecasters speak of a "cold Canadian air mass" and "moist tropical air mass."

The longer an air mass remains stationary over a region, the more definite its physical attributes become. Within each air mass there is a homogenous mix of temperature and humidity that sometimes extends through the lower half of the troposphere. Such masses of air possess all the physical characteristics of the atmosphere discussed in earlier chapters and thus link the Earth–atmosphere energy budget and water–weather systems.

Air Masses Affecting North America

Air masses generally are classified according to the moisture and temperature characteristics of their source regions:

1. *Moisture*—designated **m** for maritime (wet) and **c** for continental (dry).
2. *Temperature* (latitude)—designated **A** (arctic), **P** (polar), **T** (tropical), **E** (equatorial), and **AA** (Antarctic).

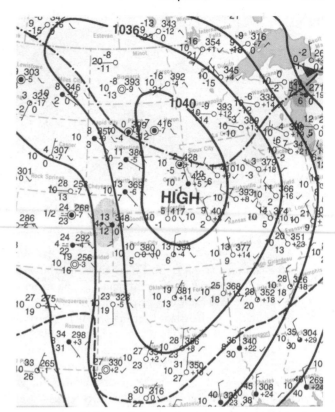

FIGURE 8.3 Winter high-pressure system.
A cP air mass with a central pressure of 1042.8 mb (30.76 in.), air temperature of −17°C, dew-point temperature of −21°C (2° and −5°F, respectively), with clear, calm, stable conditions, dominates the Midwest. Note the pattern of isobars portraying the cP air mass. The dotted lines are the −18°C (0°F) and 0°C (32°F) isotherms. [Weather map insert courtesy of National Weather Service, NOAA.]

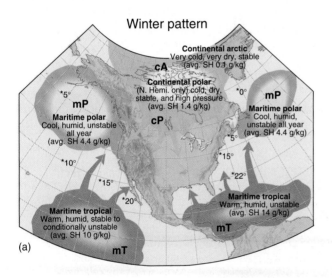

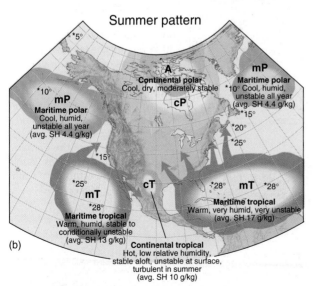

FIGURE 8.2 Principal air masses.
Air masses that influence North America in winter and summer. (*Sea-surface temperature in °C; SH = specific humidity.)

The principal air masses that affect North America in winter and summer are mapped in Figure 8.2.

Continental polar (**cP**) air masses form only in the Northern Hemisphere and are most developed in winter when they dominate cold weather conditions. These cP air masses are major players in middle- and high-latitude weather. The cold, dense cP air lifts moist, warm air in its path, producing lifting, cooling, and condensation. As we will see, the greater the temperature difference between this cold air mass and warmer air masses, the more dramatic is the weather produced. An area covered by cP air in winter experiences cold, stable air, clear skies, high pressure, and anticyclonic wind flow, all visible on the weather map in Figure 8.3. The Southern Hemisphere lacks the necessary landmasses (continentality) at high latitudes to produce continental polar characteristics.

Maritime polar (**mP**) air masses in the Northern Hemisphere exist northwest and northeast of the North American continent over the northern oceans. Within them, cool, moist, unstable conditions prevail throughout the year. The Aleutian and Icelandic subpolar low-pressure cells reside within these mP air masses, especially in their well-developed winter pattern (see the chapter-opening satellite image).

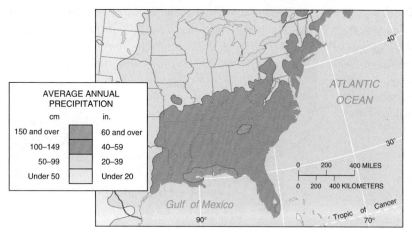

FIGURE 8.4 Influence of mT Gulf/Atlantic air mass.
The pattern of precipitation over the southeastern United States shows the influence of the warm, moist, and generally unstable mT Gulf/Atlantic air mass. Precipitation decreases with distance inland from the source region over the Gulf of Mexico and Atlantic Ocean.

Two maritime tropical (**mT**) air masses—the *mT Gulf/Atlantic* and the *mT Pacific*—influence North America. The humidity experienced in the East and Midwest is created by the mT Gulf/Atlantic air mass, which is particularly unstable and active from late spring to early fall (Figure 8.4). In contrast, the mT Pacific is stable to conditionally unstable and generally lower in moisture content and available energy. As a result, the western United States, influenced by this weaker Pacific air mass, receives lower average precipitation than the rest of the country.

Please review Figure 6.14 and the discussion of subtropical high-pressure cells off the coast of North America—the moist, unstable conditions on the western edge of the Atlantic (east coast), and the drier, stable conditions on the eastern edge of the Pacific (west coast). These conditions, coupled respectively with ocean currents that are warmer (Gulf Stream) and cooler (California Current), produce the characteristics of each source region for these different maritime air masses.

Air Mass Modification

As air masses migrate from their source regions, their temperature and moisture characteristics modify and slowly

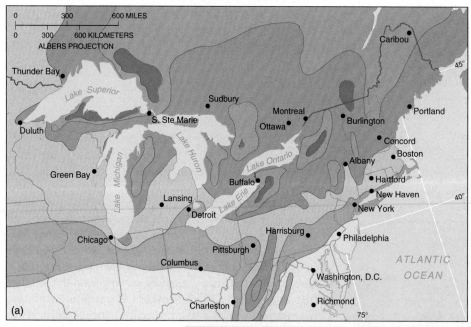

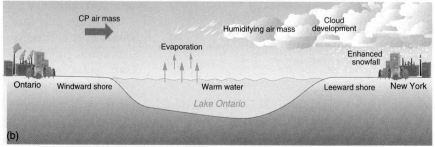

FIGURE 8.5 Lake-effect snowbelts of the Great Lakes.
(a) Locally, heavy areas of snowfall are associated with the lee side of each of the Great Lakes. In winter, cold cP and cA air masses pass from colder land surfaces across the relatively warmer water of these lakes.
(b) The air masses are warmed and humidified (water vapor is added) from the lake water. The humid, now-unstable air yields heavy snowfall as it moves onshore and becomes chilled. The strongest effect is generally limited to about 50 km (30 mi) inland up to 100 km (60 mi). [Snowfall data from the *Climatic Atlas of the United States* (Washington, DC: Department of Commerce, NOAA, 1983), p. 53.]

take on the characteristics of the land over which they pass. For example, an mT Gulf/Atlantic air mass may carry humidity to Chicago and on to Winnipeg but gradually will lose its initial characteristics of high humidity and warmth with each day's passage northward.

Similarly, below-freezing temperatures occasionally reach into southern Texas and Florida, brought by an invading winter cP air mass from the north. However, that air mass warms from the −50°C (−58°F) of its source region in central Canada. In winter, as a cP air mass moves southward over warmer land, it moderates, warming especially after it leaves areas covered by snow (the snowline).

Modification of cP air as it moves south and east produces snowbelts that lie to the east of each of the Great Lakes. As below-freezing cP air passes over the warmer Great Lakes, it absorbs heat energy and moisture from the lake surfaces (is *humidified*). This enhancement produces heavy lake-effect snowfall downwind into Ontario, Michi-

gan, Pennsylvania, and New York—some areas receiving in excess of 250 cm (100 in.) in average snowfall a year (Figure 8.5).

Atmospheric Lifting Mechanisms

For air masses to cool adiabatically (by expansion) and to reach the dew-point temperature and saturate, condense, form clouds, and perhaps precipitate, they must lift and rise in altitude. Four principal lifting mechanisms operate in the atmosphere: convergent lifting (air flows toward an area of low pressure), convectional lifting (stimulated by local surface heating), orographic lifting (air is forced over a barrier such as a mountain range), and frontal lifting (along the leading edges of contrasting air masses). Descriptions of all four mechanisms follow and are summarized in Figure 8.6.

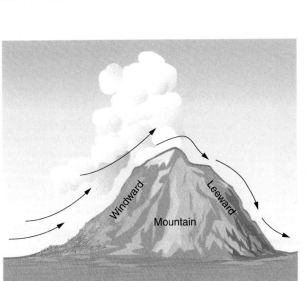

(a) Convergent

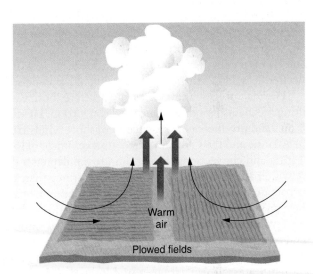

(b) Convectional (local heating)

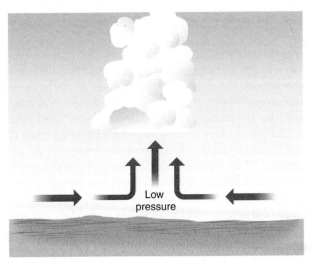

(c) Orographic (barrier)

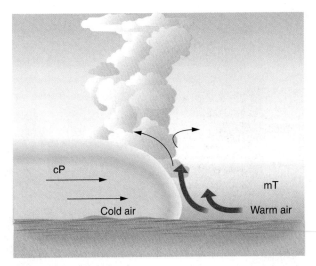

(d) Frontal (e.g. cold front)

FIGURE 8.6 Four atmospheric lifting mechanisms.
(a) Convergent lifting. (b) Convectional lifting. (c) Orographic lifting. (d) Frontal lifting.

Convergent Lifting

Air flowing from different directions into the same low-pressure area is converging, displacing air upward in **convergent lifting**. In the tropics, convergent lifting of warm, moist air produces disturbances that can lead to the development of a tropical storm. All along the equatorial region, the southeast and northeast trade winds converge, forming the intertropical convergence zone (ITCZ) and areas of extensive uplift, towering cumulonimbus cloud development, and high average annual precipitation (see Figures 6.12 and 6.13b).

Convectional Lifting

When an air mass passes from a maritime source region to a warmer continental region, heating from the warmer land causes lifting and convection in the air mass. Other sources of surface heating might include an urbanized area (heat island) or an area of dark soil in a plowed field—the warmer surfaces produce **convectional lifting**. If conditions are unstable, initial lifting sustains and clouds develop. Figure 8.7 illustrates convectional action stimulated by local heating, with unstable conditions present in the atmosphere. The rising parcel of air continues its ascent because it is warmer (less dense) than the surrounding environment.

Florida's precipitation generally illustrates both these lifting mechanisms: convergence and convection. Heating of the land produces convergence of onshore winds from the Atlantic and the Gulf of Mexico. As an example of local heating and convectional lifting, Figure 8.8 depicts a day on which the landmass of Florida was warmer than the surrounding Gulf of Mexico and Atlantic Ocean. Because the Sun's radiation gradually heats the land throughout the day and warms the air above it, convectional showers tend to form in the afternoon and early evening, causing the highest frequency of days with thunderstorms in the United States. Florida appears highlighted and painted with clouds.

Towering cumulonimbus clouds are summertime features in the regions of North America that experience the mT Gulf/Atlantic air mass and, to a lesser extent, by the weaker mT Pacific air mass. Convectional precipitation also dominates along the ITCZ, over tropical islands, and anywhere that moist, unstable air is heated from below or where the inflowing trade winds produce a dynamic convergence.

Orographic Lifting

The physical presence of a mountain acts as a topographic barrier to migrating air masses. **Orographic lifting** (*oro* means "mountain") occurs when air is forcibly lifted upslope as it is pushed against a mountain. The lifting air cools adiabatically. Stable air forced upward in this manner may produce stratiform clouds, whereas unstable or conditionally unstable air usually forms a line of cumulus and cumulonimbus clouds. An orographic barrier enhances convectional activity and causes additional lifting during the passage of weather fronts and cyclonic systems, thereby extracting more moisture from passing air masses.

Figure 8.9a illustrates the operation of orographic lifting under unstable conditions. The wetter intercepting slope is termed the *windward slope*, as opposed to the drier far-side slope, known as the *leeward slope*. Moisture is condensed from the lifting air mass on the windward side of the mountain; on the leeward side, the descending air mass is heated by compression, and any remaining water in the air evaporates. Thus, air beginning its ascent up a mountain can be warm and moist, but finishing its descent on the leeward slope it becomes hot and dry.

In North America, **chinook winds** (called *föhn* or *foehn* winds in Europe) are the warm, downslope airflows characteristic of the leeward side of mountains. Such winds can bring a 20 C° (36 F°) jump in temperature and greatly reduced relative humidity on the lee side of the mountains.

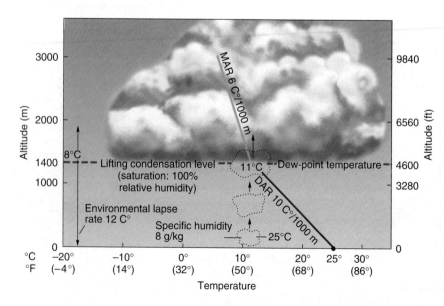

FIGURE 8.7 Local heating and convection.
Local heating and convection under unstable atmospheric conditions. Note that specific humidity is 8 g/kg and beginning temperature in the air parcel is 25°C. If you look back to Figure 7.12, you will find that air with a specific humidity of 8 g/kg must be cooled to 11°C to achieve the dew-point temperature. In the example, this temperature is reached after 14 C° of adiabatic cooling at 1400 m (4600 ft). Note that the DAR (dry adiabatic rate) is used when the air parcel is less than saturated, changing to the MAR (moist adiabatic rate) above the lifting-condensation level at 1400 m.

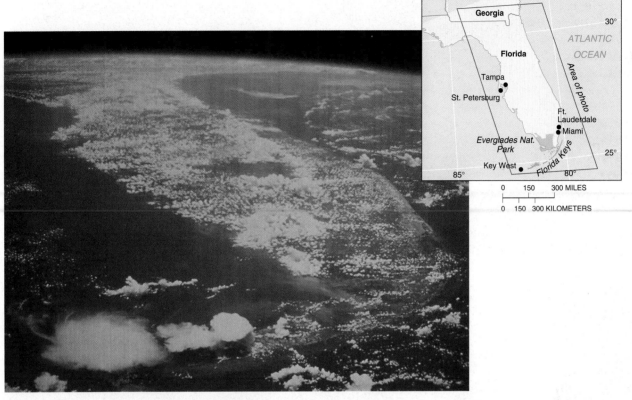

FIGURE 8.8 Convectional activity over the Florida peninsula.
Cumulus clouds cover the land, with several cells developing into cumulonimbus thunderheads. Warm, moist air from the Gulf of Mexico is lifted by local heating as it passes over the relatively warmer land. [Project Gemini photo from NASA.]

News Report 8.1

Mountains Set Precipitation Records

Because orographic lifting is limited in areal extent to locations where a topographic barrier exists, it is the least dominant lifting mechanism worldwide. But mountain ranges are the most consistent of all the precipitation-inducing mechanisms. Both the greatest average annual precipitation and the greatest maximum annual precipitation on Earth occur on the windward slopes of mountains that intercept moist tropical trade winds.

The world's greatest average annual precipitation occurs in the United States on Mount Waialeale, on the island of Kaua'i, Hawai'i. This moun-

tain rises 1569 m (5147 ft) above sea level. On its windward slope, rainfall averages 1234 cm (486 in., or 40.5 ft) a year for the years 1941–1992 (topping the previous average of 460 in. and other lower values). In contrast, the rain-shadow side of Kaua'i receives only 50 cm (20 in.) of rain annually. If no islands existed at this location, this portion of the Pacific Ocean would receive only an average 63.5 cm (25 in.) of precipitation a year.

Another place receiving world-record precipitation is Cherrapunji, India, 1313 m (4309 ft) above sea level at 25° N latitude, in the Assam Hills

south of the Himalayas. Because of the summer monsoons that pour in from the Indian Ocean and the Bay of Bengal, Cherrapunji has received 930 cm (366 in., or 30.5 ft) of rainfall in one month and a total of 2647 cm (1042 in., or 86.8 ft) in one year! Not surprisingly, Cherrapunji is the all-time precipitation record holder for a single year and for every other time interval from 15 days to 2 years. The average annual precipitation there is 1143 cm (450 in., 37.5 ft), placing it second only to Mount Waialeale.

The term **rain shadow** is applied to dry regions leeward of mountains. East of the Cascade Range, Sierra Nevada, and Rocky Mountains, such rain-shadow patterns predominate (Figure 8.9b and c). In fact, the precipitation pattern of windward and leeward slopes persists worldwide, as confirmed by the precipitation maps for North America

(Figure 9.6) and the world (Figure 10.2). See News Report 8.1 for more about the role of orographic barriers in setting precipitation records.

The state of Washington provides an excellent example of this concept, as shown in Figure 8.10. The Olympic Mountains and Cascade Mountains orographically lift

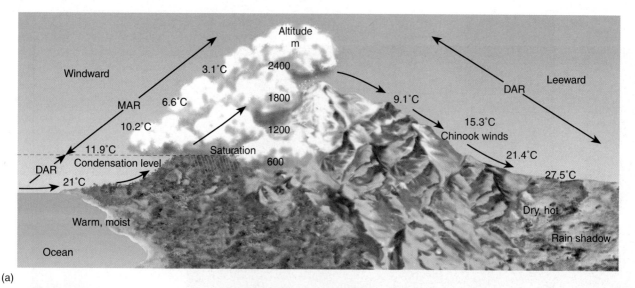

(a)

(b)

(c)

FIGURE 8.9 Orographic precipitation.

(a) Orographic barrier and precipitation patterns—unstable conditions assumed. Prevailing winds force warm, moist air upward against a mountain range, producing adiabatic cooling, eventual saturation, cloud formation, and precipitation. On the leeward slope, as the "dried" air descends, compressional heating warms it, creating the hot, relatively dry rain shadow of the mountain. (b) The windward slopes of the Sierra Nevada Mountain Range and leeward rainshadow conditions of Nevada are visible on this *Terra* satellite image. (c) Rainshadow produced by descending, warming air contrast with the clouds of the windward side. Dust is stirred up by leeward slope winds. [(b) MODIS image from *Terra* satellite, 3/12/2000 courtesy of MODIS Land Rapid Response Team, NASA/GSFC; (c) photo by author.]

FIGURE 8.10 Orographic patterns in Washington State. ➡

(a) Four stations in Washington provide examples of orographic effects: windward precipitation and leeward rain shadows. Isohyets (isolines of equal precipitation amounts) on the map indicate rainfall (in inches). (b) Vegetation and landscapes on this true-color MODIS image from *Terra* demonstrate these precipitation patterns. [(a) Data from J. W. Scott, and others, *Washington: A Centennial Atlas* (Bellingham, WA: Center for Pacific Northwest Studies, Western Washington University, 1989), p. 3; (b) satellite *Terra* image courtesy of MODIS Land Response Team, NASA/GSFC.]

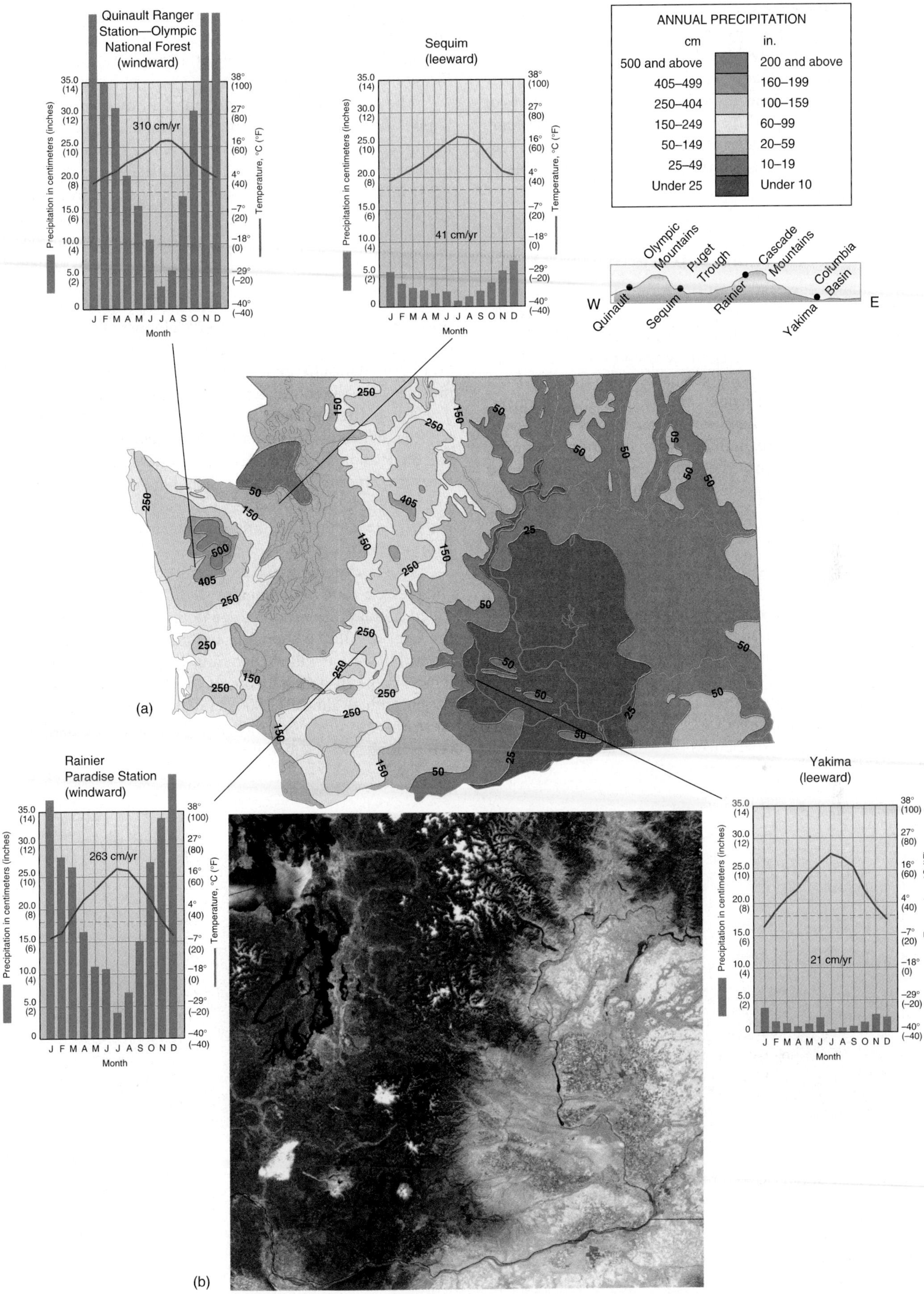

Quinault Ranger Station—Olympic National Forest (windward)
310 cm/yr
Sequim (leeward)
41 cm/yr
Rainier Paradise Station (windward)
263 cm/yr
Yakima (leeward)
21 cm/yr
Precipitation in centimeters (inches)
Temperature, °C (°F)
35.0 (14)
30.0 (12)
25.0 (10)
20.0 (8)
15.0 (6)
10.0 (4)
5.0 (2)
0
38° (100)
27° (80)
16° (60)
4° (40)
–7° (20)
–18° (0)
–29° (–20)
–40° (–40)
J F M A M J J A S O N D
Month
ANNUAL PRECIPITATION
cm
in.
500 and above
200 and above
405–499
160–199
250–404
100–159
150–249
60–99
50–149
20–59
25–49
10–19
Under 25
Under 10
Olympic Mountains
Puget Trough
Cascade Mountains
Columbia Basin
W
E
Quinault
Sequim
Rainier
Yakima
250
150
50
405
500
25
(a)
(b)

invading mP air masses from the North Pacific Ocean, squeezing out annual precipitation of more than 500 cm and 400 cm, respectively (200 in. and 160 in.).

The Quinault Ranger and Rainier Paradise weather stations demonstrate windward-slope rainfall. The cities of Sequim, in the Puget Trough, and Yakima, in the Columbia Basin, are on the leeward side of these mountain ranges and are characteristically low in annual rainfall, being in the rain shadow.

Frontal Lifting (Cold and Warm Fronts)

The leading edge of an advancing air mass is called its *front*. Vilhelm Bjerknes (1862–1951) first applied the term while working with a team of meteorologists in Norway during World War I. Weather systems seemed to them to be migrating air-mass "armies," doing battle along fronts. A front is a place of atmospheric discontinuity, a narrow zone forming a line of conflict between two air masses of different temperature, pressure, humidity, wind direction and speed, and cloud development. The leading edge of a cold air mass is a **cold front**, whereas the leading edge of a warm air mass is a **warm front**.

Cold Front   On weather maps, such as those shown in Figure 8.14 and in Figure 8.18, a cold front is a line with triangular spikes drawn that point in the direction of frontal movement along an advancing cP or mP air mass. The steep face of the cold air mass suggests its ground-hugging nature, caused by its density and uniform physical character (Figure 8.11).

Warm, moist air in advance of the cold front lifts upward abruptly and experiences the same adiabatic rates of cooling and factors of stability or instability that pertain to all lifting air parcels. A day or two ahead of the cold front's passage, high cirrus clouds appear, telling observers that a lifting mechanism is on the way.

The cold front's advance is marked by a wind shift, temperature drop, and lowering barometric pressure due to lifting along the front. Air pressure reaches a local low as the line of most intense lifting passes, usually just ahead of the front itself. Clouds may build along the cold front into characteristic cumulonimbus form and may appear as an advancing wall of clouds. Precipitation usually is heavy, containing large droplets, and can be accompanied by hail, lightning, and thunder.

The aftermath of a cold front passage usually brings northerly winds in the Northern Hemisphere as anticyclonic high-pressure advances (southerly winds in the Southern Hemisphere); lower temperatures; increasing air pressure from the cooler, denser air; and broken cloud cover.

The particular shape and size of the North American landmass and its latitudinal position present conditions where cP and mT air masses are best developed and have the most direct access to each other. The resulting contrast can lead to dramatic weather, particularly in late spring, with sizable temperature differences from one side of a cold front to the other.

A fast-advancing cold front can cause violent lifting and create a zone right along or slightly ahead of the front called a **squall line**. Along a squall line, such as the one in the Gulf of Mexico shown in Figure 8.12, wind patterns are turbulent and wildly changing, and precipitation is intense. The well-defined frontal clouds in the photograph rise abruptly to almost 17,000 m (56,000 ft), with new thunderstorms forming along the front. Tornadoes also may develop along such a squall line.

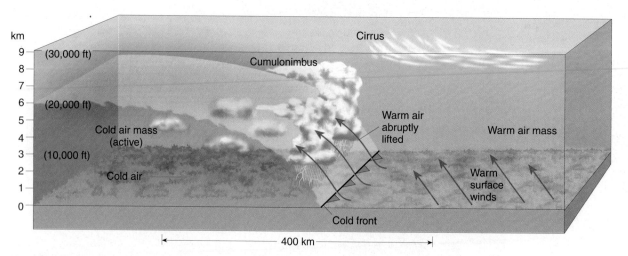

FIGURE 8.11   A typical cold front.
Denser, advancing cold air forces warm moist air to lift abruptly. As the air is lifted, it cools by expansion at the DAR, cooling to the dew-point temperature as it rises to a level of condensation and cloud formation. Cumulonimbus clouds may produce large raindrops, heavy showers, lightning and thunder, and hail.

is denser. Instead, the warm air tends to push the cooler, underlying air into a characteristic wedge shape, with the warmer air sliding up over the cooler air. Thus, in the cooler-air region a temperature inversion is present, sometimes causing poor air drainage and stagnation.

Figure 8.13 illustrates a typical warm front in which mT air is gently lifted, leading to stratiform cloud development and characteristic nimbostratus clouds and drizzly precipitation. A warm front produces a progression of cloud development: High cirrus and cirrostratus clouds announce the advancing frontal system; then clouds lower and thicken to altostratus; and finally the clouds lower and thicken to stratus within several hundred kilometers of the front.

Midlatitude Cyclonic Systems

The conflict between contrasting air masses can develop a **midlatitude cyclone**, or **wave cyclone**. This migrating low-pressure center with converging, ascending air, spirals inward counterclockwise in the Northern Hemisphere (or inward clockwise in the Southern Hemisphere). Because of the undulating nature of frontal boundaries and the steering flow of the jet streams, the term *wave* is appropriate. The combination of the *pressure gradient force*, *Coriolis force*, and *surface friction* generates this cyclonic motion (see discussion in Chapter 6).

Before World War I, weather maps displayed only pressure and wind patterns. V. Bjerknes added the concept of fronts, and his son Jacob contributed the concept of migrating centers of cyclonic low-pressure systems.

Wave cyclones dominate weather patterns in the middle and higher latitudes of both the Northern and Southern Hemispheres and act as a catalyst for air mass conflict.

FIGURE 8.12 Cold front and squall line.
Cold front and squall line are marked by a sharp line of cumulonimbus clouds in the Gulf of Mexico. The cloud formation rises to 17,000 m (56,000 ft). The passage of such a frontal system over land often produces strong winds and possibly tornadoes. [Space Shuttle photo from NASA.]

Warm Front   A line with semicircles facing in the direction of frontal movement denotes a warm front on weather maps (see Figure 8.14). The leading edge of an advancing warm air mass is unable to displace cooler, passive air, which

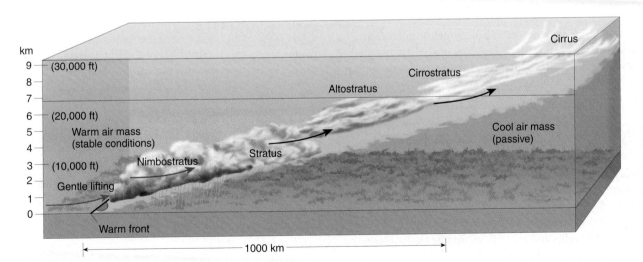

FIGURE 8.13 A typical warm front.
Note the sequence of cloud development as the warm front approaches. Warm air slides upward over a wedge of cooler, passive air near the ground. Gentle lifting of the warm, moist air produces nimbostratus and stratus clouds and drizzly rain showers, in contrast to the more dramatic precipitation associated with the passage of a cold front.

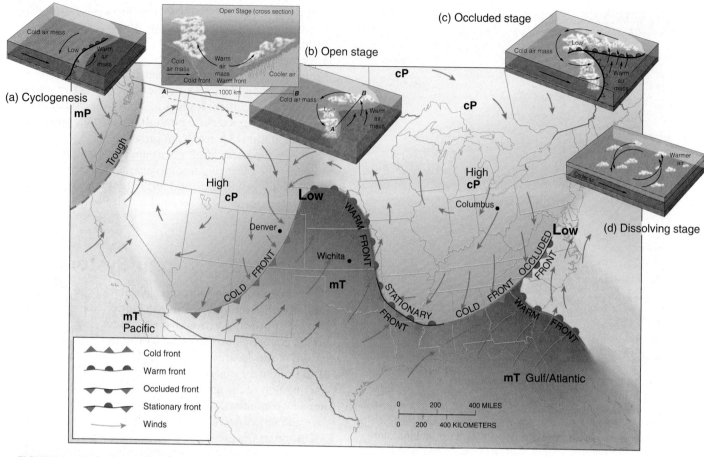

FIGURE 8.14 Idealized stages of a midlatitude wave cyclone.
(a) Cyclogenesis is where surface convergence and lifting begin. (b) The open stage. (c) The occluded stage. (d) The dissolving stage is reached at the end of the storm track as the cyclone spins down, no longer energized by the latent heat from condensing moisture. A system may take from 3–10 days to cross the continent along storm tracks that vary seasonally. After studying the text, can you describe conditions in Denver depicted on this weather map? Wichita? Columbus?

Such a midlatitude cyclone can initiate along the polar front, particularly in the region of the Icelandic and Aleutian subpolar low-pressure cells in the Northern Hemisphere. The intense high-speed winds of the jet streams guide cyclonic systems along their tracks (see Figures 6.17 and 6.18).

Life Cycle of a Midlatitude Cyclone

Figure 8.14 shows the birth, maturity, and death of a typical midlatitude cyclone in several stages, along with an idealized weather map. On the average, a midlatitude cyclone takes 3–10 days to progress through these stages from the area where it develops to the area where it finally dissolves. However, chaos rules and every day's weather map departs from the ideal in some manner.

Cyclogenesis   Cyclogenesis is the atmospheric process in which low-pressure systems develop and strengthen. Along the polar front, cold and warm air masses converge and conflict.

The polar front is a discontinuity of temperature, moisture, and winds that establishes potentially unstable conditions. For a wave cyclone to form along the polar front, a point of air *convergence* at the surface must be matched by a compensating area of air *divergence* aloft. Even a slight disturbance along the polar front, perhaps a small change in the path of the jet stream, can initiate the converging, ascending flow of air and thus a surface low-pressure system (illustrated in Figure 8.14a).

In addition to the polar front, certain other areas are associated with wave cyclone development and intensification: the eastern slope of the Rockies and other north–south mountain barriers, the Gulf Coast, and the east coasts of North America and Asia.

Open Stage   To the east of the developing low-pressure center, warm air begins to move northward along an advancing front, while cold air advances southward to the west of the center. See this movement on Figure 8.14b as a trough, or area of beginning convergence. The growing circulation system then vents into upper-level winds. As the midlatitude cyclone matures, the counterclockwise flow (in the Northern Hemisphere) draws the cold air mass from the north and west and the warm air mass from the south (here centered over western Nebraska). In the cross section, you can see the profiles of both a cold front and a warm front and each air mass segment.

On the map, Denver has just experienced the passage of a cold front. Before the front passed, you see on the map that winds were from the southwest, but now the winds have shifted and are from the northwest. Temperature and humidity have gone from the warm moist mT to colder conditions as the cP air mass moves over Denver. Wichita, Kansas, experienced the passage of a warm front and now is in the midst of the warm-air segment of the cyclone.

Such an open stage of a midlatitude cyclone occurred April 20, 2000, with low pressure centered over the Iowa–Missouri border. Figure 8.15 shows you a portion of the daily weather map for that day and an image from *SeaWiFS* sensors aboard satellite *OrbView-2*. Note how the isobars portray the cyclone and the low of 997.6 mb (29.45 in.) on the map. Note the cloud pattern stretching along the cold front and the swirling clouds around the low. Compare the cold front and warm front and overall patterns with Figure 8.14b.

Occluded Stage   Because the cP air mass is cooler in temperature and higher in pressure than the mT air mass, the cold air is denser and heavier. The cooler, more unified, air mass acts like a bulldozer blade and, therefore, moves faster than the warm front. Cold fronts can travel at an average 40 kmph (25 mph), whereas warm fronts average roughly half that at 16–24 kmph (10–15 mph). Thus, a cold front often overtakes the cyclonic warm front, wedging beneath it, producing an **occluded front** (*occlude* means "to close"). An occluded front stretches south from the center of low pressure in Virginia to the border between the Carolinas (Figure 8.14c). Precipitation may be moderate to heavy initially and then taper off as the warmer air wedge is lifted higher by the advancing cold air mass. On this idealized map a cold front and a warm front remain active to the south of the occluded area.

Also note the designation of a **stationary front** on the weather map (in Arkansas and Mississippi). This frontal symbol tells you that there is a stalemate between cooler and warmer air masses where air flow on either side is almost parallel to the front, although in opposite directions. Some gentle lifting is producing light to moderate precipitation. Eventually the stationary front will begin to move, as one of the air masses assumes dominance, evolving into a warm or a cold front.

(a)   3:00 P.M. EST

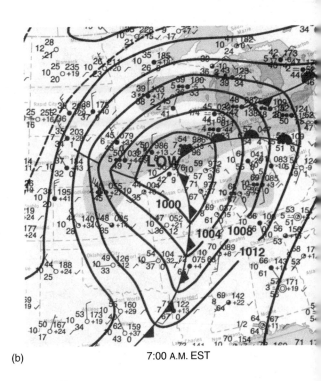

(b)   7:00 A.M. EST

FIGURE 8.15 Open stage of a midlatitude cyclone.
(a) *SeaWiFS* image of a cyclonic system over the Midwest. The cloud patterns are areas of precipitation; clear skies are behind the cold front as cP air mass covers the landscape. (b) A segment from the April 20, 2000, daily weather map, about 8 hours earlier than the image, showing the low-pressure system centered on 997.6 mb (29.45 in.). Counterclockwise winds circulate around the low. [(a) *SeaWiFS* image used with permission of ORBIMAGE. All rights reserved. (b) Segment of "Daily Weather Map," courtesy of National Weather Service, NOAA.]

Dissolving Stage   The final, dissolving, stage of the mid-latitude cyclone occurs when its lifting mechanism is completely cut off from the warm air mass, which was its source of energy and moisture. Remnants of the cyclonic system then dissipate in the atmosphere, perhaps after passage across the country (Figure 8.14d).

Storm Tracks   Cyclonic storms—1600 km (1000 mi) wide—and their attendant air masses move across the continent along **storm tracks**, which shift latitudinally with the Sun and the seasons. Typical storm tracks that cross North America are farther northward in summer and farther southward in winter (Figure 8.16). As the storm tracks begin to shift northward in the spring, cP and mT air masses are in their clearest conflict. This is the time of strongest frontal activity, featuring thunderstorms and tornadoes. Storm tracks follow the path of upper-air winds, which direct storm systems across the continent.

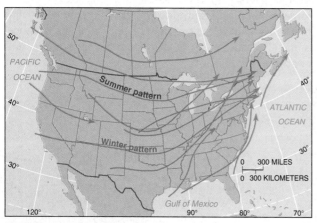

(a) Average storm tracks

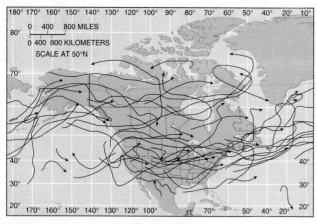

(b) Actual storm tracks in March 1991

FIGURE 8.16 Typical and actual storm tracks.
(a) Cyclonic storm tracks over North America vary seasonally. The tracks indicate several locations of cyclogenesis. (b) Actual cyclonic tracks during March 1991 over North America. [(b) From *Storm Data* 33, no. 3 (March 1991); Asheville, NC: NOAA, (NESDIS), National Climatic Data Center.]

A map of actual storm tracks for March 1991 in Figure 8.16b demonstrates several areas of cyclogenesis: the northwest over the Pacific Ocean, the Gulf of Mexico, the eastern seaboard, and the Arctic. Cyclonic circulation also frequently develops on the lee side of mountain ranges, as along the Rockies from Alberta south to Colorado. By crossing the mountains, such systems gain access to the moisture-laden, energy-rich mT air masses from the Gulf of Mexico.

Analysis of Daily Weather Maps—Forecasting

Synoptic analysis is the evaluation of weather data collected at a selected time. Building a database of wind, pressure, temperature, and moisture conditions is key to *numerical* (computer-based) *weather prediction* and the development of weather-forecasting models. Development of numerical models is a great challenge because the atmosphere operates as a nonlinear (irrational) system, tending toward chaotic behavior. Slight variations in input data or slight changes in the basic assumptions of the model's behavior can produce widely varying forecasts. As our knowledge of the interactions that produce weather and our instruments and software improve, so too will the accuracy of our forecasts.

Weather data necessary for the preparation of a synoptic map and forecast include:

- Barometric pressure (sea level and altimeter setting)
- Pressure tendency (steady, rising, falling)
- Surface air temperature
- Dew-point temperature
- Wind speed, direction, and character (gusts, squalls)
- Type and movement of clouds
- Current weather
- State of the sky (current sky conditions)
- Visibility; vision obstruction (fog, haze)
- Precipitation since last observation

For links to weather maps, current forecasts, satellite images, and the latest radar, please go to the *Geosystems* Home Page, to Chapter 8, "Destinations," and you will find many related links to the Internet. With the Internet, there is no need to wait for television or the newspaper to bring you the latest satellite image.

NOAA operates the Forecast Systems Laboratory, Boulder, Colorado, that is developing new forecasting tools (see http://www.fsl.noaa.gov). A few innovations include: wind profilers using radar to profile winds from the surface to high altitudes; a High Performance Computing System (277 CPUs networked to act as one) to model software for 3-D weather models and other computations; improved international cooperation and weather data dissemination; and a new standard Advanced Weather Interactive Processing System (AWIPS) enhanced by the installation of 883 Automated Surface Observing System, or ASOS (Figure 8.17).

(a)

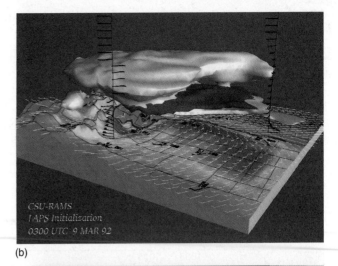

CSU-RAMS
IAPS Initialization
0300 UTC 9 MAR 92

(b)

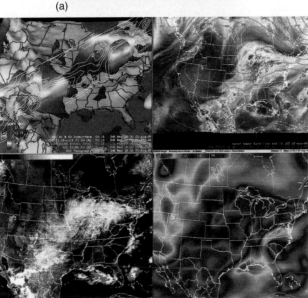

(c)

(d)

(e)

FIGURE 8.17 ASOS weather instruments and AWIPS workstation display.

(a) Automated Surface Observing System (ASOS) installation, one of 883 ASOS stations in use for data gathering as part of the U.S. primary surface weather observing network. (b) Forecast Systems Laboratory (FSL) scientists develop three-dimensional modeling programs to portray weather and simulate future weather for more accurate forecasting. On the image note the gray area that depicts significant cloud water content and the red layer within the cloud where aircraft icing is forecast. The black wind barbs off the two vertical axes denote wind profiles through the atmosphere. The view is from the southeast. (c) Powerful computers are needed to process the data and produce these 3-D virtual images on the Advanced Weather Interactive Processing System (AWIPS), employing screens with multiple frames. Of the many items that can be displayed we see in Figure 8.17c a simulated 3-D pressure analysis, water vapor and visible satellite image—including lightning strikes—and humidity. (d) An AWIPS workstation. (e) A portion of the High Performance Computing System at the FSL. [Photo (a) and (e) by Bobbé Christopherson; (b–d) images courtesy of the Forecast Systems Laboratory, National Oceanic and Atmospheric Administration, Boulder, Colorado.]

ASOS sensor instrument arrays are a primary surface weather-observing network (Figure 8.17a). An ASOS installation includes: rain gauge (tipping bucket), temperature/dew point sensor, barometer, present weather identifier, wind speed indicator, direction sensor, cloud height indicator, freezing rain sensor, thunderstorm sensor, and visibility sensor, among other items.

The sequence in Figure 8.14 illustrates an ideal midlatitude cyclone model. The actual pattern of cyclonic passage over North America is not so tidy; it is widely varied in shape and duration. Regardless, you can apply this gen-

eral model, along with your understanding of warm and cold fronts, to the actual midlatitude cyclone shown in the weather map and infrared satellite image in Figure 8.18. Preparing a weather report and forecast requires analysis of such daily weather maps and satellite images and the use of the standard weather symbols.

On the weather map in Figure 8.18, you can identify the temperatures reported by various stations, the patterns created, and the location of warm and cold fronts. The distribution of air pressure is defined by *isobars*, lines that connect points of equal pressure on the map. Although the

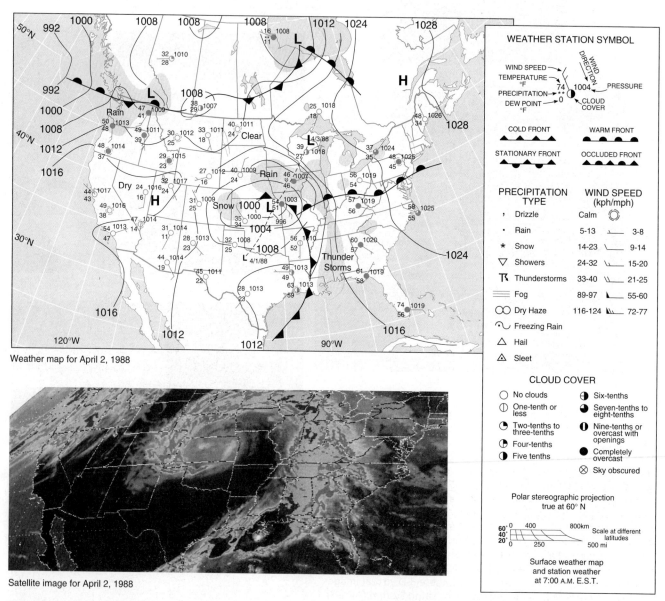

Weather map for April 2, 1988

Satellite image for April 2, 1988

FIGURE 8.18 Weather map.
Weather map and *GOES-7* satellite image for April 2, 1988. The standard weather symbols used are identified and explained. Note the track of the center of low pressure from April 1 through April 3, beginning in Texas and ending in northern Michigan (dashed line on map). The map and image also illustrate the pattern of precipitation as the warm and cold fronts sweep eastward. The mT air mass, advancing northward from the Gulf/Atlantic source region, provides the moisture for afternoon and evening convectional thunder showers. [Data for map and infrared image from National Weather Service and NOAA, NESDIS, National Climatic Data Center.]

low-pressure center identified is not intense, it is well defined over eastern Nebraska, with winds following in a counterclockwise, cyclonic path around the low pressure. This pattern is clear on the satellite image. A high-pressure cP air mass dominates Nevada and Utah. Note the dryness in Ely, Nevada, where the air temperature is −4.4°C (24°F) and the dew point is −8.8°C (16°F)! Compare this map with the weather map in Figure 8.14b.

Violent Weather

Weather provides a continuous reminder of the flow of energy across the latitudes that at times can set into motion destructive, violent weather conditions. In dollar value, weather-related damage is increasing each year, as population increases and people settle in hazardous areas. We focus on thunderstorms, tornadoes, and hurricanes.

However, we must include mention of such things as *ice storms* of **sleet** (freezing rain, ice glaze, and ice pellets), snow blizzards, and low temperatures. Sleet is caused when precipitation falls through a below-freezing layer of air near the ground (Figure 8.19). A large region was without power for weeks. Imagine a coating of ice on everything, weighing on power lines and tree limbs!

Weather-related losses exceeded $10 billion per year through the 1990s, eclipsing the previous average of less than $2 billion a year. As mentioned earlier, 1998 broke all records for weather-related damage worldwide. We now look at specific types of violent weather: thunderstorms, tornadoes, and tropical cyclones. Government research and monitoring of violent weather is centered at NOAA's National Severe Storms Laboratory in several cities (see **http://www.nssl.noaa.gov**; consult this site for each of the topics that follow).

Thunderstorms

The condensation of large quantities of water vapor in clouds liberates tremendous amounts of energy. This process locally heats the air, rapidly changing its density and buoyancy and causing violent updrafts. As rising air parcels pull surrounding air into the column, updrafts become stronger. Raindrops form and descend through the cloud, and their frictional drag pulls air toward the ground, causing violent downdrafts. The resulting giant cumulonimbus clouds signal dramatic weather moments—heavy precipitation, lightning, thunder, hail, blustery winds, and maybe tornadoes. Such thunderstorms may develop under three conditions: within an air mass (particularly in warm, moist air), in a line along a cold front or other convergent boundary, or where mountain slopes cause orographic lifting.

Thousands of thunderstorms occur on Earth at any given moment. Equatorial regions and the ITCZ experience many of them, exemplified by the city of Kampala in Uganda, East Africa (north of Lake Victoria), which sits virtually on the equator and averages a record 242 days a year

FIGURE 8.19 Aftermath of record ice storm.
For more than a week in January 1998, record blizzards and ice storms from a nor'easter storm blanketed New England and eastern Canada, causing almost a billion dollars in damage. These broken birch trees in Maine, snapped under the weight of an ice coating as thick as 8 cm (3 in.), remain as reminders of this weather disaster. Power was out for weeks. [Photo by Bobbé Christopherson.]

of thunderstorms. Figure 8.20 shows the annual distribution of days with thunderstorms across the United States and Canada. You can see that, in North America, most thunderstorms occur in areas dominated by mT air masses.

Atmospheric Turbulence   Most airplane flights experience at least some turbulence—the encountering of air of different densities, or air layers moving at different speeds and directions. This is a natural state of the atmosphere and passengers are asked to keep seat belts fastened when in their seats to avoid injury, even if the seat-belt light is turned off.

Thunderstorms can produce severe turbulence in the form of downbursts, which are exceptionally strong downdrafts. Downbursts are classified by size: a *macroburst* is at least 4.0 km (2.5 mi) wide and in excess of 210 kmph (130 mph), a *microburst* is less in size and speed. A microburst causes rapid changes in wind speed and direction, and it causes the dreaded *wind shear* that can bring down aircraft. Such turbulence events are short-lived and hard to detect, although the Forecast Systems Laboratory, among others, is making progress in developing forecasting methods.

Lightning and Thunder   An estimated 8 million lightning strikes occur each day on Earth. **Lightning** refers to flashes of light caused by enormous electrical discharges—tens of millions to hundreds of millions of volts—that briefly superheat the air to temperatures of

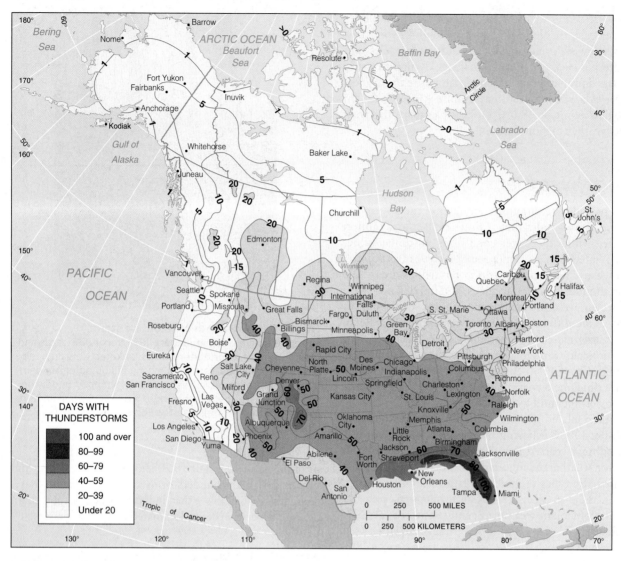

FIGURE 8.20 Thunderstorm occurrence.
Average annual number of days with thunderstorms. [Data courtesy of U.S. National Weather Service; Map Series 3, *Climatic Atlas of Canada*, Atmospheric Environment Service, Canada.]

15,000–30,000°C (27,000–54,000°F). A buildup of electrical energy between areas within a cumulonimbus cloud or between the cloud and the ground creates lightning (Figure 8.21a). The violent expansion of this abruptly heated air sends shock waves through the atmosphere as the sonic bang of **thunder.**

NASA's Lightning Imaging Sensor (LIS) aboard the *Tropical Rainfall Measuring Mission* (*TRMM*) satellite monitors lightning. The LIS can image lightning strikes day or night, within clouds, or between cloud and ground. The sensor's data show that about 90% of all strikes occur over land in response to increased convection over relatively warmer continental surfaces, with expected seasonal shifts with the high Sun as shown in Figure 8.21b and c. (See **http://thunder.msfc.nasa.gov/lis/**.)

Lightning poses a hazard to aircraft, people, animals, trees, and structures. Certain precautions are mandatory when a lightning discharge threatens, because lightning causes nearly 200 deaths and thousands of injuries each year in the United States and Canada. When lightning is imminent, agencies such as the National Weather Service issue *severe storm warnings* and caution people to remain indoors. If the hair on your head or neck begins to stand on end, get indoors, or if in the open get low to the ground, crouched on both feet and not laying down, in a low spot, if possible. Your hair is telling you that a charge is building in that area. The place *not* to seek shelter is beneath a tree, for trees are good conductors of electricity and often are hit by lightning.

Hail Ice pellets called **hail** generally form within a cumulonimbus cloud. Raindrops circulate repeatedly above and below the freezing level in the cloud, adding layers of ice until their weight no longer can be supported by the circulation in the cloud (Figure 8.22a). Hail may also grow from the addition of moisture on a snow pellet.

Pea-sized and marble-sized hail are common, although hail the size of golf balls and baseballs happens at least once

FIGURE 8.21 Seasonal images show global lightning.
(a) Multiple cloud-to-ground lightning strikes captured in a time-lapse photo. (b) A composite of three months' data derived from NASA's Lightning Imaging Sensor (LIS) records all lightning strikes between 35° N and 35° S latitudes during (b) winter (December 1999–February 2000) and (c) summer (June, July, August 2000). The LIS is aboard the *Tropical Rainfall Measuring Mission* satellite launched in 1997. [(a) Photo from C. Clark, NOAA Photo Library, NOAA Central Library, NSSL. Images (b) and (c) courtesy of NASA's Global Hydrology and Climate Center, Marshall Spaceflight Center, Huntsville, Alabama.]

(a)

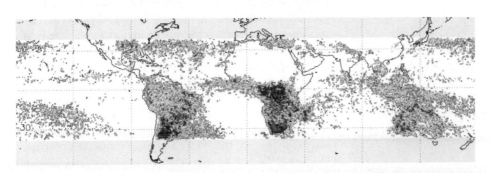

(b)   Winter (Dec. 1999, Jan. and Feb. 2000)

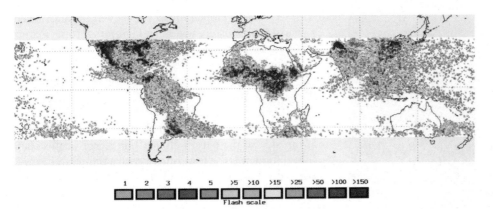

(c)   Summer (June, July, August 2000)

or twice a year somewhere in North America. For larger hail to form, the frozen pellets must stay aloft for longer periods. The largest authenticated hailstone in the world landed in Kansas in 1970; it measured 14 cm (5.6 in.) in diameter and weighed 758 g (1.67 lb)!

Hail is common in the United States and Canada, although somewhat infrequent at any given place. Hail occurs perhaps every one or two years in the highest-frequency areas. Annual hail damage in the United States tops $800 million (Figure 8.22b). The pattern of hail occurrence

(a)

(b)

FIGURE 8.22 Hailstones and hailstone damage.
(a) Irregular layers give evidence of numerous convective trips through a cumulonimbus cloud. (b) Hail can severely damage windows, automobiles, crops, livestock, and people.
[(a) © Arthur C. Smith III/Grant Heilman Photography, Inc.; (b) Photo by *The Denver Post*, 1990.]

across the United States and Canada is similar to that of thunderstorms shown in Figure 8.20.

Tornadoes

The updrafts associated with a cumulonimbus cloud appear on satellite images as pulsing bubbles of clouds (Figure 8.23). Friction with the ground slows surface winds, but

FIGURE 8.23 Thunderstorms on the Great Plains.
These boiling cloud tops signify severe thunderstorms on the Great Plains along a cold front with upper-level support, on October 9, 2001. This storm system produced nearly 100 reports of hail and strong winds and 22 tornado strikes in Nebraska and Oklahoma. [*NOAA-12* AVHRR image courtesy of NOAA.]

higher in the troposphere, winds blow faster. Thus, a body of air moves faster at altitude than at the surface, creating a rotation in the air along a horizontal axis parallel to the ground (picture a rolling pin; see Figure 8.24a). When that rotating air encounters the strong updrafts associated with frontal activity, the axis of rotation shifts to a vertical alignment perpendicular to the ground (Figure 8.24b).

According to one hypothesis for *supercell tornadoes*, this spinning, rising column of mid-troposphere-level air forms a **mesocyclone**. A mesocyclone can range up to 10 km (6 mi) in diameter and rotate vertically within a *supercell cloud* (the parent cloud) to a height of thousands of meters (Figure 8.24c). As a mesocyclone extends vertically and contracts horizontally, wind speeds accelerate in an inward vortex (as ice skaters draw their arms in to accelerate while spinning or as water speeds up when it spirals down a sink drain). A well-developed mesocyclone will produce heavy rain, large hail, blustery winds, and lightning; some mature mesocyclones generate tornado activity.

As more moisture-laden air is drawn up into the circulation of a mesocyclone, more energy is liberated and the rotation of air becomes more rapid as a *wall cloud* forms (Figure 8.25). The swirl of the mesocyclone itself is visible, as are dark gray **funnel clouds** that pulse from the bottom side of the parent cloud. The terror of this stage of development is the lowering of such a funnel to Earth—a **tornado** (Figures 8.24d and e and 8.25). When tornado circulation occurs over water, a **waterspout** forms, and the sea or lake surface water is drawn up into the funnel.

Tornado Measurement and Science   A tornado's diameter can range from a few meters to a few hundred meters. It can last from a few moments to tens of minutes. Pressures inside a tornado funnel usually are about 10% less than in the surrounding air. This disparity is similar to the pressure difference between sea level and an altitude

FIGURE 8.24 Mesocyclone and tornado formation.
(a) Strong wind aloft, establishes spinning along a horizontal axis. (b) Updraft from thunderstorm development tilts the rotating air along a vertical axis. (c) Mesocyclone forms as a rotating updraft within the thunderstorm. If a tornado forms, it will descend from the cloud base and the lower portion of the mesocyclone. (d) A fully developed tornado locks onto the ground. (e) This tornado is emerging from the base of the wall cloud near Oakfield, Wisconsin. [Photos by (d) C. Lavoie/First Light; (e) Don Lloyd/AP/Wide World Photos.]

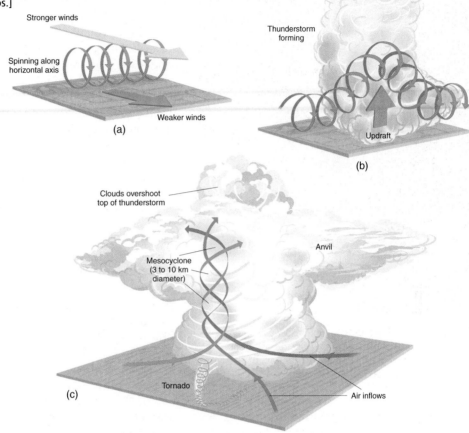

FIGURE 8.25 Supercell tornado.
A supercell tornado descends from the cloud base near Spearman, Texas. Strong hail is falling to the left of the tornado. [Photo by Howard Bluestein, all rights reserved.]

of 1000 m (3300 ft). Such a sharp horizontal-pressure gradient causes an in-rushing convergence and severe wind speeds at the funnel. Winds can exceed 485 kmph (300 mph), with at least a third of all tornadoes exceeding 181 kmph (113 mph).

The late Theodore Fujita, a noted meteorologist from the University of Chicago, designed a scale for classifying tornadoes according to wind speed as indicated by relat-

ed property damage. The Fujita Scale ranks tornadoes and is widely used (Table 8.1). Of all tornadoes, the frequency distribution on the Fujita Scale is: F0–F1 = 74%, F2–F3 = 25%, and F4–F5 = 1%. Theoretically, the scale continues from F6 to F12, although no tornado is expected to ever reach F6.

Tornadoes have struck all 50 states and all the Canadian provinces and territories. May and June are the peak months, as you see in Figure 8.26, based on 51 years of records. Other continents report a small number of tornadoes each year, but North America receives the greatest share because its latitudinal position, topography (the lay of the land), and shape permit contrasting air masses to confront one another. Looking at the spatial distribution map, can you identify the infamous "Tornado Alley"—a corridor that experiences the highest average number of tornadoes on Earth?

Using satellites, aircraft, and surface measurements, the National Center for Atmospheric Research and the University of Oklahoma completed a 2-year project in 1995 called VORTEX (Verification of the Origin of Rotation in Tornadoes). VORTEX intercepted 10 tornadoes for close study. The VORTEX Project paved the way for Subvortex, the latest cooperative research effort. In 1997 a Subvortex team equipped with a twin-Doppler radar unit on a van obtained high-resolution images of an F1 tornado southwest of Tulsa. (See **http://mrd3.nssl.ucar. edu/~vortex/** for a summary of case studies and modeling efforts.)

The National Severe Storms Forecast Center in Kansas City, Missouri, is a key forecasting center. As a result of Doppler radar, only about 15% of tornadoes strike without some public warning. Warning times of about 12 minutes are generally possible with current technology (see the Storm Prediction Center at **http://www.spc.noaa.gov/ index.shtml**).

Recent Tornado Records In the United States, 41,297 tornadoes were recorded in the 51 years from 1950 through 2000, causing 4501 deaths or about 88 deaths

Table 8.1	The Fujita Scale
F-number	**Wind Speed; Damage Specs**
F0	18–32 m/sec (64–116 kmph; 40–72 mph); *light damage*: break branches, damage chimneys.
F1	33–49 m/sec (117–180 kmph; 73–112 mph); *moderate damage*: beginning of hurricane wind speed designation, roof coverings peel off, mobile homes pushed off foundations.
F2	50–69 m/sec (181–253 kmph; 113–157 mph); *considerable damage*: roofs torn off frame houses, large trees uprooted or snapped, box cars pushed over, small missiles generated.
F3	70–92 m/sec (254–332 kmph; 158–206 mph); *severe damage*: roofs torn off well-constructed houses, trains overturned, trees uprooted, cars thrown.
F4	93–116 m/sec (333–419 kmph; 207–260 mph); *devastating damage*: well-built houses leveled, cars thrown, large missiles generated.
F5	117–142 m/sec (420–512 kmph; 261–318 mph); *incredible damage*: houses lifted and carried distance to disintegration, car-sized missiles fly farther than 100 m, bark removed from trees.

See **http://www.tornadoproject.com/fscale/fscale.htm**.

FIGURE 8.26 Tornado occurrence in the United States, 1950–2000.
(a) Average number of tornadoes per 26,000 km² (10,000 mi²) for 1950 to 2000. (b) Average number of tornadoes per month. Remarkably, the January average of 15 was exceeded by 155 tornadoes during January 1999. [Data courtesy of the National Severe Storm Forecast Center, National Weather Service, Kansas City, Missouri.]

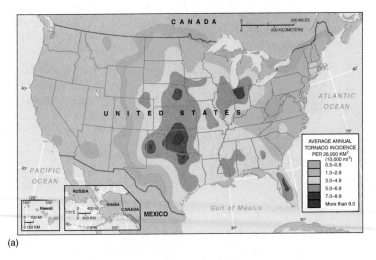

(a)

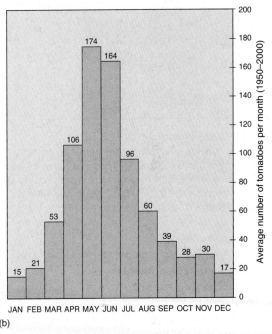

(b)

per year, although tornadoes took 130 lives in the United States in 1998 alone. Of note, these tornadoes resulted in more than 77,000 injuries and property damage of over $25 billion over this half century. The yearly average of $475 million in damage is rising at about $50 million a year. Importantly, the 1998–2000 seasons produced 4 F5 and 22 F4 tornadoes.

The annual average number of tornadoes before 1990 was 787. Interestingly, since 1990 the average per year has risen to 1172, with 1992, 1995, and 1996 surpassing 1200; 1999 more than 1300; and 1998 more than 1400. During the year 2000, 1071 tornadoes occurred. Reasons for the annual increase in tornado frequency in North America range from global climate change to better reporting by a larger population armed with videotape recorders. Researchers are pursuing many questions with an enthusiasm for the science and an awe of the subject, as summarized by Howard Bluestein, foremost tornado researcher and meteorologist:

> The source of rotation in some tornadoes appears to be preexisting vortices near the ground above where convective storms are growing; in others, the source seems to be linked to the mesocyclone. Does the mesocyclone itself descend to the ground and intensify to become the tornado, or does it trigger events near the ground that create tornadoes from other sources of rotation? Can it do both? We hope the answers to these questions are soon forthcoming. Our quest for discovery has not taken away from the respect we have for the awesome power the tornado harbors, nor the thrill for viewing the violent motions in the tornado or the beauty of the storm. We eagerly await the next act in the atmospheric play starring the tornado.[*]

[*]H. Bluestein, *Tornado Alley, Monster Storms of the Great Plains* (New York: Oxford University Press, 1999), p. 162.

Tropical Cyclones

A powerful manifestation of the Earth–atmosphere energy budget is the **tropical cyclone**, which originates entirely within tropical air masses. The tropics extend from the Tropic of Cancer at 23.5° N to the Tropic of Capricorn at 23.5° S, containing the equatorial zone between 10° N and 10° S. Approximately 80 tropical cyclones occur annually worldwide. Some 45 per year are powerful enough to be classified as hurricanes, typhoons, and cyclones (different regional names for the same type of tropical storm)—30% of these in the western North Pacific. Of note, the lowest sea-level pressure recorded on Earth was 870 mb (25.69 in.) in the center of Typhoon Tip, in October 1979, northwest of Guam.

Cyclonic systems forming in the tropics are quite different from midlatitude cyclones because the air of the tropics is essentially homogeneous, with no fronts or conflicting air masses of differing temperatures. In addition, the warm air and warm seas ensure abundant water vapor and thus the necessary latent heat to fuel these storms.

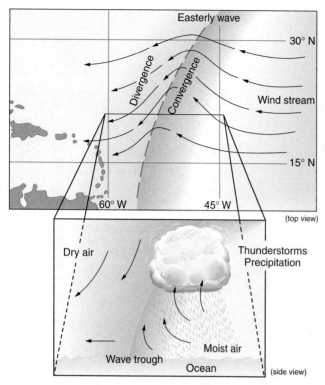

FIGURE 8.27 Easterly wave in the tropics.
Development of a low-pressure center along an easterly
(westward-moving) wave. Moist air rises in an area of convergence at the surface to the east of the wave trough. Wind
flows bend and converge before the trough and diverge
downwind from the trough.

Tropical cyclones convert heat energy from the ocean into
the mechanical energy in the wind.

What mechanism triggers the start of a tropical cyclone? Meteorologists now think that cyclonic motion
begins with slow-moving easterly waves of low pressure in
the trade-wind belt of the tropics, such as the Caribbean
area (Figure 8.27). Sea-surface temperatures must exceed
approximately 26°C (79°F). Tropical cyclones form along
the eastern (leeward) side of these migrating troughs of low
pressure, a place of convergence and rainfall. Surface airflow then spins into the low-pressure area (convergence),
ascends, and flows outward (divergence) aloft. This important divergence aloft acts as a chimney, pulling more
moisture-laden air into the developing system. To maintain and strengthen this vertical convective circulation,
there must be little or no wind shear to interrupt or block
the air flow.

The map in Figure 8.28 shows areas in which tropical
cyclones form and some of the characteristic storm tracks
traveled. The map indicates the range of months during
which tropical cyclones are most likely to appear. They
tend to occur when the equatorial low-pressure trough
(ITCZ) is farthest from the equator, during the months
following the summer solstice in each hemisphere. For
example, storms that strike the southeastern United States
do so mostly between August and October, following the
June 21 solstice. Officially, tropical cyclone season is from
June 1 to November 30 each year. (For comparison, the

average midlatitude cyclonic storm tracks are in blue on
the map.)

Relative to North and Central America, tropical depressions (low-pressure areas) intensify into tropical storms
as they cross the Atlantic. A rule of thumb has emerged: If
they mature early along their track, they tend to curve
northward toward the North Atlantic and miss the United
States. The critical position is approximately 40° W longitude. If a tropical storm matures after it reaches the longitude of the Dominican Republic (70° W), then it has a
higher probability of hitting the United States.

Since record keeping began in 1870, the second-
greatest year for number of Atlantic hurricanes was 1995.
In 1998 Hurricanes Bonnie and Georges damaged property and took lives. Hurricane Mitch (October 26–
November 4, 1998) was the deadliest Atlantic Hurricane
in 218 years, killing more than 12,000 people in Central
America. The years 1999 and 2000 had normal occurrences: 12 named storms and 8 hurricanes, and 14 storms
and 8 hurricanes, respectively.

Hurricanes, Typhoons, and Cyclones Tropical cyclones are potentially the most destructive storms experienced by humans, claiming thousands of lives each year
worldwide. This is especially true when they attain wind
speeds and low-pressure readings that upgrade their status to a full-fledged **hurricane**, **typhoon**, or *cyclone*
(>65 knots, 74 mph, 119 kmph). Across the globe, such
storms bear these three different names, among others.
Around North America, the term *hurricane* is in use,
typhoon applies in the western Pacific (Japan, Philippines),
and *cyclone* in Indonesia, Bangladesh, and India. By any
name they can be destructive killers. Worldwide, about
10% of all tropical disturbances have the right ingredients to become hurricanes or typhoons (see the National
Hurricane Center at **http://www.nhc.noaa.gov/**, or the
Joint Typhoon Warning Center at **http://www.npmoc.
navy.mil/jtwc.html**).

Table 8.2 presents the criteria for classification of tropical cyclones on the basis of wind speed, and it lists some of
their features. When you hear meteorologists speak of a
"category 4" hurricane, they are using the Saffir–Simpson
Hurricane Damage Potential Scale to estimate possible
damage from hurricane-force winds (see table). This scale,
using wind speeds and central pressure criteria, ranks hurricanes and typhoons in five categories, from smaller, category 1 storms to extremely dangerous (and rare) category
5. Damage depends on the degree of property development
at a storm's landfall site, how prepared citizens are for the
blow, the storm surge, embedded tornadoes, and the overall intensity of the storm.

The cyclone that struck Bangladesh in 1970 killed an
estimated 300,000 people, and the one in 1991 claimed more
than 200,000 lives. The Galveston, Texas, hurricane of 1900
killed 6000; Hurricane Audry (1957), 400; Hurricane Gilbert
(1988), 318; Hurricane Camille (1969), 256; and Hurricane
Mitch (1998), more than 12,000 dead with many still missing. The record holder for property damage is 1992's Hurricane Andrew, which tallied $20 billion in damage.

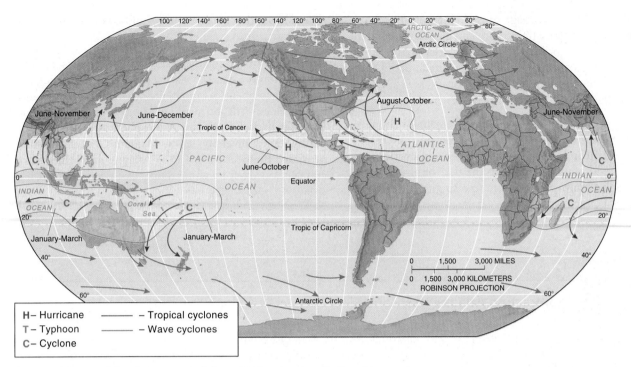

FIGURE 8.28 Worldwide pattern of the most intense tropical cyclones.
Typical tropical storm tracks with principal months of occurrence and regional names. For the
North Atlantic region from 1871 to 2000, 1084 tropical cyclones (storms and hurricanes) developed. The peak day in this Atlantic region (assuming a 9-day moving average calculation)
is September 10, with 64 cyclones for the 129-year period. Also shown for comparison are
characteristic storm tracks of midlatitude cyclones.

Table 8.2   Tropical Cyclone Classification

Designation	Winds	Features
Tropical disturbance	Variable, low	Definite area of surface low pressure; patches of clouds
Tropical depression	Up to 34 knots (63 kmph, 39 mph)	Gale force, organizing circulation; light to moderate rain
Tropical storm	35–63 knots (63–118 kmph, 39–73 mph)	Closed isobars; definite circular organization; heavy rain; assigned a name
Hurricane (Atlantic and E. Pacific) Typhoon (W. Pacific) Cyclone (Indian Ocean, Australia)	Greater than 65 knots (119 kmph, 74 mph)	Circular, closed isobars; heavy rain, storm surges; tornadoes in right-front quadrant

Saffir-Simpson Hurricane Damage Potential Scale

Category	Wind Speed	Notable Atlantic Examples
1	65–82 knots (74–95 mph)	—
2	83–95 knots (96–110 mph)	—
3	96–113 knots (111–130 mph)	1985 Elena; 1991 Bob; 1995 Roxanne, Marilyn; 1998 Bonnie
4	114–135 knots (131–155 mph)	1979 Frederic; 1985 Gloria; 1995 Felix, Luis, Opal; 1998 Georges
5	>135 knots (>155 mph)	1935 No. 2; 1938 No. 4; 1960 Donna; 1961 Carla; 1969 Camille; 1979 David; 1988 Gilbert; 1989 Hugo; 1992 Andrew; 1998 Mitch

Statistically, the damage caused by tropical cyclones is increasing substantially as more and more development occurs along susceptible coastlines, whereas loss of life is decreasing in most parts of the world owing to better forecasting and understanding of these storms. *Science* offered this "Perspective":

> The risk of human losses is likely to remain low, however, because of a well-established warning and rescue system and ongoing improvements in hurricane prediction. A main concern is the risks of high damage costs (up to $100 billion in a single event) because of ongoing population increases in coastal areas and increasing investment in buildings and extensive infrastructure in general.*

Ironically, the relatively mild hurricane seasons from the mid-1960s to mid-1990s encouraged weak zoning and rapid development of vulnerable coastal lowlands. This is problematic as Atlantic hurricanes since 1995 entered a more intensive period of activity. Atmospheric scientists at Colorado State University, led by William Gray, successfully predicted the near-record 1995 Atlantic hurricane season and developed a forecasting model (see Focus Study 8.1).

*L. Bengtsson, "Hurricane Threats," *Science* 293 (July 20, 2001): p. 441.

Focus Study 8.1

Forecasting Atlantic Hurricanes

Forecasters at the National Hurricane Center (NHC) in Miami were busy throughout the 1995–2000 hurricane seasons, for this was the most active 6-year period in the history of the NHC, with 76 named tropical storms, including 48 hurricanes (23 intense). This was a record level of activity despite the El-Niño-retarded 1997 season. In 2000 the Atlantic tropical storm season was 134% of expected Net Tropical Cyclone activity, which translates to 14 named storms (9.3 is average), including 8 hurricanes (5.8 is average, or 138% of normal). For comparison, tropical storm activity for 1995 was 235% of an average season.

An important innovation to help the NHC and state and local governments assess these storms is a new forecasting method developed by a team at Colorado State University led by William M. Gray (see **http://typhoon. atmos.colostate.edu/forecasts/**).

Their analysis of weather records for the period 1900–2000 disclosed a significant causal relationship between Atlantic tropical storms and more intense hurricanes that make landfall along the U.S. Gulf and East Coasts with several meteorological variables. Of 16 potential predictive elements used by the Colorado scientists, 5 key forecast variables emerge:

1. The presence or absence of warm water off the coast of Peru in the eastern Pacific. Sea-surface temperatures are part of the El Niño and La Niña phenomena (see Chapter 10 focus study). A strong El Niño featuring warm surface water in the eastern Pacific Ocean tends to dampen the development of Atlantic hurricanes. Associated with El Niño events are strong westerly winds in the upper troposphere (12,000 m, 40,000 ft) that tend to shear off the tops of developing tropical depressions and storms. The opposite effect occurs during La Niña episodes when cool water occurs in the eastern Pacific Ocean. These periods are associated with weak westerly winds, thus encouraging storm intensification in the Atlantic. During 1997, the Pacific El Niño was the strongest of the century, more than double the intensity of the 1982–1983 El Niño! Such an episode reduces the Atlantic storm season.

2. Rainfall, temperature, and pressure patterns in West Africa, specifically the Sahel along the southern margins of the Sahara (10°–20° N). Temperature and pressure regimes from February to May can set the stage for heavy rainfall during the normal June to September rainy season. A wet February to May stimulates storm formation off the African coast. A drought in this same region tends to retard Atlantic storm development.

3. Directional flows of equatorial stratospheric winds (20,000–23,000 m altitude, 68,000–75,000 ft), which tend to shift and reverse roughly every 12 to 16 months. These are the quasi-biennial oscillation (QBO) winds. Winds from the west (westerlies) double Atlantic tropical storm probability and intensity, whereas when the QBO easterly flow intensifies, storm tops shear and development is discouraged.

4. Sea-surface temperatures in the Atlantic Ocean, between Africa and the Caribbean, and along the North American East Coast. Warmer surface water favors tropical storm development. Important is the functioning of the Atlantic *thermohaline circulation,* or *Atlantic conveyor belt.* Imagine a river of water flowing northward from the Caribbean. Near Greenland the current sinks deep beneath the surface and flows southward to the southern Atlantic Ocean. Higher than average sea-surface temperature and salinity level are related to a more intense tropical storm/hurricane season—the trend since 1994. Lower temperatures and salinity tend to dampen storm activity—characteristic of 1900–1925 and 1970–1993 periods.

5. The status of the Azores high-pressure anticyclone (see Figure 6.14 and text discussion). If the circulation system is weak and below average in strength, then the east

(continued)

Focus Study 8.1 (continued)

Atlantic trade winds are weakened and tropical storm development is favored. A strong Azores high dampens development.

Dr. Gray's team has issued forecasts for 17 years and scored better than climatology (long-term averages), especially when it comes to predicting named storms. New discoveries followed each of these years as the forecast model was refined.

The Colorado State team issues its forecast each December (for the next year) and issues updates in April, June, and August. (To check the accuracy of the present forecast, see the URL listed previously.)

The 1995 Atlantic Hurricane Season—An Example

As Atlantic hurricane seasons go (June 1 to November 1), 1995 was the second-most active since record keeping began in 1870. Of 19 tropical storms, 11 became hurricanes, and 5 of these achieved category 3 status or higher. Only 1933 (21 tropical storms) and 1969 (12 hurricanes) exceeded this total. During the Atlantic hurricane season, the year 1995 experienced tropical storms 52% of the season and hurricanes 27% of the time—both records. Hurricanes Allison, Erin, and Opal together killed 66 and damaged $3.7 billion in U.S. property—the season totaled $8 billion in overall damage.

Figure 1 maps all 19 tropical storms, including 11 hurricanes (colored tracks). The season began with Allison on June 3–11 and ended with Tanya from October 27 through November 1. Figure 2 is a satellite image from August 30 that shows tropical storms Karen and Luis, Hurricanes Humberto and Iris, and the remnants of tropical storm Jerry off the image to the far left.

Each storm went through stages of tropical depression, tropical storm, hurricane, and extratropical depression, based on wind speed (depression, storm, hurricane) and location of the low-pressure center (tropical or *(continued)*

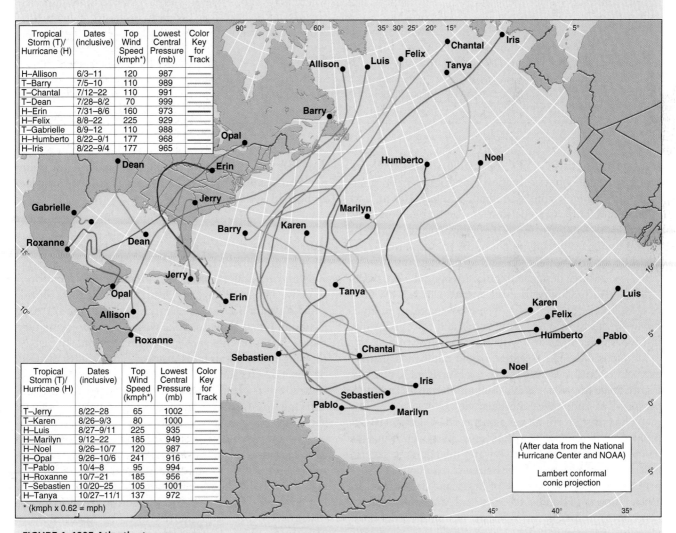

Tropical Storm (T)/ Hurricane (H)	Dates (inclusive)	Top Wind Speed (kmph*)	Lowest Central Pressure (mb)	Color Key for Track
H–Allison	6/3–11	120	987	
T–Barry	7/5–10	110	989	
T–Chantal	7/12–22	110	991	
T–Dean	7/28–8/2	70	999	
H–Erin	7/31–8/6	160	973	
H–Felix	8/8–22	225	929	
T–Gabrielle	8/9–12	110	988	
H–Humberto	8/22–9/1	177	968	
H–Iris	8/22–9/4	177	965	

Tropical Storm (T)/ Hurricane (H)	Dates (inclusive)	Top Wind Speed (kmph*)	Lowest Central Pressure (mb)	Color Key for Track
T–Jerry	8/22–28	65	1002	
T–Karen	8/26–9/3	80	1000	
H–Luis	8/27–9/11	225	935	
H–Marilyn	9/12–22	185	949	
H–Noel	9/26–10/7	120	987	
H–Opal	9/26–10/6	241	916	
T–Pablo	10/4–8	95	994	
H–Roxanne	10/7–21	185	956	
T–Sebastien	10/20–25	105	1001	
H–Tanya	10/27–11/1	137	972	

* (kmph x 0.62 = mph)

(After data from the National Hurricane Center and NOAA)

Lambert conformal conic projection

FIGURE 1 1995 Atlantic storm season.
Atlantic, Caribbean, and Gulf of Mexico hurricane season, 1995. Noted on the map are inclusive dates from each storm's birth as a tropical depression to dissipation, highest wind speed in kmph (kmph × 0.62 = mph), and lowest pressure (mb). [Data courtesy of NOAA and the National Hurricane Center, Miami.]

Focus Study 8.1 *(continued)*

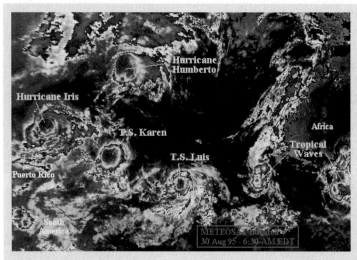

FIGURE 2 Satellite observes four storms at one time. Tropical storms Karen and Luis and Hurricanes Humberto and Iris march westward across the Atlantic Ocean in fairly tight formation during late August 1995. The remnant of tropical storm Jerry is seen at the far left. [Image from National Weather Service and NOAA, (NESDIS), National Climatic Data Center.]

extratropical). Top wind speed and lowest central pressure are noted on the map for each hurricane.

Several areas were particularly hard hit. Three hurricanes crossed the Florida panhandle within 160 km (100 mi) of each other in June, July, and September. Also, the eastern Caribbean was savaged, especially the islands of Dominica, Saint Thomas, and Saint Maarten, where Luis and Marilyn hit within a few days of each other in September. As one researcher noted, "Maybe lightning does strike the same place twice!"

No matter how accurate storm forecasts become, coastal and lowland property damage will continue to increase until better hazard zoning and development restrictions are in place. The property insurance industry appears to be taking action to promote these improvements. They are requiring tougher building standards to obtain coverage—or, in some cases, they are refusing to insure property along vulnerable coastal lowlands.

Physical Structure   Fully organized tropical cyclones have an intriguing physical appearance (Figure 8.29). They range in diameter from a compact 160 km (100 mi), to 1000 km (600 mi), to some western Pacific typhoons that attain 1300–1600 km (800–1000 mi). Vertically, these storms dominate the full height of the troposphere.

The inward-spiraling clouds form dense rain bands, with a central area designated the *eye*. Around the eye swirls a thunderstorm cloud called the *eyewall*, which is the area of most intense precipitation. The eye remains an enigma, for in the midst of devastating winds and torrential rains the eye is quiet and warm with even a glimpse of blue sky or stars possible. The structure of the rain bands, central eye, and eyewall is clearly visible for Hurricane Gilbert in Figure 8.29a (top view), 8.29b (oblique view from an artist's perspective), and 8.29c (side-radar view).

The entire storm moves along at 16–40 kmph (10–25 mph). When it makes **landfall** (moves ashore), it pushes dangerous **storm surges** of seawater inland, often several meters in depth. Storm surges often catch people by surprise and cause the majority of hurricane drownings. In 1998, Hurricanes Bonnie, Georges, and Mitch moved slowly onshore, producing damaging storm surges along North Carolina, Mississippi, and Central America, respectively. The slow progress of these storms produced much flooding from the sustained rains.

The strongest winds of a tropical cyclone are usually recorded in its right-front quadrant (relative to the storm's directional path), where dozens of fully developed tornadoes may be embedded at the time of landfall. For example, Hurricane Camille in 1969 had up to 100 tornadoes embedded in that quadrant.

Hurricane Andrew–1992   The period between August 24 and September 11, 1992, displayed nature's power and capacity to cause personal, societal, and economic tragedy. Those 19 days saw Hurricane Andrew (Florida and Louisiana), Hurricane Iniki (Kaua'i, Hawai'i), and Typhoon Omar (Guam) strike with record-breaking fury. Figure 8.30 presents an AVHRR image from August 24, showing Andrew as it moved across Florida, into the Gulf of Mexico. The locator map shows the entire track. The central eye is clearly visible, as are the tight rain bands marking the rapid inward-spiraling, counterclockwise wind flow.

Sustained winds were 225 kmph (140 mph), with gusts to 282+ kmph (175+ mph). Studies completed by meteorologist Theodore Fujita estimated that winds in the eyewall reached 320 kmph (200 mph) in small vortexes. The remnants of Hurricane Andrew continued northward, reaching Québec and eastern Canada by August 28. Locally, heavy rains were produced as the remnants of Andrew interacted with weather fronts moving through the area. In Québec, many local 24-hour precipitation records were broken.

The tragedy from Andrew is that the storm destroyed or seriously damaged 70,000 homes and left 200,000 peo-

(a)

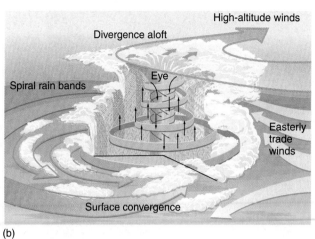

(b)

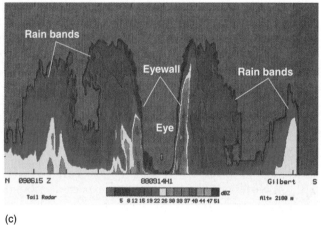

(c)

FIGURE 8.29 Profile of a hurricane.
For the Western Hemisphere, Hurricane Gilbert attained the record size (1600 km, or 1000 mi, in diameter) and record low barometric pressure (888 mb, or 26.22 in.). Gilbert's sustained winds reached 298 kmph (185 mph) with peaks exceeding 320 kmph (200 mph). Hurricane Gilbert, September 13, 1988: (a) *GOES-7* satellite image; (b) a stylized portrait of a mature hurricane, drawn from an oblique perspective (cutaway view shows the eye, rain bands, and wind flow patterns); (c) *SLAR* (side-looking airborne radar) image from an aircraft flying through the center of the storm. Rain bands of greater cloud density are false-colored in yellows and reds. The clear sky in the central eye is dramatically portrayed. [(a), (b), (c) NOAA and the National Hurricane Center, Miami.]

ple homeless between Miami and the Florida Keys. A year later, 60,000 people still were homeless and reconstruction was progressing slowly. Had Hurricane Andrew made landfall a few kilometers farther north, it would have blasted Miami; damage estimates for that missed catastrophe are more than $85 billion.

The Everglades, the coral reefs north of Key Largo, some 10,000 acres of mangrove wetlands, and coastal southern pine forests all took significant hits from Andrew. Natural recovery will take years, which offers an opportunity for study. Scientific assessments judge the Everglades to be quite resilient because the region's ecosystems have evolved naturally with periodic hurricanes over the millennia. An important fact is that storms damage human structures

more than they damage natural systems. Remember this perspective: *Urbanization, agriculture, pollution, and water diversion pose a greater ongoing threat to the Everglades than do hurricanes!*

New buildings, apartments, and government offices are opening right next to still-visible rubble and bare foundation pads. Unfortunately, sensitive planning to guide the settlement of these high-risk areas has never been policy. Public or private decision makers do not seem prone to thoughtful hazard planning, whether along coastal lowlands, river floodplains, or earthquake fault zones.

The consequence is that our entire society bears the financial cost of planning failure, whether in Florida, California, or the Midwest, not to mention those victims who

(a)

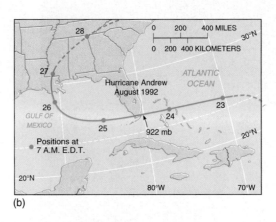

(b)

FIGURE 8.30 Hurricane Andrew.
(a) *NOAA-10* AVHRR image of Hurricane Andrew, August 24, 1992 (4:41 P.M. EDT) right after passage across southern Florida. (b) The track of Hurricane Andrew from August 23 through 28, 1992. [Image from NOAA as supplied by the EROS Data Center.]

directly shoulder the physical, emotional, and economic hardship of the event. This recurrent, yet avoidable cycle—construction, devastation, reconstruction, devastation—was reinforced ironically in *Fortune* magazine more than 30 years ago after the destruction by Hurricane Camille:

> Before long the beachfront is expected to bristle with new motels, apartments, houses, condomini-

ums, and office buildings. Gulf Coast businessmen, incurably optimistic, doubt there will ever be another hurricane like Camille, and even if there is, they vow, the Gulf Coast will rebuild bigger and better after that.*

**Fortune, October 1969, p. 62.

Summary and Review—Weather

Weather is the short-term condition of the atmosphere; **meteorology** is the scientific study of the atmosphere. The spatial implications of atmospheric phenomena and their relationship to human activities strongly link meteorology to physical geography. Analyzing and understanding patterns of wind, air pressure, temperature, and moisture conditions portrayed on daily weather maps is key to numerical (computer) weather forecasting.

weather (p. 211)
meteorology (p. 212)

● **Describe** air masses that affect North America and *relate* their qualities to source regions.

Specific conditions of humidity, stability, and cloud coverage occur in a regional, homogenous **air mass**. The longer an air mass remains stationary over a region, the more definite its physical attributes become. The homogeneity of temperature and humidity in an air mass sometimes extends through the lower half of the troposphere. Air masses are categorized by their moisture content—**m** for maritime (wetter) and **c** for

continental (drier)—and their temperature (a function of latitude)—designated **A** (arctic), **P** (polar), **T** (tropical), **E** (equatorial), and **AA** (Antarctic).

air mass (p. 212)

1. How does a source region influence the type of air mass that forms over it? Give specific examples of each basic classification.
2. Of all the air masses, which are of greatest significance to the United States and Canada? What happens to them as they migrate to locations different from their source regions? Give an example of air mass modification.

● *Identify* types of atmospheric lifting mechanisms and *describe* four principal examples.

Air masses can be lifted by **convergent lifting** (airflows conflict, forcing some of the air to lift), **convectional lifting** (air passing over warm surfaces gains buoyancy), **orographic lifting** (passage over a topographic barrier), and *frontal lifting*. In North America, **chinook winds** (called *föhn* or *foehn* winds in

Europe) are the warm, downslope airflows characteristic of the leeward side of mountains. Orographic lifting creates wetter windward slopes and drier leeward slopes situated in the **rain shadow** of the mountain. Conflicting air masses at a front produce a **cold front** (and sometimes a zone of strong wind and rain) or a **warm front**. A zone right along or slightly ahead of the front, called a **squall line**, is characterized by turbulent and wildly changing wind patterns and intense precipitation.

convergent lifting (p. 216)
convectional lifting (p. 216)
orographic lifting (p. 216)
chinook winds (p. 216)
rain shadow (p. 217)
cold front (p. 220)
warm front (p. 220)
squall line (p. 220)

3. Explain why it is necessary for an air mass to be lifted if there is to be saturation, condensation, and precipitation.
4. What are the four principal lifting mechanisms that cause air masses to ascend, cool, condense, form clouds, and perhaps produce precipitation? Briefly describe each.
5. Differentiate between the structure of a cold front and a warm front.

● *Analyze* **the pattern of orographic precipitation and** *describe* **the link between this pattern and global topography.**

The physical presence of a mountain acts as a topographic barrier to migrating air masses. *Orographic lifting* (*oro* means "mountain") occurs when air is forcibly lifted upslope as it is pushed against a mountain. It cools adiabatically. An orographic barrier enhances convectional activity and causes additional lifting during the passage of weather fronts and cyclonic systems, thereby extracting more moisture from passing air masses. The precipitation pattern of windward and leeward slopes persists worldwide. The state of Washington is presented in the chapter as an excellent example of this concept.

6. When an air mass passes across a mountain range, many things happen to it. Describe each aspect of a mountain crossing by a moist air mass. What is the pattern of precipitation that results?
7. Explain how the distribution of precipitation in the state of Washington is influenced by the principles of orographic lifting.

● *List* **the measurable elements that contribute to weather and** *describe* **the life cycle of a midlatitude cyclonic storm system.**

Synoptic analysis involves the collection of weather data at a specific time, as shown on the synoptic weather map in Figure 8.18. Building a data base of wind, pressure, temperature, and moisture conditions is key to numerical (computer-based) weather prediction and the development of weather-forecasting models. Preparing a weather report and forecast requires

analysis of a daily weather map and satellite images that depict atmospheric conditions. Weather data from these sources include the following:

- Barometric pressure (sea level and altimeter setting)
- Pressure tendency (steady, rising, falling)
- Surface air temperature
- Dew-point temperature
- Wind speed, direction, and character (gusts, squalls)
- Type and movement of clouds
- Current weather
- State of the sky (current sky conditions)
- Visibility; vision obstruction (fog, haze)
- Precipitation since last observation

A **midlatitude cyclone**, or **wave cyclone**, is a vast low-pressure system that migrates across the continent, pulling air masses into conflict along fronts. **Cyclogenesis**, the birth of the low-pressure circulation, can occur off the west coast of North America, along the polar front, along the lee slopes of the Rockies, in the Gulf of Mexico, and along the East Coast. A midlatitude cyclone can be thought of as having a life cycle of birth, maturity, old age, and dissolution. An **occluded front** is produced when a cold front overtakes a warm front in the maturing cyclone. Sometimes a **stationary front** develops between conflicting air masses, where air flow is parallel to the front on both sides. These systems are guided by the jet streams of the upper troposphere along seasonally shifting **storm tracks**.

midlatitude cyclone (p. 221)
wave cyclone (p. 221)
cyclogenesis (p. 222)
occluded front (p. 223)
stationary front (p. 223)
storm tracks (p. 224)

8. Differentiate between a cold front and a warm front as types of frontal lifting and what you would experience with each one.
9. How does a midlatitude cyclone act as a catalyst for conflict between air masses?
10. What is meant by cyclogenesis? In what areas does it occur and why? What is the role of upper-tropospheric circulation in the formation of a surface low?
11. Diagram a midlatitude cyclonic storm during its open stage. Label each of the components in your illustration, and add arrows to indicate wind patterns in the system.
12. What is your principal source of weather data, information, and forecasts? Where does your source obtain its data? Have you used the Internet and World Wide Web to obtain weather information? In what ways will you personally apply this knowledge in the future? What benefits do you see?

● *Analyze* **various forms of violent weather and the characteristics of each.**

The violent power of some weather phenomena poses a hazard to society. Severe ice storms involve **sleet** (freezing rain, ice glaze, and ice pellets), snow blizzard, and crippling ice coatings on roads, power lines, and crops. Thunderstorms produce **lightning** (electrical discharges in the atmosphere), **thunder** (sonic bangs produced by the rapid expansion of air after intense heating by lightning), and **hail** (ice pellets formed within cumulonimbus clouds). A spinning, cyclonic column rising to midtroposphere level—a **mesocyclone**—is sometimes visible as the swirling mass of a cumulonimbus cloud. Dark gray **funnel clouds** pulse from the bottom side of the parent cloud. A **tornado** is formed when the funnel connects with Earth's surface. A **waterspout** forms when a tornado circulation occurs over water.

Within tropical air masses, large low-pressure centers can form along easterly wave troughs. Under the right conditions, a tropical cyclone is produced. Depending on wind speeds and central pressure, a **tropical cyclone** can become a **hurricane**, **typhoon**, or cyclone, when winds exceed 65 knots (74 mph, 119 kmph). As forecasting and the public's perception of weather-related hazards have improved, loss of life has decreased, although property damage continues to increase. Great damage occurs to occupied coastal lands when hurricanes make **landfall** and when winds drive ocean water inland in **storm surges**.

sleet (p. 227)
lightning (p. 227)
thunder (p. 228)
hail (p. 228)
mesocyclone (p. 230)
funnel clouds (p. 230)

tornado (p. 230)
waterspout (p. 230)
tropical cyclone (p. 233)
hurricane (p. 234)
typhoon (p. 234)
landfall (p. 238)
storm surges (p. 238)

13. What constitutes a thunderstorm? What type of cloud is involved? What type of air mass would you expect in an area of thunderstorms in North America?

14. Lightning and thunder are powerful phenomena in nature. Briefly describe how they develop.

15. Describe the formation process of a mesocyclone. How is this development associated with that of a tornado?

16. Evaluate the pattern of tornado activity in the United States. Where is Tornado Alley? What generalizations can you make about the distribution and timing of tornadoes?

17. What are the different classifications for tropical cyclones? List the various names used worldwide for hurricanes.

18. What factors contributed to the incredible damage cost of Hurricane Andrew? Why have such damage figures increased, whereas loss of life has decreased over the past 30 years?

19. What forecast factors did scientists use to accurately predict the 1995 Atlantic hurricane season?

20. Relative to improving weather forecasting, what are some of the technological innovations discussed in this chapter?

NetWork

The *Geosystems* Home Page provides on-line resources for this chapter on the World Wide Web. To begin: Once on the Home Page, click on this textbook, scroll the Table of Contents menu, select this chapter, and click "Begin." You will find self-tests that are graded, short essay and review exercises, specific updates for items in the chapter, and many links to interesting related pathways on the Internet under "Destinations." *Geosystems* is at **http://www.prenhall.com/christopherson**.

Critical Thinking

A. Select "Destinations" from the left-hand frame of Chapter 8 on the *Geosystems* Home Page. Be prepared—this is a fantastic listing of many links on the Net. Any item in color and underlined will open and take you directly to a website. Sample at least five of these weather links and describe what you find. You may want to mark some of these as links with your own browser for ready reference when you need weather information.

B. Relative to coastal devastation from tropical weather, the following line appears in the text: "This recurrent, yet avoidable cycle—construction, devastation, reconstruction, devastation. . . ." This describes the ever-increasing dollar losses to property from tropical storms and hurricanes, yet at a time when improved forecasts have resulted in a significant reduction in loss of life. (These issues are discussed further in Chapter 16's Focus Study 16.1.) In your opinion, what is the solution to halt these increasing losses, this cycle of destruction? How would you implement your plan?

C. From your local newspaper (or an Internet source), collect daily weather maps for five successive days. Compare your map sequence and identify principal weather elements on the maps. Were published forecasts for your region accurate? What source is listed for these forecast maps? Government or private contractor?

Tracy Smith, Research Meteorologist

After almost two decades at the Forecast Systems Laboratory (FSL), Tracy Smith is still enthusiastic about her work. Tracy asserted, "Weather is exciting! There is such satisfaction in developing a model and forecasting something and then having it happen—predict it and then find it. We are challenged to improve weather forecasting. If we can add a few minutes to severe weather warning times and narrow the focus of the warning area, or improve aviation safety, lives can be saved."

FSL developed the Advanced Weather Interactive Processing System (AWIPS) to assist forecasters across the country (pictured in Figure 8.17). Tracy works on computer models to improve the "data ingest" of all the information acquired from a growing list of new technologies. As part of her Rapid Update Cycle (RUC) effort, she is working to increase the update frequency, accuracy, and resolution of forecasting models. Tracy said, "This is one of the reasons atmospheric science is such a wide-open field as we attack these challenges."

When I asked about her interest in weather she responded with a spirited laugh, "I didn't really decide to be a meteorologist until I was 15 or 16 years old!" I reminded her that this was an interesting thing for a teenager to say. She said, "When I was four or five my brother and sister brought a schoolbook home called *Hurricanes and Twisters.* I read it over and over." She was fascinated by the sky and, living in northeastern Ohio, she remembers those "big booming nocturnal thunderstorms in the summer."

When she was seven years old, the great Palm Sunday tornado outbreak of 1965 hit Ohio at Pittsfield (with an F5) and nearby Grafton (with an F4) where she visited her grandparents. She said, "I remember that night, wak-

FIGURE 1 Tracy Smith, Research Meteorologist. Tracy works at the Forecast Systems Laboratory, National Weather Service, NOAA. Here she shows this author the workings of an AWIPS workstation. [Photo by Bobbé Christopherson.]

ing up and hearing this great storm outside." Tracy's grandparents' house was slightly damaged but the homes around them were leveled—she walked along the street and saw the devastation. Her childhood fascination and experiences blossomed into a great career working with the atmospheric sciences she loves. Tracy admits to sometimes using vacation time to storm chase!

Tracy graduated from Bowling Green University, Ohio, with a degree in geography and a minor in math. She then earned her Master's degree in atmospheric sciences at Colorado State University, Fort Collins. For her thesis, she studied the radiation budget at the top of the atmosphere over the Indian monsoon region, using satellite data. Her committee was made up of both atmospheric science and geography professors. "Geography is important to atmospheric science because we are dealing with spatial elements of the atmosphere, hydrologic cycle, and topography. And, because of the impact on people, geography brings in the human aspect," she states.

Tracy participates in the daily weather briefings at FSL and once a month conducts the briefing. She says it is important to practice communication skills by using these weather briefings in order to be able to explain the forecast to others.

As to the future: "We want to refine the system, gain higher resolution, integrate all new data sources as they develop, and improve accuracy. We are just scratching the surface to get all this operational. The speed and capability of these computers are increasing so fast that this will be the next revolution in forecasting for us. Improving scale and accuracy are our ongoing challenges." And with her infectious laugh she asserts, "I like my job! I am paid to learn and study things and follow my instincts to solve big problems. What better way to spend your day than trying to understand the world around you while helping society!" Tracy's enthusiasm is obvious as she demonstrates the AWIPS workstation to me.

Irrigated cranberry bog near the coast at **Duxbury, Massachusetts.** [Photo by Bobbé Christopherson.]

9

Water Resources

Key Learning Concepts

After reading the chapter, you should be able to:

- *Illustrate* the hydrologic cycle with a simple sketch and *label* it with definitions for each water pathway.
- *Relate* the importance of the water-budget concept to your understanding of the hydrologic cycle, water resources, and soil moisture for a specific location.
- *Construct* the water-balance equation as a way of accounting for the expenditures of water supply and *define* each of the components in the equation and their specific operation.
- *Describe* the nature of groundwater and *define* the elements of the groundwater environment.
- *Identify* critical aspects of freshwater supplies for the future and *cite* specific issues related to sectors of use, regions and countries, and potential remedies for any shortfalls.

The physical reality of life is defined by water. Our lives are bathed and infused with water. Our own bodies are about 70% water, as are plants and animals. We use water to cook, bathe, wash clothing, and dilute our wastes. We water small gardens and vast agricultural tracts. Most industrial processes would be impossible without water. It is the essence of our existence and is therefore the most critical resource supplied by Earth systems—the essential resource for life.

Fortunately, water is a renewable resource, constantly cycling through the environment, endlessly renewed. Even so, some 80 countries face impending water shortages, either in quantity or quality, or both. One billion people lacked access to safe water in 2001; some 1.8 billion lacked adequate sanitary facilities. During the first half of the new century, water availability per person will drop by 74%, as population increases and adequate quality water decreases. Author Peter Gleick's *The World's Water 2000–2001* (Washington, DC: Island Press, 2000, p. 13), summarizes:

The overall economic and social benefits of meeting basic water requirements far outweigh any reasonable assessment of the costs of providing those needs. . . . Water-related diseases cost society on the order of $125 billion per year. . . . Yet the cost of providing new infrastructure needs for all major urban water sectors has been estimated at around $25 to $50 billion per year. While these costs are far below the costs of failing to meet these needs, they are two to three times the average rate of spending for water during the 1980s and 1990s.

In the last two chapters, we saw how the exchanges of energy between water and the atmosphere drive Earth's weather systems. The flow of water links the atmosphere, ocean, land, and living things through exchanges of energy and matter. In particular, the energy and moisture exchange between plants and the atmosphere is important to the status of the climate system and the range of responses from plant communities to climate change.

In this chapter: The hydrologic cycle and global water balance give us a model for understanding the global plumbing system, so this important cycle begins the chapter. Water spends time in the ocean, in the air, on the surface, and underground as groundwater. Water availability to plants from precipitation and from the soil is critical to water-resource issues.

We look at the water resource using a water-budget approach—similar in many ways to a money budget—in which we examine water "receipts" and "expenses" at specific locations. Precipitation provides the principal receipt of moisture, whereas evaporation and plant transpiration are the principal expenditures. This budget approach can be applied at any scale, from a small garden, to a farm, to a regional landscape.

Water is not always naturally available where and when we want it. Consequently, we rearrange surface-water resources to suit our needs. We drill wells, build cisterns and reservoirs, and dam and divert streams to redirect water either spatially (geographically, from one area to another) or temporally (over time, from one part of the calendar to another). All of this activity constitutes water-resource management.

This chapter concludes by considering the quantity and quality of the water we withdraw and consume for irrigation, industrial, and municipal uses—our specific water supply. Adequate water supplies in terms of quantity and quality loom as *the resource issue* for many parts of the world in this century. Ismail Serageldin, chair of the World Water Commission, is quoted bluntly in a recent book on water issues: ". . . the wars of the twenty-first century will be fought over water. . . . Water is the most critical issue facing human development."*

The Hydrologic Cycle

Vast currents of water, water vapor, ice, and energy are flowing about us continuously in an elaborate, open global-plumbing system. Together they form the **hydrologic cycle**, which has operated for billions of years, from the lower atmosphere to several kilometers beneath Earth's surface. The cycle involves the circulation and transformation of water throughout Earth's atmosphere, hydrosphere, lithosphere, and biosphere. (See NASA's Global Hydrology and Climate Center site, at **http://www.ghcc. msfc.nasa.gov/**, for a combined government and academic effort to study the global hydrologic cycle and related climatic effects.)

Modern study of the hydrologic cycle involves computer modeling, direct observation, and remote sensing. A better understanding of the hydrologic cycle is central to understanding water resources and global climate change. Such systems as the hydrologic cycle operate in a chaotic manner, making model building difficult.

A Hydrologic Cycle Model

Figure 9.1 is a simplified model of this complex system. Let's use the ocean as a starting point for our discussion, although we could jump into the model at any point. More than 97% of Earth's water is in the ocean, and here most evaporation and precipitation occur. We can trace 86% of all evaporation to the ocean. The other 14% is from the land, including water moving from the soil into plant roots and passing through their leaves (a process called *transpiration*, described later in this chapter).

In the figure, you can see that, of the 86% of evaporation rising from the ocean, 66% combines with 12% advected (moving horizontally) from the land to produce the 78% of all precipitation that falls back into the ocean. The remaining 20% of moisture evaporated from the ocean, plus 2% of land-derived moisture, produces the 22% of all precipitation that falls over land. Clearly, the bulk of continental precipitation comes from the oceanic portion of the cycle.

*M. De Villiers, *Water, The Fate of Our Most Precious Resource* (New York: Houghton Mifflin Co., 2000), p. 13–4.

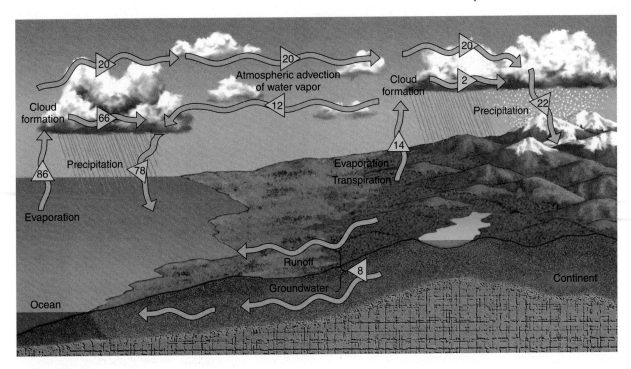

FIGURE 9.1 The hydrologic cycle model.
The model shows how water travels endlessly through the hydrosphere, atmosphere, litho-sphere, and biosphere. The triangles show global average values as percentages. Note that all evaporation (86% + 14% = 100%) equals all precipitation (78% + 22% = 100%), when all of Earth is considered. Regionally, various parts of the cycle will vary, creating imbalances and, depending on climate, surpluses in one region and shortages in another.

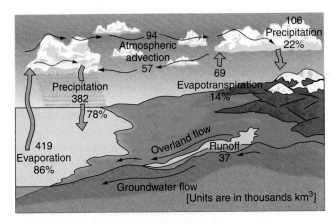

FIGURE 9.2 The global water balance in the hydrologic cycle.
The annual volume of water in all parts of the hydrologic cycle as measured in thousands of cubic kilometers. A balance exists between total evaporation and transpiration and precipitation and between advection in the atmosphere and surface runoff of water (1 km³ × 0.24 = 1 mi³). Percentages are drawn from Figure 9.1, as given in the small directional arrows.

Figure 9.2 presents a global water balance in the hydrologic cycle. The percentages from Figure 9.1 are given with the volume (in 1000 km³) of water flowing along these pathways.

Surface Water

Precipitation that reaches Earth's surface follows two basic pathways: It either flows overland or soaks into the soil. Along the way **interception** occurs when precipitation strikes vegetation or other ground cover. Intercepted water that drains across plant leaves and down their stems to the ground is *stem flow* and can be an important moisture route to the ground surface. Precipitation that falls directly to the ground, coupled with drips onto the ground from vegetation (excluding stem flow), constitutes *throughfall*. Water soaks into the subsurface through **infiltration**, or penetration of the soil surface. It further permeates soil or rock through downward movement called **percolation**. These concepts are shown in Figure 9.3.

The atmospheric advection of water vapor from sea to land and land to sea at the top of Figures 9.1 and 9.2 appears to be unbalanced: 20% (94,000 km³) moving inland but only 12% (57,000 km³) moving out to sea. However, this exchange is balanced by the 8% (37,000 km³) land runoff that flows to the sea. Most of this runoff—about 95%—comes from surface waters that wash across land as overland flow and streamflow. Only 5% of runoff is slow-moving subsurface groundwater. These percentages indicate that the small amount of water in rivers and streams is very dynamic, whereas the large quantity of subsurface water is sluggish in comparison and represents only a small portion of total runoff.

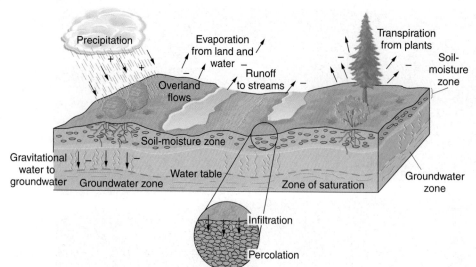

FIGURE 9.3 The soil-moisture environment.
Precipitation supplies the soil-moisture environment. The principal pathways for water include interception by plants; throughfall to the ground; collection on the surface, forming overland flow to streams; transpiration and evaporation from plants; evaporation from land and water; and gravitational water moving to subsurface groundwater. Water moves from the surface into the soil by infiltration and percolation.

The residence time for a water molecule in any part of the hydrologic cycle determines its relative importance in affecting Earth's climates. The short time spent by water in transit through the atmosphere (an average of 10 days) plays a role in temporary fluctuations in regional weather patterns. Long residence times, such as the 3000–10,000 years in deep-ocean circulation, groundwater aquifers, and glacial ice, act to moderate temperature and climatic changes. These slower parts of the cycle work as a "system memory"; the long periods over which heat energy is stored and released buffer the effects of change.

To observe and describe the hydrologic cycle and its related energy budgets, scientists established in 1988 the Global Energy and Water Cycle Experiment (GEWEX). This is part of the World Climate Research Program and represents an important focus in climate-change studies. As part of this effort the GEWEX Americas Prediction Project, a multi-scale hydrometeorological investigation of the hydrologic cycle, is underway (GAPP, see **http://www.ogp.noaa.gov/mpe/gapp/index.htm**). The goal is to correlate land surface, general circulation, regional climate, and hydrologic models to better predict seasonal and annual water resource changes. Now that you are acquainted with the hydrologic cycle, let's examine the concept of the soil-water budget as a method of assessing water resources.

Soil-Water-Budget Concept

A **soil-water budget** can be established for any area of Earth's surface—a continent, country, region, field, or front yard. Key is measuring the precipitation input and its distribution to satisfy the "demands" of plants, evaporation, and soil moisture storage in the area considered. Such a budget can examine any time frame, from minutes to years.

Think of a soil-water budget as a money budget: precipitation income must be balanced against expenditures of evaporation, transpiration, and runoff. Soil-moisture storage acts as a savings account, accepting deposits and withdrawals of water. Sometimes all expenditure demands are met, and any extra water results in a surplus. At other times, precipitation and soil moisture income are inadequate to meet demands, and a deficit, or water shortage, results.

Geographer Charles W. Thornthwaite (1899–1963) pioneered in applied water-resource analysis and worked with others to develop a water-balance methodology. They applied water-balance concepts to geographic problems, especially to irrigation, which requires accurate quantity and timing of water application to maximize crop yields.

Thornthwaite also developed methods for estimating evaporation and transpiration. He recognized the important relation between water supply and local water demand as an essential climatic element. In fact, his initial use of these techniques was to develop a climatic classification system.

The Soil-Water-Balance Equation

To understand Thornthwaite's water-balance methodology and "accounting" or "bookkeeping" procedures, we must first understand some terms and concepts. Figure 9.4 illustrates the essential aspects of a soil-water budget. Precipitation (mostly rain and snow) provides the moisture input. The object, as with a money budget, is to account for the ways in which this supply is distributed: actual water taken by evaporation and plant transpiration, extra water that exits in streams and subsurface groundwater, and recharge or utilization of soil-moisture storage.

Figure 9.4 organizes the water-balance components into an equation. As in all equations, the two sides must balance; that is, the precipitation receipt (left side) must be fully accounted for by expenditures (right side). Follow this water-balance equation as you read the following paragraphs. To help you learn these concepts I use letter abbreviations (such as PRECIP) for the components.

Precipitation (PRECIP) Input The moisture supply to Earth's surface is **precipitation** (PRECIP, or P). It arrives as rain (water drops), sleet (icy glaze and ice pellets), snow (ice crystals), and hail (balls or irregular lumps of ice). In

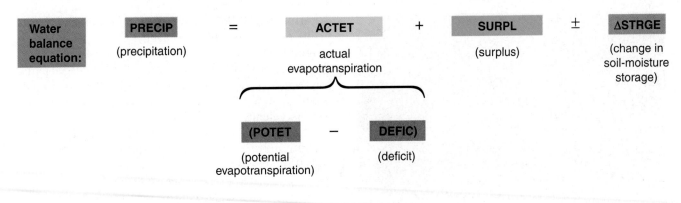

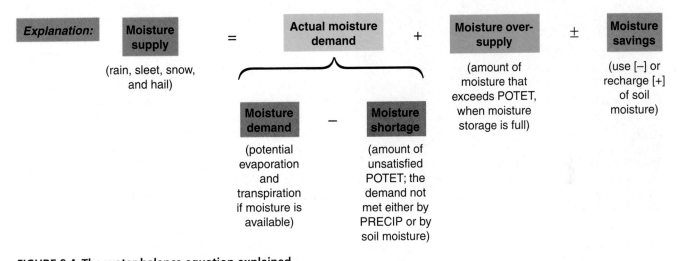

FIGURE 9.4 The water-balance equation explained.
The outputs (components to the right of the equal sign) are an accounting of expenditures
from the precipitation input (to the left).

some climates, PRECIP also deposits on Earth's surface as dew, frost, or fog. (Recall from the hydrologic cycle that 78% of Earth's PRECIP falls on the ocean and 22% falls on land. It is the land portion we examine here.)

Precipitation is measured with the **rain gauge**. A rain gauge is essentially a large measuring cup, collecting rainfall and snowfall so the water can be measured by depth, weight, or volume (Figure 9.5). Wind can cause an undercatch because the drops or snowflakes are not falling vertically. For example, a wind of 37 kmph (23 mph) produces an undercatch as great as 40%, meaning that an actual 1-in. rainfall might gauge at only 0.6 in. The *windshield* you see above the gauge's opening reduces this error by catching raindrops that arrive at an angle. According to the World Meteorological Organization (http://www.wmo.ch/), more than 40,000 weather-monitoring stations are operating worldwide, with more than 100,000 places measuring precipitation.

Figure 9.6 maps precipitation patterns for the United States and Canada. Note the generally wet East and Northwest and the drier western interiors and extreme north. (Chapter 10 presents a world precipitation map in Figure 10.2.) PRECIP is the principal input to the water-balance equation (Figure 9.4). All the components discussed next are outputs or expenditures of this water.

Actual Evapotranspiration (ACTET) **Evaporation** is the net movement of free water molecules away from a wet surface into air that is less than saturated. **Transpiration** is a cooling mechanism in plants. When a plant transpires, it moves water through small openings (stomata) in the underside of its leaves. The water evaporates, cooling the plant, much as perspiration cools humans. Transpiration is partially controlled by the plants themselves. Control cells around the stomata conserve or release water. Transpired quantities can be significant: On a hot day, a single tree can transpire hundreds of liters of water; a forest, millions of liters.

Evaporation and transpiration are important water-budget expenditures, and both respond directly to air temperature and humidity. Rates decrease when air is cold (can hold less moisture) or when there is high humidity (at or near saturation); both increase when air is hot (can hold more moisture) and dry (not near saturation). Evaporation and transpiration are combined into one term—**evapotranspiration**. (In the hydrologic cycle, Figure 9.1, 14% of

FIGURE 9.5 A rain gauge.
A standard rain gauge is cylindrical. A funnel guides water into a bucket that is sitting on an electronic weighing device. The gauge minimizes evaporation, which would cause low readings. The wind shield around the top of the gauge minimizes the undercatch produced by wind. [Photo by Bobbé Christopherson.]

evaporation and transpiration occurs from land and plants.) Now, let's examine ways to estimate evapotranspiration rates.

Potential Evapotranspiration (POTET) Evapotranspiration is an actual expenditure of water. In contrast, **potential evapotranspiration** (POTET, or PE) is the amount of water that would evaporate and transpire under optimum moisture conditions (adequate precipitation and adequate soil-moisture supply).

Filling a bowl with water and letting it evaporate illustrates this concept: When the bowl becomes dry, is there still an evaporation demand? The demand, of course, remains, regardless of whether the bowl is dry. The amount of water that would evaporate or transpire if water always were available is the POTET, or the amount that would evaporate from the bowl if it constantly were supplied with water. The amount of ultimate POTET demand that went unmet in the dry bowl is the shortage, or deficit (DEFIC). Note that in the water-balance equation, when we subtract the deficit from the potential evapotranspiration, we derive what actually happened—ACTET.

Determining POTET Although exact measurement is difficult, one method of measuring POTET employs an **evaporation pan**, or *evaporimeter*. As evaporation occurs, water in measured amounts is automatically replaced in the pan, equaling the amount that evaporated. Mesh screens over the pan protect against mismeasurement due to wind, which accelerates evaporation.

A more elaborate measurement device is a **lysimeter**. A tank, approximately a cubic meter in size or larger, is buried in a field with its upper surface left open. The lysimeter isolates a representative volume of soil, subsoil, and plant cover, thus allowing measurement of the moisture moving through the sampled area. A weighing lysimeter has this embedded tank resting on a weighing scale (Figure 9.7). A rain gauge next to the lysimeter measures the precipitation input. (See the Agricultural Research Service at http://www.ars.usda.gov/.)

Lysimeters and evaporation pans are limited somewhat in availability across North America, but they still provide a data base from which to develop ways of estimating POTET. Several methods of estimating POTET on the basis of meteorological data are widely used and easily implemented for regional applications. Thornthwaite developed one of these methods. He discovered that if you know monthly mean air temperature and daylength, you could approximate POTET. These data are readily available, so calculating POTET is easy and fairly accurate for most midlatitude locations. (Recall that daylength is a function of a station's latitude.) His method can work with data from hourly, daily, or annual time frames or with historical data to re-create water conditions in past environments.

Thornthwaite's method works better in some climates than in others. It ignores warm or cool air movements (wind) across a surface, sublimation and surface-water retention at subfreezing temperatures, poor drainage, and frozen soil (*permafrost*, Chapter 17), although there is allowance for snow accumulations. Nevertheless, we use his water-balance method here because of its great utility as a teaching tool and its overall ease of application and acceptable results for geographic studies that analyze water budgets over large areas.

Figure 9.8 presents POTET values derived by the Thornthwaite method for the United States and Canada. Note that higher values occur in the South, with highest readings in the Southwest (higher average air temperatures and lower relative humidities). Lower POTET values are found at higher latitudes and elevations (lower average temperatures).

Compare this POTET (demand) map with the PRECIP (supply) map in Figure 9.6. The relation between the two determines the remaining components of the water-balance equation in Figure 9.4. From the two maps, can you identify regions where PRECIP is greater than POTET (for example, the eastern United States)? Or where POTET is greater than PRECIP (for example, the southwestern United States)? Where you live, is the water demand usually met by the precipitation supply? Or does your area experience a natural shortage? One way to determine the

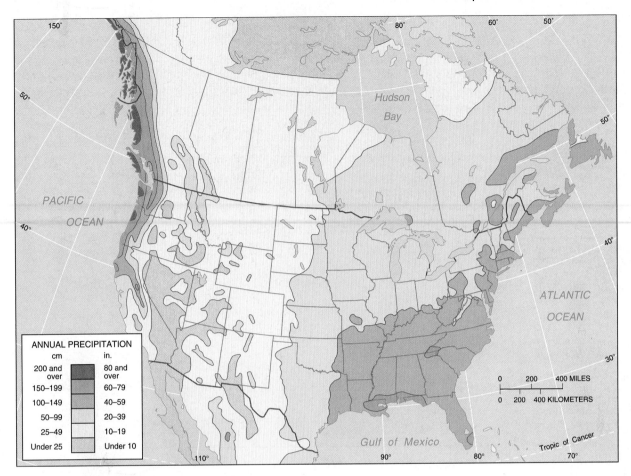

FIGURE 9.6 Precipitation in North America.
Annual precipitation (water supply, PRECIP) in the United States and Canada. [Adapted from U.S. National Weather Service, U.S. Department of Agriculture, and Environment Canada.]

FIGURE 9.7 Lysimeter.
A weighing lysimeter for measuring evaporation and transpiration. The various pathways of water are tracked: Some water remains as soil moisture, some is incorporated into plant tissues, some drains from the bottom of the lysimeter, and the remainder is credited to evapotranspiration. Given natural conditions, the lysimeter measures actual evapotranspiration. [Adapted from illustration courtesy of Lloyd Owens, Agricultural Research Service, USDA, Coshocton, Ohio.]

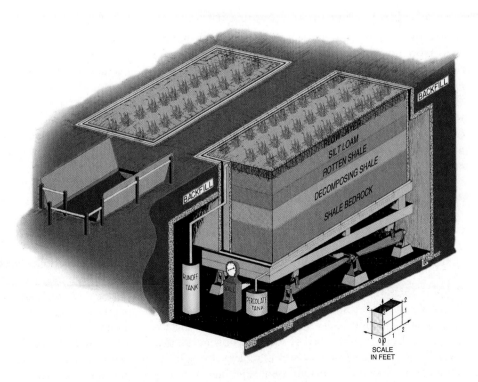

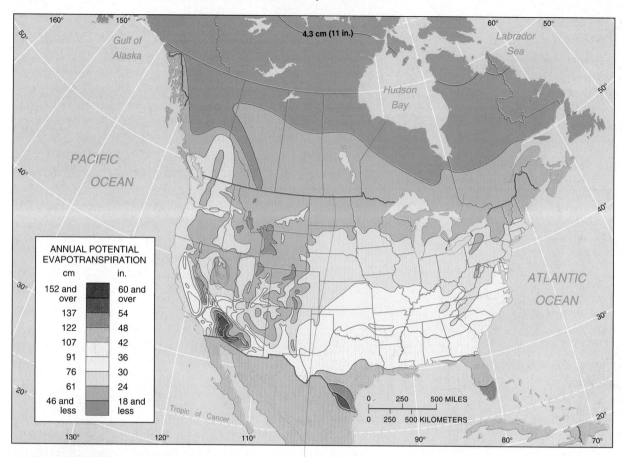

FIGURE 9.8 Potential evapotranspiration (water demand, POTET) for the United States and Canada.
[From C. W. Thornthwaite, "An approach toward a rational classification of climate," *Geographical Review* 38 (1948): 64. Adapted by permission from the American Geographical Society. Canadian data adapted from M. Sanderson, "The climates of Canada according to the new Thornthwaite classification," *Scientific Agriculture* 28 (1948): 501–517.]

answer, where applicable, is to note whether people use sprinklers for lawns and gardens during summer months and, if they do, whether the sprinklers are temporary or are permanently installed.

Deficit (DEFIC) The POTET demand can be satisfied in three ways: by PRECIP, by moisture stored in the soil, or through artificial irrigation. If these three sources are inadequate to meet ultimate demand, the location experiences a moisture shortage. This unsatisfied POTET is **deficit** (DEFIC). By subtracting DEFIC from POTET, we determine the **actual evapotranspiration**, or ACTET, that takes place. Under ideal conditions, potential and actual amounts of evapotranspiration are about the same, so plants do not experience a water shortage; droughts result from deficit conditions.

Surplus (SURPL) If POTET is satisfied and the soil is full of moisture, then additional water input becomes **surplus** (SURPL). This excess water might sit on the surface in puddles, ponds, and lakes, or it might flow across the surface toward stream channels or percolate through the soil underground. The **overland flow** combines with pre-

cipitation and groundwater flows into river channels to make up the **total runoff**. Because surplus water generates most streamflow, or runoff, the water-balance approach is useful for indirectly estimating streamflow.

Soil-Moisture Storage (ΔSTRGE) A "savings account" of water that receives recharge "deposits" and provides for "withdrawals" is **soil-moisture storage** (ΔSTRGE). This is the volume of water stored in the soil that is accessible to plant roots. The delta symbol Δ means that this component includes both *recharge* and *utilization* (use) of soil moisture; the Δ in math means "change." Soil moisture comprises two categories of water—hygroscopic and capillary—but only capillary water is accessible to plants (Figure 9.9).

Hygroscopic water is inaccessible to plants because it is a molecule-thin layer that is tightly bound to each soil particle by the hydrogen bonding of water molecules (Figure 9.9, left). Hygroscopic water exists even in the desert, but it is unavailable to meet POTET demands. Relative to plants, soil is at the **wilting point** when all that remains is this unextractable water; plants wilt and eventually die after a prolonged period of such moisture stress.

FIGURE 9.9 Types of soil moisture.
Hygroscopic and gravitational water are
unavailable to plants; only capillary water
is available. [After D. Steila, *The Geography of Soils*, © 1976, p. 45. Reprinted by
permission of Prentice Hall, Inc., Upper
Saddle River, NJ.]

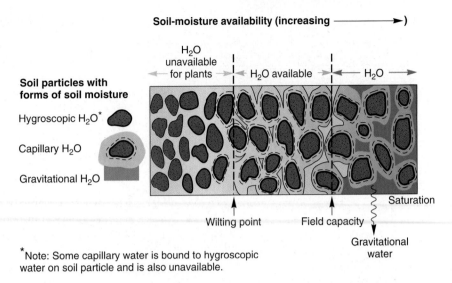

Soil-moisture availability (increasing ⟶)

Soil particles with
forms of soil moisture

Hygroscopic H_2O^*

Capillary H_2O

Gravitational H_2O

H_2O unavailable for plants | H_2O available | H_2O ⟶

Saturation

Wilting point   Field capacity

Gravitational water

*Note: Some capillary water is bound to hygroscopic
water on soil particle and is also unavailable.

FIGURE 9.10 Soil-moisture availability.
The relation between soil-moisture
availability and soil texture determines the distance between the two
curves that show field capacity and
wilting point. A loam soil (one-third
each of sand, silt, and clay) has
roughly the most available water per
vertical foot of soil exposed to plant
roots. [After U.S. Department of
Agriculture, *1955 Yearbook of Agriculture—Water*, p. 120.]

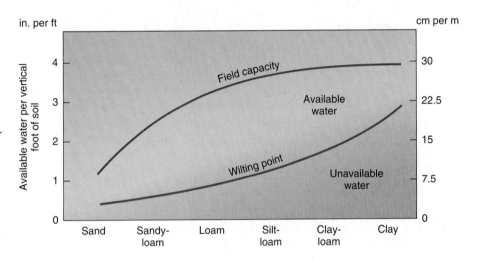

Capillary water is generally accessible to plant roots because it is held against the pull of gravity in the soil by hydrogen bonds between water molecules (surface tension) and by hydrogen bonding between water molecules and the soil. Most capillary water that remains in the soil is **available water** in soil-moisture storage. After some water drains from the larger pore spaces, the amount of available water remaining for plants is termed **field capacity**, or storage capacity. It is removable to meet POTET demands through the action of plant roots and surface evaporation (Figure 9.9, center). Field capacity is specific to each soil type, and the amount can be determined by soil surveys.

When soil becomes saturated after a precipitation event, any water surplus in the soil body becomes **gravitational water**. It percolates from the shallower capillary zone to the deeper groundwater zone (Figure 9.9, right).

Figure 9.10 shows the relation of soil texture to soil-moisture content. Different plant species send roots to different depths and therefore reach different amounts of soil moisture. A soil blend that maximizes available water is best for plants. On the basis of Figure 9.10, can you determine the soil texture with the greatest quantity of available water?

As **soil-moisture utilization** removes soil water, the plants must exert greater effort to extract the amount of moisture they need. As a result, even though a small amount of water may remain in the soil, plants may be unable to exert enough pressure to use it. The resulting *unsatisfied demand* is a *deficit*. Avoiding a deficit and reducing plant growth inefficiencies are the goals of irrigation, for the harder plants must work to get water, the less their yield and growth will be. (For more on drought, see http://www.ngdc.noaa.gov/paleo/drought/drght_what.html.)

Whether from natural precipitation or artificial irrigation, water infiltrates the soil and replenishes available water, a process called **soil-moisture recharge**. The texture and the structure of the soil dictate available pore spaces, or **porosity**. The property of the soil that determines the rate of soil-moisture recharge is its **permeability**. Permeability depends on particle sizes and the shape and packing of soil grains.

Water infiltration is rapid in the first minutes of precipitation and slows as the upper soil layers become saturated, even though the deeper soil is still dry. Agricultural practices, such as plowing and adding sand or manure to loosen soil structure, can improve both soil permeability

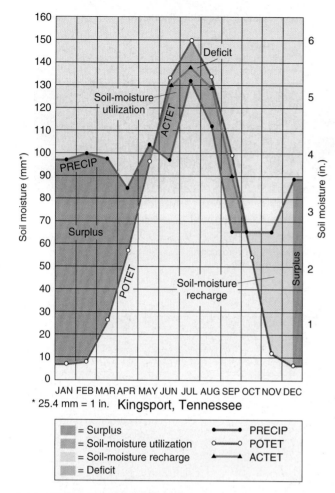

JAN FEB MAR APR MAY JUN JUL AUG SEP OCT NOV DEC
* 25.4 mm = 1 in. **Kingsport, Tennessee**

= Surplus — PRECIP
= Soil-moisture utilization ○—○ POTET
= Soil-moisture recharge ▲—▲ ACTET
= Deficit

FIGURE 9.11 Sample water budget.
Annual average water-balance components graphed for Kingsport, Tennessee. The comparison of plots for precipitation inputs and potential evapotranspiration outputs determines the condition of the soil-moisture environment. A typical pattern of spring surplus, summer soil-moisture utilization, a small summer deficit, autumn soil-moisture recharge, and ending surplus highlights the year.

and the depth to which moisture can efficiently penetrate to recharge soil-moisture storage. You may have found yourself working to improved soil permeability with a house plant or garden, working the soil to increase the rate of soil-moisture recharge.

Sample Water Budgets

Using all of these concepts, we can graph the water-balance components for several representative cities. Let's begin by looking at Kingsport, in the extreme northeastern corner of Tennessee at 36.6° N 82.5° W, elevation 390 m (1280 ft).

Figure 9.11 graphs the water-balance components for long-term supply and demand, using monthly averages. The monthly values for PRECIP and POTET smooth the actual daily and hourly variability. On the graph, a comparison of PRECIP and POTET by month determines whether there is a *net supply* or a *net demand* for water. There is a net supply (blue line) from October to May, but the warm days from June to September create a net water demand. If we assume a soil-moisture storage capacity of 100 mm (4.0 in.), typical of shallow-rooted plants, the net water-demand months are satisfied through soil-moisture utilization (green area).

Table 9.1 shows the actual data for Kingsport's water-balance equations for the year and for the months of March and September. Here, you can see the interaction of these various components. Check these three equations and perform the functions indicated to see whether we accounted for all the PRECIP received. Compare March and September in the equations with the same months in the graph in Figure 9.11.

Obviously, not all stations experience the surplus moisture patterns of this humid-continental region. Different climatic regimes experience different relations among water-balance components. Figure 9.12 presents water-balance graphs for several other cities in North America and the Caribbean Sea region. Among these examples, compare the summer minimum precipitation of Berkeley, California, with the summer maximum of Seabrook, New Jersey; the drier prairies of Saskatchewan with the more humid conditions in Ottawa; and the precipitation receipts of San Juan, Puerto Rico, with the arid desert of Phoenix, Arizona.

Water Budget and Water Resources

Water distribution is uneven over space and time. Because we require a steady supply, we build large-scale management projects intended to redistribute the water resource

Table 9.1 Annual, March, and September Water Balances for Kingsport, Tennessee

	PRECIP	=	(POTET	−	DEFIC)	+	SURPL	±	ΔSTRGE
Annual	1119.0	=	(781.0	−	16.0)	+	354.0	±	0
	(44.1)	=	(30.7	−	0.6)	+	(14.0)	±	0
March	97.0	=	(24.0	−	0)	+	73.0	±	0
	(3.8)	=	(0.9	−	0)	+	(2.9)	±	0
September	66.0	=	(99.0	−	8.0)	+	0	−	25.0
	(2.6)	=	(3.9	−	0.3)	+	0	−	(1.0)

Note. All quantities in millimeters (inches).

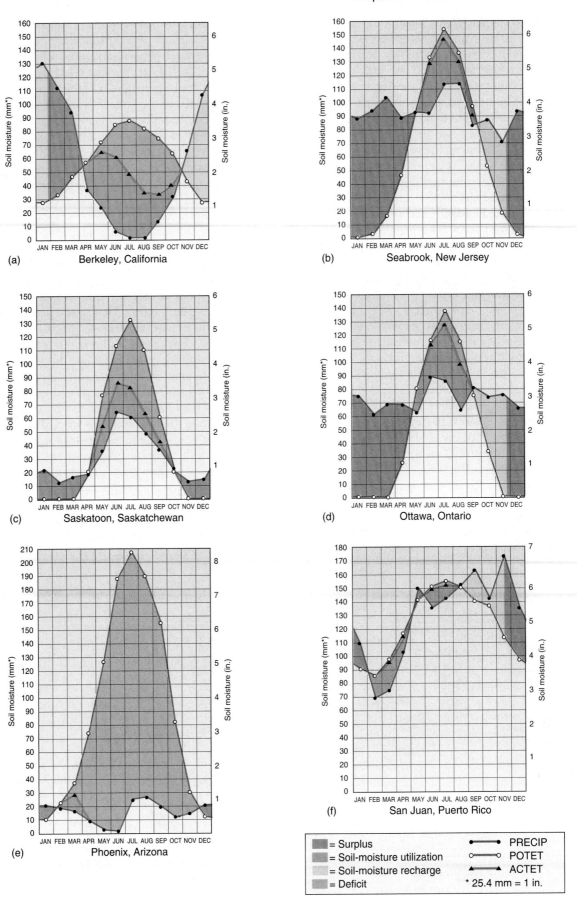

FIGURE 9.12 Sample water budgets for selected stations.
Sample water-balance regimes for selected stations in North America
and the Caribbean region.

either geographically, by moving water from one place to another, or over time, by storing water from time of receipt until it is needed. In this way deficits are reduced, surpluses are held for later release, and water availability is improved to satisfy evapotranspiration and society's demands. The water balance permits analysis of the contribution streams make to water resources.

Streams may be perennial (constantly flowing; *perennial* = "through the year") or intermittent. In either case, the total runoff that moves through them comes from surplus surface-water runoff, subsurface throughflow, and groundwater. Figure 9.13 maps annual global river runoff for the world. Highest runoff amounts are along the equator within the tropics, reflecting the continual rainfall along the Intertropical Convergence Zone (ITCZ). Southeast Asia also experiences high runoff, as do northwest coastal mountains in the Northern Hemisphere. In countries having great seasonal fluctuations in runoff, groundwater becomes an important reserve. Regions of lower runoff coincide with Earth's subtropical deserts, rain-shadow areas, and continental interiors, particularly in Asia.

In the United States, several large water-management projects already operate: the Bonneville Power Authority in Washington State, the Tennessee Valley Authority in the Southeast, the California Water Project, and the Central Arizona Project. In Canada, the Churchill Falls and Nelson River Projects of Manitoba and the proposed $12 billion Gull Island Dam and related hydroelectric projects in Québec along the lower Churchill, Saint-Jean, and Romaine Rivers are significant. The best sites for multipurpose hydroelectric projects already are taken, so battles among conflicting interests and analysis of negative environmental impacts invariably accompany new project proposals.

A particularly ambitious regional water project is the Snowy Mountains Hydroelectric Authority in Australia, where water-balance variations created the need to relocate water. In 1999 the project celebrated its 50th anniversary since the beginning of construction. In the Snowy Mountains, part of the Great Dividing Range in extreme southeastern Australia, precipitation ranges from 100 to 200 cm (40 to 80 in.) a year, while interior Australia receives less than 50 cm (20 in.) and less than 25 cm (10 in.) farther inland (see the world precipitation map in Figure 10.2). POTET values are greater and PRECIP lower throughout the Australian interior, creating water deficits, compared with conditions in higher elevations and orographic precipitation of the Snowy Mountains that produce water surpluses.

Using some of the longest tunnels ever built, vast pumping systems, numerous reservoirs, and power plants, this project included many tributary streams and drainage systems high in the mountains. The westward-flowing rivers—the Murray, Tumut, and Murrumbidgee—receive the diverted water from the Snowy Mountains, and, as a result, new acreage is now in production in what once was dry outback, generating $8.5 billion in annual income. These interior lands were formerly served only by wells that drew upon meager groundwater resources. Today, more than 2 million acre-feet of irrigation water are pumping into these farmlands from the Snowy Mountains (Figure 9.14). Despite some concerns about the economics of the project, it is a source of Australian pride. (See http://www.snowyhydro.com.au/.)

An interesting application of the water-balance approach to water resources is analyzing an event, such as the Dust Bowl, an urban flood, or a hurricane. Focus Study 9.1 presents Hurricane Camille (1969) and its generally positive effects on regional water resources.

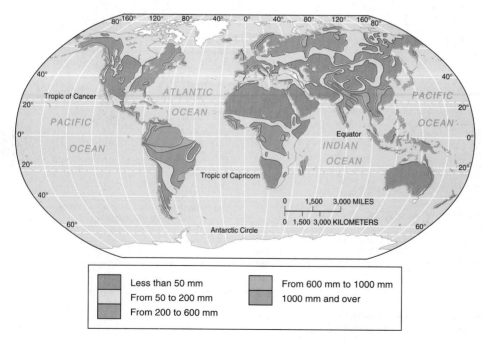

FIGURE 9.13 Annual global river runoff.
Distribution of runoff is closely correlated with climatic region, as expected. But it is poorly correlated with human population distribution and density. [Data from the Institute of Geography, Russian Academy of Sciences, Moscow, as presented by the World Resources Institute.]

Less than 50 mm	From 600 mm to 1000 mm
From 50 to 200 mm	1000 mm and over
From 200 to 600 mm	

FIGURE 9.14 The Snowy Mountains Scheme, Australia, provides water.
Snowy Mountains irrigation water makes possible the growing of rice, potatoes, tomatoes, and other crops in Australia's eastern interior. [Photo courtesy of Jenny McLeod, Policy Advisor, Murray Irrigation Limited, Deniliquin, NSW, Australia.]

Focus Study 9.1

Hurricane Camille, 1969: Water-Balance Analysis Points to Moisture Benefits

Hurricane Camille was one of the most devastating hurricanes of the 20th century. Ironically, it was significant not only for the disaster it brought (256 dead, $1.5 billion damage) but also for the drought it abated (ended).

Figure 1 shows Camille inland from the Gulf Coast north of Biloxi, Mississippi. Hurricane-force winds sharply diminished after the hurricane made landfall, leaving a vast rainstorm that traveled from the Gulf Coast through Mississippi, western Tennessee, Kentucky, and into central Virginia (Figure 2). Severe flooding drowned the Gulf Coast near landfall and the James River basin of Virginia, where torrential rains produced record floods. But, Camille actually had beneficial aspects; it ended a year-long drought along major portions of its storm track.

(continued)

FIGURE 1 *ESSA-9* **satellite image of Hurricane Camille.**
Hurricane Camille made landfall on the night of August 17, 1969, and continued inland on August 18. The storm is progressing northward through Mississippi in the image. [Image from NOAA.]

Focus Study 9.1 (continued)

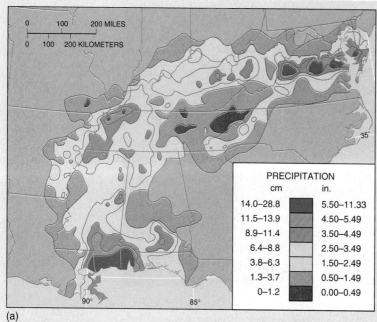

(a)

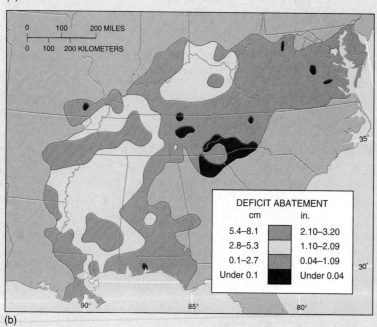

(b)

FIGURE 2 Camille affects water budgets in a beneficial way.
A water-resource view of Hurricane Camille's impact on local water budgets: (a) precipitation attributable to the storm (moisture supply), (b) resulting deficit abatement (avoided moisture shortages) attributable to Camille. [Data and maps by R. Christopherson. All rights reserved.]

PRECIPITATION

cm	in.
14.0–28.8	5.50–11.33
11.5–13.9	4.50–5.49
8.9–11.4	3.50–4.49
6.4–8.8	2.50–3.49
3.8–6.3	1.50–2.49
1.3–3.7	0.50–1.49
0–1.2	0.00–0.49

Precipitation from Camille that exceeds 1.2 cm (0.5 in.)

DEFICIT ABATEMENT

cm	in.
5.4–8.1	2.10–3.20
2.8–5.3	1.10–2.09
0.1–2.7	0.04–1.09
Under 0.1	Under 0.04

Water shortages that did not occur because of Camille's precipitation

Figure 2a maps the precipitation from Camille (moisture supply) and portrays the storm's track from landfall in Mississippi to the coast of Virginia and Delaware. By comparing actual water budgets along Camille's track with water budgets for the same three days with Camille's rainfall totals removed artificially, I analyzed the moisture impact of the storm on the region's water budget. Figure 2b maps the moisture shortages that were avoided because of Camille's rains—termed "deficit abatement." Think of this as drought that did not continue because Camille's rains occurred. Over vast portions of the affected area, Camille reduced dry-soil conditions, restored pastures, and filled low reservoirs.

Thus, Camille's monetary benefits inland outweighed its damage by an estimated 2 to 1 ratio. (Of course, the tragic loss of life does not fit into a financial equation.) Hurricanes should be viewed as normal and natural meteorological events that have terrible destructive potential, mainly to coastal lowlands, but also contribute to the precipitation regimes of the southern and eastern United States. According to weather records, about one-third of all hurricanes making landfall in the United States provide beneficial precipitation to local water budgets.

Groundwater Resources

Groundwater is an important part of the hydrologic cycle, although it lies beneath the surface, beyond the soil-moisture root zone. It is tied to surface supplies through pores in soil and rock. Groundwater is the largest potential freshwater source in the hydrologic cycle—larger than all surface lakes and streams combined. Between Earth's land surface and a depth of 4 km (13,000 ft) worldwide, some 8,340,000 km³ (2,000,000 mi³) of water resides, a volume comparable to 70 times all the freshwater lakes in the world. Despite this volume and its obvious importance, groundwater is widely abused by pollution and overconsumption in quantities beyond natural replenishment rates. Remember: *groundwater is not an independent source of water for it is tied to surface supplies for recharge.*

About 50% of the U.S. population derives a portion of its freshwater from groundwater sources. Between 1950 and 1995, annual groundwater withdrawal increased 150%. In some states, such as Nebraska, groundwater supplies 85% of water needs and as high as 100% in rural areas. In Canada, about 6 million people (two-thirds of them live in rural areas) rely on groundwater for domestic needs—1.5 billion m³/year (53 billion ft³/year). Figure 9.15 shows potential groundwater resources in the United States and Canada. An important consideration in many regions is that accumulation occurred over millions of years, so care must be taken not to exceed this long-term buildup with excessive short-term demands.

In many ways, groundwater is better than surface water. It is available in many parts of the world that lack dependable surface runoff. Whereas surface supplies are

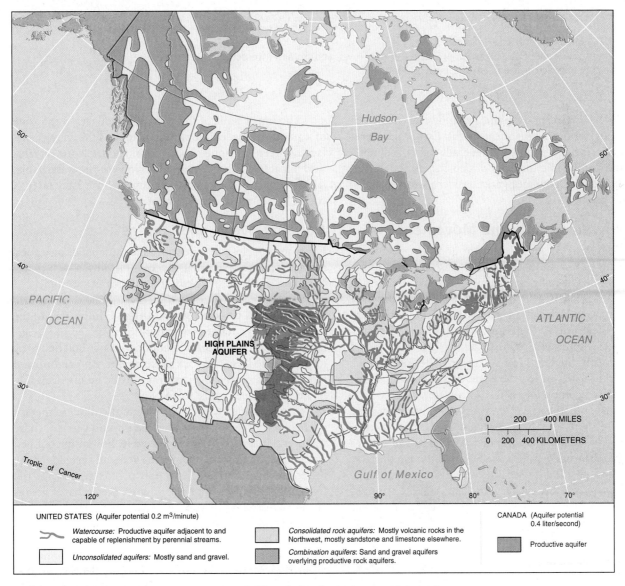

FIGURE 9.15 Groundwater resource potential for the United States and Canada.
Highlighted areas of the United States are underlain by productive aquifers capable of yielding freshwater to wells at 0.2 m³/min or more (for Canada, 0.4 l/s). [Courtesy of Water Resources Council for the United States and the Inquiry on Federal Water Policy for Canada.]

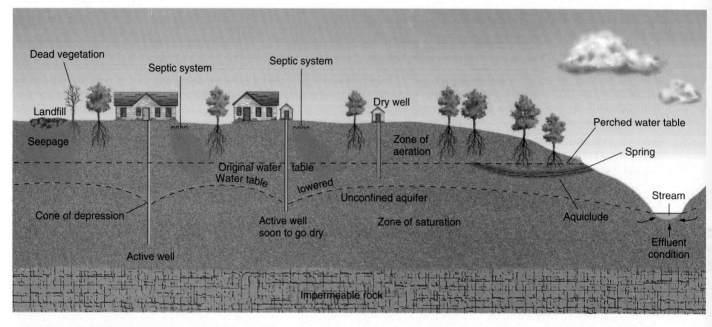

FIGURE 9.16 Groundwater characteristics.
Subsurface groundwater characteristics, processes, water flows, and human interactions.
Follow the concepts across the illustration as you read.

affected by short-term drought, groundwater is not (although long-term drought affects both). Except in severely polluted areas, groundwater is generally free of sediment, color, and pathogenic (disease) organisms, although polluted groundwater conditions are considered irreversible—you can't take it back.

Groundwater Profile and Movement

Figure 9.16 brings together many groundwater phenomena in a single illustration, and it is the basis for the following discussion. Groundwater begins as surplus water, which percolates downward as gravitational water from the zone of capillary water. This excess surface water moves through the **zone of aeration**, where soil and rock are less than saturated (some pore spaces contain air).

Eventually, the water reaches an area where subsurface water accumulates, called the **zone of saturation**. Here, the pores are completely filled with water. Like a hard sponge made of sand, gravel, and rock, the saturation zone stores water in its countless pores and voids. The saturated zone may include the saturated portion of the aquifer and a part of the underlying *aquiclude*, even though the later has such low permeability that its water is inaccessible.

The *porosity* of any rock layer depends on the arrangement, size, and shape of its individual particles, the nature of any cement between them, and their degree of compaction. Subsurface rocks are either *permeable* or *impermeable*. Their permeability depends on whether they conduct water readily (higher permeability) or tend to obstruct its flow (lower permeability). An **aquifer** is a rock layer that is permeable to groundwater flow in usable amounts. An **aquiclude** is a body of rock that does not conduct water in usable amounts (also called an *aquitard*).

The upper limit of the water that collects in the zone of saturation is called the **water table**. It is the contact surface between the zone of saturation and the zone of aeration (all across Figure 9.16). The slope of the water table, which generally follows the contours of the land surface, controls groundwater movement.

Aquifers, Wells, and Springs

For many people, adequate water supply depends on a good aquifer beneath them that is accessible with a well or exposed on a hillside where water emerges as a spring.

Confined and Unconfined Aquifers Important to the behavior of an aquifer is whether it is confined or unconfined. A **confined aquifer** is bounded above and below by impermeable layers of rock or sediment. An **unconfined aquifer** has a permeable layer on top and an impermeable one beneath (see Figure 9.16).

Confined and unconfined aquifers also differ in the size of their recharge area, which is the ground surface where water enters an aquifer to recharge it. For an unconfined aquifer, the **aquifer recharge area** generally extends above the entire aquifer; the water simply percolates down to the water table. But in a confined aquifer, the recharge area is far more restricted, as you can see in the figure.

Once identification of recharge areas of both confined and unconfined aquifers takes place, the prevailing government body should zone them to prohibit pollution discharges, septic and sewage-system installations, or hazardous-material dumping. An informed population should insist on such action because protection is cheaper than the nearly impossible cleanup or replacement of contaminated groundwater sources. In the illustration, note the improp-

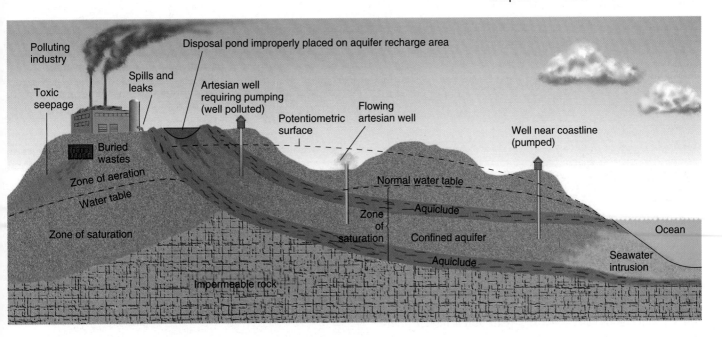

erly located disposal pond on the aquifer recharge area, contaminating wells to the right.

Confined and unconfined aquifers also differ in their water pressure. A well drilled into an unconfined aquifer (see Figure 9.16) must be pumped to make the water level rise above the water table. In contrast, the water in a confined aquifer is under the pressure of its own weight, creating a pressure level to which the water can rise on its own, called the **potentiometric surface**.

The potentiometric surface actually can be above ground level (right side in figure). Under this condition, **artesian water**, or groundwater confined under pressure, may rise in a well and even flow at the surface without pumping, if the top of the well is lower than the potentiometric surface. (These wells are called *artesian*, for the Artois area in France where they are common.) In other wells, however, pressure may be inadequate, and the artesian water must be pumped the remaining distance to the surface.

Wells, Springs, and Streamflows   The slope of the water table, which broadly follows the contour of the land surface, controls groundwater movement (see Figure 9.16). Groundwater tends to move toward areas of lower pressure and elevation.

Water wells can work only if they penetrate the water table. Too shallow a well will be a "dry well"; too deep a well will punch through the aquifer and into the impermeable layer below, also yielding little water. This is why water wells should be drilled in consultation with a *hydrogeologist*.

Where the water table intersects the surface, it creates springs (see Figure 9.16, left, and Figure 9.17). Such an intersection also occurs in lakes and riverbeds. Ultimately,

FIGURE 9.17 An active spring.
Springs provide evidence of groundwater emerging at the surface. This spring flows into Crystal Lake, Salt River Range, Wyoming. [Photo by Bobbé Christopherson.]

groundwater may enter stream channels to flow as surface water (stream near center in Figure 9.16). In fact, during dry periods, the water table may sustain river flows.

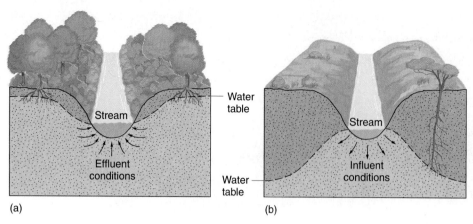

FIGURE 9.18 Groundwater interaction with streamflow. (a) Effluent stream base flow is partially supplied by a high water table, characteristic of humid regions. (b) Influent stream supplies a lower water table, characteristic of drier regions. Water table may drop below the stream channel, effectively drying it out.

Figure 9.18 illustrates the relation between the water table and surface streams in two different climatic settings. In humid climates, where the water table generally supplies a continuous base flow to a stream and is higher than the stream channel, the stream is called *effluent* because it receives the water flowing out (effluent) from the surrounding ground. The Mississippi River and countless other streams are examples. In drier climates, with lower water tables, water from *influent* streams flows into the adjacent ground, sustaining vegetation along the stream. The Colorado River and the Rio Grande of the American West are examples of influent streams.

Overuse of Groundwater

As water is pumped from a well, the surrounding water table within an unconfined aquifer may experience **drawdown**, or become lowered. Drawdown occurs if the pumping rate exceeds the replenishment flow of water into the aquifer, or the horizontal flow around the well. The resultant lowering of the water table around the well is called a **cone of depression** (see Figure 9.16, left).

Overpumping   Aquifers frequently are pumped beyond their flow and recharge capacities, a condition known as **groundwater mining**. Today, large tracts experience chronic groundwater overdrafts in the Midwest, West, lower Mississippi Valley, Florida, and the intensely farmed Palouse region of eastern Washington State. In many places, the water table or artesian water level has declined more than 12 m (40 ft). In the United States, groundwater mining is of special concern in the great High Plains Aquifer, which is the topic of Focus Study 9.2. About half of India's irrigated water and half of industrial and urban water needs are met by the groundwater reserve. And, in approximately 20% of India's agricultural districts, groundwater mining through more than 17 million wells is beyond recharge rates.

In the Middle East, conditions are even more severe, as detailed in News Report 9.1. The groundwater resource beneath Saudi Arabia accumulated over tens of thousands of years ("fossil aquifers"), but the increasing withdrawals are not being naturally recharged to any appreciable de-

FIGURE 9.19 Water desalination. Freshwater is supplied to Saudi Arabia from the Jabal water desalination plant along the Red Sea. Saudi Arabia obtains a significant amount of its water from desalination of seawater. [Photo by Liaison Agency.]

gree at present due to the desert climate—in essence, a nonrenewable groundwater resource. Some researchers suggest that groundwater in the region will be depleted in a decade, although worsening water-quality problems will no doubt arise before this date. Desalinization of seawater to augment diminishing groundwater supplies is becoming increasingly important as a freshwater source (Figure 9.19).

> Like any renewable resource, groundwater can be tapped indefinitely as long as the rate of extraction does not exceed the rate of replenishment. But just like a bank account, a groundwater reserve will dwindle if withdrawals exceed deposits. Few governments have established and enforced rules and regulations to insure that groundwater sources are exploited at a sustainable rate. . . . No government has yet adequately tackled the issue of groundwater depletion, but it is at least getting more attention.[*]

[*]L. Brown, et al., and the Worldwatch Institute, *Vital Signs, The Environmental Trends That Are Shaping Our Future* (New York: W.W. Norton & Co., 2000), p. 122, 123.

Focus Study 9.2

High Plains Aquifer Overdraft

Earth's largest known aquifer is the High Plains aquifer. It lies beneath the American High Plains, an eight-state, 450,600-km² (174,000-mi²) area from southern South Dakota to Texas (Figure 1a). Precipitation over the region varies from 30 cm in the southwest to 60 cm in the northeast (12 to 24 in.). For several hundred thousand years, the aquifer's sand and gravel were charged with meltwaters from retreating glaciers. For the past 100 years, however, High Plains groundwater has been heavily mined, and

(continued)

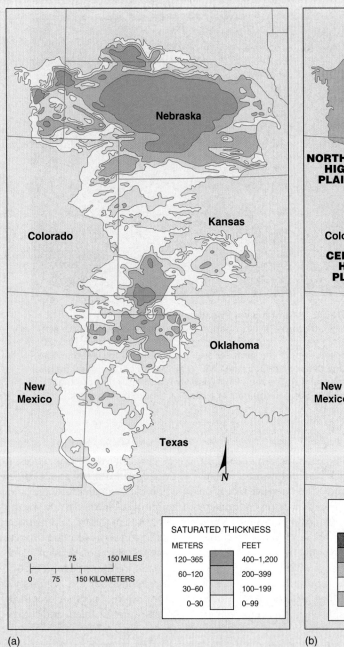

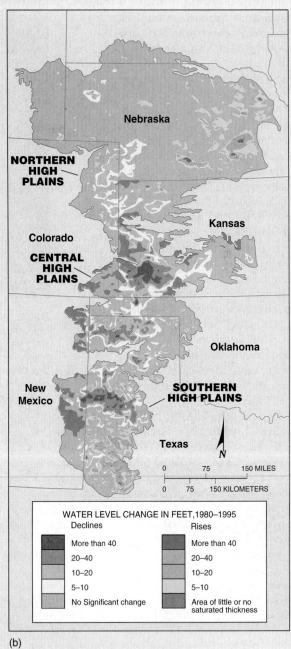

(a)   (b)

FIGURE 1 High Plains aquifer.
(a) The High Plains is the largest known aquifer in North America mapped here showing its average saturated thickness. (b) Water-level changes in the aquifer from 1980 to 1995, given in feet.
[(a) After D. E. Kromm and S. E. White, "Interstate groundwater management preference differences: The High Plains region," *Journal of Geography* 86, no. 1 (January–February 1987): 5. (b) USGS "Water-Level Changes in the High Plains Aquifer, 1980 to 1995," Fact Sheet FS-068-97, Lincoln, Nebraska: 1998.]

Focus Study 9.2 *(continued)*

mining intensified after World War II, when center-pivot irrigation devices were introduced (Figure 2). These large circular devices provide vital water to wheat, sorghums, cotton, corn, and about 40% of the grain fed to cattle in the United States. The USGS began monitoring this groundwater mining in 1988. (See http://ne.water.usgs.gov/highplains/hpactivities.html for links or http://webserver.cr.usgs.gov/nawqa/hpgw/HPGW_home.html for a study.)

The High Plains aquifer irrigates about one-fifth of all U.S. cropland: 120,000 wells provide water for 5.7 million hectares (14 million acres). This is down from the peak of 170,000 wells in 1978. In 1980, water was being pumped from the aquifer at the rate of 26 billion cubic meters (21 million acre-feet) a year, an increase of more than 300% since 1950. By 1995 withdrawals had decreased 10% due to declining well yields and increasing pumping costs.

In the past five decades, the water table in the aquifer has dropped more than 30 m (100 ft), and throughout the 1980s it has averaged a 2-m (6-ft) drop each year. The USGS estimates that recovery of the High Plains aquifer (those portions that have not been crushed or subsided) would take at least 1000 years if groundwater mining stopped today!

Figure 1b maps changes in water levels from 1980 to 1995. Declining water levels are most severe in northern Texas where the saturated thickness of the aquifer is least, through the Oklahoma panhandle and into Kansas. Rising water levels are noted in portions of south-central Nebraska and a portion of Texas owing to recharge from surface irrigation, a period of above-normal precipitation years, and downward percolation from canals and reservoirs.

Obviously, billions of dollars of agricultural activity cannot be abruptly halted, but neither can profligate

FIGURE 2 Central-pivot irrigation.
Myriad central-pivot irrigation systems water crops in north-central Nebraska. A growing season for corn requires from 10 to 20 revolutions of the sprinkler arm, depending on the weather. The arm delivers about 3 cm (1.18 in.) of water per revolution (rainfall equivalent). The High Plains aquifer is at depths of more than 76 m (250 ft) in this part of Nebraska. [Photo by Comstock.]

water mining continue. This issue raises tough questions: How best to manage cropland? Can extensive irrigation continue? Can the region continue to meet the demand to produce commodities for export? Should we continue high-volume farming of certain crops that are in chronic oversupply? Should we rethink federal policy on crop subsidies and price supports? What would be the impact on farmers and rural communities of any changes to the existing system?

Present irrigation practices, if continued, will destroy about half of the High Plains aquifer resource (and two-thirds of the Texas portion) by the year 2020. Add to this the approximate 10% loss of soil moisture due to increased evapotranspiration demand caused by climatic warming, as forecast by computer models for this region by 2050, and we have a portrait of a major regional water problem and a challenge for society.

Collapsing Aquifers   A possible effect of water removal from an aquifer is that the aquifer, which is a layer of rock or sediment, will lose its internal support. Water in the pore spaces between rock grains is not compressible, so it adds structural strength to the rock. If the water is removed through overpumping, air infiltrates the pores. Air is readily compressible, and the tremendous weight of overlying rock may crush the aquifer. On the surface, the visible re-

News Report 9.1

Middle East Water Crisis: Running on Empty

The Persian Gulf states soon may run out of freshwater. Their vast groundwater resource is overpumped to such an extent that salty seawater is encroaching into aquifers tens of kilometers inland! In this decade, groundwater on the Arabian Peninsula may become undrinkable. Imagine having the hottest issue in the Middle East become water, not oil!

Remedies for groundwater overuse are neither easy nor cheap. In the Persian Gulf area, additional freshwater is obtained by desalination of seawater, using desalination plants along the coasts. These processing plants remove salt from seawater by distillation and evaporation processes. In fact, approximately 60% of the world's 4000 desalination plants are presently operating in Saudi Arabia and other Persian Gulf states (see Figure 9.19).

A water pipeline is planned to carry water overland from Turkey in the Middle East through two branches. The Gulf water line would run southeast through Jordan and Saudi Arabia, with extensions into Kuwait, Abu Dhabi, and Oman. The other branch would run south through Syria, Jordan, and Saudi Arabia to the cities of Makkah (Mecca) and Jeddah. This pipeline would import 6 million cubic meters (1.68 billion gallons) of water a day some 1500 km (930 mi), the distance from New York City to St. Louis! (See the links listed under Middle East Water Information Network at http://water1.geol.upenn.edu.)

Other remedies are possible. Traditional agricultural practices could be modernized to use less water. Urban water use could be made more efficient to reduce groundwater demand.

Aquifers and rivers could be shared, although at present no negotiated accords exist for this purpose in the Middle East. Conflicts at various scales pose a grave threat to water pipelines and desalination facilities throughout the region.

There are severe water shortages in the Middle East. . . . The resources of the Nile, the Tigris–Euphrates, and the Jordan are overextended owing both to natural causes and to those deriving from human behavior. . . . In addition to a severe shortage in the quantity of water, there is a growing concern over water quality.*

*N. Kliot, *Water Resources and Conflict in the Middle East* (London: Routledge, 1994), p. 1.

---

sult may be land subsidence, cracked house foundations, and changes in drainage. Unfortunately, collapsed aquifers may not be rechargeable, even if surplus gravitational water becomes available, because pore spaces may be permanently collapsed. The water-well fields that serve the Tampa Bay–St. Petersburg, Florida, area are a case in point. Pond and lake levels, swamps, and wetlands in the area are declining and land surfaces subsiding as the groundwater drawdown for export to the cities increases.

Houston, Texas, provides another example. Because groundwater and crude oil were removed, land within an 80-km (50-mi) radius of Houston subsided more than 3 m (10 ft) over the years. Yet, another example is the Fresno area of California's San Joaquin Valley. After years of intensive pumping of groundwater for irrigation, land levels have dropped almost 10 m (33 ft) because of a combination of water removal and soil compaction from agricultural activity.

Surface mining ("strip mining") for coal in eastern Texas, northeastern Louisiana, Wyoming, Arizona, and other locales, also destroys aquifers. Aquifer collapse is an increasing problem in West Virginia, Kentucky, and Virginia, where low-sulfur coal is mined to meet current air-pollution requirements for coal-burning electric power plants.

Saltwater Encroachment   When aquifers are overpumped near the ocean, another problem arises. Along a coastline, fresh groundwater and salty seawater establish a natural interface (*contact surface*). But excessive withdrawal of freshwater can cause this interface to migrate inland. As a result, wells near the shore may become contaminated with saltwater, and the aquifer may become useless as a freshwater source. Figure 9.16 illustrates this seawater intrusion (far-right side). Pumping freshwater back into the aquifer may halt contamination by seawater, but once contaminated, the aquifer is difficult to reclaim.

Pollution of Groundwater

When surface water is polluted, groundwater inevitably becomes contaminated because it is recharged from surface-water supplies. Surface water flows rapidly and flushes pollution downstream, but slow-moving groundwater, once contaminated, remains polluted virtually forever.

Pollution can enter groundwater from industrial injection wells (wastes pumped into the ground), septic tank outflows, seepage from hazardous-waste disposal sites, industrial toxic waste, agricultural residues (pesticides, herbicides, fertilizers), and urban solid-waste landfills. As an example, at U.S. gasoline stations, leaking is suspected from some 10,000 underground gasoline storage tanks. And, not all underground storage tanks are yet regulated by the EPA, with local jurisdictions varying in laws and degrees of surveillance. You have probably seen the now-commonplace excavations at gas stations to remove old tanks and contaminated soil and either the abandonment of the site or the installation of new tanks. Some cancer-causing additives to

gasoline, such as the oxygenate MTBE (methyl tertiary butyl ether), are proving troublesome to thousands of local water supplies.

For management purposes, about 35% of pollution is *point source* (such as a gasoline tank or septic tank); 65% is *nonpoint source* (from a broad area, such as an agricultural field, or urban runoff). Regardless of the spatial nature of the source, pollution can spread over a great distance, illustrated in Figure 9.16. The extent of groundwater pollution is underestimated since the nature of aquifers makes them inaccessible.

In the face of government inaction, and even attempts to reverse some protection laws, serious groundwater contamination continues nationwide, adding validity to a quote from almost 20 years ago:

> One characteristic is the practical irreversibility of groundwater pollution, causing the cost of clean-up to be prohibitively high. . . . It is a questionable ethical practice to impose the potential risks associated with groundwater contamination on future generations when steps can be taken today to prevent further contamination.*

Our Water Supply

Human thirst for adequate water supplies (quantity and quality) will be the major issue in this century. Internationally, increases in per capita water use are double the rate of population growth. Since we are so dependent on water, it seems that humans should cluster where good

*J. Tripp, "Groundwater Protection Strategies," in *Groundwater Pollution, Environmental and Legal Problems* (Washington, DC: American Association for the Advancement of Science, 1984), p. 137. Reprinted by permission.

water is plentiful. But accessible water supplies are not well correlated with population distribution or the regions where population growth is greatest.

Table 9.2 shows estimated world water supplies, present world population, estimated population figures for 2025, population change forecast between 2001 and 2050 at present growth rates, and a comparison of population and global runoff percentages for each continent. The table gives an idea of the unevenness of Earth's water supply (tied to climatic variability) and water demand (tied to level of development, affluence, and per capita consumption).

For example, North America's mean annual discharge is 5960 km^3 and Asia's is 13,200 km^3. However, North America has only 4.8% of the world's population, whereas Asia has 61.4%, with a population doubling time less than half of North America's. In northern China, 550 million people living in approximately 500 cities lack adequate water supplies. For comparison, note that the 1990 floods cost China $10 billion, whereas water shortages are running at more than $35 billion a year in costs to their economy. The Yellow River, a water resource for many Chinese, runs dry every year and in 1997 it failed to reach the sea for almost 230 days! An analyst for the World Bank stated that China's water shortages pose a more serious threat than floods during this century.

Water is critical for survival. Just to produce our food requires enormous quantities of water, for growing, cleaning, processing, and waste disposal. Important to the assessment of water use is the measuring standards used for water resources. See News Report 9.2 for personal water use and water measurements. We take for granted the quality and quantity of our water supply, because it always seems to be there with little effort on our part.

At present, the world's people are withdrawing 30% of the runoff that is accessible. This idea of "accessibility"

Region (2001 population in millions)	Land Area in Thousands of km²	(mi²)	Share of Mean Annual Discharge in km³/year	(BGD)	Global Stream Runoff (%)	Global Population, 2025 (%)	Population, 2025 (millions)	Projected Population Change (+) 2001–2050 (%)
Africa (818)	30,600	(11,800)	4,220	(3,060)	11	16.2	1,268	120
Asia (3720)	44,600	(17,200)	13,200	(9,540)	36	60.2	4,714	41
Australia–Oceania (31)	8,420	(3,250)	1,960	(1,420)	5	0.5	40	49
Europe (727)	9,770	(3,770)	3,150	(2,280)	8.8	9.2	717	—
North America (416) (Canada, Mexico, U.S.)	22,100	(8,510)	5,960	(4,310)	15	6.6	513	46
Central and South America (425)	17,800	(6,880)	10,400	(7,510)	26	7.3	567	53
Global (6137) (excluding Antarctica)	134,000	(51,600)	38,900	(28,100)	—	100.00	7,818	47

Table 9.2   Estimate of Available Global Water Supply Compared by Region and Population

Note. BGD, billion gallons per day. Population data from "2001 World Population Data Sheet," Washington, DC: Population Reference Bureau, 2001.

News Report 9.2

Personal Water Use and Water Measurements

On an individual level, statistics indicate that urban dwellers directly use an average 680 liters (180 gallons) of water per person per day, whereas rural populations average only 265 liters (70 gallons) or less per person per day. However, each of us indirectly accounts for a much greater water demand, because we all consume food and other products produced with water. For the United States, dividing the total water withdrawal (402 BGD in 1995, the latest year data is available) by a population of 285 million people (2001) yields a per capita direct and indirect use of 5337 liters (1410 gallons) of water per day! As population and per capita affluence increase, greater demands are placed on the water resource base, which is essentially fixed. (This calculation is approximate since water withdrawal data is unavailable for 2001.)

Simply providing the variety of food we enjoy requires voluminous water. For example: 77 g (2.7 oz) of broccoli requires 42 l (11 gal) of water

to grow and process; producing 250 ml (8 oz) of milk requires 182 l (48 gal) of water; producing 28 g (1 oz) of cheese requires 212 l (56 gal); producing 1 egg requires 238 l (63 gal), and a 113 g (4 oz) beef patty requires 2314 l (616 gal).

And then there are the toilets, the majority of which still flush approximately four gallons of water! Imagine the spatial complexity of servicing the desert city of Las Vegas, with 125,000 hotel rooms times the number of toilets times flushes per day—in addition to the usage of a resident population of 1.3 million people.

Clearly, the average American lifestyle depends on enormous quantities of water and thus is vulnerable to shortfalls or quality problems. Demand-side water conservation and efficiency remain as potentially our best water resource.

How is all this water measured? In most of the United States, hydrologists measure streamflow in cubic feet per

second (ft^3/s); Canadians use cubic meters per second (m^3/s). In the eastern United States, or for large-scale assessments, water managers use millions of gallons a day (MGD), billions of gallons a day (BGD), or billions of liters per day (BLD). Eventually the metric system will prevail as the United States converts to the more convenient international system of measurement units.

In the western United States, where irrigated agriculture is so important, the measure frequently used is acre-feet per year. One acre-foot is an acre of water, 1 foot deep, equivalent to 325,872 gallons (43,560 ft^3, or 1234 m^3, or 1,233,429 liters). An acre is an area that is about 208 feet on a side and is 0.4047 hectares. For global measurements, 1 km^3 = 1 billion cubic meters = 810 million acre-feet; 1000 m^3 = 264,200 gallons = 0.81 acre-feet. For smaller measures, 1 m^3 = 1000 liters = 264.2 gallons.

---

is important, because about 20% of global runoff is remote and not readily available to meet water demand. Timing of flows is also important, for approximately 50% of total runoff remains beyond use as uncaptured floodwater. For example, the greatest river in terms of runoff discharge on Earth—the Amazon—is about 95% remote from population centers, or not readily available.

Water resources differ from other resources in that there is no alternative substance to water. A 1997 United Nations study found that one-third of humans were experiencing moderate to high levels of water stress; one forecast increases this ratio to two-thirds by 2025. Water shortages increase the probability for international conflict, endanger public health, reduce agricultural productivity, and damage life-supporting ecological systems. New dams and river-management schemes might increase runoff accessibility by 10% over the next 30 years, but population is projected to grow by 32% in the same time period.

Water Supply in the United States

The U.S. water supply derives from surface and groundwater sources that are fed by an average daily precipitation of 4200 BGD (billion gallons a day). That sum is based on an average annual precipitation value of 76.2 cm (30 in.)

divided evenly among the 48 contiguous states (excluding Alaska and Hawai'i).

The 4200 billion gallons of average daily precipitation mentioned is unevenly distributed across the country and unevenly distributed throughout the year. For example, New England's water supply is so abundant that only about 1% of available water is consumed each year. (The same is true in Canada, where the resource greatly exceeds that in the United States.) But in the dry Colorado River Basin mentioned earlier, the discharge is completely consumed. In fact, by treaty and compact agreements, the Colorado actually is budgeted beyond its average discharge! This paradox results from political misunderstanding of the water resource.

Figure 9.20 shows how the U.S. supply of 4200 BGD is distributed. Viewed daily, the national water budget has two general outputs: 71% actual evapotranspiration (ACTET) and 29% surplus (SURPL). The 71% actual evapotranspiration involves 2970 BGD of the daily supply. It passes through nonirrigated land, including farm crops and pasture, forest and browse (young leaves, twigs, and shoots), and nonagricultural vegetation. Eventually, it returns to the atmosphere to continue its journey through the hydrologic cycle. The remaining 29% surplus is what we directly use.

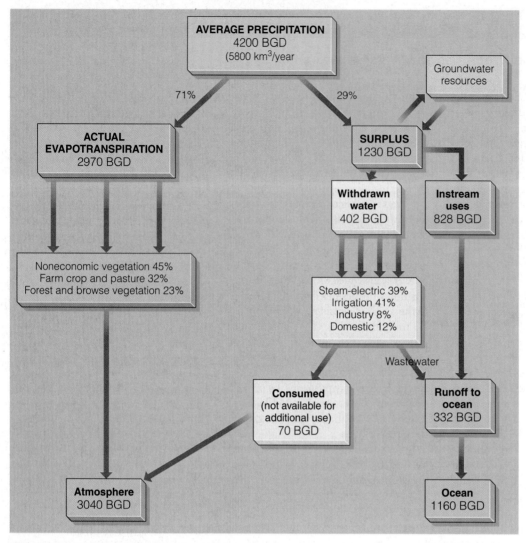

FIGURE 9.20 U.S. water budget.
Daily water budget for the contiguous 48 states in billions of gallons a day (BGD). As of 1995, approximately 30% of available surplus was withdrawn for agricultural (irrigation–livestock), municipal (domestic–commercial), and industrial (industry–mining–thermoelectric) use. [Data from Wayne B. Solley, Robert Pierce, and Howard Perlman, *Estimated Use of Water in the United States in 1995* (Denver: USGS Circular 1200, 1998)—the latest year data available.]

Instream, Nonconsumptive, and Consumptive Uses

The surplus 1230 BGD is runoff, available for withdrawal, consumption, and various instream uses.

- *Instream* uses are those that use streamwater in place: navigation, wildlife and ecosystem preservation, waste dilution and removal, hydroelectric power production, fishing resources, and recreation.
- *Nonconsumptive uses*, or **withdrawal**, remove water from the supply, use it, and then return it to the same supply. Nonconsumptive water is used by industry, agriculture, municipalities, and in steam–electric power generation. A portion of water withdrawn is consumed.
- **Consumptive uses** remove water from a stream but do not return it, so it is not available for a second or

third use. Some consumptive examples include water that evaporates or is vaporized in steam-electric plants.

When water returns to the system, water quality usually is altered—water is contaminated chemically with pollutants or waste or thermally with heat energy. In Figure 9.20, this portion of the budget is returned to runoff and eventually the ocean. This wastewater represents an opportunity to extend the resource through reuse.

Contaminated or not, returned water becomes a part of all water systems downstream. A good example is New Orleans, the last city to withdraw municipal water from the Mississippi River. New Orleans receives diluted and mixed contaminants added throughout the entire Missouri–Ohio–Mississippi River drainage! This includes: the effluent from chemical plants, runoff from

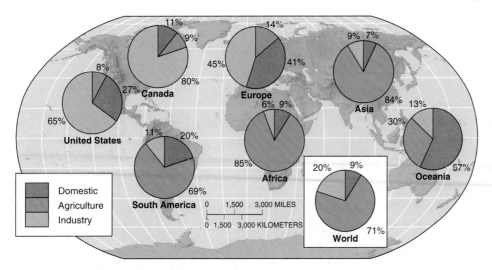

FIGURE 9.21 Water withdrawal by sector.
Compare industrial water use among the geographic areas, as well as agricultural and municipal uses. [After World Resources Institute, *World Resources 2000–2001*, Data Table FW.1, pp. 276–77.]

millions of acres of farm fields treated with fertilizer and pesticides, treated and untreated sewage, oil spills, gasoline leaks, wastewater from thousands of industries, runoff from urban streets and storm drains, and turbidity from countless construction sites and from mining, farming, and logging activities. Abnormally high cancer rates among citizens living along the Mississippi River between Baton Rouge and New Orleans have led to the ominous label "Cancer Alley" for the region.

The estimated U.S. withdrawal of water for 1995 was 402 BGD, an increase of 290% since 1940 but a decrease of 10% since the peak year of 1980 and 2% less than 1990. Total per capita use dropped from 1340 gal/day in 1990 to 1280 gal/day in 1998. A shift in emphasis from supply-side to demand-side planning, pricing, and management produced these efficiencies.

The four main uses of withdrawn water in the United States in 1995 were steam–electric power (39%), industry–mining (8%), irrigation–livestock (41%), and domestic–commercial (12%). In contrast, Canada uses only 12% of its withdrawn water for irrigation and 70% for industry. Figure 9.21 compares regions by their use of withdrawn water during 1998. It graphically illustrates the differences between more-developed and less-developed parts of the world. (For studies of water use in the United States, see **http://water.usgs.gov/public/watuse/**.)

Future Considerations

When precipitation is budgeted, the limits of the water resource become apparent. How can we satisfy the growing demand for water? Water available per person declines as population increases, and individual demand increases with economic development, affluence, and technology. Thus,

world population growth since 1970 reduced per capita water supplies by a third. Also, pollution limits the water-resource base, so that even before *quantity* constraints are felt, *quality* problems may limit the health and growth of a region. The international aspect of the problem is illustrated by the 200 major river basins in the world (basins where rivers drain into an ocean, lake, or inland sea). One hundred and forty-five countries possess territory within an international river basin—truly a global commons.

A composite index comparing available water resources to current use patterns, supplies, and national income reveals regional scarcities in Africa, the Middle East, Asia, Peru, and Mexico (Figure 9.22).

> Unlike other important commodities such as oil, copper, or wheat, freshwater has no substitutes for most of its uses. It is also impractical to transport the large quantities of water needed in agriculture and industry more than several hundred kilometers. Freshwater is now scarce in many regions of the world, resulting in severe ecological degradation, limits on agriculture and industrial production, threats to human health, and increased potential for international conflict.[*]

Clearly, cooperation is needed, yet we continue toward a water crisis without a concept of a *world water economy* as a frame of reference. When will more international coordination begin, and which country or group of countries will lead the way to sustain future water resources?

*[*]S. L. Postal, and others, "Human appropriation of renewable fresh water," *Science* 271, no. 5250 (February 9, 1996), p. 785.

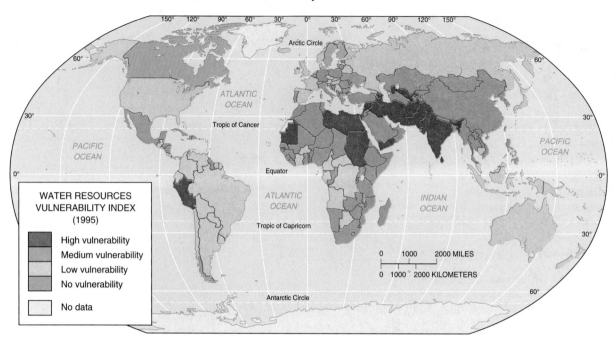

FIGURE 9.22 Global water scarcity.
Water resource vulnerability, a composite index. [Compiled by The World Resources Institute
from data gathered by the Stockholm Environment Institute, *Comprehensive Assessment of
the Freshwater Resources of the World*, 1997, appearing in *World Resources 1998–1999* (New
York: The World Resources Institute, 1998) p. 223.]

Summary and Review—Water Resources

● *Illustrate* **the hydrologic cycle with a simple sketch
and** *label* **it with definitions for each water pathway.**

The flow of water links the atmosphere, ocean, and land
through energy and matter exchanges. The **hydrologic cycle**
is a model of Earth's water system, which has operated for billions of years from the lower atmosphere to several kilometers
beneath Earth's surface. **Interception** occurs when precipitation strikes vegetation or other ground cover. Water soaks into
the subsurface through **infiltration**, or penetration of the soil
surface. It further permeates soil or rock through vertical
movement called **percolation**.

> hydrologic cycle (p. 246)
> interception (p. 247)
> infiltration (p. 247)
> percolation (p. 247)

1. Sketch and explain a simplified model of the complex
   flows of water on Earth—the hydrologic cycle.

2. What are the possible routes that a raindrop may take on
   its way to and into the soil surface?

3. Compare precipitation and evaporation volumes from
   the ocean with those over land. Describe advection flows
   of moisture and countering surface and subsurface runoff.

● *Relate* **the importance of the water-budget concept
to your understanding of the hydrologic cycle, water
resources, and soil moisture for a specific location.**

A **soil-water budget** can be established for any area of Earth's
surface by measuring the precipitation input and the output of
various water demands in the area considered.

Groundwater is a subsurface part of the hydrologic cycle,
tied to surface supplies through soil and rock. Groundwater is
the largest potential freshwater source in the hydrologic cycle.
Streams represent only a tiny fraction of all water (1250 km³,
or 300 mi³), the smallest volume of any of the freshwater categories. Yet streams are the portion of the hydrologic cycle
on which we most depend, representing four-fifths of all the
water we use.

> soil-water budget (p. 248)

4. How might an understanding of the hydrologic cycle in
   a particular locale, or a soil-moisture budget of a site,
   assist you in assessing water resources? Give some specific
   examples.

● *Construct* **the water-balance equation as a way of
accounting for the expenditures of water supply
and** *define* **each of the components in the equation
and their specific operation.**

The moisture supply to Earth's surface is **precipitation** (PRE-
CIP, or P), arriving as rain, sleet, snow, and hail. Precipitation
is measured with the **rain gauge**. **Evaporation** is the net
movement of free water molecules away from a wet surface
into air. **Transpiration** is the movement of water through
plants and back into the atmosphere; it is a cooling mecha-

nism for plants. Evaporation and transpiration are combined into one term—**evapotranspiration**. The ultimate demand for moisture is **potential evapotranspiration** (POTET, or PE), the amount of water that *would* evaporate and transpire under optimum moisture conditions (adequate precipitation and adequate soil moisture). Evapotranspiration is measured with an **evaporation pan** (*evaporimeter*) or the more elaborate **lysimeter**.

Unsatisfied POTET is **deficit** (DEFIC). By subtracting DEFIC from POTET, we determine **actual evapotranspiration**, or ACTET. Ideally, POTET and ACTET are about the same, so that plants have sufficient water. If POTET is satisfied and the soil is full of moisture, then additional water input becomes **surplus** (SURPL), which may puddle on the surface, flow across the surface toward stream channels, or percolate underground through the soil. The **overland flow** to streams includes precipitation and groundwater flows into river channels to make up the **total runoff** from the area.

A "savings account" of water that receives deposits and provides withdrawals as water-balance conditions change is the **soil-moisture storage** (ΔSTRGE). This is the volume of water stored in the soil that is accessible to plant roots. In soil, **hygroscopic water** is inaccessible because it is a molecule-thin layer that is tightly bound to each soil particle by hydrogen bonding. As available water is utilized, soil reaches the **wilting point** (all that remains is unextractable water). **Capillary water** is generally accessible to plant roots because it is held in the soil by surface tension and hydrogen bonding between water and soil. Almost all capillary water that remains in the soil is **available water** in soil-moisture storage. After water drains from the larger pore spaces, the available water remaining for plants is termed **field capacity**, or storage capacity. When soil is saturated after a precipitation event, surplus water in the soil becomes **gravitational water** and percolates to groundwater. As **soil-moisture utilization** removes soil water, the plants work harder to extract the same amount of moisture. Whereas, **soil-moisture recharge** is the rate at which needed moisture enters the soil.

The texture and the structure of the soil dictate available pore spaces, or **porosity**. The soil's **permeability** is the degree to which water can flow through it. Permeability depends on particle sizes and the shape and packing of soil grains.

precipitation (p. 248)
rain gauge (p. 249)
evaporation (p. 249)
transpiration (p. 249)
evapotranspiration (p. 249)
potential evapotranspiration (p. 250)
evaporation pan (p. 250)
lysimeter (p. 250)
deficit (p. 252)
actual evapotranspiration (p. 252)
surplus (p. 252)
overland flow (p. 252)
total runoff (p. 252)
soil-moisture storage (p. 252)
hygroscopic water (p. 252)
wilting point (p. 252)
capillary water (p. 253)
available water (p. 253)
field capacity (p. 253)
gravitational water (p. 253)
soil-moisture utilization (p. 253)
soil-moisture recharge (p. 253)
porosity (p. 253)
permeability (p. 253)

5. What does this statement mean? "The soil-water budget is an assessment of the hydrologic cycle at a specific site."
6. What are the components of the water-balance equation? Construct the equation and place each term's definition below its abbreviation in the equation.
7. Using the annual water-balance data for Kingsport, Tennessee, work the values through the water-balance "bookkeeping" method. Does the equation balance?
8. Explain how to derive actual evapotranspiration (ACTET) in the water-balance equation.
9. What is potential evapotranspiration (POTET)? How do we go about estimating this potential rate? What factors did Thornthwaite use to determine this value?
10. Explain the operation of soil-moisture storage, soil-moisture utilization, and soil-moisture recharge. Include discussion of field capacity, capillary water, and wilting point concepts.
11. In the case of silt-loam soil from Figure 9.10, roughly what is the available water capacity? How is this value derived?
12. In terms of water balance and water management, explain the logic behind the Snowy Mountains Scheme in southeastern Australia.

● *Describe* the nature of groundwater and *define* the elements of the groundwater environment.

Groundwater is a part of the hydrologic cycle, but it lies beneath the surface beyond the soil-moisture root zone. Groundwater does not exist independently because its replenishment is tied to surface surpluses. Excess surface water moves through the **zone of aeration**, where soil and rock are less than saturated. Eventually, the water reaches the **zone of saturation**, where the pores are completely filled with water.

The *permeability* of subsurface rocks depends on whether they conduct water readily (higher permeability) or tend to obstruct its flow (lower permeability). They can even be impermeable. An **aquifer** is a rock layer that is permeable to groundwater flow in usable amounts. An **aquiclude** (aquitard) is a body of rock that does not conduct water in usable amounts.

The upper limit of the water that collects in the zone of saturation is called the **water table**; it is the contact surface between the zones of saturation and aeration. A **confined aquifer** is bounded above and below by impermeable layers of rock or sediment. An **unconfined aquifer** has a permeable layer on top and an impermeable one beneath. The **aquifer recharge area** extends over an entire unconfined aquifer. Water in a confined aquifer is under the pressure of its own weight, creating a pressure level to which the water can rise on its own, called the **potentiometric surface**, which can be above ground level. Groundwater confined under pressure is **artesian water**; it may rise up in wells and even flow out at the surface without pumping, if the head of the well is below the potentiometric surface.

As water is pumped from a well, the surrounding water table within an unconfined aquifer will experience **drawdown**, or become lower, if the rate of pumping exceeds the horizontal flow of water in the aquifer around the well. This excessive pumping causes a **cone of depression**. Aquifers frequently are pumped beyond their flow and recharge capacities, a condition known as **groundwater mining**.

> groundwater (p. 259)
> zone of aeration (p. 260)
> zone of saturation (p. 260)
> aquifer (p. 260)
> aquiclude (p. 260)
> water table (p. 260)
> confined aquifer (p. 260)
> unconfined aquifer (p. 260)
> aquifer recharge area (p. 260)
> potentiometric surface (p. 261)
> artesian water (p. 261)
> drawdown (p. 262)
> cone of depression (p. 262)
> groundwater mining (p. 262)

13. Are groundwater resources independent of surface supplies, or are the two interrelated? Explain your answer.
14. Make a simple sketch of the subsurface environment, labeling zones of aeration and saturation and the water table in an unconfined aquifer. Then add a confined aquifer to the sketch.
15. At what point does groundwater utilization become groundwater mining? Use the High Plains aquifer example to explain your answer.

16. What is the nature of groundwater pollution? Can contaminated groundwater be cleaned up easily? Explain.

● *Identify* critical aspects of freshwater supplies for the future and *cite* specific issues related to sectors of use, regions and countries, and potential remedies for any shortfalls.

Nonconsumptive uses, or water **withdrawal**, remove water from the supply, use it, and then return it to the stream. **Consumptive uses** remove water from a stream but do not return it, so the water is not available for a second or third use. Americans in the 48 contiguous states withdraw approximately one-third of the available surplus runoff for irrigation, industry, and municipal uses. Water-resource planning regionally and globally, using water-budget principles, is essential if Earth's societies are to have enough water of adequate quality.

> withdrawal (p. 268)
> consumptive uses (p. 268)

17. Describe the principal pathways involved in the water budget of the contiguous 48 states. What is the difference between withdrawal and consumptive use of water resources? Compare these with instream uses.
18. Characterize each of the sectors withdrawing water: irrigation, industry, and municipalities. What are the present usage trends in more-developed and less-developed nations?
19. Briefly assess the status of world water resources. What challenges are there in meeting future needs of an expanding population and growing economies?

NetWork

The *Geosystems* Home Page provides on-line resources for this chapter on the World Wide Web. To begin: Once on the Home Page, click on this textbook, scroll the Table of Contents menu, select this chapter, and click "Begin." You will find self-tests that are graded, short essay and review exercises, specific updates for items in the chapter, and many links to interesting related pathways on the Internet under "Destinations." *Geosystems* is at **http://www.prenhall.com/christopherson**.

Critical Thinking

A. Select your campus, yard, or perhaps a house plant and apply the water-balance concepts. What is the supply of water? Estimate the ultimate water supply and demand for the area you selected. Estimate water needs and how they vary with season as components of the water budget change.

B. You have no doubt had several glasses of water between the time you awoke this morning and the time you are reading these words. Where did this water originate? Obtain the name of the water company or agency, determine whether they are using surface or groundwater to meet demands, and check how the water is metered and billed. If your state or province requires water quality reporting, obtain a copy of the analysis of your tap water.

What about the water on your campus? If the campus has its own wells, how is the quality tested? Who on campus is in charge of supervising these wells? Lastly, what is your subjective assessment of your water: taste, smell, hardness, clarity? Compare these perceptions with others in your class.

Death Valley, southeastern California, in full bloom following record rains brought on by the 1997–1998 El Niño in the distant tropics of the Pacific Ocean. [Photo by Bobbé Christopherson.]

10

Global Climate Systems

Key Learning Concepts

After reading the chapter, you should be able to:

- *Define* climate and climatology and *explain* the difference between climate and weather.
- *Review* the role of temperature, precipitation, air pressure, and air mass patterns used to establish climatic regions.
- *Review* Köppen's development of an empirical climate classification system and *compare* his with other ways of classifying climate.
- *Describe* the A, C, D, and E climate classification categories and *locate* these regions on a world map.
- *Explain* the precipitation and moisture efficiency criteria used to determine the B climates and *locate* them on a world map.
- *Outline* future climate patterns from forecasts presented and *explain* the causes and potential consequences.

Earth experiences an almost infinite variety of *weather*—conditions of the atmosphere at any given time and place. But if we consider the weather over many years, including its variability and extremes, a pattern emerges that constitutes **climate**. Think of climate patterns as dynamic rather than static. Climate is more than a consideration of simple averages of temperature and precipitation.

Today, climatologists know that intriguing global-scale linkages exist in the Earth–atmosphere–ocean system:

- Strong monsoonal rains in West Africa are correlated with the development of intense Atlantic hurricanes. In a few days, an "easterly wave" over West Africa may become a furious hurricane crossing Florida.

275

- An El Niño in the Pacific is tied to drought-breaking rains in the American West, floods in Louisiana and Northern Europe, a weak Atlantic hurricane season, and a drought in Australia and Southeast Asia. The strongest El Niño this century dominated 1997–1998; the next is expected in 2002–2003 during the life of this edition of *Geosystems*. Of significance is the recurring El Niño phenomenon, the subject of Focus Study 10.1.

- Global climate change—record-breaking global average temperatures, glacial ice melt, drying soil-moisture conditions, changing crop yields, spreading of infectious disease, changing distributions of plants and animals, declining coral reef health and fisheries, the thawing of high-latitude lands—is now the subject of important multinational treaties.

- Climate patterns are changing at a pace not evidenced in the records of the past millennia—changes in climate and natural vegetation during the next 50 years could exceed the total of all changes since the peak of the last ice-age episode, some 18,000 years ago. Scientists rush to understand and interpret the evidence and forecast what to expect in the future. The climatic regions we study in this chapter will migrate during the twenty-first century.

In this chapter: Climates are so diverse that no two places on Earth's surface experience exactly the same climatic conditions; in fact, Earth is a vast collection of microclimates. However, broad similarities among local climates permit their grouping into climatic regions.

Early climatologists faced the challenge of what to use as a basis for climate classification. Many of the physical elements of the environment that you studied in the first nine chapters of this text now link together to explain climates. One climatologist, Wladimir Köppen (pronounced KUR-pen), developed a system in the early 1900s that uses temperature and precipitation data. Though imperfect, his system is still widely used and easily understood, and we use a modified version of it in this chapter.

Climatologists use powerful computer models to simulate changing complex interactions in the atmosphere, hydrosphere, lithosphere, and biosphere. This chapter concludes with a discussion of climate change and its vital implications for society.

Earth's Climate System and Its Classification

Climatology, the study of climate, is the analysis of long-term weather patterns, including extreme weather events, over time and space to find areas of similar weather statistics, identified as **climatic regions**. Observed patterns grouped into regions are at the core of climate classification.

The climate where you live may be humid with distinct seasons, or dry with consistent warmth, or moist and cool—almost any combination is possible. There are places where it rains more than 20 cm (8 in.) each month, with monthly average temperatures remaining above 27°C (80°F) year-round. Other places may be rainless for a decade at a time. A climate may have temperatures that average above freezing every month yet still threatens severe frost problems for agriculture. Students reading *Geosystems* in Singapore experience precipitation every month ranging from 13.1 to 30.6 cm (5.1 to 12.0 in.), or 228.1 cm (89.8 in.) during an average year, whereas students at the university in Karachi, Pakistan, measure only 20.4 cm (8 in.) of rain over an entire year.

Which weather elements combine to produce Earth's climates? They include insolation, temperature, humidity, seasonal precipitation, atmospheric pressure and winds, air masses, types of weather disturbances, and cloud coverage. Note that climate cannot actually be observed and really does not exist at any particular moment. Climate is therefore a conceptual statistical construction from these measured weather elements. Figure 10.1 presents a schematic view of Earth's climate system, showing both internal and external processes and linkages that influence climate.

Climate Components: Insolation, Temperature, Pressure, Air Masses, and Precipitation

The principal elements of climate are insolation, temperature, pressure, air masses, and precipitation. The first nine chapters discussed each of these elements. We review them briefly here. Insolation is the energy input for the climate system, but it varies widely over Earth's surface by latitude (see Chapter 2 and Figures 2.9, 2.10, and 2.11). Daylength and temperature patterns vary diurnally (daily) and seasonally (yearly). The principal controls of temperature are latitude, altitude, land–water heating differences, and cloud cover. The pattern of world temperatures and their annual ranges is in Chapter 5 (see Figures 5.14, 5.16, and 5.17).

Temperature variations result from a coupling of dynamic forces in the atmosphere to Earth's pattern of atmospheric pressure and resulting global wind systems (see Figures 6.11 and 6.13). Important too are the location and physical characteristics of air masses, those vast bodies of homogeneous air that form over oceanic and continental source regions.

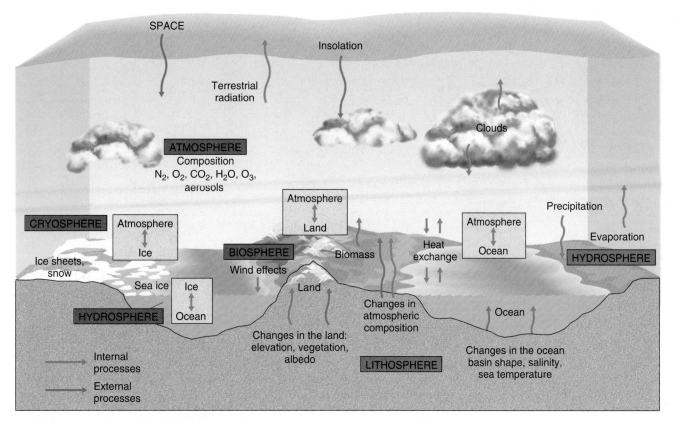

FIGURE 10.1 A schematic of Earth's climate system.
Imagine you are hired to write a computer program that simulates Earth's climates. To do this, you must consider including all the material covered in this text! *Internal processes* that influence climate involve the atmosphere, hydrosphere (streams and oceans), cryosphere (polar ice masses and glaciers), biosphere, and lithosphere (land)—all energized by insolation. All interact to produce climatic patterns. *External processes*, principally from human activity, affect this climatic balance and force climate change. [After J. Houghton, *The Global Climate* (Cambridge, UK: Cambridge University Press, 1984), and the Global Atmospheric Research Program.]

Moisture is the remaining input to climate. The hydrologic cycle transfers moisture, with its tremendous latent heat energy, through Earth's climate system (see Figure 9.1). The moisture input to climate is precipitation in all its forms. Figure 10.2 shows the worldwide distribution of precipitation, our moisture supply. Its patterns are important, for it is a key climate control factor. Average temperatures and daylength are the basic factors that help us approximate POTET (potential evapotranspiration), a measure of natural moisture demand.

Most of Earth's desert regions, areas of permanent drought, are in lands dominated by subtropical high-pressure cells, with bordering lands grading to grasslands and to forests as precipitation increases. The most consistently wet climates on Earth straddle the equator in the Amazon region of South America, the Congo region of Africa, and Indonesia and Southeast Asia, all of which are influenced by equatorial low pressure and the intertropical convergence zone (ITCZ, see Figure 6.11).

Simply relating the two principal climatic components—temperature and precipitation—reveals general climate types (Figure 10.3). Temperature and precipitation patterns, plus other weather factors, provide the key to climate classification.

Classification of Climatic Regions

The ancient Greeks simplified their view of world climates into three zones: The "torrid zone" referred to warmer areas south of the Mediterranean; the "frigid zone" was to the north; and the area where they lived was labeled the "temperate zone," which they considered the optimum climate. They believed that travel too close to the equator or too far north would surely end in death. But the world is a diverse place and Earth's myriad climatic variations are more complex than these simple views.

Classification is the process of grouping data or phenomena in related categories. Such generalizations are important tools in science and are especially useful for the spatial analysis of climatic regions. Just as there is no agreed-upon climate classification system, neither is there a single set of empirical (data) or genetic (causal) criteria to

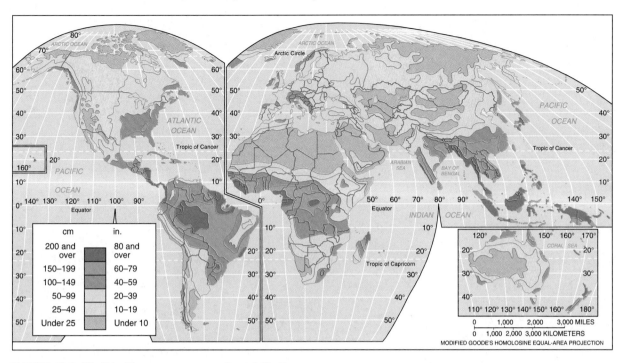

FIGURE 10.2 Worldwide average annual precipitation.
The causes that produce these patterns should be recognizable to you: temperature and pressure patterns; air mass types; convergent, convectional, orographic, and frontal lifting mechanisms; and the general energy availability that decreases toward the poles.

Focus Study 10.1

The El Niño Phenomenon—Record Intensity, Global Linkages

Climate is the consistent behavior of weather over time, but average weather conditions also include extremes that depart from normal. The El Niño–Southern Oscillation (ENSO) in the Pacific Ocean forces the greatest interannual variability of temperature and precipitation on a global scale. The two strongest ENSO events in 120 years hit in 1997–1998 and 1982–1983. Peruvians coined the name El Niño ("the boy child") because these episodes seem to occur around the traditional December celebration time of Christ's birth. Actually El Niños can occur as early as spring and summer and persist through the year.

Revisit Figure 6.22 and see that the northward-flowing Peru current dominates the region off South America's West Coast. These cold waters move toward the equator and join the westward movement of the south equatorial current.

The Peru current is part of the normal counterclockwise circulation of winds and surface ocean currents around the subtropical high-pressure cell dominating the eastern Pacific in the Southern Hemisphere. As a result, a location such as Guayaquil, Ecuador, normally receives 91.4 cm (36 in.) of precipitation each year under dominant high pressure, whereas islands in the Indonesian archipelago receive more than 254 cm (100 in.) under dominant low pressure. This normal alignment of pressure is shown in Figure 1a.

What Is ENSO?
Occasionally, for unexplained reasons, pressure patterns and surface ocean temperatures shift from their usual locations. Higher pressure than normal develops over the western Pacific, and lower pressure over the eastern Pacific. Trade winds normally moving from east

to west weaken and can be reduced or even replaced by an eastward (west-to-east) flow. The shifting of atmospheric pressure and wind patterns across the Pacific is the *Southern Oscillation*.

Sea-surface temperatures increase, sometimes more than 8 C° (14 F°) above normal in the central and eastern Pacific, replacing the normally cold, upwelling, nutrient-rich water along Peru's coastline. Such ocean-surface warming, the "warm pool," may extend to the International Date Line. This surface pool of warm water is known as El Niño. Thus, the designation ENSO is derived—El Niño–Southern Oscillation. This condition is shown in Figure 1b in illustration and satellite image.

The thermocline (boundary of colder, deep-ocean water) lowers in depth in the eastern Pacific Ocean. The change in wind direction and warmer surface water slows the normal

(continued)

Focus Study 10.1 *(continued)*

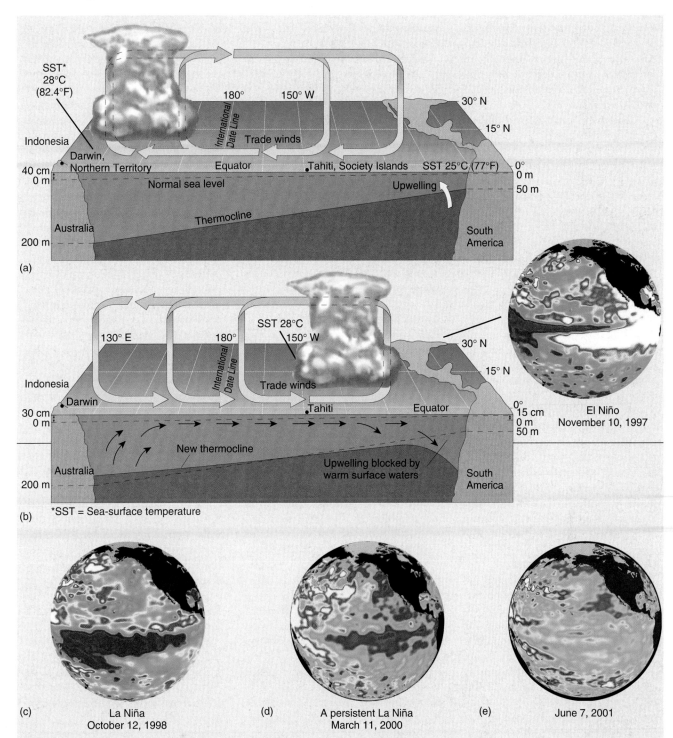

(a)

(b) *SST = Sea-surface temperature

El Niño
November 10, 1997

(c) La Niña
October 12, 1998

(d) A persistent La Niña
March 11, 2000

(e) June 7, 2001

FIGURE 1 Normal, El Niño, and La Niña changes in the Pacific.
(a) Normal patterns in the Pacific; (b) El Niño wind and weather patterns across the Pacific Ocean
and *TOPEX/Poseidon* satellite image for November 10, 1997 (white and red colors indicate warmer
surface water—a warm pool). (c) *TOPEX/Poseidon* image of La Niña conditions in transition in the
Pacific on October 12, 1998 (purple and blue colors for cooler surface water—a cool pool). (d) A
persistent La Niña in March 11, 2000 satellite image. And (e) image from June 7, 2001, showing no
El Niño as of that date, with equatorial waters slowly warming and sea-surface temperature near
normal. (Note: the *TOPEX/Poseidon* satellite accurately measures sea-surface height, not water temperature directly. The satellite can detect changes in sea level due to thermal expansion of warmer
water and contraction of cooler water relative to normal patterns (higher sea surface = warmer;
lower sea surface = cooler). [(a) and (b) Adapted and author corrected from C. S. Ramage, "El
Niño." © 1986 by *Scientific American*, Inc.; (b) (c) and (d) *TOPEX/Poseidon* images courtesy of Jet
Propulsion Laboratory, NASA.]

Focus Study 10.1 *(continued)*

upwelling currents that control nutrient availability. This loss of nutrients affects the phytoplankton and food chain, depriving many fish, marine mammals, and predator birds of nourishment.

Scientists at the National Oceanographic and Atmospheric Administration (NOAA) speculate that ENSO events occurred nine times between 1396 and 1941–1942. They are certain that ENSO events occurred in 1953, 1957–1958, 1965, 1969–1970, 1972–1973, 1976–1977, 1982–1983 (second strongest event), 1986–1987, 1991–1993 (one of the longest), and the most intense episode in 1997–1998 that disrupted global weather.

The expected interval for recurrence is 3 to 5 years, but it may range from 2 to 12 years. The frequency and intensity of ENSO events increased through the twentieth century, a topic of much research by scientists to see if there is a relation to global climate change. Recent studies suggest ENSO might be more responsive to global change than previously thought. Surface temperatures in the central tropical Pacific returned to near normal (neutral) by mid-2001 (Figure 1e). Slight warming during fall 2001 indicated a forthcoming El Niño event in 2002–2003.

La Niña—El Niño's Cousin

When surface waters in the central and eastern Pacific cool to below normal by 0.4 C° (0.7 F°) or more, the condition is dubbed La Niña, Spanish for "the girl." This is a weaker condition and less consistent than El Niño. There is no correlation in the strength or weakness of each. For instance, following the record 1997–1998 ENSO event, the subsequent La Niña was not

as strong as predicted and shared the Pacific with lingering warm water.

Between 1900 and 1998 there were 13 La Niñas of note, the latest in 1988, 1995, and late 1998 to 2000 (Figure 1 c and d). According to National Center for Atmospheric Research (NCAR) scientist Kevin Trenberth, El Niños occurred 31% and La Niñas 23% of the time between 1950 and 1997, with the remaining 46% of the time the Pacific was in a more neutral condition. Don't look for symmetry and opposite effects between the two events, for there is great variability possible, except perhaps in Indonesia where remarkable drought (El Niño) and heavy rain (La Niña) correlations seem strong.

Global Effects Related to the ENSO and La Niña

Effects related to ENSO and La Niña occur worldwide: droughts in South Africa, southern India, Australia, and the Philippines; strong hurricanes in the Pacific, including Tahiti and French Polynesia; and flooding in the southwestern United States and mountain states, Bolivia, Cuba, Ecuador, and Peru. In India, every drought between 1525 and 1900 seems linked to ENSO events. The Atlantic hurricane season weakens during El Niño years and strengthens during La Niñas. (Refer back to the discussion in Focus Study 8.1 and review William Gray's Atlantic hurricane forecasting methods based on these linkages.) Increasing ocean temperatures also are tied to the record level of coral bleaching happening across the tropics, a subject addressed in Chapter 16.

Precipitation in the southwestern United States is greater in El Niño than La Niña years. The Pacific Northwest is wetter with La Niña than

El Niño. El Niño-enhanced rains in 1998 produced the wildflower bloom in Death Valley pictured in this chapter's opening photo. The Colorado River flooding shown in Chapter 15's Focus Study 15.1 was in part attributable to the 1982–1983 El Niño. Yet, as conditions vary, other El Niños have produced drought in the very regions that flooded during a previous episode.

Since the 1982–1983 event, and with the development of remote-sensing satellites and computing capability, scientists now are able to identify the complex global interconnections among surface temperatures, pressure patterns in the Pacific, occurrences of drought in some places, excessive rainfall in others, and the disruption of fisheries and wildlife. Estimates place the overall damage from the 1982–1983 ENSO at more than $8 billion worldwide. Present estimates of weather-related costs for the ENSO in 1997–1998 exceed $80 billion, with some 300 million people displaced and 30,000 deaths.

Discovery of these truly Earth-wide relations and spatial impacts is at the heart of physical geography. The climate of one location is related to climates elsewhere, although it should be no surprise that Earth operates as a vast integrated system. "It is fascinating that what happens in one area can affect the whole world. As to why this happens, that's the question of the century. Scientists are trying to make order out of chaos," says NOAA scientist Alan Strong. (For ENSO monitoring and forecasts, see Climate Prediction Center at http://www.ncep.noaa.gov/ or the Jet Propulsion Laboratory at http://www.jpl.nasa.gov/elnino/ or NOAA's El Niño Theme Page at http://www.pmel.noaa.gov/toga-tao/el-nino/nino-home.html.)

---

which everyone agrees. Any classification system should be viewed as developmental, because it is always open to change and improvement.

A climate classification based on *causative* factors—for example, the interaction of air masses—is called a **genetic classification**. A climate classification based on *statistical data* of observed effects is an **empirical classification**. Climate classifications based on temperature and precipitation data are examples of empirical classifications. This

chapter features descriptions of climatic regions that include both genetic factors of causal elements and empirical factors of statistical elements.

Genetic classifications explain climates in terms of net radiation, thermal regimes, or air mass dominance over a region. Empirical classifications are descriptive and single out selected weather data. One empirical classification system, published by C. W. Thornthwaite in 1948, identified moisture regions using aspects of the water-budget ap-

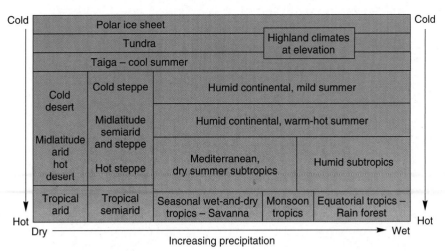

FIGURE 10.3 Climatic relationships.
Temperature and precipitation schematic reveals climatic relationships. Based on general knowledge of your college location, can you identify its approximate location on the schematic diagram? Now locate the region of your birthplace.

proach (Chapter 9) and vegetation types. Another empirical classification system, which forms the framework for our climate regions discussion, is the Köppen classification system.

The Köppen Climate Classification System

The Köppen classification system was designed by Wladimir Köppen (1846–1940), a German climatologist and botanist, and is widely used for its ease of comprehension. His classification work began with an article on heat zones in 1884. By 1900, he included plant communities in his selection of some temperature criteria, using a world vegetation map prepared by French plant physiologist Alphonse de Candolle in 1855. He added letter symbols to designate climate types (example: *Af* for *tropical rain forest*).

Later, Köppen reduced the role played by plants in setting boundaries and moved his system toward strictly climatological empiricism—using weather data. The first wall map showing world climates, co-authored with his student Rudolph Geiger, was introduced in 1928 and soon was widely adopted. Never completely satisfied with his system, Köppen continued to refine it until his death.

Classification Criteria   The basis of any empirical classification system is the choice of criteria used to draw lines on a map to designate different climates. **Köppen–Geiger climate classification** uses *average monthly temperatures*, *average monthly precipitation*, and *total annual precipitation* to devise its spatial categories and boundaries. But we must remember that boundaries really are transition zones of gradual change. The trends and overall patterns of boundary lines are more important than their precise placement, especially with the small scales generally used on world maps (News Report 10.1).

The modified Köppen–Geiger system has its drawbacks. It does not consider winds, temperature extremes, precipitation intensity, amount of sunshine, cloud cover,

---

News Report 10.1

What's in a Boundary?

Originally, Köppen proposed that the isotherm boundary between mesothermal C and microthermal D climates be a coldest month of −3°C (26.6°F) or lower. That might be an accurate criterion for Europe, but for conditions in North America, the 0°C isotherm is considered more appropriate. The difference between the 0 and −3°C isotherms covers an area about the width of the state of Ohio. Remember, these isotherm lines are really transition zones and do not mean abrupt change from one temperature to another.

A line denoting at least one month below freezing runs from New York City roughly along the Ohio River, trending westward until it meets the dry climates in the southeastern corner of Colorado. In Figure 10.5 you see this boundary used. A line marking −3°C as the coldest month would run farther north along Lake Erie and the southern tip of Lake Michigan. From year to year, the position of the 0°C isotherm for January can shift several hundred kilometers as weather conditions vary.

Climate change adds another dimension to this question of accurate placement of statistical boundaries—for they are shifting. The Intergovernmental Panel on Climate Change (IPCC, discussed later in this chapter) predicts a 150- to 550-km (90- to 350-mi) range of possible poleward shift of climatic patterns in the midlatitudes during this century. As you examine North America in Figure 10.5, use the graphic scale to get an idea of the magnitude of these potential shifts.

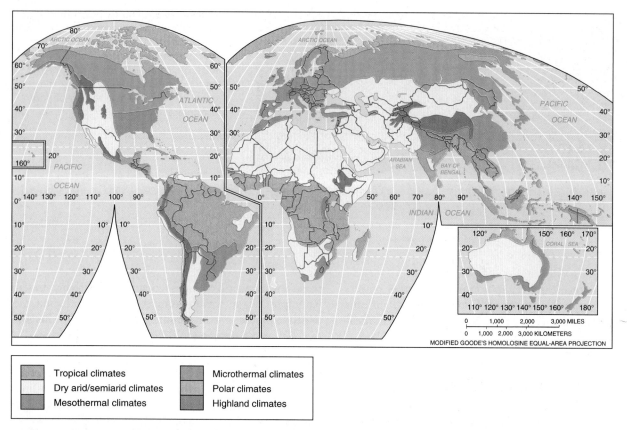

FIGURE 10.4 Climate regions generalized.
Six general categories of the Köppen climate classification system. Relative to the question asked about your campus and birthplace in the caption to Figure 10.3, locate these two places on this climate map.

or net radiation. As an empirical system, it does not consider the causes of precipitation or temperature patterns. Yet the system is important because its *correlation with the actual world is reasonable and the input data are standardized and readily available.* For our purposes of general understanding, a modified Köppen–Geiger climate classification of six climate designations is useful as an organizational tool.

Köppen's Climatic Designations Figure 10.4 shows the distribution of each of Köppen's six climate classifications on the land. This generalized map shows the spatial pattern of climate. The Köppen system uses capital letters (A, C, D, E, H, and B) to designate climatic categories by latitude, from the equator (A) to the poles (E), plus an H category for the special case of highlands. Of the six categories, all but B are based purely on temperature criteria:

A Tropical (equatorial regions)
C Mesothermal (humid subtropical, Mediterranean, and marine west coast regions)
D Microthermal (humid continental, subarctic regions)
E Polar (polar regions)
H Highland (Compared with lowlands at the same latitude, highlands have lower temperatures—recall the *normal lapse rate*—and more efficient precipitation

due to lower moisture demand at higher elevations. This difference justifies the distinction between highlands and surrounding lowland climates.)

Only one climate classification is based on moisture as well:

B Dry (deserts and steppes; arid and semiarid regions, respectively)

Within each climate classification, additional lowercase letters are used to signify temperature and moisture conditions. For example, in a *tropical rain forest Af* climate, the *A* tells us that the average coolest month is above 18°C (64.4°F, average for the month), and the *f* indicates that the weather is constantly wet (German *feucht*, for "moist"), with the driest month receiving at least 6 cm (2.4 in.) of precipitation. As you can see in the map of A climates, the *tropical rain forest Af* climate straddles the equator.

For another example, in a *Dfa* climate, the *D* means that the average warmest month is above 10°C (50°F), with at least one month falling below 0°C (32°F); the *f* says that at least 3 cm (1.2 in.) of precipitation falls during every month; and the *a* indicates a warmest summer month averaging above 22°C (71.6°F). Thus, a *Dfa* climate is a humid-continental, hot-summer climate in the microthermal D category.

Try not to get lost in the alphabet soup of this system, for it is meant to help you understand complex climates through a simplified set of symbols. To simplify the detailed criteria, look at the box placed at the end of each climate category that presents Köppen's guidelines. To help keep it straight, always say the climate's descriptive name followed by its symbol: for example, *Mediterranean dry-summer Csa, tundra ET,* or *cold midlatitude steppe BSk*. We follow this convention throughout the text.

Global Climate Patterns

World climates are presented in Figure 10.5. The following sections describe specific climates, organized around each of the main categories A, C, D, E, and B. An opening box (in the climate map color) at the beginning of each climate gives you a simple description of the climate category.

A map showing distribution appears in the introductory box of each climate and includes the cities represented in climographs. A box at the end of each section lists the specific Köppen criteria for each climate.

Climographs are presented for cities that represent particular climates. A **climograph** is a graph that shows monthly temperature and precipitation. In addition I add location coordinates, average annual temperature, total annual precipitation, elevation, the local population, annual temperature range, annual hours of sunshine (if available, as an indication of cloudiness), and a location map.

Discussions of soils, vegetation, and major terrestrial biomes that fully integrate these global climate patterns are presented in Chapters 18, 19, and 20. Table 20.1 synthesizes all this information and will enhance your understanding of this chapter, so please place a tab on the page with Table 20.1 in Chapter 20 and refer to it as you read.

Tropical Climates (A)

Tropical A climates occupy about 36% of Earth's surface, including both ocean and land areas—Earth's most extensive climate category. The tropical A climates straddle the equator from about 20° N to 20° S, roughly between the Tropics of Cancer and Capricorn, thus the designation *tropical*. Tropical A climates stretch northward to the tip of Florida and south-central Mexico, central India, and Southeast Asia. In A climates, the coolest month must be warmer than 18°C (64.4°F), making these climates truly winterless. Consistent daylength and almost perpendicular Sun angle throughout the year generate this warmth.

A climates are subdivided by annual precipitation into *tropical rain forest Af, tropical monsoon Am,* and *tropical savanna Aw*.

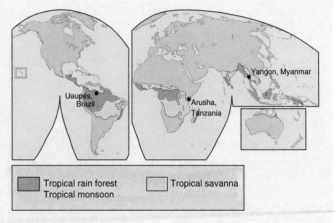

Tropical Rain Forest Climates (Af)

The *tropical rain forest Af* climate is constantly moist and warm. Convectional thunderstorms, triggered by local heating and trade-wind convergence, peak each day from midafternoon to late evening inland and earlier in the day where marine influence is strong along coastlines. Precipitation follows the migrating intertropical convergence zone (ITCZ, Chapter 6). The ITCZ shifts northward and southward with the summer Sun throughout the year, but it influences *tropical rain forest Af* regions all year long. Not surprisingly, water surpluses are enormous, creating the world's greatest stream discharges in the Amazon and Congo Rivers.

High rainfall sustains lush evergreen broadleaf tree growth, producing Earth's equatorial and tropical rain forests (Figure 10.6). Their leaf canopy is so dense that little light diffuses to the forest floor, leaving the ground surface dim and sparse in plant cover. Dense surface vegetation occurs along riverbanks, where light is abundant. (Widespread deforestation of Earth's rain forest is detailed in Chapter 20.)

High temperature promotes energetic bacterial action in the soil so that organic material is quickly consumed. Heavy precipitation washes away certain minerals and nutrients. The resulting soils are somewhat sterile and can support intensive agriculture only if supplemented by fertilizer.

Uaupés, Brazil (Figure 10.7), is characteristic of *tropical rain forest Af*. On the climograph you can see that the lowest-precipitation month receives nearly 15 cm (6 in.), and the annual temperature range is barely 2 C° (3.6 F°). In all such climates, the diurnal (day to night) temperature range exceeds the annual average minimum–maximum (coolest to warmest) range: Day–night temperatures can range more than 11 C° (20 F°), more than five times the annual monthly average range.

The only interruption of *tropical rain forest Af* climates is in the highlands of the South American Andes and in East Africa (see Figure 10.5). There, higher elevations produce lower temperatures; Mount Kilimanjaro is less than 4° south of the equator, but at 5895 m (19,340 ft) it has permanent glacial ice on its summit (although this ice is shrinking due to higher air temperatures). Köppen placed such mountainous sites in the H (highland) climate designation.

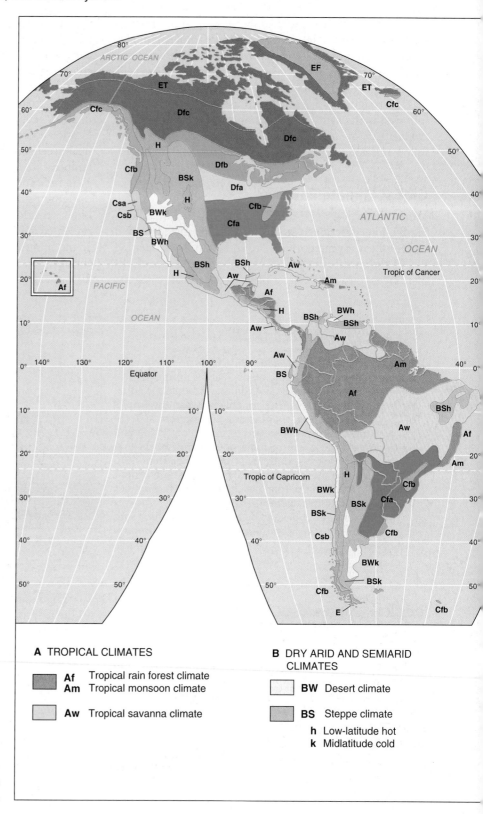

A TROPICAL CLIMATES

Af Tropical rain forest climate
Am Tropical monsoon climate

Aw Tropical savanna climate

B DRY ARID AND SEMIARID CLIMATES

BW Desert climate

BS Steppe climate

h Low-latitude hot
k Midlatitude cold

FIGURE 10.5 World climates according to the Köppen–Geiger classification system.

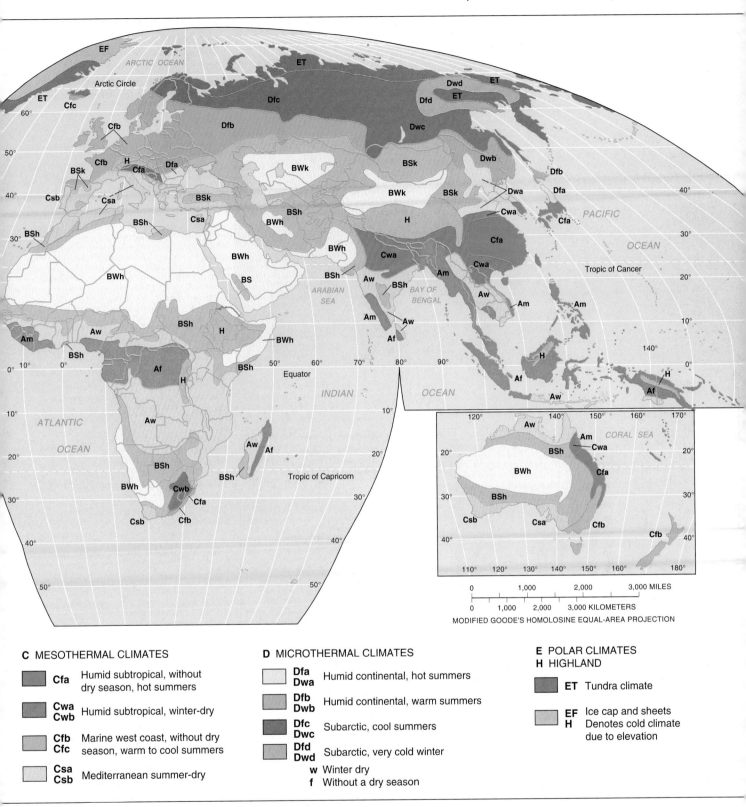

C MESOTHERMAL CLIMATES

	Cfa	Humid subtropical, without dry season, hot summers
	Cwa Cwb	Humid subtropical, winter-dry
	Cfb Cfc	Marine west coast, without dry season, warm to cool summers
	Csa Csb	Mediterranean summer-dry

D MICROTHERMAL CLIMATES

	Dfa Dwa	Humid continental, hot summers
	Dfb Dwb	Humid continental, warm summers
	Dfc Dwc	Subarctic, cool summers
	Dfd Dwd	Subarctic, very cold winter
	w	Winter dry
	f	Without a dry season

E POLAR CLIMATES
H HIGHLAND

| | ET | Tundra climate |
| | EF H | Ice cap and sheets Denotes cold climate due to elevation |

MODIFIED GOODE'S HOMOLOSINE EQUAL-AREA PROJECTION

FIGURE 10.6 The tropical rain forest.
The lush equatorial rain forest and Sangha River, near Ouesso, Congo, Africa. [Photo by BIOS (M. Gunther)/ Peter Arnold, Inc.]

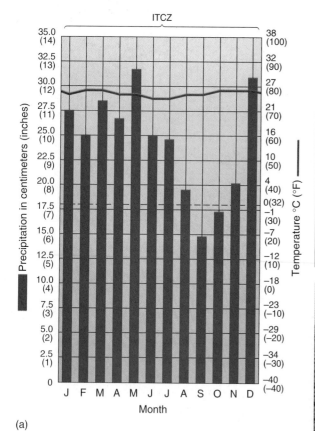

Station: Uaupés, Brazil   **Af**   **Elevation:** 86 m (282.2 ft)
Lat/long: 0°08' S 67°05' W   **Population:** 7500
Avg. Ann. Temp.: 25°C (77°F)   **Ann. Temp. Range:**
Total Ann. Precip.:   2 C° (3.6 F°)
        291.7 cm (114.8 in.)   **Ann. Hr of Sunshine:** 2018

(a)

FIGURE 10.7 Tropical rain forest climate.
(a) Climograph for Uaupés, Brazil (*tropical rain forest Af*).
(b) The rain forest near Uaupés along a tributary of the Rio Negro. [Photo by Will and Deni McIntyre/Photo Researchers, Inc.]

(b)

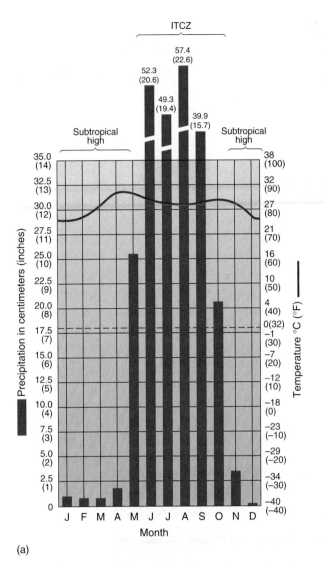

Station: Yangon, Myanmar* **Am**
Lat/long: 16°47' N 96°10' E
Avg. Ann. Temp.: 27.3°C (81.1°F)
Total Ann. Precip.:
    252.7 cm (99.5 in.)
*(Formerly Rangoon, Burma)

Elevation: 23 m (76 ft)
Population: 2,458,700
Ann. Temp. Range:
    5.5 C° (9.9 F°)

FIGURE 10.8 Tropical monsoon climate.
(a) Climograph for Yangon, Myanmar (formerly Rangoon, Burma) (*tropical monsoon Am*).
(b) The monsoonal forest near Malang, Java, at the Purwodadi Botanical Gardens. [Photo by Tom McHugh/Photo Researchers, Inc.]

Tropical Monsoon Climates (Am)

The *tropical monsoon Am* climates feature a dry season that lasts one or more months. Rainfall brought by the ITCZ affects these areas from 6 to 12 months of the year. (Remember, the ITCZ affects the *tropical rain forest Af* climate region throughout the year.) The dry season occurs when the convergence zone is not overhead. Yangon, Myanmar (formerly Rangoon, Burma), is an example of this climate type, as illustrated by the climograph and photograph in Figure 10.8.

Tropical monsoon Am climates lie principally along coastal areas within the tropical rain forest climatic realm and experience seasonal variation of wind and precipitation. Evergreen trees grade into thorn forests on the drier margins near the adjoining savanna climates.

Tropical Savanna Climates (Aw)

Tropical savanna Aw climates exist poleward of the *tropical rain forest Af* climates. The ITCZ reaches these climate regions for about six months or less of the year as it migrates with the summer Sun. These starkly seasonal convectional rains accompany the shifting ITCZ. The *w* signifies this winter-dry condition, when rains depart the area and subtropical high pressure dominates. Thus, POTET (moisture demand) exceeds PRECIP (moisture supply) in winter, causing water-budget deficits.

Temperatures vary more in tropical savanna climates than in tropical rain forest regions. The tropical savanna regime can have two temperature maximums during the year because the Sun's direct rays are overhead twice—before and after the summer solstice in each hemisphere

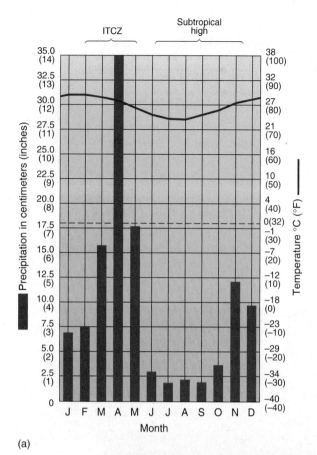

Station: Arusha, Tanzania
Lat/long: 3°24' S 36°42' E
Avg. Ann. Temp.: 26.5°C (79.7°F)
Total Ann. Precip.:
       119 cm (46.9 in.)

Elevation: 1387 m (4550 ft)
Population: 100,000
Ann. Temp. Range:
       4.1 C° (7.4 F°)
Ann. Hr of Sunshine: 2600

(a)

FIGURE 10.9 Tropical savanna climate.
(a) Climograph for Arusha, Tanzania (*tropical savanna Aw*);
note the intense dry period. (b) Characteristic landscape in
Kenya, with plants adapted to seasonally dry water budgets.
[Photo by Stephen J. Krasemann/DRK Photo.]

(b)

as the Sun moves between the equator and the tropic.
Dominant grasslands with scattered trees, drought-resistant to cope with the highly variable precipitation, characterize the *tropical savanna Aw*.

Arusha, Tanzania, is a characteristic *tropical savanna Aw*
station (Figure 10.9). This city of more than 100,000 people is east of the famous Serengeti Plains savanna grassland

and Olduvai Gorge, site of human origins, and north of
Tarangire National Park. Temperatures are consistent with
tropical A climates, despite the elevation of the station. Note
the marked dryness from June to October, which defines
changing dominant pressure systems rather than annual
changes in temperature. This region is near the transition
to the dryer hot-desert steppe climates to the northeast.

Köppen Guidelines
Tropical Climates—A

Consistently warm, with all months averaging above 18°C (64.4°F); annual PRECIP exceeds POTET.

Af—Tropical rain forest:
f = All months receive PRECIP in excess of 6 cm (2.4 in.).

Am—Tropical monsoon:
m = A marked short dry season, with 1 or more months receiving less than 6 cm (2.4 in.) PRECIP;

an otherwise excessively wet rainy season. ITCZ dominant 6–12 months.

Aw—Tropical savanna:
w = Summer wet season, winter dry season; ITCZ dominant 6 months or less, winter water-balance deficits.

Mesothermal Climates (C)

Mesothermal C climates occupy the second-largest percentage of Earth's land and sea surface, about 27%. When land area alone is considered, they rank only fourth. Together, A and C climates dominate more than half of Earth's oceans and about one-third of its land area. Approximately 55% of the world's population resides in C climates. These warm and temperate lands are designated mesothermal, meaning "middle temperature."

The mesothermal C climates, and nearby portions of the microthermal D climates, are regions of great weather variability, for these are the latitudes of greatest air mass conflict. The C climatic region marks the beginning of true seasonality; contrasts in temperature are evidenced by vegetation, soil, and human lifestyle adaptations. Subdivisions of the C classification are based on precipitation variability: *humid subtropical*, *marine west coast*, and *Mediterranean*.

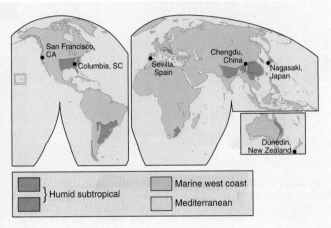

Humid Subtropical Hot-Summer Climates (Cfa, Cwa)

The *humid subtropical hot-summer climates* (*Cfa, Cwa*) are either moist all year (*f*) or have a pronounced winter-dry period (*w*) and feature a hot summer (indicated by the *a*). During summer, the *humid subtropical hot-summer Cfa* climate is influenced by maritime tropical air masses generated over warm waters along eastern coasts. This warm, moist, unstable air produces convectional showers over land. In fall, winter, and spring, maritime tropical and continental polar air masses interact, generating frontal activity and frequent midlatitude cyclonic storms. These two mechanisms produce year-round precipitation. Overall, precipitation averages 100–200 cm (40–80 in.) a year.

Nagasaki, Japan (Figure 10.10), is characteristic of an Asian *humid subtropical hot-summer Cfa* station, whereas Columbia, South Carolina, is characteristic of the North American *Cfa* climate region (Figure 10.11). Unlike the precipitation of humid subtropical hot-summer cities in the United States (Atlanta, Memphis, Norfolk, New Orleans, and Columbia), Nagasaki's winter precipitation is a bit less because of the effects of the Asian monsoon. However, the lower precipitation of winter is not quite dry enough to change its classification to a winter-dry *w*. In

comparison to higher rainfall amounts in Nagasaki (196 cm, 77 in.), Columbia's precipitation totals 126.5 cm (49.8 in.), Atlanta's 122 cm (48 in.) annually (Figure 10.11b).

Humid subtropical winter-dry Cwa climates are related to the winter-dry, seasonal pulse of the monsoons. They extend poleward from tropical savanna climates and have a summer month that receives 10 times more precipitation than their driest winter month. A representative station is Chengdu, China. Figure 10.12 demonstrates the strong correlation between precipitation and the high-summer Sun.

The habitability of the humid *subtropical hot-summer Cfa* and *humid subtropical winter-dry Cwa* climates and their ability to sustain populations are borne out by the concentration of people in north-central India, the bulk of China's 1.28 billion people, and the many who live in climatically similar portions of the United States.

The *Cwa* monsoonal climates hold several precipitation records. Cherrapunji, India, in the Assam Hills south of the Himalayas, is the all-time precipitation record holder for a single year and for every other time interval from 15 days to 2 years. Because of the summer monsoons that pour in from the Indian Ocean and the Bay of Bengal, Cherrapunji has received 930 cm (30.5 ft) of rainfall in one month and 2647 cm (86.8 ft) in one year—both records.

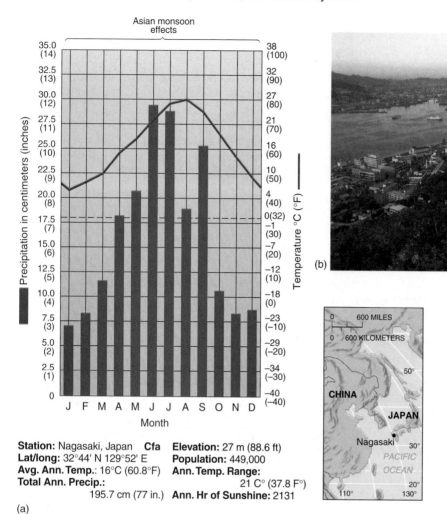

(a)

Station: Nagasaki, Japan **Cfa** **Elevation:** 27 m (88.6 ft)
Lat/long: 32°44' N 129°52' E **Population:** 449,000
Avg. Ann. Temp.: 16°C (60.8°F) **Ann. Temp. Range:**
Total Ann. Precip.: 21 C° (37.8 F°)
195.7 cm (77 in.) **Ann. Hr of Sunshine:** 2131

FIGURE 10.10 Humid subtropical climate, Asian region.
(a) Climograph for Nagasaki, Japan (*humid subtropical Cfa*). (b) Landscape near Nagasaki.
[(b) Photo by Ken Straiton/First Light.]

Marine West Coast Climates (Cfb, Cfc)

Marine west coast climates, featuring mild winters and cool summers, dominate Europe and other middle-to-high-latitude west coasts (see Figure 10.5). In the United States, these climates with their cooler summers are in contrast to the hot-summer humid climate of the southeastern United States. (The *b* means the warmest month is below 22°C, or 71.6°F, and a *c* means that the one to three warmest months are above 10°C.)

Maritime polar air masses—cool, moist, unstable—dominate *marine west coast Cfb* and *Cfc* climates. Weather systems forming along the polar front and maritime polar air masses move into these regions throughout the year, making weather quite unpredictable. Coastal fog, annually totaling 30 to 60 days, is a part of the moderating marine influence. Frosts are possible and tend to shorten the growing season.

Marine west coast climates are unusually mild for their latitude. They extend along the coastal margins of the Aleutian Islands in the North Pacific, cover the southern third of Iceland in the North Atlantic and coastal Scandinavia, and dominate the British Isles. It is hard to imagine that such high-latitude locations can have average monthly temperatures above freezing throughout the year!

Unlike the extensive influence of marine west coast regions in Europe, mountains restrict this climate to coastal environs in Canada, Alaska, Chile, and Australia (Figure 10.13). (A temperature graph for Vancouver, British Columbia, appears in Figure 5.12 with a photo of this *marine west coast Cfb* city.)

The climograph for Dunedin, New Zealand, demonstrates the moderate temperature patterns and the annual temperature range for a *marine west coast Cfb* station in the Southern Hemisphere (Figure 10.14).

Station: Columbia, South Carolina **Cfa**
Lat/long: 34° N 81° W
Avg. Ann. Temp.: 17.3°C (63.1°F)
Total Ann. Precip.:
126.5 cm (49.8 in.)

Elevation: 96 m (315 ft)
Population: 537,000
Ann. Temp. Range:
20.7 C° (37.3 F°)
Ann. Hr of Sunshine: 2800

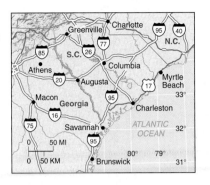

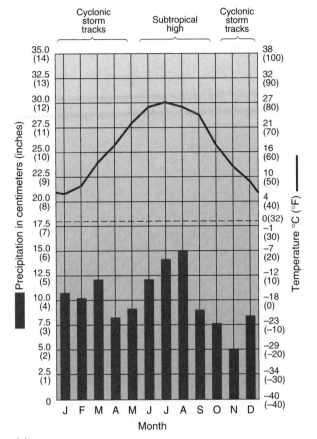

(a)

(b)

FIGURE 10.11 Humid subtropical climate, North American region.
(a) Climograph for Columbia, South Carolina (*humid subtropical Cfa*). Note the more consistent precipitation pattern compared to Nagasaki, as Columbia receives seasonal cyclonic storm activity and summer convection showers within maritime tropical air. (b) The mixed deciduous and evergreen forest in southern Georgia, typical of the humid subtropical southeastern United States. [(b) Photo by Bobbé Christopherson.]

An interesting anomaly occurs in the eastern United States. In portions of the Appalachian highlands, increased elevation lowers summer temperatures in the surrounding *humid subtropical hot summer Cfa* classification, producing a *marine west coast Cfb* climate. The climograph for Bluefield, West Virginia (Figure 10.15), reveals marine west coast temperature and precipitation patterns, despite its location in the east. Vegetation similarities between the Appalachians and the Pacific Northwest have enticed many emigrants from the East to settle in these climatically familiar environments in the Northwest.

Mediterranean Dry-Summer Climates (Csa, Csb)

Across the planet during summer months, shifting cells of subtropical high pressure block moisture-bearing winds from adjacent regions. This shifting of stable, warm to hot, dry air over an area in summer and away from these regions in the winter creates a pronounced dry-summer and wet-winter pattern. For example, the continental tropical air mass over the Sahara in Africa shifts northward in summer over the Mediterranean region and blocks

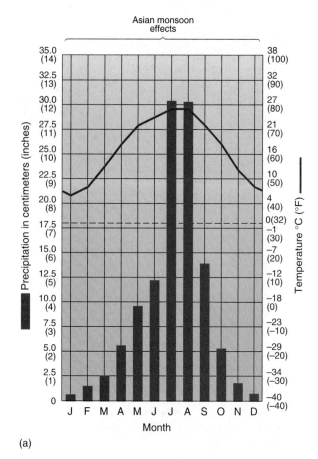

Asian monsoon
effects

(a)

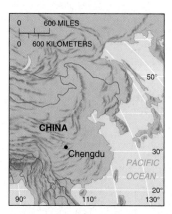

Station: Chengdu, China **Cwa**   **Elevation:** 498 m (1633.9 ft)
Lat/long: 30°40′ N 104°04′ E   **Population:** 2,260,000
Avg. Ann. Temp.: 17°C (62.6°F)   **Ann. Temp. Range:**
Total Ann. Precip.:   20 C° (36 F°)
　　114.6 cm (45.1 in.)   **Ann. Hr of Sunshine:** 1058

(b)

FIGURE 10.12 Humid subtropical winter-dry climate.
(a) Climograph for Chengdu, China (*humid subtropical Cwa*). Note the summer-wet monsoonal
precipitation. (b) Landscape of southern interior China characteristic of this winter-dry climate.
This valley is near Mount Daliang in Sichuan Province. [Photo by Jin Zuqi/Sovfoto/Eastfoto.]

FIGURE 10.13 A marine west coast climate.
The natural vegetation in the Channel Islands, Strait of Georgia, British Columbia, Canada (*marine west coast Cfb*). [Photo
by Bobbé Christopherson.]

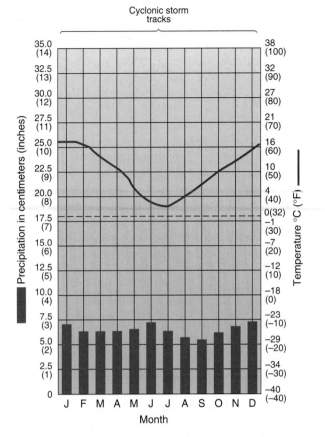

Cyclonic storm tracks

Station: Dunedin, New Zealand **Cfb**
Lat/long: 45°54' S 170°31' E
Avg. Ann. Temp.: 10.2°C (50.3°F)
Total Ann. Precip.:
          78.7 cm (31.0 in.)

Elevation: 1.5 m (5 ft)
Population: 77,500
Ann. Temp. Range:
          14.2 C° (25.5 F°)

(a)

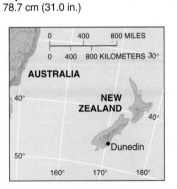

(b)

FIGURE 10.14 A Southern Hemisphere marine west coast climate.
(a) Climograph for Dunedin, New Zealand (*marine west coast Cfb*). (b) Meadow, forest, and mountains on South Island, New Zealand. [Photo by Brian Enting/Photo Researchers, Inc.]

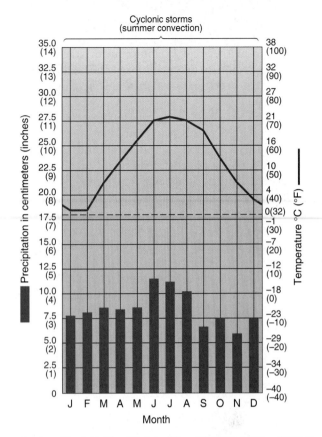

Cyclonic storms (summer convection)

Station: Bluefield, West Virginia **Cfb**
Lat/long: 37°16' N 81°13' W
Avg. Ann. Temp.:
          12°C (53.6°F)
Total Ann. Precip.:
          101.9 cm (40.1 in.)

Elevation: 780 m (2559 ft)
Population: 16,000
Ann. Temp. Range:
          21 C° (37.8 F°)

(a)

(b)

FIGURE 10.15 Marine west coast climate in the Appalachians of the East.
(a) Climograph for Bluefield, West Virginia (*marine west coast Cfb*). (b) Characteristic mixed forest of Dolly Sods Wilderness in the Appalachian highlands. [Photo by David Muench Photography, Inc.]

maritime air masses and cyclonic storm tracks. The Mediterranean climate designation specifies that at least 70% of annual precipitation occurs during the winter months.

Worldwide cool offshore currents (the California current, Canary current, Peru current, Benguela current, and West Australian current) produce stability in overlying air masses along west coasts, poleward of subtropical high pressure. The world climate map (see Figure 10.5) shows Mediterranean dry-summer climates along the western margins of North America, central Chile, and the southwestern tip of Africa, as well as across southern Australia and the Mediterranean Basin—the climate's namesake region.

Figure 10.16 compares the climographs of Mediterranean dry-summer cities San Francisco and Sevilla

(Seville), Spain. Coastal maritime effects moderate San Francisco's climate, producing a cooler summer. The transition to a hot-summer designation (*Csb* to *Csa*) occurs no more than 24–32 km (15–20 mi) inland from San Francisco. The photos in Figure 10.16 show oak-savanna landscapes near Olvera, Spain, and in California.

The *Mediterranean dry-summer* climate brings summer water-balance deficits. Winter precipitation recharges soil moisture, but water use usually exhausts soil moisture by late spring. Large-scale agriculture requires irrigation, although some subtropical fruits, nuts, and vegetables are uniquely suited to these conditions. Natural vegetation features a hard-leafed, drought-resistant variety known locally as *chaparral* in the western United States. (Chapter 20 discusses local names for this type of vegetation in other parts of the world.)

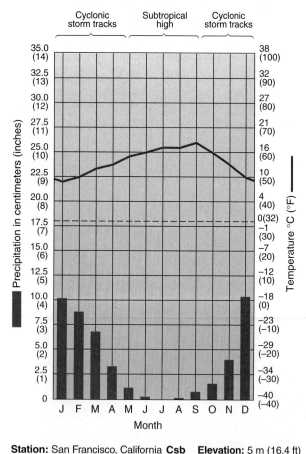

FIGURE 10.16 Mediterranean climates, California and Spain.
Climographs for (a) San Francisco, California (*Mediterranean cool-summer Csb*) and (b) Sevilla, Spain (*Mediterranean hot-summer Csa*). (c) The countryside around Olvera, Andalusia, Spain. (d) Central California landscape of oak savanna. [Photos by (c) Kaz Chiba/Liaison Agency, Inc.; (d) Bobbé Christopherson.]

FIGURE 10.16 *(continued)*

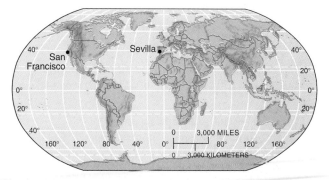

(c)

(d)

Köppen Guidelines
Mesothermal Climates—C

Warmest month above 10°C (50°F); coldest month above 0°C (32°F) but below 18°C (64.4°F); seasonal climates.

Cfa, Cwa—Humid subtropical:
a = Hot summer; warmest month above 22°C
(71.6°F).
f = Year-round PRECIP.
w = Winter drought, summer wettest month 10
times more PRECIP than driest winter month.

Cfb, Cfc—Marine west coast: Mild-to-cool summer
f = Receives year-round PRECIP.
b = Warmest month below 22°C (71.6°F) with 4
months above 10°C.
c = 1–3 months above 10°C.

Csa, Csb—Mediterranean summer dry:
s = Pronounced summer drought with 70% of
PRECIP in winter.
a = Hot summer with warmest month above 22°C
(71.6°F).
b = Mild summer; warmest month below 22°C.

Microthermal Climates (D)

Humid microthermal climates have long winters with some summer warmth. Here the term *microthermal* means cool temperate to cold. Approximately 21% of Earth's land surface is influenced by these climates, equaling about 7% of Earth's total surface. These climates occur poleward of the mesothermal climates and experience great temperature ranges related to continentality and air mass conflicts.

Because the Southern Hemisphere lacks substantial landmasses, microthermal climates develop there only in highlands. (In Figure 10.5, note the absence of microthermal D climates in the Southern Hemisphere.) Microthermal climates range from a *humid continental hot-summer Dfa* in Chicago to the formidable extremes of a frigid *subarctic Dwd* in Verkhoyansk, Siberia. (The letters *a, b, c*—hot, warm, cool—carry the same meaning as in mesothermal C climates.)

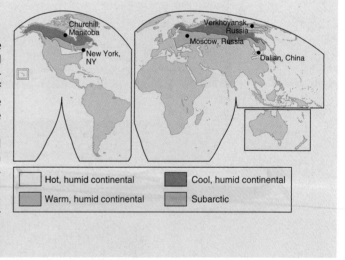

Hot, humid continental

Warm, humid continental

Cool, humid continental

Subarctic

Humid Continental Hot-Summer Climates (Dfa, Dwa)

Humid continental hot-summer Dfa climates are differentiated by their annual precipitation distribution. Both humid continental moist-all-year (*Dfa*) and winter-dry (*Dwa*) climates are influenced by maritime tropical air masses in the summer. In North America, frequent weather activity is possible between conflicting air masses—maritime tropical and continental polar—especially in winter. The climographs for New York City and Dalian,

China, illustrate these two hot-summer mesothermal climates (Figure 10.17).

Originally before European settlement, forests covered the *humid continental hot-summer Dfa* climatic region of the United States as far west as the Indiana–Illinois border. Beyond that approximate line, tall-grass prairies extended westward to about the 98th meridian (98° W in central Kansas) and the approximate location of the 51 cm (20 in.) isohyet (line of equal precipitation). Further west the short-grass prairies reflected lower precipitation receipts.

Deep sod made farming difficult for the first settlers, as did the climate. However, native grasses soon were replaced with domesticated wheat and barley. Various inventions (barbed wire, the self-scouring steel plow, well-drilling techniques, windmills, railroads, and the six-shooter) aided nonnative peoples' expansion into the region. In the United States today, the *humid continental hot-summer Dfa* climatic region is the location of corn, soybean, hog, and cattle production (Figure 10.17c).

The dry winter associated with the vast Asian landmass, specifically Siberia, is exclusively assigned *w* because of an extremely dry-winter high-pressure anticyclone. The dry monsoons of southern and eastern Asia are produced in the winter months by this system, as winds blow out of Siberia toward the Pacific and Indian Oceans. The intruding cold of continental air is a significant winter feature.

Humid Continental Mild-Summer Climates (Dfb, Dwb)

Soils are thinner and less fertile in the cooler microthermal climates, yet agricultural activity is important and includes dairy cattle, poultry, flax, sunflowers, sugar beets, wheat, and potatoes. Frost-free periods range from fewer than 90 days in the north to as many as 225 days in the south. Overall, precipitation is less than in the hot-summer regions to the south; however, notably heavier snowfall is important to soil moisture recharge when it melts. Various snow-capturing strategies are in use, including fences and tall stubble left standing after harvest in fields to create snow drifts and thus more moisture retention on the soil.

The dry-winter aspect of the mild-summer climate (*Dwb*) occurs only in Asia, in a far-eastern area poleward of the winter-dry mesothermal climates. A representative *humid continental mild-summer Dwb* climate along Russia's east coast is Vladivostok usually one of only two ice-free ports in that country.

Characteristic *Dfb* stations are Duluth, Minnesota, and Saint Petersburg, Russia. Figure 10.18 presents a climograph for Moscow, which is at 55° N, or about the same latitude as the southern shore of Hudson Bay in Canada. The photos of landscapes near Moscow, Russia, and Sebago Lake, inland from Portland, Maine, show summer and late winter scenes, respectively.

Subarctic Climates (Dfc, Dwc, Dwd)

Farther poleward, seasonal change becomes greater. The short growing season is more intense during long summer days. The three cold subarctic climates—*Dfc, Dwc, Dwd*—include vast stretches of Alaska, Canada, northern Scandinavia, and Russia. Discoveries of minerals and petroleum reserves have led to new interest in portions of these regions.

Areas that receive 25 cm (10 in.) or more of precipitation a year on the northern continental margins and are covered by the so-called snow forests of fir, spruce, larch, and birch are the *boreal forests* of Canada and the *taiga* of Russia. These forests are in transition to the more open northern woodlands and to the tundra region of the far north. Forests thin out to the north when the warmest month drops below an average temperature of 10°C (50°F).

Soils are thin in these lands once scoured by glaciers. Precipitation and potential evapotranspiration both are low, so soils are generally moist and either partially or totally frozen beneath the surface, a phenomenon known as *permafrost*.

The Churchill, Manitoba, climograph (Figure 10.19) shows average monthly temperatures below freezing for 7 months of the year, during which time light snow cover and frozen ground persist. High pressure dominates Churchill during its cold winter—this is the source region for the continental polar air mass. Churchill is representative of the *subarctic Dfc* climate: annual temperature range of 40 C° (72 F°) and low precipitation of 44.3 cm (17.4 in.).

The dry-winter *subarctic Dwc* and *subarctic Dwd* climates occur only within Russia. The intense cold of Siberia and north-central and eastern Asia is difficult to comprehend, for these areas experience a coldest month with an average temperature lower than −38°C (−36.4°F) for 7 months. And as described in Chapter 5, minimum temperatures of below −68°C (−90°F) were recorded there. Yet summer maximum temperatures in these same areas normally exceed +37°C (+98°F)!

A *subarctic Dwd* station is Verkhoyansk, Siberia (Figure 10.20). For 4 months of the year, average temperatures fall below −34°C (−29.2°F). Verkhoyansk has probably the world's greatest annual temperature range from winter to summer: a remarkable 63 C° (113.4 F°)! Winters feature brittle metals and plastics, triple-thick windowpanes, and temperatures that render straight antifreeze a solid.

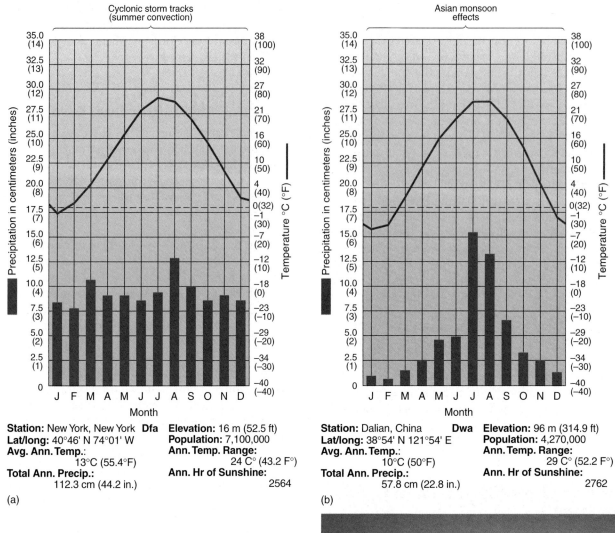

Station: New York, New York **Dfa**
Lat/long: 40°46' N 74°01' W
Avg. Ann. Temp.:
            13°C (55.4°F)
Total Ann. Precip.:
            112.3 cm (44.2 in.)

Elevation: 16 m (52.5 ft)
Population: 7,100,000
Ann. Temp. Range:
            24 C° (43.2 F°)
Ann. Hr of Sunshine:
                        2564

(a)

Station: Dalian, China **Dwa**
Lat/long: 38°54' N 121°54' E
Avg. Ann. Temp.:
            10°C (50°F)
Total Ann. Precip.:
            57.8 cm (22.8 in.)

Elevation: 96 m (314.9 ft)
Population: 4,270,000
Ann. Temp. Range:
            29 C° (52.2 F°)
Ann. Hr of Sunshine:
                        2762

(b)

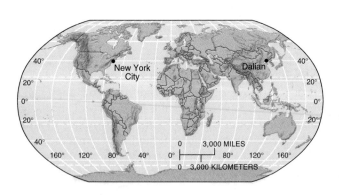

(c)

FIGURE 10.17 Humid continental hot-summer climates, New York and China.
Climographs for (a) New York City (*humid continental Dfa*) and (b) Dalian, China (*humid continental Dwa*). (c) Typical humid continental deciduous forest and ready-to-harvest soybean field in central Indiana near Zelma in the American Midwest. [Photo by Bobbé Christopherson.]

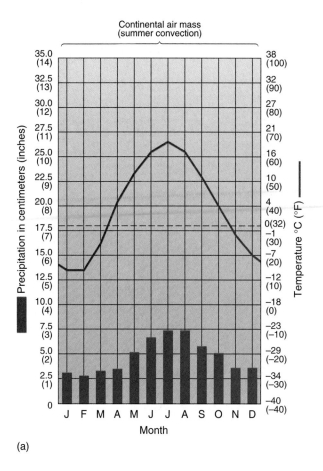

Continental air mass
(summer convection)

(a)

Station: Moscow, Russia   **Dfb**
Lat/long: 55°45' N 37°34' E
Avg. Ann. Temp.: 4°C (39.2°F)
Total Ann. Precip.:
        57.5 cm (22.6 in.)

Elevation: 156 m (511.8 ft)
Population: 9,900,000
Ann. Temp. Range:
        29 C° (52.2 F°)
Ann. Hr of Sunshine:
        1597

(b)

(c)

FIGURE 10.18 Humid continental mild-summer climate.
(a) Climograph for Moscow, Russia (*humid continental Dfb*). (b) Fields near Saratov, Russia,
during the short summer season. (c) Late-winter scene of Sebago Lake and forests inland
from Portland, Maine. [(b) Photo by Wolfgang Kaehler/Liaison Agency, Inc.; (c) photo by
Bobbé Christopherson.]

Köppen Guidelines
Microthermal Climates—D

Warmest month above 10°C (50°F); coldest month
below 0°C (32°F); cool temperate-to-cold conditions; snow climates. In Southern Hemisphere, only
in highland climates.

Dfa, Dwa—Humid continental:
a = Hot summer; warmest month above 22°C
    (71.6°F).
f = Year-round PRECIP.
w = Winter drought.

Dfb, Dwb—Humid continental:
b = Mild summer; warmest month below 22°C
    (71.6°F).
f = Year-round PRECIP.

Dfc, Dwc, Dwd—Subarctic: Cool summers, cold winters.
f = Year-round PRECIP.
w = Winter drought.
c = 1–4 months above 10°C.
d = Coldest month below −38°C (−36.4°F), in
    Siberia only.

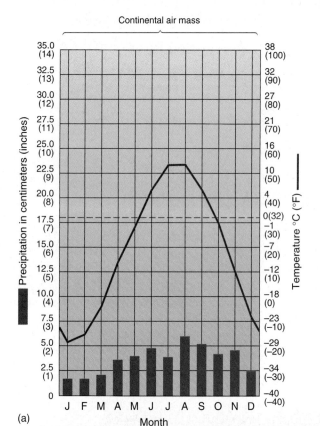

Continental air mass

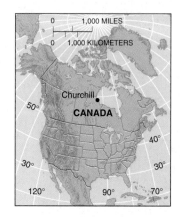

Station: Churchill, Manitoba **Dfc**
Lat/long: 58°45' N 94°04' W
Avg. Ann. Temp.: −7°C (19.4°F)
Total Ann. Precip.:
  44.3 cm (17.4 in.)

Elevation: 35 m (114.8 ft)
Population: 1400
Ann. Temp. Range:
  40 C° (72 F°)
Ann. Hr of Sunshine:
  1732

FIGURE 10.19 Subarctic climate.
(a) Climograph for Churchill, Manitoba (*subarctic Dfc*).
(b) Winter scene with foraging polar bear in the
Churchill area near Hudson Bay. (c) Summer scene as
forests give way to tundra landscapes in Churchill area.
[(b) Photo by Art Wolfe/Tony Stone Images. (c) Photo
by Norbert Rosing/Animals Animals/Earth Scenes.]

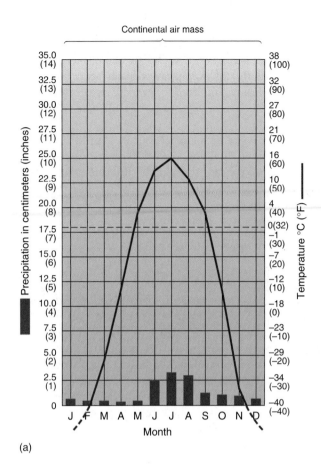

Continental air mass

Station: Verkhoyansk, Russia **Dwd** **Elevation:** 137 m (449.5 ft)
Lat/long: 67°35' N 133°27' E   **Population:** 1400
Avg. Ann. Temp.: −15°C (5°F)   **Ann. Temp. Range:**
Total Ann. Precip.:   63 C° (113.4 F°)
15.5 cm (6.1 in.)

(a)

(b)

FIGURE 10.20 Extreme subarctic winter-dry climate.
(a) Climograph for Verkhoyansk, Russia (*subarctic Dwd*). (b) Scene in the town of Verkhoyansk during the short summer. [Photo by Dean Conger/National Geographic Society.]

Polar Climates (E)

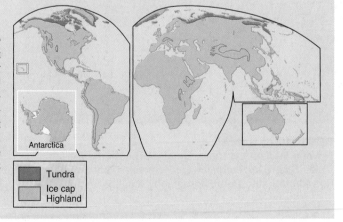

The polar climates—*tundra ET, ice cap EF*, and *polar marine EM*—cover about 19% of Earth's total surface and about 17% of its land area. These climates have no true summer like that in lower latitudes. Poleward of the Arctic and Antarctic Circles, daylength increases in summer until daylight becomes continuous, yet average monthly temperatures never rise above 10°C (50°F). Daylength, which in part determines the amount of insolation received, and low Sun altitude in the sky are the principal climatic factors in these frozen and barren regions. Yet, in winter, the Sun drops below the horizon poleward of 66.5° latitude, producing continuous night.

Tundra

Ice cap
Highland

Tundra Climate (ET)

In a *tundra ET* climate, land is under continuous snow cover for 8–10 months, but when the snow melts and spring arrives, numerous plants appear—stunted sedges, mosses, flowering plants, and lichens. Much of the area experiences permafrost and ground ice conditions. The tundra is also the summer home of mosquitoes of legend and black gnats. Global warming is bringing dramatic changes to the tundra, its plants, animals, and permafrost ground conditions. In 1998, parts of Canada registered temperatures 5 C° above average, 2.5 C° above normal for all of Canada—the warmest in the 51-year record. As organic peat deposits in the tundra thaw, vast stores of carbon are released to the atmosphere, further adding to the greenhouse gas problem. A section in Chapter 17 features this region.

Tundra ET climates are strictly a Northern Hemisphere occurrence, except for elevated mountain locations in the Southern Hemisphere and a portion of the Antarctic Peninsula. Because of elevation, the summit of Mount Washington in New Hampshire (1914 m, or 6280 ft) statistically qualifies as highland *tundra ET* climate of small scale.

Ice Cap Climate (EF)

Most of Antarctica and central Greenland fall within the *ice cap EF* climate, as does the North Pole, with all months averaging below freezing. Both regions are dominated by dry, frigid air masses, with vast expanses that never warm above freezing. The area of the North Pole is actually a sea covered by ice, whereas Antarctica is a substantial continental landmass covered by Earth's greatest ice sheet. For comparison, winter minimums at the South Pole (July) fre-quently drop below the temperature of solid carbon dioxide or "dry ice" (−78°C, or −109°F).

Antarctica is constantly snow-covered but receives less than 8 cm (3 in.) of precipitation each year. However, Antarctic ice has accumulated to several kilometers of thickness and is the largest repository of freshwater on Earth (Figure 10.21).

This ice is a vast historical record of Earth's atmosphere. Within it, thousands of past volcanic eruptions from all over the world have deposited ash layers, and ancient combinations of atmospheric gases lie trapped in frozen bubbles. Chapter 17 presents analysis of ice cores taken from Greenland and Antarctica.

Polar Marine Climate (EM)

Polar marine EM stations are more moderate than other polar climates in winter, with no month below −7°C (20°F), yet they are not as warm as *tundra ET* climates. Because of marine influences, annual temperature ranges are low. This climate exists along the Bering Sea, the southern tip of Greenland, northern Iceland, Norway, and in the Southern Hemisphere, generally over oceans between 50° S and 60° S. Precipitation, which frequently falls as sleet, is greater in these regions than in continental E climates.

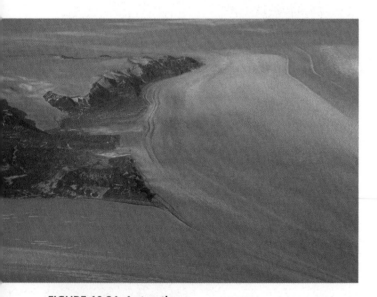

FIGURE 10.21 Antarctica.
Flowing glaciers and the mountains of the Antarctic land-scape; Earth's frozen freshwater reservoir. [Photo by Galen Rowell/Mountain Light Photography, Inc.]

Köppen Guidelines
Polar Climates—E

Warmest month below 10°C (50°F); always cold; ice climates.

ET—Tundra:
Warmest month 0–10°C (32–50°F); PRECIP moisture supply exceeds small POTET moisture demand; snow cover 8–10 months.

EF—Ice cap:
Warmest month below 0°C (32°F); PRECIP exceeds very small POTET moisture demand; the polar regions.

EM—Polar marine:
All months above −7°C (20°F), warmest month above 0°C (32°F); annual temperature range −17 C° (30 F°).

Dry Arid and Semiarid Climates (B)

The *dry arid and semiarid B* climates are the only ones that Köppen classified by precipitation rather than temperature. Dry climates are the world's arid deserts and semiarid regions, with their unique plants, animals, and physical features. They occupy more than 35% of Earth's land area and clearly are the most extensive climate over land. The mountains, bare rock strata, long vistas, and the resilient struggle for life are all magnified by the dryness. Sparse vegetation leaves the landscape exposed; POTET moisture demand exceeds PRECIP moisture supply throughout *dry arid and semiarid B* climates, creating permanent water deficits. The extent of this dryness distinguishes desert and steppe climatic regions. (In addition, refer to specific annual and daily desert temperature regimes, including the highest record temperatures, discussed in Chapter 5; surface energy budgets covered in Chapter 4; desert landscapes in Chapter 15; and desert environments in Chapter 20.)

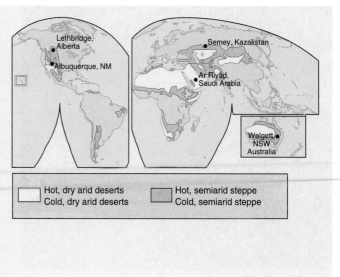

Hot, dry arid deserts
Cold, dry arid deserts

Hot, semiarid steppe
Cold, semiarid steppe

Desert Characteristics

Desert vegetation is typically *xerophytic*: drought-resistant, waxy, hard-leafed, and adapted to aridity and low transpiration loss. Along watercourses, plants called *phreatophytes*, or "water-well plants," have roots that penetrate to great depths for the water they need (Figure 10.22).

The world climate map in Figure 10.5 reveals the pattern of Earth's dry climates, which cover broad regions between 15° and 30° N and S latitudes. In these areas, subtropical high-pressure cells predominate, with subsiding, stable air and low relative humidity. Under generally cloudless skies, these subtropical deserts extend to western continental margins, where cool, stabilizing ocean currents operate offshore and summer advection fog forms. The Atacama Desert of Chile, the Namib Desert of Namibia, the Western Sahara of Morocco, and the Australian Desert each lie adjacent to such a coastline.

Orographic lifting intercepts moisture-bearing weather systems over western mountains to extend these dry regions into higher latitudes. Note these rain shadows in North and South America on the climate map. The isolated interior Asia, far distant from any moisture-bearing air masses, falls within the *dry arid and semiarid* climates as well.

Major subdivisions include *deserts BW* (PRECIP *less than* one-half of POTET) and *semiarid steppes BS* (PRECIP

more than one-half of POTET). Köppen developed simple formulas to determine the seasonal effectiveness of rainfall—whether precipitation falls principally in the winter with a dry summer, in the summer with a dry winter, or is evenly distributed. Winter rains are most effective because they fall at a time of lower moisture demand (Figure 10.23). The names of the four B-climate cities used in this section are plotted on these graphs.

Hot Low-Latitude Desert Climates (BWh)

Hot low-latitude desert BWh climates are Earth's true tropical and subtropical deserts and feature annual average temperatures above 18°C (64.4°F). They generally reside on the western sides of continents, although Egypt, Somalia, and Saudi Arabia also fall within this classification. Rainfall is from local summer convectional showers. Some regions receive nearly nothing, whereas others may receive up to 35 cm (14 in.) of precipitation a year. A representative *hot low-latitude desert BWh* station is Ar Riyāḍ (Riyadh), Saudi Arabia (Figure 10.24).

Along the Sahara's southern margin is a drought-tortured region. Human populations suffered great hardship as desert conditions gradually expanded over their homelands (Figure 10.24c). Chapter 15 presents the process of desertification (expanding desert conditions).

FIGURE 10.22 Desert landscape.
Desert plants are particularly well adapted to the harsh environment. Here the silt-laden Colorado River flows just north of Moab, Utah, cutting through beautiful red sandstone. [Photo by Bobbé Christopherson.]

Cold Midlatitude Desert Climates (BWk)

Cold midlatitude desert BWk climates cover only a small area: the countries along the southern border of Russia, the Gobi Desert, and Mongolia in Asia; the central third of Nevada and areas of the American Southwest, particularly at high elevations; and Patagonia in Argentina. Because of lower temperatures and lower moisture demand values, rainfall must be low for a station to qualify as a *cold midlatitude desert BWk* climate; consequently, total annual average rainfall is only about 15 cm (6 in.).

A representative station is Albuquerque, New Mexico, with 20.7 cm (8.1 in.) of precipitation and an annual average temperature of 14°C (57.2°F) (Figure 10.25). Across central Nevada stretches a characteristic expanse of cold midlatitude desert, greatly modified by a century of livestock grazing and other extensive uses (Figure 10.25b). Comparing *cold midlatitude desert BWk* and *hot low-latitude desert BWh* stations reveals an interesting similarity in the annual temperature range and a distinct difference in precipitation patterns. The role of summer convectional thunderstorms is evident on the climograph in Figure 10.25.

Hot Low-Latitude Steppe Climates (BSh)

Hot low-latitude steppe BSh climates generally exist around the periphery of hot deserts, where shifting subtropical high-pressure cells create a distinct summer-dry and winter-wet pattern. Average annual precipitation in *hot low-latitude steppe BSh* areas is usually below 60 cm (23.6 in.).

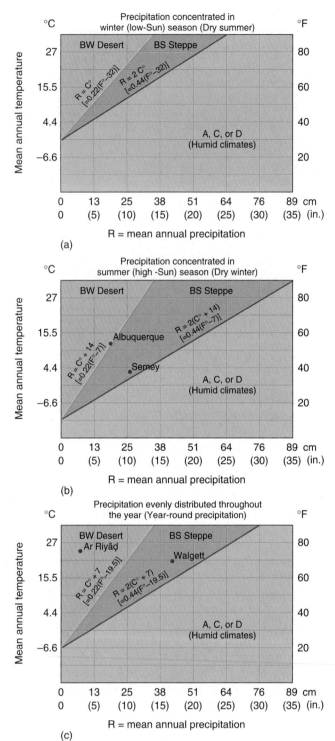

FIGURE 10.23 Determining a dry arid or semiarid climate.
Graphs demonstrate the moisture criteria used to divide the desert and steppe climates. The four B-climate cities used as examples in this section are noted on the graphs. (a) If the wettest winter month was three times wetter than the driest summer month, Köppen considered the summer dry. (b) The reverse reflects the role of potential evapotranspiration: Winter is considered dry if the wettest summer month receives 10 times the precipitation of the driest winter month. (c) B-climate classification if precipitation is evenly distributed throughout the year.

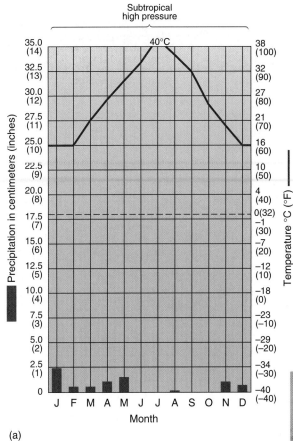

Subtropical
high pressure

(a)

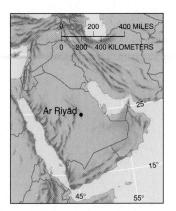

Station: Ar Riyāḍ (Riyadh), **BWh Elevation:** 609 m (1998 ft)
Saudi Arabia   **Population:** 1,308,000
Lat/long: 24°42'N 46°43' E   **Ann. Temp. Range:**
Avg. Ann. Temp.: 26°C (78.8°F)   24 C° (43.2 F°)
Total Ann. Precip.:
        8.2 cm (3.2 in.)

(b)

(c)

FIGURE 10.24 Hot low-latitude desert climate.
(a) Climograph for Ar Riyāḍ (Riyadh), Saudi Arabia (*hot low-latitude desert Bwh*). (b) Photograph of the Arabian desert sand dunes near Ar Riyāḍ. (c) Herders bring a few cattle to market near Timbuktu, Mali. Precipitation has been below normal in the region since 1966.
[(b) Photo by Ray Ellis/Photo Researchers, Inc.
(c) Photo by Betty Press/Woodfin Camp & Associates.]

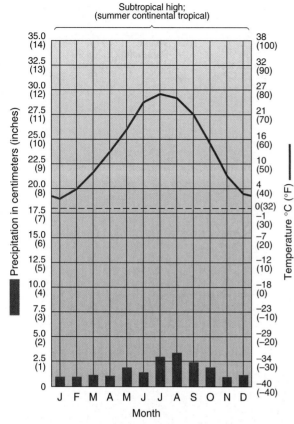

Subtropical high;
(summer continental tropical)

(a)

Station: Albuquerque, **BWk**   Elevation: 1620 m (5315 ft)
   New Mexico   Population: 370,000
Lat/long: 35°03′ N 106°37′ W   Ann. Temp. Range:
Avg. Ann. Temp.: 14°C (57.2°F)   24 C° (43.2 F°)
Total Ann. Precip.:   Ann. Hr of Sunshine:
     20.7 cm (8.1 in.)   3420

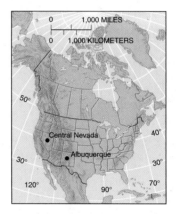

(b)

FIGURE 10.25 Cold midlatitude desert climate.
(a) Climograph for Albuquerque, New Mexico (*cold midlatitude desert BWk*). (b) Cold, high
desert landscape of the Basin and Range Province in Nevada, east of Ely along highway
U.S. 50. [Photo by Bobbé Christopherson.]

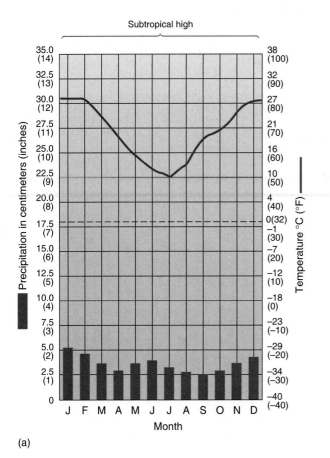

Subtropical high

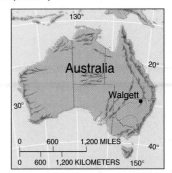

Station: Walgett, New South **BSh**   **Elevation:** 133 m (436 ft)
Wales, Australia   **Population:** 2160
Lat/long: 30°0′ S 148°07′ E   **Ann. Temp. Range:**
Avg. Ann. Temp.: 20°C (68°F)   17 C° (31 F°)
Total Ann. Precip.:
45.0 cm (17.7 in.)

(b)

FIGURE 10.26 Hot low-latitude steppe climate.
(a) Climograph for Walgett, New South Wales, Australia
(*hot low-latitude steppe BSh*). (b) Vast plains characteristic
of north-central New South Wales. [Photo by Otto
Rogge/Stock Market.]

Walgett, in interior New South Wales, Australia, provides
a Southern Hemisphere example of this climate (Figure
10.26). This climate is seen around the Sahara's periphery, and in the Iran, Afghanistan, Turkmenistan, and
Kazakstan region.

Cold Midlatitude Steppe Climates (BSk)

The *cold midlatitude steppe BSk* climates occur poleward
of about 30° latitude and the *cold midlatitude desert BWk*
climates. Such midlatitude steppes are not generally found
in the Southern Hemisphere. As with other dry climate regions, rainfall in the steppes is widely variable and undependable, ranging from 20 to 40 cm (7.9 to 15.7 in.). Not
all rainfall is convectional, for cyclonic storm tracks penetrate the continents; however, most storms produce little
precipitation.

   Figure 10.27 presents a comparison between Asian and
North American *cold midlatitude steppe BSk*. Consider
Semey (Semipalatinsk) in Kazakstan (greater temperature
range, precipitation evenly distributed) and Lethbridge,
Alberta (lesser temperature range, summer maximum con-
ventional precipitation).

Köppen Guidelines
Dry Arid and Semiarid Climates—B

POTET (moisture demand) exceeds PRECIP (moisture
supply) in all B climates. Subdivisions based on
PRECIP timing and amount and mean annual
temperature.

Earth's arid climates:

BWh—Hot low-latitude desert.
BWk—Cold midlatitude desert.
BW = PRECIP less than ½ POTET.
h = Mean annual temperature > 18°C (64.4°F).
k = Mean annual temperature < 18°C (64.4°F).

Earth's semiarid climates:

BSh—Hot low-latitude steppe.
BSk—Cold midlatitude steppe.
BS = PRECIP more than ½ POTET but not equal to it.
h = Mean annual temperature > 18°C (64.4°F).
k = Mean annual temperature < 18°C (64.4°F).

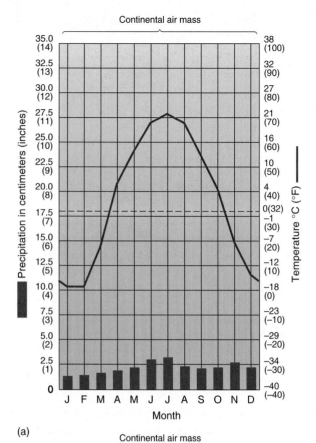

Continental air mass

(a)

Station: Semey (Semipalatinsk), **BSk**
Kazakstan
Lat/long: 50°21' N 80°15' E
Avg. Ann. Temp.: 3°C (37.4°F)
Total Ann. Precip.:
26.4 cm (10.4 in.)

Elevation: 206 m (675.9 ft)
Population: 330,000
Ann. Temp. Range:
39 C° (70.2 F°)

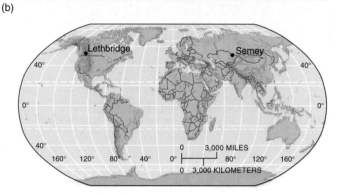

(b)

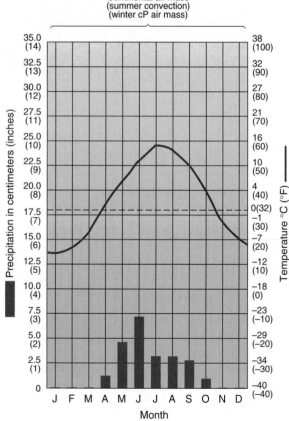

Continental air mass
(summer convection)
(winter cP air mass)

(c)

Station: Lethbridge, Alberta   **BSk**
Lat/long: 49°42' N 110°50' W
Avg. Ann. Temp.: 2.9°C (37.3°F)
Total Ann. Precip.:
25.8 cm (10.2 in.)

Elevation: 910 m (2985 ft)
Population: 58,840
Ann. Temp. Range:
24.3 C° (43.7 F°)

(d)

FIGURE 10.27 Cold midlatitude steppe climate.
Climograph for (a) Semey (Semipalatinsk) in Kazakstan; (b) herders in the region near Semey.
(c) Climograph for Lethbridge in Alberta. (d) Canadian prairies and grain elevators of southern Alberta. [(b) Photo by Sovfoto/Eastfoto; (d) photo by author.]

Global Climate Change

Significant climatic change has occurred on Earth in the past and most certainly will occur in the future. There is nothing society can do about long-term influences that cycle Earth through swings from ice ages to warmer periods. However, our global society must address short-term changes that are influencing global temperatures within the life span of present generations.

Record-high global temperatures dominated the past two decades, records for both land and ocean and for both day and night. 1998 was the all-time record year for warmth, 2001 was second and eclipsed the previous records set in 1997 and 1995. Understanding this warming and all related impacts is an important applied topic of Earth systems science and the spatial analysis ability of physical geography.

Global Warming

More than a decade ago, climatologists Richard Houghton and George Woodwell described the present climatic condition:

> The world is warming. Climatic zones are shifting. Glaciers are melting. Sea level is rising. These are not hypothetical events from a science fiction movie; these changes and others are already taking place, and we expect them to accelerate over the next years as the amounts of carbon dioxide, methane, and other trace gases accumulating in the atmosphere through human activities increase.[*]

The 2001 Intergovernmental Panel on Climate Change (IPCC) *Third Assessment Report* stated:

> . . . there is new and stronger evidence that most of the warming observed over the past 50 years is attributable to human activities. . . . Both temperature and sea level are projected to continue to rise throughout the 21st century for all scenarios studied.[†]

The IPCC, formed in 1988, is an organization operating under the United Nations Environment Programme (UNEP) and the World Meteorological Organization (WMO) and is the scientific organization coordinating global climate change research, climate forecasts, and policy formulation.

Human activities are enhancing Earth's natural greenhouse effect. There is an international scientific consensus that air temperatures are the highest since recordings were begun in earnest more than 140 years ago. This warming trend is very likely[‡] due to a buildup of green-house gases. In terms of **paleoclimatology**, the science that studies past climates (discussed in Chapter 17), proxy indicators (ice core data, sediments, coral reefs, ancient pollen, tree-ring density, among others) point to the recent past as the warmest in the last 600 years, and further, that the increase in temperature during the twentieth century is very likely the largest in any century over the past 1000 years. Earth is within 1 C° (1.8 F°) of equaling the highest average temperature of the past 125,000 years (based on ice-core data).

The rate of warming in the past 30 years exceeds any comparable period in the entire measured temperature record, according to NASA scientists. Figure 10.28a plots observed annual temperatures and 5-year mean temperatures from 1880 through 2001. The temperature map in (b) uses the same base period as the graph (1951–1980), to give you an idea of how warm the record year of 1998 was across the globe. Various organizations and agencies coordinate and conduct global climate change research. News Report 10.2 offers an overview and contact information.

Scientists are working to determine the difference between *forced fluctuations* (human-caused) and *unforced fluctuations* (natural) as a key to predicting future climate trends. Because the gases that generate temperature changes are human in origin, various management strategies are possible to reduce human-forced changes. *Radiatively active* gases are atmospheric gases, such as carbon dioxide (CO_2), methane (CH_4), nitrous oxide (N_2O), chlorofluorocarbons (CFCs), and water vapor, which absorb and radiate infrared wavelengths. Let's begin by examining the problem at its roots.

Carbon Dioxide and Global Warming   Carbon dioxide and water vapor are the principal radiatively active gases causing Earth's natural greenhouse effect. They are transparent to light but opaque to the infrared wavelengths radiated by Earth. Thus, they transmit light from the Sun to Earth but delay heat-energy loss to space. While detained, this heat energy is absorbed and reradiated over and over, warming the lower atmosphere. As concentrations of these infrared-absorbing gases increase, more heat energy remains in the atmosphere and temperatures increase.

The Industrial Revolution, which began in the mid-1700s, initiated tremendous burning of fossil fuels. This, coupled with the destruction and inadequate replacement of harvested forests, continues to increase atmospheric carbon dioxide levels. Carbon dioxide alone is responsible for 64% of the global warming trend. Table 10.1 shows the increasing percentage of carbon dioxide in the lower atmosphere from pre-industrial 1750 to the present—a 31.6% increase—and gives estimates for the future. The current rate of increase is faster than at any time in the past 20,000 years, whereas the present concentration tops anything over the last 420,000 years.

Global emissions of carbon from fossil fuels in 2000 reached 6.3 billion metric tons (combined with oxygen to form 18.9 billion metric tons of carbon dioxide)—400% more than the 1.6 billion metric tons of carbon emissions in 1950. Globally, carbon emissions increased 0.6% in the

---

[*]R. Houghton and G. Woodwell, "Global climate change," *Scientific American*, April 1989, p. 36.

[†]IPCC, Working Group I, *Climate Change 2001, The Scientific Basis* (London: Cambridge Press, 2001, p. ix).

[‡]As the standard scientific reference on climate change the IPCC uses the following words to indicate levels of confidence: *virtually certain* (greater than a 99% chance the result is true), *very likely* (90–99% chance), *likely* (66–90% chance), *medium likelihood* (33–66% chance), *unlikely* (10–33% chance), and *very unlikely* (1–10% chance).

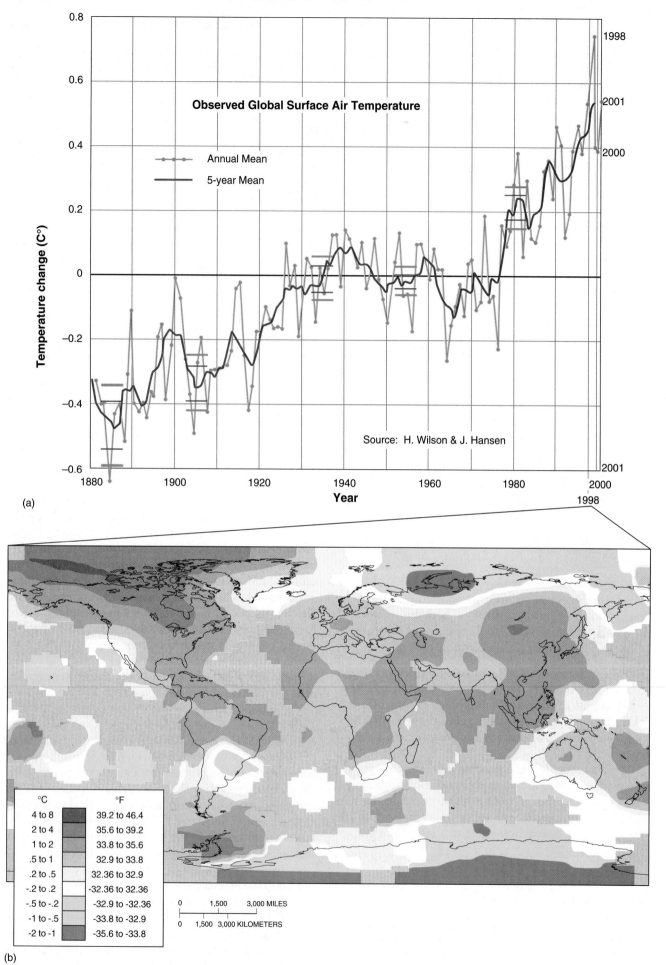

Observed Global Surface Air Temperature

Source: H. Wilson & J. Hansen

(a)

°C	°F
4 to 8	39.2 to 46.4
2 to 4	35.6 to 39.2
1 to 2	33.8 to 35.6
.5 to 1	32.9 to 33.8
.2 to .5	32.36 to 32.9
-.2 to .2	-32.36 to 32.36
-.5 to -.2	-32.9 to -32.36
-1 to -.5	-33.8 to -32.9
-2 to -1	-35.6 to -33.8

(b)

310

News Report 10.2

Coordinating Global Climate Change Research

A cooperative global network of all United Nation members participate in the United Nations Environment Programme (UNEP, http://www.unep.org) and the World Meteorological Organization (WMO, http://www.wmo.ch/). The World Climate Research Programme (WCRP, http://www.wmo.ch/web/wcrp/wcrp-home.html) and its network under the supervision of the Global Climate Observing System (GCOS, http://www.wmo.ch/web/gcos/gcoshome.html) coordinate data gathering and research. The ongoing climate assessment process within the UNEP is conducted by the Intergovernmental Panel on Climate Change (IPCC, http://www.ipcc.ch/), with completed reports issued by three Working Groups in 1990, the 1992 supplementary report, 1995, and the latest *Third Assessment Report* in 2001.

In the United States coordination is found at the U.S. Global Change Research Program (http://www.usgcrp.gov/). An overall source for information is http://globalchange.gov, which publishes an on-line monthly summary of all related developments. Also important are programs and services at NASA agencies such as Goddard Institute for Space Studies (GISS, http://www.giss.nasa.gov/), Global Hydrology and Climate Center (GHCC, http://www.ghcc.msfc.nasa.gov), and at NOAA agencies at the National Climate Data Center (NCDC, http://www.ncdc.noaa.gov/) and the National Environmental Satellite, Data, and Information Service (NESDIS, http://www.nesdis.noaa.gov/), among others. The Pew Center on Global Climate Change offers credible analysis and overview and has issued several policy reports at http://www.pewclimate.org/.

The multi-agency National Ice Center is at http://www.natice.noaa.gov/. Important research is done at the National Center for Atmospheric Research (http://www.ncar.ucar.edu/). For Canada, information and research is coordinated by Environment Canada (http://www.ec.gc.ca/climate/index.html). The effect of global warming on permafrost, which involves half of Canadian land area, is found at http://www.socc.uwaterloo.ca/permafrost/permafrost_current.cfm.

Table 10.1	Carbon Dioxide Concentration in the Lower Atmosphere	

	CO$_2$ Concentration*	
Year	**Percent (%)**	**Parts per million**
1750	0.028	280
1888	0.029	290
1970	0.032	320
1985	0.035	345.9
2001	0.037	369.7
2020 (estimate)	0.047	470
2050 (estimate)	0.053	530
2100 range (est.)	0.55–0.97	550–970

*See Chapter 3 and pp. 219–24 in IPCC Working Group I, *Climate Change 2001, The Scientific Basis* (Washington: Cambridge University Press, 2001).

FIGURE 10.28 Global temperature trends.
(a) Global temperature trends from 1880 to 2001. The 0 baseline represents the 1951–1980 global average. Comparing annual temperatures and 5-year mean temperatures gives a sense of overall trends. (b) Temperature map shows temperature anomalies for the record year 1998; the coloration represents °C departures from the base period 1951–1980. [(a) and (b) Courtesy of J. E. Hansen, Goddard Institute, NASA, and National Climate Data Center, NOAA.]

1990s, compared with 1.5% in the 1980s—a positive downturn in the increase rate. However, in the United States carbon emissions rose 13% between 1990 and 2000, to 24% of the global total.

Figure 10.29 shows sources of excessive (non-natural) carbon dioxide by country or region in 1995 and forecasted for 2025. Developing countries are clearly identified as the sector with the greatest probable growth in fossil fuel consumption and new carbon dioxide production. However, national and corporate policies could alter this forecast by actively steering developing countries toward alternative energy sources (renewable, low temperature, labor intensive) and redirecting industrial countries away from wasteful past practices and toward greater efficiencies.

Methane and Global Warming   Another radiatively active gas contributing to the overall greenhouse effect is methane (CH$_4$), which, at more than 1% per year, is increasing in concentration even faster than carbon dioxide. Air bubbles in ice show that concentrations of methane in the past, between 500 and 27,000 years ago, were approximately 0.7 ppm, whereas current atmospheric concentrations are 1.745 ppm, or more than double the preindustrial level. We are at an atmospheric concentration of methane that is higher than at any time in the past 420,000 years.

Methane is generated by such organic processes as digestion and rotting in the absence of oxygen (anaerobic processes). About 50% of the excess methane comes from bacterial action in the intestinal tracts of livestock and from organic activity in flooded rice fields. Burning of vegetation

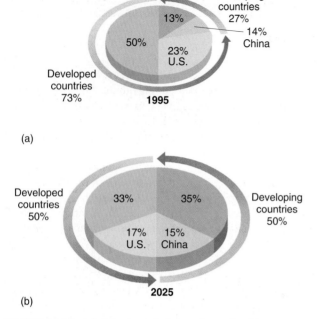

(a)

(b)

FIGURE 10.29 Origin of excessive carbon dioxide.
Countries and regions of origin (a) for excessive carbon dioxide in 1995 and (b) forecast for 2025. [Data from United Nations Environment Programme.]

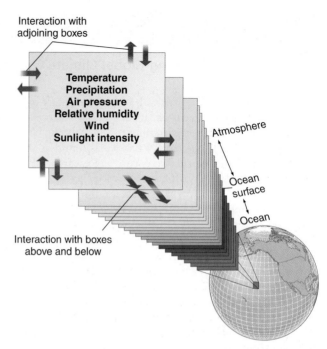

FIGURE 10.30 A general circulation model scheme.
Temperature, precipitation, air pressure, relative humidity, wind, and sunlight intensity are sampled in myriad grid boxes. In the ocean, sampling is limited, but temperature, salinity, and ocean current data are considered. The interactions within a grid layer, and between layers on all six sides, are modeled in a general circulation model program.

causes another 20% of the excess, and bacterial action inside the digestive systems of termite populations also is a significant source. Methane is thought responsible for at least 19% of the total atmospheric warming, complementing the warming caused by the buildup of carbon dioxide.

Other Greenhouse Gases Nitrous oxide (N_2O) is the third most important greenhouse gas that is forced by human activity—up 17% in atmospheric concentration since 1750, higher than at any time in the past 1000 years. Fertilizer use increases the processes in soil that emit nitrous oxide, although more research is needed to fully understand the relationships. Chlorofluorocarbons (CFCs) and other halocarbons also contribute to global warming. CFCs absorb infrared in wavelengths missed by carbon dioxide and water vapor in the lower troposphere. As radiatively active gases, CFCs enhance the greenhouse effect in the troposphere and are a cause of ozone depletion and slight cooling in the stratosphere.

Climate Models and Future Temperatures

The scientific challenge in understanding climate change is to sense climatic trends in what is essentially a nonlinear (unpredictable), chaotic natural climate system. Imagine the tremendous task of building a computer model of all climatic components and to program these linkages (shown in Figure 10.1) over different time frames and at various scales!

Using mathematical models originally established for forecasting weather, scientists developed a complex computer climate model known as a **general circulation model (GCM)**. There are at least a dozen established GCMs now operating around the world. Submodel programs for the atmosphere, ocean, land surface, cryosphere, and biosphere operate within the GCM. The most sophisticated models couple atmosphere and ocean submodels and are known as *Atmosphere-Ocean General Circulation Models (AOGCMs)*.

The first step in describing a climate is to define a manageable portion of Earth's climatic system for study. Climatologists create dimensional "grid boxes" that extend from beneath the ocean to the tropopause, in multiple layers (Figure 10.30). Resolution of these boxes in the atmosphere is about 250 km in the horizontal and 1 km (155–0.6 mi) in the vertical; in the ocean the boxes use the same horizontal resolution and a vertical resolution of about 200 to 400 m (650–1300 ft). Analysts deal not only with the climatic components within each grid layer but also with the interaction among the layers on all sides.

A comparative benchmark among the operational GCMs is *climatic sensitivity* to doubling of carbon dioxide levels in the atmosphere. GCMs do not predict specific temperatures, but they do offer various scenarios of global warming. The Goddard Institute for Space Studies GCM projects a distinct warming by 2029, using a moderate-case scenario (Figure 10.31). Consistent with other computer studies, unambiguous warming appears over some continental regions, with the greatest warming assumed to

FIGURE 10.31 July 2029 temperature forecast.
July temperature projections by the Goddard Institute for Space Studies' general circulation model for 2029 using a moderate-case scenario. [From J. E. Hansen, and others, Goddard Institute for Space Studies, NASA.]

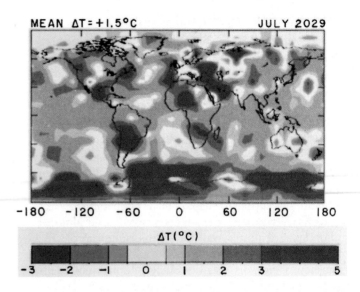

News Report 10.3

The IPCC Process

The Intergovernmental Panel on Climate Change (IPCC), established jointly in 1988 under the United Nations Environment Programme (UNEP) and the World Meteorological Organization (WMO), is the principal global scientific body researching climate change. The IPCC mission is to: (1) assess available scientific and socioeconomic information on climate change and its impacts, mitigation options, and possible adaptation to change; (2) to provide information to the Conference of the Parties (COP) and the U.N. Framework Convention on Climate Change (FCCC); and, (3) formulate response strategies.

The IPCC process produced a series of three assessment reports (1990, 1995, 2001), a supplementary report (1992), and various technical papers. For the 2001 *Third Assessment Report*, 17 lead authors, 515 contributing authors, and 420 reviewers worked over three years. In addition reviews from governments and experts expanded the peer review process to include about 2500 scientists. The process is unprecedented and considers all points of view in formulating findings and forecasts.

A sample of statements from *Climate Change 2001, The Scientific Basis*:

- An increasing body of observations gives a collective picture of a warming world and other changes in the climate system.
- Emissions of greenhouse gases and aerosols due to human activities continue to alter the atmosphere in ways that are expected to affect the climate system.
- There is new and stronger evidence that most of the warming observed over the last 50 years is attributable to human activities.
- Human influence will continue to change atmospheric composition throughout the twenty-first century.
- Global average temperatures and sea level are projected to rise under all IPCC future climate scenarios.

- A nearly worldwide decrease in mountain glacier extent and ice mass is consistent with worldwide surface temperature increases, as are decreases in snow cover, and shortening seasons of lake and river ice in the Northern Hemisphere.
- A systematic decrease in spring and summer sea-ice extent and thickness in the Arctic Ocean consistent with an increase in temperature over adjacent land and ocean surfaces—an estimated 43% decline of the Arctic Ocean sea-ice thickness happened between 1976–1996.
- Ocean heat content increased as it absorbed excess atmospheric heat, thus delaying the full extent of present global warming. Thermal expansion of the oceanic mass represents about 25% of sea level rise.
- Anthropogenic climate change will persist for many centuries.

occur over high latitudes. Note regions of forecasted cooling on the map. Remarkably, as forecast, Canada had its warmest summer and year, and second warmest winter, in history in 1998—average temperatures in some areas were 2.5 C° (4.5 F°) above normal. GCM-generated maps correlate well with the observed global warming patterns experienced in the 1990s.

The 2001 IPCC *Third Assessment Report* predicts a range of average warming between 1990 to 2100 from a "low forecast" of 1.4 C° (2.5 F°), to a "high forecast" of 5.8 C° (10.4 F°). The middle case of 3.6 C° (6.5 F°) represents a significant increase in global land and ocean temperatures and will produce consequences. News Report 10.3 offers a summary overview of the IPCC process.

Consequences of Global Warming

The consequences of uncontrolled atmospheric warming are complex. Regional climate responses are expected as temperature, precipitation, soil moisture, and air mass characteristics change. Although the ability to accurately forecast such regional changes is still evolving, some consequences of warming are forecast and in several regions are already underway.

Effects on World Food Supply and the Biosphere

Modern single-crop (monoculture) agriculture is delicate and more susceptible to changes in air temperature, water availability, irrigation, and soil chemistry than traditional multicrop agriculture. Available soil moisture is projected to be at least 10% less throughout the midlatitudes 30 years from now as hot, dry weather ensues. To maintain managed ecosystems, more energy, water, and resources will have to be expended.

Crop patterns, as well as natural habitats of plants and animals, will shift to maintain preferred temperatures. According to climate models, in the midlatitudes, climatic regions could shift poleward by 150 to 550 km (90 to 350 mi) during this century, thus improving Canadian agriculture. The possibility exists that billions of dollars in agricultural losses in one region would be countered by corresponding gains in another. The greatest food security risks are with those people that are poor, isolated, landless, or living near sea level along vulnerable coastlines and islands.

Biosphere models predict that a global average of 30% (varying regionally from 15%–65%) of the present forest cover will undergo major species redistribution—greatest at high latitudes and least in the tropics. Many plant species are already "on the move" to more favorable locations. Land dwellers also must adapt to changing forage. Several insect species are now found at unexpected higher latitudes, while disappearing from their expected lower-latitude habitats.

Studies completed by the U.S. Institute of Public Health, The Netherlands Environmental Protection Agency, The Development Research Centre in India, and the World Health Organization point to health impacts of climate change. Populations previously unaffected by malaria, dengue fever, lymphatic filariasis, and yellow fever (all mosquito vector), schistosomiasis (water snail vector), and sleeping sickness (tsetse fly vector) will be at greater risk in subtropical and midlatitude areas.

Melting Glaciers, Melting Ice Sheets, and Sea Level

Perhaps the most pervasive (widespread) climatic effect of global warming is rapid escalation of ice melt. Mount Kilimanjaro in Africa, portions of the South American Andes, and the Himalayas will very likely lose most of their glacial ice within the next two decades, affecting local water resources. Glacial ice continues its retreat in Alaska. NASA scientists determined that Greenland's ice sheet is thinning by about 1 m per year. Bill Krabill from NASA Goddard Space Flight Center said, ". . . these results pro-vide evidence that the margins of the ice sheet are in a process of change. The thinning cannot be accounted for by increased melting alone. It appears that ice must be flowing more quickly into the sea through glaciers." (See http://visibleearth.nasa.gov/cgi-bin/viewrecord?2172 for a map by the Airborne Topographic Mapper of this ice loss.)

The additional meltwater, especially from continental ice masses and glaciers, is adding to a rise in sea level worldwide. Satellite remote sensing is monitoring global sea level, sea ice, and continental ice. Worldwide measurements confirm that sea level rose during the last century.

Surrounding the margins of Antarctica, and constituting about 11% of its surface area, are numerous ice shelves, especially where sheltering inlets or bays exist. Covering many thousands of square kilometers, these ice shelves extend over the sea while still attached to continental ice. The loss of these ice shelves does not significantly raise sea level, for they already displace seawater. The concern is for the possible surge of grounded continental ice that the ice shelves hold back from the sea.

Although ice shelves constantly break up to produce icebergs, some large sections have recently broken free. In 1998 an iceberg the size of Delaware (150 km by 35 km, or 90 by 20 mi) broke off the Ronne Ice Shelf, southeast of the Antarctic Peninsula. In March 2000 an iceberg tagged B-15 broke off the Ross Ice Shelf (some 90° longitude west of the Antarctic Peninsula), measuring twice the area of Delaware, 300 km by 40 km, or 190 mi–25 mi (Figure 10.32a).

Since 1993, six ice shelves have disintegrated in Antarctica. About 8000 km^2 (3090 mi^2) of ice shelf are gone, changing maps, freeing up islands to circumnavigation, and creating thousands of icebergs. (See the multi-agency National Ice Center at http://www.natice.noaa.gov/home.htm for an update.) The Larsen Ice Shelf, along the east coast of the Antarctic Peninsula, has been retreating slowly for years. Larsen-A suddenly disintegrated in 1995. (Figure 10.32b). In only 35 days in early 2002, Larsen-B collapsed into icebergs.

This ice loss is likely a result of the 2.5 C° (4.5 F°) temperature increase in the region in the last 50 years. As researchers summarized about Larsen,

> This breakup [of Larsen-A] followed a period of steady retreat that coincided with a regional trend of atmospheric warming. The observations imply that after an ice shelf retreats beyond a critical limit, it may collapse rapidly as a result of disturbed mass balance.[*]

In response to the increasing warmth, the Antarctic Peninsula is sporting new vegetation growth, previously not seen there.

A loss of polar ice mass, augmented by melting of alpine and mountain glaciers (which experienced more than

---

[*]H. Rott, P. Skvarca, and T. Nagler, "Rapid collapse of northern Larsen Ice Shelf, Antarctica," *Science* 271 (February 9, 1996): 788.

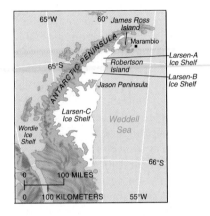

(a)

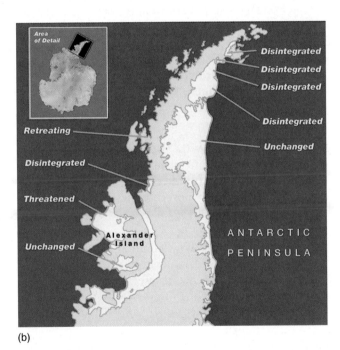

(b)

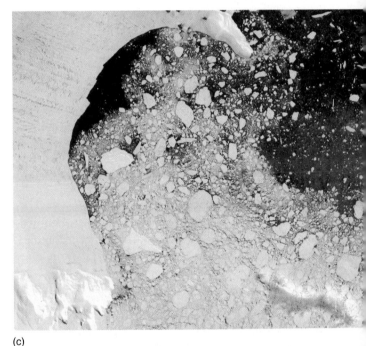

(c)

FIGURE 10.32 Disintegrating ice shelves along the Antarctic coast.
(a) March 2000 image of iceberg B-15 as it breaks off the Ross Ice Shelf, near Roosevelt Island, Antarctica. (b) Map of the Antarctic Peninsula showing the status of ice shelves between 1995–1998. Note the location of the Larsen ice shelves. (c) Continued disintegration and retreat of Larsen-B is evident in a February, 2000, *Landsat-7* image. [(a) MODIS sensor aboard satellite *Terra* courtesy of J. Descloitres, MODIS Land Science Team, NASA. (b) Map provided by the British Antarctic Survey. All rights reserved. (c) Satellite image courtesy of *Landsat-7* team, Goddard Space Flight Center, NASA.]

30% decrease in overall ice mass during the last century), will affect sea-level rise. The IPCC assessment states that "between one-third to one-half of the existing mountain glacier mass could disappear over the next hundred years."

Also, "there is conclusive evidence for a worldwide recession of mountain glaciers. . . . This is among the clearest and best evidence for a change in energy balance at the Earth's surface since the end of the 19th century."

Sea-level rise must be expressed as a range of values that are under constant reassessment. During the last century, sea-level rose 10–20 cm (4–8 in.), a rate 10 times higher than the average rate during the last 3000 years. The 2001 IPCC forecast for global mean sea-level rise this century, given regional variations, is a range from 0.11–0.88 m (4.3–34.6 in.). The median value of 0.48 m (18.9 in.) is two to four times the rate of increase over the last century. These increases would continue beyond 2100 even if greenhouse gas concentrations are stabilized.

The Scripps Institution of Oceanography in La Jolla, California, has kept ocean temperature records since 1916. Significant temperature increases are being recorded to depths of more than 300 m (1000 ft) as ocean temperature records are set. Even the warming of the ocean itself will contribute about 25% of sea-level rise, simply because of *thermal expansion* of the water. In addition, any change in ocean temperature has a profound effect on weather and, indirectly, on agriculture and soil moisture. In fact the ocean system appears to have delayed some surface global warming during the past century through absorption of excess atmospheric heat.

A quick survey of world coastlines shows that even a moderate rise could bring change of unparalleled proportions. At stake are the river deltas, lowland coastal farming valleys, and low-lying mainland areas, all contending with high water, high tides, and higher storm surges. Particularly tragic social and economic consequences will affect small island states—being able to adjust within their present country boundaries, disruption of biological systems, loss of biodiversity, reduction in water resources, among the impacts.

There could be both internal and international migration of affected human populations, spread over decades, as people move away from coastal flooding from the sea level rise. Clearly, physical geography is in an important position to synthesize all the spatial variables needed in planning to cope with these changes, whatever their mildness or severity.

Political Action to Slow Global Warming

A product of the 1992 Earth Summit in Rio de Janeiro was the United Nations Framework Convention on Climate Change (FCCC). The leading body of the Convention is the *Conference of the Parties* (*COP*) operated by the countries that ratified the FCCC, 186 countries by 2000. Meetings were held in Berlin (*COP-1*, 1995) and Geneva (*COP-2*, 1996). These meetings set the stage for *COP-3* in Kyoto, Japan, in December 1997, where 10,000 participants adopted the *Kyoto Protocol* by consensus. The latest gatherings were *COP-6* held in the Hague in late 2000 and *COP-7* in Marrakech, Morocco, in 2001. Seventeen national academies of science endorsed the Kyoto Protocol. (For updates on the status of the Kyoto Protocol, see http://www.unfccc.int/resource/kpstats.pdf.)

The Kyoto Protocol binds more-developed countries to a collective 5.2% reduction in greenhouse gas emissions as measured at 1990 levels for the period 2008 to 2012. Within this group goal, various countries promised cuts: Canada is to cut 6% and moved quickly to ratify the Protocol, the European Union 8%, and Australia 8%, among many others. The *Group of 77* countries plus China favor a 15% reduction by 2010.

Relative to the United States and its goal of a 7% cut below 1990 levels, the administration withdrew in 2001 from the climate treaty process and abandoned emission control goals, alone in its dissent. The president asked the National Research Council to assess the level of IPCC science. The NRC quickly responded in a report affirming conclusions about human-induced climate change, saying that the IPCC Third Assessment Report ". . . accurately reflects the current thinking of the scientific community on this issue" (NRC, *Climate Change Science, An Analysis of Some Key Questions*, Washington: National Academy Press, May 2001).

The Kyoto Protocol is far-reaching in scope, including calling for international cooperation in meeting goals, technology development and emission transfers, leniency for less-developed countries, "clean development" initiatives, and "emissions trading" schemes between industrialized countries and individual industries. The goal, simply and boldly stated, is to "prevent dangerous anthropogenic interference with the climate system."

Mitigation Actions with "No Regrets" The Intergovernmental Panel on Climate Change (IPCC) declared that "no regrets" opportunities to reduce carbon dioxide emissions are available in most countries. The IPCC Working Group III defines this as follows:

No regrets options are by definition greenhouse gas emissions reduction options that have negative net costs, because they generate direct and indirect benefit that are large enough to offset the costs of implementing the options.

Benefits that equal or exceed their cost to society include reduced energy cost, improved air quality and health, reduction in tanker spills and oil imports, and deployment of renewable and sustainable energy sources, among others. This holds true without even considering the benefits of slowing the rate of climate change.

One key to "no regrets" is the untapped energy-efficiency potential. For Europe, scientists determined that carbon emissions could be reduced to less than half the 1990 level by 2030, at a negative cost (reported by the International Project for Sustainable Energy Paths, http://www.ipsep.org/). In the United States, five Department of Energy national laboratories (Oak Ridge, Lawrence Berkeley, Pacific Northwest, National Renewable Energy, and Argonne) reported that the United States can meet the Kyoto carbon emission reduction targets with negative overall costs (cash benefit savings) ranging from –\$7 to –\$34 billion. (For more, see Working Group III, *Climate Change 2001*, *Mitigation*, London: Cambridge University Press, 2001, pp. 21, 474–76, and 506–07.)

Summary and Review—Global Climate Systems

● *Define* climate and climatology and *explain* the difference between climate and weather.

Climate is dynamic, not static. **Climate** is a synthesis of weather phenomena at many scales, from planetary to local, in contrast to weather, which is the condition of the atmosphere at any given time and place. Earth experiences a wide variety of climatic conditions that can be grouped by general similarities into climatic regions. **Climatology** is the study of climate and attempts to discern similar weather statistics and identify **climatic regions**.

climate (p. 275)
climatology (p. 276)
climatic regions (p. 276)

1. Define climate and compare it with weather. What is climatology?
2. Explain how a climatic region synthesizes climate statistics.
3. How does the El Niño phenomenon produce the largest interannual variability in climate? What are some of the changes and effects that occur worldwide?

● *Review* the role of temperature, precipitation, air pressure, and air mass patterns used to establish climatic regions.

Climatic inputs include insolation (pattern of solar energy in the Earth–atmosphere environment), temperature (sensible heat energy content of the air), precipitation (rain, sleet, snow, and hail; the supply of moisture), air pressure (varying patterns of atmospheric density), and air masses (regional-sized homogeneous units of air). Climate is the basic element in ecosystems, the natural, self-regulating communities of plants and animals that thrive in specific environments.

4. How do radiation receipts, temperature, air pressure inputs, and precipitation patterns interact to produce climate types? Give an example from a humid environment and one from an arid environment.
5. Evaluate the relationships among a climatic region, ecosystem, and biome.

● *Review* Köppen's development of an empirical climate classification system and *compare* his with other ways of classifying climate.

Classification is the process of ordering or grouping data in related categories. A **genetic classification** is based on causative factors, such as the interaction of air masses. An **empirical classification** is one based on statistical data, such as temperature or precipitation.

classification (p. 277)
genetic classification (p. 280)
empirical classification (p. 280)

6. What are the differences between a genetic and an empirical classification system?
7. Describe Köppen's approach to climatic classification. What are the factors used in his system?

● *Describe* the A, C, D, and E climate classification categories and *locate* these regions on a world map.

The **Köppen–Geiger climate classification** system is a modified empirical classification system and describes distinct environments on Earth: tropical A, mesothermal C, microthermal D, polar E, and arid B climates. Highland climates are assigned an H to denote that they are caused by elevation. The main climatic groups are divided into subgroups based on temperature and the seasonal timing of precipitation. The system uses average monthly temperatures, average monthly precipitation, and total annual precipitation to devise its spatial categories and boundaries. These data are plotted on a **climograph** to display the characteristics of the climate.

Köppen–Geiger climate classification (p. 281)
climograph (p. 283)

8. List and discuss each of the principal climate designations. In which one of these general types do you live? Which classification is the only type associated with the annual distribution and amount of precipitation?
9. What is a climograph, and how is it used to display climatic information?
10. Which of the major climate types occupies the most land and ocean area on Earth?
11. Characterize the tropical A climates in terms of temperature, moisture, and location.
12. Using Africa's tropical climates as an example, characterize the climates produced by the seasonal shifting of the ITCZ with the high Sun.
13. Mesothermal C climates occupy the second largest portion of Earth's entire surface. Describe their temperature, moisture, and precipitation characteristics.
14. Explain the distribution of the *humid subtropical Cfa* and *Mediterranean dry-summer Csa* climates at similar latitudes and the difference in precipitation patterns between the two types. Describe the difference in vegetation associated with these two climate types.
15. Which climates are characteristic of the Asian monsoon region?

16. Explain how a *marine west coast Cfb* climate type can occur in the Appalachian region of the eastern United States.

17. What role do offshore currents play in the distribution of the *marine west coast Cfb* climate designation? What type of fog is formed in these regions?

18. Discuss the climatic designation for the coldest places on Earth outside the poles. What do each of the letters in the Köppen classification indicate?

● *Explain* the precipitation and moisture efficiency criteria used to determine the B climates and *locate* them on a world map.

The dry and semiarid climates are the only ones that Köppen classified by precipitation rather than temperature. Dry climates are the world's arid deserts and semiarid regions, with their unique plants, animals, and physical features. The *dry arid and semiarid B* climates occupy more than 35% of Earth's land area, clearly the most extensive climate over land.

Major subdivisions are *arid deserts BW* (PRECIP less than one-half of POTET) and *semiarid steppes BS* (PRECIP more than one-half of POTET). Köppen developed simple formulas to determine the usefulness of rainfall on the basis of the season.

19. In general terms, what are the differences among the four desert classifications? How are moisture and temperature distributions used to differentiate these subtypes?

20. Relative to the distribution of dry climates, describe at least three locations where they occur across the globe and the reasons for their presence in these locations.

● *Outline* future climate patterns from forecasts presented and *explain* the causes and potential consequences.

Various activities of present-day society are producing climatic changes, particularly a global warming trend. The 1980s and 1990s were dominated by the highest average annual temperatures experienced since the advent of instrumental measurements. There is a scientific consensus building that global warming is related to the greenhouse effect.

These conditions were further supported by the 2001 *Third Assessment Report* from the Intergovernmental Panel on Climate Change. IPCC predicted surface-temperature response to a doubling of carbon dioxide with a range of increase from 1.4 C° (2.5 F°) to 5.8 C° (10.4 F°) between the present and 2100. Natural climatic variability over the span of Earth's history is the subject of **paleoclimatology**. A **general circulation model (GCM)** is used to forecast climate patterns and is evolving to greater capability and accuracy than in the past. People and their political institutions can use GCM forecasts to form policies aimed at reducing unwanted climate change.

paleoclimatology (p. 309)
general circulation model (GCM) (p. 312)

21. Explain climate forecasts. How do general circulation models (GCMs) produce such forecasts?

22. Describe the potential climatic effects of global warming on polar and high-latitude regions. What are the implications of these climatic changes for persons living at lower latitudes?

23. How is climatic change affecting agricultural and food production? Natural environments? Forests? The possible spread of disease?

24. What are the present actions being taken to delay the effects of global climate change? What is the Kyoto Protocol? The operation of the Conference of the Parties? What is the current status of U.S. and Canadian government action on the Protocol?

 NetWork

The *Geosystems* Home Page provides on-line resources for this chapter on the World Wide Web. To begin: Once on the Home Page, click on this textbook, scroll the Table of Contents menu, select this chapter, and click "Begin." You will find self-tests that are graded, review exercises, specific updates for items in the chapter, and many links to interesting related pathways on the Internet under "Destinations." *Geosystems* is at **http://www.prenhall.com/christopherson**.

Critical Thinking

A. The text asked that you find the climate conditions for your campus and your birthplace and locate these two places on Figures 10.3, 10.4, and 10.5. Briefly describe the information sources you used: library, Internet, teacher, and phone calls to state and provincial climatologists. Briefly show how you worked through the Köppen climate criteria that established the climate classification for your two cities.

B. El Niño conditions are forecast to return to the Pacific during the life of this edition of *Geosystems*. On the *Geosystems* Home Page and in Focus Study 10.1 there are URL references for El Niño and La Niña. Sample three or four sources to determine the up-to-date status of the El Niño phenomena at the present time. How are current conditions different from the record El Niño event in 1997–1998? Or La Niña in 1998–2000? What Internet links were most helpful to you in completing this status report?

C. Many external factors force climate. The chart "Global and annual mean radiative forcing for the year 2000, relative to 1750" is presented below (from *Climate Change 2001, The Scientific Basis*, Washington: Cambridge University Press, 2001, Figure 3, p. 8, and Figure 6.6, p. 392).

The estimates of radiative forcing in Watts per square meter units are given on the y-axis (vertical axis). The level of scientific understanding is noted along the x-axis (horizontal axis), arranged from "high" to "very low." Those columns above the "0" value (in red) indicate *positive forcing*, such as the greenhouse gases grouped in the far-left column. Columns that fall below the "0" value (in blue) indicate *negative forcing*, such as the haze from sulfate aerosols, fourth column from left. The vertical line between the markers on each column is an estimate of the uncertainty range. Where no column appears but there is instead a line denoting a range, there is no central estimate given present uncertainties, such as for mineral dust.

Assume you are a policy maker with a goal of reducing the rate of global warming, that is, reducing positive radiative forcing of the climate system. What strategies do you suggest to alter the height of the columns and adjust the mix of elements that cause warming? Assign priorities to each suggested strategy to denote most-to-least effective in moderating climate change. Brainstorm and discuss your strategies with others.

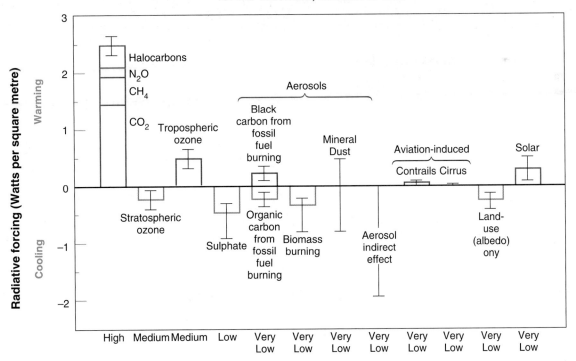

The Global Mean Radiative Forcing of the Climate System for the Year 2000, Relative to 1750

Dramatic Hawaiian landscape where active lava flows plunge into the ocean, forming the Kamoamoa bench and representing some of the newest land surface on Earth. The Kīlauea eruptions are now the longest continuous period of active eruptions in history.
[Photo by Bobbé Christopherson.]

PART THREE

The Earth–Atmosphere Interface

Earth is a dynamic planet whose surface is shaped by active physical agents of change. Two broad systems organize these agents in Part 3—endogenic and exogenic. The **endogenic system** (Chapters 11 and 12) encompasses internal processes that produce flows of heat and material from deep below Earth's crust. Radioactive decay principally powers these processes. The materials involved constitute the *solid realm* of Earth. Earth's surface responds by moving, warping, and breaking, sometimes in dramatic episodes of earthquakes and volcanic eruptions, constructing the crust.

At the same time, the **exogenic system** (Chapters 13 through 17) involves external processes that set into motion air, water, and ice, all powered by solar energy—this is the *fluid realm* of Earth's environment. These media carve, shape, and reduce the landscape. One such process, weathering, breaks up and dissolves the crust. Erosion picks up these materials; transports them in rivers, winds, coastal waves, and flowing glaciers; and deposits them along the way. Thus, Earth's surface is the interface between two vast open systems: one that builds the landscape and one that tears it down.

321

The coasts of Saudi Arabia, to the east (top), and the Sudan (bottom), to the west, illustrate a remarkable fit and geological match of lands pulling apart due to plate tectonics. The Red Sea occupies this part of the East African rift system. A dust storm is blowing off the Sudan coast at 21° N. (In this photo, north is to the left, south is to the right.) [Astronaut photo from Shuttle STS-43 courtesy of JSC Digital Image Collection, NASA.]

11

The Dynamic Planet

Key Learning Concepts

After reading the chapter, you should be able to:

- *Distinguish* between the endogenic and exogenic systems, *determine* the driving force for each, and *explain* the pace at which these systems operate.
- *Diagram* Earth's interior in cross section and *describe* each distinct layer.
- *Illustrate* the geologic cycle and *relate* the rock cycle and rock types to endogenic and exogenic processes.
- *Describe* Pangaea and its breakup and *relate* several physical proofs that crustal drifting is continuing today.
- *Portray* the pattern of Earth's major plates and *relate* this pattern to the occurrence of earthquakes, volcanic activity, and hot spots.

The twentieth century was a time of great discovery about Earth's internal structure and dynamic crust, yet much remains on the scientific frontier of Earth systems science. Discoveries have revolutionized our understanding of how continents and oceans came to be arranged as they are. A new era of understanding is emerging, combining various sciences within the study of physical geography. One task of physical geography is to explain the *spatial implications* of all this new information and its effect on the landscape.

In this chapter: Earth's interior is organized, with a core surrounded by roughly concentric shells of material. It is unevenly heated by the radioactive decay of unstable elements. A rock cycle produces three classes of rocks through igneous, sedimentary, and metamorphic processes. Internal processes coupled with the rock cycle, hydrologic cycle, and tectonic cycle result in a varied crustal surface, featuring

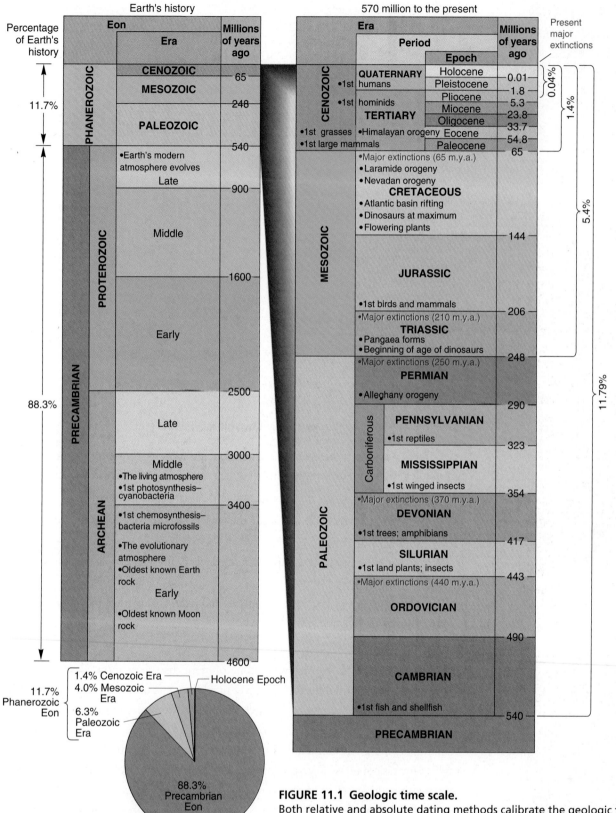

FIGURE 11.1 Geologic time scale.
Both relative and absolute dating methods calibrate the geologic time scale. Relative dating determines the sequence of events and time intervals between them. Technological means, especially radiometric dating, determine absolute dates. In the column at the left, note that 88% of geologic time occurred during the Precambrian Era. Highlights of Earth's history also are shown in the figure as bulleted items, including the six major extinctions of life forms (the sixth extinction episode is underway at the present time), and dates appear in m.y.a., million years ago. [Data from Geological Society of America.]

irregular fractures, extensive mountain ranges both on land and the ocean floor, drifting continental and oceanic crust, and frequent earthquakes and volcanic events. All of this movement of materials results from endogenic forces within Earth—the subject of this chapter.

The Pace of Change

The **geologic time scale** is a summary timeline of all Earth history, shown in Figure 11.1. It reflects currently accepted names of time intervals for each segment of Earth's history, from vast *eons* through briefer *eras*, *periods*, and *epochs*. The time scale depicts two important kinds of time: *relative* (what happened in what order) and *absolute* (actual number of years before the present). Also on the geologic time scale, see the labels denoting the six major extinctions of life forms in Earth history. These range from 440 million years ago (m.y.a.) to the present one caused by modern civilization.

Relative time is the *sequence* of events, based on the relative positions of rock strata above or below each other. Relative time is based on the important general principle of *superposition*, which states that *rock and sediment always are arranged with the youngest beds "superposed" toward the top of a rock formation and the oldest at the base, if they have not been disturbed*. The study of these sequences is called *stratigraphy*. Thus, relative time places the Precambrian at the bottom (beginning) of the time scale and the Holocene (today) at the top. Important time clues—namely *fossils*, the remains of ancient plants and animals—lie embedded within these strata. Since approximately 3.6 billion years ago, life has left its evolving imprint in the rocks.

Scientific methods such as radiometric dating determine *absolute time*, the actual "millions of years ago" (m.y.a.) shown on the time scale. These absolute ages refine the time-scale sequence and lend greater accuracy to relative dating sequences. See News Report 11.1 for more information on radioactivity and Earth's time clock. For more on the geologic time scale, see http://www.ucmp.berkeley.edu/exhibit/geology.html.

A fundamental principle of Earth science is **uniformitarianism**. Uniformitarianism assumes that *the same physical processes active in the environment today have been operating throughout geologic time*. For example, if streams carve valleys now, they must have done so 500 million years ago. The phrase "the present is the key to the past" describes this principle. Evidence from exploration and from the landscape record of volcanic eruptions, earthquakes, and exogenic processes support uniformitarianism. The concept was first proposed by James Hutton in his *Theory of the Earth* (1795) and was later amplified by Charles Lyell in *Principles of Geology* (1830). Today, the earlier Hutton version colors scientific work because of his view that rates and some processes do change, whereas essential underlying principles of biology, physics, geology, and chemistry remain uniform.

In contrast, the principle of *catastrophism* places the vastness of Earth's age and the complexity of its rocks into a shortened time span. Catastrophism holds that Earth is young and that mountains, canyons, and plains formed

---

News Report 11.1

Radioactivity: Earth's Time Clock

The age of Earth and the age of the earliest known crustal rock are astounding, for we think in terms of Earth's trips around the Sun and the pace of our own lives. We need something greater than human time to measure the vastness of geologic time. Nature has provided a way: *radiometric dating*. It is based on the steady decay of certain atoms.

An atom contains protons and neutrons in its nucleus. Certain forms of atoms (isotopes) have unstable nuclei; that is, the protons and neutrons do not remain together indefinitely. As particles break away and the nucleus disintegrates, radiation is emitted and the atom decays into a different element—this process is *radioactivity*.

Radioactivity provides the steady time clock needed to measure the age of ancient rocks. It works because the decay rates for different isotopes are determined precisely, and they do not vary beyond established uncertainties. Further refining of these geochronologies is an ongoing process. These rates are expressed as *half-life*, the time required for one-half of the unstable atoms in a sample to decay into "daughter" isotopes. Some examples of unstable elements that become stable elements and their half-lives include uranium-238 to lead-206 (4.5 billion years); thorium-232 to lead-208 (14.1 billion years); potassium-40 to argon-40 (1.3 billion years); and, in organic materials, carbon-14 to nitrogen-14 (5730 years).

The presence of these decaying elements and stable end products in sediment or rock allows scientists to read the radiometric "clock." They compare the amount of original isotope in the sample with the amount of decayed end product in the sample. If the two are in a ratio of 1:1 (equal parts), one half-life has passed. Errors can occur if the sample was disturbed or subjected to natural weathering processes that might alter its radioactivity.

To increase accuracy, investigators may check a sample using more than one radiometric measurement. Calibration of the past 10,000 years correlates through tree-ring analysis; the past 45,000 years is calibrated using fossils found in lake sediments, among several other cross checks available. The 3.96-billion-year-old date for the Acasta Gneiss, Earth's oldest known rock, was verified using several radiometric methods.

through catastrophic events that did not require eons of time. Because there is little physical evidence to support catastrophism, it is more appropriately considered a belief than a serious scientific hypothesis. Ancient landscapes, such as the Appalachian Mountains or the heartland–core regions of the continents, or the vast drifting of the continents and other fragments of crust, represent a much broader time scale of events, beyond the severe limitations of catastrophism.

Within the principle of uniformitarianism, catastrophic events such as massive landslides, earthquakes, volcanic eruptions, and sometimes cyclic episodes of mountain building punctuate geologic time. These episodes occur as interruptions in the generally uniform processes that shape the slowly evolving landscape. Here, the *punctuated equilibrium* concept studied in the life sciences and paleontology might apply to aspects of Earth's long developmental history.

To understand the phenomena seen at Earth's surface, we must have knowledge of our planet's internal structure and energy. Let us now journey deep within Earth to see its inner workings.

Earth's Structure and Internal Energy

Along with the other planets and the Sun, Earth is thought to have condensed and congealed from a nebula of dust, gas, and icy comets about 4.6 billion years ago (Chapter 2). Scientists are observing this same formation process underway elsewhere in our Milky Way Galaxy and the Universe. Previously, the oldest surface rock discovered on Earth (known as the Acasta Gneiss) was found in northwestern Canada; it was radiometrically dated to an age of 3.96 billion years. Scientists from Missouri's Saint Louis University, the Canadian Geological Survey, and Australian National University confirmed the age.

Recent research found detrital zircons (particles of pre-existing zirconium silica oxides transported and deposited, forming rock) in Western Australia dating between 4.2 and 4.4 billion years old; these are possibly the oldest materials in Earth's crust. These discoveries tell us something significant: Earth was forming continental crust at least 4 billion years ago, during the Archean Eon.

As Earth solidified, gravity sorted materials by density. Heavier substances such as iron gravitated slowly to its center, and lighter elements such as silica slowly welled upward to the surface and became concentrated in the crust. Consequently, Earth's interior is sorted into roughly concentric layers, each one distinct in either chemical composition or temperature. Heat energy migrates outward from the center by conduction and by physical convection in the more fluid or plastic layers in the mantle and nearer the surface.

Our knowledge of Earth's *internal differentiation* into these layers is acquired entirely through indirect evidence, because we are unable to drill more than a few kilometers into Earth's crust. There are several physical properties of Earth materials that enable us to approximate the nature of the interior. For example, when an earthquake or underground nuclear test sends shock waves through the planet, the cooler areas, which generally are more rigid, transmit these **seismic waves** at a higher velocity than do the hotter areas, where seismic waves are slowed to lower velocity. This is the science of *seismic tomography*, as if Earth is subjected to a kind of CAT scan with every earthquake.

Density also affects seismic-wave velocities. Plastic zones simply do not transmit some seismic waves; they absorb them. Some seismic waves are reflected as densities change, whereas others are refracted, or bent, as they travel through Earth. Thus, the distinctive ways in which seismic waves pass through Earth and the time they take to travel between two surface points help seismologists deduce the structure of Earth's interior (Figure 11.2).

Figure 11.3 illustrates the dimensions of Earth's interior compared with surface distances in North America to give you a sense of size and scale. An airplane flying from Halifax, Nova Scotia, to San Francisco would travel the same distance as that from Earth's center to its surface. Note the thinness of the crust, extending only from Vallejo to San Francisco on the far west coast.

Earth's Core

A third of Earth's entire mass, but only a sixth of its volume, lies in its dense core. The **core** is differentiated into two regions—*inner core* and *outer core*—divided by a transition zone several hundred kilometers wide (see Figure 11.2b). The inner core is thought to be solid iron and well above the melting temperature of iron at the surface, but it remains solid because of tremendous pressure. The iron in the core is impure, probably combined with silicon and possibly oxygen and sulfur. Recent research points to the conclusion that the inner core may be a single, enormous crystal of iron. The outer core is molten, metallic iron with a lighter density than the inner core.

Earth's Magnetism The fluid outer core generates at least 90% of Earth's magnetic field and the magnetosphere that surrounds and protects Earth from the solar wind. One hypothesis explains that circulation patterns in the outer core are influenced by Earth's rotation and the new discovery that Earth's solid inner core rotates slightly faster than the rest of the planet. This convective circulation generates electrical currents, which in turn induce the magnetic field.

An intriguing feature of Earth's magnetic field is that its polarity sometimes fades to zero and then returns to full strength, with north and south magnetic poles reversed! In the process, the field does not blink on and off but instead diminishes slowly to low intensity and then rapidly regains its full strength. This **geomagnetic reversal** has taken place nine times during the past 4 million years and hun-

FIGURE 11.2 Earth in cross section.
(a) Cutaway showing Earth's interior. (b) Earth's interior in cross section, from the inner core to the crust. (c) Detail of the structure of the lithosphere and its relation to the asthenosphere. (For comparison to the densities noted, the density of water is 1.0 g/cm³, and that of mercury, a liquid metal, is 13.0 g/cm³.) A revised estimate made in 2000 of Earth's mass (or weight) set it at 5.972 sextillion metric tons (5,972 followed by 18 zeros). [Photo from NASA.]

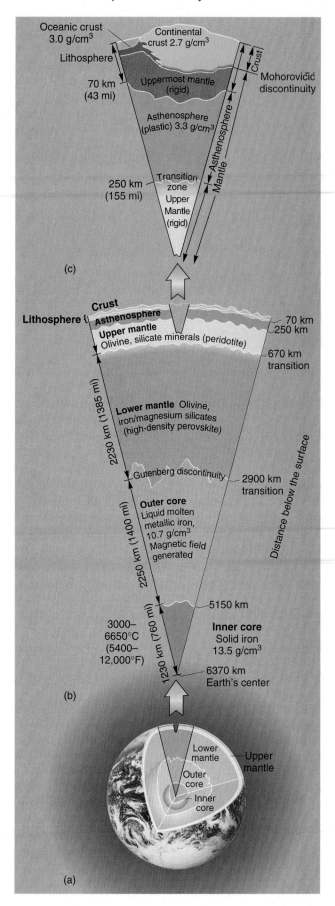

dreds of times over Earth's history. The average period of a magnetic reversal is 500,000 years; occurrences possibly vary from as short as several thousand years to as long as tens of millions of years. The obvious question for us is, Why does this happen?

The reasons for these magnetic reversals are unknown at present. However, the spatial patterns they create at Earth's surface are a key tool in understanding the evolution of landmasses and the movements of the continents. When new iron-bearing rocks solidify from molten material (lava) at Earth's surface, the small magnetic particles in the rocks align according to the orientation of the magnetic poles at that time. This alignment is then locked in place as the rocks cool and solidify.

All across Earth, rocks of the same age bear an identical record of magnetic reversals in the form of measurable magnetic "stripes" in any magnetic material they contain, such as iron particles. These stripes illustrate global patterns of changing magnetism. Matching segments allow scientists to reassemble past continental arrangements. Later in this chapter we see the importance of these magnetic reversals.

An apparent trend in recent geologic time is toward more frequent reversals. When Earth is without polarity in its magnetic field, a random pattern of magnetism in crustal rocks results. Transition periods last from 4000 to 8000 years. The effects of these low-intensity episodes on life are still speculative, but without a magnetosphere, the surface environment is unprotected from cosmic radiation and solar wind. Given present rates of magnetic field decay, we are perhaps 2000 years away from entering the next phase of field changes.

Earth's Mantle

Figure 11.2b shows a transition zone several hundred kilometers wide, at an average depth of about 2900 km (1800 mi), dividing Earth's outer core from its mantle. Scientists at the California Institute of Technology analyzed more than 25,000 earthquakes and determined that this transition area is uneven, with ragged peak-and-valley-like formations. New scientific discoveries determined that some of the motions in the mantle may be created by this rough texture and low-velocity zone (speed of seismic waves in hot material) at what is called the *Gutenberg discontinuity*. A *discontinuity* is a place where physical differences occur between adjoining regions in Earth's interior, such as between the outer core and lower mantle.

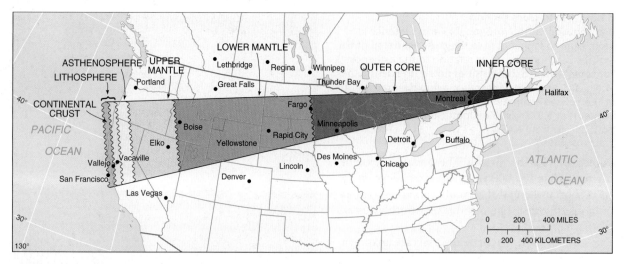

FIGURE 11.3 Distances from core to crust.
Surface map, using the distance from Halifax, Nova Scotia, to San Francisco, to compare the distance from Earth's center to the surface (6370 km, or 3963 mi). The thickness of the continental crust is the same as the distance between Vallejo (in the eastern portion of San Francisco Bay) and the city of San Francisco (roughly 20 mi, 32 km).

The **mantle** (lower and upper together) represents about 80% of Earth's total volume. The mantle is rich in oxides of iron and magnesium and silicates (FeO, MgO, and SiO_2). They are dense and tightly packed at depth, grading to lesser densities toward the surface. A broad transition zone of several hundred kilometers, centered about 670 km (415 mi) below the surface, separates the upper mantle from the lower mantle. The entire mantle experiences a gradual temperature increase with depth and a stiffening due to increased pressures. The denser lower mantle is thought to contain a mixture of iron, magnesium, and silicates, with some calcium and aluminum.

The upper mantle divides into three fairly distinct layers: upper mantle, asthenosphere, and *uppermost mantle*, shown in Figure 11.2c. Next to the crust is the uppermost mantle, a high-velocity zone just below the crust, where seismic waves transmit through a rigid, cooler layer. This uppermost mantle, along with the crust, makes up the *lithosphere*, approximately 45–70 km thick.

Below the lithosphere, from about 70 km down to 250 km, is the **asthenosphere**, or plastic layer (from the Greek *asthenos*, meaning "weak"). It contains pockets of increased heat from radioactive decay and is susceptible to slow convective currents in these hotter (and therefore less dense) materials.

Because of its dynamic condition, the asthenosphere is the least rigid region of the mantle, with densities averaging 3.3 g/cm^3. About 10% of the asthenosphere is molten in asymmetrical patterns and hot spots. The resulting slow movement in this zone disturbs the overlying crust and creates tectonic activity—the folding, faulting, and general deformation of surface rocks. In return, movement of the crust apparently influences currents throughout the mantle.

The depth affected by convection currents is the subject of much scientific research. One body of evidence

states that mixing occurs throughout the entire mantle, upwelling from great depths at the core–mantle boundary, sometimes in small blobs of material, other times "megablobs" of mantle convect toward and away from the crust. Another view states that mixing in the mantle is layered, segregated above and below the 670-km boundary. Presently, evidence indicates some truth in both positions. As an example, there are hot spots on Earth, such as those under Hawai'i and Iceland, that appear to be at the top of tall plumes of rising mantle rock that are anchored deep in the lower mantle, where the rise of warmer, less dense material begins. In other surface regions, slabs of crust descend and penetrate to the lower mantle. For basics on Earth's interior and plate tectonics, see **http://www.solarviews.com/eng/earthint.htm**.

Earth's Lithosphere and Crust

The lithosphere includes the **crust** and uppermost mantle to about 70 km (43 mi) depth (Figure 11.2c). An important internal boundary between the crust and the high-velocity portion of the uppermost mantle is another discontinuity, called the **Mohorovičić discontinuity**, or **Moho** for short. It is named for the Yugoslavian seismologist who determined that seismic waves change at this depth owing to sharp contrasts of materials and densities.

Figure 11.2c illustrates the relation of the crust to the rest of the lithosphere and the asthenosphere below. Crustal areas beneath mountain masses extend deep, perhaps to 50–60 km (31–37 mi), whereas the crust beneath continental interiors averages about 30 km (19 mi) in thickness. Oceanic crust averages only 5 km (3 mi). The crust is only a fraction of Earth's overall mass. Drilling through the crust into the uppermost mantle remains an elusive scientific goal (see News Report 11.2).

News Report 11.2

Drilling the Crust to Record Depths

Scientists wanting to sample mantle material directly have unsuccessfully tried for decades to penetrate Earth's crust to the Moho discontinuity (the crust-mantle boundary). The longest-lasting deep-drilling attempt is on the northern Kola Peninsula near Zapol-yarny, Russia, 250 km north of the Arc-tic Circle—the *Kola Borehole (KSDB)*. Twenty years of high-technology drilling (1970–1989) produced a hole 12,262 km deep (7.6 mi, or 40,230 ft), purely for exploration and science. Crystalline rock 1.4 billion years old at 180°C (356°F) was reached. The site has other active boreholes, and the fifth is underway. (See an analysis log at http://asuwlink.uwyo.edu/~seismic/kola/ or http://icdp.gfz-potsdam.de/html/kola/wellsite.html.) A record holder for depth for a gas well is in Ok-lahoma; it was stopped at 9750 m (32,000 ft) when the drill bit ran into molten sulfur.

Oceanic crust is thinner than con-tinental crust and is the object of several drilling attempts. The International Ocean Drilling Program (ODP), a cooperative effort directed by Texas A & M University, drilled a 2.5-km-

FIGURE 1 Ocean drilling ship.
A modern ocean-floor drilling ship, the *JOIDES Resolution*. The International Ocean Drilling Pro-gram operates the research ship. Since it began operations in 1984, through 2001, *JOIDES* has spent more than 5000 days at sea, trav-eled more than half a million kilo-meters, drilled 1555 holes into the seafloor, and recovered 190,049 m (623,500 ft) of core for analysis. [Photo by ODP/Texas A & M University.]

deep hole in an oceanic rift near the Galápagos Islands from 1975 to 1993, with further drilling planned at the site (Figure 1). An ocean floor of distinct layers of lava and deeper magma cham-bers was found. But the Moho and Earth's mantle remain untapped. See http://www-odp.tamu.edu/ or http://

www.oceandrilling.org/ for the ODP home page and information. Construction of a new largest deep-ocean drilling ship is ongoing in Okayama, Japan, with completion scheduled for 2006. The new ODP ship will be able to drill to 7 km, more than three times the *JOIDES Resolution*.

The composition and texture of continental and ocean-ic crusts are quite different, and this difference is a key to the concept of drifting continents. The oceanic crust is denser than continental crust. In collisions, the denser oceanic material plunges beneath the lighter, more buoy-ant continental crust.

- *Continental crust* is essentially **granite**; it is crystalline and high in silica, aluminum, potassium, calcium, and sodium. (Sometimes continental crust is called *sial*, shorthand for *si*lica and *al*uminum.) Most important, continental crust is relatively low in density, averag-ing 2.7 g/cm^3. Compare this with other densities given in Figure 11.2.
- *Oceanic crust* is **basalt**; it is granular and high in sili-ca, magnesium, and iron. (Sometimes oceanic crust is called *sima*, shorthand for *si*lica and *ma*gnesium.) It is denser than continental crust, averaging 3.0 g/cm^3.

Buoyancy is the principle that something less dense, such as wood, floats in something denser, such as water. The principles of buoyancy and balance were combined in the 1800s into the important principle of **isostasy**. Isosta-sy explains certain vertical movements of Earth's crust.

Think of Earth's crust as floating on the denser lay-ers beneath, much as a boat floats on water. Where the load is greater, owing to glaciers, sediment, or mountains, the crust tends to sink, or ride lower in the asthenosphere. Without that load (for example, when a glacier melts), the crust rides higher, in a recovery uplift known as *isostatic re-bound*. Thus, the entire crust is in a constant state of com-pensating adjustment, or isostasy, slowly rising and sinking in response to its own burdens, as it is pushed and dragged about by currents in the asthenosphere (Figure 11.4).

Earth's crust is the outermost shell: an irregular, brit-tle layer that resides restlessly on a dynamic and diverse in-terior. Let us examine the processes at work on this crust and the variety of rock types that compose the landscape.

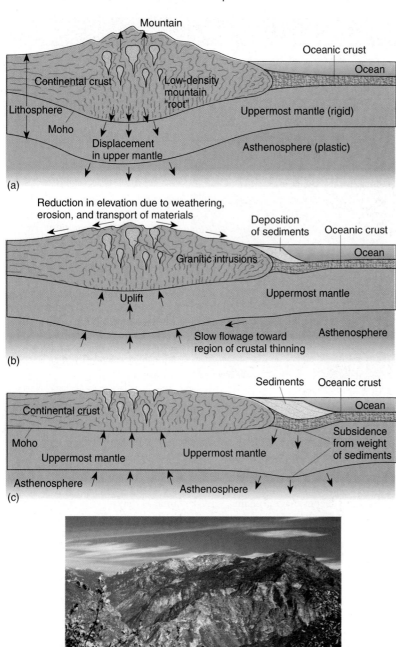

(a)

(b)

(c)

(d)

FIGURE 11.4 Isostatic adjustment of the crust.
Earth's entire crust is in a constant state of compensating adjustment, as suggested by these three sequential stages. In (a), the mountain mass slowly sinks, displacing mantle material. In (b), because of the loss of mass through erosion and transportation, the crust isostatically adjusts upward and sediments accumulate in the ocean. As the continental crust thins (c), the heavy sediment load offshore begins to deform the lithosphere beneath the ocean. (d) The melting of ice from the last ice age and losses of overlying sediments are thought to produce an ongoing isostatic uplift of portions of the Sierra Nevada batholith. [Photo by author.]

The Geologic Cycle

Earth's crust is in an ongoing state of change, being formed, deformed, moved, and broken down by physical, chemical, and biological processes. While the endogenic (internal) system is at work building landforms, the exogenic (external) system is busily wearing them down. This vast give-and-take at the Earth–atmosphere–ocean interface is called the **geologic cycle**. It is fueled from two sources—Earth's internal heat and solar energy from space—influenced by the ever-present leveling force of Earth's gravity.

Figure 11.5 illustrates the geologic cycle, combining many of the elements presented in this text. The geologic cycle is composed of three subsystems:

- The *hydrologic cycle* is the vast system that circulates water, water vapor, ice, and energy throughout the Earth–atmosphere–ocean environment. This cycle rearranges Earth materials through erosion, transportation, and deposition, and it circulates water as the critical medium that sustains life. (The hydrologic cycle and the discussion of water's properties were in Chapters 7 and 9.)

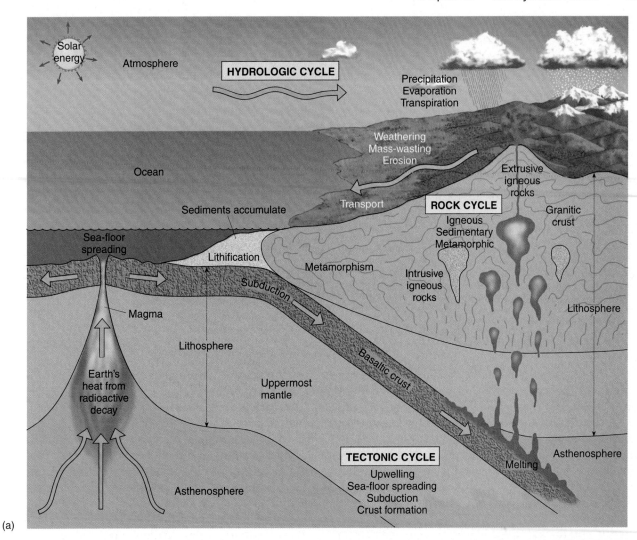

(a)

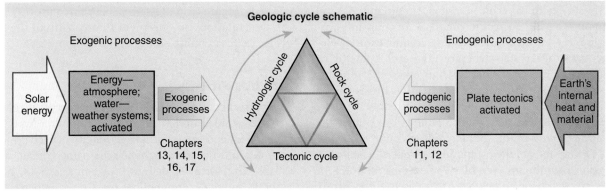

(b)

FIGURE 11.5 The geologic cycle.
(a) The geologic cycle is a model showing the interactive relation among the hydrologic cycle, rock cycle, and tectonic cycle. (b) Earth's surface is where two dynamic systems interact—the endogenic (internal) and exogenic (external).

- The *rock cycle*, through processes in the atmosphere, crust, and mantle, produces three basic rock types—igneous, sedimentary, and metamorphic. We examine this next.
- The *tectonic cycle* brings heat energy and new materials to the surface and recycles old materials to mantle depths, creating movement and deformation of

the crust. We discuss the tectonic cycle later in this chapter.

The Rock Cycle

To begin our look at the rock cycle, we see that only eight natural elements compose 99% of Earth's crust. Just two of

Table 11.1   Common Elements in Earth's Crust

Element	Percentage of Earth's Crust by Weight
Oxygen (O)	46.6
Silicon (Si)	27.7
Aluminum (Al)	8.1
Iron (Fe)	5.0
Calcium (Ca)	3.6
Sodium (Na)	2.8
Potassium (K)	2.6
Magnesium (Mg)	2.1
All others	1.5
Total	100.00

Note. A quartz crystal (SiO_2) consists of Earth's two most abundant elements, silicon (Si) and oxygen (O). Inset photo from *Laboratory Manual in Physical Geology*, 3rd ed., R. M. Busch, ed. © 1993 by Macmillan Publishing Co.

these—oxygen and silicon—account for 74.3% of the crust (Table 11.1). Oxygen, the most reactive gas in the lower atmosphere, readily combines with other elements. For this reason, the percentage of oxygen is greater in the crust (about 47%) than in the atmosphere (about 21%). The internal differentiation process explains the relatively large percentages of lightweight elements such as silicon and aluminum in the crust. These less-dense elements migrate toward the surface, as discussed earlier.

Minerals and Rocks   Earth's elements combine to form minerals. A **mineral** is an inorganic (nonliving) natural compound having a specific chemical formula and possessing a crystalline structure. The combination of elements and the crystal structure give each mineral its characteristic hardness, color, density, and other properties. For example, the common mineral *quartz* is silicon dioxide, SiO_2, and has a distinctive six-sided crystal. Of the more than 4200 minerals, about 30 minerals comprise the *rock-forming minerals* most commonly encountered. *Mineralogy* is the study of the composition, properties, and classification of minerals (see **http://webmineral.com/** or **http://www.minsocam.org/MSA/Research_Links.html**).

One of the most widespread mineral families on Earth is the *silicates*, because silicon and oxygen are so common and because they readily combine with each other and with other elements. This mineral family includes quartz, feldspar, clay minerals, and numerous gemstones. *Oxides* are a group of minerals in which oxygen combines with metallic elements, such as iron to form hematite, Fe_2O_3. And, there are the *sulfides* and *sulfates* groups in which sulfur compounds combine with metallic elements, to form pyrite (FeS_2) and anhydrite ($CaSO_4$), respectively.

Another important mineral family is the *carbonate* group, which features carbon in combination with oxygen and other elements such as calcium, magnesium, and potassium. An example is the mineral calcite ($CaCO_3$), a form of calcium carbonate.

A **rock** is an assemblage of minerals bound together (such as granite, a rock containing three minerals), or a mass of a single mineral (such as rock salt), or undifferentiated material (such as the noncrystalline glassy obsidian, volcanic glass), or even solid organic material (such as coal)—all are rocks. Thousands of different rocks have been identified. All can be identified as one of three kinds, depending on the processes that formed them: *igneous* (melted), *sedimentary* (from settling out), and *metamorphic* (altered). Figure 11.6 illustrates these three processes and the interrelations among them that constitute the **rock cycle**. Let us examine each rock-forming process.

Igneous Processes

An **igneous rock** is one that solidifies and crystallizes from a molten state. Familiar examples are granite, basalt, and rhyolite. Igneous rocks form from **magma**, which is molten rock beneath the surface (hence the name *igneous*, which means "fire-formed" in Latin). Magma is fluid, highly gaseous, and under tremendous pressure. It either *intrudes* into crustal rocks, cools, and hardens, or it *extrudes* onto the surface as **lava**.

The cooling history of an igneous rock—how fast it cooled and how steadily its temperature dropped—determines its crystalline physical characteristics, or crystallization. Igneous rocks range from coarse-grained (slower cooling, with more time for larger crystals to form) to fine-grained or glassy (faster cooling).

Igneous rocks comprise approximately 90% of Earth's crust, although they are frequently covered by sedimentary rocks (sandstone, shale, limestone), soil, or oceans. Figure 11.7 illustrates the variety of occurrences of igneous rocks, both on and beneath Earth's surface.

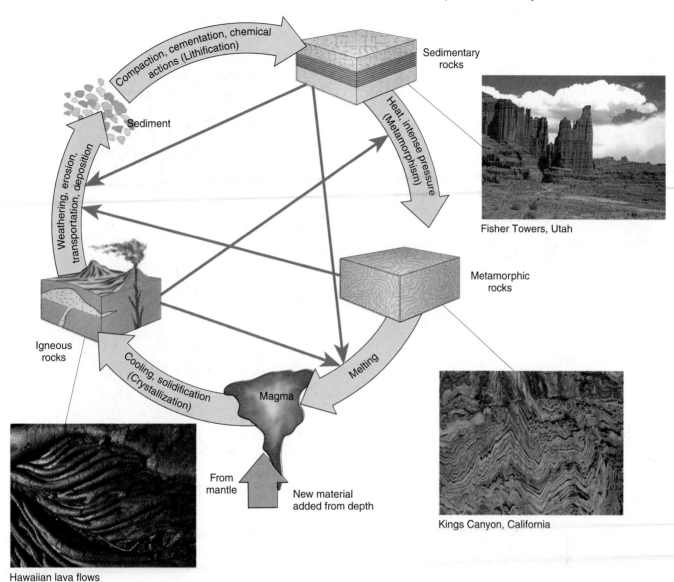

FIGURE 11.6 The rock cycle.
A rock-cycle schematic demonstrating the relation among igneous, sedimentary, and metamorphic processes. The arrows indicate that each rock type can enter the cycle at various points and be transformed into other rock types. [Adapted by permission from R. M. Busch, ed., *Laboratory Manual in Physical Geology*, 3rd ed., © 1993 by Macmillan Publishing Co. Photos by Bobbé Christopherson.]

Intrusive and Extrusive Igneous Rocks Intrusive igneous rock that cools slowly in the crust forms a **pluton**, a general term for any intrusive igneous rock body, regardless of size or shape that invaded layers of crustal rocks. The Roman god of the Underworld, Pluto, is the namesake. The largest pluton form is a **batholith**, defined as an irregular-shaped mass with a surface greater than 100 km^2, 40 mi^2 (Figure 11.8a). Batholiths form the mass of many large mountain ranges—for example, the Sierra Nevada batholith in California, the Idaho batholith, and the Coast Range batholith of British Columbia and Washington State.

Smaller plutons include the magma conduits of ancient volcanoes that have cooled and hardened. Those that form parallel to layers of sedimentary rock are *sills*; those that cross layers of the rock they invade are *dikes* (see Figure 11.7). Magma also can bulge between rock strata and produce a lens-shaped body, called *laccolith*, a type of sill. In addition, magma conduits themselves may solidify in roughly cylindrical forms that stand starkly above the landscape when finally exposed by weathering and erosion. Shiprock *volcanic neck* in New Mexico is such a feature, as suggested by the art in Figure 11.7 and shown in the inset photo. Weathering action of air, water, and ice can expose all of these intrusive forms.

Volcanic eruptions and flows produce extrusive igneous rock, such as lava that cools and forms basalt (Figure 11.8b). Chapter 12 presents volcanism.

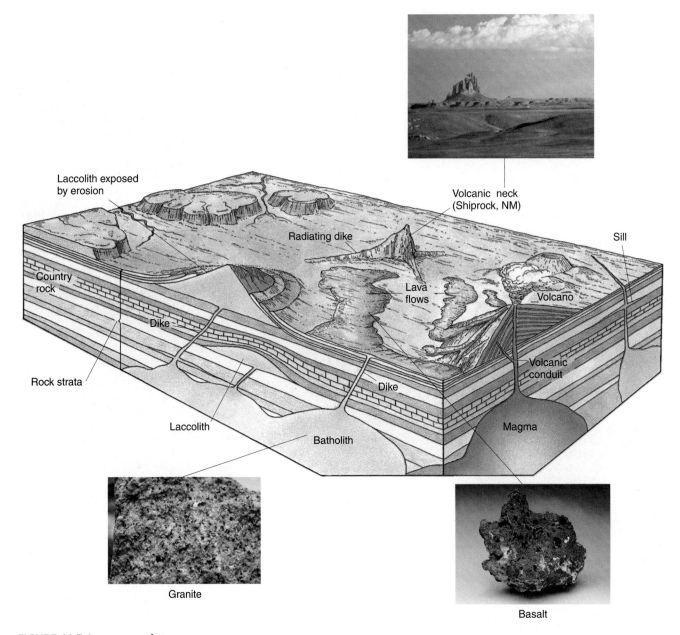

FIGURE 11.7 Igneous rock types.
The variety of occurrences of igneous rocks, both intrusive (below the surface) and extrusive
(on the surface). Inset photographs show samples of granite (intrusive) and basalt (extrusive).
[Photos from R. M. Busch, ed., *Laboratory Manual in Physical Geology*, 3rd ed., © 1993 by
Macmillan Publishing Co. Photo of Shiprock volcanic neck by author.]

Classifying Igneous Rocks   Mineral composition and
texture usually classify igneous rocks (Table 11.2). The two
broad categories are

1. *Felsic* igneous rocks—derived both in composition
   and name from *fel*dspar and *silic*a. Felsic minerals are
   generally high in silica, aluminum, potassium, and
   sodium and have low melting points. Rocks formed
   from felsic minerals generally are lighter in color and
   are less dense than mafic mineral rocks.
2. *Mafic* igneous rocks—derived both in composition
   and name from *ma*gnesium and *fer*ri*c* (Latin for

iron). Mafic minerals are low in silica, high in magnesium and iron, and have high melting points.
Rocks formed from mafic minerals are darker in
color and of greater density than felsic mineral rocks.

Table 11.2 shows that the same magma that produces
coarse-grained granite, when it slowly cools beneath the
surface, can form fine-grained *rhyolite* when it cools above
the surface (see inset photo). If it cools quickly, magma having a silica content comparable to granite and rhyolite may
form the dark, smoky, glassy-textured rock called *obsidian*,
or volcanic glass (see inset photo). Another glassy rock,

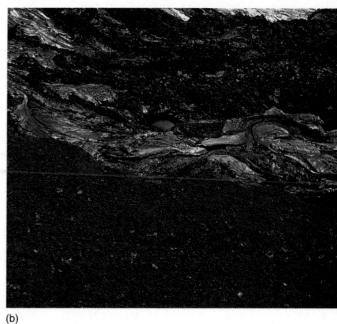

(a)

(b)

FIGURE 11.8 Intrusive and extrusive rocks.
(a) Exposed granites of the Sierra Nevada batholith. The boulders sitting on the granite are erratics left behind by glacial ice that melted and retreated thousands of years ago.
(b) Basaltic lava flows on Hawai'i; the glow is from a skylight into an active lava tube where molten lava is visible; the shiny surface is where lava recently flowed out of the skylight.
[Photos by Bobbé Christopherson.]

called *pumice*, forms when escaping gases bubble a frothy texture into the lava. Pumice is full of small holes, is light in weight, and is low enough in density to float in water (see inset photo).

On the mafic side, basalt is the most common fine-grained extrusive igneous rock. It makes up the bulk of the ocean floor, accounting for 71% of Earth's surface. It appears in lava flows such as those on the big island of Hawai'i (Figure 11.8b). An intrusive counterpart to basalt, formed by slow cooling of the parent magma, is *gabbro*.

Sedimentary Processes

Solar energy and gravity drive the process of *sedimentation*, with water as the principal transporting medium. Existing rock is digested by weathering, picked up and moved by erosion and transportation and deposited along river, beach, and ocean sites, where burial initiates the rock-forming process. The formation of **sedimentary rock** involves **lithification** processes of cementation, compaction, and a hardening of sediments.

Most sedimentary rocks derive from fragments of existing rock or organic materials. Bits and pieces of former rocks—principally quartz, feldspar, and clay minerals—erode and then mechanically transport by water (lake, stream, ocean, overland flow), ice (glacial action), wind, and gravity. They are transported from "higher-energy" sites, where the carrying medium has the energy to pick up

and move them, to "lower-energy" sites, where the material is dumped.

The common sedimentary rocks are *sandstone* (sand that became cemented together), *shale* (mud that became compacted into rock), *limestone* (bones and shells that became cemented or calcium carbonate that precipitated in ocean waters), and *coal* (ancient plant remains that became compacted into rock). Some minerals, such as calcium carbonate, dissolve into solution and form sedimentary deposits by precipitating from those solutions to form rock. This is an important process in the oceanic environment.

Characteristically, sedimentary rocks are laid down by wind, water, or ice in horizontally layered beds. Different environmental conditions produce a variety of sedimentary forms. Various cements, depending on availability, fuse rock particles together. Lime, or calcium carbonate ($CaCO_3$), is the most common, followed by iron oxides (Fe_2O_3) and silica (SiO_2). Drying (dehydration), heating, or chemical reactions can also unite particles.

The layered strata of sedimentary rocks form an important record of past ages. **Stratigraphy** is the study of the sequence (superposition), thickness, and spatial distribution of strata. These sequences yield clues to the age and origin of the rocks. Figure 11.9a shows a sandstone sedimentary rock in a desert landscape. Note the multiple layers in the formation and how differently they resist weathering processes. The climatic history of this rock's formation environment is disclosed in the stratigraphy—drier periods

Table 11.2   Igneous Rock Minerals

	Felsic Minerals (feldspars and silica)	Mafic Minerals (magnesium and iron)	Ultramafic Minerals (low silica)
General characteristics	Higher ◄——————————— Silica content ——————————► Lower		
	Higher ◄——————————— Resistance to weathering ——————————► Lower		
	◄————— Increased potassium and sodium Increased calcium, iron, and magnesium —————►		
	Lower ◄——————————— Melting temperatures ——————————► Higher		
	Lighter ◄——————————— Coloration ——————————► Darker		

	Felsic Minerals	Mafic Minerals			Ultramafic Minerals
Mineral families	Quartz Feldspars	Mica	Amphibole	Pyroxene	Olivine
	Potassium feldspars Sodium feldspars Calcium feldspars				
	SiO₂ K, Al, Si Na, Al, Si, Ca Ca, Al, Si (Orthoclase) (plagioclase) (aluminosilicates)	K, Fe, Mg, Al, Si (biotite: black; muscovite: white)	Fe, Mg, Al, Si (complex) (hornblende: black)	Fe, Mg, Si (dark)	Mg, Fe (dark green) (no quartz, no feldspars)
Coarse-grained texture (intrusive) slower cooling rate	Granite Diorite			Gabbro	Peridotite
Fine-grained texture (extrusive) faster cooling rate	Rhyolite Andesite Dacite (sodic feldspar) (Mount St. Helens)			Basalt	
Other textures	Obsidian (glassy) Pumice (vesicular)			Scoria (vesicular)	

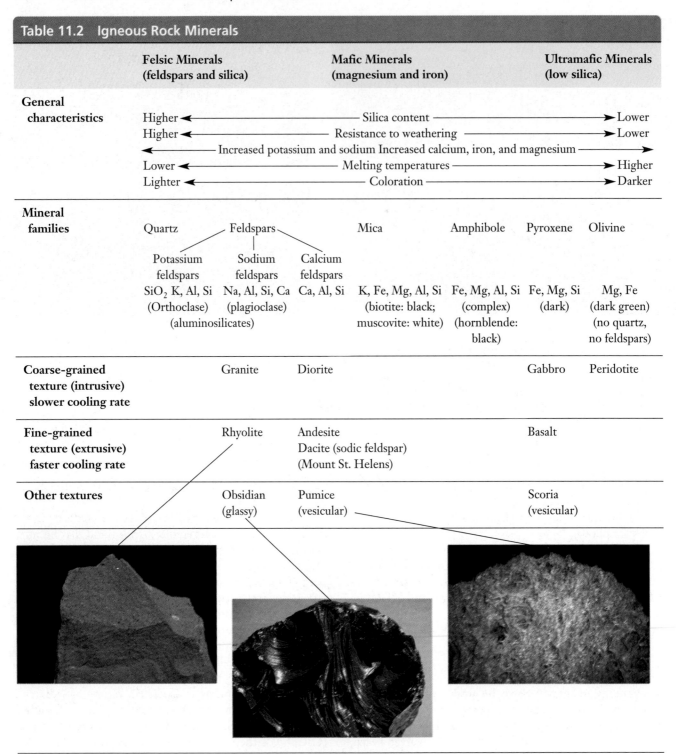

Source: Photos from R. M. Busch, ed., *Laboratory Manual in Physical Geology*, 3rd ed., © 1993 by Macmillan Publishing Co.

of former dunes (near the top), wetter periods of horizontal deposits (in the middle and lower portions).

The two primary sources of sedimentary rocks are the mechanically transported bits and pieces of former rock, called clastic sediments, and the dissolved minerals in solution, termed chemical sediments.

Clastic Sedimentary Rocks   Weathered and fragmented rocks that are further worn in transport provide *clastic sediments*. Table 11.3 lists the range of *clast* sizes—everything from boulders to microscopic clay particles—and the form they take as *lithified rock*. Common sedimentary rocks from clastic sediments include siltstone (silt) or mudstone (mud),

(a)

(b)

Sandstone

Limestone

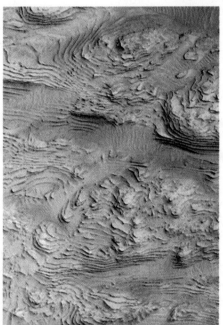

(c)   Martian sedimentary formations

FIGURE 11.9 Sedimentary rock types.
Two types of sedimentary rock. (a) Sandstone formation with sedimentary strata subjected to differential weathering; note weaker underlying siltstone. (b) A limestone landscape, formed by chemical sedimentary processes, in south-central Indiana. (c) Ancient sediment deposition through repeated cycles of erosion and deposition is evident in the Martian western Arabia Terra (8° N and 7° W). [(a) Photos by author; (b) photo by Bobbé Christopherson; inset photo by Richard M. Busch; (c) image from Mars Global Surveyor courtesy of NASA/JPL/Malin Space Science Systems.]

(a)

(b)

FIGURE 11.10 Death Valley, wet and dry.
A Death Valley landscape (a) one day after a record rainfall when the valley was covered by several square kilometers of water only a few centimeters deep. (b) One month later the water had evaporated, and the same valley is coated with evaporites (borated salts). [Photos by author.]

Table 11.3	Clastic Sediment Sizes and Related Rock Form	
Unconsolidated Sediment	**Grain Size**	**Rock Form**
Boulders, cobbles	>80 mm	Conglomerate (breccia, if pieces are angular)
Pebbles, gravel	>2 mm	
Coarse sand	0.5–2.0 mm	Sandstone
Medium-to-fine sand	0.062–0.5 mm	Sandstone
Silt	0.002–0.062 mm	Siltstone (mudstone)
Clay	<0.002 mm	Shale

Chemical sediments from inorganic sources are deposited when water evaporates and leaves behind a residue of salts. These *evaporites* may exist as common salts, such as gypsum or sodium chloride (table salt). They often appear as flat, layered deposits across a dry landscape. The pair of photographs in Figure 11.10 dramatically demonstrates this process; one was made in Death Valley National Park one day after a record 2.57-cm (1.01-in.) rainfall, and the other photo was made one month later.

Chemical deposition also occurs in the water of natural hot springs from chemical reactions between minerals and oxygen. This is a sedimentary process related to *hydrothermal activity*. An example is the massive deposit of travertine, a form of calcium carbonate, at Mammoth Hot Springs in Yellowstone National Park (Figure 11.11).

shale (up to 0.06-mm-diameter particles), and sandstone (sand particles that range from 0.06 to 2 mm in diameter).

Chemical Sedimentary Rocks   *Chemical sedimentary* rocks are not formed from physical pieces of broken rock but instead from dissolved minerals, transported in solution and chemically precipitated from solution (they are essentially nonclastic). The most common chemical sedimentary rock is **limestone**, which is lithified calcium carbonate, derived from inorganic and organic sources. A similar form is *dolomite*, which is lithified calcium–magnesium carbonate, $CaMg(CO_3)_2$. Limestone from marine organic origins is most common; it is biochemical—derived from shell and bone produced by biological activity. Once formed, these rocks are vulnerable to chemical weathering, which produces unique landforms, as discussed in the weathering section of Chapter 13 and as shown in Figure 11.9b.

FIGURE 11.11 Hydrothermal deposits.
Mammoth Hot Springs, Yellowstone National Park, is an example of a hydrothermal deposit principally composed of travertine ($CaCO_3$), deposited as a chemical precipitate from heated spring water as it evaporated. [Photo by author.]

Metamorphic Processes

Any rock, either igneous or sedimentary, may be transformed into a **metamorphic rock** by going through profound physical or chemical changes under pressure and increased temperature. (The name *metamorphic* comes from a Greek word meaning "to change form.") Metamorphic rocks generally are more compact than the original rock and therefore are harder and more resistant to weathering and erosion (Figure 11.12).

Several conditions can cause metamorphism. Most common is when subsurface rock is subjected to high tem-

(a)

(b)

FIGURE 11.12 Metamorphic rocks.
(a) A metamorphic rock outcrop in Greenland, the Amitsoq Gneiss, at 3.8 billion years old, one of the oldest rock formations on Earth. (b) Precambrian Vishnu Schist composes these rock cliffs in the inner gorge of the Grand Canyon in northern Arizona. These metamorphic rocks lie beneath many layers of sedimentary rocks deep in the continent. Despite the fact that the schist is harder than steel, the Colorado River has cut down into the uplifted Colorado Plateau, exposing and eroding these ancient rocks. [(a) Photo by Kevin Schafer/Peter Arnold, Inc.; (b) photo by author.]

Table 11.4 Metamorphic Rocks

Parent Rock	Metamorphic Equivalent	Texture
Shale (clay minerals)	Slate	Foliated
Granite, slate, shale	Gneiss	Foliated
Basalt, shale, peridotite	Schist	Foliated
Limestone, dolomite	Marble	Nonfoliated
Sandstone	Quartzite	Nonfoliated

peratures and high compressional stresses occurring over millions of years. Igneous rocks become compressed during collisions between slabs of Earth's crust (see the discussion of plate tectonics later in this chapter). Sometimes rocks simply are crushed under great weight when a crustal area is thrust beneath other crust. In another setting, igneous rocks may be sheared and stressed along earthquake fault zones, causing metamorphism.

Metamorphic rocks comprise the ancient roots of mountains. Exposed at the bottom of the inner gorge of the Grand Canyon in Arizona, a Precambrian (Archean) metamorphic rock called the Vishnu Schist is a remnant of such an ancient mountain root (Figure 11.12b).

Another metamorphic condition occurs when sediments collect in broad depressions in Earth's crust and, because of their own weight, create enough pressure in the bottommost layers to transform the sediments into metamorphic rock, or *regional metamorphism*. Also, molten magma rising within the crust may "cook" adjacent rock, a process called *contact metamorphism*.

Metamorphic rocks may be changed both physically and chemically from the original rocks. Table 11.4 lists some metamorphic rock types, their parent rocks, and resultant textures. If the mineral structure demonstrates a particular alignment after metamorphism, the rock is called *foliated*, and some minerals may appear as wavy striations (streaks or lines) in the new rock (Table 11.4, left photo). On the other hand, parent rock with a more homogeneous (evenly mixed) makeup may produce a *nonfoliated* rock (Table 11.4, right photo). Which form do you see in the inset photo of a metamorphic rock in Figure 11.6 (lower right)?

In this section, you have seen how three rock-forming processes yield the igneous, sedimentary, and metamorphic materials of Earth's crust. Next we look at how heat energy and motions originating within Earth move huge slabs of crust, causing the continents to *drift*. The fact that the continents have migrated thousands of kilometers, and continue to migrate, is a revolutionary discovery in the *tectonic cycle* that came to light only in the last century.

Plate Tectonics

Have you ever looked at a world map and noticed that a few of the continental landmasses—particularly South America and Africa—appear to have matching shapes, like pieces of a jigsaw puzzle? The incredible reality is that the continental pieces indeed once were fitted together! Continental landmasses not only migrated to their present locations but continue to move at speeds up to 6 cm (2.4 in.) per year.

We say that the continents are *adrift* because convection currents in the asthenosphere and the rest of the mantle are dragging them around. The key point is that the arrangement of continents and oceans we see today is not permanent but is in a continuing state of change.

A continent, such as North America, is actually a collage of crustal pieces that migrated from elsewhere to form the landscape we know. Through a historical chronology, let us trace the discoveries that led to the plate tectonics theory—a major revolution in geoscience.

A Brief History

As early mapping gained accuracy, some observers noticed the symmetry among continents, particularly the fit between South America and Africa. Abraham Ortelius (1527–1598), a geographer, noted the apparent fit of some continental coastlines in his *Thesaurus Geographicus* (1596). In 1620, English philosopher Sir Francis Bacon noted gross similarities between the edges of Africa and South America (although he did not suggest that they had drifted apart). Benjamin Franklin wrote in 1780 that the crust must be a shell that can break and shift by movements of fluid below! Others wrote—unscientifically—about such apparent relationships, but it was not until much later that a valid explanation came forward.

In 1912, German geophysicist and meteorologist Alfred Wegener publicly presented in a lecture his idea that Earth's landmasses migrate. His book, *Origin of the Continents and Oceans*, appeared in 1915. Today, Wegener is regarded as the father of this concept, which he first called **continental drift**. Scientists at the time were unreceptive to Wegener's revolutionary proposal. Thus, a great debate began, lasting almost 50 years until a scientific consensus developed that Wegener was right after all.

Wegener thought that all landmasses formed one supercontinent approximately 225 million years ago, during the Triassic Period. This one landmass he called **Pangaea**, meaning "all Earth" (see Figure 11.16b). Although his initial model kept the landmasses together too long and his proposal included an incorrect driving mechanism for the moving continents, Wegener's arrangement of Pangaea and its breakup was correct, Pangaea being only the latest of earlier supercontinent arrangements over the span of Earth history.

To come up with his Pangaea fit, he studied the geologic record (rock strata), the fossil record, and the climatic record for the continents. He concluded that South America and Africa correlated in many complex ways. He decided that the large midlatitude coal deposits, which date to the Permian and Carboniferous Periods (245–360 million years ago), exist because these regions once were nearer the equator and therefore covered by lush vegetation that became coal.

As modern scientific capabilities built the case for continental drift, the 1950s and 1960s saw a revival of interest in Wegener's concepts and, finally, confirmation. Aided by an avalanche of discoveries, the plate tectonics theory today is universally accepted as an accurate model of the way Earth's surface evolves. *Tectonic*, from the Greek *tektonikùs*, meaning "building" or "construction," refers to changes in the configuration of Earth's crust as a result of internal forces. **Plate tectonic** processes include upwelling of magma, crustal plate movements, sea-floor spreading and subduction of crust, earthquakes, volcanic activity, warping, folding, and faulting of the crust.

Sea-Floor Spreading and Production of New Crust

The key to establishing the theory of continental drift was a better understanding of the seafloor. The seafloor has a remarkable feature: an interconnected worldwide mountain chain (ridge) some 64,000 km (40,000 mi) in extent and averaging more than 1000 km (600 mi) in width. A striking view of this great undersea mountain chain opens Chapter 12. How did this global mountain chain get there?

In the early 1960s, geophysicists Harry H. Hess and Robert S. Dietz proposed **sea-floor spreading** as the mechanism that builds this mountain chain and drives continental movement. Hess said that these submarine mountain ranges, called the **mid-ocean ridges**, were the direct result of upwelling flows of magma from hot areas in the upper mantle and asthenosphere and perhaps from the deeper lower mantle.

When mantle convection brings magma up to the crust, the crust fractures, and the magma extrudes onto the seafloor and cools to form new seafloor. This process builds the mid-ocean ridges and spreads the seafloor laterally. The concept is illustrated in Figure 11.13, which shows how the ocean floor is rifted (moves apart) and scarred along mid-ocean ridges.

As new crust generates and the seafloor spreads, magnetic particles in the lava orient with the magnetic field in force at the time the lava cools and hardens. The particles become locked in this alignment as new seafloor forms. This alignment creates a kind of magnetic tape recording in the seafloor. The continually forming oceanic crust records each magnetic reversal and reorientation of Earth's polarity.

Figure 11.14 illustrates just such a recording from the Mid-Atlantic Ridge south of Iceland. The colors denote the alternating magnetic polarization. Note the mirror images that develop on either side of the sea-floor rift as a result of the nearly symmetrical spreading of the sea floor.

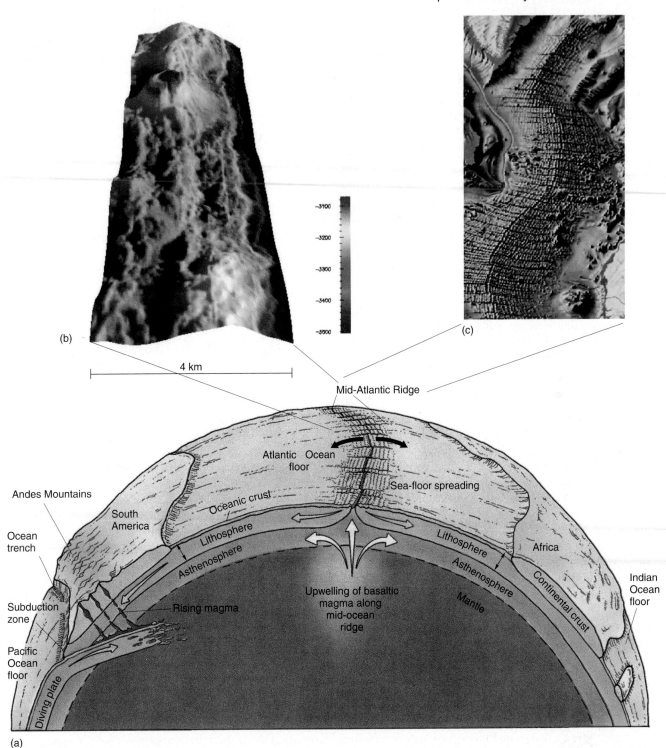

FIGURE 11.13 Crustal movements.
(a) Sea-floor spreading, upwelling currents, subduction, and plate movements, shown in cross section. Arrows indicate the direction of the spreading. (b) A 4-km-wide image of the Mid-Atlantic Ridge showing linear faults, a volcanic crater, a rift valley, and ridges. The image was taken by the TOBI (towed ocean-bottom instrument) at approximately 29° N latitude. (c) Detail from Tharp's ocean-floor map. [(a) After P. J. Wyllie, *The Way the Earth Works*, © 1976, by John Wiley & Sons, adapted by permission; (b) image courtesy of D. K. Smith, Woods Hole Oceanographic Institute, Woods Hole, Massachusetts. All rights reserved; (c) Bruce C. Heezen and Marie Tharp, courtesy Marie Tharp.]

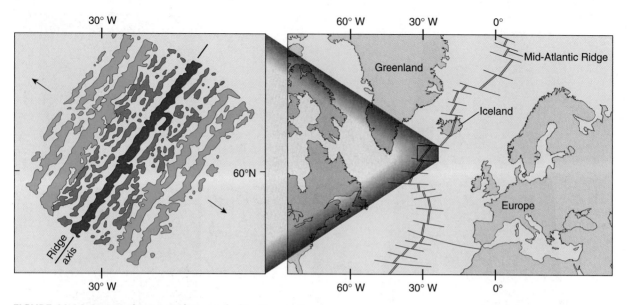

FIGURE 11.14 Magnetic reversals recorded in the seafloor.
Magnetic reversals recorded in the seafloor south of Iceland along the Mid-Atlantic Ridge.
Note rocks with similar magnetic orientation are roughly the same distance from the spreading center. [Magnetic reversals reprinted from J. R. Heirtzler, S. Le Pichon, and J. G. Baron, *Deep-Sea Research* 13, © 1966, Pergamon Press, p. 247.]

These periodic reversals of Earth's magnetic field are a valuable clue to understanding sea-floor spreading, helping scientists fit together pieces of Earth's crust.

These sea-floor recordings of Earth's magnetic-field reversals and other measurements permitted the determination of the age of the seafloor. The complex harmony of the two concepts of plate tectonics and sea-floor spreading became clearer. The youngest crust anywhere on Earth is at the spreading centers of the mid-ocean ridges, and, with increasing distance from these centers, the crust gets steadily older (Figure 11.15). The oldest seafloor is in the western Pacific near Japan, dating to the Jurassic Period. Note on the map in the figure the distance between this basin and its spreading center in the South Pacific west of South America.

Overall, the seafloor is relatively young—nowhere does it exceed 208 million years in age, remarkable when you remember that Earth's age is 4.6 billion years. The reason is that oceanic crust is short-lived—the oldest sections, farthest from the mid-ocean ridges, are slowly plunging beneath continental crust along Earth's deep oceanic trenches. The discovery that the seafloor is young demolished earlier thinking that the oldest rocks would be found there.

Subduction of the Crust

In contrast to the upwelling zones along the mid-ocean ridges are the areas of descending crust elsewhere. On the left side of Figure 11.13a, note how one plate of the crust is diving beneath another, into the mantle. Recall that the basaltic ocean crust has a density of 3.0 g/cm^3, whereas continental crust averages a lighter 2.7 g/cm^3. As a result, when

continental crust and oceanic crust slowly collide, the denser ocean floor will grind beneath the lighter continental crust, thus forming a **subduction zone**, as shown in the figure.

The world's deep ocean trenches coincide with these subduction zones and are the lowest features on Earth's surface. Deepest is the Mariana Trench near Guam, which descends below sea level to −11,030 m (−36,198 ft); next in depth are the Puerto Rico Trench at −8605 m (−28,224 ft) and, in the Indian Ocean, the Java Trench at −7125 m (−23,376 ft).

The subducted portion of crust is dragged down into the mantle, where it remelts and eventually is recycled as magma, rising again toward the surface through deep fissures and cracks in crustal rock. Volcanic mountains such as the Andes in South America and the Cascade Range from northern California to the Canadian border form inland of these subduction zones as a result of rising plumes of magma, as suggested in Figure 11.13a. The fact that spreading ridges and subduction zones are areas of earthquake and volcanic activity provides important proof of plate tectonics. (For an overview, see **http://geology.er.usgs.gov/eastern/tectonic.**) Now, using current scientific findings, let us go back and reconstruct the past, and Pangaea.

The Formation and Breakup of Pangaea

The supercontinent of Pangaea and its subsequent breakup into today's continents represent only the last 225 million years of Earth's 4.6 billion years, or only the most recent 1/23 of Earth's existence. During the other 22/23 of geologic time, other things were happening. The landmasses as we know them were unrecognizable.

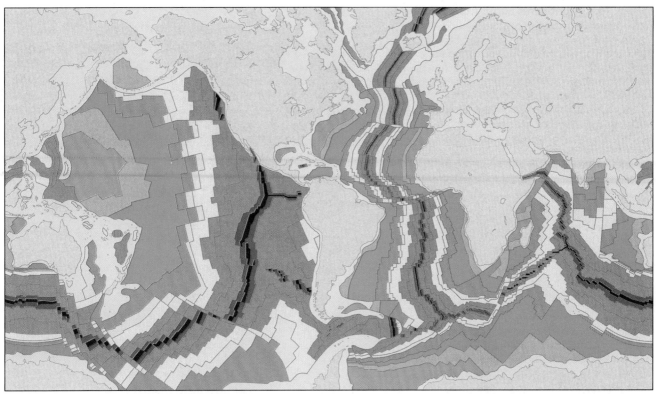

FIGURE 11.15 Relative age of the oceanic crust.
Compare the light tan color near the East Pacific rise in the eastern Pacific Ocean with the same light tan color along the Mid-Atlantic ridge. What does the difference in width tell you about the rates of plate motion in the two locations? [Adapted with the permission of W. H. Freeman and Company from *The Bedrock Geology of the World* by R. L. Larson and others, © 1985.]

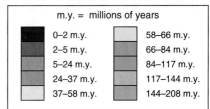

m.y. = millions of years

	0–2 m.y.		58–66 m.y.
	2–5 m.y.		66–84 m.y.
	5–24 m.y.		84–117 m.y.
	24–37 m.y.		117–144 m.y.
	37–58 m.y.		144–208 m.y.

Figure 11.16a begins with the pre-Pangaea arrangement of 465 million years ago (during the middle Ordovician Period). Figure 11.16b illustrates an updated version of Wegener's Pangaea, 225–200 million years ago (Triassic–Jurassic Periods). The movement of plates that occurred by 135 million years ago (the beginning of the Cretaceous Period) is in Figure 11.16c. Figure 11.16d presents the arrangement 65 million years ago (beginning of the Tertiary Period). Finally, the present arrangement in modern geologic time (the late Cenozoic Era) is in Figure 11.16e.

Earth's present crust is divided into at least 14 plates, of which about half are major and half are minor, in terms of area (Figure 11.17). Literally hundreds of smaller pieces and perhaps dozens of microplates that migrated together comprise these broad plates. The arrows in the figure indicate the direction in which each plate is presently moving, and the length of the arrows suggests the rate of movement during the past 20 million years.

Compare Figure 11.17 with the image of the ocean floor in Figure 11.18. In the image, satellite radar-altimeter measurements determined the sea-surface height to a remarkable accuracy of 0.03 m, or 1 in. The sea-surface elevation is far from uniform; it is higher or lower in direct response to the mountains, plains, and trenches of the ocean floor beneath it. This sea-floor topography causes slight differences in Earth's gravity.

For example, a massive mountain on the ocean floor exerts a high gravitational field and attracts water to it, producing higher sea level above the mountain; over a trench there is less gravity, resulting in a drop in sea level. A mountain 2000 m (6500 ft) high that is 20 km (12 mi) across its base creates a 2-m rise in the sea surface. Until Scripps Institution of Oceanography and NOAA developed this use of satellite technology these tiny changes in height were not visible. Now we have a comprehensive map of the ocean *floor*, derived from remote sensing of the ocean *surface*!

The illustration of the seafloor that begins Chapter 12 also may be helpful in identifying the plate boundaries in the following discussion. It is interesting to correlate the features illustrated in Figures 11.17 and 11.18 and Chapter 12's opening map. A plate motion calculator by the University of Tokyo is at **http://triton.ori.u-tokyo.ac.jp/~intridge/pmc/nuvel1.html**.

FIGURE 11.16 Continents adrift, from 465 million years ago to the present.
Observe the formation and breakup of Pangaea and the types of motions occurring at plate boundaries. [(a) from R. K. Bambach, "Before Pangaea: The Geography of the Paleozoic World," *American Scientist* 68 (1980): 26–38, reprinted by permission; (b–e) from R. S. Dietz and J. C. Holden, *Journal of Geophysical Research* 75, no. 26 (September 10, 1970): 4939–4956, © The American Geophysical Union; (f) remote sensing image courtesy of S. Tighe, University of Rhode Island, and R. Detrick, Woods Hole Oceanographic Institute, Woods Hole, Massachusetts.]

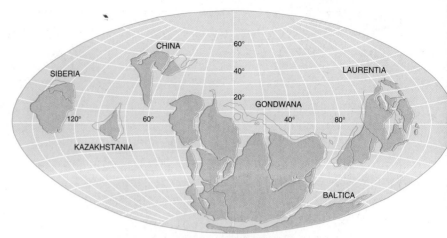

(a) 465 million years ago

Pangaea—all Earth. Panthalassa ("all seas"), became the Pacific Ocean, and the Tethys Sea (partly enclosed by the African and the Eurasian plates), became the Mediterranean Sea, and trapped portions formed the present-day Caspian Sea. The Atlantic Ocean did not exist.

Africa shared a common connection with both North and South America. Today, the Appalachian Mountains in the eastern United States and the Atlas Mountains of northwestern Africa reflect this common ancestry; they are, in fact, portions of the same mountain range, torn thousands of kilometers apart!

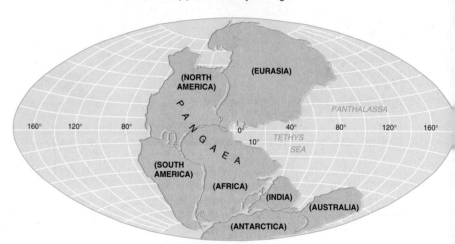

(b) 225 million years ago

New seafloor that formed since the last map is highlighted (gray tint). An active spreading center rifted North America away from landmasses to the east, shaping the coast of Labrador. India was farther along in its journey, with a spreading center to the south and a subduction zone to the north; the leading edge of the India plate was diving beneath Eurasia.

The outlines of South America, Africa, India, Australia, and southern Europe appear within the continent called *Gondwana* in the Southern Hemisphere. North America, Europe, and Asia, comprise *Laurasia*, the northern portion.

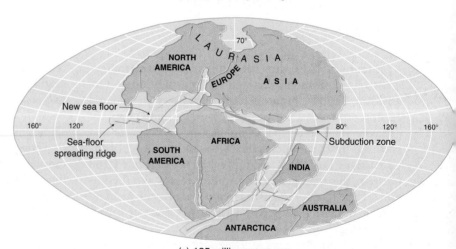

(c) 135 million years ago

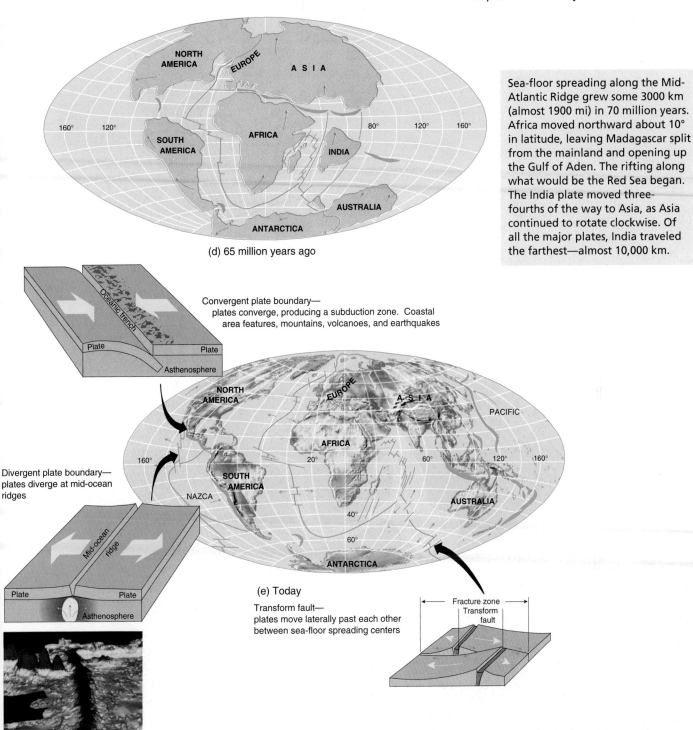

(d) 65 million years ago

Sea-floor spreading along the Mid-Atlantic Ridge grew some 3000 km (almost 1900 mi) in 70 million years. Africa moved northward about 10° in latitude, leaving Madagascar split from the mainland and opening up the Gulf of Aden. The rifting along what would be the Red Sea began. The India plate moved three-fourths of the way to Asia, as Asia continued to rotate clockwise. Of all the major plates, India traveled the farthest—almost 10,000 km.

Convergent plate boundary—
plates converge, producing a subduction zone. Coastal area features, mountains, volcanoes, and earthquakes

Divergent plate boundary—
plates diverge at mid-ocean ridges

(e) Today

Transform fault—
plates move laterally past each other between sea-floor spreading centers

Fracture zone
Transform fault

(f) East-Pacific rise,
sea-floor spreading ridge (center of image at 9° N)

From 65 million years ago until the present, more than half of the ocean floor was renewed. The northern reaches of the India plate underthrust the southern mass of Asia through subduction, forming the Himalayas in the upheaval created by the collision. Plate motions continue to this day.

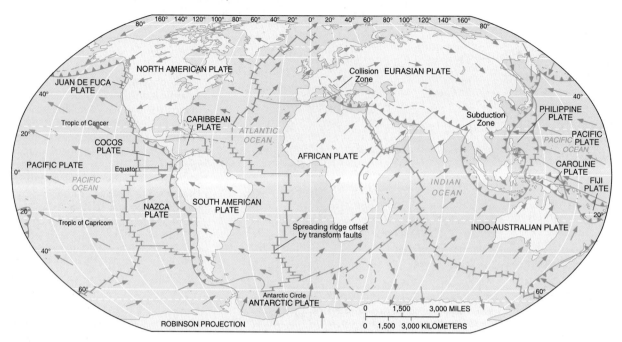

FIGURE 11.17 Earth's major lithospheric plates and their movements.
Each arrow represents 20 million years of movement. The longer arrows indicate that the
Pacific and Nazca plates are moving more rapidly than the Atlantic plates. Compare the
length of these arrows with those light-tan areas on Figure 11.15. Symbol legend is with
Figure 11.20. [Adapted from U.S. Geodynamics Committee.]

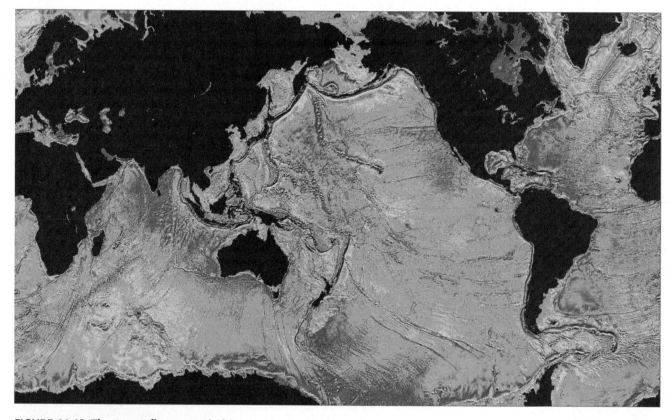

FIGURE 11.18 The ocean floor revealed.
A global gravity anomaly map derived from *Geosat* and *ERS-1* altimeter data. The radar
altimeters measured sea-surface heights. Variation in sea-surface elevation is a direct indication of the topography of the ocean floor. [Global gravity anomaly map image courtesy of
D. T. Sandwell, Scripps Institution of Oceanography. All rights reserved, 1995.]

Plate Boundaries

The boundaries where plates meet clearly are dynamic places, although slow-moving within human time frames. The block diagram inserts in Figure 11.16e show the three general types of motion and interaction that occur along the boundary areas:

- *Divergent boundaries* (lower left in figure) are characteristic of sea-floor spreading centers, where upwelling material from the mantle forms new seafloor ("constructional") and crustal plates spread apart. The spreading makes these zones of tension. An example noted in the figure is the divergent boundary along the East Pacific rise, which gives birth to the Nazca plate (moving eastward) and the Pacific plate (moving northwestward). Whereas most divergent boundaries occur at mid-ocean ridges, there are a few within continents themselves. An example is the Great Rift Valley of East Africa, where crust is rifting apart.
- *Convergent boundaries* (upper left) are characteristic of collision zones, where areas of continental and oceanic crust collide. These are zones of compression and crustal loss ("destructional"). Examples include the subduction zone off the west coast of South and Central America and the area along the Japan and Aleutian Trenches. Along the western edge of South America, the Nazca plate collides with and is subducted beneath the South American plate. This convergence creates the Andes Mountains chain and related volcanoes. The collision of India and Asia mentioned earlier is another example of a convergent boundary.
- *Transform boundaries* (lower right) occur where plates slide laterally past one another at right angles to a seafloor spreading center, neither diverging nor converging, and usually with no volcanic eruptions. These are the right-angle fractures stretching across the mid-ocean ridge system worldwide (visible in Figures 11.17 and 11.18 and Chapter 12 opening illustration).

Related to plate boundaries, another piece of the tectonic puzzle fell into place in 1965 when University of Toronto geophysicist Tuzo Wilson first described the nature of transform boundaries and their relation to earthquake activity. All the spreading centers on Earth's crust feature these perpendicular scars (Figure 11.19). Some transform faults are a few hundred kilometers long; others, such as those along the East Pacific rise, stretch out 1000 km or more (over 600 mi). Across the entire ocean floor, spreading-center mid-ocean ridges are the location of **transform faults.** The faults generally are parallel to the direction in which the plate is moving. How do they occur?

You can see in Figure 11.19 that mid-ocean ridges are not simple straight lines. When a mid-ocean rift begins, it opens at points of weakness in the crust. The fractures you see in the figure began as a series of offset breaks in the crust in each portion of the spreading center. As new ma-terial rises to the surface, building the mid-ocean ridges and spreading the plates, these offset areas slide past each other, in horizontal faulting motions. The resulting fracture zone, which follows these breaks in Earth's crust, is active only along the fault section between ridges of spreading centers, as shown in Figure 11.19.

Along transform faults (the fault section between C and D in the figure) the motion is one of horizontal displacement—no new crust is formed or old crust subducted. In contrast, beyond the spreading centers, the two sides of the fracture zones join and are inactive. In fact, the plate pieces on either side of the fracture zone are moving in the same direction, away from the spreading center (the fracture zones between A and B and between E and F in the figure).

The name *transform* was assigned because of this apparent transformation in the direction of the fault movement. The famous San Andreas fault system in California, where continental crust has overridden a transform system, relates to this type of motion. The fault that triggered the 1995 Kobe earthquake in Japan also has this type of horizontal motion; as does the North Anatolian fault in Turkey and its latest 1999 quake.

Earthquake and Volcanic Activity

Plate boundaries are the primary location of earthquake and volcanic activity, and the correlation of these phenomena is an important aspect of plate tectonics. The next chapter discusses earthquakes and volcanic activity in more detail; however, their general relationship to the tectonic plates is important to mention.

Figure 11.20 maps earthquake zones, volcanic sites, hot spots, and plate motion. The "ring of fire" surrounding the Pacific Basin, named for the frequent incidence of volcanoes, is evident. The subducting edge of the Pacific plate thrusts deep into the crust and mantle, producing molten material that makes its way back toward the surface, causing active volcanoes along the Pacific Rim. Such processes occur similarly at plate boundaries throughout the world.

Hot Spots

A dramatic aspect of Earth's internal dynamics is the estimated 50 to 100 **hot spots** across Earth's surface (also shown in Figure 11.20). These are individual sites of upwelling material arriving at the surface in tall plumes from the mantle or rising near the surface producing thermal effects in groundwater and the crust. Some of these sites can be developed for geothermal power (Focus Study 11.1). Hot spots occur beneath both oceanic and continental crust and are anchored deep in the stiff lower mantle, tending to remain fixed relative to migrating plates. Thus, the area of a plate that is above a hot spot is locally heated for the brief geologic time it is there (a few hundred thousand or million years).

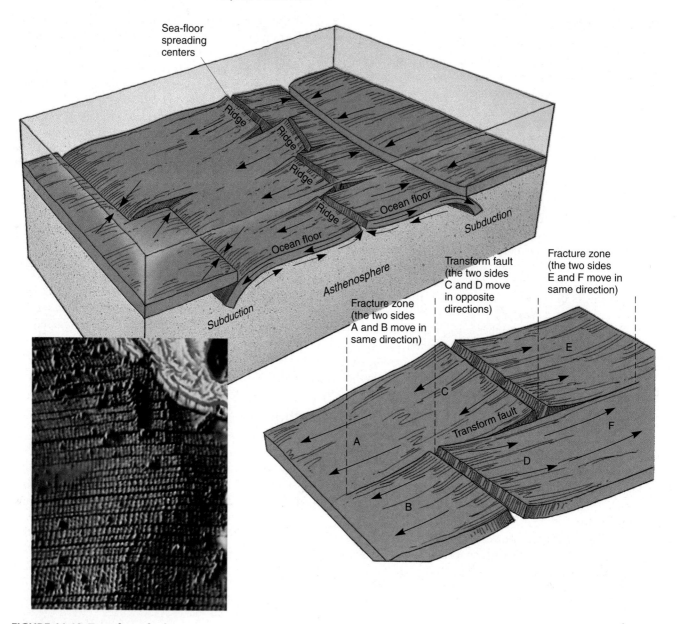

FIGURE 11.19 Transform faults.
Appearing along fracture zones, a transform fault is only that section between spreading centers where adjacent plates move in opposite directions—between C and D. [Adapted from B. Isacks, J. Oliver, and L. R. Sykes, *Journal of Geophysical Research* 73, (1968): 5855–5899, © The American Geophysical Union. Detail insert from Tharp's ocean-floor map, Bruce C. Heezen and Marie Tharp, courtesy Marie Tharp.]

The Pacific plate moved across a hot, upward-erupting plume for the last 80 million years, creating a string of volcanic islands stretching northwestward away from the hot spot. This hot spot produced, and continues to form, the Hawaiian–Emperor islands chain (Figure 11.21). Thus, the ages of the islands or seamounts in the chain increase northwestward from the island of Hawai'i, as you can see from the ages marked in the figure. The oldest island in the Hawaiian part of the chain is Kaua'i, approximately 5 mil-

lion years old; it is weathered, eroded, and deeply etched with canyons and valleys.

To the northwest of this active hot spot in Hawai'i, the island of Midway rises as a part of the same system. From there, the Emperor Seamounts follow northwestward until they reach about 40 million years of age. At that point, this linear island chain shifts direction northward. At the northernmost extreme, the seamounts that formed about 80 million years ago are now approaching

Heat from Earth—Geothermal Energy and Power

A tremendous amount of endogenic energy flows from Earth's interior toward the surface. Temperatures at the base of Earth's crust range from 200°–1000°C (392°–1800°F). Convection and conduction transports this geothermal energy in enormous quantities from the mantle to the crust, yet, geothermal power development is limited in extent to certain locations (Figure 1). Despite such site specific limitations, applications of geothermal energy for power production are a viable energy source for the present and future.

Geothermal energy is produced when pockets of magma and hot portions of the crust heat groundwater. This energy is transmitted to the surface by heated water or steam accessed through the drilling of wells. For an effective *underground thermal reservoir* to form, an aquifer strata must have high porosity, and high permeability that allows heated water to move freely through connecting pore spaces. (See Figure 9.16 and the accompanying text for illustration and definition of these terms.)

Ideally, this aquifer should heat groundwater to 180° to 350°C (355°–600°F) and be accessible to drilling within 3 km of the surface, although 6–7 km depths are workable (1.9 mi; 3.7–4.3 mi). Lower temperatures will still produce steam. With drilling costs as a major overhead expense, the shallower the well needed, the better. An impermeable rock strata above the thermal reservoir aquifer helps preserve the resource and describes a *hydrothermal reservoir*, the most common form of thermal repository.

Geothermal energy literally refers to heat from Earth's interior, whereas *geothermal power* relates to specific applied strategies of geothermal electric or geothermal direct applications. *Geothermal electric* uses steam, or hot water that flashes to steam, to drive a turbine-generator. In *geothermal direct*, hot water is used to heat buildings, to cool buildings using a heat exchange system, to heat greenhouses, to heat

FIGURE 1 Surface geothermal activity in Yellowstone.
Yellowstone is the most famous national park that contains thermal features. Approximately 10,000 geysers, mudpots, and hot springs are evidence of molten pockets of heat in an unstable crust. This geothermal resource area crosses park boundaries to the north into Montana. A controversy involving development of those thermal aquifers and the possible changes that development might produce inside the national park is being debated. [Photo by author.]

soil, and for manufacturing processes, aquaculture, and to heat swimming pools, among many uses.

For either electrical production or direct use, the end-use facilities must be fairly near the well head because of heat losses in piping the hot water or steam over distance. Of course, once electricity is generated, it can feed into the regional power grid for transmission to distant markets.

The first geothermal electrical generating station was built in Larderello, Italy, in 1904, and it has been in continuous operation since 1913 with an installed capacity of 360 MWe (megawatts electric). Today, geothermal applications are in use in 30 countries, through some 200 plants, producing an output of 8000 MWe and direct heat of 12,000 MWt (megawatts thermal). In Reykjavik, Iceland, the majority of space heating is geothermal. In the Paris Basin, France, some 25,000 residences are heated using geothermal direct 70°C (158°F) water. In the Philippines almost 30% of total electrical production is generated with geothermal energy. The top six countries for installed geothermal electrical

generation are the United States (2850 MWe), Philippines (1848 MWe), Italy (769 MWe), Mexico (743 MWe), Indonesia (590 MWe), and Japan (530 MWe). Worldwide potential is estimated to be in excess of 80,000 MWe and hundreds of thousands of MWt.

Other than the limiting factors of site-specific production and the locational aspect of having the power plant or end-use near the well, there are several other considerations with which to deal: (1) Although fuel costs are low, equipment costs and maintenance are high due to mineral deposition on piping and equipment. (2) Corrosion of metal alloys, principally by sulfide and chloride compounds, is an ongoing problem. (3) Depending on how the resource is maintained, loss of aquifer pressure and lowering groundwater levels can reduce production after a decade or so, requiring new wells to be sunk into the thermal reservoir. And (4), local pollution from water discharges, localized fog formation, and some winter icing may pose some difficulties.

An important mitigation that can extend thermal reservoir potential is

(continued)

Focus Study 11.1 (continued)

reinjection of water into the heat-bearing aquifer. This process can help dispose of surface wastewaters. In some locations reclaimed water (gray water) is used for reinjection. The Geysers in northern California, presently the world's largest-capacity geothermal electrical generation installation, uses this reinjection of wastewater.

The Geysers Geothermal Field (the name despite the lack of any geysers in the area) began production in 1960, increased to a peak in 1989 at 1967 Mwe, and lowered to a present capacity of 1070 MWe (Figure 2). By the mid-1990s some 600 wells had been drilled. The average well depth is 2500 m (8200 ft). Declining production relates to the expected problems of decreased yields. The Geysers are operating below the capacity of the field potential at the time of this writing.

Some 2800 MWe of geothermal power capacity is operating in California, Hawai'i, Nevada, and Utah. Figure 3 maps low-, medium-, and high-temperature known geothermal resource areas (KGRA) in the United States. About 300 cities are within 8 km (5 mi) of a KGRA. The Department of Energy has identified about 9000 potential development sites. In Canada, a proposed project at Mount Meager, B.C., will produce electricity; geothermal direct heats buildings at Carleton University, Ottawa; and some direct operations are in process in Nova Scotia.

As with other non-fossil fuel, potentially renewable energy resources, energy industry politics plays a role in the slow implementation of geothermal power. In times past, the price of geothermal steam was coupled with fossil fuel prices, so events completely unrelated to geothermal power could dictate its competitive stance. Recently, geothermal plant owners in California have had to sue utilities for timely payments during the energy distribution crisis of 2000–2001 in California. Apparently the potential of heat from Earth will have to wait a little longer to

FIGURE 2 The Geysers Geothermal Field, California.
The Geysers is the largest geothermal electric installation in the world. One of the 21 generating stations is Eagle Rock, Unit 11, with a 73 MWe capacity, in commercial operation since 1975. [Photo courtesy of Calpine, San Jose, California, http://www.geysers.com.]

FIGURE 3 Geothermal resources.
Geothermal resources in the conterminous United States. [Map courtesy of USGS, from W. A. Duffield, J. H. Sass, and M. L. Sorey, *Tapping the Earth's Natural Heat*, USGS Circular 1125, 1994, p. 35.]

become a more significant player. (For more information see http://www.eren.doe.gov/geothermal/, or http:// geothermal.marin.org/, or http://www.geothermal.org/, or http://www2.smu.edu/geothermal/.)

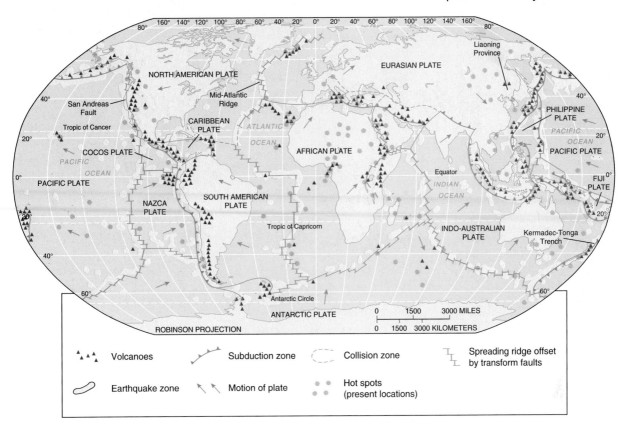

FIGURE 11.20 Earthquake and volcanic activity locations.
Earthquake and volcanic activity in relation to major tectonic plate boundaries and principal
hot spots. [Earthquake and volcano data from *Earthquakes*, B. A. Bolt, © 1988 W. H. Freeman
and Company; reprinted with permission. Hot spots adapted from USGA data.]

the Aleutian Trench, where they eventually will be subducted beneath the Eurasian plate. Perhaps some 80 million years hence, the Hawaiian Islands too will slowly disappear into the trench. Aloha!

The big island of Hawai'i, the newest, actually took less than 1 million years to build to its present stature. The island is a huge mound of lava, melded from several seafloor fissures through five volcanoes, rising from the seafloor 5800 m (19,000 ft) to the ocean surface. From sea level, its highest peak, Mauna Kea, rises to 4205 m (13,796 ft) elevation. This total height of almost 10,000 m (32,800 ft) represents the highest mountain on Earth, if measured from the seafloor. In all, the island of Hawai'i contains about 40,000 km³ of basalt, enough to cover the states of Massachusetts, Connecticut, and Rhode Island to a depth of 1 km!

The youngest island in the Hawaiian chain is still a seamount, a submarine mountain that does not reach the surface. It rises 3350 m (11,000 ft) from its base but is still 975 m (3200 ft) beneath the ocean surface. Even though this new island will not experience the tropical Sun for about 10,000 years, it is already named Lö'ihi (noted on the map).

Iceland is an example of an active hot spot sitting astride a mid-ocean ridge—visible on the different maps and images of the seafloor. It is an excellent example of a segment of mid-ocean ridge rising above sea level. This hot spot has generated enough material to form Iceland and eruptions continue from deep in the mantle. As a result, Iceland is still growing in area and volume. The youngest rocks are near the center of the island, with rock age increasing toward the eastern and western coasts. This is further evidence that, indeed, Earth is a dynamic planet!

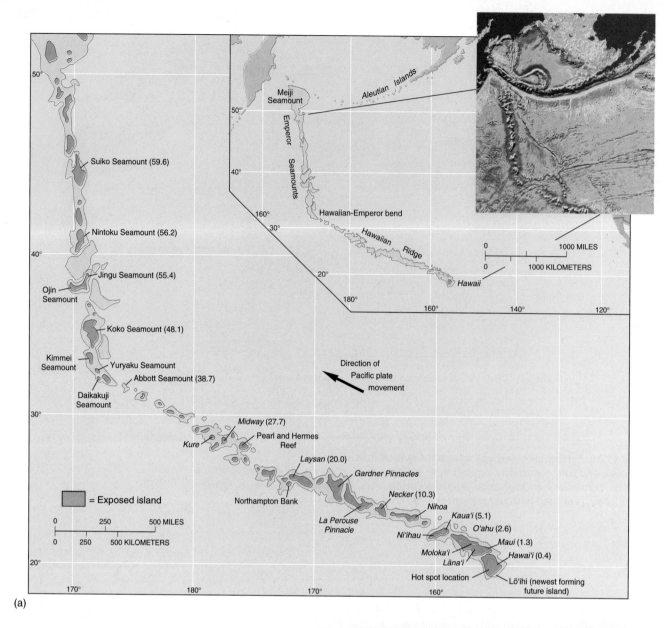

(a)

FIGURE 11.21 Hot spot tracks across the North Pacific.

Hawai'i and the linear volcanic chain of islands known as the Emperor Seamounts. (a) The islands and seamounts in the chain are progressively younger toward the southeast. Ages, in millions of years, are shown in parentheses. Note that Midway Island is 27.7 million years old, meaning that the site was over the plume 27.7 million years ago. (b) Lö'ihi is forming 975 m (3200 ft) beneath the Pacific Ocean; presently an undersea volcano (seamount), it will continue to grow into the next Hawaiian island. This view is from South Point, Hawai'i, Lö'ihi is east of this headland. [(a) After D. A. Clague, "Petrology and K–Ar (Potassium–Argon) Ages of Dredged Volcanic Rocks from the Western Hawaiian Ridge and the Southern Emperor Seamount Chain," *Geological Society of America Bulletin* 86 (1975): 991; inset from global gravity anomaly map image, Scripps Institution of Oceanography. All rights reserved. (b) photo by Bobbé Christopherson.]

(b)

Summary and Review—The Dynamic Planet

● *Distinguish* between the endogenic and exogenic systems, *determine* the driving force for each, and *explain* the pace at which these systems operate.

The Earth–atmosphere interface is where the **endogenic system** (internal), powered by heat energy from within the planet, interacts with the **exogenic system** (external), powered by insolation and influenced by gravity. These systems work together to produce Earth's diverse landscape. The **geologic time scale** is an effective device for organizing the vast span of geologic time. It depicts the sequence of Earth's events (relative time) and the approximate actual dates (absolute time).

The most fundamental principle of Earth science is **uniformitarianism**. Uniformitarianism assumes that the *same physical processes active in the environment today have been operating throughout geologic time.*

endogenic system (p. 321)
exogenic system (p. 321)
geologic time scale (p. 325)
uniformitarianism (p. 325)

1. To what extent is Earth's crust active at this time in its history?
2. Define the endogenic and the exogenic systems. Describe the driving forces that energize these systems.
3. How is the geologic time scale organized? What is the basis for the time scale in relative and absolute terms? What era, period, and epoch are we living in today?
4. Contrast uniformitarianism and catastrophism as models for Earth's development.

● *Diagram* Earth's interior in cross section and *describe* each distinct layer.

We have learned about Earth's interior from indirect evidence—the way its various layers transmit **seismic waves**. The **core** is differentiated into an inner core and an outer core—divided by a transition zone. Earth's magnetic field is generated almost entirely within the outer core. Polarity reversals in Earth's magnetism are recorded in cooling magma that contains iron minerals. The patterns of **geomagnetic reversal** frozen in rock helps scientists piece together the story of Earth's mobile crust.

Beyond Earth's core lies the **mantle**, differentiated into lower mantle and upper mantle. It experiences a gradual temperature increase with depth and a stiffening due to increased pressures. The upper mantle is divided into three fairly distinct layers. The uppermost mantle, along with the crust, makes up the *lithosphere*. Below the lithosphere is the **asthenosphere**, or *plastic layer*. It contains pockets of increased heat from radioactive decay and is susceptible to slow convective currents in these hotter materials. An important internal boundary between the **crust** and the high-velocity portion of the uppermost mantle is the **Mohorovičić discontinuity**, or **Moho**. *Continental crust* is basically **granite**; it is crystalline and high in silica, aluminum, potassium, calcium, and sodium. *Oceanic crust* is **basalt**; it is granular and high in silica, magnesium, and iron. The principles of buoyancy and balance were combined in the 1800s into the important principle of **isostasy**. Isostasy explains certain vertical movements of Earth's crust.

seismic waves (p. 326)
core (p. 326)
geomagnetic reversal (p. 326)
mantle (p. 328)
asthenosphere (p. 328)
crust (p. 328)
Mohorovičić discontinuity (Moho) (p. 328)
granite (p. 329)
basalt (p. 329)
isostasy (p. 329)

5. Make a simple sketch of Earth's interior, label each layer, and list the physical characteristics, temperature, composition, and range of size of each on your drawing.
6. What is the present thinking on how Earth generates its magnetic field? Is this field constant, or does it change? Explain the implications of your answer.
7. Describe the asthenosphere. Why is it also known as the plastic layer? What are the consequences of its convection currents?
8. What is a discontinuity? Describe the principal discontinuities within Earth.
9. Define isostasy and isostatic rebound, and explain the crustal equilibrium concept.
10. Diagram the uppermost mantle and crust. Label the density of the layers in grams per cubic centimeter. What two types of crust were described in the text in terms of rock composition?

● *Illustrate* the geologic cycle and *relate* the rock cycle and rock types to endogenic and exogenic processes.

The **geologic cycle** is a model of the internal and external interactions that shape the crust. A **mineral** is an inorganic natural compound having a specific chemical formula and possessing a crystalline structure. A **rock** is an assemblage of minerals bound together (such as granite, a rock containing three minerals), or it may be a mass of a single mineral (such as rock salt). Thousands of different rocks have been identified.

The geologic cycle comprises three cycles: the *hydrologic cycle*, the *tectonic cycle*, and the **rock cycle**. The rock cycle describes the three principal rock-forming processes and the rocks they produce. **Igneous rocks** form from **magma**, which is molten rock beneath the surface. Magma is fluid, highly gaseous, and under tremendous pressure. It either *intrudes* into crustal rocks, cools, and hardens, or it *extrudes* onto the surface as **lava**. Intrusive igneous rock that cools slowly in the crust forms a **pluton**. The largest pluton form is a **batholith**.

The cementation, compaction, and hardening of sediments into **sedimentary rocks** is called **lithification**. These layered strata form important records of past ages.

Stratigraphy is the study of the sequence (superposition), thickness, and spatial distribution of strata that yield clues to the age and origin of the rocks.

Clastic sedimentary rocks are derived from the bits and pieces of weathered rocks. Chemical sedimentary rocks are not formed from physical pieces of broken rock, but instead are dissolved minerals, transported in solution, and chemically precipitated out of solution (they are essentially nonclastic). The most common chemical sedimentary rock is **limestone**, which is lithified calcium carbonate, $CaCO_3$.

Any rock, either igneous or sedimentary, may be transformed into a **metamorphic rock**, by going through profound physical or chemical changes under pressure and increased temperature.

geologic cycle (p. 330)
mineral (p. 332)
rock (p. 332)
rock cycle (p. 332)
igneous rock (p. 332)
magma (p. 332)
lava (p. 332)
pluton (p. 333)
batholith (p. 333)
sedimentary rock (p. 335)
lithification (p. 335)
stratigraphy (p. 335)
limestone (p. 338)
metamorphic rock (p. 339)

11. Illustrate the geologic cycle and define each component: rock cycle, tectonic cycle, and hydrologic cycle.

12. What is a mineral? A mineral family? Name the most common minerals on Earth. What is a rock?

13. Describe igneous processes. What is the difference between intrusive and extrusive types of igneous rocks?

14. Characterize felsic and mafic minerals. Give examples of both coarse- and fine-grained textures.

15. Briefly describe sedimentary processes and lithification. Describe the sources and particle sizes of sedimentary rocks.

16. What is metamorphism, and how are metamorphic rocks produced? Name some original parent rocks and their metamorphic equivalents.

● *Describe* **Pangaea and its breakup and *relate* several physical proofs that crustal drifting is continuing today.**

The present configuration of the ocean basins and continents are the result of *tectonic processes* involving Earth's interior dynamics and crust. Alfred Wegener coined the phrase

continental drift to describe his idea that the crust is moved by vast forces within the planet. **Pangaea** was the name he gave to a single assemblage of continental crust some 225 million years ago that subsequently broke apart. Earth's crust is fractured into huge slabs or plates, each moving in response to flowing currents in the mantle. The all-encompassing theory of **plate tectonics** includes **sea-floor spreading** along **mid-ocean ridges** and denser oceanic crust diving beneath lighter continental crust along **subduction zones**.

continental drift (p. 340)
Pangaea (p. 340)
plate tectonics (p. 340)
sea-floor spreading (p. 340)
mid-ocean ridge (p. 340)
subduction zone (p. 342)

17. Briefly review the history of the theory of continental drift, sea-floor spreading, and the all-inclusive plate tectonics theory. What was Alfred Wegener's role?

18. Define upwelling, and describe related features on the ocean floor. Define subduction and explain the process.

19. What was Pangaea? What happened to it during the past 225 million years?

20. Characterize the three types of plate boundaries and the actions associated with each type.

● *Portray* **the pattern of Earth's major plates and *relate* this pattern to the occurrence of earthquakes, volcanic activity, and hot spots.**

Occurrences of often damaging earthquakes and volcanoes are correlated with plate boundaries. Three types of plate boundaries form: divergent, convergent, and transform. Along the offset portions of mid-ocean ridges, horizontal motions produce **transform faults**.

As many as 50 to 100 **hot spots** exist across Earth's surface, where tall plumes of magma, anchored in the lower mantle, remain fixed as drifting plates are penetrated by eruptions. **Geothermal energy** literally refers to heat from Earth's interior, whereas *geothermal power* relates to specific applied strategies of geothermal electric or geothermal direct applications.

transform faults (p. 347)
hot spots (p. 347)
geothermal energy (p. 349)

21. What is the relation between plate boundaries and volcanic and earthquake activity?

22. What is the nature of motion along a transform fault? Name a famous example of such a fault.

Floor of the Oceans, 1975, by Bruce C. Heezen and Marie Tharp.
The scarred ocean floor is clearly visible: sea-floor spreading centers marked by oceanic ridges that stretch over 64,000 km (40,000 mi), subduction zones indicated by deep oceanic trenches, and transform faults slicing across oceanic ridges.

Follow the East Pacific rise (an ocean ridge and spreading center) northward as it trends beneath the west coast of the North American plate, disappearing under earthquake-prone California. The continents, offshore submerged continental shelves, and the expanse of the sediment-covered abyssal plain are all identifiable on this illustration.

On the floor of the Indian Ocean, you can see the wide track along which the India plate traveled northward to its collision with the Eurasian plate. Vast deposits of sediment cover the Indian Ocean floor, south of the Ganges River to the east of India. Sediments derived from the Himalayan Range blanket the floor of the Bay of Bengal (south of Bangladesh) to a depth of 20 km (12.4 mi). These sediments result from centuries of soil erosion in the land of the monsoons.

Isostasy can be examined in action in central and west-central Greenland, where the weight of the ice sheet has depressed portions of the land far below sea level—now visible with the ice artificially removed by the cartographer (see the extreme upper-left corner of the map). In contrast, the region around Canada's Hudson Bay became ice-free about 8000 years ago and has isostatically rebounded 300 m (1100 ft).

In the area of the Hawaiian Islands, the hot-spot track is marked by a chain of islands and seamounts that you can follow along the Pacific plate from Hawai'i to the Aleutians. Subduction zones south and east of Alaska and Japan, as well as along the western coast of South and Central America, are visible as dark trenches. Along the left margin of the map, you can find Iceland's position on the Mid-Atlantic Ridge. [© 1980 by Marie Tharp. Reproduced by permission of Marie Tharp.]

NetWork

The *Geosystems* Home Page provides on-line resources for this chapter on the World Wide Web. To begin: Once on the Home Page, click on this textbook, scroll the Table of Contents menu, select this chapter, and click "Begin." You will find self-tests that are graded, review exercises, specific updates for items in the chapter, and many links to interesting related pathways on the Internet under "Destinations." *Geosystems* is at **http://www.prenhall.com/christopherson**.

Critical Thinking

A. Using the maps in this chapter, determine your present location relative to Earth's crustal plates. Now, using Figure 11.16b, approximately identify where your present location was 225 million years ago; express it in a rough estimate using the equator and the longitudes noted on the map.

B. Relative to the motion of the Pacific plate shown in Figure 11.21 (note the scale in lower-left corner), the island of Midway formed 27.7 million years ago over the hot spot that is active under the southeast coast of the big island of Hawai'i today. Given the scale of the map, roughly determine the average annual speed of the Pacific plate in cm per year for Midway to have traveled this distance.

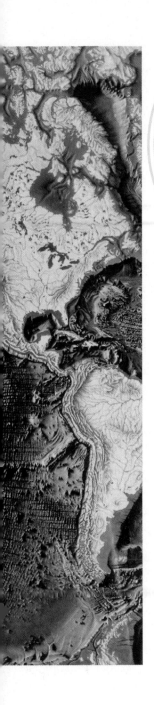

12

Tectonics, Earthquakes, and Volcanism

Key Learning Concepts

After reading the chapter, you should be able to:

- *Describe* first, second, and third orders of relief and *relate* examples of each from Earth's major topographic regions.
- *Describe* the several origins of continental crust and *define* displaced terranes.
- *Explain* compressional processes and folding; *describe* four principal types of faults and their characteristic landforms.
- *Relate* the three types of plate collisions associated with orogenesis and *identify* specific examples of each.
- *Explain* the nature of earthquakes, their measurement, and the nature of faulting.
- *Distinguish* between an effusive and an explosive volcanic eruption and *describe* related landforms, using specific examples.

Earth's physical systems move to the front pages whenever an earthquake strikes or a volcanic eruption threatens a city. Earthquakes may strike at any time on Earth, but many are in sparsely populated areas. In 1999 more than 30,000 perished in quakes in Turkey, Greece, Taiwan, and Mexico (Figure 12.1). Fifteen earthquakes greater than magnitude 7.0 struck in 2001, killing 21,400 people, compared to only 231 quake-related deaths in 2000—20,103 deaths were in Gujarat state, eastern India, alone, with 166,000 injuries, leaving 600,000 homeless. In 2002, lava lake eruptions from the composite volcano Nyiragongo caused devastation to nearby Goma, Congo—lava flows divided this town, on the East African Rift. This temporal variability in earthquake and volcanic occurrences only adds to the fear felt by many.

To varying degrees, in each country, poor planning, shoddy building standards, and weak emergency preparedness worsen the impact of such events. Well-prepared

(a)

(b)

FIGURE 12.1 Earthquakes strike Turkey and Taiwan.
(a) Earthquakes in Turkey struck Izmet (August 17, 1999)
and Düzce (November 12, 1999), bringing into question
building standards, practices, and political inaction, as
20,000 people lost their lives. (b) The damage in Chi-Chi,
Nantou County, Taiwan, which killed more than 2000
people and left 100,000 homeless, was caused by an earthquake on September 21, 1999. Because of extensive instru-
mentation across the landscape and in buildings, quality
data were collected for analysis. This preparedness permitted location and magnitude determinations in less than 2
minutes. [Photos by (a) AP Photo/Enric Marti; (b) Reuters/
Simon Kwang/Archive Photos.]

countries are not exempted. An unexpected earthquake devastated the region surrounding Kobe, Japan, in 1995. The
magnitude 6.9 quake killed 5500 people and caused more
than $100 billion in damage—the worst in Japan since
1923. Volcanic eruptions pose similar threats to life and
property along plate boundaries and hot spot locations.

In this chapter: We examine processes that construct
Earth's surface and create world structural regions. Tectonic processes deform, recycle, and reshape Earth's crust.
These processes occur sometimes in dramatic episodes but
most often in slow, deliberate motions that build the landscape. Continental crust has been forming throughout most
of Earth's 4.6-billion-year existence. The arrangement of
continents and oceans, the origin of mountain ranges, the
topography of the land and the seafloor, and the locations
of earthquake and volcanic activity are all evidence of our
dynamic Earth. Principal seismic and volcanic zones occur
along plate boundaries and hot spot locations, thus linking
plate tectonics to the devastation of major earthquakes and
local threats of major volcanic eruptions and their attendant potential global climatic impact.

We begin our look at tectonics and volcanism on the
ocean floor, hidden from direct view. The illustration that
opens this chapter is a striking representation of Earth with
its blanket of water removed, as revealed to us through
decades of direct and indirect observation. Careful examination of this portrait gives us a helpful review of the concepts learned in the previous chapter, laying the foundation
for this and subsequent chapters. Be sure to take the quick
tour presented with this chapter-opening map. Try to correlate this sea-floor illustration with the maps of crustal
plates and plate boundaries shown in Figures 11.17 and
the gravity anomaly image in Figure 11.18.

Earth's Surface Relief Features

Relief refers to vertical elevation differences in the landscape. Examples include the low relief of Nebraska and
Saskatchewan, medium relief in foothills along mountain
ranges, and high relief in the Rockies and Himalayas. The
undulating form of Earth's surface, including its relief, is
called **topography**, portrayed so effectively on topographic maps—the lay of the land.

The relief and topography of Earth's landforms played
a vital role in human history: High mountain passes both
protected and isolated societies, ridges and valleys dictated transportation routes, and vast plains necessitated developing faster methods of communication and travel.

Earth's topography has stimulated human invention and spurred adaptation.

Crustal Orders of Relief

Today's computer capabilities and tools such as the global positioning system (GPS) for determining location and elevation have enhanced our understanding of Earth's relief and topography. Scientists put elevation data in digital form; the data then are available for computer manipulation and display. The resulting *digital elevation models* (*DEMs*) aid in the scientific analysis of topography, area–altitude distributions, slopes, and local stream-drainage characteristics.

The U.S. Geological Survey has prepared a digitized shaded-relief map of the United States. The entire map is made up of 12 million spot elevations, with less than a kilometer between any two. A Web site prepared by the USGS, "A Tapestry of Time and Terrain," uses this data, colored geologic regions, and detailed topography in a useful virtual map: **http://tapestry.usgs.gov.**

For convenience of description, geographers group the landscape's topography into three *orders of relief.* These orders classify landscapes by scale, from vast ocean basins and continents down to local hills and valleys.

First Order of Relief   The first order of relief is the coarsest level of landforms, including huge continental platforms

and ocean basins. **Continental platforms** are the masses of crust that reside above or near sea level, including the undersea continental shelves along the coastlines. The **ocean basins** are entirely below sea level and are portrayed in the chapter-opening illustration. Approximately 71% of Earth is covered by water.

Second Order of Relief   The second order of relief is the intermediate level of landforms, for both continental and ocean-basin features. Continental features in the second order of relief include mountain masses, plains, and lowlands. A few examples are the Alps, Canadian and American Rockies, west Siberian lowland, and Tibetan Plateau. The great rock cores ("shields") that form the heart of each continental mass are of this second order. In the ocean basins, second order of relief includes continental rises, slopes, abyssal plains, mid-ocean ridges, submarine canyons, and subduction trenches—all visible in the sea-floor illustration that opens this chapter.

Third Order of Relief   The third and most-detailed order of relief includes individual mountains, cliffs, valleys, hills, and other landforms of smaller scale. These features are identifiable as local landscapes.

Hypsometry   Figure 12.2 is a *hypsographic curve* (from the Greek *hypsos* meaning "height") that shows the distribution

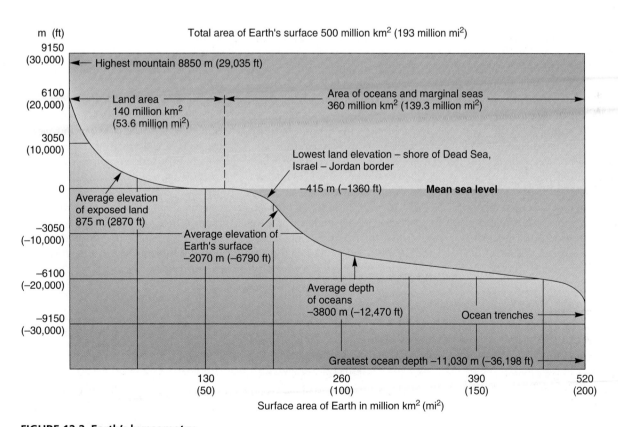

FIGURE 12.2 Earth's hypsometry.
Hypsographic curve of Earth's surface, charting area and elevation as related to mean sea level. From the highest point above sea level (Mount Everest) to the deepest oceanic trench (Mariana Trench), Earth's overall relief is almost 20 km (12.5 mi). The new height given for Mount Everest was announced in November 1999.

News Report 12.1

Mount Everest at New Heights

The announcement came November 11, 1999, in Washington, DC, at the opening reception of the 87th annual meeting of the American Alpine Club. The press conference was held at National Geographic Society headquarters. Mount Everest has a new revised elevation determined by direct Global Positioning System (GPS) placement on the mountain's icy summit!

In May 1999, mountaineers Pete Athans and Bill Crouse reached the summit with five Sherpas. The climbers operated Global Positioning System satellite equipment on the very top of Mount Everest and determined the precise height of the world's tallest mountain. In Chapter 1, the photograph in News Report 1.1 shows a GPS unit placed by climber Wally Berg 18 m below the summit in 1998. This site is still Earth's highest benchmark with a permanent metal marker installed in rock. Other GPS instruments in the region permit differential calibration for exact measurement (Figure 1). This is a continuation of a measurement effort begun in 1995 by Bradford Washburn, renowned mountain photographer/explorer and honorary director of Boston's Museum of Science.

Washburn announced the new measurement of 8,850 meters (29,035 feet). Everest's new elevation is close to the previous official measure of 8,848 meters (29,028 feet) set in 1954 by the Survey of India after picking the average measurements from 12 different survey points around the mountain. The margin of error in this new GPS

FIGURE 1 GPS installation measures Mount Everest.
A global positioning system (GPS) installation near Namche Bazar in the Khumbu (Everest) region of Nepal. The station is at latitude 27.8° N, longitude 86.7° E, at 3523 m (11,558 ft). The summit of Mount Everest is 30 km (18.6 mi) distant—visible above the right side of the weather dome. A network of such GPS installations correlated with the unit the climbers took to Everest's summit. [Photo by Charles Corfield, science manager to the 1998 and 1999 expeditions.]

measurement exceeds by many times the 1954 effort.

The National Geographic Society press release stated: "This latest measurement stemmed from Washburn's desire to use lightweight GPS receivers, along with lithium batteries that work in severe temperature conditions, to establish the highest bedrock survey station. Then, running two receivers simultaneously, one at the South Col and the other at the summit, an extremely accurate altitude was established for the top of the mountain." Wally Berg's preliminary place-

ment of a GPS unit was an important step in meeting this goal.

Washburn stated, "the reading of 29,035 feet (8,850 meters) showed no measurable change in the height of Everest calculated since GPS observations began four years ago. But from these GPS readings it appears that the horizontal position of Everest seems to be moving steadily and slightly northeastward; between 3 and 6 mm a year (up to 0.25 in. a year)." The mountain range is being driven further into Asia by plate tectonics—the continuing collision of Indian and Asian landmasses.

of Earth's surface by area and elevation in relation to sea level. Relative to Earth's diameter of 12,756 km (7926 mi), the surface is of low relief, only about 20 km (12.5 mi) from highest peak to lowest oceanic trench. For perspective, Mount Everest is 8.8 km (5.5 mi) above sea level and the Mariana Trench is 11 km (6.8 mi) below sea level. Note the new GPS measurement for the summit of Mount Everest announced in November 1999—8850 m

(29,035 ft). See News Report 12.1 for more on this geographic milestone.

The average elevation of Earth's solid surface is actually under water: −2070 m (−6790 ft) below mean sea level. The average elevation for exposed land is only +875 m (+2870 ft). For the ocean depths, average elevation is −3800 m (−12,470 ft). From this description you can see that, on the average, the oceans are much deeper than con-

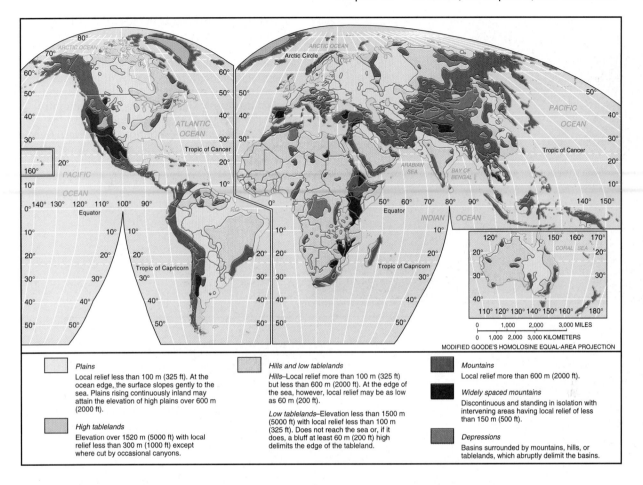

FIGURE 12.3 Earth's topographic regions.
Earth's topography is characterized as plains, high and low tablelands, hills, mountains, and depressions. [After R. E. Murphy, "Landforms of the World," *Annals of the Association of American Geographers* 58, no. 1 (March 1968). Adapted by permission.]

tinental regions are high. Overall, the underwater ocean basins, ocean floor, and submarine mountain ranges form Earth's largest "landscape."

Earth's Topographic Regions

The three orders of relief can be further generalized into six topographic regions: plains, high tablelands, hills and low tablelands, mountains, widely spaced mountains, and depressions (Figure 12.3). An arbitrary elevation or descriptive limit that is in common use defines each type of topography (see the figure's legend).

Four of the continents possess extensive *plains*, areas with local relief of less than 100 m (325 ft) and slope angles of 5° or less. Some plains have high elevations of more than 600 m (2000 ft); in the United States, the high plains attain elevations above 1220 m (4000 ft). The Colorado Plateau, Greenland, and Antarctica are notable *high tablelands*, with elevations exceeding 1520 m (5000 ft). *Hills* and *low tablelands* dominate Africa.

Mountain ranges, characterized by local relief exceeding 600 m (2000 ft), occur on each continent. Earth's relief

and topography are undergoing constant change as a result of processes that form crust.

Crustal Formation Processes

How did Earth's continental crust form? What gave rise to the three orders of relief just discussed? Ultimately, the answers to both questions are the combined effects of tectonic activity, which is driven by our planet's internal energy, and the exogenic processes of weathering and erosion, powered by the Sun through the actions of air, water, ice, and waves.

Tectonic activity generally is slow, requiring millions of years. Endogenic (internal) processes result in gradual uplift and new landforms, with major mountain-building occurring along plate boundaries. These uplifted crustal regions are quite varied, but we think of them in three general categories, all discussed in this chapter:

- Residual mountains and stable continental cores, formed from inactive remnants of ancient tectonic activity;

FIGURE 12.4 Continental shields.
(a) Portions of major continental shields that have been exposed by erosion. Adjacent portions of these shields remain covered. (b) Canadian shield landscape in central Labrador interior, stable for hundreds of millions of years, stripped by past glaciations, and marked by intrusive igneous dikes (magmatic intrusions). [(a) After R. E. Murphy, "Landforms of the World," *Annals of the Association of American Geographers* 58, no. 1 (March 1968), adapted by permission. (b) Photo by John Eastcott/Yva Momatiuk/The Image Works.]

(b)

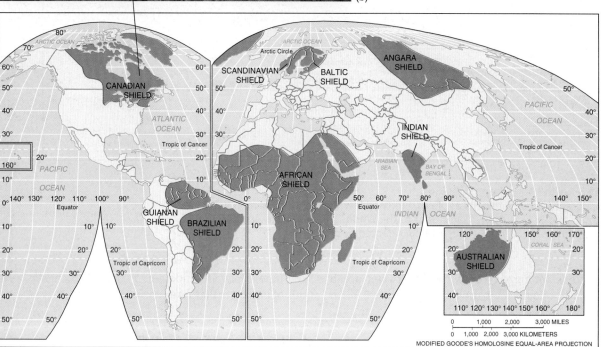

(a)

- Tectonic mountains and landforms, produced by active folding, faulting, and crustal movements;
- Volcanic features, formed by the surface accumulation of molten rock from eruptions of subsurface materials.

Thus, several distinct processes operate in concert to produce the continental crust we see around us.

Continental Shields

All continents have a nucleus of ancient crystalline rock on which the continent "grows" with the addition of crustal fragments and sediments. This nucleus is the *craton*, or heartland region, of the continental crust. Cratons generally have been eroded to a low elevation and relief. Most date to the Archean and exceed 2 billion years of age. The lack of basaltic components in these cratons offers a clue to their stability. The lithosphere in cratonic regions (crust and lithospheric uppermost mantle) is thicker than the lithosphere beneath younger portions of continents and oceanic crust.

A **continental shield** is a region where a craton is exposed at the surface. Figure 12.4 shows the principal areas of exposed shields and a photo of the Canadian shield. Layers of sedimentary rock surround these shields and appear quite stable over time. An example of such a stable *platform* is the region that stretches from east of the Rockies to the Appalachians and northward into central and eastern Canada.

Building Continental Crust and Terranes

The formation of continental crust is very complex and takes hundreds of millions of years. It involves the entire sequence of sea-floor spreading and formation of oceanic crust, its later subduction and remelting, and its subsequent rise as new magma, all summarized in Figure 12.5.

To understand this process, study Figure 12.5 (and Figure 11.13). Begin with the magma that originates in the asthenosphere and wells up along the mid-ocean ridges. Basaltic magma is formed from minerals in the upper mantle that are rich in iron and magnesium. Such magma has less than 50% silica and has a low-viscosity (thin) texture—it tends to flow. This mafic material rises to erupt at spreading centers and cools to form new basaltic seafloor, which spreads outward to collide with continental crust along its far edges. This denser oceanic crust plunges beneath the lighter continental crust, into the mantle, where it remelts. The new magma then rises and cools, forming more continental crust, in the form of intrusive granitic igneous rock.

As the subducting oceanic plate works its way under a continental plate, it takes with it trapped seawater and sediment from eroded continental crust. The remelting incorporates the seawater, sediments, and surrounding crust into the mixture. As a result, the magma, generally called a *melt*, that migrates upward from a subducted plate contains 50%–75% silica and aluminum (called andesitic or silicic depending on silica content). The melt has a high-viscosity (thick) texture—it tends to block and plug conduits to the surface.

Bodies of such silica-rich magma may reach the surface in explosive volcanic eruptions, or they may stop short and become subsurface intrusive bodies in the crust, cooling slowly to form granitic crystalline plutons such as batholiths (see Figures 11.7 and 11.8a). Note that this composition is quite different from the magma that rises directly from the asthenosphere at spreading centers. In these processes of crustal formation, you can literally follow the cycling of materials in the tectonic cycle.

Each of Earth's major lithospheric plates actually is a collage of many crustal pieces acquired from a variety of sources. Crustal fragments of ocean floor, curving chains (or arcs) of volcanic islands, and other pieces of continental crust all have been forced against the edges of continental shields and platforms. These slowly migrating crustal pieces, which have become attached or accreted to the plates, are called **terranes** (not to be confused with "terrain," which refers to the topography of a tract of land).

These displaced terranes, sometimes called *microplate* or *foreign terranes*, have histories different from those of the continents that capture them. They are usually framed by fractures and differ in rock composition and structure from their new continental homes.

In the region surrounding the Pacific, accreted terranes are particularly prevalent. At least 25% of the growth of western North America can be attributed to the accretion of terranes since the early Jurassic Period (190 million years ago). A good example is the Wrangell Mountains, which lie just east of Prince William Sound and the

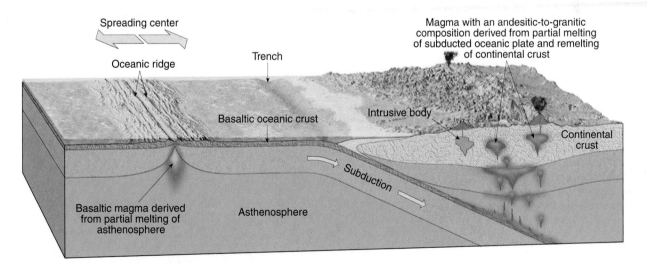

FIGURE 12.5 Crustal formation.
Material from the asthenosphere upwells along sea-floor spreading centers. Basaltic ocean floor is subducted beneath lighter continental crust, where it melts, along with its cargo of sediments, water, and minerals. This melting generates magma, which makes its way up through the crust to form igneous intrusions and extrusive eruptions. [After E. J. Tarbuck and F. K. Lutgens, *Earth, An Introduction to Physical Geology*, 5th ed., © Prentice Hall, 1996, Figure 20.20, p. 502.]

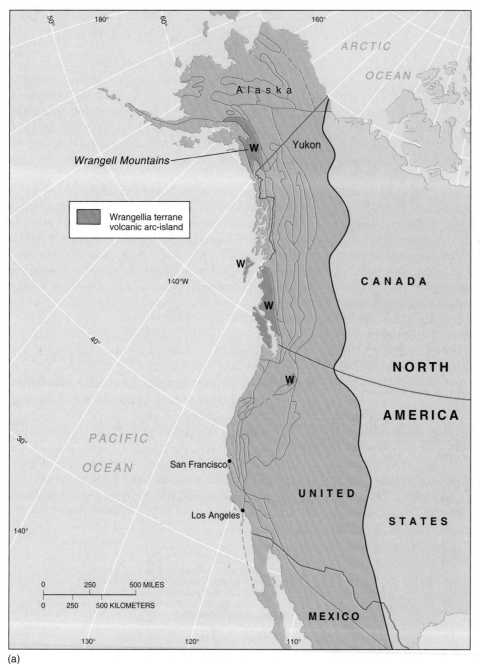

(a)

(b) Wrangell Mountains, Alaska

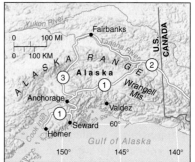

FIGURE 12.6 North American terranes.
(a) Wrangellia terranes, highlighted among the other terranes along the western margin of
North America, occur in four segments. (b) Snow-covered Wrangell Mountains north of the
Chugach Mountains in this November 7, 2001, image, in east-central Alaska and across the
Canadian border. (Note Mount McKinley (Denali) at 6194 m, 20,322 ft, in upper-left part of
the image, north of Cook Inlet, highest elevation in North America.) [(a) Based on data from
U.S. Geological Survey. (b) *Terra* image from MODIS sensor, courtesy of MODIS Land Rapid
Response Team, NASA/GSFC.]

city of Valdez, Alaska. The *Wrangellia terranes*—a former
volcanic island arc and associated marine sediments from
near the equator—migrated approximately 10,000 km
(6200 mi) to form the Wrangell Mountains and three other
distinct formations along the western margin of the continent (Figure 12.6).

The Appalachian Mountains, extending from Alabama to the Maritime Provinces of Canada, possess bits of
land once attached to ancient Europe, Africa, South America, Antarctica, and various oceanic islands. The discovery
of terranes, made only in the 1980s, demonstrates one of
the ways continents are assembled.

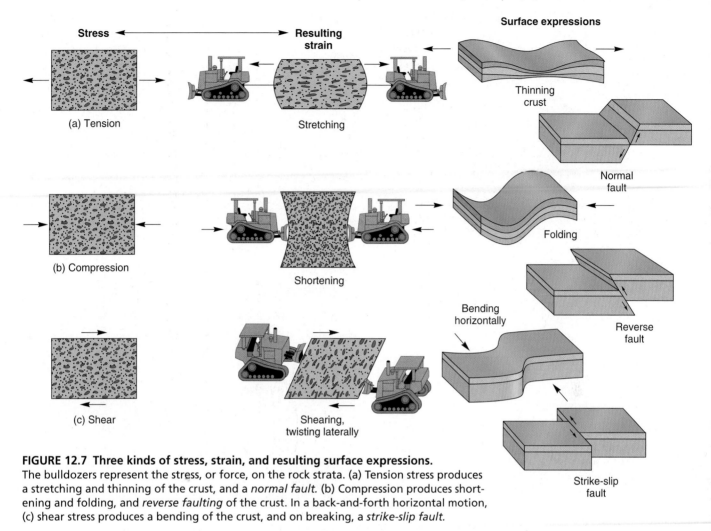

FIGURE 12.7 Three kinds of stress, strain, and resulting surface expressions.
The bulldozers represent the stress, or force, on the rock strata. (a) Tension stress produces
a stretching and thinning of the crust, and a *normal fault.* (b) Compression produces shortening and folding, and *reverse faulting* of the crust. In a back-and-forth horizontal motion,
(c) shear stress produces a bending of the crust, and on breaking, a *strike-slip fault.*

Crustal Deformation Processes

Rocks, whether igneous, sedimentary, or metamorphic, are
subjected to powerful stress by tectonic forces, gravity, and
the weight of overlying rocks. There are three types of
stress: *tension* (stretching), *compression* (shortening), and
shear (twisting or tearing), as shown in Figure 12.7.

Strain is how rocks respond to stress. Strain is expressed in rocks by *folding* (bending) or *faulting* (breaking).
Think of stress as a force and resulting strain as the deformation in the rock. Whether a rock bends or breaks depends on several factors, including composition and how
much pressure is on the rock. An important quality is
whether the rock is *brittle* or *ductile.* The patterns created
by these processes are clearly visible in the landforms we see
today, especially in mountain areas. Figure 12.7 illustrates
each type of stress and its resulting strain and the surface expressions that develop.

Folding and Broad Warping

When rock strata that are layered and flat are subjected to
compressional forces, they become deformed (Figure 12.8).

Convergent plate boundaries intensely compress rocks, de-
forming them in a process known as **folding.** As an analogy, if we take sections of thick fabric, stack them flat on a
table, and then slowly push on opposite ends of the stack,
the cloth layers will bend and rumple into folds similar to
those shown in Figure 12.8a.

If we then draw a line down the center axis of a resulting ridge, and a line down the center of a resulting
trough, we are able to see how the names of the folds are
assigned. Along the *ridge* of a fold, layers *slope downward
away from the axis,* which is called an **anticline.** In the
trough of a fold, however, layers *slope downward toward the
axis,* called a **syncline.**

If the axis of either type of fold is not "level" (horizontal, or parallel to Earth's surface), the fold–axis layers
then *plunge* (dip down) at an angle. Knowledge of how folds
are angled to Earth's surface and where they are located is
important for the petroleum industry. For example, petroleum geologists know that oil and natural gas collect in the
upper portions of anticlinal folds in permeable rock layers
such as sandstone. This is the "anticlinal theory of petroleum accumulation" attributed to I. C. White, founder of
the West Virginia Geological Survey.

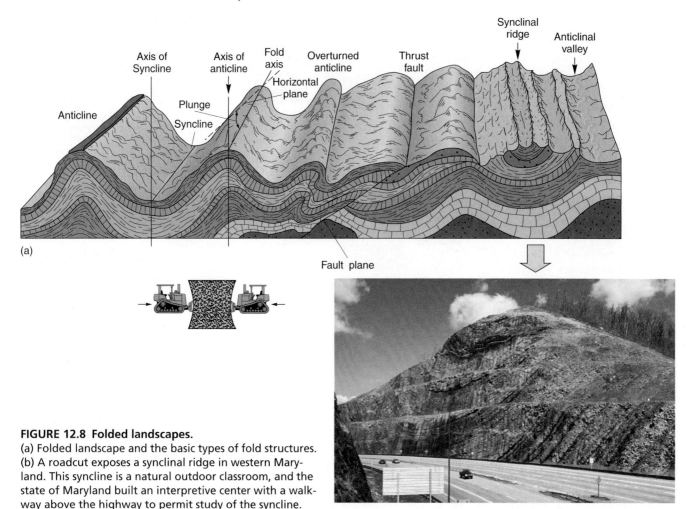

FIGURE 12.8 Folded landscapes.
(a) Folded landscape and the basic types of fold structures.
(b) A roadcut exposes a synclinal ridge in western Maryland. This syncline is a natural outdoor classroom, and the state of Maryland built an interpretive center with a walkway above the highway to permit study of the syncline.
[Photo by Mike Boroff/Photri-Microstock.]

Figure 12.8a further illustrates various folds that have been weathered and reduced by exogenic processes:

- A residual "synclinal ridge" may form within a syncline, because different rock strata offer greater resistance to weathering processes. An actual synclinal ridge is exposed dramatically in an interstate highway roadcut in Figure 12.8b.
- Compressional forces often push folds far enough that they actually overturn upon their own strata ("overturned anticline").
- Further stress eventually fractures the rock strata along distinct lines, and some overturned folds are thrust upward, causing a considerable shortening of the original strata ("thrust fault").

The Canadian Rocky Mountains, the Appalachian Mountains, and areas of the Middle East illustrate well the complexity of folded landscapes. Satellites allow us to view many of these structures from an orbital perspective, as in Figure 12.9. Just north of the Persian Gulf are the Zagros Mountains of Iran. This area was a dispersed terrane that

separated from the Eurasian plate. However, the collision produced by the northward push of the Arabian block is now shoving this terrane back into Eurasia and forming an active margin known as the Zagros crush zone, a zone of continuing collision more than 400 km (250 mi) wide. In the satellite image, anticlines form the parallel ridges; active weathering and erosion processes are exposing the underlying strata.

In addition to the rumpling of rock strata just discussed, broad warping actions also affect Earth's continental crust. These actions produce similar up-and-down bending of strata, but the bends are far greater in extent than those produced by folding. Warping forces include mantle convection, isostatic adjustment such as the weight of previous ice loads across northern Canada, and crustal swelling above an underlying hot spot. Warping features can be small, individual, foldlike structures called basins and domes (Figure 12.10a, b). They can also range up to regional features the size of the Ozark Mountain complex in Arkansas and Missouri, the Colorado Plateau in the West, the Richat dome in Mauritania, or the Black Hills of South Dakota (Figure 12.10 c, d).

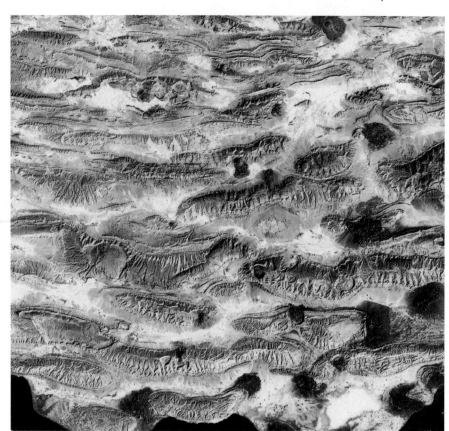

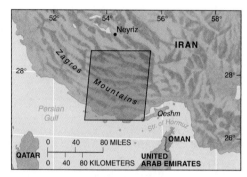

FIGURE 12.9 Folding in the Zagros crush zone, Iran.
The Zagros Mountains are a product of the Zagros crush zone between the Arabian and Eurasian plates. This area was a dispersed terrane (migrating crustal piece) that separated from the Eurasian plate. However, the collision produced by the northward push of the Arabian block is now shoving this terrane back into Eurasia and forming the folded mountains shown. [NASA image.]

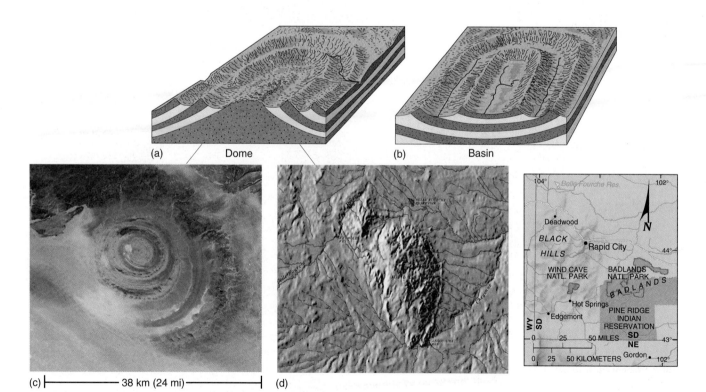

(a)   Dome   (b)   Basin

(c) |—————— 38 km (24 mi) ——————| (d)

FIGURE 12.10 Domes and basins.
(a) An upwarped dome. (b) A structural basin. (c) The Richat dome structure of Mauritania.
(d) The Black Hills of South Dakota is a dome structure, shown here on a digitized relief map.
[(c) NASA/Mark Marten/Science Source/Photo Researchers, Inc.; (d) from USGS digital terrain map I-2206, 1992.]

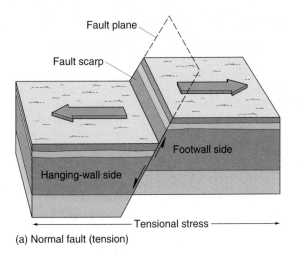

Fault plane

Fault scarp

Footwall side

Hanging-wall side

Tensional stress

(a) Normal fault (tension)

(a)

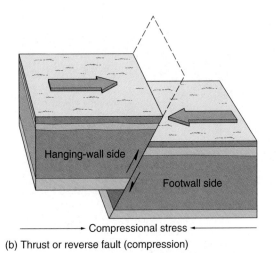

Hanging-wall side

Footwall side

Compressional stress

(b) Thrust or reverse fault (compression)

(b)

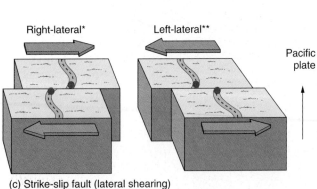

Right-lateral*   Left-lateral**

(c) Strike-slip fault (lateral shearing)

* Viewed from either dot on each road, movement to opposite side is *to the right*.
** Viewed from either dot on each road, movement to opposite side is *to the left*.

Pacific plate

North American plate

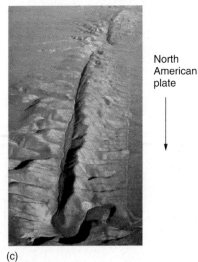

(c)

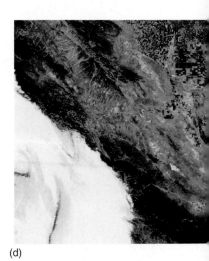

(d)

FIGURE 12.11 Types of faults.
(a) A normal fault produced by tension in the crust, visible along the edge in Tashkent, Uzbekistan. (b) A thrust, or reverse fault, produced by compression in the crust, visible in these offset strata in coal seams and volcanic ash in British Columbia. (c) A strike-slip fault produced by lateral shearing, clearly seen looking north along the San Andreas fault rift zone on the eastern edge of the Coast Ranges in California—Pacific plate to the left, North American plate to the right. (d) Can you find the San Andreas fault on the satellite image? Look for a linear rift stretching from southeast to northwest, at the edge of the coastal mountains and San Joaquin Valley. [(a) Photo by Fred McConnaughey/Photo Researchers, Inc.; (b) photo by Fletcher and Baylis/Photo Researchers, Inc.; (c) Kevin Schafer/Peter Arnold, Inc. (d) *Terra* MISR image courtesy of MISR Team, NASA/GSFC/JPL.]

Faulting

A freshly poured concrete sidewalk is smooth and strong. But stress the sidewalk by driving heavy equipment over it, and the resulting strain might cause a fracture. Pieces on either side of the fracture may move up, down, or horizontally, depending on the direction of stress. Similarly, when rock strata are stressed beyond their ability to remain a solid unit, they express the strain as a fracture. Rocks on either side of the fracture displace relative to the other side in a process known as **faulting**. Thus, *fault zones* are areas where fractures in the rock demonstrate crustal movement. At the moment of fracture, a sharp release of energy occurs, called an **earthquake** or *quake*.

The fracture surface along which the two sides of a fault move is the *fault plane*. The names of the three basic types of faults, illustrated in Figure 12.11, are based on the tilt and orientation of the fault plane. A normal fault forms when rocks are pulled apart by tensional stress. A thrust or reverse fault results when rocks are forced together by compressional stress. And, a strike-slip fault forms when rocks are torn by lateral-shearing stress.

Normal Fault When forces pull rocks apart, the tension causes a **normal fault**, or tension fault. When the break occurs, rock on one side moves vertically along an inclined fault plane (Figure 12.11a). The downward-shifting side is the *hanging wall*; it drops relative to the *footwall block*. The exposed fault plane sometimes is visible along the base of faulted mountains, where individual ridges are truncated by the movements of the fault and appear as triangular facets at the ends of the ridges. A cliff formed by faulting is commonly called a *fault scarp*, or *escarpment*.

Reverse (Thrust) Fault Compressional forces associated with converging plates force rocks to move *upward* along the fault plane. This is a **reverse fault**, or compression fault (Figure 12.11b). On the surface, it appears similar to a normal fault, although more collapse and landslides may occur from the hanging wall component. In England, when miners worked along a reverse fault, they would stand on the lower side (footwall) and hang their lanterns on the upper side (hanging wall), giving rise to these terms.

If the fault plane forms a low angle relative to the horizontal, the fault is termed a **thrust fault**, or *overthrust fault*, indicating that the overlying block has shifted far over the underlying block (see Figure 12.8, "thrust fault"). Place your hands palms-down on your desk, with fingertips together, and slide one hand up over the other—this is the motion of a low-angle thrust fault, with one side pushing over the other.

In the Alps, several such overthrusts result from compressional forces of the ongoing collision between the African and Eurasian plates. Beneath the Los Angeles Basin, overthrust faults produce a high risk of earthquakes and caused many quakes in the 20th century, including the $30 billion 1994 Northridge earthquake. These blind (unknown until they rupture) thrust faults beneath the Los Angeles region are a major earthquake threat in the future.

Strike-Slip Fault If movement along a fault plane is horizontal, such as produced along a transform fault, it forms a **strike-slip fault** (Figure 12.11c; refer to Figure 11.19 for review). The movement is *right-lateral* or *left-lateral*, depending on the motion perceived when you observe movement on one side of the fault relative to the other side.

Although strike-slip faults do not produce cliffs (scarps), as do the other types of faults, they can create linear rift valleys. This is the case with the San Andreas fault system of California. The rift valley is clearly visible in Figure 12.11c, where the edges of the North American and Pacific plates are grinding past one another as a result of transform-fault movement. The evolution of this fault system is shown in Figure 12.12.

Note in the figure how the East Pacific rise developed as a spreading center with associated transform faults (1), while the North American plate was progressing westward after the breakup of Pangaea. Forces then shifted the transform faults toward a northwest–southeast alignment along a weaving axis (2). Finally, the western margin of North America overrode those shifting transform faults (3). Note the convergence rate is a rapid 4 cm (1.6 in.) per year!

In *relative* terms, the motion along this series of transform faults is right-lateral, whereas in *absolute* terms the North American plate is still moving westward. Consequently, the San Andreas system is a series of faults that are *transform* (associated with a former spreading center), *strike-slip* (horizontal in motion), and *right-lateral* (one side is moving to the right relative to the other side).

The North Anatolian fault system in Turkey is a strike-slip fault and has a right-lateral motion similar to the San Andreas system (Figure 12.13). A progression of earthquakes have hit along the entire extent of the fault system in Turkey since 1939; the latest devastation in August and November 1999, and again in February 2002.

Faults in Concert Combinations of faults can produce distinctive landscapes. In the U.S. interior west, the *Basin and Range Province* (featuring a parallel series of mountains and valleys) experienced tensional forces caused by uplifting and thinning of the crust (illustrated later in Chapter 15, Figure 15.23). This movement cracked the surface to form aligned pairs of normal faults and a distinctive landscape (Figure 12.14a).

The term **horst** applies to upward-faulted blocks; **graben** refers to downward-faulted blocks. One example of a horst and graben landscape is the Great Rift Valley of East Africa (associated with crustal spreading); it extends northward to the Red Sea, which fills the rift formed by parallel normal faults (Figure 12.14b). Another example is the Rhine graben, through which the Rhine River flows in Europe.

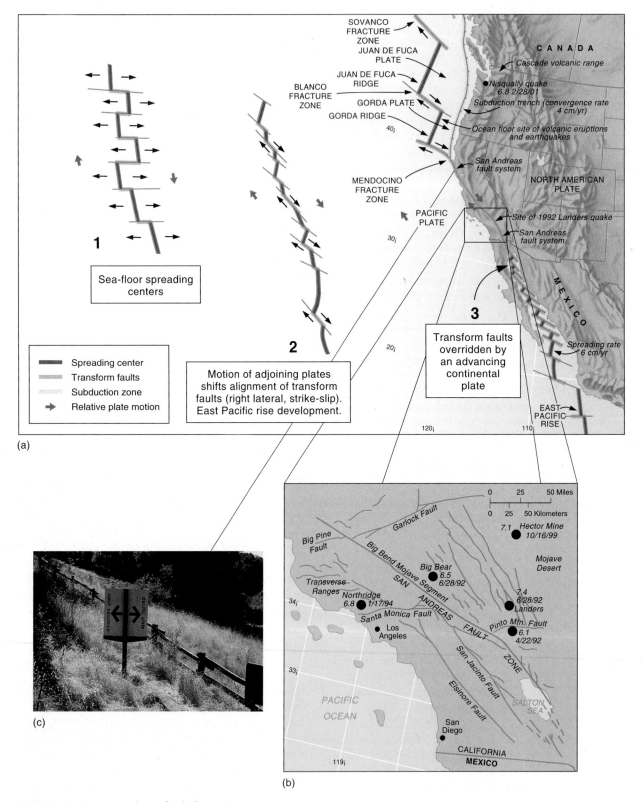

FIGURE 12.12 San Andreas fault formation.
(a) Formation of the San Andreas fault system as a series of transform faults in three successive stages. (b) Enlargement shows the portion of southern California where the 1992 Landers, the 1994 Northridge (Reseda), and 1999 Hector Mine earthquakes occurred. Magnitude ratings are shown for six quakes. (Note the epicenter location of the January 2001 magnitude 6.8 Nisqually quake, Washington state, related to the subduction zone offshore.) (c) Trail sign near the epicenter of the 1906 San Francisco earthquake roughly marks the Pacific–North American plate boundary, Marin County, Point Reyes National Seashore, California. [(c) Photo by author.]

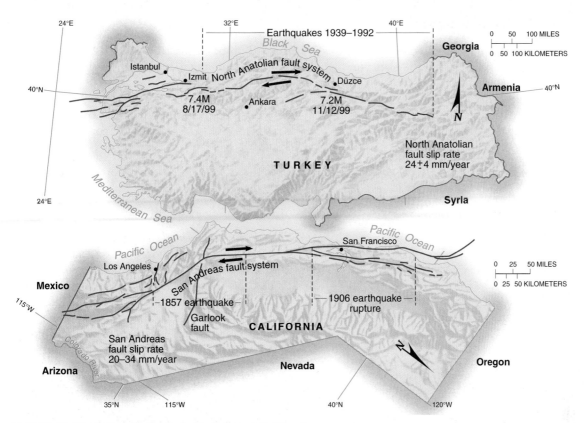

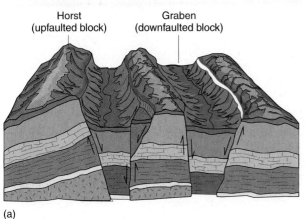

FIGURE 12.13 Strike-slip faults in Turkey and California.
Note these two strike-slip, right-lateral faults and the dates of activity along each—the North Anatolian (Turkey) and San Andreas (California). California is oriented with north to the right for comparison purposes. The San Andreas system is more complex than the North Anatolian because it is a dense network of active faults. [After USGS map by Ross S. Stein.]

FIGURE 12.14 Faulted landscapes.
(a) Pairs of faults produce a horst and graben landscape characteristic of the Basin and Range Province in the western United States. (b) The Red Sea occupies a down-dropped block that is part of the rift system that runs through East Africa. [(b) NASA photo from *Gemini.*]

Orogenesis (Mountain Building)

The processes that work to produce crust and the tectonic forces that bend, warp, and break it brings us to orogenesis. Let us now look at specific mountain-building processes.

Orogenesis literally means the birth of mountains (*oros* comes from the Greek for "mountain"). An *orogeny* is a mountain-building episode, occurring over millions of years. It can occur through large-scale deformation and uplift of the crust. It also may include the capture of migrating terranes and cementation of them to the continental margins, and the intrusion of granitic magmas to form plutons. The net result of this accumulating material is a thickening of the crust. The granite plutons often become exposed by erosion following uplift. Uplift is the final act of an orogenic cycle of mountain building. Earth's

371

major chains of folded and faulted mountains, called *orogens*, are remarkably well correlated with the plate tectonics model.

No orogeny is a simple event; many involve previous developmental stages dating back millions of years, and the processes are ongoing today. Major mountain ranges, and the orogens that caused them, include

- Rocky Mountains of North America (*Laramide orogeny*, 40–80 million years ago)
- Sierra Nevada of California (*Sierra Nevadan orogeny*, 35 million years ago, with older batholithic intrusions dating back 130–160 million years)
- Appalachian Mountains and the Ridge and Valley Province (nearly parallel ridges and valleys) of the eastern United States (*Alleghany orogeny*, 250–300 million years ago, preceded by at least two earlier orogenies; in Europe this is contemporary with the *Hercynian orogeny*)

- Alps of Europe (*Alpine orogeny*, 20–120 million years ago and continuing to the present, with many earlier episodes, Figure 12.15)
- Himalayas of Asia (*Himalayan orogeny*, 45–54 million years ago, beginning with the collision of the India plate and Eurasia plate and continuing to the present)

Types of Orogenies

Figure 12.16 illustrates three types of convergent plate collisions that cause orogenesis:

(a) *Oceanic plate–continental plate collision orogenesis*. This type of convergence is now occurring along the Pacific coast of the Americas and has formed the Andes, the Sierra of Central America, the Rockies, and other western mountains. We see folded sedimentary formations, with intrusions of magma forming granitic plutons at the heart of these moun-

FIGURE 12.15 European Alps.
Western (France), Central (Italy), and Eastern (Austria) segments comprise the crescent shape of the Alps. Complex overturned faults and crustal shortening due to compressional forces occur along convergent plates. The Alps are some 1200 km (750 mi) in length, occupying 207,000 km² (80,000 mi²). Note the aerosols and other pollutants concentrated south of the Alps in northern Italy. [*Terra* MODIS image courtesy of MODIS Land Rapid Response Team, NASA/GSFC.]

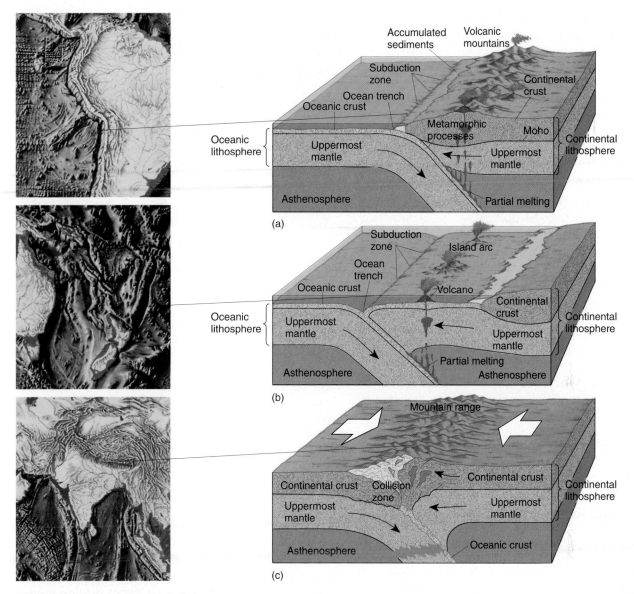

FIGURE 12.16 Three types of plate convergence.
Real-world examples illustrate three types of crustal collisions. (a) Oceanic–continental
(example: Nazca plate–South American plate collision and subduction). (b) Oceanic–oceanic
(example: New Hebrides Trench near Vanuatu, 16° S 168° E). (c) Continental–continental (example: India plate and Eurasian landmass collision and resulting Himalayan Mountains). Can
you identify more of these plate convergence areas on the chapter-opening map?
[Illustrations on left from *Floor of the Oceans,* 1975, by Bruce C. Heezen and Marie Tharp
© 1980. Reproduced by permission of Marie Tharp.]

tains. Their buildup has been augmented by capturing of displaced terranes, cemented during their
collision with the continental mass. Also, note the
associated volcanic activity inland from the subduction zone.

(b) *Oceanic plate–oceanic plate collision orogenesis.* Such collisions can produce either simple volcanic island arcs
or more complex arcs, such as Indonesia and Japan,
which include deformation and metamorphism of
rocks, and granitic intrusions. These processes
formed the chains of island arcs and volcanoes that
continue from the southwestern Pacific to the west-

ern Pacific, the Philippines, the Kurils, and on
through portions of the Aleutians.

Both collision types, (a) oceanic–continental and
(b) oceanic–oceanic, are active around the Pacific Rim. Both
are thermal in nature, because the diving plate melts and
migrates back toward the surface as molten rock. The region of active volcanoes and earthquakes around the Pacific
is known as the **circum-Pacific belt** or, more popularly,
the **ring of fire**.

(c) *Continental plate–continental plate collision orogenesis.*
Here the orogenesis is mechanical; large masses of

continental crust are subjected to intense folding, overthrusting, faulting, and uplifting. The converging plates crush and deform both marine sediments and basaltic oceanic crust. The formation of the European Alps is a result of such compression forces and includes considerable crustal shortening, forming great overturned folds, called *nappes* (see Figure 12.15).

As mentioned earlier, the collision of India with the Eurasian landmass produced the Himalayan Mountains. That collision is estimated to have shortened the overall continental crust by as much as 1000 km (about 600 mi) and to have produced telescoping sequences of thrust faults at depths of 40 km (25 mi). The Himalayas feature the tallest above-sea-level mountains on Earth, including Mount Everest at 8850 m elevation (29,035 ft; a 1999 GPS-based measurement) and all 10 of Earth's highest peaks.

The disruption created by this plate collision has reached far under China, and frequent earthquakes there signal the continuation of this rapid-paced collision. As evidence of this ongoing strain, the January 2001 quake in Gujarat, India, was along a shallow, east–west tending thrust fault that gave way under the pressure of the northward-pushing India plate. More than one million buildings were destroyed or damaged!

The Grand Tetons and the Sierra Nevada

The Sierra Nevada of California and the Grand Tetons of Wyoming are examples of recent stages of mountain building. Each is a *tilted-fault block* mountain range, in which a normal fault on one side of the range has produced a tilted landscape of dramatic relief (Figure 12.17). Magma intruded into those blocks, slowly cooling to form granitic cores of coarsely crystalline rock. After tremendous tectonic uplift and the removal of overlying material through weathering, erosion, and transport, those granitic masses are now exposed in each mountain range. In some areas, overlying material previously covered these batholiths by more than 7500 m (25,000 ft).

Recent research in the Sierra Nevada disclosed that some of the uplift was isostatic, caused by the erosion of overburden and loss of melting ice mass following the last ice age some 18,000 years ago. The accumulation of sediments in the adjoining valley depressed the crust, thus enhancing relief in the landscape.

The Appalachian Mountains

The old, eroded, fold-and-thrust belt of the eastern United States and southeastern Canada (250–300 million years old) contrasts with the younger, higher mountains of western North America (35–80 million years old). As noted, the *Alleghany orogeny* followed at least two earlier orogenic cycles of uplift and the accretion of several captured terranes.

The original material for the Appalachian Mountains resulted from the collisions that produced Pangaea. In fact, the Atlas Mountains of northwestern Africa were connected to the Appalachians at some time in the past, but the

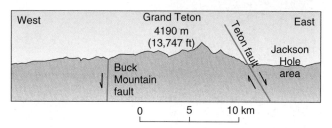

FIGURE 12.17 Tilted-fault block.
The Teton Range in Wyoming is an example of a tilted-fault block, a range of scenic beauty featuring 2130 m (7,000 ft) rugged relief between Jackson Hole and the summits. The Grand Teton is the highest peak in the range at 4190 m (13,747 ft). [Photo by author.]

Atlas Mountains, embedded in the African plate, rafted apart from the Appalachians.

The Appalachian Mountain region comprises several landscape subregions (refer to the locator map in Figure 12.18): the Ridge and Valley Province (elongated sequences of folded sedimentary rock); the Blue Ridge Province (principally of crystalline rock, highest where North Carolina, Virginia, and Tennessee converge); the Piedmont (hilly to gentle terrain along most of the eastern and southern margins of the mountains); and the east coastal plain (from gentle hills to flat plains that extend to the coast).

Figure 12.18 displays the linear folds of the Appalachian system. Note how these dissected ridges are cut through by rivers, forming *water gaps*. These important breaks in the rugged ridges greatly influenced migration, settlement patterns, and the diffusion of cultural traits in the 1700s. The initial flow of people, goods, and ideas was guided by this topography.

World Structural Regions

Examine the first two maps in this chapter (the chapter opener and Figure 12.3) and you will note two vast alpine systems on the continents. In the Western Hemisphere, the *Cordilleran system* stretches from Tierra del Fuego at the southern tip of South America to the massive peaks of Alaska, including the relatively young Rocky and Andes Mountains along the western margins of the North and South American plates. In the Eastern Hemisphere, the

FIGURE 12.18 The Appalachian Mountains.
Appalachian Mountain region of Pennsylvania south through Maryland, Virginia, and West Virginia in a *Terra* MODIS sensor image from October 11, 2000, showing true fall colors. The highly folded nature of this entire region is clearly visible in the satellite image. [*Terra* image courtesy of MODIS Land Science Team, NASA/GSFC.]

Eurasian–Himalayan system stretches from the European Alps across Asia to the Pacific Ocean and contains younger and older components.

These mountain systems also are shown on the structural region map as the *Alpine system* (Figure 12.19). The map defines seven fundamental structural regions that possess distinctive types of landscapes, grouped because of their shared physical characteristics. Looking at the distribution of these regions helps summarize the three rock-forming processes (igneous, sedimentary, and metamorphic), plate tectonics, landform origins and construction, and overall orogenesis.

As you examine the map, identify the continental shields at the heart of each landmass. Continental platforms composed of sedimentary deposits surround these areas. Various mountain chains, rifted regions, and isolated volcanic areas are noted on the map. On the continent of Australia, you see older mountain sequences to the east, sedimentary layers covering basement rocks west of these ranges, and portions of the original Gondwana, an ancient landscape in the central and western region. (Remember that Gondwana was a landmass that included Antarctica, Australia, South America, Africa, and the southern portion of India; it broke away from Pangaea some 200 million years ago.)

Earthquakes

Crustal plates do not glide smoothly past one another. Instead, tremendous friction exists along plate boundaries. The *stress* (a force) of plate motion builds *strain* (a deformation) in the rocks until friction is overcome and the sides along plate boundaries suddenly break loose. The two sides of the fault plane then lurch into new positions, moving from centimeters to several meters, and release enormous amounts of seismic energy into the surrounding crust. This energy radiates throughout the planet, diminishing with distance, but sufficient enough to register on instruments worldwide. The 1906 earthquake that devastated San Francisco and the 1999 earthquake in Armenia, Colombia, that took more than 1000 lives, remind us of the danger and unpredictability of these tectonic forces of nature (Figure 12.20).

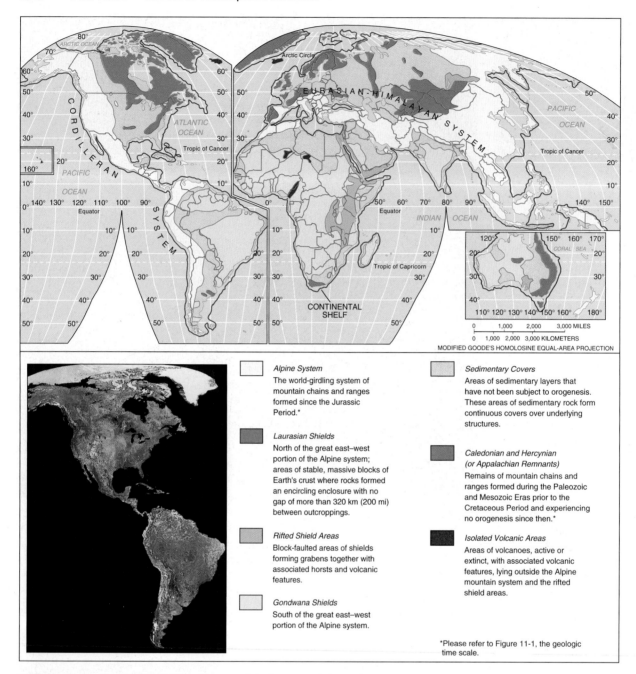

FIGURE 12.19 World structural regions and major mountain systems.
Some of the regions appear larger than the structures themselves because each region includes related landforms adjacent to the central feature. (Correlate aspects of this map with Figures 12.3 and 12.4.) Structural regions in the Western Hemisphere are visible on the composite false-color *Landsat* image inset (vegetation is portrayed in red). [After R. E. Murphy, "Landforms of the World," *Annals of the Association of American Geographers* 58, no. 1 (March 1968). Adapted by permission. Inset image courtesy of EROS Data Center and the National Geographic Society.]

Expected Quakes and Those of Deadly Surprise

In the Liaoning Province of northeastern China, ominous indications of tectonic activity began in 1970. Troubling symptoms included land uplift and tilting, increasing minor tremors, and changes in the region's magnetic field—after almost 120 years of quiet. These precursors of tectonic events continued for almost 5 years before Chinese scientists took the bold step of forecasting an earthquake. In February 1975, some 3 million people evacuated in what turned out to be a timely manner; the quake struck 6 hours later, within the predicted time frame. Ninety percent of the buildings in the city of Haicheng fell. Thousands of lives were claimed saved—an earthquake had been forecast and preparatory action taken for the first time in history.

(a)   (b)

FIGURE 12.20 Destructive earthquakes 1906 and 1999.
(a) A view of San Francisco, devastated by the 1906 earthquake and subsequent fire. The view is to the east toward Nob Hill and the still-standing Fairmont Hotel. (b) Massive destruction in Armenia, Colombia, and surrounding region, from a January 1999 magnitude 6.0 quake, leaving more than 1000 people dead and 200,000 homeless. [Photos by (a) Corbis-Bettmann; and, (b) Ricardo Mazalon, AP/Wide World Photos.]

In contrast, only 17 months later, at Tangshan in the northeastern province of Hebei (Hopei), 145 km (90 mi) southeast of Beijing, China's capital city, a severe earthquake occurred without warning. No precursors (warning signs) were sensed for scientists to build a forecast. Consequently, this magnitude 7.4 quake killed about 250,000 people! (This is the official death toll; other estimates range as high as 650,000 fatalities.) The earthquake also destroyed 95% of the buildings and 80% of the industrial structures, and it severely damaged more than half the bridges and highways in the area. The jolt and ground acceleration threw people against the ceilings of their homes. An old, previously undetected fault ruptured, and the rocks shifted 1.5 m (5 ft) along 8 km (5 mi) in the heart of Tangshan. What are the mechanisms that produce such different tectonic events? Why are some expected, whereas most strike in total surprise?

Students and faculty at California State University–Northridge, in the San Fernando Valley of southern California, need no reminder of the power of earthquakes. Their campus was near the epicenter of the most devastating earthquake in U.S. history in terms of property damage—$30 billion in destruction across the region. The quake was caused by one of the many low-angle thrust faults that underlie the Los Angeles region. The January 17, 1994, Northridge (Reseda) earthquake, a magnitude 6.8, and more than 10,000 aftershocks caused approximately $350 million damage to campus buildings two weeks before the beginning of spring classes. The semester began just three weeks late in 450 temporary trailers. Geography professors taught some class sessions outdoors for a brief time. Amazingly, graduation was still held in May!

Focus, Epicenter, Foreshock, and Aftershock

The subsurface area along a fault plane, where the motion of seismic waves is initiated, is the *focus*, or hypocenter, of an earthquake (see labels in Figure 12.21). The area at the surface directly above the focus is the *epicenter*. Shock waves produced by an earthquake radiate outward through the crust from the focus and epicenter. Some of the seismic waves are conducted throughout the planet to distant instruments. From these seismic wave patterns and the nature of their transmission through the layers of the planet, Earth's interior is explored by scientists.

An *aftershock* may occur after the main shock, sharing the same general area of the epicenter; some aftershocks rival the main tremor in magnitude. A *foreshock* also is possible, preceding the main shock. The pattern of foreshocks is now regarded as an important consideration in the forecast effort. (Before the 1992 Landers earthquake in southern California, at least two dozen foreshocks occurred along the soon-to-erupt portion of the fault.)

Earthquake Intensity and Magnitude

Tectonic earthquakes are those quakes associated with faulting. A worldwide network of more than 4000 **seismograph** instruments record vibrations transmitted as waves of energy throughout Earth's interior and in the crust. Using this device and actual observations, scientists rate earthquakes on two kinds of scales: a *qualitative* damage intensity scale and a *quantitative* magnitude-of-energy-released scale.

A damage *intensity scale* is useful in classifying and describing damage to terrain and structures after an

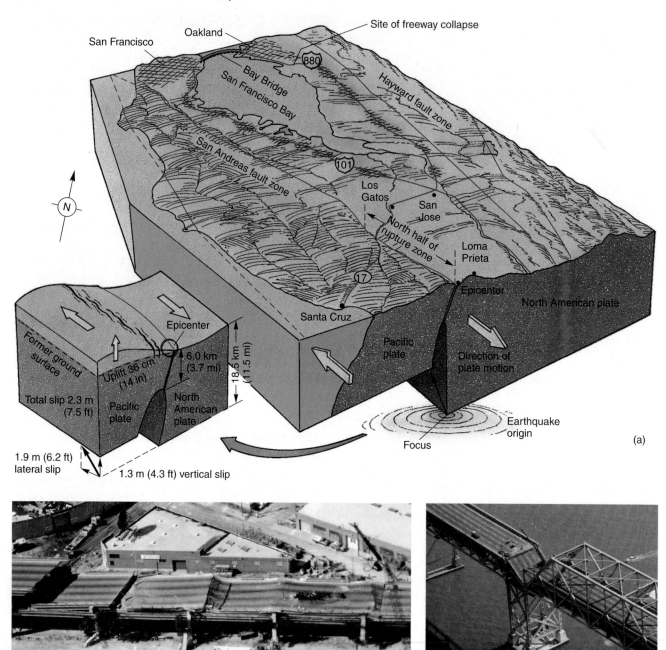

FIGURE 12.21 Anatomy of an earthquake.
(a) The fault-plane solution for the 1989 Loma Prieta, California, earthquake shows the lateral and vertical (thrust) movements occurring at depth. There was no surface expression of this fault plane. Damage totaled $8 billion, 14,000 people were displaced from their homes, 4000 were injured, and 67 killed. (b) In only 15 seconds, more than 2 km (1.2 mi) of the Route 880 Cypress Freeway collapsed. (c) Section failure in the San Francisco–Oakland Bay Bridge actually is the way the bridge handles strain without completely failing. [(a) After P. J. Ward and R. A. Page, *The Loma Prieta Earthquake of October 17, 1989* (Washington, DC: U.S. Geological Survey, November 1989), p. 1; (b) and (c) courtesy of California Department of Transportation.]

earthquake. Earthquake intensity is rated on the arbitrary *Mercalli scale*, a Roman numeral scale from I to XII, representing "barely felt" to "catastrophic total destruction." It was designed in 1902 and modified in 1931. Table 12.1 shows this scale and the number of quakes in each category that are expected each year.

Moment Magnitude Scale Revises the Richter Scale
In 1935 Charles Richter designed a system to estimate earthquake magnitude. In this method, a seismograph located at least 100 km (62 mi) from the epicenter of the quake records the amplitude of seismic waves. That measurement is then charted on the **Richter scale**. The rela-

Table 12.1   Earthquake Characteristics and Frequency Expected Each Year

Effects in Populated Areas	Approximate Intensity (modified Mercalli scale)	Approximate Magnitude (Moment Magnitude scale)	Number per year
Nearly total damage	XII	>8.0	1 every few years
Great damage	X–XI	7–7.9	18
Considerable-to-serious damage to buildings: railroad tracks bent	VIII–IX	6–6.9	120
Felt-by-all, with slight damage to buildings	V–VII	5–5.9	800
Felt-by-some to felt-by-many	III–IV	4–4.9	6,200
Not felt, but recorded	I–II	2–3.9	500,000

Source: USGS, Earthquake Information Center.

tion of magnitude to energy released is still a useful feature of his scale.

The Richter scale is logarithmic: Each whole number on it represents a 10-fold increase in the measured wave amplitude. Translated into energy, each whole number signifies a 31.5-fold increase in energy released. Thus, a magnitude of 3.0 on the Richter scale represents 31.5 times more energy than a 2.0, and 992 times more energy than a 1.0. It is difficult to imagine the power released by the Tangshan quake, which was rated a 7.6 on the Richter scale.

Today, the Richter scale is improved and made more quantitative. Revision was needed because the scale did not properly measure or differentiate between quakes of high intensity. Seismologists wanted to know more about what they call the *seismic moment* to understand a broader range of possible motions during an earthquake.

The **moment magnitude scale**, in use since 1993, is more accurate for large earthquakes than Richter's *amplitude magnitude* scale. Moment magnitude considers the amount of fault slippage produced by the earthquake, the size of the surface (or subsurface) area that ruptured, and the nature of the materials that faulted, including how resistant they were to failure. The new scale rates the 1994 Northridge quake, mentioned earlier, at magnitude 6.8 and considers extreme ground acceleration (movement upward), which the Richter amplitude magnitude method underestimated. Richter was familiar in everyday usage, although moment magnitude is the correct scale and is the scale used in this book unless stated otherwise.

A reassessment of past quakes has increased the rating of some and decreased that of others. As an example, the 1964 earthquake at Prince William Sound in Alaska had an amplitude magnitude of 8.6, but on the moment magnitude scale it increases to a magnitude 9.2. Please note that Table 12.2 includes more earthquakes for the period following 1960 than for the years before 1960. This is not because earthquake frequency has increased; rather, it reflects an effort to include recent events that affected increased population densities in vulnerable areas.

The National Earthquake Information Center in Golden, Colorado, reports earthquake epicenters. For reference, see http://wwwneic.cr.usgs.gov/ and the National Geophysical Data Center at http://www.ngdc. noaa.gov/seg/. As part of the "Learning from Earthquakes Project," reconnaissance teams examine the sites of major quakes; see http://www.eqnet.org. For a real time listing of earthquakes worldwide, go to http://wwwneic.cr.usgs. gov/neis/bulletin/bulletin.html.

The Nature of Faulting

We earlier described types of faults and faulting motions. The specific mechanics of how a fault breaks remain under study, but **elastic-rebound theory** describes the basic process. Generally, two sides along a fault appear to be locked by friction, resisting any movement despite the powerful forces acting on the adjoining pieces of crust. Stress continues to build strain along the *fault plane* surfaces, storing elastic energy like a wound-up spring. When the strain buildup finally exceeds the frictional lock, both sides of the fault abruptly move to a condition of less strain, releasing a burst of mechanical energy. This type of sudden movement rocked the region around Kobe, Japan, in 1995, with a magnitude 6.9 quake. See News Report 12.2 for an account of this significant event.

Think of the fault plane as a surface with irregularities that act as sticking points, preventing movement, similar to two pieces of wood held together by drops of glue of different sizes rather than an even coating of glue. Research scientists at the USGS and the University of California identify these small areas of high strain as *asperities*. They are the points that break and release the sides of the fault.

If the fracture along the fault line is isolated to a small asperity break, the quake will be small in magnitude. Clearly, as some asperities break (perhaps recorded as small foreshocks), the strain increases on surrounding asperities that remain intact. Thus, small earthquakes in an area may be precursors to a major quake. However, if the break involves the release of strain along several asperities, the quake will be greater in extent and will involve the shifting of massive amounts of crust. The latest evidence points to a wavelike pattern, as rupturing spreads along the fault plane, rather than the entire fault surface giving way at once.

Table 12.2 A Sampling of Significant Earthquakes**

Year	Date	Location	Number of Deaths	Mercalli Intensity	Moment Magnitude (Richter)
1556	Jan. 23	Shaanxi Province, China	830,000	*	*
1737	Oct. 11	Calcutta, India	300,000	*	*
1812	Feb. 7	New Madrid, Missouri	Several	XI–XII	*
1857	Jan. 9	Fort Tejon, California	*	X–XI	*
1870	Oct. 21	Montreal to Québec, Canada	*	IX	*
1886	Aug. 31	Charleston, South Carolina	*	IX	6.7
1906	Apr. 18	San Francisco, California	3,000	XI	7.7 (8.25)
1923	Sept. 1	Kwanto, Japan	143,000	XII	7.9 (8.2)
1939	Dec. 27	Erzincan, Turkey	40,000	XII	7.6 (8.0)
1960	May 22	Southern Chile	5,700	XII	9.5 (8.6)
1964	Mar. 28	Southern Alaska	131	X–XII	9.2 (8.6)
1970	May 31	Northern Peru	66,000	*	7.9 (7.8)
1971	Feb. 9	San Fernando, California	65	VII–IX	6.7 (6.5)
1972	Dec. 23	Managua, Nicaragua	5,000	X–XII	6.2 (6.2)
1976	Jul. 28	Tangshan, China	250,000	XI–XII	7.4 (7.6)
1978	Sept. 16	Iran	25,000	X–XII	7.8 (7.7)
1985	Sept. 19	Mexico City, Mexico	7,000	IX–XII	8.1 (8.1)
1988	Dec. 7	Armenia–Turkey border	30,000	XII	6.8 (6.9)
1989	Oct. 17	Loma Prieta (near Santa Cruz, California)	66	VII–IX	7.0 (7.1)
1991	Oct. 20	Uttar Pradesh, India	1,700	IX–XI	6.2 (6.1)
1994	Jan. 17	Northridge (Reseda), California	66	VII–IX	6.8
1995	Jan. 17	Kobe, Japan	5,500	XII	6.9
1996	Feb. 17	Indonesia	110	X	8.1
1997	Feb. 28	Armenia–Azerbaijan	1,100	XII	6.1
1997	May 10	Northern Iran	1,600	XII	7.3
1998	May 30	Afghanistan–Tajikistan	4,000	XII	6.9
1998	Jul. 17	Papua, New Guinea	2,200	X	7.1
1999	Jan. 26	Armenia, Colombia	1,000	VIII–IX	6.0
1999	Aug. 17	Izmit, Turkey	17,100	VIII–XI	7.4
1999	Sept. 7	Athens, Greece	150	VI–VIII	5.9
1999	Sept. 20	Chi-Chi, Taiwan	2,500	VI–X	7.6
1999	Sept. 30	Oaxaca, Mexico	33	VI	7.5
1999	Oct. 16	Hector Mine, California	0	*	7.1
1999	Nov. 12	Düzce, Turkey	700	VI–X	7.2
2001	Jan. 26	Gujarat state, India	19,998	X–XII	7.7

*Data not available.

**There is not a recent increase in earthquakes; this table merely reflects more detail on the recent record.

Earthquakes and the San Andreas Fault

In 1906, San Francisco was a city of 400,000 people (there are several million in the metropolitan region today). In that year, a magnitude 7.7 (8.25 Richter) tremor rocked the city, felling buildings and initiating a firestorm fed by broken gas pipes. After the quake, movement along a fault was evident over a stretch of 435 km (270 mi), prompting intensive research to discover the nature of such faulting. The elastic-rebound theory developed as a result of research along the rupture. (Realize that this quake occurred 6 years before Wegener proposed his continental drift hypothesis!)

The fault system that devastated San Francisco in 1906, and coastal and inland California throughout recorded history, is the San Andreas. The San Andreas fault system provides a good example of a spreading center overridden by an advancing continental plate (refer to Figure 12.12 for its evolution).

The tectonic history of the San Andreas was brought home to hundreds of millions of television viewers as they watched the 1989 baseball World Series played in Candlestick Park near San Francisco. Less than half an hour before the call to "play ball," a powerful earthquake rocked the region, turning sportscasters into newscasters and sports fans into disaster witnesses. This earthquake involved a

News Report 12.2

A Tragedy in Kobe, Japan—The Hyogo-ken Nanbu Earthquake

Early on the morning of January 17, 1995, a year to the day after the Northridge quake, a section of the Nojima fault zone, approximately 16 km (10 mi) below the surface, became the focus of a devastating earthquake. On the moment magnitude scale the quake registered 6.9; it killed 5500 people, injured 26,000, and caused more than $100 billion in damage. About 200,000 homes, or 10% of all the housing in the metropolitan area, were damaged; some 80,000 buildings completely collapsed; major transportation arteries were severed; and the busy port was brought to a standstill (Figure 1).

For Japan, this was the worst quake loss since the 1923 temblor that hit Tokyo. The Kobe region had large earthquakes in A.D. 868, 1596, and

1916, but after almost 80 years of relative calm, the people did not expect this surprise—this despite the fact the fault and the earthquake mechanism is well understood. The fault moved in a right-lateral, strike-slip motion, similar to the San Andreas fault in California. This event moved the land along three segments, where displacement of 1.0 to 1.5 m occurred in a rupture zone that stretched over 30 km (18 mi) in length.

The lessons are many, especially the occurrence of liquefaction and the failure of filled land. Reclaimed areas of Osaka Bay taken from the sea by landfill liquefied quickly in the shaking. Soil surfaces failed, collapsing the structures they supported (Figure 1b). Importantly, these human-made land-

fills and islands are characteristic of the San Francisco Bay region, where, since the early 1900s, about half the original surface area of the bay is reclaimed with fill and occupied with buildings.

A positive note in Kobe and Osaka was the good performance of landfill sites where drains were installed and where deeply anchored pilings were in place to support structures. The cost of such retrofits in the United States could easily exceed several hundred billion dollars. In the meantime, the resemblance between the geology of the Kobe region and the San Francisco Bay region, specifically the menacing Hayward fault system, is troubling. The Kobe lesson raises great concern and a call to action.

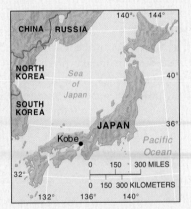

FIGURE 1 1995 earthquake in Kobe, Japan.
(a) Catastrophic failure to an elevated freeway; note the failure of the supporting pillars. (b) Ground failure, mainly liquefaction, tilts a building on the failing foundation soils to near collapse; note the space between the building and the ground on the right side. [Photos by Haruyoshi Yamaguchi/Sygma.]

portion of the San Andreas fault, near Loma Prieta, approximately 16 km (9.9 mi) east of Santa Cruz and 95 km (59 mi) south of San Francisco (Figure 12.21). The moment magnitude was 7.0. A fault had ruptured at a focus unusually deep for the San Andreas system, more than 18 km (11.5 mi) below the surface.

Unlike previous earthquakes—such as the one in 1906, when the plates shifted a maximum of 6.4 m (21 ft) relative to each other—there was no evidence of a fault plane or rifting at the surface in the Loma Prieta area. Instead, the fault plane suggested in Figure 12.21 shows the two plates moving horizontally approximately 2 m (6 ft) past each other deep below the surface, with the Pacific plate thrusting 1.3 m (4.3 ft) upward. This vertical motion is unusual for the San Andreas fault and perhaps is a clue that this

portion of the San Andreas system is more complex than previously thought.

Los Angeles Region

Since the mid-1980s, an area east of Los Angeles has experienced seven earthquakes greater than 6.0 magnitude. After a magnitude 6.1 quake in 1992, a magnitude 7.4 tremor rocked the lightly populated area near Landers, California (see the enlargement map of southern California in Figure 12.12). Although it was the single largest quake in California in 30 years, the remote location kept injuries and damage slight. Involved in this series were four different faults and some unknown segments. The main faulting caused a displacement of 6.1 m (20 ft).

For yet unexplained reasons, related earthquakes over the next few weeks struck in Mammoth Lakes, about 645 km (400 mi) to the north; at Mount Shasta in northern California; in southern Nevada and Utah; and 1810 km (1125 mi) distant in Yellowstone National Park, Wyoming.

The 1971 San Fernando, 1987 Whittier, 1988 Pasadena, 1991 Sierra Madre, and 1994 Northridge (Reseda) earthquakes are a few of the many quakes associated with deeply buried thrust faults. Earthquakes roughly originate at foci 18 km (11 mi) deep in the crust. Although not directly aligned with the San Andreas, scientists think that southern California will be affected by more of these thrust-fault actions as strain continues to build along the nearby San Andreas system of faults. The fault line that marks the contact between the Pacific and the North American plates is indeed restless.

Earthquake Forecasting and Planning

The map in Figure 12.22 plots the epicenters of earthquakes in the United States and southern Canada between 1899 and 1990. These occurrences give some indication of relative risk by region. The challenge is to discover how to predict the *specific time and place* for a quake in the short term. See News Report 12.3 for more on forecasting.

Actual implementation of an action plan to reduce death, injury, and property damage from earthquakes is difficult. The political environment adds complexity; sadly, an accurate earthquake prediction would be viewed as a threat to a region's economy (Figure 12.23). If we examine the potential socioeconomic impact of earthquake prediction on an urban community, we find surprising negative economic impacts in the period before the quake hits. Imagine a chamber of commerce, bank, real estate agent, tax assessor, or politician who would privately welcome an earthquake prediction and such negative publicity for their city.

Long-range planning is a complex subject. After the Loma Prieta earthquake in 1989, a committee report by the National Research Council concluded that,

> One of the most jarring lessons from these quakes may be that earthquake professionals have long known many of the things that could have been done to reduce devastation. . . . The cost-effectiveness of mitigation and the importance of closing the knowledge gap among researchers, building professionals, government officials, and the public are only two of the many lessons from the Loma Prieta that need immediate action.

A valid and applicable generalization is that *humans and their institutions are unable or unwilling to perceive hazards in a familiar environment.* In other words, we tend to feel secure in our homes and communities, even if they are sitting on a quiet fault zone. Such an axiom of human behavior certainly helps explain why large populations continue to live and work in earthquake-prone settings. Similar questions also can be raised about populations in areas vulnerable to floods, droughts, hurricanes and coastal storm surge, and settlements on barrier islands. (See the Natural Hazards Observer at http://www.colorado.edu/hazards/o/o.html as part of the Natural Hazards Center at the University of Colorado at http://www.colorado.edu/hazards/index.html.)

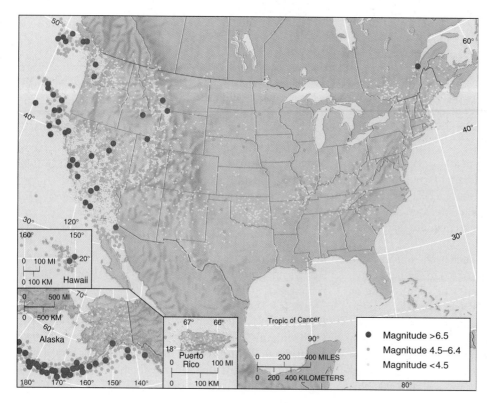

FIGURE 12.22 Seismicity of the United States and southern Canada, 1899–1990.
Earthquake occurrences of magnitude 4.5, or greater, indicate areas of greatest seismic risk. Note the contrast between stable and unstable regions, related to tectonic activity. Active seismic regions include the West Coast, the Wasatch Front of Utah northward into Canada, the central Mississippi Valley, southern Appalachians, and portions of South Carolina, upstate New York, and Ontario. [From the U.S. Geological Survey, National Earthquake Information Center. For more detailed checking on your locale, see http://geohazards.cr.usgs.gov/eq/index.html.]

News Report 12.3

Seismic Gaps, Nervous Animals, Dilitancy, and Radon Gas

How do you forecast an earthquake? One approach is to examine the history of each plate boundary and determine the frequency of past earthquakes, a study called *paleoseismology*. Paleoseismologists construct maps that provide an estimate of expected earthquake activity based on past performance. An area that is quiet and overdue for an earthquake is a *seismic gap*; such an area forms a gap in the earthquake occurrence record and is therefore a place that possesses accumulated strain. The area along the Aleutian Trench subduction zone had three such gaps until the great 1964 Alaskan earthquake filled one of them.

The areas around San Francisco and northeast of Los Angeles represent other such gaps where the fault system appears to be locked by friction and stress is producing accumulating strain. The U.S. Geological Survey in 1988 made a prediction that there was a 30% chance of an earthquake occurring with a magnitude 6.5 within 30 years in the Loma Prieta portion of the fault system. The actual 1989 quake dramatically filled a portion of the seismic gap in that region. The Nojima fault in the area of Kobe occurred in another gap.

One potentially positive discovery from the Loma Prieta disaster was that Stanford University scientists found unusually great changes in Earth's magnetic field—about 30 times more than normal—three hours before the main shock. These measurements were made 7 km (4.3 mi) from the epicenter, raising hopes that short-term warnings might be possible in the future. A valid research question is whether animals have the ability to detect these minute changes in magnetic fields, an ability that was perhaps lost in humans. If so, strange prequake animal behavior, often reported, might provide forecasting clues.

Dilitancy refers to the slight increase in volume of rock produced by small cracks that form under stress and accumulated strain. The affected region may tilt and swell in response to strain. Tiltmeters measure these changes in suspect areas. Another indicator of dilitancy is an increase in radon dissolved in groundwater, a nat-

urally occurring, slightly radioactive gas. Presently, earthquake hazard zones have thousands of radon monitors taking samples in test wells.

A seismic network in operation in Mexico City is providing 70-second warnings of arriving seismic-wave energy from distant epicenters. Following the Loma Prieta earthquake, freeway-repair workers were linked to a broadcast system that gave them 20-second alerts of aftershocks from distant epicenters. In southern California the USGS is coordinating a Southern California Seismographic Network (SCSN at http://www.trinet.org/scsn/scsn.html) to correlate 350 instruments for immediate earthquake location analysis and disaster coordination. Also, for the same region, see http://www-socal.wr.usgs.gov/.

Someday accurate earthquake forecasting may be a reality, but several questions remain. How will humans respond to a forecast? Can a major metropolitan region be evacuated for short periods of time? Can cities relocate after a disaster to areas of lesser risk?

Volcanism

A partial sampling of recent volcanic eruptions across the globe reminds us of Earth's internal energy. For more volcanic eruption listings, see http://volcano.und.nodak.edu/vwdocs/current_volcs/current.html. (Note that volcanic activity is not on the increase; this list reflects an effort to include greater detail for recent events. Years given are when recent action started; many continue for years.)

- 1995: Kīlauea (Hawai'i, since 1983: see Figures 12.27 and Part 3's opening photo), Soufriere Hills on Montserrat (Lesser Antilles/Caribbean Sea, continuing through 2001), Mount Etna (Italy, continuing through 2001), Metis Shoal (Tonga Islands);
- 1996: Mount Ruapehu (New Zealand, continuing through 2000), Karymsky (Kamchatka Peninsula, Russia, continuing), Canlaon (Philippines), Maderas (Nicaragua), Grímsvötn (Iceland), Pavlof (Alaska);
- 1997: Mount Karangetang (Indonesia), Semeru (Java, Indonesia, active since 1967), Rabaul (Papua New Guinea);
- 1998: Popocatépetl, (Mexico, continuing), Cerro Negro (Nicaragua), White Island (New Zealand), Axial Summit (sea-floor spreading center off the coast of Oregon), Stromboli (Italy);
- 1999: in addition to continuing eruptions from 1998, Tungurahua and Guagua (Ecuador), Colima (Mexico), Villarrica (Chile), Ruapehu (New Zealand), and Cerro Negro (Nicaragua);
- 2000: Shishaldin (Unimak Island, Alaska; see Figure 12.34), Lascar (Chile), White Island (New Zealand; see Figure 12.30), Tavurvur (Papua New Guinea), Sakura-Jima (Japan);
- 2001: Nyamuragira (Congo, Africa), Karangetang (Siau Island, Indonesia), Pacaya (Guatemala), Kliuchevskoi (Kamchatka, Russia; see Figure 1.25), Piton de la Fournaise (Island of Reunion), Jackson Segment Northern Gorda Ridge (seafloor off the Oregon coast), Masaya (Nicaragua), Popocatépetl (Mexico), Colima (Mexico), Tungurahua (Ecuador), Mayon (Philippines).
- 2002: Nyiragongo (Congo, Africa).

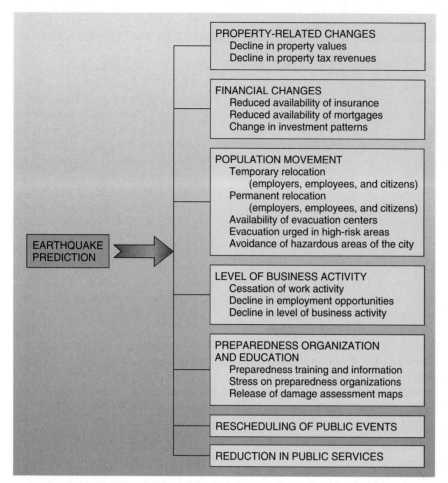

FIGURE 12.23 Socioeconomic impacts of an earthquake prediction.
Business, political, and monetary interests are at stake as scientists get closer to being able to predict earthquakes. [After J. E. Haas and D. S. Mileti, *Socioeconomic Impact of Earthquake Prediction on Government, Business, and Community*, Boulder: University of Colorado, Institute of Behavioral Sciences. Used by permission.]

More than 1300 identifiable volcanic cones and mountains exist on Earth, although fewer than 600 are active—having at least one eruption in recorded history. (An index to the world's volcanoes is at http://vulcan.wr.usgs.gov/Volcanoes/framework.html and the National Museum of Natural History Global Volcanism Program at http://www.nmnh.si.edu/gvp/.) In an average year, about 50 volcanoes erupt worldwide, varying from modest activity to major explosions. Eruptions in remote locations and at depths on the seafloor go largely unnoticed, but the occasional eruption of great magnitude near a population center makes headlines.

North America has about 70 volcanoes (mostly inactive) along the western margin of the continent. Mount St. Helens in Washington State is a famous active example, and more than 1 million visitors a year travel to Mount St. Helens Volcanic National Monument to see volcanism for themselves. A source of information about volcanoes on the Web is at Volcano World http://volcano.und.nodak.edu/, or check http://www.geo.mtu.edu/volcanoes/links/observatories.html for a listing of volcano observatories in the world.

Volcanoes produce some benefits. These include fertile soils, which develop from weathering of basaltic lava, as in Hawai'i; geothermal energy such as in Iceland, Italy, New Zealand, and California; and even new, albeit risky, real estate in Iceland, Japan, Hawai'i, and elsewhere, as lava extends shorelines seaward.

Volcanic Features

A **volcano** forms at the end of a central vent or pipe that rises from the asthenosphere and upper mantle through the crust into a volcanic mountain. A **crater**, or circular surface depression, usually forms at or near the summit. Magma rises and collects in a magma chamber deep below the volcano until conditions are right for an eruption. This subsurface magma emits tremendous heat, and in some areas it boils groundwater, producing *geothermal energy*, as seen in the thermal springs and geysers of Yellowstone National Park, or elsewhere in the world where it is harnessed for geothermal power production.

Lava (molten rock), gases, and **pyroclastics**, or *tephra* (pulverized rock and clastic materials of various sizes ejected violently during an eruption) pass through the vent to the surface and build a volcanic landform. The fact that lava can occur in many different textures and forms accounts for the varied behavior of volcanoes and the different landforms they build. In this section we look at five volcanic landforms and their origins: cinder cones, calderas, shield volcanoes, plateau basalts, and composite volcanoes.

A **cinder cone** is a small cone-shaped hill usually less than 450 m (1500 ft) high, with a truncated top formed from cinders that accumulate during moderately explosive eruptions. Cinder cones are made of pyroclastic material and *scoria* (cindery rock, full of air bubbles).

Another distinctive landform is a large basin-shaped depression called a **caldera** (Spanish for "kettle"). It forms when summit material on a volcanic mountain collapses inward after an eruption or other loss of magma. In the Cape Verde Islands (15° N 24.5° W), Fogo Island has such a caldera in the midst of a volcanic cone, opening to the sea on the eastern side. The best farmland is in the caldera, so the people choose to live and work on the floor of an active volcano. Unfortunately, eruptions began in April 1995, destroying farmland and forcing the evacuation of 5000 people. A cinder cone 600 m (1800 ft) high formed in the former farmlands. The area experienced eruptions between A.D. 1500 and 1750, followed by six more in the next century, each destroying houses and crops. The previous eruption was in 1951.

An example of a caldera in North America is beautiful Crater Lake in southern Oregon. News Report 12.4 discusses the Long Valley Caldera in California.

Location and Types of Volcanic Activity

The location of volcanic mountains on Earth is a function of plate tectonics and hot-spot activity. Volcanic activity occurs in three settings:

1. Along subduction boundaries at continental plate–oceanic plate convergence (such as Mount St. Helens and Kliuchevskoi, Siberia) or oceanic plate–oceanic plate convergence (Philippines and Japan).
2. Along sea-floor spreading centers on the ocean floor (Iceland on the Mid-Atlantic ridge, or off the coast of Oregon and Washington) and areas of rifting on continental plates (the rift zone in east Africa).
3. At hot spots, where individual plumes of magma rise to the crust (such as Hawai'i and Yellowstone National Park).

Figure 12.24 illustrates the three types of volcanic activity, which you can compare with the active volcano sites and plate boundaries shown in Figure 11.20. Figures 12.25 to 12.30 illustrate the various aspects of volcanism presented in Figure 12.24.

The variety of forms among volcanoes makes them difficult to classify; most fall in transition between one type and another. Even during a single eruption, a volcano may behave in several different ways. The primary factors in determining an eruption type are: (1) the magma's chemistry,

---

News Report 12.4

Is the Long Valley Caldera Next?

Trees are dying in the forests on Mammoth Mountain, in a portion of the Long Valley Caldera, near the California–Nevada border. The oval caldera is about 15 by 30 km (9 by 19 mi) long in a north–south direction and sits at 2000 m (6500 ft) elevation, rising to 2600 m (8500 ft) along the western side.

About 1200 tons of carbon dioxide is coming up through the soil in the old caldera each day. Carbon dioxide levels reach 30% to 96% of gases in some soil samples. The source: active and moving magma at some 3 km depth. Elsewhere in the world, this and other gas emissions assist forecasts of potential volcanic activity. At Mam-

moth Mountain, these gases signify magmatic activity and may be a portent of things to come (Figure 1).

A powerful volcanic eruption 730,000 years ago formed the Long Valley Caldera. This ancient eruption exceeded the volume output of the 1980 Mount St. Helens event by more than 500 times! Volcanic activity of a lesser extent has occurred in the area over the past several hundred thousand years, most recently between A.D. 1720 and 1850 in the Mono Lake area north of Long Valley.

In the late 1970s and again in 1996, continuing through 1998, swarms of earthquakes shook Long Valley—thousands of quakes overall,

some three dozen greater than magnitude 3.0. The subsurface foci for these quakes is at about 11 km beneath the region. As of this writing, the surface has not been lifted or deformed by this activity. The combination of volcanic gas production and earthquakes is drawing scientific and public attention to this area because of the potential for a future massive eruption somewhere along the Mono-Inyo Craters stretching north of Long Valley. The region is a popular tourist mecca and recreational center, with second-home residential development and a growing year-round population. For updates, see http://quake.wr.usgs.gov/VOLCANOES/LongValley/.

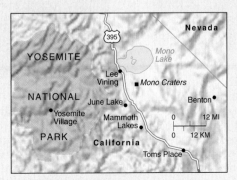

FIGURE 1 Trees are dying. Forest killed by carbon dioxide and a warning sign about the danger. The hazard has killed one person to date. The CO_2 is gassing up through soils from subsurface magma and related volcanic activity. [Photo by Bobbé Christopherson.]

FIGURE 12.24 Tectonic settings of volcanic activity.
Magma rises and lava erupts from rifts, through crust above subduction zones, and where thermal plumes at hot spots break through the crust. [After U.S. Geological Survey, *The Dynamic Planet* (Washington, DC: Government Printing Office, 1989).]

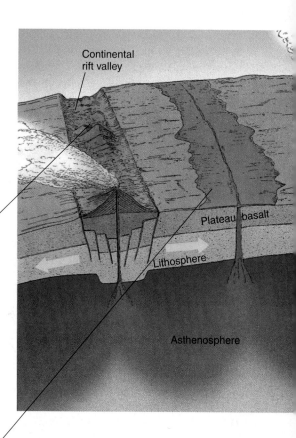

FIGURE 12.25 Rift Valley.
This rift is in Iceland, near Gullfoss, where the mid-Atlantic ridge system is seen above sea level, spreading as new material increases the size of this island country. [Photo by Nada Pecnik/Visuals Unlimited.]

FIGURE 12.26 Two expressions of volcanic activity.
In the foreground are plateau (flood) basalts characteristic of the Columbia Plateau in Oregon. Mount Hood, in the background to the west of the plateau, is a composite volcano, part of the Cascade Range of active and dormant volcanoes that result from subduction of the Juan de Fuca plate beneath the North American plate. [Photo by author.]

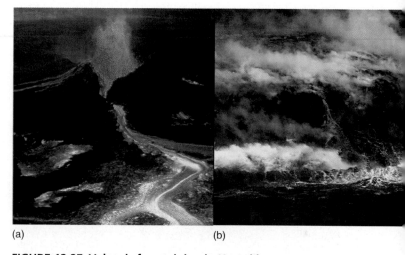

(a)   (b)

FIGURE 12.27 Volcanic fountaining in Hawai'i.
(a) Kīlauea erupts a fountain of lava from its East Rift spatter cone. (b) Lava flows reach the sea, adding new land to the big island of Hawai'i, in continuing eruptions from Earth's most active volcano. [(a) Photo by Kepa Maly/U.S. Geological Survey; (b) photo by Bobbé Christopherson.]

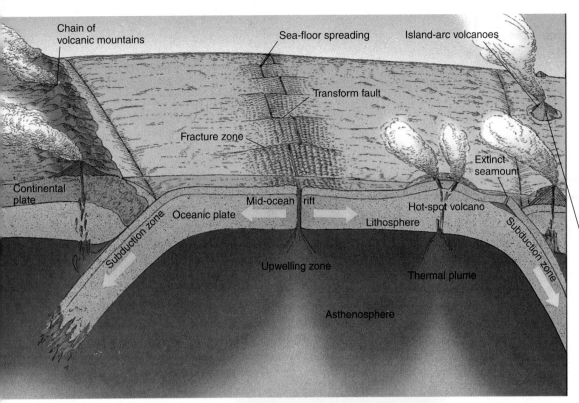

Chain of volcanic mountains

Sea-floor spreading

Island-arc volcanoes

Transform fault

Fracture zone

Continental plate

Extinct seamount

Mid-ocean rift

Subduction zone   Oceanic plate

Hot-spot volcano

Lithosphere

Subduction zone

Upwelling zone

Thermal plume

Asthenosphere

FIGURE 12.29 Extrusive igneous rocks compared.
A dacite volcanic bomb (lighter rock on left) from Mount St. Helens and a basalt lava rock (darker rock on right) from Hawai'i. What does this coloration tell you about the rocks' history and composition? Find these rock types in Table 11.2. [Photos by Bobbé Christopherson.]

FIGURE 12.28 Kīlauea landscape.
Aerial photo of the newest land on the planet produced by the massive flows of basaltic lavas from the Kīlauea volcano, Hawai'i Volcano National Park. In the distance, ocean water quenches the intense 1200°C (2200°F) lava on contact, producing steam and hydrochloric acid mist. [Photo by Bobbé Christopherson.]

FIGURE 12.30 Volcanic island off the coast of New Zealand.
White's Isle, New Zealand (37.5° S 177° E) actively erupted through 2001, sometimes gas and steam, other times bursts of pyroclastics and ash without warning. [Photo by Harvey Lloyd/Stock Market.]

which is related to its source, and (2) the magma's viscosity. Viscosity is the magma's resistance to flow ("thickness"), ranging from low viscosity (very fluid) to high viscosity (thick and flowing slowly). We consider two types of eruptions—effusive and explosive—and the characteristic landforms they build.

Effusive Eruptions

Effusive eruptions are the relatively gentle ones that produce enormous volumes of lava annually on the seafloor and in places such as Hawai'i and Iceland. These direct eruptions from the asthenosphere and upper mantle produce a low-viscosity magma that is very fluid and cools to form a dark, basaltic rock (less than 50% silica and rich in iron and magnesium). Gases readily escape from this magma because of its low viscosity, causing a relatively gentle **effusive eruption** that pours out on the surface, with relatively small explosions and little pyroclastics. However, dramatic fountains of basaltic lava sometimes shoot upward, powered by jets of rapidly expanding gases.

An effusive eruption may come from a single vent or from the flank of a volcano, through a side vent. If such vents form a linear opening, they are called *fissures*; these sometimes erupt in a dramatic *curtain of fire* (sheets of molten rock spraying into the air). In Iceland, active fissures are spread throughout the plateau landscape. Rift zones capable of erupting tend to converge on the central crater, or vent, as they do in Hawai'i. The interior of such a crater, often a sunken caldera, may fill with low-viscosity magma during an eruption, forming a molten lake, which then may overflow lava downslope in dramatic rivers and falls of molten rock.

On the island of Hawai'i, the continuing Kīlauea eruption is the longest in recorded history, active since January 3, 1983—the 55th episode of this series began in February 1997! More than 60 separate eruptions occurred at Kīlauea's summit between 1778 (the arrival of westerners) and 1982. During 1989–1990, lava flows from Kīlauea actually consumed several visitor buildings in the Hawai'i Volcanoes National Park and two housing subdivisions in Kalapana, Kapa'ahu, and Kaimu. Kīlauea has produced more lava than any other vent on Earth in recorded history (Part 3's opening photo, Figures 12.27 and 12.28). Pu'u O'o is the active crater at Kīlauea and is presently continuing to send streams of lava to the ocean (Figure 12.31).

FIGURE 12.31 Scientists monitor the active Pu'u O'o, at Kīlauea, Hawai'i.
The cracked and faulted crater of Pu'u O'o, part of the Kīlauea volcano, Hawai'i, is a risky environment for volcanic research; note the helicopter and two people at the crater rim edge! Direct measurement and monitoring give scientists data to forecast volcanic eruptions. Active lava from the crater continues to make its way to the ocean. [Photo by Bobbé Christopherson.]

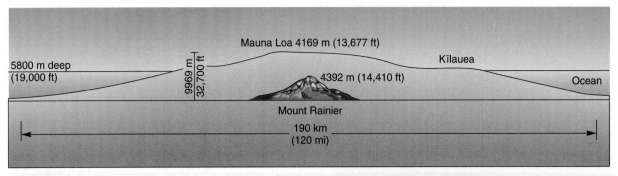

FIGURE 12.32 Shield and composite volcanoes compared.
Comparison of Mauna Loa in Hawai'i, a shield volcano, and Mount Rainier in Washington
State, a composite volcano. Their strikingly different profiles signify their different tectonic
origins. Note the gently sloped shield shape in the inset photo. [After U.S. Geological Survey,
Eruption of Hawaiian Volcanoes (Washington, DC: Government Printing Office, 1986). Inset
photo by Bobbé Christopherson.]

A typical mountain landform built from effusive eruptions is gently sloped, gradually rising from the surrounding landscape to a summit crater. The shape is similar in
outline to a shield of armor lying face up on the ground
and therefore is called a **shield volcano**. The shield shape
and size of Mauna Loa in Hawai'i is distinctive when compared with Mount Rainier in Washington, which is a different type of volcano (explained shortly) and the largest
in the Cascade Range (Figure 12.32).

The height of the Mauna Loa shield is the result of
successive eruptions, flowing one on top of another. Mauna
Loa is one of five shield volcanoes that make up the island
of Hawai'i. At least 1 million years were needed to accumulate its mass. Mauna Kea is slightly taller, but Mauna
Loa is the most massive single mountain on Earth.

Effusive eruptions send material out through hot spots
or elongated fissures, such as a continental rift valley, forming extensive sheets of basaltic lava on the surface (see
Figure 12.25). The Columbian Plateau of the northwestern United States, some 2–3 km thick, is the result of the
eruption of **plateau basalts**, or *flood basalts* (foreground in
Figure 12.26).

More than double the size of the Columbian Plateau
is the Deccan Traps, which dominates west-central India.
(*Trap* is Dutch for "staircase," referring to the typical step-
like eroded lavas.) The Siberian Traps is more than twice
the area of that in India and is exceeded only by the Ontong
Java Plateau, which covers an area of the seafloor in the
Pacific (Figure 12.33). These regions are sometimes referred to as *plateau basalt provinces*. Of these extensive igneous provinces, no presently active sites come close in size
to the largest of the extinct igneous provinces, some of
which formed more than 200 million years ago.

Explosive Eruptions

Volcanic activity inland from subduction zones produces
the infamous explosive volcanoes. Magma produced by the
melting of subducted oceanic plate and other materials is
thicker (more viscous) than magma from effusive volcanoes; it is 50%–75% silica and high in aluminum. Consequently, it tends to block the magma conduit inside the
volcano; the blockage traps and compresses gases, causing
pressure to build and conditions for a possible **explosive
eruption**. From this magma, a lighter dacitic rock forms at
the surface, as illustrated in the comparison in Figure 12.29
(see *dacite* in Table 11.2). Unlike the volcanoes in Hawai'i
Volcanoes National Park, where tourists gather at observation platforms or hike out across the cooling lava flows
to watch the relatively calm effusive eruptions, these explosive eruptions do not invite close inspection and can ex-
plode with little warning.

The term **composite volcano** describes these explosively formed mountains. (They are sometimes called
stratovolcanoes because they are built up in alternating layers of ash, rock, and lava, but shield volcanoes also can exhibit a stratified structure, so *composite* is the preferred
term.) Composite volcanoes tend to have steep sides, are
more conical in shape than shield volcanoes, and therefore
are also known as *composite cones*. If a single summit vent
erupts repeatedly, a remarkable symmetry may develop as
the mountain grows in size, as demonstrated by Mount
Orizaba in Mexico, Mount Shishaldin in Alaska (Figure
12.34), Mount Fuji in Japan, the pre-1980-eruption shape
of Mount St. Helens in Washington, and Mount Mayon
in the Philippines.

Owing to its chemical makeup and high viscosity, rising magma in a composite volcano forms a plug near the

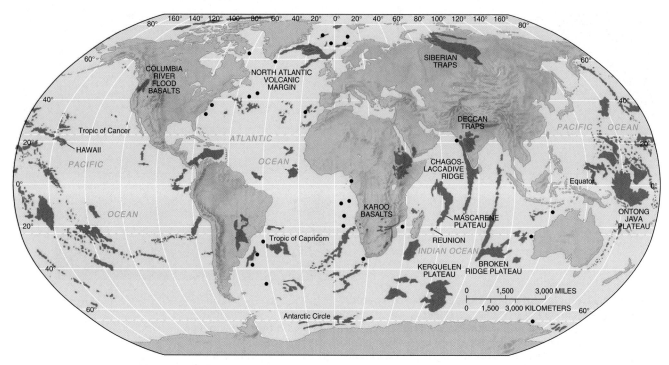

FIGURE 12.33 Earth's extensive igneous provinces.
Plateau basalts form large igneous provinces across the globe. Many are ancient and inactive. The largest known is the Ontong Java Plateau, which covers nearly 2 million square kilometers (0.8 million square miles) on the Pacific Ocean floor. [After M. F. Coffin and O. Eldholm, "Large Igneous Provinces," *Scientific American* (October 1993): 42–43. © Scientific American, Inc.]

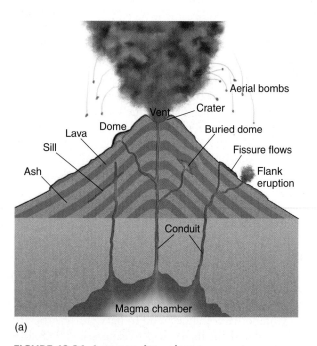

(a)

(b)

FIGURE 12.34 A composite volcano.
(a) A typical composite volcano with its cone-shaped form. (b) An eruption of Mount Shishaldin, Alaska, a composite volcano, December 23, 1995. It was active through 2001. [(b) Photo courtesy of AeroMap U.S., Inc., Anchorage, Alaska.]

surface. The blockage causes tremendous pressure to build, keeping the trapped gases compressed and liquefied. When the blockage can no longer hold back this pressurized inferno, explosions equivalent to megatons of TNT blast the tops and sides off these mountains. This type of eruption produces much less lava than effusive eruptions, but larger amounts of *pyroclastics*, which include volcanic *ash* (<2 mm in diameter), dust, cinders, *lapilli* (up to 32 mm in

Table 12.3	Notable Composite Volcano Eruptions*		
Date	Location	Number of Deaths	Amount Extruded (mostly pyroclastics) in km³ (mi³)
Prehistoric	Yellowstone, Wyoming	Unknown	2400 (576)
4600 B.C.	Mount Mazama (Crater Lake, Oregon)	Unknown	50–70 (12–17)
1900 B.C.	Mount St. Helens	Unknown	4 (0.95)
A.D. 79	Mount Vesuvius, Italy	20,000	3 (0.7)
1500	Mount St. Helens	Unknown	1 (0.24)
1815	Tambora, Indonesia	66,000	80–100 (19–24)
1883	Krakatau, Indonesia	36,000	18 (4.3)
1902	Mont Pelée, Martinique	29,000	Unknown
1912	Mount Katmai, Alaska	Unknown	12 (2.9)
1943–1952	Paricutín, Mexico	0	1.3 (0.30)
1980	Mount St. Helens	54	4 (0.95)
1985	Nevado del Ruiz, Colombia	23,000	1 (0.24)
1991	Mount Unzen, Japan	10	2 (0.5)
1991	Mount Pinatubo, Philippines	800	12 (3.0)
1992	Mount Spurr/Mount Shishaldin, Alaska	0	1 (0.24)
1993	Galeras Volcano, Colombia	5	1 (0.24)
1993	Mount Mayon, Philippines	0	1 (0.24)
1994	Kliuchevskoi, Russia	0	1 (0.24)

*See opening section under Volcanism for listings of other volcanic activity since 1995.

diameter), *scoria* (volcanic slag), pumice, and *aerial bombs* (explosively ejected blobs of incandescent lava). Table 12.3 presents some notable composite volcano eruptions. Focus Study 12.1 details the highly publicized 1980 eruption of Mount St. Helens.

Mount Pinatubo Eruption   In June 1991, after 600 years of dormancy, Mount Pinatubo in the Philippines erupted. The summit of the 1460-m (4795-ft) volcano ex-

ploded, devastating many surrounding villages and permanently closing Clark Air Force Base operated by the United States. Fortunately, scientists from the U.S. Geological Survey and local scientists accurately predicted the eruption. A resulting timely evacuation of the surrounding countryside saved thousands of lives, but 800 people were killed.

Although volcanoes are regional events, their spatial implications can be worldwide. The single volcanic eruption

Focus Study 12.1

The 1980 Eruption of Mount St. Helens

Probably the most studied and photographed composite (explosive) volcano on Earth is Mount St. Helens, located 70 km (45 mi) northeast of Portland, Oregon, and 130 km (80 mi) south of the Tacoma–Seattle area of Washington. Mount St. Helens is the youngest and most active of the Cascade Range of volcanoes, which form a line from Mount Lassen in California to Mount Baker in Washington (Figure 1). The Cascade Range is the product of the Juan de Fuca sea-floor spreading center off the coast of northern

California, Oregon, Washington, and British Columbia and the plate subduction that occurs offshore, as identified in the upper right of Figure 12.12.

The mountain had been quiet since 1857. New activity began in March 1980, with a sharp earthquake registering a magnitude 4.1. The first eruptive outburst occurred one week later, beginning with a magnitude 4.5 quake and continuing with a thick black plume of ash and the development of a small summit crater. Ten days later, the first volcanic earthquake,

called a *harmonic tremor*, registered on the many instruments that had been hurriedly placed around the volcano. Harmonic tremors are slow, steady vibrations, unlike the sharp releases of energy associated with tectonic earthquakes and faulting. Harmonic tremors told scientists that magma was on the move within the mountain.

Also developing was a massive bulge on the north side of the mountain. This bulge indicated the direction of the magma flow within the volcano.

(continued)

Focus Study 12.1 *(continued)*

(a)

(b)

(c)

FIGURE 1 Mount St. Helens before and after the eruption—days and years later.
(a) Mount St. Helens prior to the 1980 eruption. (b) The devastated and scorched land shortly after the eruption in 1980. The scorched earth and tree blowdown area covered some 38,950 hectares (95,000 acres). (c) A landscape in recovery as life moves back in and takes hold to establish new ecosystems in 1999 along the Toutle River and debris flow. [(a) Photo by Pat and Tom Lesson/Photo Researchers; (b) photo by Krafft-Explorer/Photo Researchers, Inc.; (c) photo by Bobbé Christopherson.]

A bulge represents the greatest risk from a composite volcano, for it could signal a potential lateral burst through the bulge and across the landscape.

Early on Sunday, May 18, the area north of the mountain was rocked by a magnitude 5.0 quake, the strongest to date. The mountain, with its distended 245-m (800-ft) bulge, was shaken, but nothing happened. Then a second quake (magnitude 5.1) hit at 8:32 A.M., loosening the bulge and launching the eruption. David Johnston, a volcanologist with the U.S. Geological Survey, was only 8 km (5 mi) from the mountain, servicing instruments, when he saw the eruption begin. He radioed headquarters in Vancouver, Washington, saying, "Vancouver, Vancouver, this is it!" He perished in the eruption. For a continuously updated live picture of Mount St. Helens from Johnston's approximate observation point and other specifics, go to http://www.fs.fed.us./gpnf/mshnvm/volcanocam/. The camera is mounted below the roof line at the Johnston Ridge Observatory 8 km from the mountain. The observatory is open to the public and is named for this brave scientist.

As the contents of the mountain exploded, a surge of hot gas (about 300°C, or 570°F), steam-filled ash, pyroclastics, and a *nuée ardente* (rapidly moving, very hot, explosive ash and incandescent gases) moved northward, hugging the ground and traveling at speeds up to 400 kmph (250 mph) for a distance of 28 km (17 mi).

The slumping north face of the mountain produced the greatest landslide witnessed in recorded history; about 2.75 km³ (0.67 mi³) of rock, ice, and trapped air, all fluidized with steam, surged at speeds approaching 250 kmph (155 mph). Landslide materials traveled for 21 km (13 mi) into the valley, blanketing the forest, covering a lake, and filling the rivers below. A series of photographs, taken at 10-second intervals from the east looking west, records this sequence (Figure 2). The eruption continued with intensity for nine hours, first clearing out old rock from the throat of the volcano and then blasting new material.

As destructive as such eruptions are, they also are constructive, for this is the way in which a volcano eventually builds its height. Before the eruption, Mount St. Helens was 2950 m (9677 ft) tall; the eruption blew away 418 m (1370 ft). Today, Mount St. Helens is building a lava dome within its crater. The thick lava rapidly and repeatedly plugs and breaks in a series of lesser dome eruptions that may continue for several decades. The dome already is more than 300 m (1000 ft) high, so a new mountain is being born from the eruption of the old. The dome is covered with a dozen survey benchmarks and several tiltmeters to monitor the status of the volcano. Swarms of minor earthquakes along the north flank of the mountain began in November 2001, marking this mountain as still unstable.

An ever-resilient ecology is recovering as plants and animals reclaim the devastated landscape. Please refer to the dramatic comparison photos from 1983 and 1999 in Figure 19.27. More than two decades have passed, yet strong interest in the area continues; scientists and more than 1 million tourists visited the Mount St. Helens Volcanic National Monument each year. For more information, see the Cascades Volcano Observatory at http://vulcan.wr.usgs.gov/ and specifically Mount St. Helens at http://vulcan.wr.usgs.gov/Volcanoes/MSH/framework.html.

Focus Study 12.1 *(continued)*

FIGURE 2 The Mount St. Helens eruption sequence and corresponding schematics.
[Photo sequence by Keith Ronnholm. All rights reserved.]

of Mount Pinatubo was significant to the global environment and energy budget, as discussed in Chapters 1, 3, 4, 5, and 10:

- 15–20 million tons of ash and sulfuric acid mist were blasted into the atmosphere, concentrating at 16–25 km (10–15.5 mi) altitude
- 12 km^3 (3.0 mi^3) of material was ejected and extruded by the eruption (12 times the volume from Mount St. Helens)
- 60 days after the eruption, about 42% of the globe was affected (from 20° S to 30° N) by the thin, spreading aerosol cloud in the atmosphere
- colorful twilight and dawn skies were observed worldwide
- an increase in atmospheric albedo of 1.5% (4.3 W/m^2) occurred
- an increase in the atmospheric absorption of insolation followed (2.5 W/m^2)
- a decrease in net radiation at the surface and a lowering of Northern Hemisphere average temperatures of 0.5 C° (0.9 F°) were measured
- atmospheric scientists and volcanologists were able to study the eruption aftermath using satellite-borne orbiting sensors and general circulation model computer simulations.

Volcano Forecasting and Planning

The USGS and the Office of Foreign Disaster Assistance (U. S. AID) operate the Volcano Disaster Assistance Program (VDAP; see **http://vulcan.wr.usgs.gov/Vdap/framework.html**). The need for such a program is evident in that, over the past 15 years, volcanic activity has killed 29,000 people, forced more than 800,000 to evacuate their homes, and caused more than $3 billion in damage. The program was established after 23,000 people died in the eruption of Nevado del Ruiz, Colombia, in 1985.

As an example of the need for accurate forecasting, a volcano watch continues at Soufriere Hills, an erupting composite volcano on the island of Montserrat in the Lesser Antilles (17° N 62° W). The Montserrat Volcano Observatory reports a dozen evacuations since volcanic activity began in 1995 and strengthened through 1998. Are these precursors to a main event? Or just routine rumblings that will grow quiet again? Although the U.S. VDAP is in place to help local scientists with eruption forecasts, residents remain skeptical after many false alerts, which only heightens the risk. VDAP is at work at more than two dozen volcanoes in the world.

The key to the effort is to set up mobile volcano-monitoring systems at the most-threatened sites. An effort such as this led to the life-saving evacuation of 60,000 people hours before Mount Pinatubo exploded. In addition, satellite remote sensing is helping VDAP to monitor eruption cloud dynamics, atmospheric emissions and climatic effects, and lava and thermal measurements; to make topographic measurements; to estimate volcanic hazard potential; and to enhance geologic mapping—all in an effort to better understand our dynamic planet. Integrated seismographic networks and monitoring are making possible early warning systems.

In this era of the Internet you can access "volcano cams" positioned around the world giving you 24-hour surveillance of many volcanoes. For an exciting visual adventure, go to the following URL and add it to your bookmarks (note whether it is day or night for the location you are checking): **http://vulcan.wr.usgs.gov/LivingWith/volcano_cams.html#world**.

Summary and Review—Tectonics, Earthquakes, and Volcanism

● *Describe* first, second, and third orders of relief and *relate* examples of each from Earth's major topographic regions.

Earth's surface is dramatically shaped by tectonic forces generated within the planet. **Relief** is the vertical elevation difference in a local landscape. The undulating physical surface of Earth, including relief, is called **topography**. Convenient descriptive categories are termed *orders of relief*. The coarsest level of landforms includes the **continental platforms** and **ocean basins**; the finest comprises local hills and valleys.

relief (p. 358)
topography (p. 358)
continental platforms (p. 359)
ocean basins (p. 359)

1. How does the map of the ocean floor (chapter-opening illustration) exhibit the principles of plate tectonics? Briefly analyze.
2. What is meant by an "order of relief"? Give an example from each order.

3. Explain the difference between relief and topography.

● *Describe* the several origins of continental crust and *define* displaced terranes.

The continents are formed as a result of several processes, including upwelling material from below the crust and migrating portions of crust. The source of material generated by plate tectonic processes determines the behavior and the composition of the crust that results.

A continent has a nucleus of ancient crystalline rock called a *craton*. A region where a craton is exposed is termed a **continental shield**. As continental crust forms, it is enlarged through accretion of dispersed **terranes**. An example is the *Wrangellia terrane* of the Pacific Northwest and Alaska.

continental shield (p. 362)
terranes (p. 363)

4. What is a craton? Relate this structure to continental shields and platforms, and describe these regions in North America.

5. What is a migrating terrane, and how does it add to the formation of continental masses?

6. Briefly describe the journey and destination of the Wrangellia Terrane.

● *Explain* **compressional processes and folding;** *describe* **four principal types of faults and their characteristic landforms.**

Folding, broad warping, and faulting, deform the crust and produce characteristic landforms. Compression causes rocks to deform in a process known as **folding**, during which rock strata bend and may overturn. Along the *ridge* of a fold, layers *slope downward away from the axis*, which is an **anticline**. In the *trough* of a fold, however, layers *slope downward toward the axis*, called a **syncline**.

When rock strata are stressed beyond their ability to remain a solid unit, they express the strain as a fracture. Rocks on either side of the fracture are displaced relative to the other side in a process known as **faulting**. Thus, *fault zones* are areas where fractures in the rock demonstrate crustal movement. At the moment of fracture, a sharp release of energy, called an **earthquake**, or *quake*, occurs.

When forces pull rocks apart, the tension causes a **normal fault**, sometimes visible on the landscape as a scarp, or escarpment. Compressional forces associated with converging plates force rocks to move upward, producing a **reverse fault**. A low-angle fault plane is referred to as a **thrust fault**. Horizontal movement along a fault plane that produces a linear rift valley is a **strike-slip fault**. In the U.S. interior west, the Basin and Range Province is an example of aligned pairs of normal faults and a distinctive horst-and-graben landscape. The term **horst** is applied to upward-faulted blocks; **graben** refers to downward-faulted blocks.

folding (p. 365)
anticline (p. 365)
syncline (p. 365)
faulting (p. 369)
earthquake (p. 369)
normal fault (p. 369)
reverse fault (p. 369)
thrust fault (p. 369)
strike-slip fault (p. 369)
horst (p. 369)
graben (p. 369)

7. Diagram a simple folded landscape in cross section, and identify the features created by the folded strata.

8. Define the four basic types of faults. How are faults related to earthquakes and seismic activity?

9. How did the Basin and Range Province evolve in the western United States? What other examples exist of this type of landscape?

● *Relate* **the three types of plate collisions associated with orogenesis and** *identify* **specific examples of each.**

Orogenesis is the birth of mountains. An orogeny is a mountain-building episode, occurring over millions of years, that thickens continental crust. It can occur through large-scale deformation and uplift of the crust. It also may include the capture of migrating terranes and cementation of them to the continental margins and the intrusion of granitic magmas to form plutons.

Oceanic plate–continental plate collision orogenesis is now occurring along the Pacific coast of the Americas and has formed the Andes, the Sierra of Central America, the Rockies, and other western mountains. *Oceanic plate–oceanic plate* collision orogenesis produces either simple volcanic island arcs or more complex arcs such as Japan, the Philippines, the Kurils, and portions of the Aleutians. The region around the Pacific contains expressions of each type of collision in the **circum-Pacific belt**, or the **ring of fire**.

Continental plate–continental plate collision orogenesis is quite mechanical; large masses of continental crust, such as the Himalayan Range, are subjected to intense folding, overthrusting, faulting, and uplifting.

orogenesis (p. 371)
circum-Pacific belt (p. 373)
ring of fire (p. 373)

10. Define orogenesis. What is meant by the birth of mountain chains?

11. Name some significant orogenies.

12. Identify on a map Earth's two large mountain chains. What processes contributed to their development?

13. How are plate boundaries related to episodes of mountain building? Explain how different types of plate boundaries produce differing orogenic episodes and different landscapes.

14. Relate tectonic processes to the formation of the Appalachians and the Alleghany orogeny.

● *Explain* **the nature of earthquakes, their measurement, and the nature of faulting.**

Earthquakes generally occur along plate boundaries; major ones can be disastrous. Earthquakes result from faults, which are under continuing study to learn the nature of faulting, stress and the buildup of strain, irregularities along fault plane surfaces, the way faults rupture, and the relationship among active faults. Earthquake prediction and improved planning are active concerns of *seismology*, the study of earthquake waves and Earth's interior. Seismic motions are measured with a **seismograph**.

Charles Richter developed the **Richter scale**, a measure of earthquake magnitude. A more precise and quantitative scale that assesses the *seismic moment* is now used, especially for larger quakes; this is the **moment magnitude scale**. The specific mechanics of how a fault breaks are under study, but the **elastic-rebound theory** describes the basic process. In general, two sides along a fault appear to be locked by friction, resisting any movement. This stress continues to build strain along the fault surfaces, storing elastic energy like a wound-up spring. When energy is released abruptly as the rock breaks, both sides of the fault return to a condition of less strain.

seismograph (p. 377)
Richter scale (p. 378)
moment magnitude scale (p. 379)
elastic-rebound theory (p. 379)

15. Describe the differences in human response between the two earthquakes that occurred in China in 1975 and 1976.

16. Differentiate between the Mercalli and moment magnitude and amplitude scales. How are these used to describe an earthquake? Why has the Richter scale been updated and modified?

17. What is the relationship between an epicenter and the focus of an earthquake? Give examples from the Loma Prieta, California, and Kobe, Japan, earthquakes.

18. What local soil and surface conditions in San Francisco severely magnifies the energy felt in earthquakes?

19. How do the elastic-rebound theory and asperities help explain the nature of faulting?

20. Describe the San Andreas fault and its relationship to ancient sea-floor spreading movements along transform faults.

21. How is the seismic gap concept related to expected earthquake occurrences? Are any gaps correlated with earthquake events in the recent past? Explain.

22. What do you see as the biggest barrier to effective earthquake prediction?

● *Distinguish* between an effusive and an explosive volcanic eruption and *describe* related landforms, using specific examples.

Volcanoes offer direct evidence of the makeup of the asthenosphere and uppermost mantle. A **volcano** forms at the end of a central vent or pipe that rises from the asthenosphere through the crust into a volcanic mountain. A **crater**, or circular surface depression, usually forms at the summit. Areas where magma is near the surface may heat groundwater, producing *geothermal energy*.

Eruptions produce **lava** (molten rock), gases, and **pyroclastics** (pulverized rock and clastic materials ejected violently during an eruption) that pass through the vent to openings and fissures at the surface and build volcanic landforms, such as a **cinder cone**, a small hill, or a large basin-shaped depression sometimes caused by collapse of a volcano's summit and called a **caldera**.

Volcanoes are of two general types, based on the chemistry and the viscosity of the magma involved. **Effusive eruption** produces a **shield volcano** (such as Kīlauea in Hawai'i) and extensive deposits of **plateau basalts**, or flood basalts. Magma of higher viscosity leads to an **explosive eruption** (such as Mount Pinatubo in the Philippines), producing a **composite volcano**. Volcanic activity has produced some destructive moments in history but constantly creates new seafloor, land, and soils.

  volcano (p. 384)
  crater (p. 384)
  geothermal energy (p. 384)
  lava (p. 384)
  pyroclastics (p. 384)
  cinder cone (p. 384)
  caldera (p. 384)
  effusive eruption (p. 388)
  shield volcano (p. 389)
  plateau basalts (p. 389)
  explosive eruption (p. 389)
  composite volcano (p. 389)

23. What is a volcano? In general terms, describe some related features.

24. Where do you expect to find volcanic activity in the world? Why?

25. Compare effusive and explosive eruptions. Why are they different? What distinct landforms are produced by each type? Give examples of each.

26. Describe several recent volcanic eruptions.

NetWork

The *Geosystems* Home Page provides on-line resources for this chapter on the World Wide Web. To begin: Once on the Home Page, click on this textbook, scroll the Table of Contents menu, select this chapter, and click "Begin." You will find self-tests that are graded, review exercises, specific updates for items in the chapter, and many links to interesting related pathways on the Internet under "Destinations." *Geosystems* is at **http://www.prenhall.com/christopherson**.

Critical Thinking

A. Using the topographic regions map in Figure 12.3, assess the region within 100 km (62 mi) and 1000 km (620 mi) of your campus. Describe the topographic character and the variety of relief within these two regional scales. You may want to consult local maps and atlases in your analysis. Do you perceive that this type of topographic region influences lifestyles? Economic activities? Transportation? History of the region?

B. Determine from library, Internet, and local agencies, the seismic potential of the region in which your campus is sited. If a hazard exists, does it influence regional planning? Availability of property insurance? If appropriate, are there existing disaster or emergency plans for your campus? Community, state or province? Explain any other significant issues you found in working on this critical thinking challenge.

Spheroidal weathering of granite rocks in the Alabama Hills (in foreground) of eastern California. [Photo by Bobbé Christopherson.]

13

Weathering, Karst Landscapes, and Mass Movement

Key Learning Concepts

After reading the chapter, you should be able to:

- *Define* the science of geomorphology.
- *Illustrate* the forces at work on materials residing on a slope.
- *Define* weathering and *explain* the importance of the parent rock and joints and fractures in rock.
- *Describe* frost action, crystallization, hydration, pressure-release jointing, and the role of freezing water as physical weathering processes.
- *Describe* the susceptibility of different minerals to the chemical weathering processes called hydrolysis, oxidation, carbonation, and solution.
- *Review* the processes and features associated with karst topography.
- *Portray* the various types of mass movements and *identify* examples of each in relation to moisture content and speed of movement.

A benefit you receive from a physical geography course is a new appreciation of the scenery. Whether you go by foot, bicycle, car, train, or plane, travel is an opportunity to experience Earth's varied landscapes and to witness the active processes that produce them.

In the last two chapters, we discussed the endogenic (internal) processes of our planet and how they produce landforms. However, as the landscape is formed, several exogenic (external) processes simultaneously wear and waste it. Perhaps you noticed rough and broken highways in areas that experience freezing temperatures. The pavement breaks into chunks each winter and develops potholes. Or maybe you have seen older marble structures such as tombstones that rainwater has etched and dissolved. Perhaps after a rainstorm you noticed mud flowing from a hillslope or maybe large segments of a hillside ready to let loose in a slide or flow of debris.

Here we begin a five-chapter examination of exogenic processes at work on the landscape. This chapter examines weathering and mass movement of the lithosphere. The next four chapters look at specific exogenic agents and their handiwork—river systems, wind-influenced landscapes, deserts in water-deficit regions, coastal processes and landforms, and regions worked by ice and glaciers. Whether you enjoy time along a river, love the desert or the waves along a coastline, or live in a place where glaciers once carved the land, you will find something of interest in these chapters.

In this chapter: We look at physical (mechanical) and chemical weathering processes that break up, dissolve, and generally reduce the landscape. Such weathering releases essential minerals from bedrock for soil formation or enrichment. Perhaps you live in a region that has caves and caverns. Water has dissolved enormous underground worlds of mystery and darkness, yet many caves remain undiscovered. In addition, we examine mass movement processes that continually operate in and upon the land-scape. A news broadcast may describe an avalanche in Colombia, mudflows in Los Angeles or Peru, a landslide in Turkey, a massive rockfall in Yosemite, or lava flows on the slopes of a volcano in Italy.

Landmass Denudation

Geomorphology is the science of landforms—their origin, evolution, form, and spatial distribution. This science is an important aspect of physical geography. **Denudation** is any process that wears away or rearranges landforms. The principal denudation processes affecting surface materials include *weathering, mass movement, erosion, transportation,* and *deposition,* as produced by the agents of moving water, air, waves, and ice, all influenced by the pull of gravity.

Interactions between the structural elements of the land and denudation processes are complex. They represent a continuing struggle between Earth's internal and external processes, between the resistance of materials and weathering and erosional processes. The 15-story-tall Delicate Arch in Utah is dramatic evidence of this struggle (Figure 13.1). Differing resistances of the rocks, coupled with variations in the processes at work on the rock, are carving this delicate sculpture—an example of **differential weathering**, where a more resistant cap rock protects supporting strata below.

FIGURE 13.1 Weathered landform.
Delicate Arch—a dramatic example of differential weathering in Arches National Park, Utah. Resistant rock strata at the top of the structure have helped preserve the arch beneath them as surrounding rock was eroded away. Note the person standing at the base for a sense of scale. [Photo by author.]

Ideally, endogenic processes build *initial landscapes*, whereas exogenic processes develop *sequential landscapes* of low relief, gradual change, and stability. Several hypotheses have been proposed to model denudation processes and to account for the appearance of the landscape in its developmental stages.

Geomorphic Models of Landform Development

Famed American geographer and geomorphologist William Morris Davis (1850–1934) proposed the erosion cycle, or *geomorphic cycle model*. Davis theorized that a landscape undergoes initial uplift that is accompanied by erosion or removal of materials. From the uplifted landscape, streams begin flowing more rapidly, cutting more energetically, both headward (upstream) and deeper. Davis believed that slope angle is gradually reduced and that ridges and divides become rounded and lowered over time. According to Davis's cyclic model, the landscape eventually evolves into an old erosional surface (discussed further in Chapter 14).

Davis's theory, which helped launch the science of geomorphology and was innovative at the time, did not account for the processes being observed in modern geomorphology studies. Academic support for his cyclic and evolutionary landform development model declined by 1960. Although not generally accepted today, his thinking about the evolution of landscapes is influential and some of his terminology is still used.

Dynamic Equilibrium View of Landforms

Most geomorphologists prefer the **dynamic equilibrium model**, which emphasizes a balance among force, form, and process. The dynamic equilibrium model summarizes this balancing act between tectonic uplift and reduction by weathering and erosion, between the resistance of rocks and the ceaseless attack of weathering and erosion. A dynamic equilibrium demonstrates a trend over time. According to current thinking, landscapes in a dynamic equilibrium feature ongoing adaptations to the ever-changing conditions of rock structure, climate, local relief, and elevation.

A landscape is an open system, with highly variable inputs of energy and materials: Uplift creates the *potential energy of position* above sea level and therefore a disequilibrium, an imbalance, between relief and energy. The Sun provides radiant energy that converts into *heat energy*. The hydrologic cycle imparts *kinetic energy* through mechanical motion. *Chemical energy* is made available from the atmosphere and various reactions within the crust. In response to this input of energy and materials, landforms constantly adjust toward *equilibrium*.

Endogenic events (such as earthquakes and volcanic eruptions), or exogenic events (such as heavy rainfall or forest fire), may provide new sets of relationships for the landscape. Following such destabilizing events, a landform system arrives at a **geomorphic threshold**—the point at which there is enough energy to overcome resistance against movement. At this threshold, the system breaks

FIGURE 13.2 A slope in disequilibrium.
Unstable, saturated soils gave way, leaving a debris dam partially blocking the river. [Photo by author.]

through to a new equilibrium as the landform adjusts. The pattern over time follows a sequence: (1) equilibrium stability (fluctuating around some average), (2) a destabilizing event, (3) a period of adjustment, and (4) development of a new and different condition of equilibrium stability.

The disturbed hillslope in Figure 13.2 is in the midst of compensating adjustment. The failure of saturated slopes caused a landslide into the river and set a disequilibrium condition. As a consequence, the new dam of material threw the stream into a disequilibrium between its flow and sediment load.

Slow, continuous-change events, such as soil development and erosion, tend to maintain an approximate equilibrium condition. Dramatic events such as a major landslide or dam collapse require longer recovery times before an equilibrium is reestablished. (Figure 1.5 graphically illustrates both the distinction between a steady-state equilibrium and a dynamic equilibrium, and the occurrence of a geomorphic threshold.)

Slopes   Material loosened by weathering is susceptible to erosion and transportation. However, if gravity is to move loosened material downslope, the agents of erosion must overcome the forces of friction, inertia (the tendency of objects at rest to remain at rest), and the cohesion of particles to each other (Figure 13.3a). If the slope angle is steep enough for gravity to overcome frictional forces, or if material is dislodged by the impact of raindrops, hail, falling branches, moving animals, wind, trail bikes, off-road vehicles, or construction equipment, then particle erosion and transport downslope occur.

Slopes or *hillslopes* are curved, inclined surfaces that form the boundaries of landforms. Figure 13.3b illustrates basic slope components that vary among slopes with conditions of rock structure and climate. Slopes generally

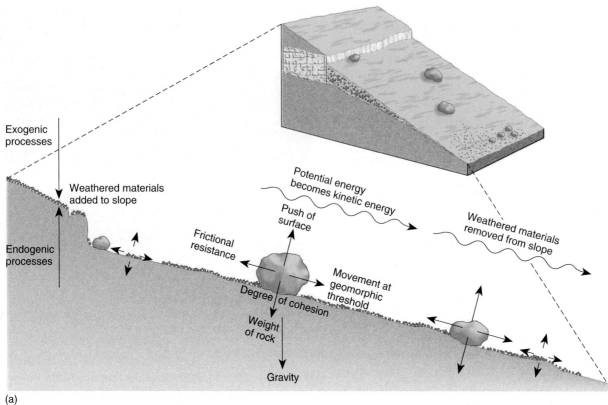

(a)

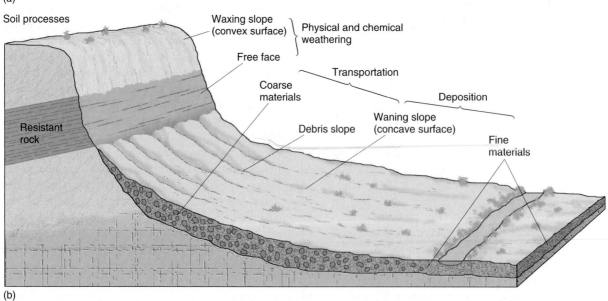

(b)

FIGURE 13.3 Slope mechanics and form.
(a) Directional forces (noted by arrows) act on materials along an inclined slope. (b) The principal elements of a slope.

feature an upper *waxing slope* near the top (*waxing* means "increasing"). This convex surface curves downward and grades into the *free face* below. The presence of a free face indicates an outcrop of resistant rock that forms a steep scarp or cliff.

Downslope from the free face is a *debris slope*, which receives rock fragments and materials from above. The condition of a debris slope reflects the local climate. In humid climates, continually moving water carries materi-

al away, lowering the debris slope. But in arid climates, debris slopes persist and accumulate. A debris slope transitions into a *waning slope*, a concave surface along the base of the slope. This surface of erosional materials gently slopes at a continuously decreasing angle to the valley floor.

A slope is an open system seeking an *angle of equilibrium*. Conflicting forces work simultaneously on slopes to establish an optimum compromise incline that balances these forces. You can identify these slope components and condi-

FIGURE 13.4 A hillslope example.
Compare this slope with the components mentioned in Figure 13.3. Note the role of the rock outcrop in interrupting the slope and the rock fragments that are loosened from the outcrop by frost action. [Photo by author.]

tions on the actual hillslope shown in Figure 13.4. When any condition in the balance is altered, all forces on the slope compensate by adjusting to a new dynamic equilibrium.

The relation between rates of weathering and breakup of slope materials, coupled with the rates of mass movement and material erosion, shape slopes. A slope is *stable* if its strength exceeds these denudation processes and *unstable* if materials are weaker than these processes. Why are hillslopes shaped in certain ways? How do slope elements evolve? How do hillslopes behave during rapid, moderate, or slow uplift? These are topics of active scientific study and research.

Now, with the concepts of landmass denudation, dynamic equilibrium, and slope development in mind, let us examine specific processes that operate to wear away landforms.

Weathering Processes

Weathering processes attack rocks at Earth's surface and to some depth below the surface. **Weathering** processes either disintegrate rock into mineral particles or dissolve them in water. Weathering processes are both physical (mechanical) and chemical.

Weathering does not transport the materials; it simply generates them for erosion and transport by the agents of water, wind, waves, and ice—all influenced by gravity. In most areas, the upper surface of bedrock undergoes continual weathering, creating broken-up rock called **regolith**. Loose surface material comes from further weathering of regolith and from transported and deposited regolith (Figure 13.5a). In some areas, regolith may be missing or undeveloped, thus exposing an outcrop of unweathered bedrock. The thickness of the soil cover depends on a bal-

ance between soil production rates and erosion (removal of soil particles)—an equilibrium between competing processes.

Bedrock is the *parent rock* from which weathered regolith and soils develop. While a soil is relatively youthful, its parent rock is traceable through similarities in composition. For example, the sand in Figure 13.5c derives its color and character from the parent rock in the background. This sandy unconsolidated fragmental material, known as **sediment**, combines with weathered rock to form the **parent material** from which soil evolves.

Factors Influencing Weathering Processes

Weathering is greatly influenced by the character of the bedrock: hard or soft, soluble or insoluble, broken or unbroken. *Jointing* in rock is important for weathering processes. **Joints** are fractures or separations in rock that occur without displacement of the sides (as would be the case in faulting). The presence of these usually plane (flat) surfaces increases the surface area of rock exposed to both physical and chemical weathering.

Important controls on weathering rates are climatic elements—precipitation, temperature, and freeze–thaw cycles. Also significant is the position of the water table and water movement (hydraulics).

Another control over weathering rates is the *geographic orientation* of a slope—whether it faces north, south, east, or west. Orientation controls the slope's exposure to Sun, wind, and precipitation. Slopes facing away from the Sun's rays tend to be cooler, moister, and more vegetated than slopes in direct sunlight. This effect of orientation is especially noticeable in the middle and higher latitudes.

Vegetation is also a factor in weathering. Although vegetative cover can protect rock by shielding it from raindrop impact and providing roots to stabilize soil, it also produces organic acids from the partial decay of organic matter; these acids contribute to chemical weathering. Plant roots can enter crevices and break up a rock, exerting enough pressure to drive rock segments apart, thereby exposing greater surface area to other weathering processes (Figure 13.6). You may have observed how tree roots can heave the sections of a sidewalk or driveway sufficiently to raise and crack the concrete.

The role of the climatic factor warrants further discussion. There is a relation among climate (annual precipitation and temperature), physical weathering, and chemical weathering processes. In general, physical weathering dominates in drier, cooler climates, whereas chemical weathering dominates in wetter, warmer climates. Extreme dryness reduces weathering rates, as is experienced in desert climates (*low-latitude arid desert BW*). In the hot, wet tropical and equatorial rain forest climates (*tropical rain forest Af*), most rocks weather rapidly, and the weathering extends deep below the surface.

The scale at which we analyze weathering processes is important. Research at *microscale* levels reveals greater complexity in the relation of climate and weathering. At the

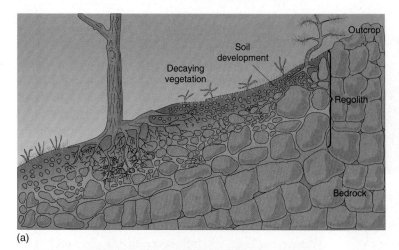

(a)

(b)

(c)

FIGURE 13.5 Regolith, soil, and parent materials.
(a) A cross section of a typical hillside. (b) A cliff exposes hillside components. (c) These reddish-colored dunes in the Navajo Tribal Park, near the Utah–Arizona border, derive their color from the red-sandstone parent materials in the background. [Photos by author.]

small scale of actual reaction sites on the rock surface, both physical and chemical weathering processes can occur across varied climate types. Hygroscopic water (a molecule-thin water layer on soil particles) and capillary water (soil water) activates chemical weathering processes, even in the driest landscape. (Review sections in Chapter 9 for these water types.)

Imagine all the factors that influence weathering rates as operating in concert: climatic influence (precipitation and temperature), soil water and groundwater, rock composition and structure (jointing), slope orientation, vegetation, and microscopic boundary-layer conditions at reaction sites. We separate these processes here for convenience of study. Of course in all this, *time* is the crucial factor, for these processes require long periods of time to operate.

Physical Weathering Processes

When rock is broken and disintegrated without any chemical alteration, the process is called **physical weathering** or *mechanical weathering*. By breaking up rock, physical weathering produces more surface area on which chemical weathering may operate. A single rock that becomes broken into eight pieces has doubled its surface area susceptible to weathering processes. We look briefly at four physical weathering processes: frost action, crystallization, hydration, and pressure-release jointing.

Frost Action   When water freezes, its volume expands as much as 9% (see Chapter 7). Such expansion creates a powerful mechanical force called **frost action**, or *freeze–thaw*

(a)

FIGURE 13.6 Organic physical weathering.

(a) Can you find the tree root growing in and on the jointing fracture in the rock? (b) Roots exert a force on the sides of this joint in the rock. [Photos by author.]

(b)

FIGURE 13.7 Climate and weathering.

Frost action shattered this granite rock; ice expansion forced the rock segments apart. [Photo by Bobbé Christopherson.]

(a)

(b)

FIGURE 13.8 Physical weathering examples.

(a) Physical weathering along joints in rock produces discrete blocks in the back country of Canyonlands National Park, Utah. (b) Frost action loosens and frees rock that falls to the base of the cliff, accumulating as a talus slope, near Wheeler Peak in Great Basin National Park, Nevada. [Photos by author.]

action, which can exceed the tensional strength of rock. Repeated freezing (expanding) and thawing (contracting) of water breaks rocks apart (Figure 13.7). Freezing actions are important in the humid microthermal D climates (*humid continental Df* and *subarctic Dfc and Dw climates*), and polar climates (*polar E climates*), and they occur at higher elevations in mountains worldwide. In arctic and subarctic climates, frost action dominates soil conditions (discussed in further detail in Chapter 17).

The work of ice begins in small openings, gradually expanding until rocks are cleaved (split). Figure 13.8 shows blocks of rock on which this *joint-block separation* occurs along existing joints and fractures. This weathering action, called *frost-wedging*, pushes portions of the rock apart. Cracking and breaking create varied shapes in the rocks, depending on the rock structure. In the photo, the softer

FIGURE 13.9 Rockfall.
Shattered rock debris from a large rockfall in Yosemite National Park. A larger rockfall involving 162,000 tons of granite shocked Yosemite in July 1996, and another in 1999. [Photo by author.]

drill-hole patterns. In spring, the pioneers hauled the blocks to town for construction material. Frost action also produces unwanted fractures that damage road pavement and burst water pipes.

Spring can be a risky time to venture into mountainous terrain. As rising temperatures melt the winter's ice, newly fractured rock pieces fall without warning and may even start rock slides. The falling rock pieces may physically shatter on impact—another form of physical weathering (Figure 13.9). One such rockfall in Yosemite National Park (1996) involved a 670-m (2200-ft) crashing drop of a 162,000-ton granite slab at 260 kmph (160 mph). The impact shattered the rock, covering 50 acres in powdered rock and felling 500 trees.

Crystallization   Especially in arid climates, dry weather draws moisture to the surface of rocks. As the water evaporates, dissolved minerals in the water grow crystals. Over time, as the crystals grow and enlarge, they exert a force great enough to spread apart individual mineral grains and begin breaking up the rock. Such *crystallization*, or *salt-crystal growth*, is a form of physical weathering.

In the Colorado Plateau of the Southwest, salty water slowly flows from rock strata. As this salty water evaporates, crystallization loosens the sand grains. Subsequent erosion and transportation by water and wind complete the sculpturing process. Deep indentations develop in sandstone cliffs, especially above impervious layers. More than 1000 years ago, Native Americans built entire villages in these weathered niches at several locations, including Mesa Verde in Colorado and Arizona's Canyon de Chelly (pronounced "canyon duh shay," Figure 13.10).

Hydration   Another process involving water, but little chemical change, is **hydration** (meaning "combination with water"). Water becomes part of the chemical composition of the mineral (such a hydrate is gypsum, which is hydrous calcium sulfate: $CaSO_4 \cdot 2H_2O$). We list this as a physical weathering process because, when some minerals absorb water, they expand, creating a strong mechanical effect that stresses the rock, forcing grains apart. A cycle of hydration and dehydration can lead to granular disintegration and further susceptibility of the rock to chemical weathering. Hydration works together with carbonation and oxidation to convert feldspar, a common mineral in many rocks, to clay minerals and silica. The hydration process is also at work on the sandstone niches shown in the cliff-dwelling photo.

Pressure-Release Jointing   Recall from Chapter 11 how rising magma that is deeply buried and subjected to high pressure forms intrusive igneous rocks called plutons. These plutons cool slowly and produce coarse-grained, crystalline granitic rocks. As the landscape is subjected to uplift, the regolith overburden is weathered, eroded, and transported away, eventually exposing the pluton as a mountainous batholith (a plutonic batholith is illustrated in Figure 11.7).

supporting rock underneath the slabs is weathering at a faster pace in *differential weathering*.

In Figure 13.8b, frost wedging loosened rock from these cliffs. The angular pieces of rock fragments cascade down and form a **talus slope**, which is a poorly sorted, cone-shaped deposit of debris at the base of a steep slope. Several talus cones constitute such a *talus slope* in the photograph.

Frost action was important to various cultures as a force to quarry rock. Pioneers in the early American West drilled holes in rock, poured water in the holes, and then plugged them. During the cold winter months, expanding ice broke off large blocks along lines determined by the

(a)

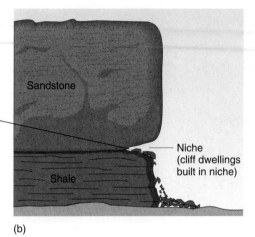

(b)

FIGURE 13.10 Physical weathering in sandstone.
(a) Cliff dwelling site in Canyon de Chelly, Arizona, occupied by the Anasazi people until about 900 years ago. The niche in the rock was formed partially by crystallization, which forced apart mineral grains and broke up the rock. The dark streaks on the rock are thin coatings of desert varnish, composed of iron oxides with traces of manganese and silica. (b) Water and impervious sandstone layer helped concentrate weathering processes in the niche. [Photo by author.]

As the tremendous weight of overburden is removed from the granite, the pressure of deep burial is relieved. Over millions of years, the granite slowly responds with an enormous physical heave. In a process known as *pressure-release jointing*, layer after layer of rock peels off in curved slabs or plates, thinner at the top of the rock structure and thicker at the sides. As these slabs weather, they slip off in a process called **sheeting**. This *exfoliation process* creates arch-shaped and dome-shaped features on the exposed landscape, sometimes forming an **exfoliation dome** (Figure 13.11). Such domes are probably the largest weathering features on Earth (in areal extent).

Chemical Weathering Processes

Chemical weathering is the actual decomposition (chemical change) of minerals in rock. Chemical weathering involves reactions between air and water and minerals in rocks. Minerals may combine with oxygen or carbon dioxide from the atmosphere, or they may dissolve and combine with water. Water is important because of its general availability and its tremendous ability to dissolve. Chemi-

cal breakdown hastens as both temperature and precipitation increase.

Although individual minerals vary in susceptibility, all rock-forming minerals are at least slightly responsive to chemical weathering. A familiar example is the eating away of cathedral façades and the etching of tombstones by acidic precipitation. Figure 13.12a illustrates chemical weathering along joints in granite as chemical processes dissolve its component minerals. This dissolution in turn opens spaces for frost action.

As an example of the way chemical weathering attacks rock, consider **spheroidal weathering**. The sharp edges and corners of rocks are rounded as the alteration of minerals progresses through the rock. Joints in the rock offer more surfaces of opportunity for more weathering. Water penetrates joints and fractures and dissolves the rock's weaker minerals or cementing materials. A boulder can be attacked from all sides, shedding spherical shells of decayed rock, like the layers of an onion. The resulting rounded edges are the basis for the name *spheroidal*.

Spheroidal weathering of rock resembles exfoliation, but it does not result from pressure-release jointing.

(a)

(b)

(c)

Examples of spheroidal weathering include the chapter-opening photo and the photos in Figure 13.12 from the Alabama Hills area of eastern California.

We now look at three chemical weathering processes: hydrolysis, oxidation, and carbonation and solution.

Hydrolysis When minerals chemically combine with water, the process is **hydrolysis**. Hydrolysis is a decomposition process that breaks down silicate minerals in rocks. Compared with hydration, a physical process in which water is simply attached to minerals in the rock with no chemical reaction, the hydrolysis process involves active participation of water in chemical reactions to produce different compounds.

For example, the weathering of feldspar minerals in granite can be caused by a reaction to the normal mild acids dissolved in precipitation:

feldspar (K, Al, Si, O) + carbonic acid and water →
residual clays + dissolved minerals + silica

So the by-products of chemical weathering of feldspar in granite include clay (such as kaolinite) and silica. As clay forms from some minerals in the granite, quartz (SiO_2) particles are left behind. The resistant quartz may wash downstream, eventually becoming sand on some distant beach. Clay minerals become a major component in soil and in shale, a common sedimentary rock.

When weaker minerals in rock are changed by hydrolysis, the interlocking crystal network breaks down, so the rock fails, and *granular disintegration* takes place. Such disintegration in granite may make the rock appear etched, corroded, and softened (Figure 13.12b).

In Table 11.2, the second line shows resistance to chemical weathering in igneous rocks. On the ultramafic end of the table (right side), the low-silica minerals olivine and peridotite are most susceptible to chemical weathering. Stability gradually increases toward the high-silica minerals such as feldspar. On the far left side of the table, quartz is resistant to chemical weathering. You can see that, because of the nature of the constituent minerals, basalt weathers faster chemically than does granite.

FIGURE 13.11 Exfoliation in granite.
Exfoliation processes loosen slabs of granite, freeing them for further weathering and downslope movement: (a) Great arches form in the White Mountains of New Hampshire. (b) Exfoliated layers of rock are visible in characteristic dome formations in granites. The loosened slabs of rock are susceptible to further weathering and downslope movement. This view is from the east side of Half Dome in Yosemite National Park, California. (c) Half Dome perspective from the west; relief is approximately 1500 m (5000 ft) from the top of the dome to the glaciated valley below. [Photos by (a) Bobbé Christopherson, (b) and (c) by author.]

(a)

(b)

(c)

FIGURE 13.12 Chemical weathering and spheroidal weathering.
(a) Chemical weathering processes act on the joints in granite to dissolve weaker minerals, leading to a rounding of the edges of the cracks. (b) Rounded granite outcrop demonstrates spheroidal weathering and the disintegration of rock. (c) The rugged, weathered Alabama Hills, Mount Whitney in the background, provides scenic backdrops for many commercials and movies. You can imagine a movie chase scene raising clouds of dust! [Photos by Bobbé Christopherson.]

Oxidation   Another example of chemical weathering occurs when certain metallic elements combine with oxygen to form oxides. This is a chemical weathering process known as **oxidation**. Perhaps the most familiar oxidation form is the "rusting" of iron in rocks or soil that produces a reddish brown stain of iron oxide (Fe_2O_3). We have all left a tool or nails outside only to find them weeks later, coated with iron oxide. The rusty color is visible on the surfaces of rock and in heavily oxidized soils such as those in the southeastern United States, southwestern deserts, or the tropics (Figure 13.13). Here is a simple oxidation reaction in iron:

iron (Fe) + oxygen (O_2) → iron oxide (hematite; Fe_2O_3)

As iron is removed from the minerals in a rock, the disruption of the crystal structures in the rock's minerals makes the rock more susceptible to further chemical weathering and disintegration.

Carbonation and Solution   The third form of chemical weathering occurs when a mineral dissolves into *solution*—for example, when sodium chloride (common table salt) dissolves in water. Water is the universal solvent because it is capable of dissolving at least 57 of the natural elements and many of their compounds.

Water vapor readily dissolves carbon dioxide, thereby yielding precipitation containing carbonic acid (H_2CO_3). This acid is strong enough to react with many minerals, especially limestone, in a process called **carbonation**. *Carbonation* simply means reactions whereby carbon combines with minerals.

Such carbonation chemical weathering transforms minerals that contain calcium, magnesium, potassium, and sodium. When rainwater attacks formations of limestone (which is calcium carbonate, $CaCO_3$), the constituent minerals dissolve and wash away with the mildly acidic rainwater:

calcium carbonate + carbonic acid and water →
calcium bicarbonate ($Ca_2^{+}CO_2H_2O$)

Walk through an old cemetery and you can observe the carbonation of marble, a metamorphic form of limestone. Weathered limestone and marble, in tombstones or in rock formations, appear pitted and weathered wherever adequate water is available for carbonation. In this era of human-induced increases of acid precipitation, carbonation processes are greatly enhanced (see Focus Study 3.2, "Acid Deposition: A Continuing Blight on the Landscape").

Chemical weathering process of carbonation dominate entire landscapes composed of limestone. These are the regions of karst topography, which we examine next.

(a)

(b)

FIGURE 13.13 Oxidation processes in rock and soil.
(a) Oxidation of iron minerals produces these brilliant red colors in the sandstone formations of Red Rock Canyon, Nevada. (b) Ultisols, soils produced by warm, moist conditions, in Sumter County, Georgia, bear the color of iron and aluminum oxides. [Photos by Bobbé Christopherson.]

Karst Topography and Landscapes

Limestone is so abundant on Earth that many landscapes are composed of it (Figure 13.14). These areas are quite susceptible to chemical weathering. Such weathering creates a specific landscape of pitted, bumpy surface topography, poor surface drainage, and well-developed solution channels (dissolved openings and conduits) underground. Remarkable mazes of underworld caverns also may develop, owing to weathering and erosion caused by groundwater.

These are the hallmarks of **karst topography**, named for the Krš Plateau in Yugoslavia, where karst processes were first studied. Approximately 15% of Earth's land area has some karst features, with outstanding examples found in southern China, Japan, Puerto Rico, Cuba, the Yucatán of Mexico, Kentucky, Indiana, New Mexico, and Florida. As an example, approximately 38% of Kentucky has sinkholes and related karst features noted on topographic maps.

Formation of Karst

For a limestone landscape to develop into karst topography, there are several necessary conditions:

- The limestone formation must contain 80% or more calcium carbonate for solution processes to proceed effectively.
- Complex patterns of joints in the otherwise impermeable limestone are needed for water to form routes to subsurface drainage channels.
- There must be an aerated (containing air) zone between the ground surface and the water table.
- Vegetation cover supplies varying amounts of organic acids that enhance the solution process.

The role of climate in providing optimum conditions for karst processes remains in debate, although the amount and distribution of rainfall appears important. Karst occurs in arid regions, but it is primarily due to former climatic conditions of greater humidity. Karst is rare in the Arctic and Antarctic because the water, although present, is generally frozen.

As with all weathering processes, time is a factor. Early in the last century, karst landscapes were thought to progress through evolutionary stages of development, as if they were aging. Today, these landscapes are thought to be locally unique, a result of specific conditions, and there is little evidence that different regions evolve sequentially along similar lines. Nonetheless, mature karst landscapes do display certain characteristic forms.

Lands Covered with Sinkholes

The weathering of limestone landscapes creates many **sinkholes**, which form in circular depressions. (Traditional studies may call a sinkhole a *doline*.) A *collapse sinkhole* forms if a solution sinkhole collapses through the roof of an underground cavern. A gently rolling limestone plain might be pockmarked by slow subsidence of surface materials in *solution sinkholes* with depths of 2–100 m (7–330 ft) and diameters of 10–1000 m (33–3300 ft), as shown in Figure 13.15a. Through continuing solution and collapse, sinkholes may coalesce to form a *karst valley*—an elongated depression up to several kilometers long.

The area southwest of Orleans, Indiana, has more than 1000 sinkholes in just 2.6 km² (1 mi²). In this area, the Lost River, a "disappearing stream," flows more than 13 km (8 mi) underground before it resurfaces at its Lost River rise (near the Orangeville rise shown in Figure 13.15e). The Lost River flow diverts from the surface through sinkholes and solution channels. Its dry bed can

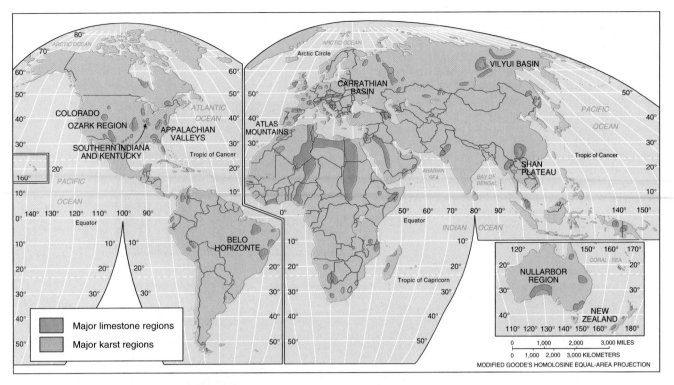

FIGURE 13.14 Karst landscapes and limestone regions.
Distribution of karst landscapes and limestone regions worldwide. [After R. E. Snead, *Atlas of the World Physical Features*, p. 76. © 1972 by John Wiley & Sons. Adapted by permission.]

be seen on the lower left of the topographic map in Figure 13.15b.

In Florida, several sinkholes have made news because lowered water tables (lowered by pumping from municipal wells) caused their collapse into underground solution caves, taking with them homes, businesses, and even new cars from an auto dealership. One such sinkhole collapsed in a suburban area in 1981, and others in 1993 and 1998 (Figure 13.16).

A complex landscape in which sinkholes intersect is a *cockpit karst*. The sinkholes can be symmetrically shaped in certain circumstances; one at Arecibo, Puerto Rico, is shaped perfectly for a radio telescope installation (Figure 13.17).

Another type of karst topography forms in the wet tropics, where deeply jointed, thick limestone beds are weathered into gorges, leaving isolated resistant blocks standing. These resistant cones and towers are most remarkable in several areas of China where *tower karst* up to 200 m (660 ft) high interrupts an otherwise lower-level plain (Figure 13.18).

Caves and Caverns

Caves form in limestone rock, because it is so easily dissolved by carbonation. The largest limestone caverns in the United States are Mammoth Cave in Kentucky (also the longest surveyed cave in the world at 560 km, 350 mi), Carlsbad Caverns in New Mexico, and Lehman Cave in Nevada.

Carlsbad Caverns are in 200-million-year-old limestone formations deposited when shallow seas covered the region. Regional uplifts associated with building of the Rockies (the Laramide orogeny, 40–80 million years ago) elevated the region above sea level, subsequently leading to active cave formation (Figure 13.19).

Caves generally form just beneath the water table, where later lowering of the water level exposes them to further development. *Dripstones* form as water containing dissolved minerals slowly drips from the cave ceiling. Calcium carbonate precipitates out of the evaporating solution, literally one molecular layer at a time, and accumulates at a point below on the cave floor. Forming depositional features, *stalactites* grow from the ceiling and *stalagmites* build from the floor; sometimes the two grow until they connect and form a continuous *column* (Figure 13.19b). A dramatic subterranean world is thus created. This is an aspect of geomorphology where amateur cavers make important discoveries about these unique habitats (see News Report 13.1). (For more on caves and related formations, see **http://www.goodearthgraphics.com/virtcave/virtcave. html.**)

Mass Movement Processes

Nevado del Ruiz, northernmost of two dozen dormant (not extinct, sometimes active) volcanic peaks in the Cordilleran Central of Colombia, had erupted six times during the

(continued on page 415)

FIGURE 13.15 Features of karst topography in Indiana.
(a) Idealized features of karst topography in southern Indiana. (b) Karst topography southwest of the town of Orleans, Indiana. On average, 1022 sinkholes occur per 1 mi² in this area. On the map, note the contour lines: Depressions are indicated with small hachures (tick marks) on the downslope side of contour lines. (c) Gently rolling karst landscape and cornfields near Orleans, Indiana. (d) This pond is in a sinkhole depression near Palmyra, Indiana. (e) This is the Orangeville rise, near Orangeville, Indiana, just north of the Lost River rise. During periods of high rainfall, this rise is almost filled with water. [(a) Adapted from W. D. Thornbury, *Principles of Geomorphology,* illustration by W. J. Wayne, p. 326; © 1954 by John Wiley & Sons. (b) Mitchell, Indiana quadrangle, USGS. Photos c, d, and e by Bobbé Christopherson.]

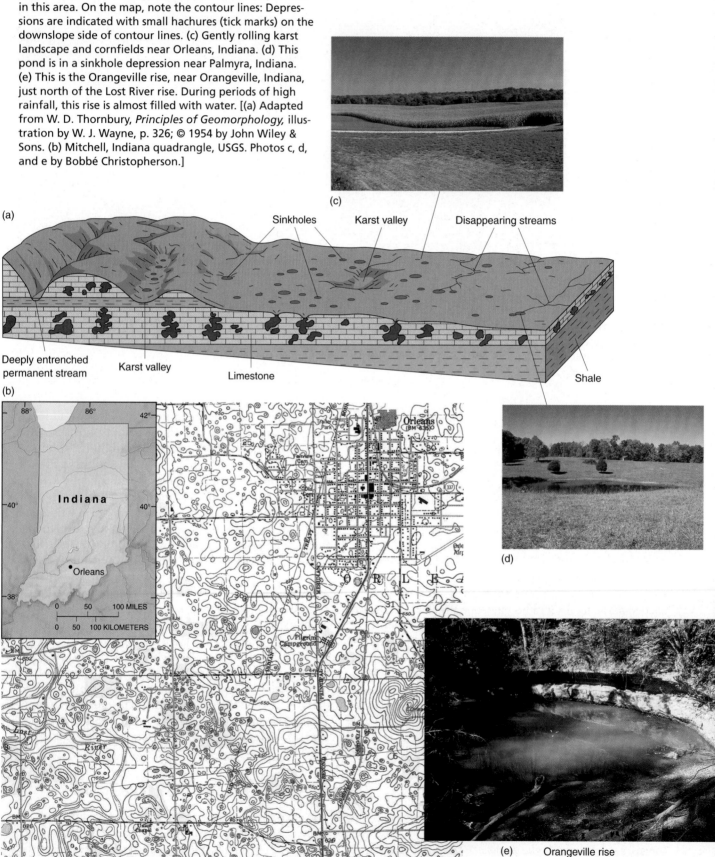

(c)

(a)

Sinkholes   Karst valley   Disappearing streams

Deeply entrenched permanent stream   Karst valley   Limestone   Shale

(b)

Indiana

Orleans

(d)

(e)   Orangeville rise

(a)

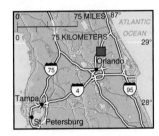

(b)

FIGURE 13.16 Sinkholes.
(a) Florida sinkhole, formed in 1981 in Winter Park, a suburb of Orlando. (b) Karst area 25 km (15.5 mi) north of Winter Park depicted on a topographic map. Note that the depressions are marked by small hachures (tick marks). [(a) Photo by Jim Tuten/Black Star; (b) Orange City quadrangle, USGS.]

FIGURE 13.17 Deep-space research from a sinkhole.
Cockpit karst topography near Arecibo, Puerto Rico, provides a natural depression for the dish antenna of a giant radio telescope. The Arecibo Observatory is part of the National Astronomy and Ionosphere Center, which is operated by Cornell University under contract with the National Science Foundation. [Photo courtesy of Cornell University.]

FIGURE 13.18 Tower karst of the Guangxi (Kwangsi) Province, China.
Resistant strata protect each tower as weathering removes the surrounding limestone. [Photo by Wolfgang Kaehler/Wolfgang Kaehler Photography.]

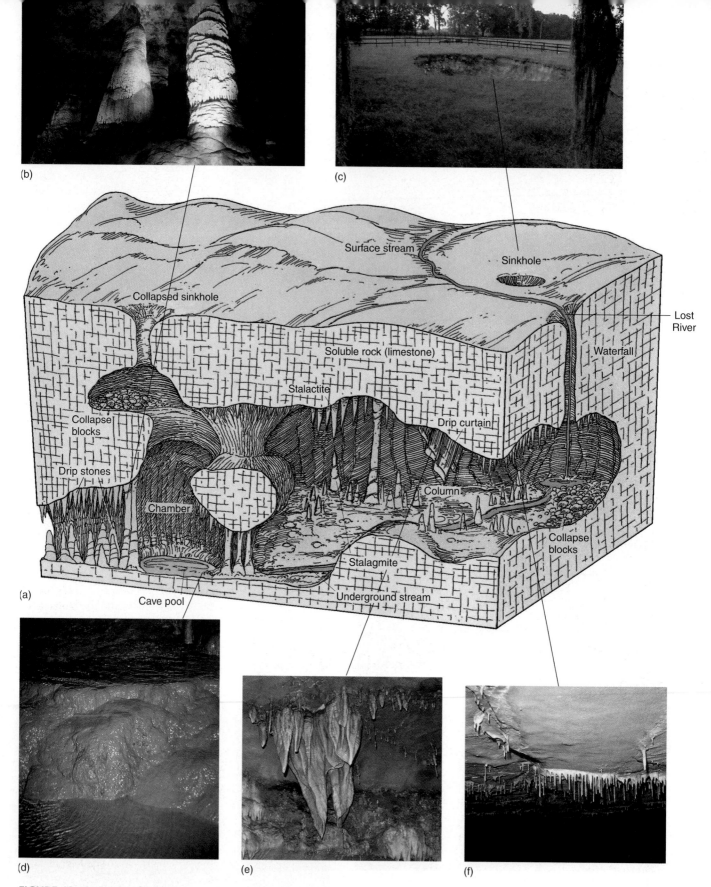

FIGURE 13.19 Cavern features.

(a) An underground cavern and related forms in limestone. (b) A column in Carlsbad Caverns, New Mexico, where a series of underground caverns includes rooms more than 1200 m (4000 ft) long and 190 m (625 ft) wide. Although a national park since 1930, unexplored portions remain. (c) A sinkhole in a Florida pasture. From Marengo Caves, Marengo, Indiana: (d) A flowstone and pool of water. (e) Dripstone drapery formations. (f) Soda straws hanging from ceiling cracks, forming one molecular layer at a time. [(b) Photo by author; (c) photo by Thomas M. Scott, Florida Geological Survey; photos (d, e, f) by Bobbé Christopherson.]

Amateurs Make Cave Discoveries

The exploration and scientific study of caves is *speleology*. Although professional physical and biological scientists carry on investigations, amateur cavers, or "spelunkers," have made many important discoveries. As an example, in the early 1940s George Colglazier, a farmer southwest of Bedford, Indiana, awoke to find his farm pond at the bottom of a deep collapsed sinkhole. This sinkhole is now the entrance to an extensive cave system that includes a subterranean navigable stream.

Cave habitats are unique. They are nearly closed, self-contained ecosystems with simple food chains and great stability. In total darkness, bacteria synthesize inorganic elements and produce organic compounds that sustain many types of cave life, including algae, small invertebrates, amphibians, and fish.

In a cave discovered in 1986, near Movile in southeastern Romania, cave-adapted invertebrates were discovered after millions of years of sunless isolation. Thirty-one of these organisms were previously unknown. Without sunlight, the ecosystem in Movile sustains on sulfur-metabolizing bacteria that synthesize organic matter using energy from oxidation processes. These chemosynthetic bacteria feed other bacteria and fungi that in turn support cave animals. The sulfur bacteria produce sulfuric acid compounds that may prove to be important in the chemical weathering of some caves.

The mystery, intrigue, and excitement of cave exploration lie in the variety of dark passageways, enormous chambers that narrow to tiny crawl spaces, strange formations, and underwater worlds that can be accessed only by cave-diving. Private-property owners and amateur adventurers discovered many of the major caves, a fact that keeps this popular science/sport very much alive. (For nearly a thousand worldwide links and information, see the **http://hum.amu.edu.pl/~sgp/spec/links.html** Web site.)

---

past 3000 years, killing 1000 people during its last eruption in 1845. On November 13, 1985, at 11 P.M. after a year of earthquakes and harmonic tremors, a growing bulge on its northeast flank, and months of small summit eruptions, Nevado del Ruiz violently erupted in a lateral explosion. The mountain was back in action.

On this night, the familiar pyroclastics, lava, and blast were not the worst problem. The hot eruption quickly melted ice on the mountain's snowy peak, liquefying mud and volcanic ash, sending a hot mudflow downslope. Such a flow is a *lahar*, an Indonesian word referring to mudflows of volcanic origin. This lahar moved rapidly down the Lagunilla River toward the villages below. The wall of mud was at least 40 m (130 ft) high as it approached Armero, a regional center with a population of 25,000. The city slept as the lahar buried its homes: 23,000 people were killed there and in other afflicted river valleys; thousands were injured; 60,000 were left homeless. The volcanic debris flow is now a permanent grave for its victims. Not all mass movements are this destructive, but such processes play a major role in the denudation of the landscape.

For more on mass movement hazards, including landslides, see the Web site of the Natural Hazards Center at the University of Colorado, Boulder, at **http://www.Colorado.edu/hazards/** or the USGS Geologic Hazards page at **http://landslides.usgs.gov/index.html**.

Mass Movement Mechanics

Physical and chemical weathering processes create an overall weakening of surface rock, which makes it more susceptible to the pull of gravity. The term **mass movement** applies to any unit movement of a body of material, propelled and controlled by gravity, such as the lahar just described. Mass movements can be surface processes or they can be submarine landslides beneath the ocean. Mass movement content can range from dry to wet, slow to fast, or small to large, and from free-falling to gradual or intermittent (see Figure 13.21).

The term *mass movement* is sometimes used interchangeably with **mass wasting**, which is the general process involved in mass movements and erosion of the landscape. To combine the concepts, we can say that the mass movement of material works to waste slopes and provide raw material for erosion, transportation, and deposition.

The Role of Slopes All mass movements occur on slopes under the influence of gravitational stress. If we pile dry sand on a beach, the grains will flow downslope until an equilibrium is achieved. The steepness of the resulting slope depends on the size and texture of the grains; this steepness is called the **angle of repose**. This angle represents a balance of the driving force (gravity) and resisting force (friction and shearing). The angle of repose for various materials commonly ranges between 33° and 37° (from horizontal), and 30° to 50° for snow avalanche slopes.

The *driving force* in mass movement is gravity. It works in conjunction with the weight, size, and shape of the surface material; the degree to which the slope is oversteepened (how far it exceeds the angle of repose); and the amount and form of moisture available (frozen or fluid). The greater the slope angle, the more susceptible the surface material is to mass wasting processes.

The *resisting force* is the shearing strength of slope material, that is, its cohesiveness and internal friction, which work against gravity and mass wasting. To reduce shearing strength is to increase shearing stress, which eventually reaches the point at which gravity overcomes friction, initiating slope failure.

Clays, shales, and mudstones are highly susceptible to hydration (physical swelling in response to the presence of water). If such materials underlay rock strata in a slope, the strata will move with less driving force energy. When clay surfaces are wet, they deform slowly in the direction of movement, and when saturated they form a viscous fluid with little shearing strength (resistance to movement) to hold back the slope. However, if the rock strata are such that material is held back from slipping, then more driving force energy may be required, such as that generated by an earthquake.

Madison River Canyon Landslide In the Madison River Canyon near West Yellowstone, Montana, a blockade of dolomite (a magnesium-rich carbonate rock) held back a deeply weathered and *oversteepened slope* (40° to 60° slope angle) for untold centuries (white area in Figure 13.20). Then, shortly after midnight on August 17, 1959, a magnitude 7.5 earthquake broke the dolomite structure along the foot of the slope. The break released 32 million cubic meters (1.13 billion cubic feet) of mountainside, which moved downslope at 95 kmph (60 mph), causing gale force

winds through the canyon. Momentum carried the material more than 120 m (about 400 ft) up the opposite canyon slope, trapping several hundred campers with about 80 m (260 ft) of rock, killing 28 people.

The mass of material also effectively dammed the Madison River and thus created a new lake, dubbed Quake Lake. The landslide debris dam established a new temporary equilibrium for the river and canyon. A channel was quickly excavated by the U.S. Army Corps of Engineers to prevent a disaster below the dam, for if Quake Lake overflowed the landslide dam, the water would quickly erode a channel and thereby release the entire contents of the new lake onto farmland downstream. This event conveys a dramatic example of the role of slopes and tectonic forces in creating massive land movements.

Classes of Mass Movements

In any mass movement, gravity pulls on a mass until the critical shear-failure point is reached—*a geomorphic threshold.* The material then can *fall, slide, flow,* or *creep*—the four classes of mass movement. Figure 13.21 summarizes these classes. Note the temperature and moisture gradients in the margins of the illustration, which show the relation

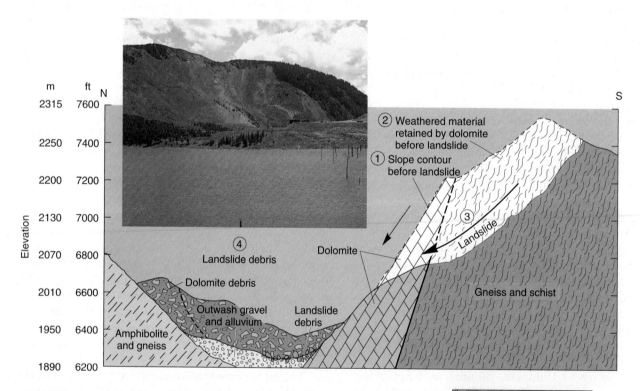

FIGURE 13.20 Madison River Landslide.
Cross section showing geologic structure of the Madison River Canyon, in Montana, where an earthquake triggered a landslide in 1959. (1) Prequake slope contour. (2) Weathered rock that failed. (3) Direction of landslide. (4) Landslide debris blocking the canyon and damming the Madison River. [After J. B. Hadley, *Landslides and Related Phenomena Accompanying the Hebgen Lake Earthquake of 17 August 1959,* U.S. Geological Survey Professional Paper 435-K, p. 115. Inset photo by Bobbé Christopherson.]

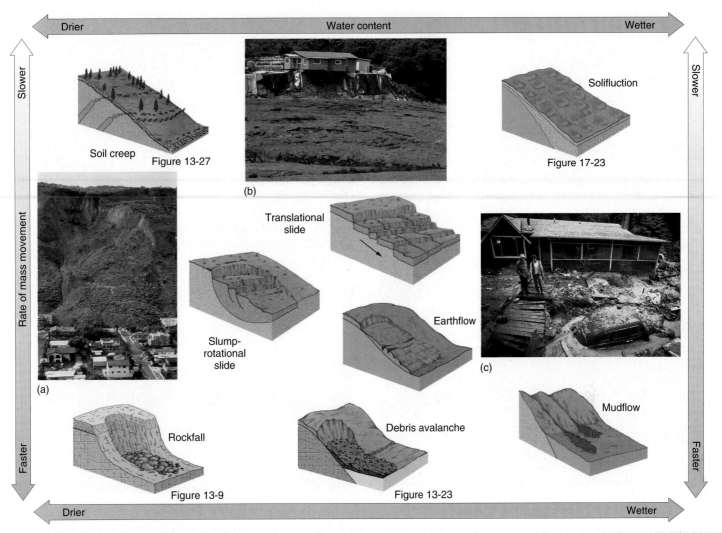

FIGURE 13.21 Mass movement classes.
Principal types of mass movement and mass wasting events. Variations in water content and rates of movement produce a variety of forms. (a) A 1995 slide in La Conchita, California, (b) saturated hillsides fail; (c) mudflows 2 m deep, both in Santa Cruz County, California. [Photos by (a) Robert L. Schuster/USGS; (b) Alexander Lowry/Photo Researchers, Inc.; (c) James A. Sugar.]

between water content and movement rate. The Madison River Canyon event was a type of slide, whereas the Nevado del Ruiz lahar mentioned earlier was a flow. We now look at specific mass movement classes.

Falls and Avalanches This class of mass movement includes rockfalls and debris avalanches. A **rockfall** is simply a volume of rock that falls through the air and hits a surface. During a rockfall, individual pieces fall independently and characteristically form a cone-shaped pile of irregular broken rocks called a talus slope at the base of a steep incline (Figure 13.22).

A **debris avalanche** is a mass of falling and tumbling rock, debris, and soil. It is differentiated from a slower debris slide or landslide by the velocity of onrushing material. This speed often results from ice and water that fluidize the debris. The extreme danger of a debris avalanche

results from its tremendous speed and consequent lack of warning.

In 1962 and again in 1970, debris avalanches roared down the west face of Nevado Huascarán, the highest peak in the Peruvian Andes. The 1962 debris avalanche contained an estimated 13 million cubic meters (460 million cubic feet) of material, burying the city of Ranrahirca and eight other towns, killing 4000 people.

An earthquake initiated the 1970 event. Upward of 100 million cubic meters (3.53 billion cubic feet) of debris buried the city of Yungay, where 18,000 people perished (Figure 13.23). This avalanche attained velocities of 300 kmph (185 mph), which is especially incredible when you consider the quantity of material involved and the fact that some boulders weighed thousands of tons. The avalanche covered a vertical drop of 4144 m (13,600 ft) and a horizontal distance of 16 km (10 mi) in just a few minutes.

FIGURE 13.22 Talus slope.
Rockfall and talus deposits at the base of a steep slope.
[Photo by author.]

FIGURE 13.23 Debris avalanche, Peru.
A 1970 debris avalanche falls more than 4100 m (2.5 mi)
down the west face of Nevado Huascarán, burying the city
of Yungay, Peru. The same area was devastated by a similar
avalanche in 1962 and by others in pre-Columbian times. A
great danger remains for the cities and towns in the valley
from possible future mass movements. [Photo by George
Plafker.]

FIGURE 13.24 Debris avalanche, Alaska.
A debris avalanche covers portions of the Cascade Glacier,
west of the Saint Elias Range in Alaska. The one pictured was
the largest of several dozen triggered by a 1979 earthquake.
[Photo by George Plafker.]

Another dramatic example of a debris avalanche, but
one that did not kill, occurred west of the Saint Elias
Range, north of Yakutat Bay in Alaska. A magnitude 7.1
earthquake triggered several dozen rockfalls, large debris
avalanches, and snow avalanches. The largest single
avalanche covered approximately 4.7 km² (1.8 mi²) of the
Cascade Glacier (Figure 13.24). The photo reveals characteristic grooves, lobes, and large rocks associated with
these fluid avalanches.

Landslides   A sudden rapid movement of a cohesive mass
of regolith or bedrock that is not saturated with moisture
is a **landslide**—a large amount of material failing simultaneously. Surprise creates the danger, for the downward pull
of gravity wins the struggle for equilibrium in an instant.
Focus Study 13.1 describes one such surprise event that
struck near Longarone, Italy, in 1963.

To eliminate the surprise element, scientists are using the global positioning system (GPS) to monitor landslide movement. With GPS, scientists measure slight land shifts in suspect areas for clues to possible mass wasting. GPS was applied in two cases in Japan and effectively identified pre-landslide movements of 2–5 cm per year, providing information to expand the area of hazard concern and warning.

Slides occur in one of two basic forms: translational or rotational (see Figure 13.21 for an idealized view of each). *Translational slides* involve movement along a planar (flat) surface roughly parallel to the angle of the slope, with no rotation. The Madison Canyon landslide described earlier was a translational slide. Flow and creep patterns also are considered translational in nature.

Rotational slides occur when surface material moves along a concave surface. Frequently, underlying clay pre-sents an impervious surface to percolating water. As a result, water flows along the clay surface, undermining the overlying block and lubricating the contact, thereby reducing the friction force. The simplest form of rotational slide is a rotational slump, in which a small block of land shifts downward. The upper surface of the slide appears to rotate backward and often remains intact. The surface may rotate as a single unit, or it may present a stepped appearance.

On Good Friday in 1964, the Pacific crustal plate plunged a bit further beneath the North American plate near Anchorage, Alaska, causing a massive earthquake. The resulting earthquake triggered more than 80 mass-wasting events. One particular housing subdivision in Anchorage experienced a sequence of translational movements accompanied by rotational slumping. Figure 13.25 shows one view of the segmented ground failure.

FIGURE 13.25 Suburban destruction in Alaska.
Anchorage, Alaska, landslide, illustrates mass movement patterns throughout a housing subdivision. [Courtesy of U.S. Geological Survey.]

Focus Study 13.1

Vaiont Reservoir Landslide Disaster

Place Northeastern Italy, Vaiont Canyon in the Italian Alps, rugged scenery, 680 m (2200 ft) above sea level.

Location 46.3° N 12.3° E, near the border of the Veneto and Friuli-Venezia Giulia regions.

Situation Centrally located in the region, an ideal situation for hydroelectric power production.

Site Steep-sided, narrow glaciated canyon, opening to populated lowlands to the west; ideal site for a narrow-crested, high dam, and deep reservoir.

The Plan Build the second highest dam in the world, using a new thin-arch design, 262 m (860 ft) high, 190 m (623 ft) crest length, impounding a reservoir capacity of 150 million cubic meters (5.3 billion cubic feet) of water (third largest reservoir in the world), and generate hydroelectric power for distribution.

Geologic Analysis
1. Steep canyon walls composed of interbedded limestone and shale; badly cracked and deformed structures; open fractures in the shale inclined toward the future reservoir body. The steepness of the canyon walls enhance the strong driving forces (gravitational) at work on rock structures.
2. High potential for bank storage (water absorption by canyon walls) into the groundwater system that will increase water pressure on all rocks in contact with the reservoir.
3. The nature of the shale beds is such that cohesion will be reduced as its clay minerals become saturated.
4. Evidence of ancient rockfalls and landslides on the north side of the canyon.
5. Evidence of creep activity along south side of the canyon.

(continued)

Focus Study 13.1 (continued)

Political/Engineering Decision Begin design and construction of a thin-arch dam at this site.

Events During Construction and Filling Large volumes of concrete were injected into the bedrock as "dental work" in an attempt to strengthen the fractured rock. During reservoir filling in 1960, 700,000 cubic meters (2.5 million cubic feet) of rock and soil slid into the reservoir from the south side. A slow creep of slope materials along the entire south side began shortly thereafter, increasing to about 1 cm per week by January 1963. Creep rate increased as the level of the reservoir rose. By mid-September 1963, the creep rate exceeded 40 cm (16 in.) a day.

This sequence of events set the stage for one of the worst dam disasters in history. Heavy rains began on September 28, 1963. Alarmed at the runoff from the rain into the reservoir and the increase in creep rate along the south wall, engineers opened the outlet tunnels on October 8 in an attempt to bring down the reservoir level—but it was too late.

The next evening, it took only 30 seconds for a landslide of 240 million cubic meters (8.5 billion cubic feet) to crash into the reservoir, shaking seismographs across Europe. A 150-m-thick (500-ft) slab of mountainside gave way (2 km by 1.6 km in area; 1.2 mi by 1 mi), sending shock waves of wind and water up the canyon 2 km (1.2 mi) and splashing a 100-m (330-ft) wave over the dam. The former reservoir was effectively filled with bedrock, regolith, and soil that almost entirely displaced its water content (Figure 1). Amazingly, the experimental dam design held.

Downstream, in the unsuspecting town of Longarone, near the mouth of Vaiont Canyon on the Piave River, people heard the distant rumble and were quickly drowned by the 69-m (226-ft) wave of water that came out of the canyon. That night 3000 people perished. As for a lesson or recommendation, we need only to reread the preconstruction geologic analysis that began this Focus Study. The Italian courts eventually prosecuted the persons responsible.

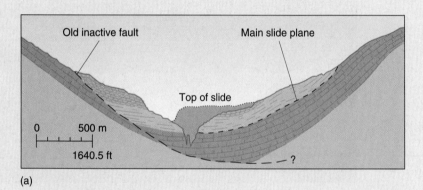

(a)

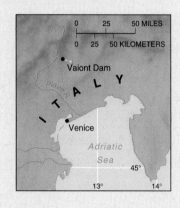

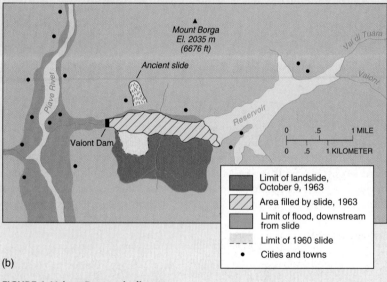

(b)

FIGURE 1 Vaiont Reservoir disaster
(a) Cross section of the Vaiont River Valley. (b) Map of the disaster site noted with evidence of previous activity. [After G. A. Kiersch, "The Vaiont Reservoir Disaster," California Division of Mines and Geology, Mineral Information Service, vol. 18, no. 7, pp. 129–138.]

— This forested mass is the main portion of the landslide.

FIGURE 13.26 The Gros Ventre slide near Jackson, Wyoming.
[Photo by Steven K. Huhtala.]

A landslide slide show and solution diagrams may be viewed at http://www.kingston.ac.uk/~ce_s011/slides.htm. For a comprehensive look at an ongoing landslide situation in southern California, see the Anaheim Hills Landslide Update at http://anaheim-landslide.com/. This site includes general background on mass movements, including many diagrams, and before and after photos.

Flows   Flows include *earthflows* and more fluid **mudflows**. When the moisture content of moving material is high, the suffix *-flow* is used (see Figure 13.21). Heavy rains can saturate barren mountain slopes and set them moving, as was the case east of Jackson Hole, Wyoming, in the spring of 1925. Material above the Gros Ventre River (pronounced "grow vaunt") broke loose and slid downslope as a unit. The slide occurred because sandstone formations rested on weak shale and siltstone, which became moistened and soft, offering little resistance to the overlying strata. The slide is still visible after 75 years, as you can see in Figure 13.26.

Because of melted snow and rain, the water content of the Gros Ventre landslide was great enough to classify it as an *earthflow*. About 37 million cubic meters (1.3 billion cubic feet) of wet soil and rock moved down one side of the canyon and surged 30 m (100 ft) up the other side. The earthflow dammed the river and formed a lake, as did the landslide across the Madison River in the Hebgen Lake area in 1959. However, in 1925 equipment was not available to excavate a channel, so the new lake filled. Two years later, the lake water broke through the temporary earthflow dam, transporting a tremendous quantity of debris over the region downstream.

Creep   A persistent, gradual mass movement of surface soil is called **soil creep**. In creep, individual soil particles are lifted and disturbed by the expansion of soil moisture as it freezes, by cycles of moistness and dryness, diurnal temperature variations, or grazing livestock or digging animals.

In the freeze–thaw cycle, particles lift at right angles to the slope by freezing soil moisture, as shown in Figure 13.27. When the ice melts, however, the particles fall straight downward in response to gravity. As the process repeats, the surface soil gradually creeps its way downslope.

The overall wasting of a slope may cover a wide area and may cause fence posts, utility poles, and even trees to lean downslope. Various strategies are used to arrest the mass movement of slope material—grading the terrain, building terraces and retaining walls, planting ground cover—but the persistence of creep nearly always wins.

Human-Induced Mass Movements (Scarification)

Every human disturbance of a slope—highway roadcut, surface-mining, or building of a shopping mall, housing development, or home—can hasten mass wasting. The newly destabilized and oversteepened surfaces are thrust into a search for a new equilibrium. Imagine the disequilibrium in slope relations created by the highway roadcut pictured in Figure 12.8b.

Large open-pit surface mines—such as the Bingham Copper Mine west of Salt Lake City, the Berkeley Pit in Butte, Montana, and numerous large coal surface mines in the U.S. East and West—are examples of human-induced mass movements, generally called **scarification** (Figure 13.28a).

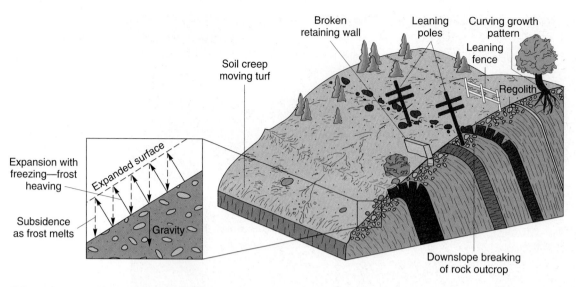

FIGURE 13.27 Soil creep and its effects.

(a)

(b)

(c)

(d)

FIGURE 13.28 Scarification.
(a) Black Mesa, Arizona, strip-mining for coal. (b) Bingham Canyon, Utah, west of Salt Lake City, strip-mining for copper and other minerals. (c) Abandoned Berkeley Pit copper mine, Butte, Montana. (d) Spoil banks in a West Virginia coal-mining area. [Photos by author.]

At the Bingham Copper Mine, a mountain literally was removed, forming a pit 4-km wide and 1-km deep (Figure 13.28b). This is easily the largest human-made excavation on Earth. The disposal of tailings (mined ore of little value) and waste material is a significant problem at any surface mine. Such large excavations produce tailing piles that are unstable and susceptible to further weathering, mass wasting, or wind dispersal.

At the abandoned Berkeley Pit, toxic drainage water is threatening the regional aquifers, the Clark Fork River, and local freshwater supply (Figure 13.28c). Residues of copper, zinc, lead, and arsenic lace one creek, now devoid of life, and are in the soil on Butte Hill, where children play. The leaching of toxic materials from tailings and waste piles poses an ever-increasing problem to streams, aquifers, and public health (Figure 13.28d). Wind dispersal is a particular problem with uranium tailings in the West, because of their radioactivity.

Where underground mining is common, particularly for coal in the Appalachians, land subsidence and collapse produce further mass movements. Homes, highways, streams, wells, and property values are severely affected.

Scientists can informally quantify the scale of human-induced scarification for comparison with natural denudation processes. R. L. Hooke, Department of Geology and Geophysics, University of Minnesota, used estimates of U.S. excavations for new housing, mineral production

(including the three largest—stone, sand and gravel, and coal), and highway construction. He then prorated these quantities of moved earth for all countries, using their gross domestic product (GDP), energy consumption, and agriculture's effect on river sediment loads. From these he calculated a global estimate for human earth moving.

Hooke estimated that humans, as a geomorphic agent, annually move 40–45 billion tons (40–45 Gt/yr) of the planet's surface. Compare this quantity with natural river sediment transfer (14 Gt/yr), movement through stream meandering (39 Gt/yr), haulage by glaciers (4.3 Gt/yr), movement due to wave action and erosion (1.25 Gt/yr), wind transport (1 Gt/yr), sediment movement by continental and oceanic mountain building (34 Gt/yr), or deep-ocean sedimentation (7 Gt/yr). As Hooke stated,

Homo sapiens has become an impressive geomorphic agent. Coupling our earth-moving prowess with our inadvertent adding of sediment load to rivers and the visual impact of our activities on the landscape, one is compelled to acknowledge that, for better or for worse, this biogeomorphic agent may be the premier geomorphic agent of our time.[*]

[*]R. L. Hooke, "On the efficacy of humans as geomorphic agents," *GSA Today*, The Geological Society of America, 4, no. 9 (September 1994): 217–226.

Summary and Review—Weathering, Karst Landscapes, and Mass Movement

● *Define* **the science of geomorphology.**

Geomorphology is the science that analyzes and describes the origin, evolution, form, and spatial distribution of landforms. The exogenic system, powered by solar energy and gravity, tears down the landscape through processes of landmass **denudation** involving weathering, mass movement, erosion, transportation, and deposition. Different rocks offer differing resistance to these weathering processes and produce a pattern on the landscape of **differential weathering**. Agents of change include moving air, water, waves, and ice. W. M. Davis's *geomorphic cycle model* characterized landscapes as evolving through stages from youth to old age. Since the 1960s, research and understanding of the processes of denudation have moved toward the **dynamic equilibrium model**, which considers slope and landform stability to be consequences of the resistance of rock materials to the attack of denudation processes.

geomorphology (p. 400)
denudation (p. 400)
differential weathering (p. 400)
dynamic equilibrium model (p. 401)

1. Define geomorphology, and describe its relationship to physical geography.

2. Define landmass denudation. What processes are included in the concept?

3. What is the interplay between the resistance of rock structures and weathering variabilities?

4. Give a brief overview of the geomorphic cycle model. What was W. M. Davis's principal contribution to the models of landmass denudation?

5. What are the principal considerations in the dynamic equilibrium model?

● *Illustrate* **the forces at work on materials residing on a slope.**

Slopes are shaped by the relation between rate of weathering and breakup of slope materials and the rate of mass movement and erosion of those materials. A slope is considered stable if it is stronger than these denudation processes; it is unstable if it is weaker. In this struggle against gravity, a slope may reach a **geomorphic threshold**—the point at which there is enough energy to overcome resistance against movement. **Slopes** that form the boundaries of landforms have several general components: *waxing slope, free face, debris slope,* and *waning slope*. Slopes seek an angle of equilibrium among the operating forces.

geomorphic threshold (p. 401)
slopes (p. 401)

6. Describe conditions on a hillslope that is right at the geomorphic threshold. What factors might push the slope beyond this point?

7. Given all the interacting variables, do you think a landscape ever reaches a stable, old-age condition? Explain.

8. What are the general components of an ideal slope?

9. Relative to slopes, what is meant by an "angle of equilibrium"? Can you apply this concept to the photograph in Figure 13.2?

● *Define* weathering and *explain* the importance of parent rock and joints and fractures in rock.

Weathering processes disintegrate both surface and subsurface rock into mineral particles or dissolve them in water. The upper layers of surface material undergo continual weathering and create broken-up rock called **regolith**. Weathered **bedrock** is the *parent rock* from which regolith forms. The unconsolidated, fragmented material that develops after weathering is **sediment**, which along with weathered rock forms the **parent material** from which soil evolves.

Important in weathering processes are **joints**, the fractures and separations in the rock. Jointing opens up rock surfaces on which weathering processes operate. Factors that influence weathering include character of the bedrock (hard or soft, soluble or insoluble, broken or unbroken), climatic elements (temperature, precipitation, freeze–thaw cycles), position of the water table, slope orientation, surface vegetation and its subsurface roots, and time.

weathering (p. 403)
regolith (p. 403)
bedrock (p. 403)
sediment (p. 403)
parent material (p. 403)
joints (p. 403)

10. Describe weathering processes operating on an open expanse of bedrock. How does regolith develop? How is sediment derived?

11. Describe the relationship between mesoscale climatic conditions and rates of weathering activities.

12. What is the relation among parent rock, parent material, regolith, and soil?

13. What role do joints play in the weathering process? Give an example from one of the illustrations in this chapter.

● *Describe* frost action, crystallization, hydration, pressure-release jointing, and the role of freezing water as physical weathering processes.

Physical weathering refers to the breakup of rock into smaller pieces with no alteration of mineral identity. The physical action of water when it freezes (expands) and thaws (contracts) is a powerful agent in shaping the landscape. This **frost action** may break apart any rock. Working in joints, expanded ice can produce *joint-block separation* through the process of *frost-wedging*. Frost action loosens rock that falls from a steep cliff, producing a **talus slope** of poorly sorted debris at the base of the slope.

Another process of physical weathering is *crystallization*; as crystals in rock grow and enlarge over time, they force apart mineral grains and break up rock. **Hydration** occurs when a mineral absorbs water and expands, thus creating a strong mechanical force that stresses rocks.

As overburden is removed from a granitic batholith, the pressure of deep burial is relieved. The granite slowly responds with *pressure-release jointing*, with layer after layer of rock peeling off in curved slabs or plates. As these slabs weather, they slip off in a process called **sheeting**. This *exfoliation process* creates an arch-shaped or dome-shaped feature on the exposed landscape, forming an **exfoliation dome**.

physical weathering (p. 404)
frost action (p. 404)
talus slope (p. 406)
hydration (p. 406)
sheeting (p. 407)
exfoliation dome (p. 407)

14. What is physical weathering? Give an example.

15. Why is freezing water such an effective physical weathering agent?

16. What weathering processes produce a granite dome? Describe the sequence of events.

● *Describe* the susceptibility of different minerals to the chemical weathering processes called hydrolysis, oxidation, carbonation, and solution.

Chemical weathering is the chemical decomposition of minerals in rock. It can cause **spheroidal weathering**, in which chemical weathering occurs in cracks in the rock. As cementing and binding materials are removed, the rock begins to disintegrate, and sharp edges and corners become rounded.

Hydrolysis breaks down silicate minerals in rock, as in the chemical weathering of feldspar into clays and silica. Water is not just absorbed, as in hydration, but actively participates in chemical reactions. **Oxidation** is the reaction of oxygen with certain metallic elements, the most familiar example being the rusting of iron, producing iron oxide. *Solution* is chemical weathering. For instance, a mild acid such as carbonic acid in rainwater will cause **carbonation**, wherein carbon combines with certain minerals, such as calcium, magnesium, potassium, and sodium.

chemical weathering (p. 407)
spheroidal weathering (p. 407)
hydrolysis (p. 408)
oxidation (p. 409)
carbonation (p. 409)

17. What is chemical weathering? Contrast this set of processes to physical weathering.

18. What is meant by the term spheroidal weathering? How is spheroidal weathering formed?

19. What is hydrolysis? How does it affect rocks?

20. Iron minerals in rock are susceptible to which form of chemical weathering? What characteristic color is associated with this type of weathering?

21. With what kind of minerals do carbon compounds react, and under what circumstances does this reaction occur? What is this weathering process called?

● *Review* the processes and features associated with karst topography.

Karst topography refers to distinctively pitted and weathered limestone landscapes. Surface circular **sinkholes** form and may extend to form a *karst valley*. A sinkhole may collapse through the roof of an underground cavern, forming a *collapse sinkhole*. The formation of caverns is part of karst processes and groundwater erosion. Limestone caves feature many unique erosional and depositional features, producing a dramatic subterranean world.

karst topography (p. 410)
sinkholes (p. 410)

22. Describe the development of limestone topography. What is the name applied to such landscapes? From what area was this name derived?

23. Differentiate among sinkholes, karst valleys, and cockpit karst. Within which form is the radio telescope at Arecibo, Puerto Rico?

24. In general, how would you characterize the region southwest of Orleans, Indiana?

25. What are some of the unique erosional and depositional features you find in a limestone cavern?

● *Portray* the various types of mass movements and *identify* examples of each in relation to moisture content and speed of movement.

Any movement of a body of material, propelled and controlled by gravity, is **mass movement**, also called **mass wasting**. The **angle of repose** of loose sediment grains represents a balance of driving and resisting forces on a slope. Mass movement of Earth's surface produces some dramatic incidents, including: **rockfalls** (a volume of rock that falls); **debris avalanches** (a mass of tumbling, falling rock, debris, and soil at high speed); **landslides** (a large amount of material failing simultaneously); **mudflows** (material in motion with a high moisture content); and, **soil creep** (a persistent movement of individual soil particles that are lifted by the expansion of soil moisture as it freezes, by cycles of wetness and dryness, and temperature variations, or the impact of grazing animals). In addition, human mining and construction activities have created massive **scarification** of landscapes.

mass movement (p. 415)
mass wasting (p. 415)
angle of repose (p. 415)
rockfall (p. 417)
debris avalanche (p. 417)
landslide (p. 418)
mudflows (p. 421)
soil creep (p. 421)
scarification (p. 421)

26. Define the role of slopes in mass movements, using the terms angle of repose, driving force, resisting force, and geomorphic threshold.

27. What events occurred in the Madison River Canyon in 1959?

28. What are the classes of mass movement? Describe each briefly and differentiate among these classes.

29. Name and describe the type of mudflow associated with a volcanic eruption.

30. Describe the difference between a landslide and what happened on the slopes of Nevado Huascarán.

31. What is scarification, and why is it considered a type of mass movement? Give several examples of scarification. Why are humans a significant geomorphic agent?

NetWork

The *Geosystems* Home Page provides on-line resources for this chapter on the World Wide Web. To begin: Once on the Home Page, click on this textbook, scroll the Table of Contents menu, select this chapter, and click "Begin." You will find self-tests that are graded, review exercises, specific updates for items in the chapter, and many links to interesting related pathways on the Internet under "Destinations." *Geosystems* is at **http://www.prenhall.com/christopherson**.

Critical Thinking

A. Locate a slope, possibly near campus, near your home, or a local roadcut. Using Figure 13.3a and b, can you identify the forces and forms of a hillslope at your site? How would you go about assessing the stability of the slope? Is there any evidence of the mass wasting of materials, soil creep, or other processes discussed in this chapter?

B. The USGS has completed a landslide hazards potential map for the conterminous United States. You can check out this resource at **http://landslides.usgs.gov/ html_files/landslides/nationalmap/national.html**. Note the zoom-in capability to look at specific areas in greater detail. The map legend rates landslide incidence and susceptibility. How does the map portray landslide incidences? In what way does the map rate landslide susceptibility to an area? Are you able to find a location that you have visited and determine its vulnerability?

C. Go to the *Geosystems* Home Page and to Critical Thinking in Chapter 13. Study the two remote sensing images. The Himalayan Mountains are the primary source of the sediment transported by the rivers. Imagine yourself as a piece of rock at the very top of Mount Everest. Describe a scenario by which you join a collection of sediments in the Bay of Bengal. Discuss the changes in weathering processes acting upon you between the high, cold reaches of the Himalayas and the hot, humid mud flats of the Ganges Delta.

Career Link 13.1

Gregory A. Pope, Assistant Geography Professor

Gregory Pope lists physical geography, weather, geomorphology, weathering processes, GIS, and global environmental change as areas of special interest. He assumed his teaching post at Montclair State in 1996 and is the advisor for students in the geography program. Greg's Master's and Ph.D. degrees were completed at Arizona State University. His dissertation developed a boundary-layer model for interpreting spatial variation in weathering. He has published several articles derived from his dissertation.

Earlier in this chapter you can see Greg's influence in my mention of the importance of scale in analyzing weathering processes. He found both chemical and physical weathering at small-scale reaction sites on rock, even in the driest landscapes.

Greg did his undergraduate work at the University of Colorado in his hometown of Colorado Springs. As a child, he camped with his family in Colorado and other parts of the West. Especially vivid to him are camping trips to the Rockies: "I didn't know what geomorphology or geology were, but the rocks and mountain formations fascinated me." Beginning when he was only 4 or 5, during automobile trips to Illinois, he was the family navigator, using his own highway map collection.

Greg remembers, "We had a geographic encyclopedia set that I read and used for homework throughout my lower grades. I also read all the ge-

FIGURE 1 Gregory A. Pope, Assistant Geography Professor Montclair State University, Upper Montclair, New Jersey. Greg, along with geography professor Patricia Beyer, led a 2001 field trip through New York's Central Park analyzing the diverse types of stones used in the park's many bridges. [Photo by Bobbé Christopherson.]

ography and geology books in my junior high before I finished eighth grade; they called it social studies, not geography, but it was really geography. I convinced my teacher to let me do an urban planning project about the unplanned growth in Colorado Springs. I asked the teacher to let me design a new city plan, as if Colorado Springs wasn't there. I did this one whole semester instead of what the rest of the class did."

"Following graduation from UCCS, I received a National Geographic Society internship and worked in Washington, DC, for half a year. I received a lot of valuable research experience working on *World*, the NGS children's magazine." He also worked as a contributing writer for the NGS National Geography Bee.

As a teaching assistant in graduate school at Arizona State, Greg, along with his advisor Ronald Dorn, began studying the White Mountains of eastern California, which fostered his interest in desert geomorphology. He knew he wanted to teach. "When I taught the labs, something all teaching assistants do, I felt comfortable. I knew teaching had to be part of my future."

(continued)

Career Link 13.1 *(continued)*

Greg is studying the weathering of ancient to modern architectural structures in Portugal and collaborating with geoarcheologists who study weathering of petroglyphs. Also, he is looking at weathering in some stone work, specifically isolating pre–Clean Air Act (1970) contamination and weathering, examining tombstones near pollution sources. With his students he examines the rates of building-stone weathering around New York City—his take on urban geology—and then maps the spatial results. In addition, Greg is doing research on the tectonic geomorphology of the Rocky Mountains, a lifelong subject of interest.

What lies ahead for geography: global change studies will be important; the integration of environmental studies is another possible avenue. Greg feels that, "Some discipline must emerge to pull all the disparate parts together—geography should be it. In the direction I am going—weathering studies, geoarcheology, and work in building my department's program—I am having fun! I am inventing stuff and combining subjects that interest me. And being a geographer, I get to synthesize all this into coherent patterns! I get to work with people from so many different fields. My advice is to have fun with it, have joy in what you choose to do."

Hogs fight to stay alive as the Neuse River floodwaters rise around a hog farm a few miles from New Bern, North Carolina. Those animals, too tired to swim, drowned in the polluted water. Many of the factory farm sewage lagoons were located on the inundated floodplain. [Photo by Chris Seward, *Raleigh News & Observer*.]

14

River Systems and Landforms

Key Learning Concepts

After reading the chapter, you should be able to:

- *Define* the term fluvial and *outline* the fluvial processes: erosion, transportation, and deposition.
- *Construct* a basic drainage basin model and *identify* different types of drainage patterns and internal drainage, with examples.
- *Describe* the relation among velocity, depth, width, and discharge and *explain* the various ways that a stream erodes and transports its load.
- *Develop* a model of a meandering stream, including point bar, undercut bank, and cutoff, and *explain* the role of stream gradient in these flow characteristics.
- *Define* a floodplain and *analyze* the behavior of a stream channel during a flood.
- *Differentiate* the several types of river deltas and *detail* each.
- *Explain* flood probability estimates and *review* strategies for mitigating flood hazards.

arth's rivers and waterways form vast arterial networks that drain the continents. They also shape the landscape by removing the products of weathering, mass movement, and erosion and transporting them downstream. To call rivers "Earth's lifeblood" is no exaggeration, inasmuch as rivers redistribute mineral nutrients important for soil formation and plant growth and serve society in many ways.

Rivers not only provide essential water supplies, they also process waste (diluting and transporting it), provide critical cooling water for manufacturing and power generation, and form essential transportation networks. Rivers have been important in the geography of human history, influencing where settlements were built, where livelihoods were made, and where borders were drawn. This chapter discusses the dynamics of river systems and their landforms.

At any moment, approximately 1250 km³ (300 mi³) of water is flowing through Earth's waterways. Even though this volume is only 0.003% of all freshwater, the work performed by this energetic flow makes it a dominant agent of landmass denudation. Of the world's rivers, those with the greatest *discharge* (stream flow rate) are the Amazon of South America (Figure 14.1), the Congo (Zaire) of Africa, the Chang Jiang (Yangtze) of Asia, and the Orinoco of South America (Table 14.1). In North America, the greatest discharges are from the Missouri–Ohio–Mississippi, Saint Lawrence, and Mackenzie River systems.

Hydrology is the science of water, its global circulation, distribution, and properties, specifically water at and below Earth's surface. For hydrology links on the Web, see **http://terrassa.pnl.gov:2080/hydroweb.html** or the Global Hydrology and Climate Center at **http://www. ghcc.msfc.nasa.gov/**; also see the Amazon River at **http://boto.ocean.washington.edu/eos/index.html**.

In this chapter: We begin with a look at the largest rivers on Earth. Essential fluvial concepts of base level, drainage basin, drainage density and patterns follows. Factors that affect streamflow characteristics and the work performed by flowing water, including erosion, transport, and deposition, are discussed. Human response to floods and floodplain management are important aspects of river management and are concluded in the chapter.

FIGURE 14.1 Mouth of the Amazon River.
The mouth of the Amazon River discharges a fifth of all the freshwater that enters the world's oceans. The mouth of the Amazon is 160 km (100 mi) wide. Millions of tons of sediments are derived from the Amazon's drainage basin, which is as large as the Australian continent. Large islands of sediment are left where the river's discharge leaves the mouth and flows into the Atlantic Ocean. [*Terra* MISR sensor image courtesy of NASA/GSFC/JPL and the MISR Team.]

Table 14.1   Largest Rivers on Earth Ranked by Discharge Volume

Rank by Volume	Average Discharge at Mouth in Thousands of m³/s (cfs)	River (with Tributaries)	Outflow/Location	Length km (mi)	Rank by Length
1	180 (6350)	Amazon (Ucayali, Tambo, Ene, Apurimac)	Atlantic Ocean/ Amapá-Pará, Brazil	6570 (4080)	2
2	41 (1460)	Congo, also known as the Zaire (Lualaba)	Atlantic Ocean/Angola, Congo	4630 (2880)	10
3	34 (1201)	Yangtze (Chàng Jiang)	East China Sea/Kiangsu, China	6300 (3915)	3
4	30 (1060)	Orinoco	Atlantic Ocean/Venezuela	2737 (1700)	27
5	21.8 (779)	La Plata estuary (Paraná)	Atlantic Ocean/Argentina	3945 (2450)	16
6	19.6 (699)	Ganges (Brahmaputra)	Bay of Bengal/India	2510 (1560)	23
7	19.4 (692)	Yenisey (Angara, Selenga or Selenge, Ider)	Gulf of Kara Sea/Siberia	5870 (3650)	5
8	18.2 (650)	Mississippi (Missouri, Ohio, Tennessee, Jefferson, Beaverhead, Red Rock)	Gulf of Mexico/Louisiana	6020 (3740)	4
9	16.0 (568)	Lena	Laptev Sea/Siberia	4400 (2730)	11
17	9.7 (348)	St. Lawrence	Gulf of St. Lawrence/Canada and United States	3060 (1900)	21
36	2.83 (100)	Nile (Kagera, Ruvuvu, Luvironza)	Mediterranean Sea/Egypt	6690 (4160)	1

Fluvial Processes and Landscapes

Stream-related processes are termed **fluvial** (from the Latin *fluvius*, meaning "river"). Geographers seek to describe stream patterns and the fluvial processes that created them. Fluvial systems, like all natural systems, have characteristic processes and produce predictable landforms. Yet a stream system can behave with randomness and disorder. The term *river* is applied to a trunk stream or an entire river system. *Stream* is a more general term not necessarily related to size. There is some overlap in usage between the two terms.

Insolation and gravity power the hydrologic cycle and are the driving forces of fluvial systems. Individual streams vary greatly, depending on the climate in which they operate, the composition of the surface, topography over which they flow, the nature of vegetation and plant cover, and the length of time they have been operating in a specific setting.

Water dislodges, dissolves, or removes surface material in a process called **erosion**. Streams produce *fluvial erosion*, in which weathered sediment is picked up for transport to new locations. Thus, a stream is a mixture of water and solids; the solids are carried in solution, suspension, and by mechanical **transport** (movement of material). Materials are laid down by another process, **deposition**. **Alluvium** is the general term for the clay, silt, and sand deposited by running water.

Base Level of Streams

American geologist and ethnologist John Wesley Powell (1834–1902) was a director of the U.S. Geological Survey, first director of the U.S. Bureau of Ethnology, explorer of the Colorado River, and a pioneer in understanding the landscape (Figure 14.2). In 1875 he put forward the idea of **base level**, or a level below which a stream cannot erode its valley. In general, the *ultimate base level* is sea level (the average level between high and low tides). As shown in Figure 14.3, you can imagine the base level as a surface extending inland from sea level, inclined gently upward under the continents. Ideally, this is the lowest practical level for all denudation processes.

Of course, Powell recognized that not every landscape has degraded all the way to sea level; clearly, other intermediate base levels are in operation. A *local base level*, or temporary one, may control the lower limit of local streams. The local base level may be a river, a lake, hard and resistant rock, or a human-made dam (see Figure 14.3). In arid landscapes, with their intermittent precipitation, valleys, plains, or other low points provide local control.

Over time, the work of streams modifies the landscape dramatically. Landforms are produced by two basic processes: (1) erosive action of flowing water, and (2) deposition of stream-transported materials. Let us begin our study by examining a basic fluvial unit—the drainage basin.

FIGURE 14.2 Powell's journey to base level.
John Wesley Powell developed his base level concept during extensive exploration of the West. He first ventured down the Colorado River in 1869 in heavy oak boats, drifting by Dead Horse Point and the bend in the river visible in the canyon. [Photo by author.]

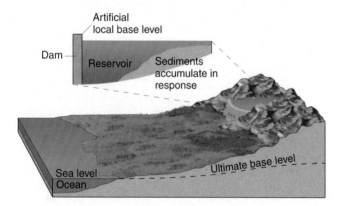

FIGURE 14.3 Ultimate and local base levels.
The concepts of ultimate base level (sea level) and local base level (natural, such as a lake, or artificial, such as a dam). Note how base level curves gently upward from the sea as it is traced inland; this is the theoretical limit for stream erosion.

Drainage Basins

Figure 14.4 illustrates drainage basin concepts. Every stream has a **drainage basin**, ranging in size from tiny to vast. Every drainage basin is defined by ridges that form *drainage divides;* that is, the ridges are the dividing lines that control into which basin precipitation drains. Drainage divides define a **watershed**, the catchment (water-receiving) area of the drainage basin. In any drainage basin, water initially moves downslope in a thin film called **sheetflow**, or *overland flow*. High ground that separates one valley from another and directs sheet flow is termed an *interfluve*. Surface runoff concentrates in *rills*, or small-scale downhill grooves, which may develop into deeper *gullies* and then on to a stream course in the valley.

Drainage Divides and Basins   Several high drainage divides, called **continental divides**, are situated in the United States and Canada. These are extensive mountain and highland regions that separate drainage basins, sending flows to the Pacific, the Gulf of Mexico, the Atlantic, Hudson Bay, or the Arctic Ocean. The principal drainage divides and drainage basins in the United States and Canada are mapped in Figure 14.5. These divides form water-resource regions and provide a spatial framework for water-management planning.

A major drainage basin system is made up of many smaller drainage basins. Each drainage basin gathers and delivers its precipitation and sediment to a larger basin, concentrating the volume into the main stream. A good example is the great Mississippi–Missouri–Ohio river system (see Figure 14.5).

Consider the travels of rainfall in north-central Pennsylvania. This water feeds hundreds of small streams that flow into the Allegheny River. At the same time, rainfall in southern Pennsylvania feeds hundreds of streams that flow into the Monongahela River. The two rivers then join at Pittsburgh to form the Ohio River. The Ohio flows southwestward and at Cairo, Illinois, connects with the Mississippi River, which eventually flows on past New Orleans into the Gulf of Mexico. Each contributing tributary, large or small, adds its discharge and sediment load to the larger river.

In our example, sediment weathered and eroded in north-central and southern Pennsylvania is transported thousands of kilometers and accumulates on the floor of the Gulf of Mexico, where it forms the Mississippi River delta.

Drainage Basins As Open Systems   Drainage basins are open systems in which inputs include precipitation and

FIGURE 14.4 A drainage basin.
A drainage divide separates the drainage
basin and its watershed from other basins.

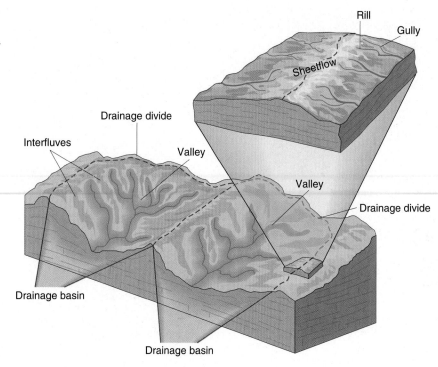

the minerals and rocks of the regional geology. Energy and materials are redistributed as the stream constantly adjusts with its landscape. System outputs of water and sediment exit through the mouth of the river, into a lake, another river, or the ocean.

Change that occurs in any portion of a drainage basin can affect the entire system, because the stream adjusts to carry the appropriate load of sediment relative to its discharge and velocity. If a river system is brought to a threshold where it can no longer maintain its present form, the relations within the drainage basin system are destabilized, initiating a transition period to a more stable condition. A stream drainage system constantly struggles toward equilibrium among the interacting variables of discharge, transported load, channel shape, and channel steepness.

Delaware River Basin Let us look at the Delaware River basin, within the Atlantic Ocean drainage region (Figure 14.6). The Delaware River headwaters are in the Catskill Mountains of New York. This basin encompasses 33,060 km^2 (12,890 mi^2) and includes parts of five states in the river's length, 595 km (370 mi) from headwaters to the mouth. The river system ends at Delaware Bay, which eventually enters the Atlantic Ocean. Topography varies from low-relief coastal plains to the Appalachian Mountains in the north. The entire basin lies within a humid, temperate climate and receives an average annual precipitation of 120 cm (47.2 in.).

The river provides water for an estimated 20 million people, not only within the basin but to cities outside the basin as well. Several major conduits export water from the Delaware River. Note on the map the Delaware Aqueduct

to New York City (in the north) and the Delaware & Raritan Canal (near Trenton). Several reservoirs in the drainage basin allow some control over water flow and storage for dry periods. The sustainability of this water resource is critical to the entire region. Clearly, planning for a drainage basin requires regional cooperation and careful spatial analysis of all variables.

Internal Drainage Most streamwater finds its way to progressively larger rivers and eventually into the ocean. In some regions, however, stream drainage does not reach the ocean. Instead, the water leaves the drainage basin by means of evaporation or subsurface gravitational flow. Such streams terminate in areas of **internal drainage**.

Portions of Asia, Africa, Australia, Mexico, and the western United States have regions with such internal drainage patterns. Internal drainage dominates the region of the Great Basin of the western United States, as shown on maps in Figures 14.5 and 15.23a. For example, the Humboldt River flows across Nevada to the west, where evaporation and seepage losses make it disappear into the Humboldt "sink." Many streams and creeks flow into the Great Salt Lake, but its only outlet is evaporation, so it exemplifies a region of internal drainage. In the Middle East, the Dead Sea region, and in Asia, the areas around the Aral Sea and Caspian Sea have no outlet to the ocean.

Drainage Density and Patterns

A primary feature of any drainage basin is its drainage density. **Drainage density** is determined by dividing the total length of all stream channels in the basin by the area of the

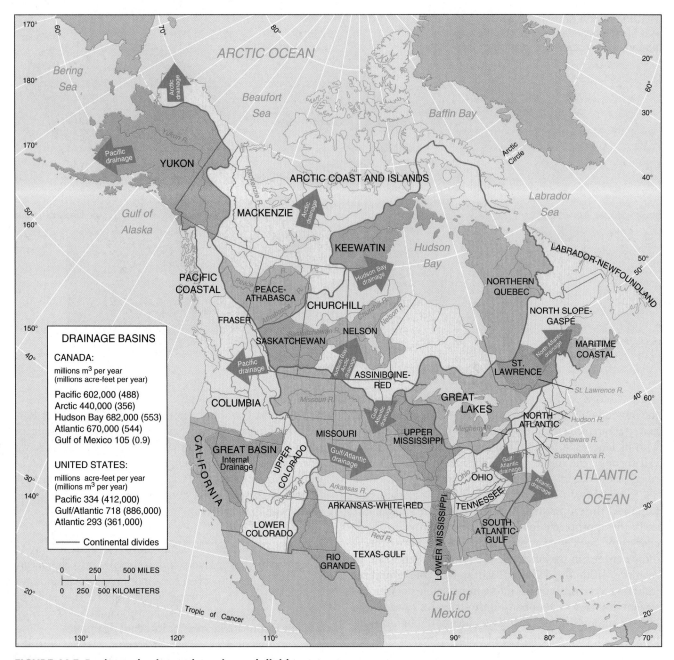

FIGURE 14.5 Drainage basins and continental divides.
Continental divides (blue lines) separate the major drainage basins that empty into the Pacific, Atlantic, Gulf of Mexico, and to the north through Canada into Hudson Bay and the Arctic Ocean. Subdividing these large-scale basins are major river basins. [After U.S. Geological Survey; *The National Atlas of Canada*, 1985, Energy, Mines, and Resources Canada; and Environment Canada, *Currents of Change—Inquiry on Federal Water Policy—Final Report 1986*.]

basin. The number and length of channels in a given area express the landscape's regional topography and surface appearance. For example, Figure 14.7 reveals a very high drainage density in a humid climate. In contrast, the typical desert has a very low drainage density.

The **drainage pattern** is the arrangement of channels in an area. Patterns are quite distinctive, for they are determined by the combination of regional steepness, variable rock resistance, variable climate, variable hydrology,

relief of the land, and structural controls imposed by the underlying rocks. Consequently, the drainage pattern of any land area on Earth is a remarkable visual summary of every characteristic—geologic and climatic—of that region.

The *Landsat* image and topographic map in Figure 14.7 are of the Ohio River drainage near the junction of West Virginia, Ohio, and Kentucky. The high-density drainage pattern and intricate dissection of the land occur because the region has generally level, easily eroded sandstone, silt-

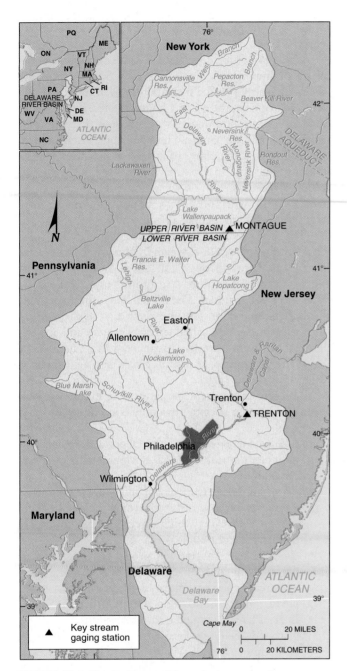

FIGURE 14.6 The Delaware River drainage basin.
This basin is under study by the USGS to assess the potential impact of global warming on a river system. [After U.S. Geological Survey, 1986, "Hydrologic events and surface water resources," *National Water Summary 1985*, Water Supply Paper 2300 (Washington, DC: Government Printing Office), p. 30.]

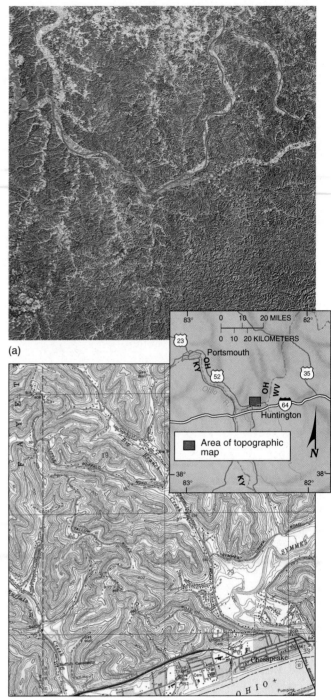

FIGURE 14.7 A stream-dissected landscape.
(a) A highly dissected topography (featuring a dendritic drainage pattern shown in Figure 14.8) around the junction of West Virginia, Ohio, and Kentucky. All of these borders are formed by rivers. (b) A portion of the USGS topographic map covering the area north of Huntington, West Virginia, reveals the intricate complexity of dissected landscapes. (The image and map are not at the same scale.) [(a) *Landsat* image from NASA; (b) Huntington Quadrangle, USGS.]

stone, and shale strata and a humid mesothermal climate. Fluvial action and other denudation processes are responsible for this dissected topography.

Common Drainage Patterns   The seven most common drainage patterns are shown in Figure 14.8. A most familiar pattern is *dendritic drainage* (Figure 14.8a). This tree-like pattern (Greek *dendron* tree) is similar to that of many natural systems, such as capillaries in the human circulatory system, the vein patterns in leaves, and tree roots. Energy expended by this drainage system is efficient because the overall length of the branches is minimized. Figure 14.7 shows dendritic drainage; on the satellite image and topographic map, you can trace the branching pattern of streams.

The *trellis drainage* pattern (Figure 14.8b) is characteristic of dipping or folded topography. Such drainage exists in the nearly parallel mountain folds of the Ridge and Valley Province in the eastern United States. Refer to Figure 12.18, a satellite image of this region that shows its distinctive drainage pattern. Here drainage patterns are influenced by rock structures of variable resistance and folded strata. The principal streams are directed by the parallel folded structures, whereas smaller dendritic tributary streams are at work on nearby slopes, joining the main streams at right angles, like a plant trellis.

The inset sketch to Figure 14.8b suggests that a headward-eroding part of one stream (to the lower right of the inset) could break through a drainage divide and *capture* the headwaters of another stream in the next valley, and indeed this does happen. The dotted line is the abandoned former channel. The sharp bends in two of the streams in the illustration are called *elbows of capture* and are evidence that one stream has breached a drainage divide. This type of capture, or *stream piracy*, can also occur in other drainage patterns.

The remaining drainage patterns in Figure 14.8 are responses to other specific structural conditions:

- A *radial* drainage pattern (c) results when streams flow off a central peak or dome, such as occurs on a volcanic mountain.
- *Parallel* drainage (d) is associated with steep slopes.
- A *rectangular* pattern (e) is formed by a faulted and jointed landscape, which directs stream courses in patterns of right-angle turns.
- *Annular* patterns (f) are produced by structural domes, with concentric patterns of rock strata guiding stream courses. Figure 12.10c provides an example of annular drainage on a dome structure.
- In areas having disrupted surface patterns, such as the glaciated shield regions of Canada, northern Europe, and some parts of Michigan and other states, a *deranged* pattern (g) is in evidence, with no clear geometry in the drainage and no true stream valley pattern.

The structure and relief of the land dictate these seven drainage patterns. Drainage patterns also occur that are discordant with the landscape through which they flow. For example, a drainage system may flow in apparent conflict with older, buried structures that have been uncovered by erosion, so that the streams appear to be *superimposed*. Where an existing stream flows as rocks are uplifted, the stream keeps its original course, cutting into the rock in a pattern contrary to its structure. Such a stream is a *superposed stream* (the stream cuts across weak and resistant rocks alike). A few examples include Wills Creek, cutting a water gap through Haystack Mountain at Cumberland, Maryland; the Columbia River through the Cascade Mountains of Washington; and the River Arun that cuts across the Himalayas.

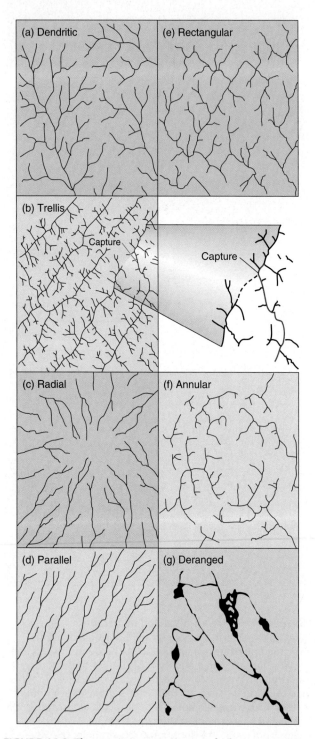

FIGURE 14.8 The seven most common drainage patterns. Each pattern is a visual summary of all the geologic and climatic conditions of its region. [After A. D. Howard, "Drainage analysis in geological interpretation: A summation," *Bulletin of American Association of Petroleum Geologists* 51 (1967), p. 2248. Adapted by permission.]

Streamflow Characteristics

A mass of water positioned above base level in a stream has potential energy. As the water flows downslope (downstream) under the influence of gravity, this energy becomes kinetic energy. The rate of this conversion from potential to kinetic energy depends on the steepness of the stream channel.

Stream channels vary in *width* and *depth*. The streams that flow in them vary in *velocity* and in the *sediment load* they carry. All of these factors increase with increasing discharge. **Discharge**, or a stream's flow rate, is calculated by multiplying the velocity of the stream by its width and depth for a specific cross section of the channel, as stated in the simple expression:

$$Q = wdv$$

where Q = discharge, w = channel width, d = channel depth, and v = stream velocity. As Q increases, some combination of channel width, depth, and stream velocity increases. Discharge is expressed either in cubic meters per second (m^3/s) or cubic feet per second (cfs).

Figure 14.9 illustrates the relation of discharge to width, depth, and velocity. The graphs show that mean velocity increases with greater discharge, despite the common misperception that downstream flow becomes more sluggish. (The increased velocity downstream often is masked by the apparent smooth, quiet flow of the water.) Not surprisingly, discharge also increases with width and depth.

Given the interplay of channel width and depth and stream velocity, the cross section of a stream varies over time, especially during heavy floods. Figure 14.10 shows changes in the San Juan River channel in Utah that occurred during a flood. Greater discharge increases the velocity and therefore the carrying capacity of the river as the flood progresses. As a result, the river's ability to scour materials from its bed is enhanced. Such scouring represents a powerful clearing action, especially in the excavation of alluvium. Scouring also can provide new recreational areas and wildlife habitats as demonstrated by the Grand Canyon experiment in News Report 14.1.

You can see in Figure 14.10 that the San Juan River's channel was deepest on October 14, when floodwaters were highest (blue line). Then, as the discharge returned to normal, the kinetic energy of the river was reduced, and the bed again filled as sediments were redeposited. You can see this process in the progression of the red, green, and purple lines. The flood and scouring process graphed in Figure 14.10 moved a depth of about 3 m (10 ft) of sediment from this cross section of the stream channel. Such adjustments in a stream channel occur as the stream system continuously works toward equilibrium, in an effort to balance discharge, velocity, and sediment load.

Exotic Streams

Most streamflows increase discharge downstream because the area being drained increases. The Mississippi River is typical. It starts as many small brooks and grows to a mighty

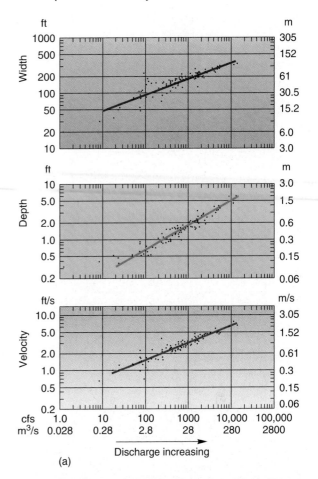

(a)

(b)

FIGURE 14.9 Effects of stream discharge.
(a) Relation of stream width, depth, and velocity to stream discharge of the Powder River at Locate, Montana. Discharge is shown in cubic meters per second (m^3/s) and cubic feet per second (cfs). (b) The Powder River area near the Wyoming border in southeastern Montana. [(a) After L. Leopold and T. Maddock, Jr., *The Hydraulic Geometry of Stream Channels and Some Physiographic Implications*, U.S. Geological Survey Professional Paper 252 (Washington, DC: Government Printing Office, 1953), p.7; (b) photo by Joyce Wilson.]

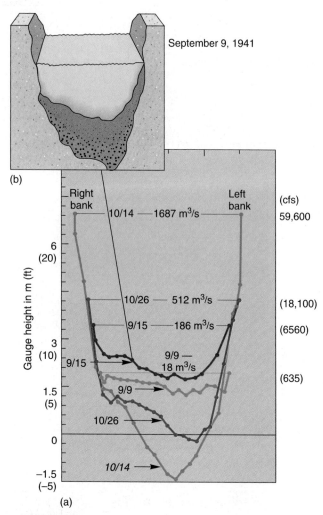

FIGURE 14.10 A flood affects a stream channel.
(a) Stream channel cross sections showing the progress of a 1941 flood on the San Juan River near Bluff, Utah. (b) Detail of stream-channel profile on September 9, 1941. [After L. Leopold and T. Maddock, Jr., *The Hydraulic Geometry of Stream Channels and Some Physiographic Implications*, U.S. Geological Survey Professional Paper 252 (Washington, DC: Government Printing Office, 1953), p. 32.]

charge to reach its mouth in the Gulf of California! The exotic Colorado River is depleted not only by passage across dry, desert lands but also by upstream removal of water for agriculture and municipal uses—see Focus Study 15.1 in Chapter 15 for a satellite image of the river's former mouth.

Stream Erosion

A stream's erosional turbulence and abrasion carve and shape the landscape through which it flows. **Hydraulic action** is the work of flowing water alone. Running water causes hydraulic squeeze-and-release action that loosens and lifts rocks. As this debris moves along, it mechanically erodes the streambed further, through the process of **abrasion**, with rock particles grinding and carving the streambed like liquid sandpaper.

The upstream tributaries in a drainage basin usually have small and irregular discharges, and most of the stream's energy is expended in turbulent eddies. As a result, hydraulic action in these upstream sections is at maximum, whereas the coarse-textured load of such a stream is small. The downstream portions of a river, however, move much larger volumes of water past a given point and carry larger suspended loads of sediment (Figure 14.11). Stream velocity determines rates of erosion and deposition. Sediment particles are deposited onto the streambed at slower velocities, whereas they are eroded at higher velocities.

Stream Transport

You may have watched a river or creek after a rainfall, the water colored brown by the heavy sediment load being transported. The amount of material available to a stream depends on topographic relief, the nature of rock and soil through which the stream flows, climate, vegetation, and human activity in a drainage basin. *Competence*, which is a stream's ability to move particles of a specific size, is a function of stream velocity. *Capacity* is the total possible load that a stream can transport. Four processes transport eroded materials: solution, suspension, saltation, and traction; each is shown in action in Figure 14.12.

Solution refers to the **dissolved load** of a stream, especially the chemical solution derived from minerals such as limestone or dolomite or from soluble salts. The main contributor of material in solution is chemical weathering. Sometimes the undesirable salt content that hinders human use of some rivers comes from dissolved rock formations and from springs in the stream channel; as an example, the San Juan and Little Colorado Rivers that flow into the Colorado River near the Utah–Arizona border add dissolved salts to the system.

The **suspended load** consists of fine-grained, clastic particles (bits and pieces of rock). They are held aloft in the stream, with the finest particles not deposited until the stream velocity slows nearly to zero. Turbulence in the water, with random upward motion, is an important mechanical factor in holding a load of sediment in suspension.

river pouring into the Gulf of Mexico. In contrast, a stream can originate in a humid region and subsequently flow through an arid region. In that case, the *discharge usually decreases with distance*, because of high potential evapotranspiration (POTET) rates in the arid area. Such a stream is called an **exotic stream** (exotic means "of foreign origin").

The Nile River exemplifies exotic streams. This great river, Earth's longest, drains much of northeastern Africa. But as it courses through the deserts of Sudan and Egypt, it loses water instead of gaining it, because of evaporation and withdrawal for agriculture. By the time it empties into the Mediterranean Sea, the Nile's flow has dwindled so much that it ranks only 36th in discharge.

The United States has exotic streams too, notably the Colorado River. Its flow decreases with distance from its source; in fact, the river no longer produces enough dis-

News Report 14.1

Scouring the Grand Canyon for New Beaches and Habitats

Glen Canyon Dam sealed the Colorado River gorge north of the Grand Canyon, near Lees Ferry and the Utah–Arizona border (see the map in Focus Study 15.1, Figure 1). Impoundment of water began in 1963, as did the inundation of many canyons upstream from the dam of the new Lake Powell.

The lake began collecting the tremendous sediment load of the Colorado River, reducing the volume of water in the river below the dam and eliminating the natural seasonal fluctuations in river discharge. The production of hydroelectricity at Glen Canyon further affected the Grand Canyon with highly variable water releases keyed to electrical turbine operations. Water in the canyon rose and fell as much as 4.3 m (14 ft) as the dis-

tant lights and air conditioning of Las Vegas and Phoenix went on and off.

All of these changes affected the canyon. Over the years, the beaches were starved for sand, and channels filled with sediment. Fisheries were disrupted and backwater channels were depleted of nutrients. In 1996 an unprecedented experiment took place. The Grand Canyon was artificially flooded with a 1260-m³/s (45,000-cfs) release from Glen Canyon. The flow lasted for 7 days and then was reduced to 224 m³/s (8000 cfs). Flows were decreased gradually to allow the newly formed beaches to drain, in contrast to the rapid opening of the pipes that initiated the flood. Lake Powell dropped 1 m (3.3 ft), and Lake Mead, downstream behind Hoover Dam, rose 0.8 m (2.6 ft).

The results were initially positive: 35% more beach area was created by the scouring of sand and sediments from the channel. Approximately 80% of the aggradation (building up of beaches) took place in the first 40 hours and was completed by the 100-hour mark. Numerous backwater channels were created, flush with fresh nutrients for the humpback chub and other endangered fish species. In contrast to these benefits were a few negatives in the form of existing ecosystem disruption. A full assessment of this experiment is ongoing. Present thinking is that its success was limited. Another scouring test is likely. For more on this experiment, see http://water.usgs.gov/pubs/FS/FS-060-99/.

(a)

(b)

FIGURE 14.11 Stream velocity and discharge increase together.
(a) A low-volume, low-velocity (but turbulent) mountain stream in the White Mountains, New Hampshire. (b) The high-discharge, high-velocity (but laminar, or smooth water flow) portion of the Ocmulgee River near Jacksonville, Georgia. [Photos by Bobbé Christopherson.]

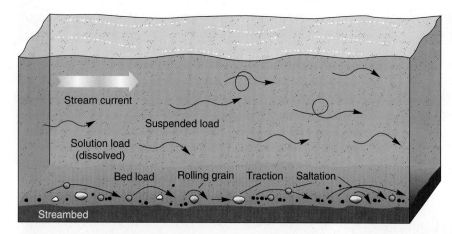

FIGURE 14.12 Fluvial transport.
Fluvial transportation of eroded materials through saltation, traction, suspension, and solution.

(a)

(b)

FIGURE 14.13 A braided stream.
(a) Braided stream pattern in Chitina River, Wrangell-Saint Elias National Park, Alaska. (b) A braided Brahmaputra River channel some 35 km (22 mi) south of Lhasa, Tibet, in a narrow valley south of the Tibetan Plateau. These streams reflect excessive sediment load associated with glacial meltwaters filled with fine sediments, or "glacial flour." [(a) Photo by Tom Bean. (b) October 13, 2001, photo by International Space Station astronaut courtesy of Earth Science and Image Analysis Lab, JSC/NASA.]

Bed load refers to coarser materials that are dragged along the streambed by **traction** or are rolled and bounced along by **saltation** (from the Latin *saltim*, which means "by leaps or jumps"). At times, it is difficult to distinguish traction from saltation and their effects on bed load. Particles transported by saltation are too large to remain in suspension, a distinction directly related to a stream's velocity and its ability to retain particles in suspension. With increased kinetic energy, parts of the bed load are rafted upward and become suspended load. Saltation is also a process in wind transport of materials (see Chapter 15).

The first Spanish explorers to visit the Grand Canyon reported in their journals that they were kept awake at night

by the thundering sound of boulders tumbling along the Colorado River bed (a combination of traction and saltation). Such sounds today are substantially lessened because of the reduced velocity and discharge of the Colorado resulting from the many dams and control facilities that now trap sediments and reduce bed load capacity.

If the load (bed and suspended) exceeds a stream's capacity, sediments accumulate as **aggradation** (the opposite of degradation) and the stream channel builds up through deposition. With excess sediment, a stream becomes a maze of interconnected channels that form a **braided stream** pattern (Figure 14.13). Braiding often occurs when reduced discharge lowers a stream's transporting ability, such as after flooding, or when a landslide occurs upstream, or from increased load where weak banks of sand or gravel exist. Locally, braiding also may result from a new sediment load from glacial meltwaters, as in Alaska's Chitina River in the photo. Glacial materials also exceed stream capacity in the 15-km (9.3-mi) stretch of the Brahmaputra River about 35 km (22 mi) south of Lhasa, Tibet (Figure 14.13b).

Flow and Channel Characteristics

Streams have two general types of flow, laminar and turbulent. *Laminar flow* is a streamlined flow of water in which individual clay and other fine particles move along evenly in generally parallel paths. Natural streams have stretches of laminar flow only when the water is deep and the channel surfaces are smooth.

Flow becomes *turbulent* in shallow streams, or where the channel is rough, as in a section of rapids. Small eddies are caused by friction between streamflows and the channel sides and bed. Complex turbulent flows propel sand, pebbles, and even boulders, increasing the suspension, traction, and saltation.

Flow characteristics of a stream are best seen in a cross-sectional view. The greatest velocities in a stream are near the surface at center channel (Figure 14.14), corresponding to the deepest part of the stream channel. Velocities de-

crease closer to the sides and bottom of the channel because of the frictional drag on the water flow. The portion of the stream flowing at maximum velocity moves diagonally across the stream from bend to bend.

Laminar and turbulent flow form three types of stream channels: *braided, straight,* and *meandering.* Where slope is gradual, stream channels develop a sinuous (snakelike) form, weaving across the landscape. This action produces a **meandering stream**. The term *meander* comes from the ancient Greek Maiandros River in Asia Minor (the present-day Menderes River in Turkey), which had what came to be known as a meandering channel pattern. The tendency to meander is evidence of a river system's struggle to operate with least effort, between self-organizing order (equilibrium) and chaotic disorder in nature.

The outer portion of each meandering curve is subject to the fastest water velocity and therefore the greatest scouring erosive action; it can be the site of a steep bank called an **undercut bank**, or *cutbank* (Figure 14.15). On the other hand, the inner portion of a meander experiences the slowest water velocity and thus receives sediment fill, forming a deposit called a **point bar**. As meanders develop, these scour-and-fill features gradually work their way downstream. As a result, the landscape near a meandering river bears meander scars of residual deposits from previous river channels (see Figure 14.20). The photograph of the Itkillik River in Alaska, Figure 14.15a, shows both meanders and meander scars.

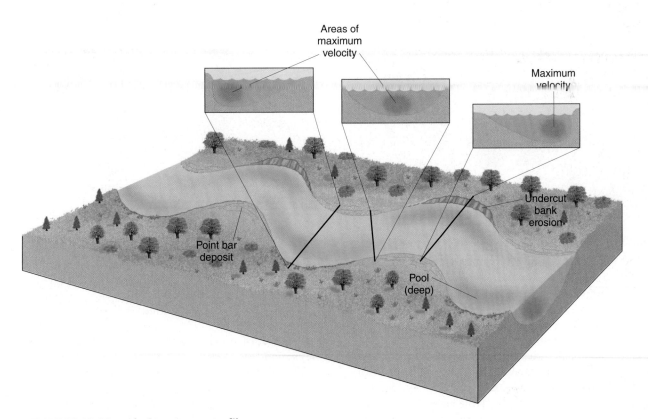

FIGURE 14.14 Meandering stream profile.
Aerial view and cross sections of a meandering stream, showing the location of maximum flow velocity, point-bar deposits, and areas of undercut bank erosion.

(a)

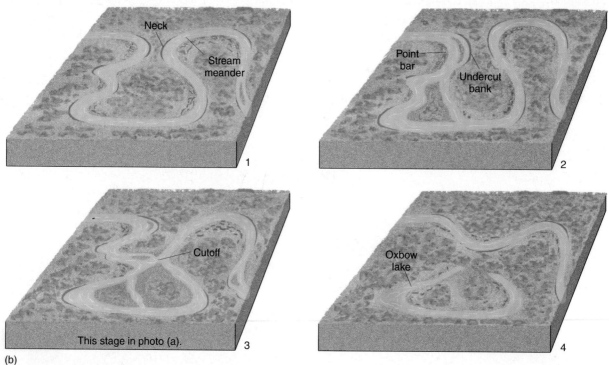

(b)

FIGURE 14.15 Meandering stream development.
(a) Itkillik River in Alaska. (b) Development of a river meander and oxbow lake simplified in four stages. [(a) U.S. Geological Survey photo.]

Meandering streams create a remarkable looping pattern on the landscape. Meanders gradually form loops, as shown in the four-part sequence in Figure 14.15b. In (1), the stream erodes its outside bank as the curve migrates downstream, forming a neck. In (2), the narrowing neck of land created by the looping meander eventually erodes through and forms a *cutoff* in (3). A cutoff marks an abrupt change in the stream's lateral movements—the stream becomes straighter.

When the former meander becomes isolated from the rest of the river, the resulting **oxbow lake** (4) may gradually fill with silt or may again become part of the river when it floods. The Mississippi River is many miles shorter today than it was in the 1830s because of artificial cutoffs that

News Report 14.2

Rivers Make Poor Political Boundaries

Streams often serve the role of natural political boundaries. Commonly, where a stream forms a boundary, the boundary line is drawn down the middle of the stream. It is easy to see how boundary disputes might arise when they are based on river channels that shift their positions quite rapidly during times of flood. The boundaries based on them are not changed, and existing land ownership remains set.

Carter Lake, Iowa, provides a fascinating example (Figure 1). The Nebraska–Iowa border originally was placed mid-channel in the Missouri River. But in 1877 the meander loop that curved around the town of Carter Lake in Iowa was cut off by the river, leaving the town "captured" by Nebraska! The old boundary marked along the former meander bend still is used as the state line. The oxbow lake created by the cutoff was named Carter Lake.

This event illustrates why boundaries always should be fixed by surveys independent of river locations. Such surveys have been completed along the Rio Grande near El Paso, Texas, and

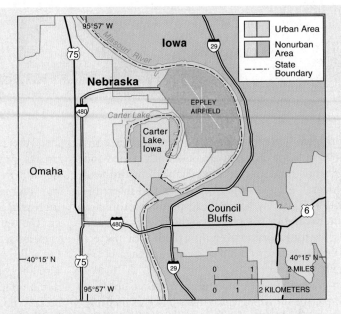

FIGURE 1   A stranded town.
Carter Lake, Iowa, sits within the curve of a former meander that was cut off by the Missouri River. The city and oxbow lake remain part of Iowa even though they are stranded within Nebraska.

along the Colorado River between Arizona and California, permanently establishing political boundaries separate from shifting river channels. In the latter example, a midpoint between the bluffs on either side of the floodplain is used as the state line. Meanwhile, Carter Lake is still in Iowa.

---

were dredged across meander necks to improve navigation and safety. News Report 14.2 looks at the difficulty of using a meandering stream as a political boundary.

Stream Gradient

Every stream runs downhill under the pull of gravity. In the course of this process, each stream develops its own **gradient**, which is the rate of elevation decline from its headwaters to its mouth. This decline is far from linear; characteristically, the *longitudinal profile* (side view) of a stream features a steeper slope upstream and a gentler slope downstream, forming an uneven concave shape (Figure 14.16). This gradient is concave for complex reasons related to the stream's having just enough energy to transport the sediment load it receives. The longitudinal profile of streams can be expressed mathematically, enabling scientists to categorize and predict stream behavior.

An important fluvial concept is that of the graded stream, nicely defined by J. H. Mackin, a geomorphologist:

A graded stream is one in which, over a period of years, slope is delicately adjusted to provide, with available discharge and with prevailing channel characteristics, just the velocity required for transportation of the load supplied from the drainage basin.*

In other words, a **graded stream** is one that attains a graded condition and does not mean that the stream is at its lowest gradient. Rather, graded represents a present balance (a dynamic equilibrium) among erosion, transportation, and deposition over time along a portion of the stream. Both high-gradient and low-gradient streams can achieve a graded condition. The longitudinal profile of a graded stream is called a *profile of equilibrium*—a parabolic curve, gently flattening toward the mouth. Stream dynamics can then be compared against this ideal balance.

*J. H. Mackin, "Concept of the graded river," *Geological Society of America Bulletin*, 59 (1948): 463.

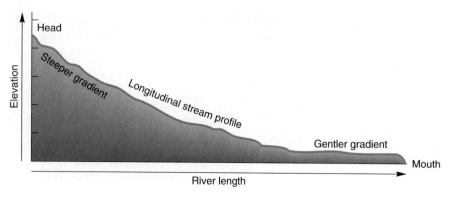

FIGURE 14.16 An ideal longitudinal profile.
Idealized cross section of the longitudinal profile of a stream, showing its gradient. Upstream segments have a steeper gradient; downstream, the gradient is gentler. The middle and lower portions in the illustration appear graded, or in dynamic equilibrium.

One problem with applying the graded stream concept is that an individual stream can have both graded and ungraded portions and may have graded sections without having an overall graded slope. Variations and interruptions are the rule rather than the exception. A profile of equilibrium may not be smooth throughout its course and cannot exist for long, for it represents a theoretical perfect balance. With streams, as in all of nature, change is the only constant.

Stream gradient may be affected by tectonic uplift of the landscape, which changes the base level. If tectonic forces slowly lift the landscape, the stream gradient will increase, stimulating renewed erosional activity. Imagine this occurring to the landscape in Figure 14.2. The meandering stream flowing through the uplifted landscape becomes *rejuvenated*; that is, the river actively returns to downcutting and eventually forms *entrenched meanders* in the landscape. Figure 14.17 depicts an actual rejuvenated landscape.

Nickpoints When the longitudinal profile of a stream shows an abrupt change in gradient, such as at a waterfall or an area of rapids, the point of interruption is termed a **nickpoint** (also spelled *knickpoint*). At a nickpoint, the conversion of potential energy to concentrated kinetic energy works to eliminate the nickpoint. Figure 14.18 shows a stream with two such interruptions.

Nickpoints can result when a stream flows across a zone of hard, resistant rock or from various tectonic uplift episodes, such as might occur along a fault line. Temporary blockage in a channel, caused by a landslide or a logjam, also could be considered a nickpoint; when the logjam breaks, the stream quickly readjusts its channel to its former grade.

Waterfalls are interesting and beautiful gradient breaks. At the edge of a fall, a stream is free-falling, moving at high velocity under the acceleration of gravity, causing increased abrasion on the channel below. The increased abrasion and hydraulic action generally undercut the waterfall. Eventually the excavation will cause the rock ledge at the lip of the fall to collapse, and the waterfall will shift a bit farther upstream. The height of the waterfall is gradually reduced as debris accumulates at its base. Thus, a nickpoint migrates upstream, sometimes for kilometers, until it becomes a series of rapids and is eventually eliminated.

At Niagara Falls on the Ontario–New York border, glaciers advanced over the region and then receded. In doing so, they exposed resistant rock strata that are underlain by less-resistant shales. As this less-resistant material continues to weather away, the overlying rock strata collapse, allowing the falls to erode farther upstream toward Lake Erie (Figure 14.19a).

Niagara Falls is a place where natural processes labor to eliminate a nickpoint, reducing this portion of the river to a series of mere rapids. In fact, the falls have retreated more than 11 km (6.8 mi) from the steep face of the Niagara escarpment (cliff) during the last 12,000 years. In the past engineers have used control facilities upstream to reduce flows over the American Falls at Niagara for inspection of the cliff (Figure 14.19b). A nickpoint is a relatively temporary and mobile feature on the landscape.

Stream Gradient and Landscape Forms In Chapter 13, we introduced William Morris Davis, a geomorphologist who founded the Association of American Geographers

FIGURE 14.17 Entrenched meanders.
The San Juan River near Mexican Hat, Utah, cuts down into the uplifted Colorado Plateau landscape producing entrenched meanders.
[(a) Photo by Betty Crowell; (b) photo by Randall M. Christopherson.]

FIGURE 14.18 Nickpoints interrupt a stream profile.
(a) Longitudinal stream profile showing nickpoints produced by resistant rock strata. Potential energy is converted into kinetic energy and concentrated at the nickpoint, accelerating erosion, which will eventually eliminate the feature. (b) A nickpoint and granite pothole interrupts stream gradient on the Pemigewassey River, Franconia Notch Park, New Hampshire. [Photo by Bobbé Christopherson.]

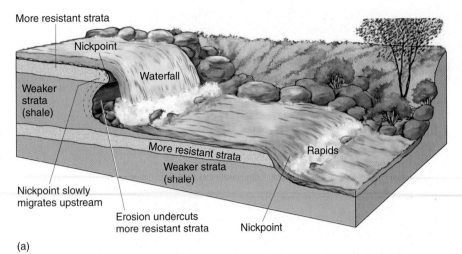

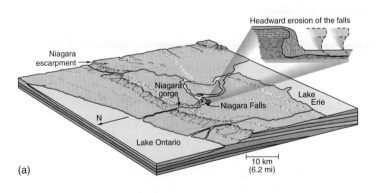

FIGURE 14.19 Retreat of Niagara Falls.
(a) Headward retreat of Niagara Falls from the Niagara escarpment. It has taken the falls about 12,000 years to reach this position at a pace of about 1.3 m (4.3 ft) per year. (b) Niagara Falls, with the American Falls portion almost completely shut off by upstream controls for inspection. Horseshoe Falls in the background is still flowing. Such inspections allow engineers to assess the progress of natural processes that are working to eliminate the Niagara Falls nickpoint. [(a) After W. K. Hamblin, *Earth's Dynamic Systems*, 6th ed. (Upper Saddle River, NJ: Macmillan Publishing, an imprint of Prentice Hall, Inc. © 1992), Figure 12.15, p. 246. Used by permission. (b) Photo courtesy of the New York Power Authority.]

in 1904. His evolutionary concepts of erosion also included fluvial processes. Drawing from G. K. Gilbert, an important geomorphologist and his contemporary, Davis incorporated the graded-stream concept into his model, identifying erosion stages in a cyclic model he called *youth*, *maturity*, and *old age*—old age is approached as the floodplain broadens and a low stream gradient produces a wide, meandering flow pattern.

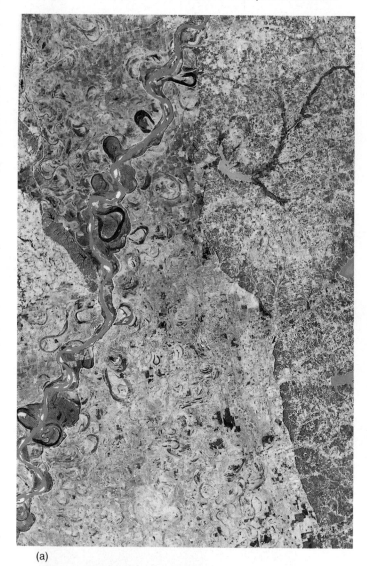

(a)

(b)

FIGURE 14.20 Meander scars.
The Mississippi River forms a portion of the Mississippi-Arkansas border near Senatobia, Mississippi. Characteristic meander patterns and scars of former channels are visible in the *Landsat* image. The image of northwestern Mississippi shows the portion of the Mississippi River that forms the Mississippi–Arkansas border. [Image by GEOPIC, Earth Satellite Corporation/Photo Researchers, Inc.]

Today, a *functional model of dynamic equilibrium* is supported by geomorphologists. The dynamic equilibrium model emphasizes the effects of individual processes interacting on streams and hillslope systems. Stream form and behavior result from complex interactions of slope, discharge, and load, all of which are variable within different climates and with different rock types. Landscapes simply do not provide enough clear evidence to support a cyclic model of evolution. Regardless, Davis's work was a breakthrough in understanding landscapes, and many of his terms are still in use.

As suggested by S. A. Schumm and R. W. Lichty, two modern geomorphologists, the validity of cyclic or functional landscape models may depend on the *time frame*. Let us consider three time frames: geologic time, graded time, and steady time. Over the long span of *geologic time*, cyclic models of evolutionary development might explain the disappearance of entire mountain ranges through denudation, for example. At the other extreme of time, *steady time* applies to short-term adjustments ongoing in a drainage basin. *Graded time* is between the two time frames, and within it lies the realm of dynamic equilibrium conditions.

Stream Deposition

After weathering, mass movement, erosion, and transportation, deposition is the next logical event in a sequence. In *deposition*, a stream deposits alluvium, or unconsolidated sediments, thereby creating depositional landforms, such as floodplains, terraces, or deltas.

As discussed earlier, stream meanders tend to migrate downstream through the landscape. Over time, the landscape near a meandering river comes to bear meander scars of residual deposits from former, abandoned channels. Former point-bar deposits leave low-lying ridges, creating a *bar-and-swale relief* (a swale is a gentle low area). The *Landsat* image in Figure 14.20 exhibits characteristic meandering scars: meander bends, oxbow lakes, natural levees, point bars, and undercut banks.

Floodplains The flat low-lying area flanking a stream channel that is subjected to recurrent flooding is a **floodplain**. It is formed when the river overflows its channel during times of high flow. Thus, when floods occur, the floodplain is inundated. When the water recedes, it leaves behind alluvial deposits that generally mask the underlying rock with their accumulating thickness. The present river channel is embedded in these alluvial deposits. Figure 14.21 illustrates a characteristic floodplain and a representative topographic map of an area near Philipp, Mississippi.

On either bank of most streams, **natural levees** develop as by-products of flooding. When floodwater arrives, the river overflows its banks, loses velocity as it spreads out, and drops a portion of its sediment load to form the

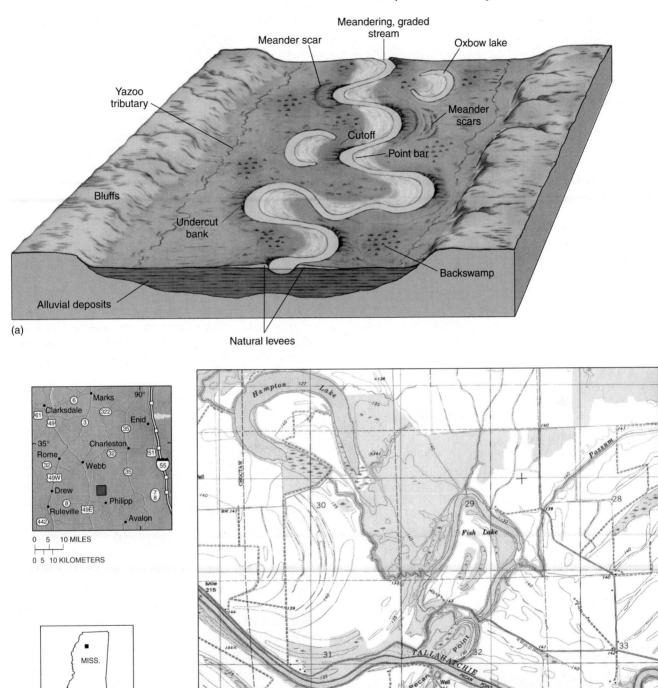

FIGURE 14.21 A floodplain.
(a) Typical floodplain landscape and related landscape features. (b) A portion of the Philipp, Mississippi, topographic map quadrangle. [(b) Topographic map from USGS.]

levees. Larger sand-sized particles drop out first, forming the principal component of the levees, with finer silts and clays deposited farther from the river. Successive floods increase the height of the levees (*levée* is French for "raising"). The levees may grow in height until the river channel becomes elevated, or *perched* above the surrounding floodplain.

On the topographic map (Figure 14.21b), you can see the natural levees represented by several contour lines that run immediately adjacent to the Tallahatchie River. These contour lines (5-ft interval) denote a height of 10–15 ft (3–4.5 m) above the river and the adjoining floodplain. Next time you have an opportunity to see a river and its floodplain, look for levees (they may be low and subtle).

Notice in Figure 14.21 an area labeled backswamp and a stream called a yazoo tributary. The natural levees and elevated channel of the river prevent this **yazoo tributary** from joining the main channel, so it flows parallel to the river and through the **backswamp** area. (The name comes from the Yazoo River in the southern part of the Mississippi floodplain.)

People build cities on floodplains despite the threat of flooding because floodplains are nearly level and they are next to water. People often are encouraged by government assurances of artificial protection from floods and disaster assistance if floods occur. Government assistance may provide the building of artificial levees on top of natural levees. Artificial levees do increase the capacity in the channel, but they also lead to even greater floods when they are overtopped by flood waters or when they fail (Figure 14.22). Are there river floodplains where you live? If so, what is your impression of present land-use patterns, local planning and zoning, and people's hazard perception overall?

The catastrophic floods along the Mississippi River and its tributaries in 1993 illustrate the risk of building settlements on floodplains. Damage estimates from those floods, detailed in News Report 14.3, exceeded $30 billion.

In 2001 some two-thirds of disaster losses were attributable to floods. Tropical storm Allison left $6 billion in damage in its wandering visit to Texas in June 2001, most damage occurred from flooding along occupied floodplains.

Stream Terraces   As explained earlier, several factors may rejuvenate stream energy and stream–landscape relations so that a stream can scour downward with renewed vigor and increased erosion. The resulting entrenchment of the river deeper into its own floodplain produces **alluvial terraces** on either side of the valley, which look like topographic steps above the river. Alluvial terraces generally appear paired at similar elevations on each side of the valley (Figure 14.23). If more than one set of paired terraces is present, the valley probably has undergone more than one episode of rejuvenation. The flat terrace areas, above the lowest section of floodplain along the river, have always been a location for settlement.

If the terraces on either side of the valley do not match in elevation, then entrenchment actions must have been continuous as the river meandered from side to side, with each meander cutting a terrace slightly lower in elevation—

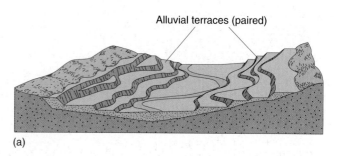

Alluvial terraces (paired)

(a)

(b)

FIGURE 14.23 Alluvial stream terraces.
(a) Alluvial terraces are formed as a stream cuts into a valley.
(b) Alluvial terraces along the Rakaia River, in New Zealand.
[(a) After W. M. Davis, *Geographical Essays* (New York: Dover, 1964 [1909]), p. 515; (b) photo by Bill Bachman/Photo Researchers, Inc.]

FIGURE 14.22 Flooding river and threatened levees.
Emergency levee repairs to prevent further erosion by raindrop impact and seepage from the river. The floodplain to the left is below the level of the Sacramento River. [Photo by California Department of Water Resources.]

News Report 14.3

The 1993 Midwest Floods

The Mississippi and Missouri Rivers are no strangers to flooding. The damage a flood might cause increases as more people settle on the vulnerable floodplains. The widespread floods of 1993 in the upper Mississippi and lower Missouri River basins exceeded peak discharge records at nearly 100 gaging (stream-measuring) stations, making it one of the greatest floods in U.S. history.

In spring 1993, a series of low-pressure systems (with their counter-clockwise winds) stalled in the West, and high pressure (with its clockwise winds) dominated the Eastern Seaboard. These systems combined to produce a region of sustained convergence, instability, and thundershowers over the Midwest. Precipitation was 150%–200% above normal for most cities in the flooded area (Figure 1a).

By late June, many reservoirs were filled, and soils were saturated by record rainfall over the region. Then came July, when some stations received 75 cm (30 in.) of precipitation in the first 3 weeks! Three remarkable aspects of "The Great Flood of 1993" were that the flood stage along some rivers was sustained for more than a month, flood crests set new historical records, and many locations experienced multiple crests.

As 10,000 km (6200 mi) of levees were overtopped, more than 1000 levees were breached (broken through). Water reached almost 10 m (32 ft) above flood stage in some areas, flooding many cities and towns, and 6 million hectares (14 million acres) of forest and farmland in the Missouri and Mississippi River drainage basins (Figure 1b). The result was a Presidential Dis-

aster Declaration for 534 counties in Illinois, Iowa, Kansas, Minnesota, Missouri, the Dakotas, Nebraska, and Wisconsin. Overall, damage estimates topped $30 billion.

Some cities had prepared for the disaster through improved levee construction and hazard zoning of susceptible floodplains. Many others had postponed raising local taxes for such action and had done little to prevent flood damage. This unevenness between areas created political friction. This flood event was a painful reminder of the power of nature in our lives and the need for improved hazard perception, preparation, and avoidance strategies. For imagery of these floods, see **http://rsd.gsfc.nasa.gov/rsd/images/ Flood_cp.html**, and for a brief paper, see **http://www.nwrfc.noaa.gov/ floods/papers/oh_2/great.htm**.

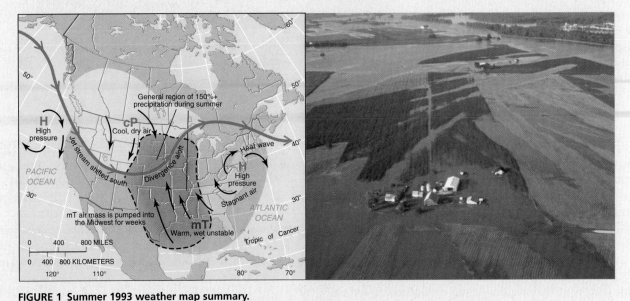

FIGURE 1 Summer 1993 weather map summary.
(a) Weather pattern summarized for the summer of 1993. Atmospheric conditions produced a steady flow of moisture-rich, unstable, maritime tropical air from the Gulf of Mexico for a prolonged period of time. (b) A farm 32 km (20 mi) south of Des Moines, Iowa, is inundated during the Midwest floods of 1993. [Photo by Les Stone/Sygma.]

a condition of *unpaired terraces.* Thus, alluvial terraces represent what originally was a depositional feature (a flood-plain) that subsequently has been eroded by its own stream because the stream experienced changes in stream load and capacity.

River Deltas   The mouth of a river is where it reaches a base level. The river's forward velocity rapidly decelerates as it enters a larger body of water. The reduced velocity causes the transported load to quickly exceed the river's carrying capacity. Coarse sediments such as sand and gravel drop

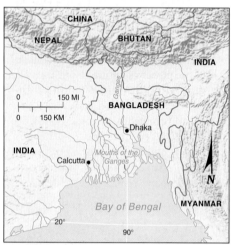

FIGURE 14.24 The Ganges River enters the Bay of Bengal.
The complex distributary pattern in the "many mouths" of the Ganges River delta in Bangladesh and extreme eastern India from the *Terra* satellite. [*Terra* MODIS sensor image courtesy of MODIS Land Team, NASA.]

out first and are deposited closest to the river's mouth. Finer clays are carried farther and form the extreme end of the deposit. The depositional plain that forms at the mouth of a river is called a **delta** for its characteristic triangular shape, named after the Greek letter delta (Δ). The significance of the Nile River delta to food production was perceived by Herodotus in ancient times (see News Report 14.4).

Each flood stage deposits a new layer of alluvium over the surface of the delta so that it grows outward. At the same time, river channels divide into smaller courses known as *distributaries*, which appear as a reverse of the dendritic drainage pattern of tributary streams discussed earlier. Here are a few examples:

- The Ganges River delta features an intricate pattern of distributaries in a *braided delta*. Bountiful alluvium carried from deforested slopes upstream provides excess sediment that is deposited to form many deltaic islands (Figure 14.24). The combined Ganges–Brahmaputra River delta complex is the largest in the world.

- The Nile River delta is an *arcuate* (arc-shaped) delta (Figure 14.25). Also arcuate are the Danube River delta in Romania where it enters the Black Sea and the Indus River delta. (See News Report 14.4 for an update on the condition of the disappearing Nile Delta.)
- The Tiber River in Italy has an *estuarine delta*, one that is in the process of filling an **estuary**, which is the seaward mouth of a river where the river's freshwater encounters seawater.

Mississippi River Delta The Mississippi River delta has an interesting history. Over the past 120 million years, the Mississippi has collected sediments throughout its vast basin and deposited them into the Gulf of Mexico. During the past 5000 years, the river has formed a succession of seven distinct deltaic complexes along the Louisiana coast (Figure 14.26).

Each generalized lobe in the illustration reflects distinct course changes in the Mississippi River, probably where the river broke through its natural levees during

News Report 14.4

The Nile Delta Is Disappearing

People along the Nile River have depended on its regular flow and annual floods for millennia. Herodotus noted in the 5th century B.C. how the people farmed the fields in the floodplain and delta regions. After harvesting their crops, they retreated from the area to their homes. They would await the annual floods that brought fresh silt and nutrients for next year's planting. This cycle of fertility continued until the completion of the Aswan High Dam in 1964. This structure caused a partial interruption in the supply of sediment to the delta, and as a result the delta

coastline continues to actively recede. Herodotus stated in *The History*, Book Two, "in the part called the Delta, it seems to me that if the Nile no longer floods . . . for all time to come, the Egyptians will suffer."

J. Stanley, an oceanographer at the Smithsonian Institution, has proposed an intriguing answer to what is happening in addition to the impact of the Aswan High Dam. Over the centuries, more than 9000 km (5500 mi) of canals were built in the delta to augment the natural distributary system. As the river discharge enters the network of canals,

flow velocity is reduced, stream competence and capacity are lost, and sediment load is deposited far short of where the delta touches the Mediterranean Sea. River flows no longer effectively reach the sea.

The Nile Delta is receding from the coast at an alarming 50 to 100 m (165 ft to 330 ft) per year. Seawater is intruding farther inland in both surface water and groundwater. Human action and reaction to this evolving situation will no doubt determine the delta's future. If Herodotus could only see the delta he described now!

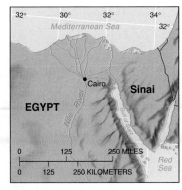

FIGURE 14.25 The Nile River delta.
The arcuate Nile River delta. Intensive agricultural activity and small settlements are visible on the delta and along the Nile River floodplain in this true color image. Cairo is at the apex of the delta. You can see the two main distributaries: Damietta to the east and Rosetta to the west. [January 30, 2001, *Terra* MISR sensor image courtesy of MISR Team, NASA/GSFC/JPL.]

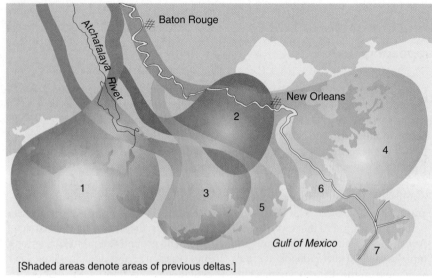

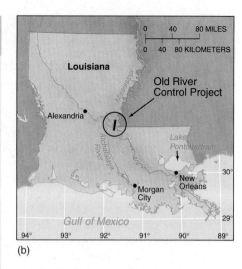

[Shaded areas denote areas of previous deltas.]

(a)

(b)

(c)

FIGURE 14.26 The Mississippi River delta.
(a) Evolution of the present delta, from 5000 years ago (1) to present (7). (b) Location map of the Old Control Structure and potential capture point (arrow) where the Atchafalaya River may one day divert the present channel. (c) The bird-foot delta of the Mississippi River receives a continuous supply of sediments, focused by controlling levees. The delta extends ever farther into the Gulf of Mexico, although subsidence of the delta and rising sea level have diminished the overall surface area. [(a) Adapted from C. R. Kolb and J. R. Van Lopik, "Depositional environments of the Mississippi River deltaic plain," in *Deltas in Their Geologic Framework* (Houston: Houston Geological Society, 1966). Adapted by permission. (c) March 5, 2001, *Terra* MODIS sensor image courtesy of Liam Gumley, Space Science and Engineering Center, University of Wisconsin, and the MODIS Science Team, NASA.]

episodes of severe flooding, thus changing the configuration of the delta (Figure 14.26a). The seventh and current delta has been building for at least 500 years and is a classic example of a *bird-foot delta*—a long channel with many distributaries and sediments carried beyond the tip of the delta into the Gulf of Mexico.

The Mississippi River delta clearly is dynamic over time as sediments accumulate on the floor of the Gulf of Mexico and the distributaries shift (Figure 14.26c). The main channel persists because of much effort and expense directed at maintaining the artificial levee system. The 3.25 million km² (1.25 million mi²) Mississippi drainage basin produces enough sediment to extend the Louisiana coast 90 m (295 ft) a year—550 million metric tonnes a year.

To further complicate this situation, compaction and the tremendous weight of the sediments in the Mississippi River create isostatic adjustments in Earth's crust. These adjustments are causing the entire region of the delta to subside, thereby placing ever-increasing stress on natural and artificial levees and other structures along the lower Mississippi.

The city of New Orleans is now almost entirely below river level, with some sections of the city below sea level. Severe flooding is a certainty for existing and planned settlements unless further intervention or urban relocation occurs. The building of multiple flood-control structures and extensive reclamation efforts by the U.S. Army Corps of Engineers apparently have only delayed the peril, as demonstrated by recent flooding.

An additional problem for the lower Mississippi Valley is the possibility, in a worst-case flood, that the river could break from its existing channel and seek a new route to the Gulf of Mexico. If you examine the map in Figure 14.26b and look at the sediment plume to the west of the main delta in (c), an obvious alternative to the Mississippi's present channel is the Atchafalaya River.

The Atchafalaya would provide a much shorter route to the Gulf of Mexico, less than one-half the present distance, and it has a steeper gradient than the Mississippi. Presently this alternative-route distributary carries about 30% of the Mississippi's total discharge. For the Mississippi to bypass New Orleans entirely would be a blessing, for it would remove the flood threat. However, this shift would be a financial disaster, as a major U.S. port would silt in and seawater would intrude into freshwater resources.

At present, artificial barriers block the Atchafalaya from reaching the Mississippi at the point shown; without the floodgates the two rivers do connect. The Old River Control Project (1963) maintains three structures and a lock about 320 km (200 mi) from the Mississippi's mouth to keep these rivers in their channels. A major flood is only a matter of time, and residents should prepare for the river-channel change.

Rivers Without Deltas   The Amazon River, Earth's highest-discharge stream, exceeds 175,000 m³/s (6.2 million cfs) discharge and carries sediments far into the deep Atlantic offshore. Yet the Amazon lacks a true delta. Its mouth, 160 km (100 mi) wide, has formed an underwater deltaic plain deposited on a sloping continental shelf. As a result, the Amazon's mouth is braided into a broad maze of islands and channels (see Figure 14.1).

Other rivers also lack deltaic formations if they do not produce significant sediment or if they discharge into strong erosive currents. The Columbia River of the U.S. Northwest lacks a delta because offshore currents remove sediment before it can accumulate into a delta.

Floods and River Management

Throughout history, civilizations have settled floodplains and deltas, especially since the agricultural revolution of 10,000 years ago, when the fertility of floodplain soils was discovered. Early villages generally were built away from the area of flooding, or on stream terraces, because the floodplain was dedicated exclusively to farming. However, as commerce grew, competition for sites near rivers grew, because these locations were important for transportation. Port and dock facilities were built, as were river bridges. Because water is a basic industrial raw material used for cooling and for diluting and removing wastes, waterside industrial sites became desirable. These competing human activities on vulnerable flood-prone lands place lives and property at risk during floods.

The abuse and misuse of river floodplains brought catastrophe to North Carolina in 1999. In short succession during September and October, Hurricanes Dennis, Floyd, and Irene delivered several feet of precipitation to the state; each storm falling on already saturated ground. About 50,000 people were left homeless and at least 50 died, while more than 4000 homes were lost and an equal amount were badly damaged (Figure 14.27a). The dollar estimate for the ongoing disaster now exceeds $10 billion. However, the real tragedy will unfold for years to come.

Hogs, in factory farms, outnumber humans in North Carolina. More than 10 million hogs, each producing two tons of waste per year, were located in about 3000 agricultural factories. These generally unregulated operations collect millions of tons of manure into open lagoons, many set on river floodplains. The hurricane downpour flushed out these waste lagoons, spewing hundreds of millions of gallons of untreated sewage into wetlands, streams, and eventually Pamlico Sound and the ocean—a spreading "dead zone." Add to this waste, hundreds of thousands of hog, poultry, and other livestock carcasses, industrial toxins, floodplain junkyard oil, and municipal waste, and you have an environmental catastrophe (Figure 14.27b). Final assessment of overall damage will take a decade, if not longer.

Catastrophic floods continue to be a threat, especially in poor nations. Bangladesh is perhaps the most persistent example. Bangladesh is one of the most densely populated countries on Earth, and more than *three-fourths* of its land area is a floodplain! The country's vast alluvial plain sprawls over an area the size of Alabama (130,000 km², or 50,000 mi²).

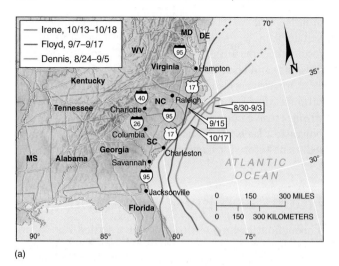

(a)

(b)

FIGURE 14.27 North Carolina 1999 floodplain disaster.
(a) Three hurricanes deluged North Carolina with several feet of rain during September and October 1999, Hurricane Floyd being the worst. (b) Hundreds of thousands of livestock were killed in the floods, which also washed out hundreds of animal sewage lagoons into wetlands, streams, and the ocean. Many of these factory farms and lagoons were sited on floodplains. [(b) Photo by Mel Nathanson, *Raleigh News & Observer*.]

The flooding severity is magnified as a consequence of human economic activities. Excessive forest harvesting in the upstream portions of the Ganges–Brahmaputra River watersheds increased runoff. Over time, the increased load carried by the river was deposited in the Bay of Bengal, creating new islands (see Figure 14.24). These islands, barely above sea level, became sites for new farming villages. As a result, about 150,000 people perished in the 1988 and 1991 floods. (For information on worldwide floods, see **http://www.dartmouth.edu/artsci/geog/floods/** or a daily flood summary at **http://www.nws.noaa.gov/oh/hod/handbook_products.htm**.)

Rating Floodplain Risk

A **flood** is a high water level that overflows the natural (or artificial) levees along any portion of a stream. Both floods and the floodplains they might occupy are rated statistically for the expected time intervals between floods. Thus, you hear about "10-year floods," "50-year floods," and so on. A *10-year flood* is the greatest level of flooding that is likely to occur once every 10 years. This also means that such flooding has only a 10% likelihood of occurring in any one year and is likely to occur about 10 times each century. For any given floodplain, such a frequency indicates a moderate threat.

A 50-year or 100-year flood is of greater and perhaps catastrophic consequence, but it is also less likely to occur in a given year. These probability ratings of flood levels are mapped for an area, and the defined floodplains that result are then labeled as a "50-year floodplain" or a "100-year floodplain."

These statistical estimates are probabilities that events will occur randomly during any single year of the specified period. Of course, two decades might pass without a 50-year

flood, or a 50-year level of flooding could occur 3 years in a row. The record-breaking Mississippi River Valley floods in 1993 easily exceeded a 1000-year flood probability. See Focus Study 14.1 for more about floodplain hazards and management strategies.

Streamflow Measurement

Flood patterns in a drainage basin are as complex as the weather, for floods and weather are equally variable, and both include a level of unpredictability. Measuring and analyzing the behavior of each large watershed and stream enables engineers and concerned parties to develop the best possible flood-management strategy. Unfortunately, reliable data often are not available for small basins or for the changing landscapes of urban areas.

The key to flood avoidance or management is to possess extensive measurements of *streamflow*, a stream's discharge and its flow pattern (Figure 14.28a). Once the cross section of a stream is fully measured, only the stream level is needed to determine discharge (using the calculation: discharge = width × depth × velocity). A *staff gauge* (a pole marked with water levels) is placed in a stream, as shown in the figure, to measure stream level. Another method involves a *stilling well* on the stream bank with a gauge mounted in it to measure stream level. A movable current meter can be used to sample stream velocity at various locations.

Approximately 11,000 stream gaging stations (hydrologists use this spelling) are in use in the United States

Floodplain Strategies

Detailed measurements of streamflows and flood records have been kept rigorously in the United States for only about 100 years, in particular since the 1940s. At any selected location along any given stream, the *probable maximum flood* (PMF) is a hypothetical flood of such a magnitude that there is virtually no possibility it will be exceeded. Because the collection and concentration of rainfall produce floods, hydrologists speak of a corollary, the *probable maximum precipitation* (PMP) for a given drainage basin, which is an amount of rainfall so great that it will never be exceeded.

These parameters are used by hydrologic engineers to establish a design flood against which to take protective measures. For urban areas near creeks, planning maps often include survey lines for a 50-year or a 100-year floodplain; such maps have been completed for most U.S. urban areas. The design flood usually is used to enforce planning restrictions and special insurance requirements. Restrictive zoning using these floodplain designations is an effective way of avoiding potential damage. (The floodplain managers' organization at http://www.floods.org/ is a respected source of information.)

Unfortunately, such political action is not generally implemented, and the scenario all too often goes like this: (1) Minimal zoning precautions are not carefully supervised; (2) a flooding disaster occurs; (3) the public is outraged at being caught off guard; (4) businesses and homeowners are surprisingly resistant to stricter laws and enforcement; (5) eventually another flood refreshes the memory and promotes more knee-jerk planning. As strange as it seems, there is little indication that our risk perception improves as the risk increases.

A Few Planning Strategies

A planning strategy used in some large river systems is to develop artificial floodplains; this is done by constructing bypass channels to accept seasonal or occasional floods. When not flood-

ed, the bypass channel can serve as farmland, often benefiting from the occasional soil-replenishing inundation. When the river reaches flood stage, large gates called weirs are opened, allowing the water to enter the bypass channel. This alternate route relieves the main channel of the burden of carrying the entire discharge.

Dams (an artificial structure placed in a river channel) and reservoirs (an impoundment of water behind a dam; a human-made lake) are common streamflow-control methods within a watershed. For conservation purposes, a dam holds back seasonal peak flows for distribution during low-water periods. In this way, streamflows

are regulated to assure year-round water supplies. Dams also are constructed for flood control, to hold back excess flows for later release at more moderate discharge levels. Adding hydroelectric power production to these functions of conservation and flood control can define a modern multipurpose reclamation project.

The function of reservoir impoundment is to provide flexible storage capacity within a watershed to regulate river flows, especially in a region with variable precipitation. Figure 1 shows one reservoir during drought conditions and during a time of wetter weather 6 years later.

(continued)

(a)

(b)

FIGURE 1 Reservoir extremes.
Comparative photographs of the New Hogan reservoir, central California, during (a) dry and (b) wet weather conditions. Reservoirs help regulate runoff variability. [Photos by author.]

Focus Study 14.1 *(continued)*

Reservoir Considerations

Unfortunately, the multipurpose benefits of reservoir construction are countered by some negative consequences. The area upstream from a dam becomes permanently drowned. In mountainous regions, this may mean loss of white-water rapids and recreational sections of a river. In agricultural areas, the ironic end result may be that a hectare of farmland is inundated upstream to preserve a hectare of farmland downstream.

Furthermore, dams built in warm and arid climates lose substantial water to evaporation, compared with the free-flowing streams they replace. Reservoirs in the southwestern United States can lose 3–4 meters (10–13 ft) of water a year. Also, sedimentation can reduce the effective capacity of a reservoir and can shorten a dam's life span, as mentioned earlier regarding Glen Canyon Dam.

Vast environmental disruption for the sake of economic gain no longer appears popular with the public. The James Bay Project in central and northern Québec is a case in point. Launched in 1970 by Hydro-Québec and only one-third complete at this time, the project might eventually include 215 dams, 25 power stations, and 20 river diversions. Many unexpected environmental problems have arisen because no environmental impact studies were completed at the outset. The early stages remain a huge experiment with fragile ecosystems. Well into the planning phase, the public learned that corporate interests sought inexpensive, publicly generated electric power, and that much of it was for export to the United States—all at public expense.

The second major phase of the James Bay Project, the Great Whale project (named for a local river), may never be completed because of the success of conservation programs begun by utilities in the northeastern United States and because of court challenges to assess impacts before further construction. In addition, the state of New York in 1992 withdrew its offer to buy 1000 megawatts of power from the Hydro-Québec project. For more on this project and other links, see http://www.edf. org/programs/International/Dams/ NAmerica/b_GreatWhale.html, and for the First Nations 1999 resolution, see http://www.afn.ca/resolutions/ 1990/sca/res14.htm.

A Final Thought About Floods

The benefit of any levee, bypass, or other project intended to prevent flood destruction is measured in avoided damage and is used to justify the cost of the protection facility. Thus, ever-increasing damage leads to the justification of ever-increasing flood-control structures. All such strategies are subjected to cost-benefit analysis, but bias is a serious drawback because such an analysis usually is prepared by an agency or bureau with a vested interest in building more flood-control projects.

As suggested in an article titled "Settlement Control Beats Flood Control,"* published 50 years ago, there are other ways to protect populations than with enormous, expensive, sometimes environmentally disruptive projects. Strictly zoning the floodplain is one approach. However, the flat, easily developed floodplains near pleasant rivers are desirable for housing, and thus weaken political resolve. A reasoned zoning strategy would set aside the floodplain for farming or passive recreation, such as a riverine park, golf course, or plant and wildlife sanctuary, or for other uses that are not hurt by natural floods. This study concludes that "urban and industrial losses would be largely obviated by set-back levees and zoning and thus cancel the biggest share of the assessed benefits which justify big dams."

*Walter Kollmorgen, *Economic Geography* 29, no. 3 (July 1953): 215.

(an average of more than 200 per state). Of these, 7000 are operated by the U.S. Geological Survey and have continuous recorders for stage (level) and discharge (see http:// water.usgs.gov/pubs/circ/circ1123/). Many of these stations automatically send telemetry data to satellites, from which information is retransmitted to regional centers (Figure 14.28b). Environment Canada's Water Survey of Canada maintains more than 3000 gaging stations (see http://www.watersurvey.sk.ca/). In the political climate surrounding hazards and planning, such hydrologic monitoring by the stream gaging network is under continual threats of budget cuts.

Hydrographs A graph of stream discharge over time for a specific place is a **hydrograph**. The hydrograph in Figure 14.29a shows the relation between precipitation input (the bar graph) and stream discharge (the curves). During dry periods, at low-water stages, the flow is described as base flow and is largely maintained by input from local groundwater (dark blue line).

When rainfall occurs in some portion of the watershed, the runoff collects and is concentrated in streams and tributaries. The amount, location, and duration of the rainfall episode determine the *peak flow*. Also important is the nature of the surface in a watershed; for example, a hydrograph for a specific portion of a stream changes after a forest fire or following urbanization of the watershed.

Human activities have enormous impact on water flow in a basin. The effects of urbanization are quite dramatic, both increasing and hastening peak flow, as you can see by comparing preurban stream flow (purple curve) and urbanized stream flow (light blue) in Figure 14.29a. In fact, urban areas produce runoff patterns quite similar to those of deserts. The sealed surfaces of the city drastically reduce infiltration and soil-moisture recharge; their effect is similar to that of the hard, nearly barren surfaces of the desert. A significant part of urban flooding occurs because of the alteration in surfaces. Shortened concentration times for peak flows strike with little warning (Figure 14.30). These issues will intensify as urbanization of vulnerable areas continues.

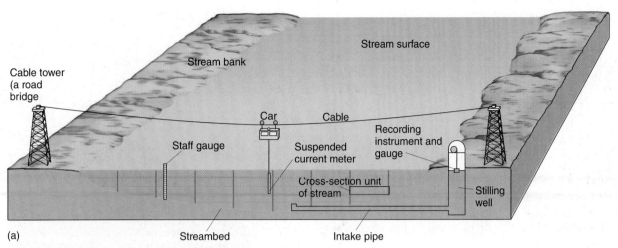

(a)

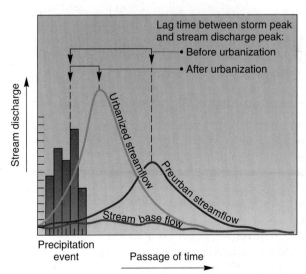

(b)

FIGURE 14.28 Streamflow measurement.
(a) A typical streamflow measurement installation may use a variety of devices: staff gauge, stilling well with recording instrument, and suspended current meter. (b) An automated hydrographic station sends telemetry to a satellite for collection by the USGS. [Photo courtesy of California Department of Water Resources.]

FIGURE 14.29 Urban flooding.
(a) Effect of urbanization on a typical stream hydrograph. Normal base flow is indicated with a dark blue line. The purple line indicates discharge after a storm, before urbanization. Following urbanization, stream discharge dramatically increases, as shown by the light blue line. (b) Severe flooding of an urban area in Linda, California, after a levee break on the Sacramento River in 1986. [(b) Photo from California Department of Water Resources.]

(b)

(a)

(b)

FIGURE 14.30 Urban creek flooding hits fast and hard.
(a) Urbanization increases runoff and hastens flooding, catching floodplain homes and businesses by surprise. (b) Even a small creek can cause great damage, such as this destroyed lumberyard—lumber was carried more than a mile downstream. [Photos by author.]

Summary and Review—River Systems and Landforms

● *Define* the term fluvial and *outline* the fluvial processes: erosion, transportation, and deposition.

River systems, fluvial processes and landscapes, floodplains, and river control strategies are important to human populations as demands for limited water resources increase. Stream-related processes are called **fluvial**. Water dislodges, dissolves, or removes surface material in the process called **erosion**. Streams produce *fluvial erosion*, in which weathered sediment is picked up for **transport**, movement to new locations. Sediments are laid down by another process, **deposition**. **Alluvium** is the general term for the clay, silt, and sand deposited by running water.

Base level is the lowest elevation limit of stream erosion. A *local base level* occurs when something interrupts the stream's ability to achieve base level, such as is created by a dam or a landslide that blocks a stream channel.

> fluvial (p. 431)
> erosion (p. 431)
> transport (p. 431)
> deposition (p. 431)
> alluvium (p. 431)
> base level (p. 431)

1. What role is played by rivers in the hydrologic cycle?
2. What are the five largest rivers on Earth in terms of discharge? Relate these to the weather patterns in each area and to regional POTET and PRECIP.
3. Define the term fluvial. What is a fluvial process?
4. What is the sequence of events that takes place as a stream dislodges material?
5. Explain the base-level concept. What happens to a local base level when a reservoir is constructed?

● *Construct* a basic drainage basin model and *identify* different types of drainage patterns and internal drainage, with examples.

The basic fluvial system is a **drainage basin**, which is an open system. *Drainage divides* define the **watershed** catchment (water receiving) area of the drainage basin. In any drainage basin, water initially moves downslope in a thin film called **sheetflow**, or *overland flow*. This surface runoff concentrates in *rills*, or small-scale downhill grooves, which may develop into deeper *gullies* and a stream course in a valley. High ground that separates one valley from another and directs sheet flow is termed an *interfluve*. Extensive mountain and highland regions act as **continental divides** that separate major drainage basins. Some regions, such as the Great Salt Lake Basin, have **internal drainage** that does not reach the ocean, the only outlets being evaporation and subsurface gravitational flow.

Drainage density is determined by the number and length of channels in a given area and is an expression of a landscape's topographic surface appearance. **Drainage pattern** refers to the arrangement of channels in an area as determined by the steepness, variable rock resistance, variable climate, hydrology, relief of the land, and structural controls imposed by the landscape. Seven basic drainage patterns are generally found in nature: dendritic, trellis, radial, parallel, rectangular, annular, and deranged.

> drainage basin (p. 432)
> watershed (p. 432)
> sheetflow (p. 432)
> continental divides (p. 432)
> internal drainage (p. 433)
> drainage density (p. 433)
> drainage pattern (p. 434)

6. What is the spatial geomorphic unit of an individual river system? How is it determined on the landscape? Define the several relevant key terms used.

7. In Figure 14.5, follow the Allegheny–Ohio–Mississippi River systems to the Gulf of Mexico. Analyze the pattern of tributaries, and describe the channel. What role do continental divides play in this drainage?

8. Describe drainage patterns. Define the various patterns that commonly appear in nature. What drainage patterns exist in your hometown? Where you attend school?

● *Describe* the relation among velocity, depth, width, and discharge and *explain* the various ways that a stream erodes and transports its load.

Stream channels vary in *width* and *depth*. The streams that flow in them vary in velocity and in the *sediment load* they carry. All of these factors may increase with increasing discharge. **Discharge** is calculated by multiplying the velocity of the stream by its width and depth for a specific cross section of the channel. Most streams increase discharge downstream. But, some streams originate in a humid region and flow through an arid region, such that discharge decreases with distance. Such an **exotic stream** is exemplified by the Nile River or the Colorado River.

Hydraulic action is the work of *turbulence* in the water. Running water causes hydraulic squeeze-and-release action to loosen and lift rocks and sediment. As this debris moves along, it mechanically erodes the streambed further, through a process of **abrasion**.

Solution refers to the **dissolved load** of a stream, especially the chemical solution derived from minerals such as limestone or dolomite or from soluble salts. The **suspended load** consists of fine-grained, clastic particles held aloft in the stream, with the finest particles not deposited until the stream velocity slows nearly to zero. **Bed load** refers to coarser materials that are dragged along the stream bed by **traction** or are rolled and bounced along by **saltation**. If the load in a stream exceeds its capacity, sediments accumulate as **aggradation** as the stream channel builds through deposition. With excess sediment, a stream becomes a maze of interconnected channels that form a **braided stream** pattern.

discharge (p. 437)
exotic stream (p. 438)
hydraulic action (p. 438)
abrasion (p. 438)
dissolved load (p. 438)
suspended load (p. 438)
bed load (p. 440)
traction (p. 440)
saltation (p. 440)
aggradation (p. 440)
braided stream (p. 440)

9. What was the impact of flood discharge on the channel of the San Juan River near Bluff, Utah? Why did these changes take place?

10. How does stream discharge do its erosive work? What are the processes at work in the channel?

11. Differentiate between stream competence and stream capacity.

12. How does a stream transport its sediment load? What processes are at work?

● *Develop* a model of a meandering stream, including point bar, undercut bank, and cutoff, and *explain* the role of stream gradient in these flow characteristics.

Where the slope is gradual, stream channels develop a sinuous form called a **meandering stream**. The outer portion of each meandering curve is subject to the fastest water velocity and can be the site of a steep **undercut bank**. On the other hand, the inner portion of a meander experiences the slowest water velocity and forms a **point bar** deposit. When a meander neck is cut off as two undercut banks merge, the meander becomes isolated and forms an **oxbow lake**.

Every stream develops its own **gradient** and establishes a longitudinal profile. A portion of the stream is designated a **graded stream** when the stream is adjusted among available discharge, channel characteristics, its velocity, and the load supplied from the drainage basin. An interruption in a stream's longitudinal profile is called a **nickpoint**. A nickpoint can occur as the stream flows across hard resistant rock or after tectonic uplift episodes.

meandering stream (p. 441)
undercut bank (p. 441)
point bar (p. 441)
oxbow lake (p. 442)
gradient (p. 443)
graded stream (p. 443)
nickpoint (p. 444)

13. Describe the flow characteristics of a meandering stream. What is the pattern of flow in the channel? What are the erosional and depositional features and the typical landforms created?

14. Explain these statements: (a) All streams have a gradient, but not all streams are graded. (b) Graded streams may have ungraded segments.

15. Why is Niagara Falls an example of a nickpoint? Without human intervention, what do you think would eventually take place at Niagara Falls?

16. What is meant by "the validity of cyclic or equilibrium models depends on which of three time frames is being considered"? Explain and discuss.

● *Define* a floodplain and *analyze* the behavior of a stream channel during a flood.

Floodplains have been an important site of human activity throughout history. Rich soils, bathed in fresh nutrients by floodwaters, attract agricultural activity and urbanization. Despite our knowledge of historical devastation by floods, floodplains are settled, raising issues of human hazard perception. The flat low-lying area along a stream channel that is subjected to recurrent flooding is a **floodplain**. It is formed when the river overflows its channel during times of high flow. On either bank of most streams, **natural levees** develop as by-products of flooding. On the floodplain, backswamps and yazoo tributaries may develop. The natural levees and elevated channel of the river prevent a **yazoo tributary** from joining the main channel, so it flows parallel to the river and through the **backswamp** area. **Alluvial terraces** are formed by the entrenchment of a river into its own floodplain.

> floodplain (p. 446)
> natural levees (p. 446)
> yazoo tributary (p. 448)
> backswamp (p. 448)
> alluvial terraces (p. 448)

17. Describe the formation of a floodplain. How are natural levees, oxbow lakes, backswamps, and yazoo tributaries produced?

18. Identify any of the features listed in question 17 on the Philipp, Mississippi, topographic quadrangle in Figure 14.21b.

19. Describe any floodplains near where you live or where you go to college. Have you seen any of the floodplain features discussed in this chapter? If so, which ones?

● *Differentiate* **the several types of river deltas and** *detail* **each.**

A depositional plain formed at the mouth of a river is called a **delta**. When the mouth of a river enters the sea and is inundated by the sea in a mix with freshwater, it is called an **estuary**.

> delta (p. 450)
> estuary (p. 450)

20. What is a river delta? What are the various deltaic forms? Give some examples.

21. How might life in New Orleans change in the next century? Explain.

22. Describe the Ganges River delta. What factors upstream explain its form and pattern? Assess the consequences of settlement on this delta.

23. What is meant by the statement "the Nile River delta is disappearing"?

● *Explain* **flood probability estimates and** *review* **strategies for mitigating flood hazards.**

A **flood** occurs when high water overflows the natural or artificial levees of a stream. Both floods and the floodplains they occupy are rated statistically for the expected time interval between floods. A 10-year flood is the greatest level of flooding that is likely once every 10 years. A graph of stream discharge over time for a specific place is called a **hydrograph**.

Collective efforts by government agencies undertake to reduce flood probability. Such management attempts include the construction of artificial levees, bypasses, straightened channels, diversions, dams, and reservoirs. Society is still learning how to live in a sustainable way with Earth's dynamic river systems.

> flood (p. 454)
> hydrograph (p. 456)

24. Specifically, what is a flood? How are such flows measured and tracked?

25. Differentiate between a hydrograph from a natural terrain and one from an urbanized area.

26. What do you see as the major consideration regarding floodplain management? How would you describe the general attitude of society toward natural hazards and disasters?

27. What do you think the author of the article "Settlement Control Beats Flood Control" meant by the title? Explain your answer, using information presented in the chapter.

NetWork

The *Geosystems* Home Page provides on-line resources for this chapter on the World Wide Web. To begin: Once on the Home Page, click on this textbook, scroll the Table of Contents menu, select this chapter, and click "Begin." You will find self-tests that are graded, review exercises, specific updates for items in the chapter, and many links to interesting related pathways on the Internet under "Destinations." *Geosystems* is at **http://www.prenhall.com/christopherson**.

Critical Thinking

A. Determine the name of the river drainage basin within which your campus is located. Where are its headwaters? Where is the river's mouth? If you are in the United States or Canada, use Figure 14.5 to locate the larger drainage basins and divides for your region. Is there any regulatory organization that oversees planning and coordination for this drainage basin?

B. Relative to the drainage basin you determined in (A), see if there is a topographic map on file in the library, geography department, or at a local outdoor recreation store that covers the portion of the basin near campus. After examining the map, can you discern a prominent drainage pattern for the area (Figure 14.8)?

C. Under "Destinations" in Chapter 14 of the *Geosystems* Home Page, there are links to many flood sources. Relative to the discussion of the 1993 Midwest flood in News Report 14.3 in the chapter, what related information do you find? Examine the sites and describe the available information.

Career Link 14.1

Julie Dian-Reed, Service Hydrologist and Weather Forecaster

Julie Dian-Reed reminisces, "Growing up, I lived in northeastern Indiana. When I was in high school, I would go sailing on Lake Michigan with my older sister and her husband. I was fascinated by the winds, clouds, and thunderstorms over the lake and how they affected sailing. Even a trip to the beach for the day allowed me to watch the thunderstorms over the lake."

Julie did her undergraduate work at the University of Indiana in geography with a major in meteorology, although she entered college intent on studying biology and music. She moved to the University of Illinois for her Master's work in climatology, also in the geography department. Her advisor was in agricultural meteorology. "I studied climatological aspects of integrated pest management, the theory that by using a variety of methods, the reliance on chemical pesticides can be reduced. I found that knowledge of weather forecasting and weather patterns aided farmers in knowing when to use pesticides. Wind and weather conditions dictate migration patterns of pests and I tried to track down these relationships."

After graduate school, she accepted a position at the Illinois State Water Survey and later worked at the Midwest Regional Climate Center, where she compiled climate data and studied climate trends. In 1992 she joined the National Weather Service in Cincinnati as a meteorological intern. She gathered surface and radar observations, issued severe storm warnings, and honed her forecasting skills. Julie was also heavily involved in public education and NWS outreach programs. The great Midwest floods hit in 1993 (see News Report 14.3), so her first full year with NWS was eventful.

Julie's next job in the NWS was at the Ohio River Forecast Center (OHRFC) as a hydrology intern. There she completed additional hydrology

FIGURE 1 Julie Dian-Reed, Service Hydrologist and Weather Forecaster
National Weather Service Wilmington, Ohio. [Photo by National Weather Service.]

coursework, pulling together various specialties into her expertise as a weather forecasting hydrologist.

She became the Service Hydrologist at the Weather Forecast Office, Wilmington, Ohio, in May 1995. "I try to get everyone thinking like a hydrologist. Once the rain hits the ground, it shouldn't pass from one area of responsibility to another, especially during flash flood and river flood situations. The geographer in me sees the spatial relations among rainfall events, hydrology, and watershed characteristics. Our staff is good at this and everyone is tuned into this approach."

Her office is co-located with the River Forecast Center, one of 13 such combined centers in the United States: "They are just across the hall, so it is nice to have that interaction. I can see what goes on regarding river forecasting and integrate this with the weather forecast. Then, from this mix, we put out a comprehensive warning or advisories to the public." The Wilmington Office was awarded a National Weather Service Modernization Award in 1996 and the Department of Commerce Silver Medal for public service during the Ohio River flooding in 1997.

Julie works with the Advanced Weather Interactive Processing System (AWIPS), installed in the office in 1998, to put out forecasts (Figure 1). "It is a nice system, where all the data are accessed in multi-frame screens on several monitors. This kind of integration is so much better than the old running-around-the-office style of trying to pull together diverse elements," she says. She also operates the Doppler radar at Wilmington on some work shifts: "Once the input data arrive, we work at the terminal to develop the derived data product."

Careers in this field are bright and involve the GIS revolution. As Julie says, "My background in geography helps me bring together diverse subjects to complement our weather and river forecasting." The future for hydrology and river forecasting will feature further integration of the GIS software in the use of weather forecasts to augment and improve river forecasting. Julie states, "I like weather. It is kind of a hobby. My husband is hooked on weather now, although he is an electrical engineer. Even after I get off work, I will take pictures if there is a flood event, sometimes in flyovers from helicopters."

The floor of Canyon de Chelly is home to Navajo farmers and shepherds. In niches in the sandstone walls, cliff dwelling ruins are signs of ancient occupation and livelihoods a thousand years ago by the Anasazi peoples. [Photo by author.]

15

Eolian Processes and Arid Landscapes

Key Learning Concepts

After reading the chapter, you should be able to:

• *Characterize* the unique work accomplished by wind and eolian processes.

• *Describe* eolian erosion, including deflation, abrasion, and the resultant landforms.

• *Describe* eolian transportation and *explain* saltation and surface creep.

• *Identify* the major classes of sand dunes and *present* examples within each class.

• *Define* loess deposits, their origins, locations, and landforms.

• *Portray* desert landscapes and *locate* these regions on a world map.

Wind is an agent of geomorphic change. Like moving water, moving air (wind) causes erosion, transportation, and deposition of materials. Like moving water, moving air is a fluid and it behaves similarly, although it has a lower viscosity (it is "thinner") than water. The polar regions are deserts as well, so there should be no surprise that Yuma, Arizona, and weather stations in Antarctica receive the same amount of annual precipitation.

Wind processes modify and move sediment in deserts and along coastlines in a variety of climates. Wind may contribute to soil formation in places far distant from the point of origin, bringing fine material from regions where glaciers deposited it. Elsewhere, fallow fields (those not planted) give up their soil resource to destructive wind erosion. Scientists are only now getting an accurate picture of the amount of windblown dust that fills the atmosphere and crosses the oceans between continents. Chemical fingerprints and satellites trace windblown dust from African soils to South America and Asian landscapes to Europe.

Arid landscapes display unique landforms and life forms: "Instead of finding chaos and disorder the observer never fails to be amazed at a simplicity of form, an exactitude of repetition and a geometric order in the desert."*

In this chapter: We examine the work of wind, associated erosion, transport, and depositional processes and resulting landforms. Most deserts are rocky and covered with desert pavement, whereas other dry landscapes are covered in sand dunes. Windblown fine particles form vast loess deposits, the basis for rich agricultural soils. For convenience of organization, we include the discussion of arid lands in this chapter. Earth's dry lands stand out in stark contrast on the water planet. In desert environments, an overall lack of moisture and stabilizing vegetation sets the landscapes in sharp, sun-baked relief. Water is the major erosional agent in the desert, yet water is the limiting resource for human development. The Colorado River is the subject of an important focus study.

The Work of Wind

The work of the wind—erosion, transportation, and deposition—is called **eolian** (also spelled *aeolian*). The word comes from Aeolus, ruler of the winds in Greek mythology. British Army major, Ralph Bagnold, while stationed in Egypt in 1925, accomplished much eolian research. An engineering officer who spent much of his time in the deserts west of the Nile, Bagnold measured, sketched, and developed hypotheses about the wind and desert forms. Imagine the scene in the 1920s: A Model-T Ford chugs across the desert west of Cairo taking Bagnold to his research area. He laid rolls of chicken wire over treacherous stretches of sand to prevent getting stuck! Bagnold's often-cited work, *The Physics of Blown Sand and Desert Dunes*, was published in 1941.

The actual ability of wind to move materials is small compared with that of other transporting agents such as water and ice, because air is so much less dense than those other media. Yet, over time, wind accomplishes enormous work. Bagnold studied the ability of wind to transport sand over the surface of a dune. Figure 15.1 shows that a steady wind of 50 kmph (30 mph) can move approximately one-half ton of sand per day over a square meter section of dune. The graph also demonstrates how rapidly the amount of transported sand increases with wind speed.

Grain size is important in wind erosion. Intermediate-sized grains move (bounced along) most easily. It is the largest and the smallest sand particles that require the strongest winds to move. The large particles are heavier, and thus require stronger winds to move. Small particles are diffi-

*R. A. Bagnold, *The Physics of Blown Sand and Desert Dunes* (London: Methuen, 1941).

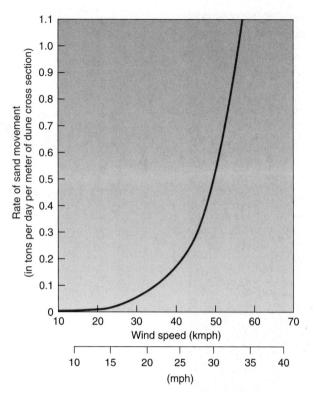

FIGURE 15.1 Sand movement and wind velocity.
Sand movement relative to wind velocity, as measured over a square meter cross section of ground surface. [After R. A. Bagnold, *The Physics of Blown Sand and Desert Dunes* (London: Methuen, 1941). Adapted by permission.]

cult to move because they exhibit a mutual cohesiveness and because they usually present a smooth surface (aerodynamic) to the wind, yet the finest dust aloft is carried from continent to continent. In addition, wind can prune and shape vegetation, especially where winds are strong in a consistent direction (Figure 15.2a).

Eolian Erosion

Two principal wind-erosion processes are **deflation**, the removal and lifting of individual loose particles, and **abrasion**, the grinding of rock surfaces with a "sandblasting" action by particles captured in the air (Figure 15.2b). Deflation and abrasion produce a variety of distinctive landforms and landscapes.

Deflation   Deflation literally blows away loose or noncohesive sediment. Fine materials are eroded away by wind deflation and moving water, leaving behind a concentration of pebbles and gravel called **desert pavement**. Resembling a cobblestone street, desert pavement protects underlying sediment from further deflation and water erosion (Figure 15.3). Desert pavements are so common that many provincial names are used for them—for example, *gibber plain* in Australia; *gobi* in China; and in Africa, *lag gravels* or *serir*, or *reg* desert if some fine particles remain. Water is an important factor in the formation of desert pavement, washing away fine materials and concentrating and cementing the remaining rock pieces.

(a)

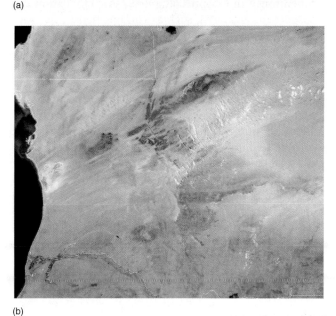

(b)

◄ **FIGURE 15.2 The work of the wind.**
(a) Wind-sculpted tree near South Point, Hawai'i. Nearly constant trade winds keep this tree pruned. (b) Streaks of wind-blown sands cover bedrock south of Atar, in central Mauritania. Note the circular Richat dome structure, just east of the Western Sahara political border corner (refer to Figure 12.10c). [(a) Photo by Bobbé Christopherson. (b) *Terra* MODIS sensor image courtesy of NASA Earth Observatory, R. Stockli, R. Simmon, and B. Montgomery.]

(a)

(b)

FIGURE 15.3 Desert pavement.
(a) Desert pavement is formed from larger rocks and fragments left after deflation and sheetwash (flowing water across the surface). (b) A typical desert pavement. [Photo by Bobbé Christopherson.]

Heavy recreational activity damages fragile desert landscapes, especially in the arid lands of the United States, where more than 15 million off-road vehicles (ORVs) are now in use. Such vehicles crush plants and animals; disrupt desert pavement, leading to greater deflation; and create ruts that easily concentrate sheetwash (water flowing across surfaces) to form gullies. Measures to restrict ORV use to specific areas, preserving the remaining desert, are controversial.

Military activities can also threaten delicate desert landscapes. A serious environmental impact of the 1991 Persian Gulf War was the disruption of desert pavement. Thousands of square kilometers of stable desert pavement were shattered by bombardment with more than 68,000 tons of explosives and were disrupted by the movement of heavy vehicles. The resulting loosened sand and silt was made available for deflation, plaguing cities and farms with increased dust and sand accumulations.

Wherever wind encounters loose sediment, deflation may remove enough material to form basins called **blowout depressions**. These depressions range from small indentations less than a meter wide up to areas hundreds of meters wide and many meters deep. Chemical weathering, although slow in the desert owing to the lack of water, is important in the formation of a blowout, for it removes the cementing materials that give particles their cohesiveness. (In arid climates chemical weathering is active on the surfaces of particles at the microscopic level, utilizing hygroscopic and capillary water in reactions.)

Large depressions in the Sahara Desert are at least partially formed by deflation. The enormous Munkhafad el Qaṭṭâra (Qaṭṭâra Depression) just inland from the Mediterranean Sea in the Western Desert of Egypt, which covers 18,000 km^2 (7000 mi^2), is now about 130 m (427 ft) below sea level at its lowest point.

Abrasion You may have seen work crews sandblasting surfaces on buildings and bridges to clean them or on streets to remove unwanted markings. Sandblasting uses a stream of compressed air filled with sand grains to quickly and smoothly abrade a surface. Abrasion by windblown particles is nature's slower version of sandblasting, and it is especially effective at polishing exposed rocks when the abrading particles are hard and angular.

Variables that affect the rate of abrasion include the hardness of surface rocks, wind velocity, and wind constancy. Abrasive action is restricted to the area immediately above the ground, usually no more than a meter or two in height, because sand grains are lifted only a short distance.

Rocks exposed to eolian abrasion appear pitted, grooved, or polished. They usually are aerodynamically shaped in a specific direction, according to the consistent flow of airborne particles carried by prevailing winds. Rocks that bear such evidence of eolian erosion are **ventifacts** (literally, "artifacts of the wind").

On a larger scale, deflation and abrasion are capable of streamlining rock structures that are aligned parallel to the most effective wind direction, leaving behind distinctive, elongated ridges or formations called **yardangs**. These wind-sculpted features can range from meters to kilometers in length and up to many meters in height. On Earth, some yardangs are large enough to be detected on satellite imagery. The Ica Valley of southern Peru contains yardangs reaching 100 m (330 ft) in height and several kilometers in length, and yardangs in the Lūt Desert of Iran attain 150 m (490 ft) in height. Abrasion is concentrated on the windward end of each yardang, with deflation operating on the leeward portions (Figure 15.4). The Sphinx in Egypt was perhaps partially formed as a yardang, suggesting a head and body. Some scientists think this shape led the ancients to complete the bulk of the sculpture artificially with masonry.

Eolian Transportation

As mentioned in Chapter 6, atmospheric circulation can transport fine material, such as volcanic debris, fire soot and smoke, and dust, worldwide within days. Wind exerts a drag, or frictional pull, on surface particles until they become airborne, just as water in a stream picks up sediment (again, think of air as a fluid). The distance that wind is capable of transporting particles in suspension varies greatly with particle size. Only the finest dust particles travel significant distances, so the finer material suspended in a dust storm is lifted much higher than the coarser particles of a sandstorm, which may be lifted only about 2 m (6.5 ft).

People living in areas of frequent dust storms are faced with infiltration of very fine particles into their homes and businesses through even the smallest cracks. (Figure 3.8 illustrates such dust storms in the Nevada desert, the blowing alkali dust in the Andes, the reddish dust of an Australian storm, and alkali dust near Mono Lake, California.) People living in desert regions and along sandy beaches,

FIGURE 15.4 A yardang.
A small wind-sculpted rock formation in Snow Canyon outside St. George, Utah. [Photo by Bobbé Christopherson.]

FIGURE 15.5 Preventing sand transport.
Popham Beach State Park, Maine. Further erosion and transport of coastal dunes can be controlled by stabilizing strategies, planting native plants, and by confining pedestrian traffic to walkways. [Photo by Bobbé Christopherson.]

FIGURE 15.6 How the wind moves sand.
(a) Eolian suspension, saltation, and surface creep are mechanisms of sediment transportation. Compare with the saltation and traction that occur in another fluid, water, in Figure 14.12. (b) Sand grains saltating along the surface in the Stovepipe Wells dune field, Death Valley. [Photo by author.]

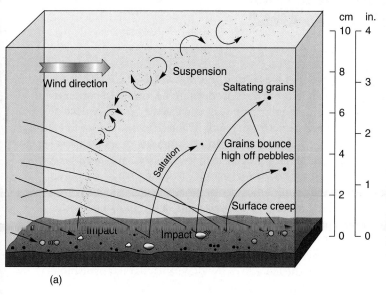

(a)

(b)

where frequent sandstorms occur, contend with the sandblasting of painted surfaces and etched window glass.

Both human and natural sand erosion and transport from a beach are slowed by conservation measures such as the introduction of stabilizing native plants, the use of fences, and the restriction of pedestrian traffic to walkways (Figure 15.5). As human settlement further encroaches on coastal dunes, sand transport becomes problematic. As an example at Nags Head, North Carolina, the Jocky's Ridge dune, some 43 m (140 ft) in height, is actively migrating over roads and yards.

The term *saltation* was used in Chapter 14 to describe movement of particles by water. The term also describes the wind transport of grains along the ground, grains usually larger than 0.2 mm (0.008 in.). About 80% of wind transport of particles is accomplished by this skipping and bouncing action (Figure 15.6a). Compared with fluvial transport, in which saltation is accomplished by hydraulic

lift, eolian saltation is executed by aerodynamic lift, elastic bounce, and impact (compare Figure 15.6a with Figure 14.12). Grains lift much higher in air than in water because of air's lower viscosity.

On impact, grains hit other grains and knock them into the air. Saltating particles crash into other particles, knocking them both loose and forward (Figure 15.6a). This type of movement is **surface creep**, which slides and rolls particles too large for saltation and affects about 20% of the material being transported. Once in motion, particles continue to be transported by lower wind velocities. In a desert or along a beach, you can hear a slight hissing sound produced by the myriad saltating grains of sand, almost like steam escaping, as they bounce along and collide with surface particles.

Through processes of weathering, erosion, and transportation, mineral grains are removed from parent rock and redistributed elsewhere. In Figure 13.5c, you can see

the relation between the composition and color of the sandstone in the background and the derived sandy surface in the foreground. Wind action is not significant in the weathering process that frees individual grains of sand from the parent rock, but it is active in relocating the weathered grains.

Eolian Depositional Landforms

The smallest features shaped by individual saltating grains are ripples (Figure 15.7). Ripples form in crests and troughs, positioned transversely (at a right angle) to the direction of the wind. Their formation is influenced by the length of time particles are airborne. Eolian ripples are similar to fluvial ripples, although the impact of saltating grains is slight in water.

Many people have never been in a desert; they know this arid landscape only from the movies, which leave the impression that most deserts are covered by sand. Instead, desert pavements predominate across most subtropical arid landscapes; only about 10% of desert areas are covered with sand. Sand grains generally are deposited as transient ridges or hills called dunes.

A **dune** is a wind-sculpted accumulation of sand. An extensive area of dunes, such as that found in North Africa, is characteristic of an **erg desert, or sand sea.** The Grand Erg Oriental in the central Sahara exceeds 1200 m (4000 ft) in depth and covers 192,000 km² (75,000 mi²), comparable to the area of Nebraska. This sand sea has been active for more than 1.3 million years and has average dune heights of 120 m (400 ft). Sahara Marzūq sand sea shown in Figure 15.8a is wider than 300 km (185 mi). Eolian processes are at work on the Martian surface forming sim-

ilar parallel ridges along the floor of Melas Chasma in Figure 15.8b (note the dust devil in the lower left corner of the image). Similar sand seas, such as the Grand Ar Rub'al Khālī Erg, are active in Saudi Arabia.

Dune fields, whether in arid regions or along coastlines, tend to migrate in the direction of strong prevailing winds. Strong seasonal winds or winds from a passing storm may sometimes prove more effective than average prevailing winds. When saltating sand grains encounter small patches of sand, their kinetic energy (motion) is dissipated and they start to accumulate; a dune is born. As height increases above 30 cm (12 in.), a steeply sloping slipface on the lee side of the dune and characteristic dune features form.

Study the dune model in Figure 15.9 and you can see that winds characteristically create a gently sloping *windward side* (stoss side), with a more steeply sloped **slipface** on the *leeward side*. A dune usually is asymmetrical in one or more directions. The angle of a slipface is the steepest angle at which loose material is stable—its *angle of repose*. Thus, the constant flow of new material makes a slipface a type of *avalanche slope*. Sand builds up as it moves over the crest of the dune to the brink; then it avalanches (falls and cascades) as the slipface continually adjusts, seeking its angle of repose (usually 30° to 34°). In this way, a dune migrates downwind, as suggested by the successive dune profiles in Figure 15.9.

Dunes have many wind-shaped styles that make classification difficult. We can simplify dune forms into three classes—*crescentic* (crescent, curved shape), *linear* (straight forms), and *star dunes* (each is summarized in Figure 15.10). The ever-changing form of these eolian deposits is part of their beauty, eloquently described by one author: "I see

(a)

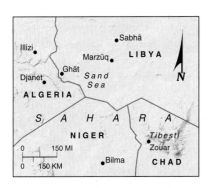

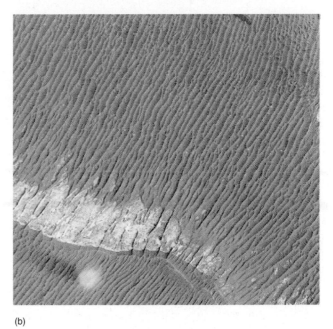

(b)

FIGURE 15.8 A sand sea.
(a) The Sahara Marzūq, an erg desert that dominates southwestern Libya. Effective northwesterly winds shape the pattern and direction of the transverse and barchanoid (series of connected barchans) dunes. This sand sea exceeds 300 km across. (b) A similar pattern of dunes appear on the Martian surface in the southern area of Melas Chasma in Valles Marineris. Note the dust devil in the lower left. Area covered is about 2 km wide. [(a) *Terra* MODIS sensor image courtesy of MODIS Land Rapid Response Team, NASA/GSFC, November 9, 2001. (b) Mars Global Surveyor, Mars Orbiter Camera image courtesy of NASA/JPL/Malin Space Science Systems, July 11, 1999.]

Effective wind direction

Previous slipfaces

Windward (stoss) slope

Slipface

Leeward slope

Direction of dune movement

Successive slipfaces created as dune migrates

FIGURE 15.9 Dune cross section.
Successive slipfaces exhibit a distinctive pattern as the dune migrates in the direction of the effective wind.

469

Class	Type	Description
Crescentic	Barchan	Crescent-shaped dune with horns pointed downwind. Winds are constant with little directional variability. Limited sand available. Only one slipface. Can be scattered over bare rock or desert pavement or commonly in dune fields.
	Transverse	Asymmetrical ridge, transverse to wind direction (right angle). Only one slipface. Results from relatively ineffective wind and abundant sand supply.
	Parabolic	Role of anchoring vegetation important. Open end faces upwind with U-shaped "blow-out" and arms anchored by vegetation. Multiple slipfaces, partially stabilized.
	Barchanoid ridge	A wavy, asymmetrical dune ridge aligned transverse to effective winds. Formed from coalesced barchans; look like connected crescents in rows with open areas between them.

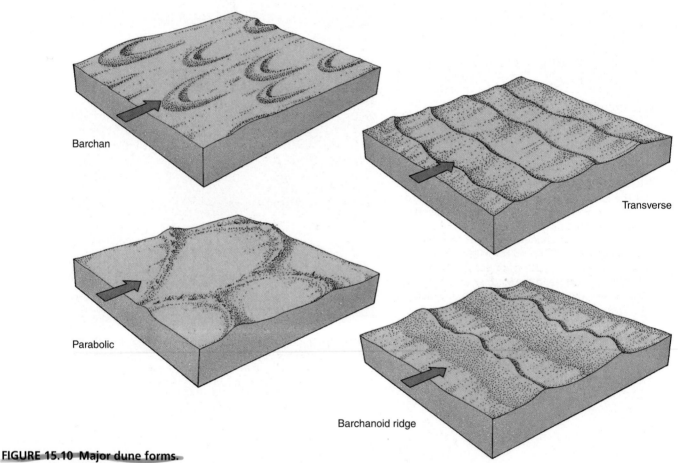

FIGURE 15.10 Major dune forms.
Arrows show wind direction. [Adapted from E. D. McKee, *A Study of Global Sand Seas*, U.S. Geological Survey Professional Paper 1052 (Washington, DC: U.S. Government Printing Office, 1979).]

hills and hollows of sand like rising and falling waves. Now at midmorning, they appear paper white. At dawn they were fog gray. This evening they will be eggshell brown."*

*J. E. Bowers, *Seasons of the Wind* (Flagstaff, AZ: Northland Press, 1985), p. 1.

Star dunes are the mountainous giants of the sandy desert. They form in response to complicated, changing wind patterns and have multiple slipfaces. They are pinwheel-shaped, with several radiating arms rising and joining to form a common central peak. The best examples of star dunes are in the Sahara and the Namib Desert, where they approach 200 m (650 ft) in height (Figure 15.11).

Class	Type	Description
Linear	Longitudinal	Long, slightly sinuous, ridge-shaped dune, aligned parallel with the wind direction; two slipfaces. Average 100 m high and 100 km long and can reach to 400 m high. Results from strong effective winds varying in one direction.
	Seif	After Arabic word for "sword"; a more sinuous crest and shorter than longitudinal dunes. Rounded toward upwind direction and pointed downwind. (Not illustrated.)
Star dune		The giant of dunes. Pyramidal or star-shaped with three or more sinuous radiating arms extending outward from a central peak. Slipfaces in multiple directions. Results from effective winds shifting in all directions. Tend to form isolated mounds in high effective winds and connected sinuous arms in low effective winds.
Other	Dome	Circular or elliptical mound with no slipface. Can be modified into barchanoid forms.
	Reversing	Asymmetrical ridge form intermediate between star dune and transverse dune. Wind variability can alter shape between forms.

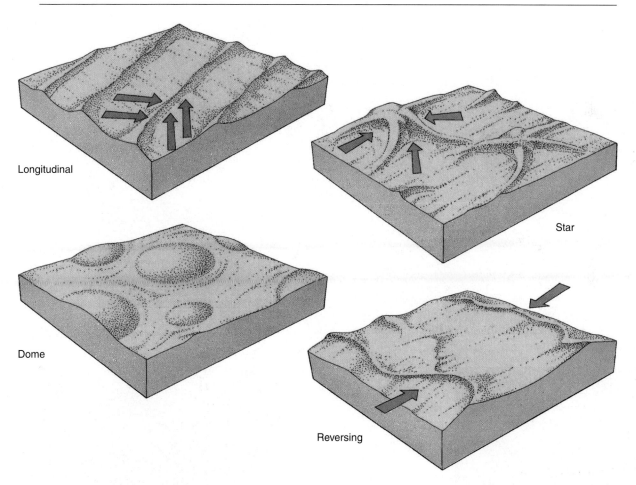

The map in Figure 15.12 shows the correlation of active sand regions with deserts (tropical, continental interior, and coastal). Important is the limited extent of desert area covered by active sand dunes—only about 10% of all continental land between 30° N and 30° S. Also noted on the map are dune fields in humid climates such as along coastal Oregon, the south shore of Lake Michigan (Figure 15.12b), along the Gulf and Atlantic coastlines, in Europe, and elsewhere. The remarkable coastal desert sands of Namibia are pictured in Figure 15.12c.

These same dune-forming principles and terms (for example, dune and slipface) apply to snow-covered landscapes. *Snow dunes* are formed as wind deposits snow in drifts. In semiarid farming areas, capturing drifting snow

FIGURE 15.11 Mountains of the desert.
Star dune in the Namib Desert in southwestern Africa.
[Photo by Comstock.]

with fences and tall stubble left in fields contributes significantly to soil moisture when the snow melts.

Loess Deposits

Approximately 15,000 years ago, in several episodes, Pleistocene glaciers retreated in many parts of the world, leaving behind large glacial outwash deposits of fine-grained clays and silts. These materials were blown great distances by the wind and redeposited in unstratified, homogeneous (evenly mixed) deposits. Peasants working along the Rhine River Valley in Germany gave these deposits the name **loess** (pronounced "luss"). No specific landforms were created; instead, loess covered existing landforms with a thick blanket of material that assumed the general topography of the existing landscape.

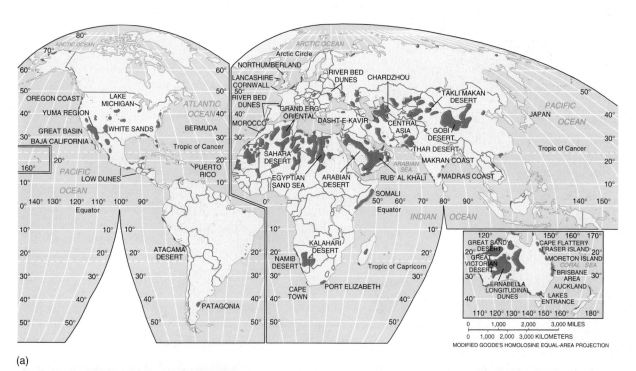

(a)

(b)

(c)

FIGURE 15.12 Sandy regions of the world.
(a) Worldwide distribution of active and stable sand regions. (b) Sand dunes along the shore of Lake Michigan in Indiana Dunes State Park, Indiana. (c) Sandy area in the Namib-Naukluft Park, Namib Desert, Namibia, Africa. [(a) After R. E. Snead, *Atlas of World Physical Features*, p. 134, ©1972 by John Wiley & Sons. Adapted by permission of John Wiley & Sons, Inc. (b) Photo by Bobbé Christopherson; (c) photo by Nigel J. Dennis/Photo Researchers, Inc.]

(a)

(b)

FIGURE 15.13 Example of loess deposits.
(a) Loess formation in Xi'an, Shaanxi Province, China, has sufficient structural strength to permit excavation for dwelling rooms. (b) A loess bluff along the Arikaree River in extreme northwestern Cheyenne County, Kansas. [(a) Photo by Betty Crowell; (b) photo by Steve Mulligan Photography.]

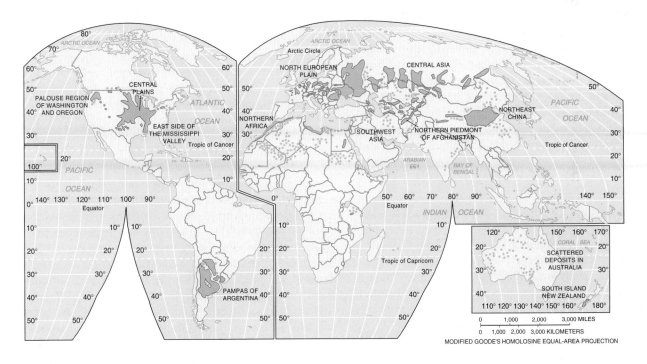

FIGURE 15.14 Worldwide loess deposits.
Dots represent small scattered loess formations. [After R. E. Snead, *Atlas of World Physical Features*, p. 138, ©1972 by John Wiley & Sons. Adapted by permission of John Wiley & Sons, Inc.]

Because of its own binding strength, loess weathers and erodes into steep bluffs, or vertical faces. At Xi'an, Shaanxi Province, China, a loess wall has been excavated for dwelling space (Figure 15.13a). When a bank is cut into a loess deposit, it generally will stand vertically, although it can fail if saturated (Figure 15.13b).

Figure 15.14 shows the worldwide distribution of loess deposits. Significant accumulations throughout the Mississippi and Missouri River valleys form continuous deposits 15–30 m (50–100 ft) thick. Loess deposits also occur in eastern Washington State and Idaho. This silt explains the fertility of the soils in these regions, for loess deposits are well drained and deep and have excellent moisture retention. Loess deposits also cover much of Ukraine, central Europe, China, the Pampas–Patagonia regions of Argentina, and lowland New Zealand. The soils derived from loess are some of Earth's "breadbasket" farming regions. Transport of loess soils occurred in the catastrophic Dust Bowl in the 1930s in the United States; see News Report 15.1.

In Europe and North America, loess is thought to be derived mainly from glacial and periglacial sources. The vast deposits of loess in China, covering more than 300,000 km^2 (116,000 mi^2) are thought to be derived from windblown desert sediment rather than glacial sources. Accumulations in the Loess Plateau of China exceed 300 m (1000 ft) thickness, forming complex weathered badlands and some good agricultural land. These windblown deposits are interwoven with much of Chinese history and society.

Overview of Desert Landscapes

Dry climates occupy about 26% of Earth's land surface. If all semiarid climates are considered, perhaps as much as 35% of all land area, constituting the largest single climatic region on Earth (see Figures 10.4 and 10.5 for the location of these *arid deserts BW* and *semiarid steppe BS* climate regions and Figure 20.3 for the distribution of these desert environments).

Deserts occur worldwide as topographic plains, such as the Great Sandy and Simpson Deserts of Australia, the Arabian and Kalahari Deserts of Africa, and portions of the extensive Taklimakan Desert, which cover some 270,000 km^2 (105,000 mi^2) in the central Tarim Basin of China. Deserts also are found in mountainous regions: interior Asia, from Iran to Pakistan, and in China and Mongolia. In South America, lying between the ocean and the Andes, is the rugged Atacama Desert. And Earth's deserts are expanding, as we will discuss shortly. Now, let us look at the link between climate and Earth's deserts.

Desert Climates

The spatial distribution of dry lands is related to three climatological settings: to subtropical high-pressure cells between 15° and 35°, both N and S latitudes (see Figures 6.11 and 6.13), to the rain shadow on the lee side of mountain ranges (see Figure 8.9), and to areas at great distance from moisture-bearing air masses, such as central Asia. Figure 15.15 portrays this distribution according to the modified Köppen climate classification used in this text and presents photographs of four major desert regions.

Desert areas possess unique landscapes created by the interaction of intermittent precipitation events, weathering processes, and wind. Rugged, hard-edged desert landscapes of cliffs and scarps contrast sharply with the vegetation-covered, rounded and smoothed slopes characteristic of humid regions.

The daily surface energy balance for El Mirage, California, presented in Figure 4.21a and b, highlights the high sensible heat conditions and intense ground heating in the desert. Such areas receive a high input of insolation through generally clear skies, and they experience high radiative heat losses at night. A typical desert water balance experiences high potential evapotranspiration (POTET) demand, low precipitation supply, and prolonged summer water deficits (see, for example, Figure 9.12e for Phoenix, Arizona). Fluvial processes in the desert generally are characterized by intermittent running water, with hard, poorly vegetated surfaces yielding high runoff during rainstorms (Figure 15.16).

Desert Fluvial Processes

Precipitation events in a desert may be rare indeed, a year or two apart, but when they do occur, a dry streambed can fill with a torrent called a **flash flood**. These channels may fill in a few minutes and surge briefly during and after a storm. Depending on the region, such a dry streambed is known as a **wash**, an *arroyo* (Spanish), or a *wadi* (Arabic). A desert highway that crosses a wash usually has signs posted

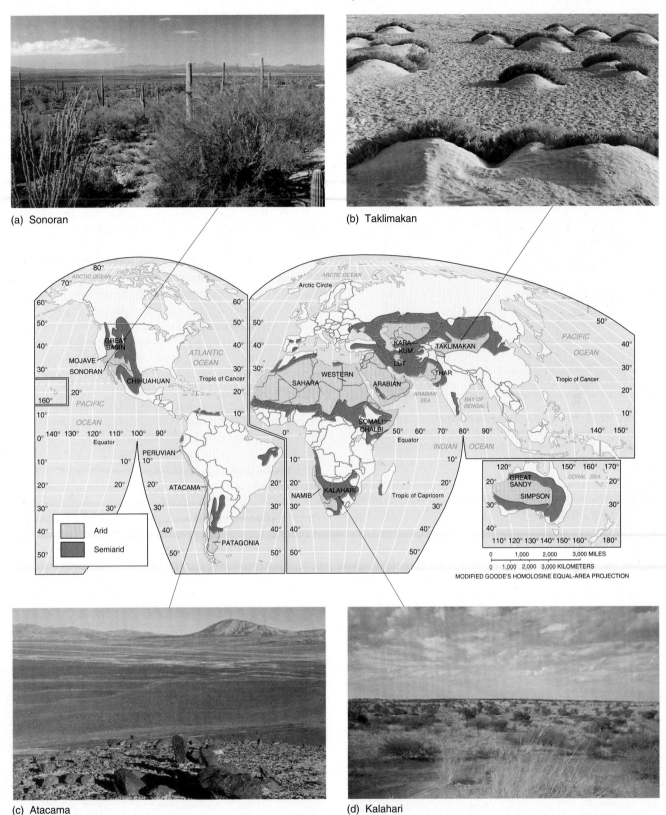

(a) Sonoran

(b) Taklimakan

(c) Atacama

(d) Kalahari

FIGURE 15.15 The world's dry regions.
Worldwide distribution of arid lands (*arid desert BW* climates) and semiarid lands (*semiarid steppe BS* climates) based on the Köppen climatic classification system. (a) Sonoran Desert in the American southwest. (b) Taklimakan Desert in central Asia. (c) Atacama Desert in subtropical Chile near Baquedano. (d) Kalahari Desert in south-central Africa. [(a) Photo by author; (b) photo by Wu Chunzhan/New China Pictures Co./East Photo; (c) photo by Jacques Jangoux/ Photo Researchers, Inc.; (d) photo by Nigel J. Dennis/Photo Researchers, Inc.]

FIGURE 15.16 The land in Canyon de Chelly.
In this marginal land, undependable water flows between the towering sandstone walls surrounding Chinle Wash in Canyon de Chelly, Arizona. [Photo by author.]

to warn drivers not to proceed if rain is in the vicinity, for a flash flood can suddenly arise and sweep away anything in its path.

When washes fill with surging flash flood waters, a unique set of ecological relationships quickly develops. Crashing rocks and boulders break open seeds that respond

to the timely moisture and germinate. Other plants and animals also spring into brief life cycles as the water irrigates their limited habitats.

At times of intense rainfall, remarkable scenes fill the desert. Figure 15.17 shows two photographs taken just one month apart in a sand dune field in Death Valley, California. A rainfall event produced 2.57 cm (1.01 in.) of precipitation in one day, in a place that receives only 4.6 cm (1.83 in.) in an average year. The stream in the photograph continued to run for hours and then collected in low spots on hard, underlying clay surfaces. The water was quickly consumed by the high evaporation demand so that, in just a month, these short-lived watercourses were dry and covered with accumulations of alluvial materials.

As runoff water evaporates, salt crusts may be left behind on the desert floor. This intermittently wet and dry low area in a region of closed drainage is called a **playa**, site of an *ephemeral lake* when water is present. Accompanying our earlier discussion of evaporites, Figure 11.10 shows such a playa in Death Valley, covered with salt precipitate just one month after this record rainfall event.

Permanent lakes and continuously flowing rivers are uncommon features in the desert, although the Nile River and the Colorado River are notable exceptions. Both these rivers are *exotic streams*, having their headwaters in a wetter region and the bulk of their course through arid regions. Focus Study 15.1 discusses the Colorado River and its problem of overuse in an arid land.

In arid climates, a prominent landform is the **alluvial fan**, which occurs at the mouth of a canyon where it exits into a valley. The fan is produced by flowing water that abruptly loses velocity as it leaves the constricted channel of the canyon and therefore drops layer upon layer of sediment along the base of the mountain block. Water then flows over the surface of the fan and produces a braided drainage pattern, shifting from channel to channel with

(a)

(b)

FIGURE 15.17 An improbable river in Death Valley.
The Stovepipe Wells dune field of Death Valley, California, shown (a) the day after a 2.57 cm (1.01 in.) rainfall and (b) one month later (identical location). [Photos by author.]

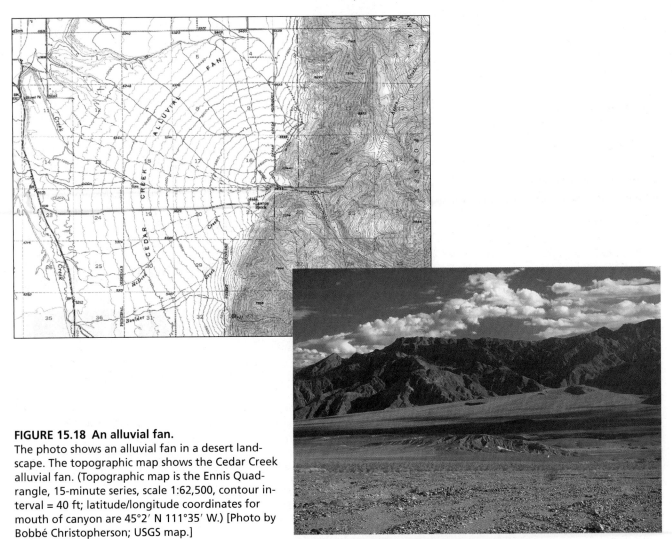

FIGURE 15.18 An alluvial fan.
The photo shows an alluvial fan in a desert landscape. The topographic map shows the Cedar Creek alluvial fan. (Topographic map is the Ennis Quadrangle, 15-minute series, scale 1:62,500, contour interval = 40 ft; latitude/longitude coordinates for mouth of canyon are 45°2′ N 111°35′ W.) [Photo by Bobbé Christopherson; USGS map.]

each precipitation event (Figure 15.18). A continuous apron, or **bajada** (Spanish for "slope"), may form if individual alluvial fans coalesce into one sloping surface (see Figure 15.23c). Fan formation of any sort is reduced in humid climates because perennial streams constantly carry away sediment, preventing its deposition.

An interesting aspect of an alluvial fan is the natural sorting of materials by size. Near the mouth of the canyon at the apex of the fan, coarse materials are deposited, grading slowly to pebbles and finer gravels with distance out from the mouth. Then sands and silts are deposited, with the finest clays and salts carried in suspension and solution all the way to the valley floor. Dissolved minerals accumulate as evaporite deposits on the valley floor left after evaporation of the water from the playa.

Well-developed alluvial fans also can be a major source of groundwater. Some cities—San Bernardino, California, for example—are built on alluvial fans and extract their municipal water supplies from them. In other parts of the world, such water-bearing alluvial fans are known as *qanat* (Iran), *karex* (Pakistan), and *foggara* (western Sahara).

Desert Landscapes

Contrary to popular belief, deserts are not wastelands, for they abound in specially adapted plants and animals. Moreover, the limited vegetation, intermittent rainfall, intense insolation, and distant vistas produce starkly beautiful landscapes (see Chapter 10's opening photo). And all deserts are not the same: For example, North American deserts have more vegetation cover than do the generally barren Asian desert expanses, as we saw in Figure 15.15.

The shimmering heat waves and related mirage effects in the desert are products of light refraction through layers of air that have developed a temperature gradient near the hot ground. The desert's enchantment is captured in the book *Desert Solitaire*:

Around noon the heat waves begin flowing upward from the expanses of sand and bare rock. They shimmer like transparent, filmy veils between my sanctuary in the shade and all the sun-dazzled world beyond. Objects and forms viewed through this tremulous flow appear somewhat displaced or distorted. . . . The great Balanced Rock floats a few inches above its

The Colorado River: A System Out of Balance

The headwaters of an exotic stream rise in a humid region of water surpluses but then flows mostly through arid lands for the rest of its journey to the sea. Exotic streams have few incoming tributaries. Consequently, an exotic stream has a discharge pattern that is opposite that of a typical stream: Instead of discharge increasing downstream, it decreases (see Chapter 14). The Nile and Colorado Rivers are prominent examples.

In the case of the Nile, the East African mountains and plateaus provide a humid source area. The Nile first rises as the remote headwaters of the Kagera River in the eastern portion of the Lake Plateau country of East Africa. On its way to Lake Victoria, it forms the partial boundary of Tanzania, Rwanda, and Uganda. The Nile itself then rises out of the lake and continues on its 6650 km (4132 mi) course to its mouth on the Mediterranean Sea near Cairo.

The Colorado River Basin

The Colorado rises on the high slopes of Mount Richthofen (3962 m, or 13,000 ft) in Rocky Mountain National Park, Colorado (Figure 1a) and flows almost 2317 km (1440 mi) to where a trickle of water disappears in the sand, kilometers short of its former mouth in the Gulf of California.

Orographic precipitation totaling 102 cm (40 in.) per year (mostly snow) falls in the Rockies, feeding the Colorado headwaters. But at Yuma, Arizona, near the river's end, annual

precipitation is a scant 8.9 cm (3.5 in.), an extremely small amount when compared with the annual potential evapotranspiration demand in the Yuma region of 140 cm (55 in.).

From its source region, the Colorado River quickly leaves the humid Rockies and spills out into the arid desert of western Colorado and eastern Utah. At Grand Junction, Colorado, near the Utah border, annual precipitation is only 20 cm (8 in.) (Figure 1b). After carving its way through the intricate labyrinth of canyonlands in Utah, the river enters Lake Powell, 945 m (3100 ft) lower in elevation than the river's source area upstream in the Rockies some 982 km (610 mi) away (Figure 1c). The Colorado then flows through the Grand Canyon chasm, formed by its own erosive power.

West of the Grand Canyon, the river turns southward, tracing its final 644 km (400 mi) as the Arizona–California border. Along this stretch sits Hoover Dam, just east of Las Vegas (Figure 1d); Davis Dam, built to control the releases from Hoover (Figure 1e); Parker Dam for the water needs of Los Angeles; three more dams for irrigation water (Palo Verde, Imperial, and Laguna, Figure 1f), and finally, Morelos Dam at the Mexican border (Figure 1g). Mexico owns the end of the river and whatever water is left, although the river no longer reaches its mouth into the Gulf of California (Figure 1i).

Figure 1j shows the annual water discharge and suspended sediment load for the Colorado River at Yuma, Ari-

zona, from 1905 to 1964. The completion of Hoover Dam in the 1930s dramatically reduced suspended sediment. Addition of Glen Canyon Dam upstream from Hoover Dam in 1963 further reduced streamflows. In fact, Lake Powell, which formed upstream behind the artificial base level of Glen Canyon Dam, is forecast to fill with sediment over the next 100 years.

Overall, the drainage basin encompasses 641,025 km^2 (247,500 mi^2) of mountain, basin-and-range, plateau, canyon, and desert landscapes, in parts of seven states and two countries. A discussion of the Colorado River is included in this chapter because of its crucial part in the history of the Southwest and its role in the future of this region.

Dividing Up the Colorado's Dammed Water

John Wesley Powell (1834–1902), the first person of record to successfully navigate the Colorado River through the Grand Canyon, was the first director of the U.S. Bureau of Ethnology and later director of the U.S. Geological Survey (1881–1892). Powell perceived that the challenge of the West was too great for individual efforts and believed that solutions to problems such as water availability could be met only through private cooperative efforts. His 1878 study (reprinted 1962), *Report of the Lands of the Arid Region of the United States*, is a conservation landmark.

(continued)

FIGURE 1 The Colorado River drainage basin.
The Colorado River basin, showing division of the upper and lower basins near Lees Ferry in northern Arizona. (a) Headwaters of the Colorado River near Mount Richthofen in the Colorado Rockies. (b) The river near Moab, Utah. (c) Glen Canyon Dam, a regulatory, administrative facility near Lees Ferry, Arizona. (d) Hoover Dam spillways in rare operation during 1983 floods. (e) Davis Dam in full release during flood. (f) Irrigation canal and cropland irrigated with Colorado River water in the Palo Verde district. (g) Morelos Dam at the Mexican border is the final stop as the river dwindles to a mere canal. (h) Central Arizona Project aqueduct west of Phoenix. (i) The Colorado River (far upper left) stops short of its former delta. The bluish-purple water in the former channel is actually an inlet for Gulf of California water; the gray deposits are sediments in mud flats along this inlet. (j) Dam construction affects river discharge and sediment yields. [(a) Photo by Bobbé Christopherson; (b through g) photos by author; (h) photo by Tom Bean/DRK Photo; (i) *Terra* ASTER (thermal emission and reflection) image courtesy of NASA/GSFC/MITI/ERSDAC/JAROS and the U.S./Japan ASTER Science Team, September 8, 2000; and, (j) data from USGS, 1985, *National Water Summary 1984*, Water Supply Paper 2275 (Washington, DC: Government Printing Office), p. 55).]

Focus Study 15.1 *(continued)*

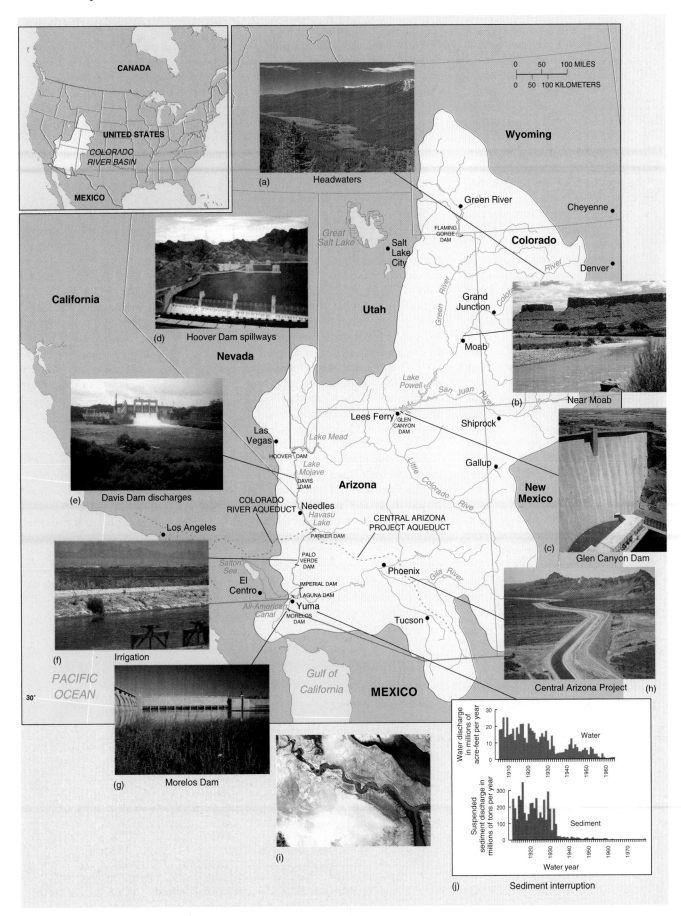

(a) Headwaters

(d) Hoover Dam spillways

(b) Near Moab

(c) Glen Canyon Dam

(e) Davis Dam discharges

(f) Irrigation

(g) Morelos Dam

(h) Central Arizona Project

(i)

(j) Sediment interruption

Focus Study 15.1 *(continued)*

Today, Powell probably would be skeptical of the intervention by government agencies in building large-scale reclamation projects. Lake Powell is named after him despite his probable opposition were he alive. An anecdote in Wallace Stegner's book *Beyond the Hundredth Meridian* relates that, at an 1893 international irrigation conference held in Los Angeles, development-minded delegates bragged that the entire West could be conquered and reclaimed from nature and that "rain will certainly follow the plow." Powell spoke against that sentiment: "I tell you, gentlemen, you are piling up a heritage of conflict and litigation over water rights, for there is not sufficient water to supply the land."* He was booed from the hall, but history has shown Powell to be correct.

The Colorado River Compact was signed by six of the seven basin states in 1923. (The seventh, Arizona, signed in 1944, the same year as the Mexican Water Treaty.) With this compact, the Colorado River basin was divided into an upper basin and a lower basin, arbitrarily separated for administrative purposes at Lees Ferry near the Utah–Arizona border (noted on the map in Figure 1). Congress adopted the Boulder Canyon Act in 1928, authorizing Hoover Dam as the first major reclamation project on the river. Also authorized was the All-American Canal into the Imperial Valley, which required an additional dam. Los Angeles then began its project to bring Colorado River water 390 km (240 mi) from still another dam and reservoir on the river to their city.

Shortly after Hoover Dam was finished and downstream enterprises were thus offered flood protection, the other projects were quickly completed. There are now eight major dams on the river and many irrigation works. The latest effort to redistribute Colorado River water is the Central Arizona Project, which carries water to the Phoenix area (Figure 1h).

Highly Variable River Flows

The flaw in all this planning and water distribution is that exotic streamflows are highly variable, and the Colorado is no exception. In 1917 the discharge measured at Lees Ferry totaled 24 million acre-feet (maf), whereas in 1934 it dropped by nearly 80 percent, to only 5.03 maf. In 1977 the discharge dropped again to 5.02 maf, but in 1984 it rose to an all-time high of 24.5 maf. The 1998 flows were above average at 17.5 maf and 16.6 maf in 1999. In addition to this variability, approximately 70% of the year-to-year discharge occurs between April and July; the other 30% is spread over the balance of the year.

The average flows between 1906 and 1930 were almost 18 maf a year, but averages dropped to 14.171 maf during the past 70 years (1930 to 1999). As a planning basis for the Colorado River Compact, the government used average river discharges from 1914 up to the treaty signing in 1923, an exceptionally high average of 18.8 maf. That amount was perceived as more than enough for the upper and lower basins each to receive 7.5 maf and, later, for Mexico to receive 1.5 maf in the 1944 Mexican Water Treaty.

We might question whether proper long-range planning should rely on the providence of high variability. Tree-ring analyses of past climates have disclosed that the only other time Colorado discharges were at the high 1914–1923 level was between A.D. 1606 and 1625! The dependable flows of the river have been consistently overestimated. This shortfall problem is shown in an estimated budget of 20 million acre-feet for the river (Table 1): Clearly the situation is out of balance, for there is not enough discharge to meet budgeted demands. A few wet years in the late 20th century sparked more confidence and triggered more disputes on how to divvy up the surpluses—only delaying the inevitable deficit crisis.

Presently, the seven states ideally want rights to a total as high as 25.0 maf. When added to the guarantee for Mexico, this comes up to 26.5 maf of wants. And six states share one common opinion: California's right to the water must be limited to a court-ordered 4.4 maf, which it exceeds every year—1998 saw California take 4.9 maf.

Water Loss at Glen Canyon Dam and Lake Powell

Glen Canyon Dam was completed and began water impoundment (Lake Powell) in 1963, 27.4 km (17 mi) north of the basin division point at Lees Ferry. The advancing water slowly flooded the deep, fluted, inner gorges of many canyons, including the Glen, Navajo, Labyrinth, and Cathedral. Glen Canyon Dam's primary purpose, according to the Bureau of Reclamation, is to regulate flows between the upper and lower basins. An additional benefit is the production of hydroelectric power, which is sold wholesale at a low rate to utilities across the Southwest. Also, there is a growing recreation and tourism industry around Lake Powell, as previously inaccessible desert scenery now can be reached by boat.

(continued)

Table 1   Estimated Colorado River Budget

Water Demand	Quantity (maf)[a]
Upper Basin (7.5) Lower Basin (7.5)[b]	15.0
Central Arizona Project (rising to 2.8 maf)	1.0
Mexican allotment (1944 Treaty)	1.5
Evaporation from reservoirs	1.5
Bank storage at Lake Powell	0.5
Phreatophytic losses (water-demanding plants)	0.5
Budgeted total demand	20.0 maf
1930–1999 average flow of the river	14.8 maf

Source: Bureau of Reclamation states of Arizona and California.
[a]1 million acre-feet = 325,872 gallons; 1.24 million liters.
[b]In-basin consumptive uses 75% agricultural.

Focu Study 15.1 *(continued)*

Many thought that Lake Mead, behind Hoover Dam downstream, could have served the primary administrative function of flow regulation, especially considering the serious water-loss problems with the Glen Canyon Dam and the absorbent rocks that contain its reservoir. Porous Navajo sandstone underlies most of the Lake Powell reservoir at Glen Canyon Dam. This sandstone absorbs an estimated 0.5 maf of the river's overall annual discharge as bank storage. The higher the lake level, the greater the loss into the sandstone. Second, Lake Powell is an open body of water in an arid desert, where hot, dry winds accelerate evaporative losses. Another 0.5 maf of the Colorado's overall discharge is lost annually from the reservoir in this manner, about a third of Colorado River reservoir evaporation losses. Third, now-permanent sand bars and

banks have stabilized along the regulated river, allowing water-demanding plants called phreatophytes to establish and extract an additional 0.5 maf of the river flow.

Intense precipitation and heavy snowpack in the Rockies, attributable to the 1982–1983 El Niño (see Focus Study 10.1), led to record-high discharge rates on the Colorado, testing the controllability of one of the most regulated rivers in the world. Federal reservoir managers were not prepared for the high discharge, since they had set aside Lake Mead's primary purpose (flood control) in favor of competing water and power interests. What followed was a human-caused flood on the most engineered river in the world!

The only time the spillways at Hoover Dam had ever operated was more than 40 years earlier, when the reservoir capacity was artificially raised

for a test; now they were opened to release the floodwaters (Figure 1d). Davis Dam, which regulates releases from Hoover Dam, was within 30 cm (1 ft) of overflow, a real problem for a structure made partially of earth fill (Figure 1e). In addition, Glen Canyon Dam was over capacity and at risk and was damaged by the volume of discharge tearing through its spillways. The decision to increase releases doomed towns and homeowners along the river, especially in subdivisions near Needles, California.

We might wonder what John Wesley Powell would think if he were alive today to witness such errant attempts to control the mighty and variable Colorado. He foretold such a "heritage of conflict and litigation."

*W. Stegner, *Beyond the Hundredth Meridian* (Boston: Houghton Mifflin, 1954), p. 343.

pedestal, supported by a layer of superheated air. The buttes, pinnacles, and fins in the windows area bend and undulate beyond the middle ground like a painted backdrop stirred by a draft of air.*

The buttes, pinnacles, and mesas of arid landscapes are resistant horizontal rock strata that have eroded differentially. Removal of the less-resistant sandstone strata produces unusual desert sculptures—arches, windows, pedestals, and delicately balanced rocks (Figure 15.19). Specifically, the upper layers of sandstone along the top of an arch or butte are more resistant to weathering and protect the sandstone rock beneath.

The removal of all surrounding rock through differential weathering leaves enormous buttes as residuals on the landscape. If you imagine a line intersecting the tops of the Mitten Buttes shown in Figure 15.20, you can gain some idea of the quantity of material that has been removed. These buttes exceed 300 m (1000 ft) in height, similar to the Chrysler Building in New York City or First Canadian Place in Toronto.

*E. Abbey, *Desert Solitaire* (New York: McGraw-Hill, 1968), p. 154. Copyright © 1968 by Edward Abbey.

FIGURE 15.19 A balanced rock—differential weathering.
Balanced Rock in Arches National Park, Utah, where writer–naturalist Edward Abbey (quoted in text) worked as a ranger years before it became a park. The overall feature is 39 m (128 ft) tall and composed of Entrada sandstone. The balanced-rock portion is 17 m (55 ft) tall and weighs 3255 metric (3577 short) tons. [Photo by author.]

(a)

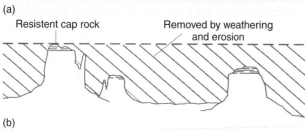

(b)

FIGURE 15.20 Monument Valley landscape.
(a) Mitten Buttes, Merrick Butte, and rainbow in Monument Valley, Navajo Tribal Park, along the Utah–Arizona border.
(b) A schematic of the tremendous removal of material by weathering, erosion, and transport. [Photo by author.]

Desert landscapes are places where stark erosional remnants can stand above the surrounding terrain as knobs or hills. Such a bare, exposed rock, called an *inselberg* (island mountain), is exemplified by Uluru (Ayers) Rock in Australia (Figure 15.21).

In a desert area, weak surface material may weather to a complex, rugged topography, usually of relatively low and varied relief. Such a landscape is called a *badland*, probably so named because in the American West it offered little economic value and was difficult to traverse in 19th-century wagons. The Badlands region of the Dakotas and northcentral Arizona (the Painted Desert) are of this form.

Sand dunes that existed in some ancient deserts have lithified, forming sandstone structures that bear the imprint of *cross-stratification*. When such a dune was accumu-

lating, sand cascaded down its slipface, and distinct bedding planes (layers) were established that remained after the dune lithified (Figure 15.22). Ripple marks, animal tracks, and fossils also are found preserved in these sandstones, which originally were eolian-deposited sand dunes.

Basin and Range Province

A province is a large region that is characterized by several geologic or physiographic traits. Characterizing the **Basin and Range Province** of the western United States are alternating basins (valleys) and mountain ranges that lie in the rain shadow of mountains to the west (Figure 15.23). The physiography and geography combine to give the province a dry climate, few permanent streams, and *internal*

FIGURE 15.21 Australian landmark.
Uluru (Ayers) Rock in Northern Territory, Australia, is an isolated mass of weathered rock. The formation is 348 m (1145 ft) high and is 2.5 km long by 1.6 km wide. Uluru Rock is sacred to Aboriginal peoples and has been protected in Uluru National Park since 1950. [Photo by Porterfield/Chickering.]

FIGURE 15.22 Cross-bedding in sedimentary rocks.
The bedding pattern, called cross-stratification, in these sandstone rocks tells us about patterns that were established in the dunes before lithification (hardening into rock). [Photo by Bobbé Christopherson.]

drainage patterns—drainage basins that lack any outlet to the ocean (see Figure 14.5, "Great Basin" and internal drainage).

The vast Basin and Range Province—almost 800,000 km² (300,000 mi²)—was a major barrier to early settlers in their migration westward. The combination of desert climate and north–south trending mountain ranges presented harsh challenges. Today, when you traverse U.S. Highway 50 across Nevada, you cross five passes (horsts) of more than 1950 m (6400 ft) and numerous basins (grabens). Throughout the drive you are reminded where you are by prideful signs that plainly state: "The Loneliest Road in America." It is difficult to imagine crossing this topography with wagons and oxen (Figure 15.23e).

How the Basin and Range formed is interesting. As the North American plate lumbered westward, it overrode former oceanic crust and hot spots at such a rapid pace that

slabs of subducted material literally were run over. The crust was stretched (tensional forces), creating a landscape fractured by many faults. The present landscape consists of nearly parallel sequences of *horsts* (upward-faulted blocks, which are the "ranges") and *grabens* (downward-faulted blocks, which are the "basins," or valleys). Figure 15.23c shows this pattern of normal faults.

John McPhee captured the feel of this desert province in his book *Basin and Range*:

> Supreme over all is silence. Discounting the cry of the occasional bird, the wailing of a pack of coyotes, silence—a great spatial silence—is pure in the Basin and Range. It is a soundless immensity with mountains in it. You stand . . . and look up at a high mountain front, and turn your head and look fifty miles down the valley, and there is utter silence.*

Basin-and-range relief is abrupt, and rock structures are angular and rugged. As the ranges erode, transported materials accumulate to great depths in the basins, gradually producing extensive desert plains. The basin's elevation averages roughly 1200–1500 m (4000–5000 ft) above sea level, with mountain crests rising higher by some 900–1500 m (3000–5000 ft). Death Valley, California, is the lowest of these basins, with an elevation of −86 m (−282 ft). However, to the west of the valley, the Panamint Range rises to 3368 m (11,050 ft) at Telescope Peak—almost 3.5 vertical kilometers (2.2 mi) of desert mountain relief!

In Figure 15.23c and d, note the **bolson**, a slope-and-basin area between the crests of two adjacent ridges in a dry region of internal drainage. Death Valley provides a dramatic example of these arid-land features. Figure 15.23c also identifies a *playa* (central salt pan), a *bajada* (coalesced alluvial fans), and a mountain front in retreat from weathering and erosional attack. A *pediment* is an area of bedrock that is layered with a thin veneer, or coating, of alluvium. It is an erosional surface, as opposed to the depositional surface of the bajada.

Vast arid and semiarid lands remain an enigma on the water planet. They challenge our technology, courage, and personal need for water. These lands hold a mysterious fascination, perhaps because they are so lacking in the moisture that infuses our lives.

Desertification

We are witnessing an unwanted expansion of the Earth's desert lands in a process known as **desertification**. This now is a worldwide phenomenon along the margins of semiarid and arid lands. Desertification is due principally to poor agricultural practices (overgrazing and agricultural activities that abuse soil structure and fertility), improper soil-moisture management, erosion and salinization,

*J. McPhee, *Basin and Range* (New York: Farrar, Straus, Giroux, 1981), p. 46.

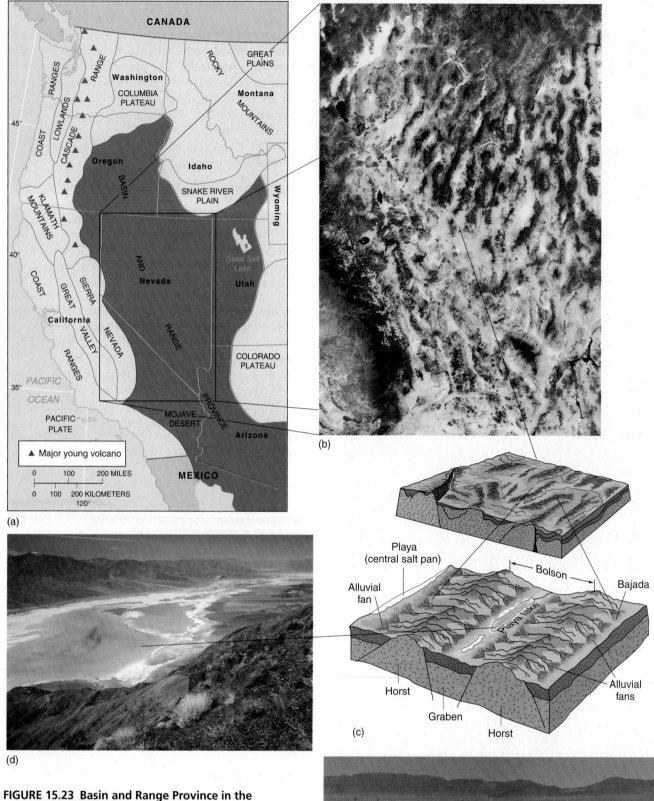

FIGURE 15.23 Basin and Range Province in the western United States.
(a) Map and (b) *Landsat* image of the area. Recent scientific discoveries demonstrate that this province extends south through northern and central Mexico. (c) A bolson in the mountainous desert landscape of the Basin and Range Province. Parallel normal faults produce a series of horsts (ranges) and grabens (basins). (d) Death Valley features a central playa, parallel mountain ranges, alluvial fans, and bajada along the base of the ranges. (e) "The Loneliest Road in America," Highway U.S. 50, slashes westward through the province. [(b) Image from NASA; (d) photo by author; (e) photo by Bobbé Christopherson.]

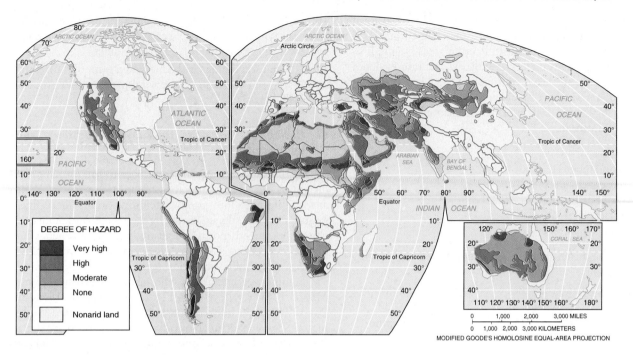

FIGURE 15.24 The desertification hazard.
Worldwide desertification estimates by the United Nations. [Data from U.N. Food and Agri-
cultural Organization (FAO), World Meteorological Organization (WMO), United Nations Edu-
cational, Scientific, and Cultural Organization (UNESCO), Nairobi, Kenya; as printed in J. M.
Rubenstein, *An Introduction to Human Geography*, Figure 14-16, p. 509. ©1999 Prentice Hall,
Inc. Used by permission.]

deforestation, and the ongoing global climatic change, which is shifting temperature and precipitation patterns.

The southward expansion of Saharan conditions through portions of the *Sahel region* has left many African peoples on land that no longer experiences the rainfall of just two decades ago. Other regions at risk of desertification stretch from Asia and central Australia to portions of North and South America. The United Nations estimates that degraded lands have covered some 800 million hectares (2 billion acres) since 1930; many millions of additional hectares are added each year. An immediate need is to improve the database for a more accurate accounting of the problem and a better understanding of what is occurring.

Figure 15.24 is drawn from a map prepared for a U.N. Conference on Desertification. Desertification areas are ranked: A moderate hazard area has an average 10%–25% drop in agricultural productivity; a high hazard area has a 25%–50% drop; and a very high hazard area has more than a 50% decrease. Because human activities and economies, especially unwise grazing practices, appear to be the major cause of desertification, actions to slow the process are readily available. The severity of this problem is magnified by the poverty in many of the affected regions. For more on these global arid lands, see the U.N. site at **http://ceps. nasm.edu:1995/drylands.html**.

Summary and Review—Eolian Processes and Arid Landscapes

● **Characterize** the unique work accomplished by wind and eolian processes.

Winds are produced by the movement of the atmosphere in response to pressure differences. Wind is a geomorphic agent of erosion, transportation, and deposition. **Eolian** processes modify and move sand accumulations along coastal beaches and deserts. Wind's ability to move materials is small compared with that of water and ice.

eolian (p. 464)

1. Who was Ralph Bagnold? What was his contribution to eolian studies?

2. Explain the term *eolian* and its application in this chapter. How would you characterize the ability of the wind to move material?

● **Describe** eolian erosion, including deflation, abrasion, and the resultant landforms.

Two principal wind-erosion processes are **deflation**, the removal and lifting of individual loose particles, and **abrasion**, the "sandblasting" of rock surfaces with particles captured in the air. Fine materials are eroded by wind deflation and moving water, leaving behind concentrations of pebbles and gravel called **desert pavement**. Wherever wind encounters loose sediment, deflation may remove enough material to form basins. Called **blowout depressions**, they range from small indentations less than a meter wide up to areas hundreds of meters wide and many meters deep. Rocks that bear evidence of eolian erosion are called **ventifacts**. On a larger scale, deflation and abrasion are capable of streamlining rock structures, leaving behind distinctive rock formations or elongated ridges called **yardangs**.

> deflation (p. 464)
> abrasion (p. 464)
> desert pavement (p. 464)
> blowout depressions (p. 465)
> ventifacts (p. 466)
> yardangs (p. 466)

3. Describe the erosional processes associated with moving air.
4. Explain deflation and the evolutionary sequence that produces desert pavement.
5. How are ventifacts and yardangs formed by the wind?

● *Describe* eolian transportation and *explain* saltation and surface creep.

Wind exerts a drag or frictional pull on surface particles until they become airborne. Only the finest dust particles travel significant distances, so the finer material suspended in a dust storm is lifted much higher than the coarser particles of a sandstorm. Saltating particles crash into other particles, knocking them both loose and forward. The motion called **surface creep** slides and rolls particles too large for saltation.

> surface creep (p. 467)

6. Differentiate between a dust storm and a sandstorm.
7. What is the difference between eolian saltation and fluvial saltation?
8. Explain the concept of surface creep.

● *Identify* the major classes of sand dunes and *present* examples within each class.

In arid and semiarid climates and along some coastlines where sand is available, dunes accumulate. A **dune** is a wind-sculpted accumulation of sand. An extensive area of dunes, such as that found in North Africa, is characteristic of an **erg desert**, or **sand sea**. When saltating sand grains encounter small patches of sand, their kinetic energy (motion) is dissipated and they start to accumulate into a dune. As height increases above 30 cm (12 in.), a steeply sloping **slipface** on the lee side and characteristic dune features are formed. Dune forms are broadly classified as *crescentic*, *linear*, and *star*.

> dune (p. 468)
> erg desert (p. 468)
> sand sea (p. 468)
> slipface (p. 468)

9. What is the difference between an erg and a reg desert? Which type is a sand sea? Are all deserts covered by sand? Explain.

10. What are the three classes of dune forms? Describe the basic types of dunes within each class. What do you think is the major shaping force for sand dunes?
11. Which form of dune is the mountain giant of the desert? What are the characteristic wind patterns that produce such dunes?

● *Define* loess deposits, their origins, locations, and landforms.

Eolian-transported materials contribute to soil formation in distant places. Windblown **loess** deposits occur worldwide and can develop into good agricultural soils. These fine-grained clays and silts are moved by the wind many kilometers, where they are redeposited in unstratified, homogeneous deposits. The binding strength of loess causes it to weather and erode in steep bluffs, or vertical faces.

Significant accumulations throughout the Mississippi and Missouri River valleys form continuous deposits 15–30 m (50–100 ft) thick. Loess deposits also occur in eastern Washington State, Idaho, much of Ukraine, central Europe, China, the Pampas–Patagonia regions of Argentina, and lowland New Zealand.

> loess (p. 472)

12. How are loess materials generated? What form do they assume when deposited?
13. Name a few examples of significant loess deposits on Earth.

● *Portray* desert landscapes and *locate* these regions on a world map.

Dry and semiarid climates occupy about 35% of Earth's land surface. The spatial distribution of these dry lands is related to subtropical high-pressure cells between 15° and 35° N and S, to rain shadows on the lee side of mountain ranges, or to areas at great distance from moisture-bearing air masses, such as central Asia.

Precipitation events are rare, yet running water is still the major erosional agent in deserts. Precipitation events may be rare, but when they do occur, a dry streambed fills with a torrent called a **flash flood**. Depending on the region, such a dry streambed is known as a **wash**, an *arroyo* (Spanish), or a *wadi* (Arabic). As runoff water evaporates, salt crusts may be left behind on the desert floor. This intermittently wet and dry low area in a region of closed drainage is called a **playa**, site of an *ephemeral lake* when water is present.

In arid climates, a prominent landform is the **alluvial fan** at the mouth of a canyon where it exits into a valley. The fan is produced by flowing water that abruptly loses velocity as it leaves the constricted channel of the canyon and deposits a layer of sediment along the mountain block. A continuous apron, or **bajada**, may form if individual alluvial fans coalesce. A *province* is a large region that is characterized by several geologic or physiographic traits. The **Basin and Range Province** of the western United States consists of alternating basins and mountain ranges. A slope-and-basin area between the crests of two adjacent ridges in a dry region of internal drainage is termed a **bolson**. **Desertification** is the process that leads to an unwanted expansion of the Earth's desert lands.

flash flood (p. 474)
wash (p. 474)
playa (p. 476)
alluvial fan (p. 476)
bajada (p. 477)
Basin and Range Province (p. 482)
bolson (p. 483)
desertification (p. 483)

14. Characterize desert energy and water balance regimes. What are the significant patterns of occurrence for arid landscapes in the world?

15. How would you describe the water budget of the Colorado River? What was the basis for agreements regarding distribution of the river's water? Why has thinking about the river's discharge been so optimistic?

16. Describe a desert bolson from crest to crest. Draw a simple sketch with the components of the landscape labeled.

17. Where is the Basin and Range Province? Briefly describe its appearance and character.

18. What is meant by desertification? Using the maps in Figures 15.15 and 15.24, and the text description, locate several of the affected regions of desert expansion.

NetWork

The *Geosystems* Home Page provides on-line resources for this chapter on the World Wide Web. To begin: Once on the Home Page, click on this textbook, scroll the Table of Contents menu, select this chapter, and click "Begin." You will find self-tests that are graded, review exercises, specific updates for items in the chapter, and many links to interesting related pathways on the Internet under "Destinations." *Geosystems* is at **http://www.prenhall.com/christopherson**.

Critical Thinking

A. "Water is the major erosion and transport medium in the desert." Respond to this quotation. How is this possible? What factors have you learned from this chapter that prove this statement true?

B. Where are the nearest eolian features (coastal, lakeshore, or desert dunes, or loess deposits) to your present location? Which causative factors discussed in this chapter explain the features you identified?

The Bahamas are the largest fringing reef in the world, comprising 700 islands and many coral reefs. The lighter areas feature ocean depths to 10 m (33 ft). The darker blues indicate depths dropping off to 4000 m (13,100 ft). The name Bahamas is from the Spanish *baja mar*, for "shallow sea." [*Terra* MODIS sensor image courtesy of NASA/GSFC and MODIS Land Response Team, April, 18, 2000.]

16

The Oceans, Coastal Processes, and Landforms

Key Learning Concepts

After reading the chapter, you should be able to:

- *Describe* the chemical composition of seawater and the physical structure of the ocean.
- *Identify* the components of the coastal environment and *list* the physical inputs to the coastal system, including tides and mean sea level.
- *Describe* wave motion at sea and near shore and *explain* coastal straightening as a product of wave refraction.
- *Identify* characteristic coastal erosional and depositional landforms.
- *Describe* barrier islands and their hazards as they relate to human settlement.
- *Assess* living coastal environments: corals, wetlands, salt marshes, and mangroves.
- *Construct* an environmentally sensitive model for settlement and land use along the coast.

Walk along a shoreline and you witness the dramatic interaction of Earth's vast oceanic, atmospheric, and lithospheric systems. At times, the ocean attacks the coast in a stormy rage of erosive power; at other times, the moist sea breeze, salty mist, and repetitive motion of the water are gentle and calming. Few have captured this confrontation between land and sea as well as biologist Rachel Carson:

> The edge of the sea is a strange and beautiful place. All through the long history of Earth it has been an area of unrest where waves have broken heavily against the land, where the tides have pressed forward over the continents, receded, and then returned. For no two successive days is the shoreline precisely the same. Not

only do the tides advance and retreat in their eternal rhythms, but the level of the sea itself is never at rest. It rises or falls as the glaciers melt or grow, as the floors of the deep ocean basins shift under its increasing load of sediments, or as the earth's crust along the continental margins warps up or down in adjustment to strain and tension. Today a little more land may belong to the sea, tomorrow a little less. Always the edge of the sea remains an elusive and indefinable boundary.[*]

Despite such variability between sea and land, many people live and work near the ocean because of commerce, shipping, fishing, and tourism. A 1995 scientific assessment estimates that about 40% of Earth's population lives within 100 km (62 mi), and 49% within 200 km (145 mi), of coastlines. In the United States, about 50% of the people live in areas designated as *coastal* (this includes the Great Lakes). Therefore, an understanding of coastal processes and landforms is important to humanity. And because these processes along coastlines often produce dramatic change, they are essential to consider in planning and development. A World Resources Institute study found as much as 50% of the world's coastlines at some risk of loss or disruption (see "Coastlines at Risk" at http://www.wri.org/wri/indictrs/coastrsk.htm).

The ocean is a vast ecosystem, intricately linked to life on the planet and to life-sustaining systems in the atmosphere, the hydrosphere, and the lithosphere. In the recent *Atlas of the Oceans—The Deep Frontier*, Jean-Michel Cousteau on "The Future of the Ocean" states,

Today, we are coming to better appreciate the extent to which our actions affect an ecosystem—and the people who depend on it—thousands of miles away. The reef fisherman in Fiji is not undone by the local poacher, but by global warming intensified by the driving of a car in downtown Toronto, Canada. Yet these connections are not all bad news. The web of interdependence is built with strands of responsibility and hope. Our ever-expanding ability to communicate across borders and oceans is helping to drive a truly global dialogue about the planet's most pressing environmental challenges.[†]

In this chapter: We begin the chapter with a brief look at our global oceans and seas—1998 was celebrated as The International Year of the Ocean by all United Nations countries (see http://www.yoto98.noaa.gov/). The physical and chemical properties of the sea distinguish it from the waters of the continent. The coastlines are areas of dynamic change and beauty, where oceans and seas confront the land. Coverage includes discussions about tides, waves, coastal erosional and depositional landforms, beaches, barrier islands, and organic processes, including corals, wetlands, salt marshes, and mangroves. A systems framework of specific inputs (components and driving forces), actions (movements and processes), and outputs (results and consequences) organize our discussion of coastal processes. We conclude with a look at the considerable human impact on coastal environments.

Global Oceans and Seas

The ocean is one of Earth's last great scientific frontiers and is of great interest to geographers. Remote sensing from orbit, aircraft, surface vessels, and submersibles is providing a wealth of data and a new capability to understand the oceanic system. The pattern of sea-surface temperatures is presented in Figure 5.11 and ocean currents in Figure 6.22. The world's oceans, their area, volume, and depth, are listed in a table in Figure 7.3. The locations of oceans and major seas are shown and listed alphabetically in Figure 16.1.

Chemical Composition of Seawater

Water is called the "universal solvent," dissolving at least 57 of the 92 elements found in nature. In fact, most natural elements and the compounds they form are found in the seas as dissolved solids, or *solutes*. Thus, seawater is a solution, and the concentration of dissolved solids is called **salinity**.

The ocean remains a remarkably homogeneous mixture. The ratio of individual salts does not change, despite minor fluctuations in overall salinity. In 1874 the British HMS *Challenger* sailed around the world, taking surface and depth measurements and collecting samples of seawater. Analyses of those samples first demonstrated the uniform composition of seawater.

Although ocean chemistry was thought to be fairly constant during the Phanerozoic Eon (the past 540 million

---

[*]"The Marginal World," in *The Edge of the Sea* by Rachel Carson. © 1955 by Rachel Carson, © renewed 1983 by R. Christie (Boston: Houghton Mifflin), p. 11.

[†]*Atlas of the Oceans—The Deep Frontier* by Sylvia Earle. © 2001 by National Geographic Society, text © 2001 by Sylvia Earle (Washington: National Geographic Society), p. 171.

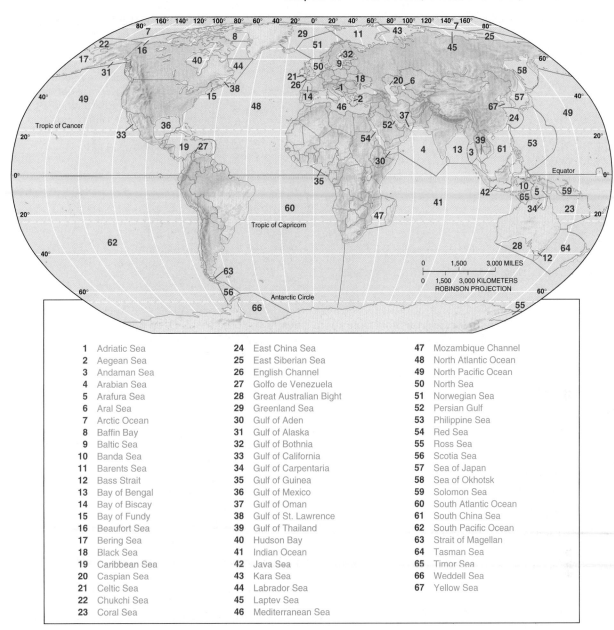

FIGURE 16.1 Principal oceans and seas of the world.
A sea is generally smaller than an ocean and is near a landmass; sometimes the term refers to
a large, inland, salty body of water. Match the alphabetized name and number with its location on the map.

1	Adriatic Sea	**24**	East China Sea	**47**	Mozambique Channel
2	Aegean Sea	**25**	East Siberian Sea	**48**	North Atlantic Ocean
3	Andaman Sea	**26**	English Channel	**49**	North Pacific Ocean
4	Arabian Sea	**27**	Golfo de Venezuela	**50**	North Sea
5	Arafura Sea	**28**	Great Australian Bight	**51**	Norwegian Sea
6	Aral Sea	**29**	Greenland Sea	**52**	Persian Gulf
7	Arctic Ocean	**30**	Gulf of Aden	**53**	Philippine Sea
8	Baffin Bay	**31**	Gulf of Alaska	**54**	Red Sea
9	Baltic Sea	**32**	Gulf of Bothnia	**55**	Ross Sea
10	Banda Sea	**33**	Gulf of California	**56**	Scotia Sea
11	Barents Sea	**34**	Gulf of Carpentaria	**57**	Sea of Japan
12	Bass Strait	**35**	Gulf of Guinea	**58**	Sea of Okhotsk
13	Bay of Bengal	**36**	Gulf of Mexico	**59**	Solomon Sea
14	Bay of Biscay	**37**	Gulf of Oman	**60**	South Atlantic Ocean
15	Bay of Fundy	**38**	Gulf of St. Lawrence	**61**	South China Sea
16	Beaufort Sea	**39**	Gulf of Thailand	**62**	South Pacific Ocean
17	Bering Sea	**40**	Hudson Bay	**63**	Strait of Magellan
18	Black Sea	**41**	Indian Ocean	**64**	Tasman Sea
19	Caribbean Sea	**42**	Java Sea	**65**	Timor Sea
20	Caspian Sea	**43**	Kara Sea	**66**	Weddell Sea
21	Celtic Sea	**44**	Labrador Sea	**67**	Yellow Sea
22	Chukchi Sea	**45**	Laptev Sea		
23	Coral Sea	**46**	Mediterranean Sea		

years), recent evidence suggests that seawater chemistry
has varied over time within a narrow range. The variations
are consistent with changes in sea-floor spreading rates,
volcanism, and sea level. Evidence is gathered from fluid in-
clusions in marine formations such as in limestones and
evaporites, which contain ancient seawaters. The ocean reflects conditions in Earth's environment.

Ocean Chemistry   Ocean chemistry is a result of complex
exchanges among seawater, the atmosphere, minerals, bottom sediments, and living organisms. In addition, significant flows of mineral-rich water enter the ocean through
hydrothermal (hot water) vents in the ocean floor. (These

vents are called "black smokers" for the dense, black
mineral-laden water that spews from them.) The uniformity of seawater results from complementary chemical reactions and continuous mixing—after all, the ocean basins
interconnect, and water circulates among them.

Seven elements account for more than 99% of the dissolved solids in seawater. They are (with their ionic form)
chlorine (as chloride Cl^-), sodium (as Na^+), magnesium
(as Mg^{2+}), sulfur (as sulfate SO_4^{2-}), calcium (as Ca^{2+}),
potassium (as K^+), and bromine (as bromide Br^-). Seawater also contains dissolved gases (such as carbon dioxide,
nitrogen, and oxygen), suspended and dissolved organic
matter, and a multitude of trace elements.

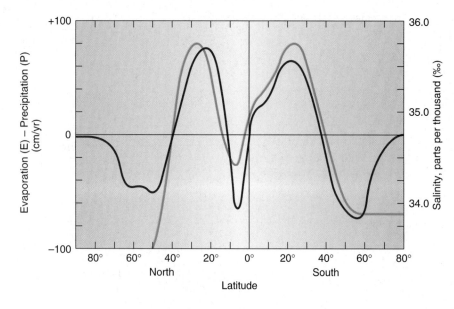

FIGURE 16.2 Variation in ocean salinity by latitude.
Salinity (green line) is principally a function of climatic conditions. Specifically important is the moisture relation expressed by the difference between evaporation and precipitation (E–P) (purple line). Why is salinity higher in the subtropics and lower along the equator? [G. Wüst, 1936.]

Commercially, only sodium chloride (common table salt), magnesium, and bromine are extracted in any significant amount from the ocean. Future mining of minerals from the seafloor is technically feasible, although it remains uneconomical.

Average Salinity: 35‰ There are several ways to express salinity (dissolved solids by volume) in seawater, using the worldwide average value:

- 3.5% (% parts per hundred)
- 35,000 ppm (parts per million)
- 35,000 mg per liter
- 35 g/kg
- 35‰ (‰ parts per thousand), the most common notation

Salinity worldwide normally varies between 34‰ and 37‰; variations are attributable to atmospheric conditions above the water and to the volume of freshwater inflows. In equatorial water, precipitation is great throughout the year, diluting salinity values to slightly lower than average (34.5‰). In subtropical oceans—where evaporation rates are greatest because of the influence of hot, dry subtropical high-pressure cells—salinity is more concentrated, increasing to 36‰. Figure 16.2 plots the difference between evaporation and precipitation and salinity by latitude to illustrate this slight spatial variability. Can you answer the question in the caption?

The term **brine** is applied to water that exceeds the average of 35‰ salinity. **Brackish** applies to water that is less than 35‰ salts. In general, oceans are lower in salinity near landmasses because of freshwater runoff and river discharges. Extreme examples include the Baltic Sea (north of Poland and Germany) and the Gulf of Bothnia (between Sweden and Finland), which average 10‰ or less salinity because of heavy freshwater runoff and low evaporation rates.

On the other hand, the Sargasso Sea, within the North Atlantic subtropical gyre, averages 38‰. The Persian Gulf has a salinity of 40‰ as a result of high evaporation rates

in a nearly enclosed basin. Deep pockets, or "brine lakes," along the floor of the Red Sea and the Mediterranean Sea register up to a salty 225‰.

Physical Structure of the Ocean

The basic physical structure of the ocean is layered, as shown in Figure 16.3. The figure also graphs four key aspects of the ocean, each of which varies with increasing depth: average temperature, salinity, dissolved carbon dioxide, and dissolved oxygen level.

The ocean's surface layer is warmed by the Sun and is wind-driven. Variations in water temperature and solutes are blended rapidly in a *mixing zone* that represents only 2% of the oceanic mass. Below the mixing zone is the *thermocline transition* zone, a more than 1-kilometer-deep region of decreasing temperature gradient that lacks the motion of the surface. Friction at these depths dampens the effect of surface currents. In addition, colder water temperatures at the lower margin tend to inhibit any convective movements.

From a depth of 1–1.5 km (0.6–0.9 mi) to the ocean floor, temperature and salinity values are quite uniform. Temperatures in this *deep cold zone* are near 0°C (32°F). Water in the deep cold zone does not freeze, however, because of its salinity and intense pressures at those depths; seawater freezes at about −2°C (28.4°F) at the surface. The coldest water is along the bottom, except near the poles, where the coldest water may be near or at the surface. Now let us shift to the edge of the sea and examine Earth's coastlines. (Many scientific activities related to the ocean are coordinated and conducted by the National Ocean Service. You can find information about these activities at http://www.nos.noaa.gov/.)

Coastal System Components

We know that the continents were formed over many millions of years. However, most of Earth's coastlines are relatively new, existing in their present state as the setting for

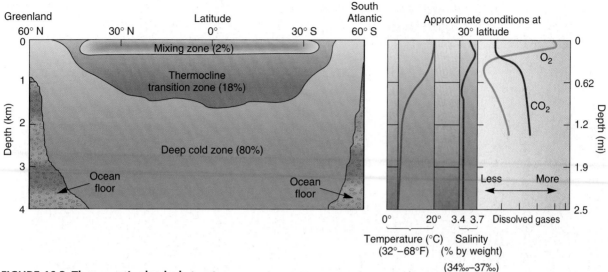

FIGURE 16.3 The ocean's physical structure.
Schematic of average physical structure observed throughout the ocean's vertical profile as sampled along a line from Greenland to the South Atlantic. Temperature, salinity, and dissolved gases are shown plotted by depth.

continuous change. A dynamic equilibrium exists among the energy of waves, tides, wind, and currents, the supply of materials, the slope of the coastal terrain, and the fluctuation of relative sea level. These interactions produce an infinite variety of erosional and depositional features and coastlines of diverse beauty.

Inputs to the Coastal System

Inputs to the coastal environment include many elements we have already discussed:

- *Solar energy* input drives the atmosphere and the hydrosphere. The conversion of insolation to kinetic energy produces prevailing winds, weather systems, and climate.
- *Atmospheric winds*, in turn, generate ocean currents and waves, key inputs to the coastal environment.
- *Climatic regimes*, which result from insolation and moisture, strongly influence coastal geomorphic processes.
- The nature of *coastal rock* is important in determining rates of erosion and sediment production.
- *Human activities* are an increasingly significant input to coastal change.

All of these inputs occur within the ever-present influence of gravity's pull, not only from Earth but also from the Moon and Sun. Gravity provides the potential energy of position for materials in motion and generates the tides.

The Coastal Environment and Sea Level

The coastal environment is called the **littoral zone**, from the Latin word for "shore." Figure 16.4 illustrates the littoral zone and includes specific components discussed later in the chapter. The littoral zone spans some land as well as water. Landward, it extends to the highest water line that occurs on shore during a storm. Seaward, it extends to where water is too deep for storm waves to move sediments on the seafloor–usually around 60 m, or 200 ft, in depth. The specific contact line between the sea and the land is the *shoreline*, although this line shifts with tides, storms, and sea-level adjustments. The *coast* continues inland from high tide to the first major landform change and may include areas considered to be part of the coast in local usage.

Because the level of the ocean varies, the littoral zone naturally shifts position from time to time. A rise in sea level causes submergence of land, whereas a drop in sea level exposes new coastal areas. In addition, uplift and subsidence of the land itself initiates changes in the littoral zone.

Sea level is an important concept. Every elevation you see in an atlas or on a map is referenced to mean sea level. Yet this average sea level changes daily with the tides and over the long term with changes in climate, tectonic plate movements, and glaciation. Thus, *sea level* is a relative term. At present, there exists no international system to determine exact sea level over time. The Global Sea Level Observing System (GLOSS) is an international group actively working on sea level issues. (For copies of their newsletters and other scientific discussion, see http://www. pol.ac.uk/psmsl/gb.html. GLOSS is part of the larger Permanent Service for Mean Sea Level, which you can find at http://www.pol.ac.uk/psmsl/programmes/.)

Mean sea level (MSL) is a value based on average tidal levels recorded hourly at a given site over many years. MSL varies spatially because of ocean currents and waves, tidal variations, air temperature and pressure differences, ocean temperature variations, slight variations in Earth's gravity, and changes in oceanic volume. Presently sea level is rising at a historical rate related to global climate change. For more information, see News Report 16.1.

At present, the overall U.S. MSL is calculated at approximately 40 locations along the coastal margins of the

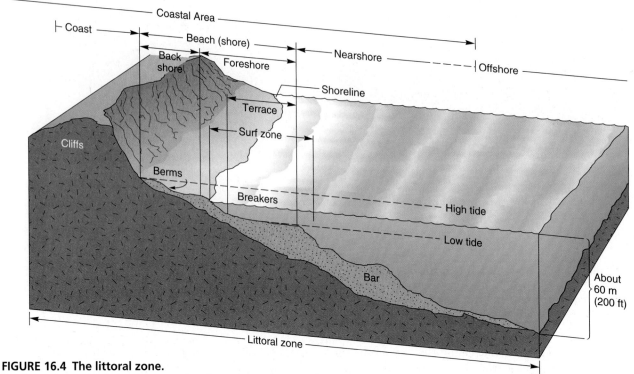

FIGURE 16.4 The littoral zone.
The littoral zone includes the coast, beach, and nearshore environments.

 News Report

News Report 16.1

Sea Level Variations and the Present MSL Increase

Sea level varies along the full extent of North American shorelines. The mean sea level (MSL) of the U.S. Gulf Coast is about 25 cm (10 in.) higher than that of Florida's east coast, which is the lowest in North America. MSL rises northward along the eastern coast, to 38 cm (15 in.) higher in Maine than in Florida. Along the U.S. western coast, MSL is higher than Florida by about 58 cm (23 in.) in San Diego and by about 86 cm (34 in.) in Oregon.

Overall, North America's Pacific coast MSL averages about 66 cm (26 in.) higher than the Atlantic coast MSL. MSL is affected by differences in ocean currents, air pressure and wind patterns, water density, and water temperature.

Over the long term, sea-level fluctuations expose a range of coastal landforms to tidal and wave processes. As average global temperatures cycle through cold or warm climatic spells, the quantity of ice locked up in the ice sheets of Antarctica and Greenland and in hundreds of mountain glaciers can increase or decrease and result in sea level changes accordingly. At the peak of the most recent Pleistocene glaciation about 18,000 b.p. (years before the present), sea level was about 130 m (430 ft) lower than it is today. On the other hand, if Antarctica and Greenland ever became ice-free (ice sheets fully melted), sea level would rise at least 65 m (215 ft) worldwide.

Just 100 years ago sea level was 38 cm (15 in.) lower along the coast of southern Florida. Venice, Italy, has experienced a rise of 25 cm (10 in.) since 1890. During the last century, sea level rose 10–20 cm (4–8 in.), a rate 10 times higher than the average rate during the last 3000 years.

Given these trends and the predicted climatic change, sea level will continue to rise and be potentially devastating for many coastal locations. A rise of only 0.3 m (1 ft) would cause shorelines worldwide to move inland an average of 30 m (100 ft)! This elevated sea level would inundate some valuable real estate along coastlines worldwide. Some 20,000 km² (7800 mi²) of land along North American shores alone would be drowned, at a staggering loss of $650 billion. A 95-cm (3.1-ft) sea-level rise could inundate 15% of Egypt's arable land, 17% of Bangladesh, and many island nations and communities. However, uncertainty exists in these forecasts, and the pace of the rise should be slow.

In 2001 the Intergovernmental Panel on Climate Change (IPCC, discussed in Chapter 10) forecast global mean sea-level rise for this century, given regional variations, as a range from 0.11 to 0.88 m (4.3 to 34.6 in.). The median value of 0.48 m (18.9 in.) is two to four times the rate of increase over the last century. These increases would continue beyond 2100 even if greenhouse gas concentrations are stabilized. Despite any uncertainty, planning should start now along coastlines worldwide because preventive strategies are cheaper than recovery costs from possible destruction. Insurance underwriters have begun the process by refusing coverage for shoreline properties vulnerable to rising sea level.

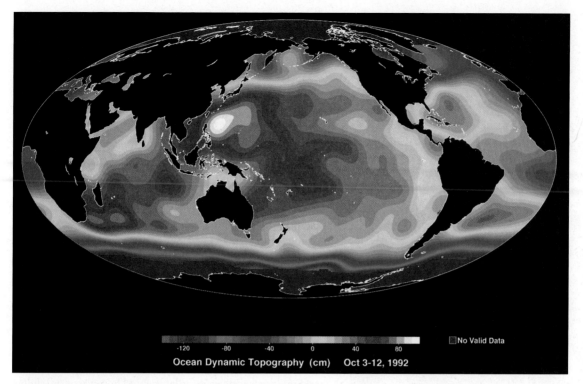

FIGURE 16.5 Ocean topography as revealed by satellite.
Sea-level data recorded by the radar altimeter aboard the *TOPEX/Poseidon* satellite,
October 3–12, 1992. The color scale is given in centimeters above or below Earth's geoid.
The overall relief portrayed in the image is about 2 m (6.6 ft). The maximum sea level is in
the western Pacific Ocean (white). The minimum is around Antarctica (blue and purple).
[Image from JPL and NASA's Goddard Space Flight Center. *TOPEX/Poseidon* is a joint
U.S.–French Mission.]

continent. These sites are being upgraded with new equipment in the Next Generation Water Level Measurement System, using next-generation tide gauges, specifically along the U.S. and Canadian Atlantic coasts, Bermuda, and the Hawaiian Islands. The *NAVSTAR* satellites that make up the Global Positioning System (GPS) make possible the correlation of data within a network of ground- and ocean-based measurements.

Remote-sensing technology augments these measurements, including the *TOPEX/Poseidon* satellite launched in 1992 (see http://topex-www.jpl.nasa.gov/). This satellite has two radar altimeters that measure changes in mean sea level at any one location every 10 days between 66° N and 66° S latitudes. These measurements are made to an astonishing precision of 4.2 cm, 1.7 in. (Figure 16.5)! Spectacular *TOPEX/Poseidon* portraits of El Niño and La Niña in the Pacific are in Focus Study 10.1. Another ocean surface topography satellite launched in December 2001, named *Jason-1*, is extending the science of determining mean sea level, ocean topography, and ocean circulation.

Coastal System Actions

The coastal system is the scene of complex tidal fluctuation, winds, waves, ocean currents, and the occasional impact of storms. These forces shape landforms ranging from gentle beaches to steep cliffs, and they sustain delicate ecosystems.

Tides

Tides are complex daily oscillations in sea level, ranging worldwide from barely noticeable to several meters. They are experienced to varying degrees along every ocean shore around the world. Tidal action is a relentless energy agent for geomorphic change. As tides flood (rise) and ebb (fall), the daily migration of the shoreline landward and seaward causes significant changes that affect sediment erosion and transportation.

Tides also are important in human activities, including navigation, fishing, and recreation. Tides are especially important to ships because the entrance to many ports is limited by shallow water, and thus high tide is required for passage. Tall-masted ships may need a low tide to clear overhead bridges. Tides also exist in large lakes, but because the tidal range is small, tides are difficult to distinguish from changes caused by wind. Lake Superior, for instance, has a tidal variation of only about 5 cm (2 in.).

Causes of Tides Tides are produced by the gravitational pull of both the Sun and the Moon. (Earth's astronomical relation to the Sun and the Moon is discussed in Chapter 2.) The Sun's influence is only about half that of

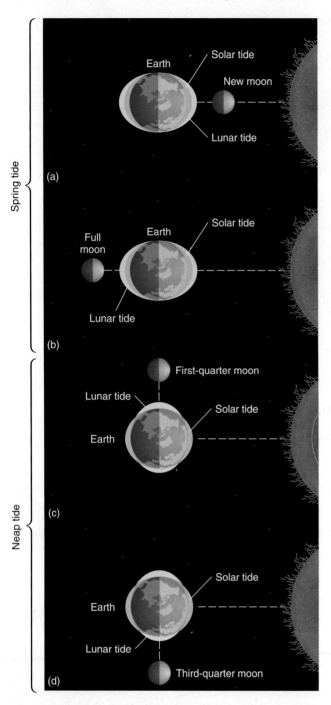

FIGURE 16.6 The cause of tides.
Gravitational relations of Sun, Moon, and Earth combine to produce spring tides (a, b) and neap tides (c, d). (Tides are greatly exaggerated for illustration.)

all experience some stretching as a result of this gravitational pull. The stretching raises large *tidal bulges* in the atmosphere (which we can't see), smaller tidal bulges in the ocean, and very slight bulges in Earth's rigid crust. Our concern here is the tidal bulges in the ocean.

Gravity and inertia are essential elements in understanding tides. *Gravity* is the force of attraction between two bodies. *Inertia* is the tendency of objects to stay still if motionless or to keep moving in the same direction if in motion. The gravitational effect on the side of Earth facing the Moon or Sun is greater than that experienced by the far side, where inertial forces are slightly greater. This difference exists because gravitational influences decrease with distance. It is this difference in the net force of gravitational attraction and inertia that generates the tides.

Figure 16.6a shows the Moon and the Sun in *conjunction* (lined up with Earth), a position in which the sum of their gravitational forces produces a large tidal bulge. The corresponding tidal bulge on Earth's opposite side is primarily the result of the farside water's remaining in position (being left behind) because its inertia exceeds the gravitational pull of the Moon and Sun. In effect, from this inertial point of view, as the nearside water and Earth are drawn toward the Moon and Sun, the farside water is left behind because of the slightly weaker gravitational pull. This arrangement produces the two opposing tidal bulges on opposite sides of Earth.

Tides appear to move in and out along the shoreline, but they do not actually do so. Instead, Earth's surface rotates into and out of the relatively "fixed" tidal bulges as Earth changes its position in relation to the Moon and Sun. Every 24 hours and 50 minutes, any given point on Earth rotates through two bulges as a direct result of this rotational positioning. Thus, every day, most coastal locations experience two high (rising) tides, known as **flood tides**, and two low (falling) tides, known as **ebb tides**. The difference between consecutive high and low tides is considered the *tidal range*.

Spring and Neap Tides   The combined gravitational effect of the Sun and Moon is strongest in the conjunction alignment and results in the greatest tidal range between high and low tides, known as **spring tides**. (*Spring* means to "spring forth"; it has no relation to the season of the year.) Figure 16.6b shows the other alignment that gives rise to spring tides, when the Moon and Sun are at *opposition*. In this arrangement, the Moon and Sun cause separate tidal bulges, affecting the water nearest to each of them. In addition, the left-behind water resulting from the pull of the body on the opposite side augments each bulge.

When the Moon and the Sun are neither in conjunction nor in opposition but are more or less in the positions shown in Figure 16.6c and d, their gravitational influences are offset and counteract each other, producing a lesser tidal range known as **neap tide**. (*Neap* means "without the power of advancing.")

Tides also are influenced by other factors, including ocean basin characteristics (size, depth, and topography), latitude, and shoreline shape. These factors cause a great variety of tidal ranges. For example, some locations may

the Moon's because of the Sun's greater distance from Earth, although it is a significant force. Figure 16.6 illustrates the relation among the Moon, the Sun, and Earth and the generation of variable tidal bulges on opposite sides of the planet.

The gravitational pull of the Moon tugs on Earth's atmosphere, oceans, and lithosphere. The same is true for the Sun, to a lesser extent. Earth's solid and fluid surfaces

(a)

(b)

(c)

FIGURE 16.7 Tidal range and tidal power.
Tidal range is great in some bays and estuaries, such as Halls Harbor near the Bay of Fundy at flood tide (a) and ebb tide (b). The flow of water between high and low tide is ideal for turning turbines and generating electricity, as is done at the Annapolis Tidal Generating Station (c), near the Bay of Fundy in Nova Scotia, in operation since 1984. [Photos by Jeff Newbery.]

experience almost no difference between high and low tides. The highest tides occur when open water is forced into partially enclosed gulfs or bays. The Bay of Fundy in Nova Scotia records the greatest tidal range on Earth, a difference of 16 m (52.5 ft) (Figure 16.7a, b). (For more on tides and tide prediction, contact the Scripps Institution of Oceanography library at **http://scilib.ucsd.edu/sio/tide/**.)

Tidal Power   The fact that sea level changes daily with the tides suggests an opportunity: Could these predictable flows be harnessed to generate electricity? The answer is yes, given the right conditions. Bays and estuaries tend to focus tidal energy, concentrating it in a smaller area than in the open ocean. This circumstance provides an opportunity to construct a dam with water gates, locks to let ships through, and turbines to generate power.

Only 30 locations in the world are suited for tidal power generation. At present, only three are producing electricity. Two are outside North America—a 4-megawatt capacity station in Russia in operation since 1968 (at Kislaya-Guba Bay on the White Sea) and a facility in France operating since 1967 (on the Rance River estuary on the Brittany coast). The tides in the Rance estuary fluctuate up to 13 m (43 ft), and power production has been almost continuous there, providing an electrical generating capacity of a moderate 240 megawatts (about 20% of the capacity of Hoover Dam).

The third area is the Bay of Fundy in Nova Scotia, Canada. At one of several favorable sites on the Bay, the Annapolis Tidal Generating Station was built in 1984. Nova Scotia Power Incorporated operates this 20-megawatt plant (Figure 16.7c). According to the Canadian government, tidal power generation at ideal sites is economically competitive with fossil fuel plants.

Waves

Friction between moving air (wind) and the ocean surface generates undulations of water called **waves**. Waves travel in groups, called *wave trains*. Waves vary widely in scale;

on a small scale, a moving boat creates a wake of small waves; at a larger scale, storms generate large groups of wave trains. At the extreme is the wind wake produced by the presence of the Hawaiian Islands, traceable westward across the Pacific Ocean surface for 3000 km (1865 mi). The islands disrupt the steady trade winds and produce related surface temperature and wind changes.

A stormy area at sea is called a *generating region* for large wave trains, which radiate outward in all directions. The ocean is crisscrossed with intricate patterns of waves traveling in all directions. The waves seen along a coast may be the product of a storm center thousands of kilometers away.

Regular patterns of smooth, rounded waves, called **swells**, are the mature undulations of the open ocean. As waves leave the generating region, wave energy continues

to run in these swells, which can range from small ripples to very large flat-crested waves. A wave leaving a deep-water generating region tends to extend its wavelength horizontally for many meters. Tremendous energy occasionally accumulates to form unusually large waves. One moonlit night in 1933, the U.S. Navy tanker *Ramapo* reported a wave in the Pacific higher than their mainmast, at about 34 m (112 ft)!

As you watch waves in open water, it appears that water is migrating in the direction of wave travel, but only a slight amount of water is actually advancing. It is the *wave energy* that is moving through the flexible medium of water. Water within a wave in the open ocean is simply transferring energy from molecule to molecule in simple cyclic undulations, called *waves of transition* (Figure 16.8). Individual

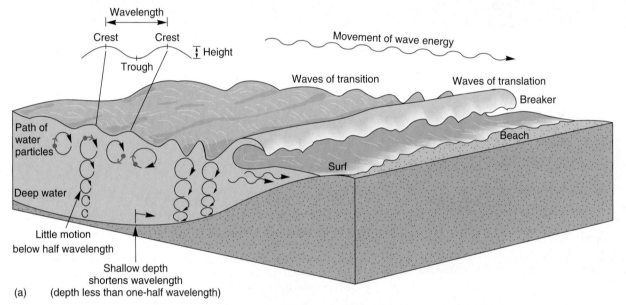

FIGURE 16.8 Wave formation and breakers.
(a) The orbiting tracks of water particles change from circular motions and swells in deep water (waves of transition) to more elliptical orbits near the bottom in shallow water (waves of translation). (b) Cascades of waves attack the shore near Pacific Grove, California. These storm-driven wave heights range from 3 to 6 m (10 to 20 ft). [Photo by Bobbé Christopherson.]

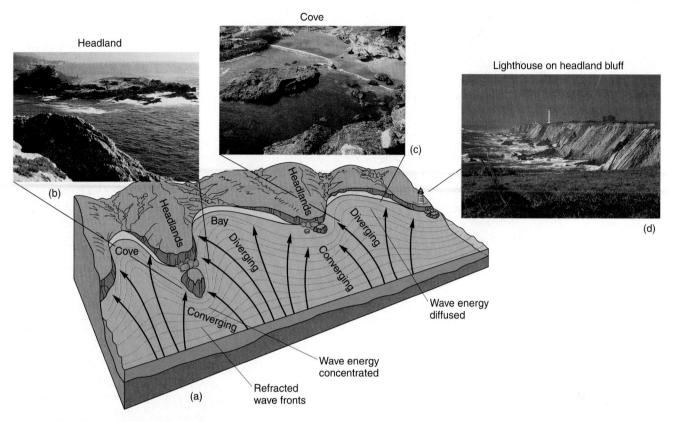

FIGURE 16.9 Coastal straightening.
(a) The process of coastal straightening is brought about by wave refraction. Wave energy is concentrated as it converges on headlands (b) and is diffused as it diverges in coves and bays (c). Headlands are frequent sites for lighthouses such as the light at Point Arena, California (d). [(b), (c) Photos by author; (d) photo by Bobbé Christopherson.]

water particles move forward only slightly, forming a vertically circular pattern.

The diameter of the paths formed by the orbiting water particles decreases with depth. As a deep-ocean wave approaches the shoreline and enters shallower water (10–20 m or 30–65 ft), the orbiting water particles are vertically restricted. This restriction causes more-elliptical, flattened orbits to form near the bottom. This change from circular to elliptical orbits slows the entire wave, although more waves continue arriving. The resultant effects are closer-spaced waves, growing in height and steepness, with sharper wave crests. As the crest of each wave rises, a point is reached when its height exceeds its vertical stability and the wave falls into a characteristic **breaker**, crashing onto the beach (Figure 16.8b).

In a breaker, the orbital motion of transition gives way to elliptical *waves of translation* in which both energy and water move toward shore. The slope of the shore determines wave style. Plunging breakers indicate a steep bottom profile, whereas spilling breakers indicate a gentle, shallow bottom profile. In some areas, unexpected high waves can arise suddenly. It is a good idea to learn to recognize severe wave conditions before you venture along the shore in these areas.

As various wave trains move along in the open sea, they interact by *interference*. These interfering waves sometimes align so that the wave crests and troughs from one wave train are in phase with those of another. When this in-phase condition occurs, the height of the waves is increased, sometimes dramatically. The resulting "killer waves" or "sleeper waves" can sweep in unannounced and overtake unsuspecting victims. Signs along portions of the California, Oregon, Washington, and British Columbia coastline warn beachcombers to watch for "killer waves." On the other hand, out-of-phase wave trains will dampen wave energy at the shore. When you observe the breakers along a beach, the changing beat of the surf actually is produced by the patterns of *wave interference* that occurred in far-distant areas of the ocean.

Wave Refraction   In general, wave action tends to straighten a coastline. Where waves approach an irregular coast, they bend around headlands, which are protruding landforms generally composed of resistant rocks (Figure 16.9). The submarine topography refracts (bends) approaching waves. The refracted energy is focused around headlands and dissipates energy in coves, bays, and the submerged coastal valleys between headlands. Thus, headlands receive the brunt of wave attack along a coastline. This **wave refraction** redistributes wave energy so that different sections of the coastline vary in erosion potential, with the long-term effect of straightening the coast.

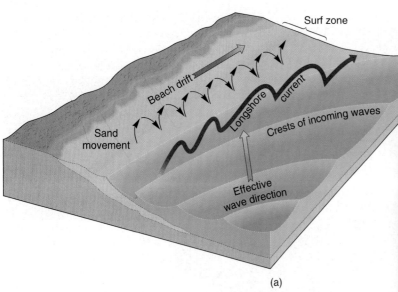

Surf zone

Beach drift

Sand movement

Longshore current

Crests of incoming waves

Effective wave direction

(a)

(b)

FIGURE 16.10 Longshore current and beach drift.
(a) Longshore currents are produced as waves approach the surf zone and shallower water. Longshore and beach drift results as substantial volumes of material are moved along the shore. (b) Processes at work along Point Reyes Beach, Point Reyes National Seashore, California. [(b) Photo by author.]

Figure 16.10 shows waves approaching the coast at an angle, for they usually arrive at some angle other than parallel. As the waves enter shallow water, they are refracted as the shoreline end of the wave slows. The wave portion in deeper water moves faster in comparison, thus producing a current parallel to the coast, zigzagging in the prevalent direction of the incoming waves. This **longshore current**, or *littoral current*, depends on wind direction and wave direction. A longshore current is generated only in the surf zone and works in combination with wave action to transport large amounts of sand, gravel, sediment, and debris along the shore as *longshore drift*.

Particles on the beach also are moved along as **beach drift**, shifting back and forth between water and land with each *swash* and *backwash* of surf. Individual sediment grains trace arched paths along the beach. You have perhaps stood on a beach and heard the sound of myriad sand grains and seawater in the backwash of surf. These dislodged materials are available for transport and eventual deposition in coves and inlets and can represent a significant volume.

Tsunami, or Seismic Sea Wave   An occasional wave that momentarily but powerfully influences coastlines is the tsunami. **Tsunami** is Japanese for "harbor wave," named for its devastating effect where its energy is focused in harbors. Often tsunami are reported incorrectly as "tidal waves," but they have no relation to the tides. They are formed by sudden, sharp motions in the seafloor, caused by earthquakes, submarine landslides, or eruptions of undersea volcanoes. Thus, they properly are called *seismic sea waves*.

A tsunami typically begins when a large undersea disturbance generates a solitary wave of great wavelength, or sometimes a group of two or three long waves. These waves generally exceed 100 km (60 mi) in wavelength, but they are only a meter or so in height. Because of their great wavelength, tsunami are affected by the topography of the deep-ocean floor and are refracted by rises and ridges. They travel at great speeds in deep-ocean water—velocities of 600 to 800 kmph (375 to 500 mph) are not uncommon—but they often pass unnoticed on the open sea because their great length makes the slow rise and fall of water hard to observe.

As a tsunami approaches a coast, however, the shallow water forces the wavelength to shorten. As a result, the wave height may increase up to 15 m (50 ft) or more. Such a wave has the potential to devastate a coastal area, causing property damage and death. For example, in 1992 the citizens of Casares, Nicaragua, were surprised by a 12-m (39-ft) tsunami that took 270 lives, and another hit Indonesia and killed 1000. Two tsunami in 1994 killed 250 in Java, Indonesia, and 62 people in the Philippines; earthquakes triggered all these waves.

Hawai'i is vulnerable to tsunami because of its position in the open central Pacific, surrounded by the ring of fire that outlines the Pacific Basin. The U.S. Army Corps of Engineers has reported 41 damaging tsunami in Hawai'i during the past 142 years—statistically, 1 every 3.5 years.

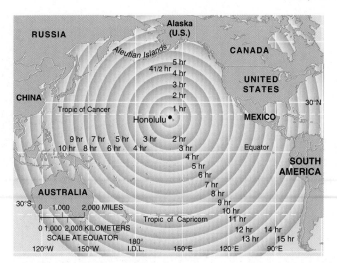

FIGURE 16.11 Tsunami travel times to Honolulu, Hawai'i. [After NOAA.]

Figure 16.11 illustrates tsunami travel times across the Pacific Ocean to Honolulu; for instance, it would take a tsunami 10 hours to reach Hawai'i from the Philippines.

Because tsunami travel such great distances so quickly and are undetectable in open ocean, accurate forecasts are difficult and occurrences are often unexpected. A warning system now is in operation for nations surrounding the Pacific, where the majority of tsunami occur. A warning is issued whenever seismic stations detect a significant quake under water, where it might generate a tsunami. Warnings always should be heeded, despite the many false alarms, for the causes lie hidden beneath the ocean and are difficult to monitor in any consistent manner. (For the tsunami research program, see **http://www.pmel.noaa.gov/ tsunami/**. The tsunami home page is at **http://www. geophys.washington.edu/tsunami/welcome.html**.)

Coastal System Outputs

As you can see, coastlines are very active places, with energy and sediment being continuously delivered to a narrow environment. The action of tides, currents, wind, waves, and changing sea level produces a variety of erosional and depositional landforms. We look first at erosional coastlines, then at depositional coastlines.

Erosional Coastal Processes and Landforms

The active margin of the Pacific Ocean along North and South America is a typical erosional coastline. *Erosional coastlines* tend to be rugged, of high relief, and tectonically active, as expected from their association with the leading edge of drifting lithospheric plates (review the plate tectonics discussion in Chapter 11). Figure 16.12 presents features commonly observed along an erosional coast.

Sea cliffs are formed by the undercutting action of the sea. As indentations slowly grow at water level, a sea cliff becomes notched and eventually will collapse and retreat. Other erosional forms evolve along cliff-dominated coastlines, including *sea caves*, *sea arches*, and *sea stacks*. As erosion continues, arches may collapse, leaving isolated stacks in the water (Figure 16.12b, c). The coasts of southern England and Oregon are prime examples of such erosional landscapes.

Wave action can cut a horizontal bench in the tidal zone, extending from a sea cliff out into the sea. Such a structure is called a **wave-cut platform**, or *wave-cut terrace*. If the relation between the land and sea level has changed over time, multiple platforms or terraces may rise like stairsteps back from the coast. These marine terraces are remarkable indicators of a changing relation between the land and sea, with some terraces more than 370 m (1200 ft) above sea level. A tectonically active region, such as the California coast, has many examples of multiple wave-cut platforms, which at times can be unstable and vulnerable to failure (Figure 16.12d, e).

Depositional Coastal Processes and Landforms

Depositional coasts generally are along land of gentle relief, where sediments from many sources are available. Such is the case with the Atlantic and Gulf coastal plains of the United States, which lie along the relatively passive, trailing edge of the North American lithospheric plate. Erosional processes and inundation, particularly during storm activity, influence depositional coasts.

Figure 16.13 illustrates characteristic landforms deposited by waves and currents. One notable deposition landform is the **barrier spit**, which consists of material deposited in a long ridge extending out from a coast. It partially crosses and blocks the mouth of a bay. Classic examples of a barrier spit include Sandy Hook, New Jersey (south of New York City), and Cape Cod, Massachusetts. The Little Sur River in California (16.13a) and Prion Bay in Southwestern National Park in Tasmania (16.13b) show barrier spits forming partway across the mouths of their respective rivers.

If a spit grows to completely cut off the bay from the ocean and form an inland lagoon, it is called a **bay barrier**, or *baymouth bar*. Spits and barriers are made up of materials that have been eroded and transported by *littoral drift* (beach and longshore drift combined). For much sediment to accumulate, offshore currents must be weak; strong currents carry material away before it can be deposited. Tidal flats and salt marshes are characteristic low-relief features wherever tidal influence is greater than wave action. If these deposits completely cut off the bay from the ocean, an inland **lagoon** is formed.

A **tombolo** occurs when sediment deposits connect the shoreline with an offshore island or sea stack (Figure 16.13c). The tombolo forms when sediments accumulate on an underwater wave-built terrace.

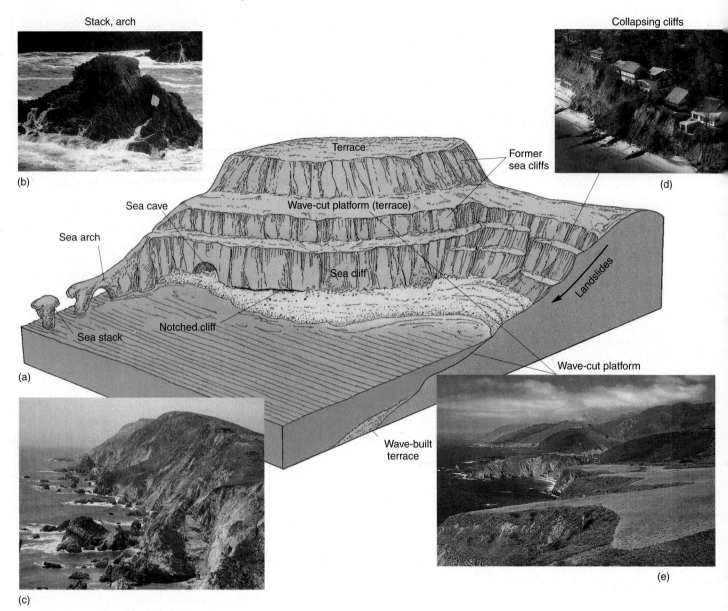

FIGURE 16.12 Erosional coastal features.
(a) Characteristic coastal erosional landforms: caves, arches (b), stacks (c), collapsing cliffs (d), and wave-cut platforms near Bixby Bridge along the Cabrillo Highway (e). [(b), (c) Photos by author; (d) photo by Lowell Georgia/Photo Researchers, Inc.; (e) photo by Bobbé Christopherson.]

Beaches Of all the features associated with a depositional coastline, beaches probably are the most familiar. Beaches vary in type and permanence, especially along coastlines dominated by wave action. Technically, a **beach** is that place along a coast where sediment is in motion, deposited by waves and currents. Material from the land temporarily resides on the beach while it is in active transit along the shore. You may have experienced a beach at some time, along a sea coast, a lake shore, or even a stream. Perhaps you have even built your own "landforms" in the sand, only to see them washed away by the waves: a lesson in erosion.

On average, the beach zone spans from about 5 m (16 ft) above high tide to 10 m (33 ft) below low tide (see

Figure 16.4). However, the specific definition varies greatly along individual shorelines. Worldwide, quartz (SiO_2) dominates beach sands because it resists weathering and therefore remains after other minerals are removed. In volcanic areas, beaches are derived from wave-processed lava. Hawai'i and Iceland, for example, feature some black-sand beaches.

Many beaches, such as those in southern France and western Italy, lack sand and are composed of pebbles and cobbles—a type of "shingle beach." Some shores have no beaches at all; scrambling across boulders and rocks may be the only way to move along the coast. The coasts of Maine and portions of Canada's Atlantic provinces are clas-

FIGURE 16.13 Depositional coastal features.
Characteristic coastal depositional landforms: beaches, spits, barriers, and lagoons. (a) The Little Sur River enters the Pacific Ocean, a barrier spit nearly blocks its way. (b) A barrier spit developing across the mouth of the New River, forming the New River Lagoon and Prion Bay at Southwestern National Park in Tasmania. (c) A tombolo at Point Sur along the central California coast, where sediment deposits connect the shore with an island. [(a), (c) Photos by Bobbé Christopherson; (b) photo by Reg Morrison/Auscape International Pty. Ltd.]

sic examples. These coasts, composed of resistant granite rock, are scenically rugged and have few beaches.

A beach acts to stabilize a shoreline by absorbing wave energy, as is evident by the amount of material that is in almost constant motion (see "sand movement" in Figure 16.10). Some beaches are stable. Others cycle seasonally: They accumulate during the summer, are moved offshore by winter storm waves, forming a submerged bar, and are redeposited onshore the following summer. Protected areas along a coastline tend to accumulate sediment, which can lead to large coastal sand dunes. Prevailing winds often drag such coastal dunes inland, sometimes burying trees and highways.

Maintaining Beaches   Changes in coastal sediment transport can fight against human activities—beaches are lost, harbors are closed, and coastal highways and beach houses can be inundated with sediment. Thus, people use various strategies to interrupt longshore drift and beach drift. The goal is either to halt sand accumulation or to force accumulation in a desired way through construction of engineered structures, called "hard" shoreline protection.

Figure 16.14 illustrates common approaches: a *jetty* to block material from harbor entrances, a *groin* to slow drift action along the coast, and a *breakwater* to create a zone of still water near the coastline. However, interrupting the littoral drift that is the natural replenishment for beaches may

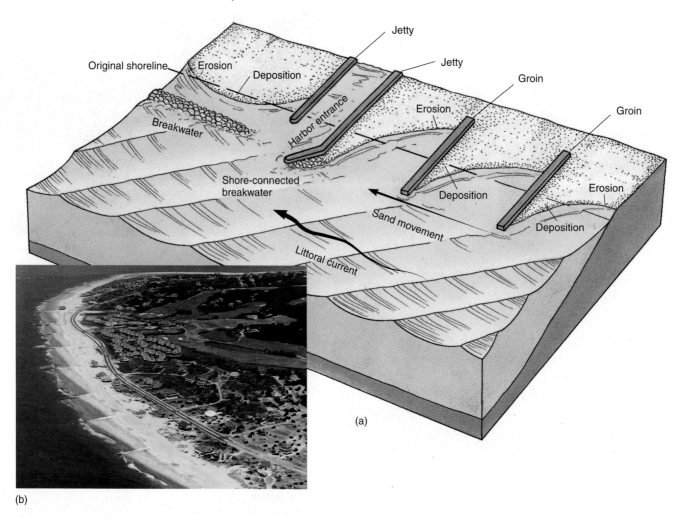

FIGURE 16.14 Interfering with the littoral drift of sand.
(a) Various constructions attempt to control littoral drift (beach drift and longshore drift)
along a coast: breakwater, jetty, and groin. (b) Aerial photograph of coastal constructions.
Note the disruption of sediment movement along the coast at Bald Head Island, North Carolina, looking west across the Cape Fear River mouth. [Photo by Robert H. Goslee.]

News Report |▭▭|

News Report 16.2

Engineers Nourish a Beach

The city of Miami, Florida, and surrounding Dade County have spent almost $70 million since the 1970s in a continuing effort to rebuild their beaches. Sand is transported to the replenishment area from a source area. To maintain a 200-m-wide beach (660-ft), planners determine net sand loss per year and set a schedule for replenishment. In Miami Beach, an 8-year replenishment cycle is maintained. During Hurricane Andrew in 1992, the replenished Miami Beach is thought to have prevented millions of dollars in shoreline structural damage.

Unforeseen environmental impact may accompany the addition of sand to a beach, especially if the sand is from an unmatched source, in terms of its ecological traits. If the new sands do not match (physically and chemically) the existing varieties, disruption of coastal marine life is possible. The U.S. Army Corps of Engineers, which operates the Miami replenishment program, is running out of "borrowing areas" for sand that matches the natural sand of the beach. A proposal to haul a different type of sand from the Bahamas is being studied as to possible environmental consequences.

In contrast to the term *hard structures*, this hauling of sand to replenish a beach is considered "soft" shoreline protection. Enormous energy and material must be committed to counteract the relentless energy that nature expends along the coast. For more on beach nourishment through Duke University, see http://www.eos.duke.edu/Research/psds/psds.htm.

(b)

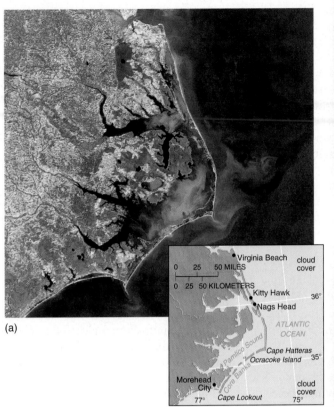

(a)

(c)

FIGURE 16.15 Barrier island chain.
(a) *Landsat* image of barrier island chain along the North Carolina coast. Hurricane Emily swept past Cape Hatteras in August 1993, causing damage and beach erosion. You can see key depositional forms: spit, island, beach, lagoon, and inlet. A sound is a large inlet of the ocean; Pamlico Sound forms an ideal example. (b) View from Cape Hatteras lighthouse illustrates the narrow strand of sand that stands between the ocean and the mainland. (c) Cape Hatteras lighthouse is moved inland in 1999 to safer ground, away from the receding shoreline. [(a) *Terra* MODIS image courtesy of NASA/GSFC; (b) photo by Eric Horan/Liaison Agency, Inc.; (c) © AP/Wide World Photos.]

lead to unwanted changes in sediment distribution down-current. Careful planning and impact assessment should be part of any strategy for preserving or altering a beach.

Beach nourishment refers to the artificial replacement of sand along a beach (see News Report 16.2). Through such efforts, a beach that normally experiences a net loss of sediment will instead show a net gain. Years of human effort and expense to build beaches can be erased by a single storm, as occurred when Hurricane Camille completely eliminated offshore islands along the Gulf Coast in 1969.

Barrier Formations Barrier chains are long, narrow, depositional features, generally of sand, that form offshore roughly parallel to the coast. Common forms are **barrier beaches**, or the broader, more extensive landform, **barrier islands**. Tidal variation in the area usually is moderate to low, with adequate sediment supplies coming from nearby coastal plains. Figure 16.15 illustrates the many features of barrier chains, using North Carolina's famed Outer Banks as an example, including Cape Hatteras, across Pamlico Sound from the mainland. The area presently is designated as one of three national seashore reserves supervised by the National Park Service.

On the landward side of a barrier formation are tidal flats, marshes, swamps, lagoons, coastal dunes, and beaches, visible in Figure 16.15. Barrier beaches appear to adjust to sea level and may naturally shift position from time to time in response to wave action and longshore currents. A break in a barrier forms an inlet that connects a bay with the ocean. The name *barrier* is appropriate, for these formations take the brunt of storm energy and actually shield the mainland. Because of the continuing loss to a barrier island, the famous Cape Hatteras lighthouse was moved inland in 1999 to safer ground (Figure 16.15c). Its new position is about 488 m (1600 ft) from the ocean, or approximately the distance it was in 1870 when it was built—that much sand was lost back to the sea. (See http://www.ncsu.edu/coast/chl/ for details of the extraordinary effort to save this landmark.)

Barrier beaches and islands are quite common worldwide, lying offshore of nearly 10% of Earth's coastlines. Examples are found offshore of Africa, India's eastern coast, Sri Lanka, Australia, Alaska's northern slope, and the shores of the Baltic and Mediterranean Seas. Earth's most extensive chain of barrier islands is along the U.S. Atlantic and Gulf Coast, extending some 5000 km (3100 mi) from Long Island to Texas and Mexico.

Barrier Island Origin and Hazards   Various hypotheses have been proposed to explain the formation of barrier islands. They may begin as offshore bars or low ridges of submerged sediment near shore and then gradually migrate toward shore as sea level rises.

Because many barrier islands seem to be migrating landward, they are an unwise choice for homesites or commercial building. Nonetheless, they are a common choice, even though they take the brunt of storm energy. The hazard represented by the settlement of barrier islands was made graphically clear when Hurricane Hugo assaulted South Carolina in 1989. The storm attacked the Grand Strand barrier islands off the northern half of South Carolina's coastline, most affecting Charleston and the South Strand portion of the islands. The storm made landfall as the worst hurricane to strike there in 35 years.

In the Charleston area, beachfront houses, barrier-island developments, and millions of tons of sand were swept away; up to 95% of the single-family homes in one community were destroyed. The southern portion of one island was torn away, and one of every four homes was destroyed. With increased development and continuing real estate appreciation, each future storm can be expected to cause ever-increasing capital losses.

The barrier islands off the Louisiana shore are disappearing at rates approaching 20 m (65 ft) per year. Hurricanes have taken their toll on these barrier islands. Also, they are affected by subsidence through compaction of Mississippi delta sediments and a changing sea level that is rising at 1 cm (0.4 in.) per year in the region. Hurricane Andrew eroded much sand from these barrier islands in 1992. In 1998, Hurricane Georges destroyed large tracts of the Chandeleur Islands (30 to 40 km, 19 to 25 mi, from the Louisiana and Mississippi Gulf Coast), leaving the mainland with reduced protection (Figure 16.16). The regional office of the USGS predicts that in a few decades the barrier islands may be gone. Louisiana's increasingly exposed wetlands are disappearing at rates of 40 km^2 (15 mi^2) per year, made worse by regional subsidence and a rising sea level.

Biological Processes: Coral Formations

Not all coastlines form by purely physical processes. Some form as the result of biological processes, such as coral growth. A **coral** is a simple marine animal with a small, cylindrical, saclike body (polyp); it is related to other marine invertebrates, such as anemones and jellyfish. Corals secrete calcium carbonate ($CaCO_3$) from the lower half of their bodies, forming a hard calcified external skeleton.

Corals live in a *symbiotic* relationship with algae: They live together in a mutually helpful arrangement, each dependent on the other for survival. Corals cannot photosynthesize, but they do ingest some of their own nourishment. Algae perform photosynthesis and convert solar energy to chemical energy in the system, providing the coral with about 60% of its nutrition and assisting the coral with the calcification process. In return, corals provide the algae with nutrients.

Coral reefs are the most diverse among marine ecosystems. Preliminary estimates of coral species range upward to nine million worldwide, yet, as in most ecosystems in water or on land, biodiversity is declining in these communities.

Figure 16.17 shows the distribution of currently living coral formations. Corals thrive in warm tropical oceans, so the difference in ocean temperature between the western coasts and eastern coasts of continents is critical to their distribution. Western coastal waters tend to be cooler, thereby discouraging coral activity, whereas eastern coastal currents are warmer and thus enhance coral growth. Living colonial corals range in distribution from about 30° N to 30° S. Corals occupy a very specific ecological zone: 10–55 m (30–180 ft) depth, 27‰–40‰ (parts per thousand) salinity, and 18° to 29°C (64°–85°F) water temperature—30°C (86°F) is the upper threshold; above that temperature the corals begin to bleach and die. Corals require clear, sediment-free water and consequently do not locate near the mouths of sediment-charged freshwater streams. For example, note the lack of these structures along the U.S. Gulf Coast.

Coral Reefs   Both solitary and colonial corals exist. It is the colonial corals that produce enormous structures. Their skeletons accumulate, forming a coral rock. Through many generations, live corals near the ocean's surface build on the foundation of older coral skeletons, which in turn may rest upon a volcanic seamount or some other submarine feature built up from the ocean floor. *Coral reefs* form by this process. Thus, a coral reef is a biologically derived sedimentary rock. It can assume one of several distinctive shapes.

In 1842, Charles Darwin hypothesized an evolution of reef formation. He suggested that, as reefs developed

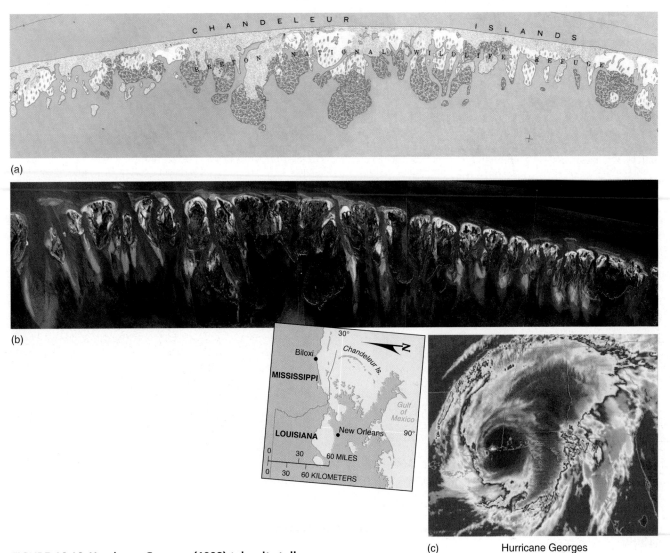

(a)

(b)

(c)   Hurricane Georges

FIGURE 16.16 Hurricane Georges (1998) takes its toll.
This 1998 hurricane eroded huge amounts of sand from the Chandeleur
Islands, just off the Louisiana and Mississippi Gulf Coast. Compare the *before*
topographic map (a) with the *after* aerial photo composite (b). *GOES-8* image
of Hurricane Georges; note the islands just south of the hurricane's eye (c).
[(b) Photo by Aerial Data Service, Earth Imaging, a USGS service company;
topographic map provided by USGS; (c) Image courtesy of NOAA.]

**FIGURE 16.17 Worldwide distribution
of living coral formations.**
Yellow areas include prolific reef growth
and atoll formation. The red dotted line
marks the geographical limits of coral activity. Colonial corals range in distribution
from about 30° N to 30° S. [After J. L.
Davies, *Geographical Variation in Coastal
Development*, Essex, England: Longman
House, 1973. Adapted by permission.]

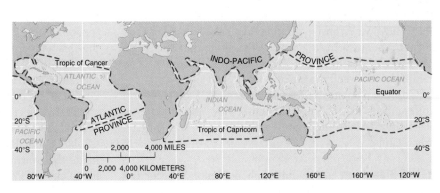

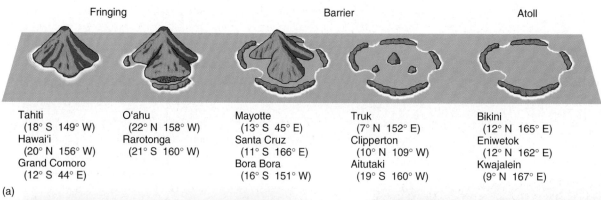

Fringing   Barrier   Atoll

Tahiti
(18° S 149° W)
Hawai'i
(20° N 156° W)
Grand Comoro
(12° S 44° W)

O'ahu
(22° N 158° W)
Rarotonga
(21° S 160° W)

Mayotte
(13° S 45° E)
Santa Cruz
(11° S 166° E)
Bora Bora
(16° S 151° W)

Truk
(7° N 152° E)
Clipperton
(10° N 109° W)
Aitutaki
(19° S 160° W)

Bikini
(12° N 165° E)
Eniwetok
(12° N 162° E)
Kwajalein
(9° N 167° E)

(a)

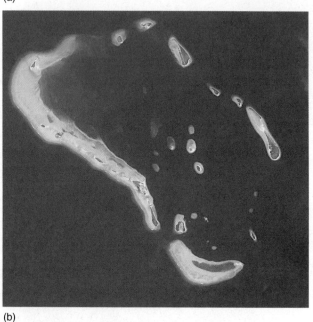

(b)

(c)

FIGURE 16.18 Coral forms.
(a) Common coral formations in a sequence of reef growth formed around a subsiding volcanic island: fringing reefs, barrier reefs, and an atoll. (b) Satellite image of a portion of the Maldive Islands, in the Indian Ocean (5° N 75° W). (c) Aerial photograph of the atolls in Bora Bora, Society Islands (16° S 152° W). [(a) After D. R. Stoddart, *The Geographical Magazine* 63 (1971): 610; (b) *Landsat-7* image courtesy of NASA; (c) photo by Harvey Lloyd/Stock market.]

around a volcanic island and the island itself gradually subsided, an equilibrium between the subsidence of the island and the upward growth of the corals is maintained. This idea, generally accepted today, is portrayed in Figure 16.18. Note the specific examples of each reef stage: *fringing reefs* (platforms of surrounding coral rock), *barrier reefs* (forming enclosed lagoons), and *atolls* (circular, ring-shaped).

The most extensive fringing reef is the Bahamian platform in the western Atlantic (see the photo that opens this chapter), covering some 96,000 km² (37,000 mi²). The largest barrier reef, the Great Barrier Reef along the shore of Queensland State in Australia, exceeds 2025 km (1260 mi) in length, is 16–145 km (10–90 mi) wide, and includes at least 200 coral-formed islands and keys (coral islets or barrier islands).

Coral Bleaching A troubling phenomenon is occurring among corals around the world as normally colorful corals turn stark white by expelling their own nutrient-supplying, colorful algae (red-brown to green), a phenomenon known as *bleaching*. Exactly why the corals eject their symbiotic partner is unknown, for without algae the corals die. Scientist are tracking their unprecedented bleaching and dying worldwide. Locations in the Caribbean Sea and the Indian Ocean, as well as off the shores of Australia, Indonesia, Japan, Kenya, Florida, Texas, and Hawai'i, are experiencing this phenomenon.

Possible causes include local pollution, disease, sedimentation, and changes in salinity. Another cause is the 1 to 2 C° (1.8 to 3.6 F°) warming of sea-surface temperatures, as stimulated by greenhouse warming of the atmosphere. In a report, the *Status of Coral Reefs of the World: 2000* from the Global Coral Reef Monitoring Network, a finding was made that warmer water is a greater threat to corals than local pollution or other environmental assaults (see http://coral.aoml.noaa.gov/gcrmn/).

Coral bleaching worldwide is continuing as average ocean temperatures climb higher, thus linking the issue of climate change to the health of all living coral formations. By the end of 2000, approximately 30% of reefs were lost, especially following the record El Niño event of 1998. (For more information and Internet links, see **http://www. usgs.gov/coralreef.html.**) The *World Atlas of Coral Reefs* in 2001 summarized:

> Humans are thus bringing new pressures to bear on the world's coral reefs and driving more profound changes, more rapidly, than any natural impact has ever done. Overfishing has become so widespread that there are few, if any, reefs in the world which are not threatened. . . . From onshore a much greater suite of damaging activities is taking place. Often remote from reefs, deforestation, urban development, and intensive agriculture are now producing vast quantities of sediments and pollutants which are pouring into the sea and rapidly degrading coral reefs. . . . A further specter overshadowing the world of coral reefs is that of global climate change.*

Wetlands, Salt Marshes, and Mangrove Swamps

Some coastal areas have great *biological productivity* (plant growth, spawning grounds for fish, shellfish, and other organisms) stemming from trapped organic matter and sed-

---

*M. D. Spalding, C. Ravilious, and E. P. Green, and UNEP/WCMC, *World Atlas of Coral Reefs* (Berkeley: University of California Press, 2001), p. 11.

iments. Such a rich coastal marsh environment can greatly outproduce a wheat field in raw vegetation per acre. Thus, coastal marshes can support rich wildlife habitats. Unfortunately, these wetland ecosystems are quite fragile and are threatened by human development.

Wetlands are saturated with water enough of the time to support *hydrophytic vegetation* (plants that grow in water or very wet soil). Wetlands usually occur on poorly drained soils. Geographically, they occur not only along coastlines but also as northern bogs (peatlands with high water tables), as potholes in prairie lands, as cypress swamps (with standing or gently flowing water), as river bottomlands and floodplains, and as arctic and subarctic environments that experience permafrost during the year.

Coastal Wetlands

Coastal wetlands are of two general types—salt marshes and mangrove swamps. In the Northern Hemisphere, **salt marshes** tend to form north of the 30th parallel, whereas **mangrove swamps** form equatorward of that line. This distribution is dictated by the occurrence of freezing conditions, which control the survival of mangrove seedlings. Roughly the same latitudinal limits apply in the Southern Hemisphere.

Salt marshes usually form in estuaries and behind barrier beaches and spits. An accumulation of mud produces a site for the growth of *halophytic* (salt-tolerant) plants. This vegetation then traps additional alluvial sediments and adds to the salt marsh area. Because salt marshes are in the intertidal zone (between the farthest reaches of high and low tides), sinuous, branching channels are produced as tidal waters flood into and ebb from the marsh (Figure 16.19).

Sediment accumulation on tropical coastlines provides the site for mangrove trees, shrubs, and other small trees.

FIGURE 16.19 Coastal salt marsh.
Salt marshes are productive ecosystems commonly occurring poleward of 30° latitude in both hemispheres. This is Gearheart Marsh, part of the Arcata Marsh system on the Pacific Coast in northern California. [Photo by Bobbé Christopherson.]

(a)

(b)

(c)

FIGURE 16.20 Mangroves.
Mangroves tend to grow equatorward of 30° latitude. (a) Mangroves along the East Alligator River (12° S latitude) in Kakadu National Park, Northern Territory, Australia. (b) Mangroves retain sediments and can form anchors for island formations, such as Aldabra Island (9° S), Seychelles. (c) Mangroves in the Florida Keys. [(a) Photo by Belinda Wright/ DRK Photo; (b) photo by Wolfgang Kaehler Photography; (c) photo by author.]

The prop roots of the mangrove are constantly finding new anchorages. The roots are visible above the water line but reach below the water surface, providing a habitat for a multitude of specialized life forms. Mangrove swamps often secure and fix enough material to form islands (Figure 16.20).

Marco Island   The development of Florida's Marco Island is an example of mangrove loss due to urbanization. Marco Island is on the southwestern Florida coast about 24 km (15 mi) south of Naples. Figure 16.21a shows Marco Island as it was in 1952: approximately 5300 acres of subtropical mangrove habitat and barrier island terrain, formed from river sediments and old shell and reef fragments.

Development began in 1962, despite tropical storms that occasionally assault the area, as Hurricane Donna did in 1960. Development included artificial landfill for housing sites and general urbanization of the entire island. Scientists regarded the island and estuarine habitat as highly productive and unique in its mixture of mangrove species, bird species, and a productive aquatic environment. Consequently, numerous challenges and court actions were initiated in attempts to halt development. But construction continued, and by 1984 Marco Island was completely developed (Figure 16.21b). Compare the 1952 digitized inventory and 1984 color-infrared aerial photograph to see the impact of urban development on this island. Note that the only remaining natural community is restricted to a very limited portion along the extreme perimeter. Today, geographic information system models greatly facilitate such comparative analyses. (See Career Link 16.1.)

The World Resources Institute and the U.N. Environment Programme estimate that from pre-agricultural times to today, mangrove losses are running between 40% (examples, Cameroon and Indonesia) and nearly 80% (examples, Bangladesh and Philippines). Deliberate removal was a common practice by many governments in the early days of settlement because of a falsely conceived fear of disease or pestilence in these swamplands.

Human Impact on Coastal Environments

Development of modern societies is closely linked to estuaries, wetlands, barrier beaches, and coastlines. Estuaries are important sites of human settlement because they pro-

(a)

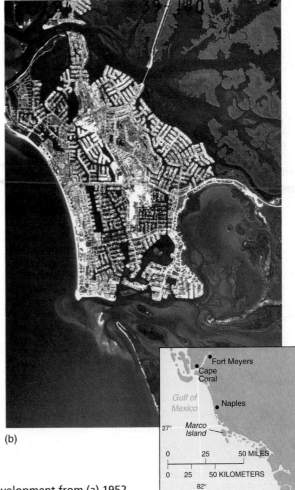

(b)

FIGURE 16.21 Mangrove loss at Marco Island.
Digitized inventory of Marco Island, Florida, showing dramatic development from (a) 1952 survey map and (b) 1984 color-infrared photograph of the island. This type of analysis is similar to GIS (Geographic Information System) studies and is invaluable in assessing human impacts. For more on GIS see Figure 1.28 and the Career Link at the end of this chapter. [From S. Patterson, *Mangrove Community Boundary Interpretation and Detection of Areal Changes on Marco Island, Florida*, Biological Report 86 (10), for National Wetlands Research Center, U.S. Fish and Wildlife Service, August 1986, pp. 23, 49.]

FIGURE 16.22 Beach erosion dooms houses on Long Island, New York.
[Photo by Mark Wexler/Woodfin Camp & Associates.]

vide natural harbors, a food source, and convenient sewage and waste disposal. Society depends on daily tidal flushing of estuaries to dilute the pollution created by waste disposal. Thus, the estuarine and coastal environment is vulnerable to abuse and destruction if development is not carefully planned. We are approaching a time when most barrier islands will be developed and occupied.

Barrier islands and coastal beaches do migrate over time. Houses that were more than a mile from the sea in the Hamptons along the southeastern shore of Long Island, New York, are now within only 30 m (100 ft) of it. The time remaining for these homes can be quickly shortened by a single hurricane or by a sequence of storms (Figure 16.22).

Despite our understanding of beach and barrier-island migration, the effect of storms, and warnings from scientists and government agencies, coastal development proceeds. Society behaves as though beaches and barrier islands are stable, fixed features, or as though they can be engineered to be permanent. Experience has shown that severe erosion generally cannot be prevented, as we have seen with the cliffs of southern California, the shores of the Great Lakes, the New Jersey shore, the devastation of Hurricane Hugo, or the devastation along the Gulf Coast and the disappearances of islands caused by Hurricane Georges. Shoreline planning is the topic of Focus Study 16.1.

The key to protective environmental planning and zoning is to *allocate responsibility and cost in the event of a disaster*. An ideal system places a hazard tax on land, based on assessed risk and restricts the government's responsibility to fund reconstruction or an individual's right to reconstruct on frequently damaged sites. Comprehensive mapping of erosion-hazard areas would help avoid the ever-increasing costs from recurring disasters.

Focus Study 16.1

An Environmental Approach to Shoreline Planning

Coastlines are places of wonderful opportunity. They also are zones of specific constraints. Poor understanding of this resource and a lack of environmental analysis often go hand in hand. The late ecologist and landscape architect Ian McHarg, in *Design with Nature*, discusses the New Jersey shore. He shows how proper understanding of a coastal environment could have avoided problems from major storms and coastal development. Much of what is presented here applies to coastal areas elsewhere. Figure 1 illustrates the New Jersey shore from ocean to back bay. Let us walk across this landscape and discover how it should be treated under ideal conditions.

Beaches and Dunes—
Where to Build?

We begin our walk at the water's edge and proceed across the beach. Sand beaches are the primary natural defense against the ocean; they act as bulwarks against the pounding of a stormy sea. The shoreline tolerates recreation, but not construction, because of its shifting, changing nature during storms, daily tidal fluctuations, and the potential effects of rising sea level. Beaches are susceptible to pollution and require environmental protection to control nearshore dumping of dangerous materials. In recent years, high water levels in the Great Lakes and along the southern California coast attest to the vulnerability of shorelines to erosion

and inundation. Wave erosion attacks cliff formations and undermines, bit by bit, the foundations of houses and structures built too close to the edge (Figure 2).

Walking inland, we encounter the primary dune. It is even more sensitive than the beach: It is fragile, easily disturbed, and vulnerable to erosion, and it cannot tolerate the passage of people trekking to the beach. Delicate plants struggle to hold the sand in place. Primary dunes are like human-made dikes in the Netherlands: They are the primary defense against the sea, so development or heavy traffic should not disturb the primary dunes. Carefully controlled access points to the beach should be designated, and re-

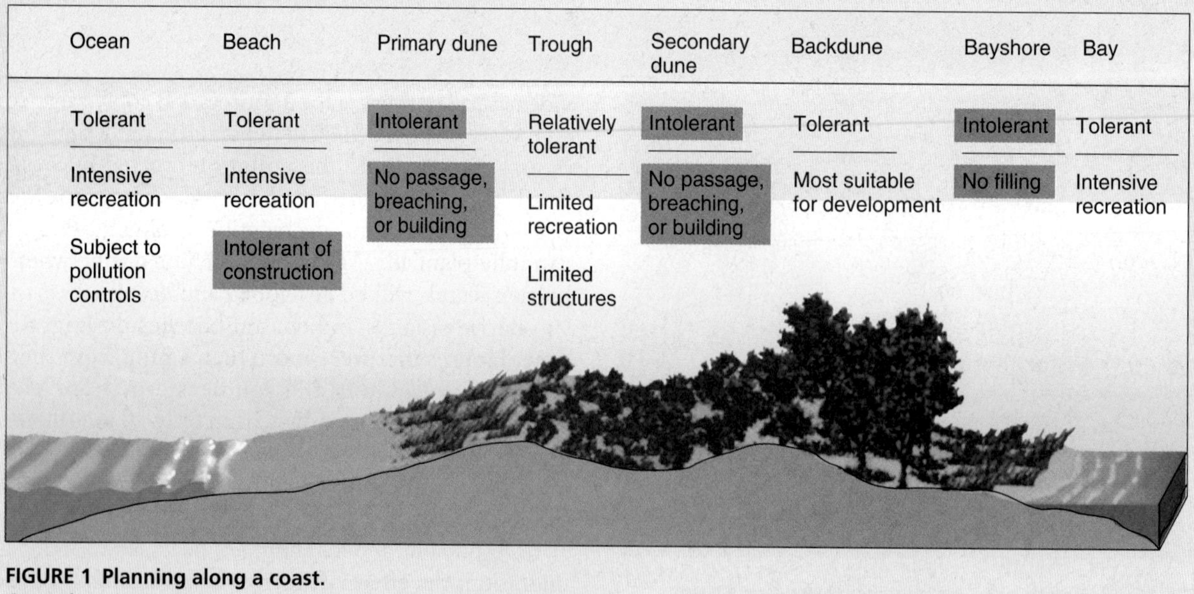

Ocean	Beach	Primary dune	Trough	Secondary dune	Backdune	Bayshore	Bay
Tolerant	Tolerant	Intolerant	Relatively tolerant	Intolerant	Tolerant	Intolerant	Tolerant
Intensive recreation	Intensive recreation	No passage, breaching, or building	Limited recreation	No passage, breaching, or building	Most suitable for development	No filling	Intensive recreation
Subject to pollution controls	Intolerant of construction		Limited structures				

FIGURE 1 Planning along a coast.
Coastal environment—a planning perspective from ocean to bay. [After *Design with Nature* by Ian McHarg. © 1969 by Ian L. McHarg. Adapted by permission of Ian McHarg.]

(continued)

Focus Study 16.1 *(continued)*

(a)

(b)

FIGURE 2 Examples of poor planning.
(a) Coastal erosion and a failed house foundation. (b) Seemingly futile efforts to halt collapse of coastal cliffs. [Photos by (a) author; (b) Bobbé Christopherson.]

stricted access should be enforced, for even foot traffic can cause destruction. (See the beach stabilization example in Figure 15.5.)

The trough behind the primary dune is relatively tolerant of limited recreation and building. The plants that fix themselves to the surface send roots down to fresh groundwater reserves and anchor the sand in the process. Thus, if construction should inhibit the surface recharge of that water supply, the natural protective ground cover could die and destabilize the environment. Or subsequent saltwater intrusion might contaminate well water. Clearly, groundwater resources and the location of recharge aquifers must be considered in planning.

Behind the trough is the secondary dune, a second line of defense against the sea. It, too, is tolerant of some use yet is vulnerable to destruction. Next is the backdune, more suitable for development than any zone between it and the sea. Further inland are the bayshore and the bay, where no dredging or filling and only limited dumping of treated wastes and toxics should

be permitted. This zone is tolerant to intensive recreation. In reality, of course, the opposite of such careful assessment and planning prevails.

A Scientific View and a Political Reality

Before an intensive government study of the New Jersey shore was completed in 1962, no analysis of coastal hazards had been done, outside of academic circles. What is common knowledge to geographers, botanists, biologists, and ecologists in the classroom and laboratory still has not filtered through to the general planning and political processes. As a result, on the New Jersey shore and along much of the Atlantic and Gulf Coasts, improper development of the fragile coastal zone (primary dunes, trough, and secondary dunes) led to extensive destruction during storms in 1962, 1992, and numerous other times.

Scientists estimate that the coastlines will continue their retreat in some cases tens of meters in a few decades. Such estimates persist along the entire East Coast around to the Gulf Coast. Society must reconcile ecology and

economics if these coastal environments are to be sustained.

South Carolina enacted the Beach Management Act of 1988 (modified 1990) to apply some of McHarg's principles. In its first 2 years, more than 70 lawsuits protested the act as an invalid seizure of private property without compensation.

Destructive or threatening hurricane activity is helping to drive the importance of such cooperative planning and proactive laws. Similar measures in other states have faced the same difficult path. Implementation of any planning process is problematic, political pressure is intense, and results are mixed. Ian McHarg voiced an optimistic hope: "May it be that these simple ecological lessons will become known and incorporated into ordinance [law] so that people can continue to enjoy the special delights of life by the sea."*

*From *Design with Nature* by Ian McHarg, p. 17. © 1969 by Ian L. McHarg. Published by Bantam Doubleday Dell Publishing Group, Inc.

Summary and Review—The Oceans, Coastal Processes, and Landforms

● *Describe* **the chemical composition of seawater and the physical structure of the ocean.**

Water is called the "universal solvent," dissolving at least 57 of the 92 elements found in nature. Most natural elements and the compounds they form are found in the seas as dissolved solids. Seawater is a solution, and the concentration of dissolved solids is called **salinity**. **Brine** exceeds the average 35‰ (parts per thousand) salinity; **brackish** applies to water that is less than 35‰. The ocean is divided by depth into a narrow mixing zone at the surface, a thermocline transition zone, and the deep cold zone.

salinity (p. 490)
brine (p. 492)
brackish (p. 492)

1. Describe the salinity of seawater: its composition, amount, and distribution.
2. Analyze the latitudinal distribution of salinity as shown in Figure 16.2. Why is salinity less along the equator and greater in the subtropics?
3. What are the three general zones relative to physical structure within the ocean? Characterize each by temperature, salinity, dissolved oxygen, and dissolved carbon dioxide.

● *Identify* **the components of the coastal environment and** *list* **the physical inputs to the coastal system, including tides and mean sea level.**

The coastal environment is called the **littoral zone** and exists where the tide-driven, wave-driven sea confronts the land. Inputs to the coastal environment include solar energy, wind and weather, ocean currents and waves, climatic variation, and the nature of coastal rock. **Mean sea level (MSL)** is based on average tidal levels recorded hourly at a given site over many years. MSL varies spatially because of ocean currents and waves, tidal variations, air temperature and pressure differences, ocean temperature variations, slight variations in Earth's gravity, and changes in oceanic volume.

Tides are complex daily oscillations in sea level, ranging worldwide from barely noticeable to many meters. Tides are produced by the gravitational pull of both the Moon and the Sun. Most coastal locations experience two high (rising) **flood tides**, and two low (falling) **ebb tides** every day. The difference between consecutive high and low tides is the tidal range. **Spring tides** exhibit the greatest tidal range, when the Moon and Sun are either in conjunction or opposition. **Neap tides** produce a lesser tidal range.

littoral zone (p. 493)
mean sea level (MSL) (p. 493)
tide (p. 495)
flood tide (p. 496)
ebb tide (p. 496)
spring tide (p. 496)
neap tide (p. 496)

4. What are the key terms used to describe the coastal environment?
5. Define mean sea level. How is this value determined? Is it constant or variable around the world? Explain.
6. What interacting forces generate the pattern of tides?
7. What characteristic tides are expected during a new Moon or a full Moon? During the first-quarter and third-quarter phases of the Moon? What is meant by a flood tide? An ebb tide?
8. Is tidal power being used anywhere to generate electricity? Explain briefly how such a plant would utilize the tides to produce electricity. Are there any sites in North America? Where are they?

● *Describe* **wave motion at sea and near shore and** *explain* **coastal straightening as a product of wave refraction.**

Friction between moving air (wind) and the ocean surface generates undulations of water that we call **waves**. Wave energy in the open sea travels through water, but the water itself stays in place. Regular patterns of smooth, rounded waves, called **swells**, are the mature undulations of the open ocean. Near shore, the restricted depth of water slows the wave, forming *waves of translation*, in which both energy and water actually move forward toward shore. As the crest of each wave rises, the wave falls into a characteristic **breaker**.

Wave refraction redistributes wave energy so that different sections of the coastline vary in erosion potential. Headlands are eroded, whereas coves and bays receive materials, with the long-term effect of straightening the coast. As waves approach a shore at an angle, refraction produces a **longshore current** of water moving parallel to the shore. This current produces the *longshore drift* of sand, sediment, and gravel and assorted materials moving along the beach as **beach drift**—together these materials make up the overall *littoral drift*.

A **tsunami** is a seismic sea wave triggered by an undersea landslide or earthquake. It travels at great speeds in the open sea and gains height as it comes ashore, posing a coastal hazard.

wave (p. 497)
swell (p. 498)
breaker (p. 499)
wave refraction (p. 499)
longshore current (p. 500)
beach drift (p. 500)
tsunami (p. 500)

9. What is a wave? How are waves generated, and how do they travel across the ocean? Does the water travel with

the wave? Discuss the process of wave formation and transmission.

10. Describe the refraction process that occurs when waves reach an irregular coastline. Why is the coastline straightened?

11. Define the components of beach drift and the longshore current and longshore drift.

12. Explain how a seismic sea wave attains such tremendous velocities. Why is it given a Japanese name?

● *Identify* **characteristic coastal erosional and depositional landforms.**

An *erosional coast* features wave action that cuts a horizontal bench in the tidal zone, extending from a sea cliff out into the sea. Such a structure is called a **wave-cut platform**, or *wave-cut terrace*. In contrast, *depositional coasts* generally are located along land of gentle relief, where depositional sediments are available from many sources. Characteristic landforms deposited by waves and currents are a **barrier spit** (material deposited in a long ridge extending out from a coast); a **bay barrier**, or *baymouth bar* (a spit that cuts off the bay from the ocean and forms an inland **lagoon**); a **tombolo** (where sediment deposits connect the shoreline with an offshore island or sea stack); and a **beach** (land along the shore where sediment is in motion, deposited by waves and currents). A beach helps to stabilize the shoreline, although it may be unstable seasonally.

> wave-cut platform (p. 501)
> barrier spit (p. 501)
> bay barrier (p. 501)
> lagoon (p. 501)
> tombolo (p. 501)
> beach (p. 502)

13. What is meant by an erosional coast? What are the expected features of such a coast?

14. What is meant by a depositional coast? What are the expected features of such a coast?

15. How do people attempt to modify littoral drift? What strategies are used? What are the positive and negative impacts of these actions?

16. Describe a beach—its form, composition, function, and evolution.

17. What success has Miami had with beach replenishment? Is it a practical strategy?

● *Describe* **barrier islands and their hazards as they relate to human settlement.**

Barrier chains are long, narrow, depositional features, generally of sand, that form offshore roughly parallel to the coast. Common forms are **barrier beaches**, and the broader, more extensive **barrier islands**. Barrier formations are transient coastal features, constantly on the move, and they are a poor, but common, choice for development.

> barrier beach (p. 505)
> barrier island (p. 505)

18. On the basis of the information in the text and any other sources at your disposal, do you think barrier islands and beaches should be used for development? If so, under what conditions? If not, why not?

19. After the Grand Strand off South Carolina was destroyed by Hurricane Hazel in 1954, settlements were rebuilt, only to be hit by Hurricane Hugo 35 years later, in 1989. Why do these recurring events happen to human populations?

● *Assess* **living coastal environments: corals, wetlands, salt marshes, and mangroves.**

A **coral** is a simple marine invertebrate that forms a hard calcified external skeleton. Over generations, corals accumulate in large reef structures. Corals live in a *symbiotic* (mutually helpful) relationship with algae; each is dependent on the other for survival.

Wetlands are lands saturated with water that support specific plants adapted to wet conditions. They occur along coastlands and inland in bogs, swamps, and river bottomlands. Coastal wetlands form as **salt marshes** poleward of the 30th parallel in each hemisphere and **mangrove swamps** equatorward of these parallels.

> coral (p. 506)
> wetlands (p. 509)
> salt marsh (p. 509)
> mangrove swamp (p. 509)

20. How are corals able to construct reefs and islands?

21. Describe a trend in corals that is troubling scientists, and discuss some possible causes.

22. Why are the coastal wetlands poleward of 30° N and S latitude different from those that are equatorward? Describe the differences.

● *Construct* **an environmentally sensitive model for settlement and land use along the coast.**

Coastlines are zones of specific constraints. Poor understanding of this resource and a lack of environmental analysis often go hand in hand, producing frequent disasters to coastal ecosystems and real property losses. Society must reconcile ecology and economics if these coastal environments are to be sustained.

23. Describe the condition of Marco Island, Florida. Was a rational model used to assess the environment prior to development? What economic and political forces were involved?

24. What type of environmental analysis is needed for rational development and growth in a region like the New Jersey shore? Evaluate South Carolina's approach to coastal hazards and protection.

The *Geosystems* Home Page provides on-line resources for this chapter on the World Wide Web. To begin: Once on the Home Page, click on this textbook, scroll the Table of Contents menu, select this chapter, and click "Begin." You will find self-tests that are graded, review exercises, specific updates for items in the chapter, and many links to interesting related pathways on the Internet under "Destinations." *Geosystems* is at **http://www.prenhall.com/christopherson**.

Critical Thinking

A. This chapter includes the following statement, "The key to protective environmental planning and zoning is to *allocate responsibility and cost in the event of a disaster*. An ideal system places a hazard tax on land, based on assessed risk, and restricts the government's responsibility to fund reconstruction or an individual's right to reconstruct at frequently damaged sites." What do you think about this as a policy statement? How would you approach implementing such a strategy? In what way could you use geographic information systems (GIS), as described in Chapter 1, to survey, assess, list owners, and follow taxation status for a vulnerable stretch of coastline?

B. Under "Destinations" in Chapter 16 of the *Geosystems* Home Page there is a link called "Coral Reefs." Sample some of the links on this page. Do you find any information about the damage to and bleaching of coral reefs reported in 1998? Which places in the world? Are there some suspect causes presented in "Bleaching Hot Spots" or any of the "Coral Reef Alliance" references?

Fran E. Evanisko, Chief Cartographer and GIS Adjunct Faculty

In Figure 16.21 we saw how layers of information can be compared so that the nature of landscape changes and human impacts can be analyzed. In this case, a digitized inventory base map was compared to urbanization and development over a 32-year period. Today, geographic information systems (GIS) is a critical tool for model building of complex spatial systems and the development of integrated maps showing a composite overlay of information. GIS enhances our ability to find relationships and linkages in such studies. See Figure 1.28 and News Report 1.2, both in Chapter 1 for more. Let's take a brief look at someone actively working in the GIS field.

Watching Fran at the computer is a lively experience. Using current GIS software from ESRI as an example, he quickly keyboards lines of instruction, accessing original programs of his own creation that pull data from various sources. In a matter of minutes he creates a detailed map showing roads, hydrologic features (streams and canals), township and section lines, and named towns. Fran is working with maps such as these to identify land ownership patterns.

As he creates, he spots something that isn't right and attacks the problem, talking and keyboarding rapidly. Each step is an exciting challenge as he snaps through lines of instruction. He doesn't quit until all inconsistencies are resolved—his powerful determination is key to his professional success.

Fran Evanisko discovered geography at Alan Hancock Community College in 1969 and was hooked when he first looked into stereoscopic lenses and saw 3-D stereopair landscape photos. He continued his education in geography at the University of California, Santa Barbara (UCSB), focusing on remote sensing, satellite imagery, and maps.

Fran graduated in geography and went to work for the Klamath National Forest as a cartographic technician, directing their geographic information

FIGURE 1 Fran E. Evanisko, Chief Cartographer and GIS Adjunct. Fran Evanisko is Chief Cartographer for the U.S. Bureau of Land Management and a GIS Adjunct Faculty member at American River College, both positions in Sacramento, California. [Photo by author.]

systems (GIS) team, when GIS was just emerging. For graduate work he returned to UCSB. He used the work experience with remote sensing in his Masters' thesis to establish methods for analysis of diverse forest vegetation. Fran, along with other graduate students and professors, worked on the original grant proposal that led to UCSB becoming a national center for geographic information analysis.

In 1992, he began a career at the U.S. Bureau of Reclamation (BLM) in California. His responsibilities include the design, development, and maintenance of large spatial data bases for California. In this capacity Fran has written many programs that convert a variety of data bases into forms that can be imported into GIS software. He has worked to automate map production at the BLM and frequently provides tech support for GIS to other field and state offices. GIS software and technology is rapidly evolving and Fran is looked to by many to assist with these changes.

Fran expresses carefully considered and strong opinions about most aspects of GIS and cartography: "Cartography is going through a revolution. Historically, maps provided a place where spatial data were recorded and served as a communication device. GIS

has changed all this. Today, in the digital world, spatial data bases are separate from maps. Cartography is the interaction with and utilization of spatial data. Rather than a map being static and old when printed, GIS maps are a dynamic, changeable, georelational medium."

When American River College (ARC), Sacramento, began the process of designing a new GIS Certificate Program, Fran was asked to serve on the GIS Advisory Committee. According to Fran, "To merely produce graduates that can push buttons of automated software is bad practice. Students need to know conceptual basics, cartography, major issues in the field, and enough to track errors and uncertainty that can develop in a GIS." Fran now teaches several of the eleven courses comprising the ARC program, now featuring two different certificates.

Fran is a geographer who emphasizes the spatial aspects of the discipline. He pushes the limits of GIS to develop new utilities and is actively working with the new programming challenge of ESRI's ARCGIS. Further, he translates his energy and drive into geographic science at the BLM and dynamic classroom teaching in the American River College GIS program.

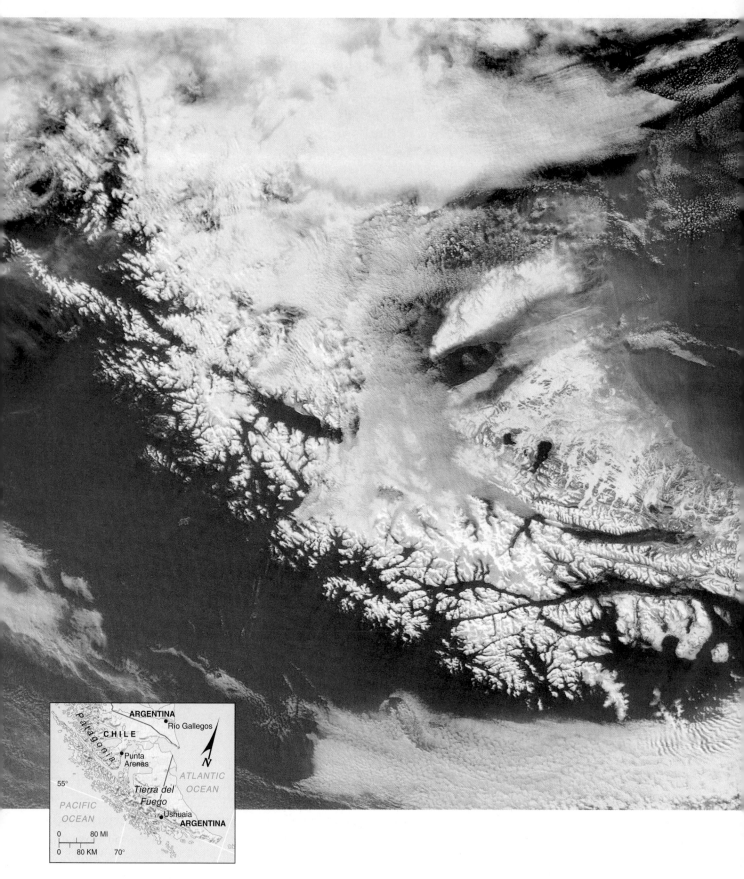

Tierra del Fuego ("Land of Fire") is the southern tip of South America and the terminus of the Andes Mountains. Peaks exceed 2400 m (7875 ft) and feature alpine glaciers and glacial topography first charted by navigator Ferdinand Magellan in 1520. [*Terra* MODIS sensor image courtesy of MODIS Land Rapid Response Team, NASA/GSFC, June 12, 2001.]

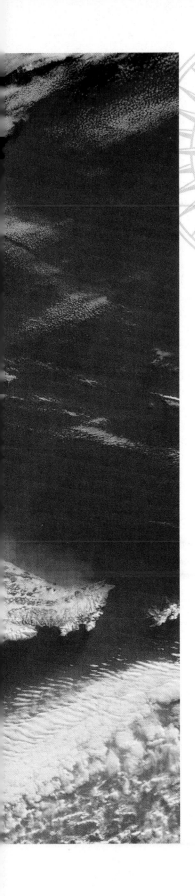

17

Glacial and Periglacial Processes and Landforms

Key Learning Concepts

After reading the chapter, you should be able to:

- *Differentiate* between alpine and continental glaciers and *describe* their principal features.
- *Describe* the process of glacial ice formation and *portray* the mechanics of glacial movement.
- *Describe* characteristic erosional and depositional landforms created by alpine glaciation and continental glaciation.
- *Analyze* the spatial distribution of periglacial processes and *describe* several unique landforms and topographic features related to permafrost and frozen ground phenomena.
- *Explain* the Pleistocene ice age epoch and related glacials and interglacials and *describe* some of the methods used to study paleoclimatology.

About 77% of Earth's freshwater is frozen, with the bulk of that ice sitting restlessly in just two places—Greenland and Antarctica. The remaining ice covers various mountains and fills some alpine valleys. More than 29 million cubic kilometers (7 million cubic miles) of water is tied up as ice. These deposits provide an extensive frozen record of Earth's climatic history over the past several million years and perhaps some clues to its climatic future. Worldwide glacial ice is in retreat, melting at rates exceeding anything in the ice record. In the European Alps alone some 75% of the glaciers have receded in the past 50 years, losing more than 50% of their ice mass since 1850. At this rate, the European Alps will have only 20% of their pre-industrial glacial ice left by 2050.

Significant to towns and cities is that many mountain cliffs and slopes are held together by permanently frozen conditions (permafrost). With the meltdown underway these slopes are thawing, making rockfalls and landslides an increasing hazard probability.

In this chapter: We focus on Earth's extensive ice deposits—their formation, movement, and the ways in which they produce various erosional and depositional landforms. Glaciers, transient landforms themselves, leave in their wake a variety of landscape features. The fate of glaciers is intricately tied to change in global temperature, which ultimately concerns us all. We discuss the methods used to decipher past climates—the science of *paleoclimatology*—and the clues for understanding the future.

We examine the cold, near-glacial world of permafrost and periglacial processes. Approximately 20% of Earth's land area is subject to freezing conditions and frost action characteristic of periglacial regions. Increasing global temperatures are more pronounced at higher latitudes. These areas near former and existing glaciers are reminders of the last ice age.

Rivers of Ice

A **glacier** is a large mass of ice, resting on land or floating as an ice shelf in the sea adjacent to land. Glaciers are not frozen lakes or groundwater ice. Instead, they form by the continual accumulation of snow that recrystallizes under its own weight into an ice mass. Glaciers are not stationary; they move slowly under the pressure of their own great weight and the pull of gravity. In fact, they move slowly in streamlike patterns, merging as tributaries into large rivers of ice, as you can see in Figure 17.1.

Today, about 11% of Earth's land area is dominated by these slowly flowing rivers of ice. But during colder episodes in the past, as much as 30% of continental land was covered by glacial ice. Through these "ice ages," below-freezing temperatures prevailed at lower latitudes than they do today, allowing snow to accumulate year after year.

Glaciers form in areas of permanent snow, both at high latitudes and at high elevations at any latitude. A **snowline** is the lowest elevation where snow can survive year-round; specifically, it is the lowest line where winter snow accumulation persists throughout the summer. Glaciers form on some high mountains along the equator, such as in the Andes Mountains of South America and on Mount Kilimanjaro in Tanzania, Africa. On equatorial mountains, the snowline is around 5000 m (16,400 ft); on midlatitude mountains, such as the European Alps, snowlines average 2700 m (8850 ft); and in southern Greenland, snowlines are as low as 600 m (1970 ft). (For Internet links to glaciers, see the Glacier Page at **http://southpole.rice.edu/**, Global Land Ice Measurements from Space at **http://wwwflag.wr.usgs.gov/GLIMS/glimshome.html**, or the National Snow and Ice Data Center at **http://www-nsidc.colorado.edu/**. *Landsat*-7 images are listed at **http://www.emporia.edu/earthsci/gage/glacier7.htm**.)

Glaciers are as varied as the landscape itself. They fall within two general groups, based on their form, size, and flow characteristics: alpine glaciers and continental glaciers.

Alpine Glaciers

With few exceptions, a glacier in a mountain range is called an **alpine glacier**, or *mountain glacier*. The name comes from the Alps mountains of central Europe, where such glaciers abound. Alpine glaciers form in several subtypes. One prominent type is a *valley glacier*, literally a river of ice confined within a valley that originally was formed by stream action. Such glaciers range in length from only 100 m (325 ft) to more than 100 km (60 mi).

FIGURE 17.1 Rivers of ice.
Apline glaciers merge from adjoining glacial valleys in the northeast region of Ellesmere Island (80° N 75° W) in the Canadian Arctic. [*Terra* ASTER image courtesy of University of Alberta, NASA/GSFC/ERS-DAC/JAROS and the U.S./Japan ASTER Science Team, July 31, 2000.]

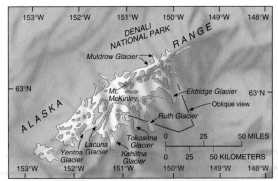

FIGURE 17.2 Glaciers in south-central Alaska.
Oblique infrared (false-color) image of Eldridge and Ruth Glaciers, with Mount McKinley at upper left, in the Alaska Range of Denali National Park. Photo made at 18,300 m (60,000 ft). [Alaska High Altitude Aerial Photography from EROS Data Center, USGS.]

In Figure 17.2, at least a half dozen valley glaciers are identifiable in the high-altitude photograph of the Alaska Range. Several are named on the map, specifically the Eldridge and Ruth Glaciers, which fill valleys as they flow from source areas near Mount McKinley.

As a valley glacier flows slowly downhill, the mountains, canyons, and river valleys beneath its mass are profoundly altered by its erosive passage. Some of the debris created by the glacier's excavation is transported on the ice, visible as dark streaks and bands being transported for deposition elsewhere; other portions of its debris load are carried within or along its base (see Figures 17.1 and 17.7c).

Most alpine glaciers originate in a mountain *snowfield* that is confined in a bowl-shaped recess. This scooped-out erosional landform at the head of a valley is called a **cirque**. A glacier that forms in a cirque is called a *cirque glacier*. Several cirque glaciers may jointly feed a valley glacier, as shown on the topographic map and photograph in Figure 17.3. The Kuskulana Glacier, which flows off the map to the west, is supplied by numerous cirque glaciers feeding several tributary valley glaciers. In the satellite image in Figure 17.1, how many valley glaciers join the main glacier?

Wherever several valley glaciers pour out of their confining valleys and coalesce at the base of a mountain range, a *piedmont glacier* is formed and spreads freely over the lowlands. Malaspina Glacier is an excellent example of a piedmont glacier (Figure 17.4). The debris deposits on the surface of the ice form beautiful streaked patterns as the glacier fans out over the coastal plain and into Yakutat Bay. A *tidal glacier*, such as the Columbia Glacier on Prince William Sound in Alaska, ends in the sea, calving (breaking off) to form floating ice called **icebergs** (Figure

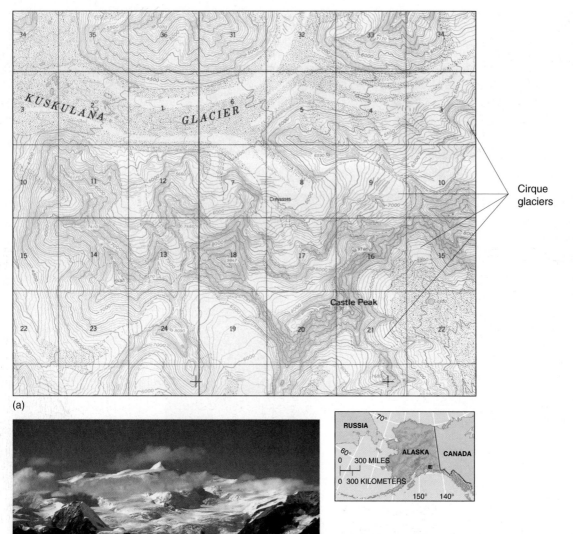

(a)

(b)

FIGURE 17.3 Topographic map from southeastern Alaska.
(a) Numerous cirque glaciers depicted on a topographic map of southeastern Alaska (McCarthy C-7 Quadrangle, 1:63,360 scale, 100-ft contour interval). The blue areas represent active glaciers, and brown-striped areas on the ice are moraines. (b) A dramatic scene depicts the active Kuskulana glacier in the Wrangell Range. [(a) Courtesy of U.S. Geological Survey; (b) photo by Steve McCutcheon/Visuals Unlimited.]

17.4b and c). Icebergs usually form wherever glaciers meet the ocean. (See Glaciers of Prince William Sound at http://www.alaska.net/~sea/glacier.html.)

Continental Glaciers

On a much larger scale than individual alpine glaciers, a continuous mass of ice is called a **continental glacier**. In its most extensive form, it is an **ice sheet**. Most of Earth's glacial ice exists in the ice sheets that blanket 80% of Greenland (1.8 million cubic kilometers, or 0.43 million cubic miles) and 90% of Antarctica (13.9 million cubic kilometers, or 3.3 million cubic miles). Antarctica alone has 91% of all the glacial ice on the planet.

The Antarctic and Greenland ice sheets have such enormous mass that large portions of each landmass beneath the ice are isostatically depressed (pressed down by weight) below sea level. Each ice sheet is more than 3000 m (10,000 ft) deep, burying all but the highest peaks.

Two additional types of continuous ice cover associated with mountain locations are *ice caps* and *ice fields*. An ice cap is roughly circular and, by definition, covers an area of less than 50,000 km² (19,300 mi²). An **ice cap** completely buries the underlying landscape. The volcanic island of Iceland features several ice caps, such as the Vatnajökull Ice Cap in Figure 17.5a's *Landsat* image. Volcanoes lie beneath these icy surfaces. Iceland's Grímsvötn Volcano erupted in 1996, producing large quantities of

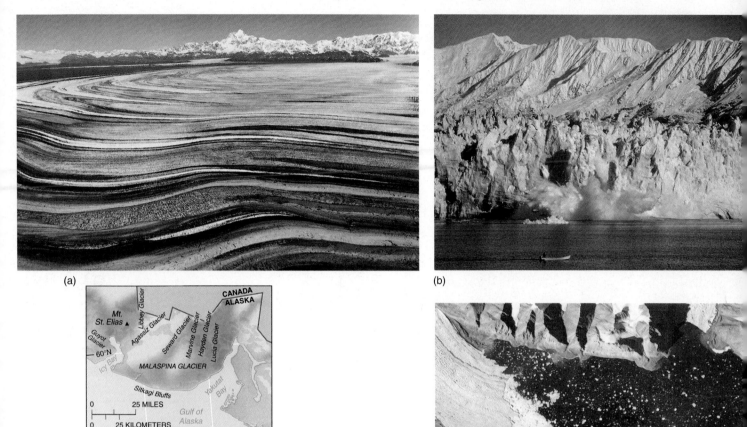

FIGURE 17.4 Piedmont glacier and ice calving into the sea.
(a) Malaspina Glacier in southeastern Alaska. This piedmont glacier is nearly the size of Rhode Island and covers some 5000 km² (1950 mi²). About 70% of Malaspina's ice comes from the Seward Glacier (upper center), which is fed by ice fields in the St. Elias Mountain Range.
(b) When glaciers reach the sea, large blocks begin calving off, forming icebergs, such as from Hubbard Glacier into Disenchantment Bay, Alaska. (c) Eugenie Glacier floats out on the water and breaks up into icebergs in Dobbin Bay, Ellesmere Island, Canada. [(a) High-altitude photo courtesy of AeroMap U.S., Inc., Anchorage, Alaska; (b) photo by Steve McCutcheon/Visuals Unlimited. (c) *Terra* ASTER sensor image courtesy of University of Alberta, NASA/GSFC/MITI/ERSDAC/JAROS, and U.S./Japan ASTER Science Team.]

melted glacial water and floods, a flow called *jökulhlaup* by Icelanders.

An **ice field** is not extensive enough to form the characteristic dome of an ice cap; instead, it extends in a characteristic elongated pattern in a mountainous region. A fine example is the Patagonian ice field of Argentina and Chile, one of Earth's largest. It attains only 90 km (56 mi) width, but stretches 360 km (224 mi), from 46° to 51° S latitude (equivalent in latitude from South Dakota to central Manitoba in the Northern Hemisphere). In an ice field, ridges and peaks are visible above the buried terrain; the term *nunatak* refers to these peaks, visible in Figure 17.5b. Note that the chapter-opening image is south of this ice field.

Continuous ice sheets or ice caps are drained by rapidly moving, solid *ice streams* that form around their periphery, moving to the sea or to lowlands. Such frozen ice streams flow from the edges of Greenland and Antarctica, through stationary and slower moving ice. An *outlet glacier* flows out from an ice sheet or ice cap but is constrained by a mountain valley or pass (Figure 17.6).

(a)

(b)

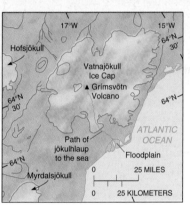

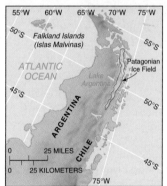

FIGURE 17.5 Ice cap and ice field.
(a) The Vatnajökull ice cap in southeastern Iceland (*jökull* means "ice cap" in Danish). Note the location of the Grímsvötn Volcano on the map and the path of the floods to the sea. (b) The southern Patagonian ice field of Argentina. [(a) *Landsat* image from NASA; (b) photo by Cosmonauts G. M. Greshko and Yu V. Romanenko, *Salyut 6*.]

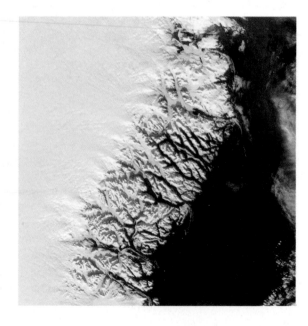

FIGURE 17.6 Greenland's east coast.
Ice streams from the ice sheet, outlet glaciers flowing to the sea, and fjords dominate the east coast of Greenland. Center of the image is near the Arctic Circle (66.5° N). [*Terra* MODIS sensor image courtesy of MODIS Land Rapid Response Team, NASA/GSFC, November 2, 2001.]

Glacial Processes

A glacier is a dynamic body, moving relentlessly downslope at rates that vary within its mass, excavating the landscape through which it flows. The mass is dense ice that is formed from snow and water through a process of compaction, recrystallization, and growth. A glacier's *mass budget* consists of net gains or losses of this glacial ice, which determine whether the glacier expands or retreats. Let us now look at glacial ice formation, mass balance, movement, and erosion before we discuss the fascinating landforms produced by these processes.

Formation of Glacial Ice

Consider for a moment the nature of ice. You may be surprised to learn that ice is both a mineral (an inorganic natural compound of specific chemical makeup and crystalline structure) and a rock (a mass of one or more minerals). Ice is a frozen fluid, a trait that it shares with igneous rocks. The accumulation of snow in layered deposits is similar to sedimentary rock formations. To give birth to a glacier, snow and ice are transformed under pressure, recrystallizing into a type of metamorphic rock. Glacial ice is a remarkable material!

The essential input to a glacier is snow that accumulates in a snowfield, a glacier's accumulation zone (Figure 17.7a). Snowfields typically are at the highest elevation of an ice sheet, ice cap, or head of a valley glacier, usually in a cirque. Avalanches from surrounding mountain slopes can add to the snowfield.

As the snow accumulation deepens in sedimentary-like layers, the increasing thickness results in increased weight and pressure on underlying ice. Rain and summer snowmelt then contribute water, which stimulates further melting, and that meltwater seeps down into the snowfield and refreezes.

Snow that survives the summer and into the following winter begins a slow transformation into glacial ice. Air spaces among ice crystals are pressed out as snow packs to a greater density. The ice recrystallizes and consolidates under pressure. In a transition step to glacial ice, snow becomes **firn**, which has a compact, granular texture.

As this process continues, many years pass before dense glacial ice is produced. Formation of **glacial ice** is analogous to metamorphic processes: Sediments (snow and firn) are pressured and recrystallized into a dense metamorphic rock (glacial ice). In Antarctica, glacial ice formation may take 1000 years because of the dryness of the climate (minimal snow input), whereas in wet climates the time is reduced to just a few years because of rapid, constant snow input to the system.

Glacial Mass Balance

A glacier is an open system, with *inputs* of snow and *outputs* of ice, meltwater, and water vapor. At its upper end, a glacier is fed by snowfall and other moisture in the *accumulation zone* (Figure 17.7a). This area ends at the **firn line**, indicating where the winter snow and ice accumulation survived the summer melting season. Toward a glacier's lower end, it is wasted (reduced) through several processes: melting on the surface, internally, and at its base; ice removal by deflation (wind); the calving of ice blocks; and sublimation (recall from Chapter 7 that this is the direct evaporation of ice). Collectively, these losses are called **ablation**.

The zone where accumulation gain balances ablation loss is the *equilibrium line* (Figure 17.7b). This area of a glacier generally coincides with the firn line. A glacier achieves *positive net balance* of mass—grows larger—during cold periods with adequate precipitation. In warmer times, the equilibrium line migrates up-glacier, and the glacier retreats—grows smaller—because of its *negative net balance*. Internally, gravity continues to move a glacier forward even though its lower terminus might be in retreat owing to ablation. The mass-balance losses from the South Cascade glacier provides a case in point (News Report 17.1).

Highways of ice that flow from accumulation areas high in the mountains are marked by trails of transported debris called *moraines* (Figure 17.7c, d). A lateral moraine accumulates along the sides as rock is loosened by abrasion and plucking from the valley walls. A medial moraine forms down the middle when two glaciers merge and their lateral moraines combine, forming an elongated deposit on and within the combined glacier.

Tributary valley glaciers merge to form a *compound valley glacier*. The flowing movement of a *compound valley glacier* is different from that of a river with tributaries. Tributary glaciers flow into a compound glacier and merge alongside one another by extending and thinning rather than by blending, as do rivers. Each tributary maintains its own patterns of transported debris (dark streaks, visible on the photo and image in Figure 17.7).

Glacial Movement

Like all minerals, ice has specific properties of hardness, color, melting point (quite low in the case of ice), and brittleness. We know the properties of ice best from those brittle little cubes in the freezer. But glacial ice has different properties, depending on its location in a glacier. In a glacier's depths, glacial ice behaves in a plastic manner, distorting and flowing in response to weight and pressure from above and the degree of slope below. In contrast, the glacier's upper portion is more like the everyday ice we know, quite brittle.

A glacier's rate of flow ranges from almost nothing to a kilometer or two per year on a steep slope. The rate of snow accumulation in the formation area is critical to the pace of glacial movement.

Glaciers are not rigid blocks that simply slide downhill. The greatest movement within a valley glacier occurs *internally*, below the rigid surface layer, where the underlying zone moves plastically forward (Figure 17.8a). At the same time, the base creeps and slides along, varying its speed with temperature and the presence of any lubricating water or saturated sediment beneath the ice. This *basal*

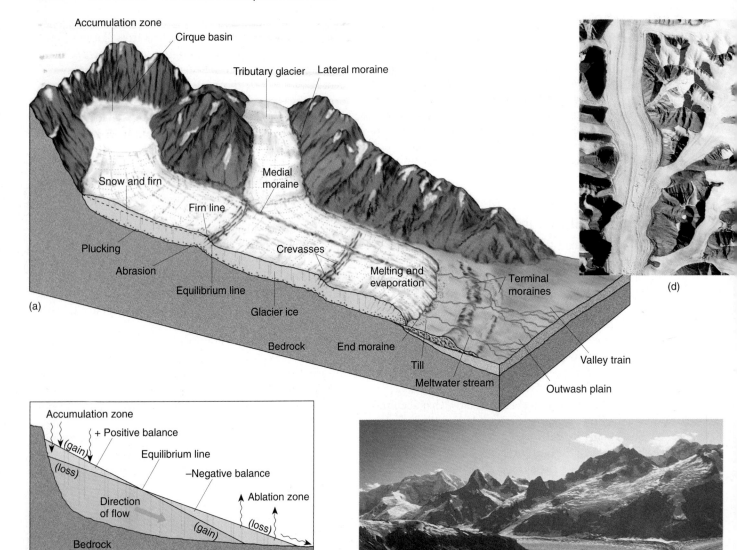

FIGURE 17.7 A retreating alpine glacier and mass balance.
(a) Cross section of a typical retreating alpine glacier. (b) Annual mass balance of a glacial system, showing how the relation between accumulation and ablation controls the location of the equilibrium line. (c) Johns Hopkins Glacier, Glacier Bay National Park, Alaska, and, (d) a valley glacier off the Agassiz ice cap, Ellesmere Island, Canada, demonstrates many of the features in the illustration. Note the formation of the medial moraine where two glaciers merge. [(c) Photo by Frank S. Balthis. (d) *Terra* ASTER sensor image courtesy of University of Alberta, NASA/GSFC/MITI/ERSDAC/JAROS, and U.S./Japan ASTER Science Team.]

slip usually is much slower than the internal plastic flow of the glacier, so the upper portion of the glacier flows ahead of the lower portion. The difference in speed stretches the glacier's brittle surface ice.

In addition, the pressure may vary in response to unevenness in the landscape beneath the ice. Basal ice may be melted by compression at one moment, only to refreeze later. This process is called *ice regelation*, meaning to re-

freeze, or re-gel. Regelation is important because it facilitates downslope movement and because the process incorporates rock debris into the glacier. Consequently, a glacier's basal ice layer, which can extend tens of meters above its base, has a much greater debris content than the ice above.

A flowing glacier can develop vertical cracks known as **crevasses** (Figure 17.8b and c). Crevasses result from fric-

News Report 17.1

South Cascade Glacier Loses Mass

The net mass balance of the South Cascade Glacier in Washington State demonstrated significant losses between 1955 and 2001. In just one year (September 1991 to October 1992), the terminus of the glacier retreated 38 m (125 ft), resulting in major changes to the surface and sides of the glacier. There was more than a 2% loss of the glacier's mass in just that one year. The net accumulation and net wastage are illustrated in Figure 1.

The reasons for these losses are complex. Possible causes are increasing average air temperature and decreasing precipitation. A comparison of the trend of this glacier's mass balance with that of others in the world shows that temperature changes apparently are causing widespread reductions in middle- and lower-elevation glacial ice.

The heavy snowfall in the Northwest associated with the La Niña episode is reflected in the positive mass balance figure for 1999 and 2000. The present wastage (ice loss) from alpine glaciers worldwide is thought by some to contribute more than 25% to the rise in sea level.

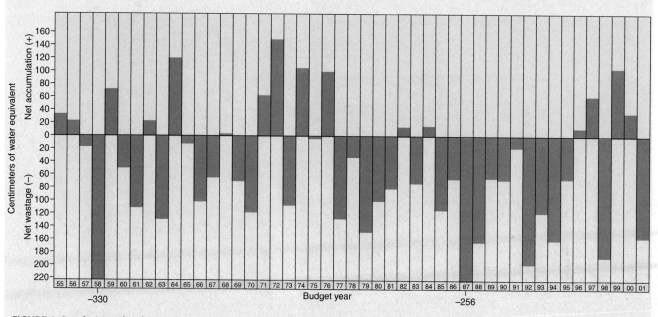

FIGURE 1 South Cascade Glacier net mass balance, 1955–2001.
A negative net mass balance has dominated this shrinking glacier since 1955. Data in centimeters (2.54 cm per in.). [Data from R. M. Krimmel, *Water, Ice, Meteorological, and Speed Measurements at South Cascade Glacier, Washington, 2001 Balance Year*. USGS Water Resources Report, Tacoma, Washington, 2001; and personal communication.]

tion with valley walls, or tension from stretching as the glacier passes over convex slopes, or compression as the glacier passes over concave slopes. Traversing a glacier, whether an alpine glacier or an ice sheet, is dangerous because a thin veneer of snow sometimes masks the presence of a crevasse. *where topography is Bad.*

Glacier Surges   Although glaciers flow plastically and predictably most of the time, some will lurch forward with little or no warning in a **glacier surge**. A surge is not quite as abrupt as it sounds; in glacial terms, a surge can be tens of meters per day. The Jakobshavn Glacier on the western Greenland coast, for example, is known to move between 7 and 12 km (4.3 and 7.5 mi) a year.

In the spring of 1986, Hubbard Glacier surged across the mouth of Russell Fjord in Alaska, cutting it off from

contact with Yakutat Bay. This area, the St. Elias Mountain Range in southeastern Alaska, is fed by annual snowfall that averages more than 850 cm (335 in.) a year, so the surge event had been predicted. But the rapidity of the surge was surprising. The glacier's movement exceeded 34 m (112 ft) per day during the peak surge, an enormous increase over its normal rate of 15 cm (6 in.) per day.

The exact cause of such a glacier surge is being studied. Some surge events result from a buildup of water pressure under the glacier, sometimes enough to actually float the glacier slightly, detaching it from its bed, during the surge. As a surge begins, icequakes are detectable, and ice faults are visible. Surges can occur in dry conditions as well, as the glacier plucks (picks up) rock from its bed and moves forward. Another cause of glacier surges is the presence of a water-saturated layer of sediment, a so-called *soft*

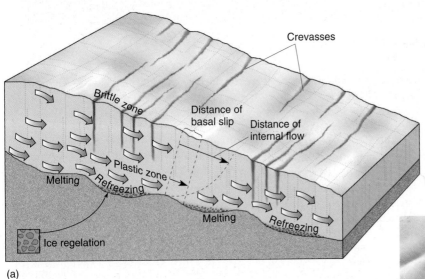

(a)

FIGURE 17.8 Glacial movement.
(a) Cross section of a glacier, showing its forward motion and brittle cracking at the surface and flow along its basal layer. (b) Surface crevasses and cracks are evidence of a glacier's forward motion (near the Don Sheldon Amphitheater, Denali National Park, Alaska). (c) Numerous crevasses in the Pine Island Glacier, off the West Antarctic Ice Sheet; the image is near 75° S 100° W. [(b) Photo by Michael Collier; (c) *Terra* ASTER sensor image courtesy of NASA/GSFC/MITI/ERSDAC/JAROS, and the U.S./Japan ASTER Science Team.]

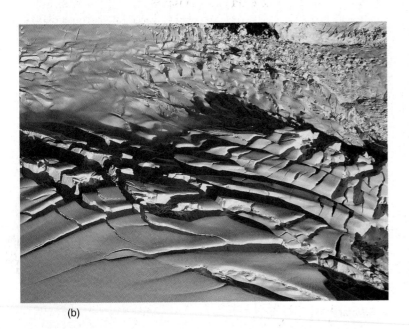

(b)

(c)

bed, beneath the glacier. This is a deformable layer that cannot resist the tremendous sheer stress produced by the moving ice of the glacier. Scientists examining cores taken from several ice streams now accelerating through the West Antarctic Ice Sheet think they have identified this cause—although water pressure is still important.

Glacial Erosion   The way in which a glacier erodes the land is similar to a large excavation project, with the glacier hauling debris from one site to another for deposition. The passing glacier mechanically plucks rock material and carries it away. Debris is carried on its surface and is also transported internally, or *englacially*, embedded within the glacier itself. There is evidence that rock pieces actually freeze to the basal layers of the glacier in a *glacial plucking*, or picking up, process and, once embedded, enable the glacier to scour and sandpaper the landscape as it moves—a process

called **abrasion**. This abrasion and gouging produce a smooth surface on exposed rock, which shines with *glacial polish* when the glacier retreats. Larger rocks in the glacier act much like chisels, gouging the underlying surface and producing glacial striations parallel to the flow direction (Figure 17.9).

Glacial Landforms

Glacial erosion and deposition produce distinctive landforms that differ greatly from the way the land looked before the ice came and went. You might expect all glaciers to create the same landforms, but alpine and continental glaciers each generate their own characteristic landscapes. We look first at erosional landforms created by alpine glaciers, then at their depositional landforms. Finally, we examine the landscape that results from continental glaciation.

FIGURE 17.9 Glacial sandpapering polishes rock.
Glacial polish and striations are examples of glacial abrasion and erosion. The polished, marked surface is seen beneath a glacial erratic—rock left behind by a retreating glacier. [Photo by Bobbé Christopherson.]

Erosional Landforms Created by Alpine Glaciation

Alpine glaciers create spectacular, dramatic landforms that bring to mind the Canadian Rockies, the Swiss Alps, or vaulted Himalayan peaks. Geomorphologist William Morris Davis depicted the stages of a valley glacier in drawings published in 1906 and redrawn here in Figure 17.10. Study of these figures reveals the handiwork of ice as sculptor:

- In (a), you see typical stream-cut valleys as they exist before glaciation. Note the prominent **V**-shape valley.
- In (b), you see the same landscape during subsequent glaciation. Glacial erosion and transport actively remove much of the regolith (weathered bedrock) and the soils that covered the stream valley landscape. As the cirque walls erode away, sharp ridges form, dividing adjacent cirque basins. These **arêtes** ("knife-edge" in French) become the sawtooth, serrated ridges in glaciated mountains. Two eroding cirques may reduce an arête to a saddle-like depression or pass, called a **col**. A **horn** (pyramidal peak) results when several cirque glaciers gouge an individual mountain summit from all sides. Most famous is the Matterhorn in the Swiss Alps, but many others occur worldwide.
- In (c), you see the same landscape at a time of warmer climate when the ice has retreated. The glaciated valleys now are **U**-shaped, greatly changed from their previous stream-cut **V** form. You can see the steep sides and the straightened course of the valleys. Physical weathering from the freeze–thaw cycle has loosened rock along the steep cliffs, where it has fallen to form *talus slopes* along the valley sides.

Note that in the cirques where the valley glaciers originated, small mountain lakes called **tarns** have formed. One cirque contains small, circular, stair-stepped lakes called

paternoster ("our father") **lakes** for their resemblance to rosary (religious) beads. Paternoster lakes may have formed from the differing resistance of rock to glacial processes or from damming by glacial deposits.

The valleys carved by tributary glaciers are left stranded high above the valley floor, because the primary glacier eroded the valley floor so deeply. These *hanging valleys* are the sites of spectacular waterfalls.

In Figure 17.11 a portion of the Bernese Alps of Switzerland is shown as it appeared in late summer of 2001. The tallest peaks are slightly higher than 4000 m (13,100 ft). Keep in mind as you examine this photo that more than 50% of the glacial ice has melted since 1850, exposing many of the glaciated features. How many of the erosional forms (arêtes, col, horn, cirque, cirque glacier, **U**-shaped valleys, tarn, truncated spurs, among others) from Figure 17.10 can you identify in this photograph?

Where a glacial trough intersects the ocean, the glacier can continue to erode the landscape, even below sea level. As the glacier retreats, the trough floods and forms a deep fjord in which the sea extends inland, filling the lower reaches of the steep-sided valley (Figure 17.12). The fjord may be flooded further by rising sea level or by changes in the elevation of the coastal region. All along the glaciated coast of Alaska, glaciers now are in retreat, thus opening many new fjords that previously were blocked by ice. Coastlines with notable fjords include those of Norway, Greenland (see Figure 17.6), Chile, the South Island of New Zealand, Alaska, and British Columbia.

Depositional Landforms Created by Alpine Glaciation

You have just seen how glaciers excavate tremendous amounts of material and create fascinating landforms in the process. Glaciers produce a different set of distinctive landforms when they melt and deposit their debris cargo at the glacier's *terminus* (end). Figure 17.7a shows that after the glacier melts, debris accumulates to mark the former margins of the glacier, both its end and sides.

Glacial Drift   Glacial drift is the general term for all glacial deposits, both unsorted and sorted. Sediments deposited by glacial meltwater are sorted by size and are termed **stratified drift**. Direct ice deposits leave unstratified and unsorted debris called **till**.

As a glacier flows to a lower elevation, a wide assortment of rock fragments become *entrained* (carried along) on its surface or embedded within its mass or in its base. As the glacier melts, this unsorted cargo is deposited on the ground surface. Such till is poorly sorted and is difficult to cultivate for farming, but the clays and finer particles can provide a basis for soil development.

Retreating glaciers leave behind large rocks (sometimes house-sized), boulders, and cobbles that are "foreign" in composition and origin from the ground on which they were deposited. These *glacial erratics*, lying in strange locations with no obvious means of transport, were an early

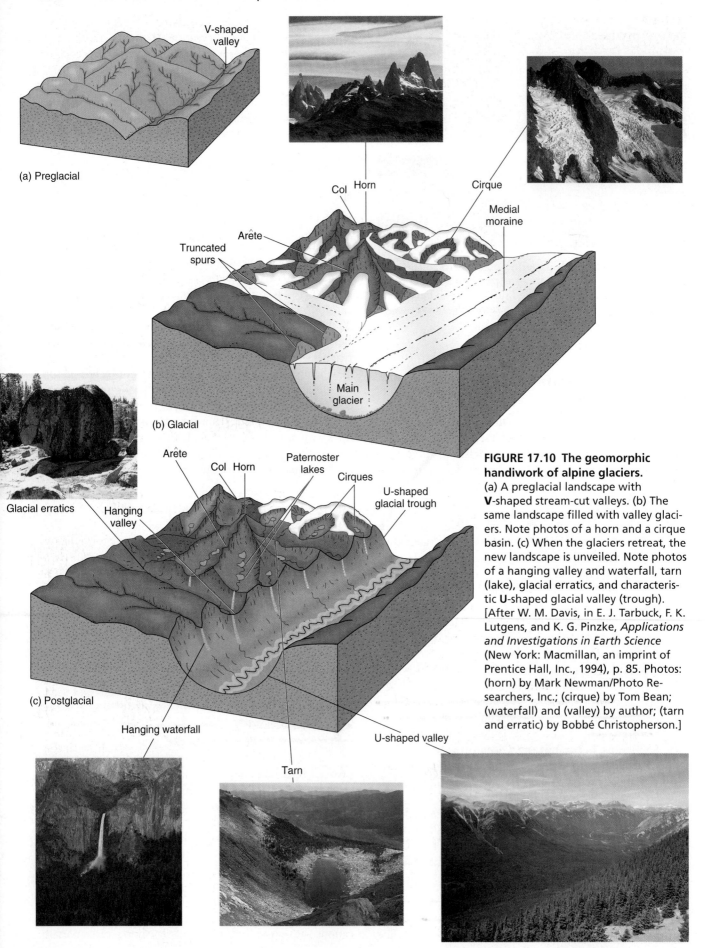

(a) Preglacial

V-shaped valley

Col Horn

Cirque

Arête

Medial moraine

Truncated spurs

Main glacier

(b) Glacial

Glacial erratics

Arête

Col Horn

Paternoster lakes

Cirques

U-shaped glacial trough

Hanging valley

(c) Postglacial

Hanging waterfall

Tarn

U-shaped valley

FIGURE 17.10 The geomorphic handiwork of alpine glaciers.
(a) A preglacial landscape with **V**-shaped stream-cut valleys. (b) The same landscape filled with valley glaciers. Note photos of a horn and a cirque basin. (c) When the glaciers retreat, the new landscape is unveiled. Note photos of a hanging valley and waterfall, tarn (lake), glacial erratics, and characteristic **U**-shaped glacial valley (trough). [After W. M. Davis, in E. J. Tarbuck, F. K. Lutgens, and K. G. Pinzke, *Applications and Investigations in Earth Science* (New York: Macmillan, an imprint of Prentice Hall, Inc., 1994), p. 85. Photos: (horn) by Mark Newman/Photo Researchers, Inc.; (cirque) by Tom Bean; (waterfall) and (valley) by author; (tarn and erratic) by Bobbé Christopherson.]

FIGURE 17.11 Erosional features of Alpine glaciation. Many of the glacial features illustrated in Figure 17.10 appear just south of Interlaken, in the Bernese Alps, Switzerland, with the Rhone Valley along the lower part of the photo. [Astronaut photo courtesy of STS-106 and NASA, Earth Science and Image Analysis Lab.]

FIGURE 17.12 Norwegian fjord.
A cruise ship sails along a coastal fjord in Norway. [Photo by Desjardins/Rapho/ Photo Researchers, Inc.]

clue that blankets of ice once had covered the land (pictured in Figure 17.10c).

Moraines **Moraine** is the name for specific landforms produced by the deposition of glacial sediments. Several types of moraines are exhibited by Hole-in-the-Wall Glacier, shown in Figure 17.13. A **lateral moraine** forms along each side of a glacier. If two glaciers with lateral moraines join, a **medial moraine** may form (see Figures 17.1 and 17.7c and d). A deposition of till that is generally spread across a surface is called a *ground moraine*, or *till plain*, and may hide the former landscape. Such plains are found in portions of the U.S. Midwest.

Eroded debris that is dropped at the glacier's farthest extent is called a **terminal moraine**. However, there also may be *end moraines*, formed at other points where a glacier paused after reaching a new equilibrium between growth and ablation. Both terminal and end moraine forms are clearly visible in Figure 17.13. Lakes may form behind terminal and end moraines after a glacier's retreat, with the moraine acting as a dam.

All of these till types are unsorted and unstratified. In contrast, streams of glacial meltwater can carry and deposit sorted and stratified glacial drift beyond a terminal moraine. Meltwater-deposited material downvalley from a glacier is called a *valley train deposit*. Peyto Glacier in Alberta, Canada, produces such a valley train that continues into Peyto Lake (Figure 17.14). Distributary stream channels appear braided across its surface. The picture also shows the milky meltwater associated with glaciers, laden with finely ground "rock flour." Meltwater is produced by glaciers at all times, not just when they are retreating.

FIGURE 17.13 Depositional features of Alpine glaciation.
Medial, lateral, terminal, and ground moraine deposits are
evident in this photo. These features were produced by the
Hole-in-the-Wall Glacier, Wrangell–St. Elias National Park,
Alaska. Also, note the kettle pond in the foreground. [Photo
by Tom Bean.]

FIGURE 17.14 A valley train deposit.
Peyto Glacier in Alberta, Canada. Note valley train, braided
stream, and milky-colored glacial meltwater. [Photo by
author.]

Erosional and Depositional Landforms Created by Continental Glaciation

The extent of the most recent continental glaciation in
North America and Europe, 18,000 years ago, is portrayed
several pages ahead in Figure 17.27. When these huge
sheets of ice advanced and retreated, they produced some,
but not all, of the erosional and depositional features characteristic of alpine glaciation. Because continental glaciers
form under different circumstances—not in mountains, but
across broad, open landscapes—the intricately carved alpine
features, lateral moraines, and medial moraines all are lacking in continental glaciation. Table 17.1 compares the erosional and depositional features of alpine (valley) and
continental glaciers.

Figure 17.15 illustrates some of the most common erosional and depositional features associated with the retreat
of a continental glacier. A **till plain** forms behind an end

Table 17.1	**Features of Valley Glaciation and Continental Glaciation Compared**	
Features and Landforms	**Alpine (Valley) Glacier**	**Continental Glaciation**
Erosional		
Striations, polish, etc.	Common	Common
Cirques	Common	Absent
Horns, arêtes, cols	Common	Absent
U-shaped valleys, truncated spurs, hanging valleys	Common	Rare
Fjords	Common	Absent
Depositional		
Till	Common	Common
Terminal moraines	Common	Common
Recessional moraines	Common	Common
Ground moraines	Common	Common
Lateral moraines	Common	Absent
Medial moraines	Common, easily destroyed	Absent
Drumlins	Rare or absent	Locally common
Erratics	Common	Common
Stratified drift	Common	Common
Kettles	Common	Common
Eskers, crevasse fillings	Rare	Common
Kames	Common	Common
Kame terraces	Common	Present in hilly country

Source: Adapted from L. D. Leet, S. Judson, and M. Kauffman,
Physical Geology, 5th ed., © 1978, p. 317. Reprinted by permission of Prentice Hall, Inc., Upper Saddle River, NJ.

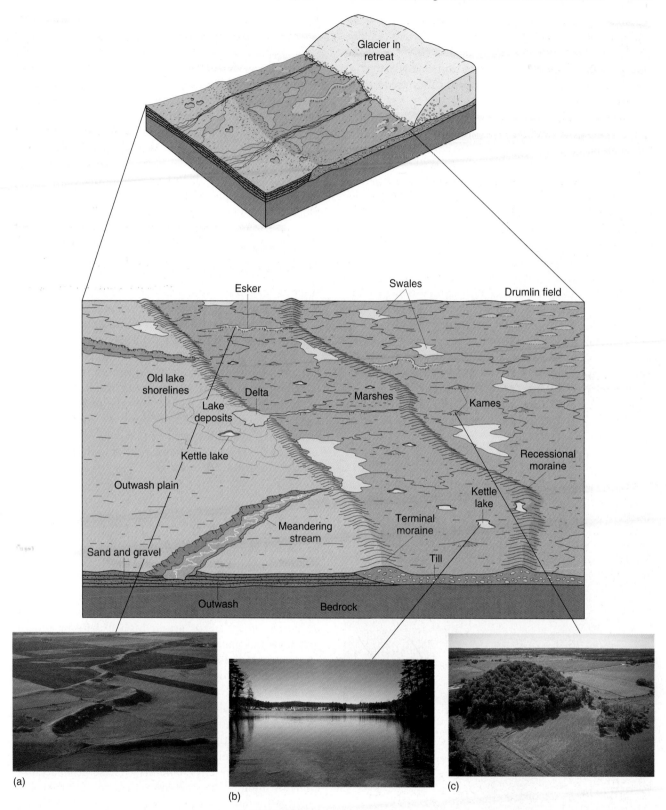

FIGURE 17.15 Continental glacier depositional features.
Common depositional landforms produced by glaciers. (a) An esker through farmland near
Dahlen, North Dakota. (b) Walden Pond is a kettle surrounded by mixed forest near Boston,
Massachusetts. (c) A kame covered by a woodlot near Campbellsport, Wisconsin. [Illustration
from R. M. Busch, ed., *Laboratory Manual in Physical Geology*, 3rd ed. (New York: Macmillan,
an imprint of Prentice Hall, Inc., 1993), p. 188. (a) Photo by Tom Bean/DRK Photo; (b) photo by
Bobbé Christopherson; (c) photo by Tom Bean.]

moraine; it features *unstratified* coarse till, has low and rolling relief, and has a deranged drainage pattern (see Figure 14.8). Beyond the morainal deposits lie the **outwash plains** of *stratified drift* featuring stream channels that are meltwater-fed, braided, and overloaded with sorted and deposited materials.

Figure 17.15a shows a sinuously curving, narrow ridge of coarse sand and gravel called an **esker**. It forms along the channel of a meltwater stream that flows beneath a glacier, in an ice tunnel, or between ice walls. As a glacier retreats, the steep-sided esker is left behind in a pattern roughly parallel to the path of the glacier. The ridge may not be continuous and in places may even appear to be branched, following the path set by the subglacial watercourse. Commercially valuable deposits of sand and gravel are quarried from some eskers.

Sometimes an isolated block of ice, perhaps more than a kilometer across, remains in a ground moraine, an outwash plain, or valley floor after a glacier has retreated. As much as 20 to 30 years is required for it to melt. In the interim, material continues to accumulate around the melting ice block. When the block finally melts, it leaves behind a steep-sided hole. Such a feature then frequently fills with water. This feature is called a **kettle**. Thoreau's famous Walden Pond, mentioned in the quotation in Chapter 7, is such a glacial kettle and is pictured in Figure 17.15b.

Another feature of outwash plains is a **kame**, a small hill, knob, or mound of poorly sorted sand and gravel that is deposited directly by water, by ice in crevasses, or in ice-caused indentations in the surface (Figure 17.15c). Kames also can be found in deltaic forms and in terraces along valley walls.

Glacial action also forms two types of streamlined hills. One is erosional, called a roche moutonnée, and the other is depositional, called a drumlin. A **roche moutonnée** ("sheep rock" in French) is an asymmetrical hill of exposed bedrock. Its gently sloping upstream side (stoss side) has been polished smooth by glacial action, whereas its downstream side (lee side) is abrupt and steep where the glacier plucked rock pieces (Figure 17.16).

A **drumlin** is deposited till that has been streamlined in the direction of continental ice movement, blunt end upstream and tapered end downstream (the opposite of a roche moutonnée). "Swarms" of drumlins occur across the landscape in portions of New York and Wisconsin, among other areas. Sometimes their shape is that of an elongated teaspoon bowl, lying face down. They attain lengths of 100–5000 m (300 ft to more than 3 mi) and heights up to 200 m (650 ft).

Figure 17.17 shows a portion of a topographic map for the area south of Williamson, New York, which experienced continental glaciation during the last ice age. In studying the map, can you identify the numerous drumlins? In what direction do you think the continental glaciers moved across this region? (Look for gentle, tapered slopes in the downstream direction of each drumlin, left behind as the glacier retreated.)

Periglacial Landscapes

In 1909, geologist W. Lozinski coined the term **periglacial** to describe cold-climate processes, landforms, and topographic features that exist along the margins of glaciers, past and present. Periglacial regions occupy more than 20% of Earth's land surface. These areas have either near-permanent ice or are at high elevation, and the ground is seasonally snow-free. Under these conditions, a unique combination of periglacial processes operates, including permafrost (Figure 17.18), frost action, and ground ice.

Climatologically, these regions are in *subarctic Dfc, Dfd,* and *polar E* climates (especially *tundra ET* climate). Such climates occur either at high latitude (tundra and boreal forest environments) or at high elevation in lower-latitude mountains (alpine environments). These periglacial regions are dominated by processes that are related to physical weathering, mass movement (Chapter 13), climate (Chapter 10), and soil (Chapter 18).

(a)

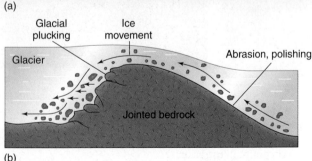

(b)

FIGURE 17.16 Glacial erosion streamlined rock.
Roche moutonnée, as exemplified by Lembert Dome in the Tuolumne Meadows area of Yosemite National Park, California. [Photo by author.]

Geography of Permafrost

When soil or rock temperatures remain below 0°C (32°F) for at least 2 years, **permafrost** ("permanent frost") develops. An area of permafrost that is not covered by glaciers is considered periglacial. Note that this criterion is based

Focus Study 16.1 *(continued)*

(a)   (b)

FIGURE 2 Examples of poor planning.
(a) Coastal erosion and a failed house foundation. (b) Seemingly futile efforts to halt collapse of
coastal cliffs. [Photos by (a) author; (b) Bobbé Christopherson.]

stricted access should be enforced, for even foot traffic can cause destruction. (See the beach stabilization example in Figure 15.5.)

The trough behind the primary dune is relatively tolerant of limited recreation and building. The plants that fix themselves to the surface send roots down to fresh groundwater reserves and anchor the sand in the process. Thus, if construction should inhibit the surface recharge of that water supply, the natural protective ground cover could die and destabilize the environment. Or subsequent saltwater intrusion might contaminate well water. Clearly, groundwater resources and the location of recharge aquifers must be considered in planning.

Behind the trough is the secondary dune, a second line of defense against the sea. It, too, is tolerant of some use yet is vulnerable to destruction. Next is the backdune, more suitable for development than any zone between it and the sea. Further inland are the bayshore and the bay, where no dredging or filling and only limited dumping of treated wastes and toxics should

be permitted. This zone is tolerant to intensive recreation. In reality, of course, the opposite of such careful assessment and planning prevails.

A Scientific View and a Political Reality

Before an intensive government study of the New Jersey shore was completed in 1962, no analysis of coastal hazards had been done, outside of academic circles. What is common knowledge to geographers, botanists, biologists, and ecologists in the classroom and laboratory still has not filtered through to the general planning and political processes. As a result, on the New Jersey shore and along much of the Atlantic and Gulf Coasts, improper development of the fragile coastal zone (primary dunes, trough, and secondary dunes) led to extensive destruction during storms in 1962, 1992, and numerous other times.

Scientists estimate that the coastlines will continue their retreat in some cases tens of meters in a few decades. Such estimates persist along the entire East Coast around to the Gulf Coast. Society must reconcile ecology and

economics if these coastal environments are to be sustained.

South Carolina enacted the Beach Management Act of 1988 (modified 1990) to apply some of McHarg's principles. In its first 2 years, more than 70 lawsuits protested the act as an invalid seizure of private property without compensation.

Destructive or threatening hurricane activity is helping to drive the importance of such cooperative planning and proactive laws. Similar measures in other states have faced the same difficult path. Implementation of any planning process is problematic, political pressure is intense, and results are mixed. Ian McHarg voiced an optimistic hope: "May it be that these simple ecological lessons will become known and incorporated into ordinance [law] so that people can continue to enjoy the special delights of life by the sea."*

*From *Design with Nature* by Ian McHarg, p. 17. © 1969 by Ian L. McHarg. Published by Bantam Doubleday Dell Publishing Group, Inc.

● *Describe* the chemical composition of seawater and the physical structure of the ocean.

Water is called the "universal solvent," dissolving at least 57 of the 92 elements found in nature. Most natural elements and the compounds they form are found in the seas as dissolved solids. Seawater is a solution, and the concentration of dissolved solids is called **salinity**. **Brine** exceeds the average 35‰ (parts per thousand) salinity; **brackish** applies to water that is less than 35‰. The ocean is divided by depth into a narrow mixing zone at the surface, a thermocline transition zone, and the deep cold zone.

salinity (p. 490)
brine (p. 492)
brackish (p. 492)

1. Describe the salinity of seawater: its composition, amount, and distribution.
2. Analyze the latitudinal distribution of salinity as shown in Figure 16.2. Why is salinity less along the equator and greater in the subtropics?
3. What are the three general zones relative to physical structure within the ocean? Characterize each by temperature, salinity, dissolved oxygen, and dissolved carbon dioxide.

● *Identify* the components of the coastal environment and *list* the physical inputs to the coastal system, including tides and mean sea level.

The coastal environment is called the **littoral zone** and exists where the tide-driven, wave-driven sea confronts the land. Inputs to the coastal environment include solar energy, wind and weather, ocean currents and waves, climatic variation, and the nature of coastal rock. **Mean sea level (MSL)** is based on average tidal levels recorded hourly at a given site over many years. MSL varies spatially because of ocean currents and waves, tidal variations, air temperature and pressure differences, ocean temperature variations, slight variations in Earth's gravity, and changes in oceanic volume.

Tides are complex daily oscillations in sea level, ranging worldwide from barely noticeable to many meters. Tides are produced by the gravitational pull of both the Moon and the Sun. Most coastal locations experience two high (rising) **flood tides**, and two low (falling) **ebb tides** every day. The difference between consecutive high and low tides is the tidal range. **Spring tides** exhibit the greatest tidal range, when the Moon and Sun are either in conjunction or opposition. **Neap tides** produce a lesser tidal range.

littoral zone (p. 493)
mean sea level (MSL) (p. 493)
tide (p. 495)
flood tide (p. 496)

ebb tide (p. 496)
spring tide (p. 496)
neap tide (p. 496)

4. What are the key terms used to describe the coastal environment?
5. Define mean sea level. How is this value determined? Is it constant or variable around the world? Explain.
6. What interacting forces generate the pattern of tides?
7. What characteristic tides are expected during a new Moon or a full Moon? During the first-quarter and third-quarter phases of the Moon? What is meant by a flood tide? An ebb tide?
8. Is tidal power being used anywhere to generate electricity? Explain briefly how such a plant would utilize the tides to produce electricity. Are there any sites in North America? Where are they?

● *Describe* wave motion at sea and near shore and *explain* coastal straightening as a product of wave refraction.

Friction between moving air (wind) and the ocean surface generates undulations of water that we call **waves**. Wave energy in the open sea travels through water, but the water itself stays in place. Regular patterns of smooth, rounded waves, called **swells**, are the mature undulations of the open ocean. Near shore, the restricted depth of water slows the wave, forming *waves of translation*, in which both energy and water actually move forward toward shore. As the crest of each wave rises, the wave falls into a characteristic **breaker**.

Wave refraction redistributes wave energy so that different sections of the coastline vary in erosion potential. Headlands are eroded, whereas coves and bays receive materials, with the long-term effect of straightening the coast. As waves approach a shore at an angle, refraction produces a **longshore current** of water moving parallel to the shore. This current produces the *longshore drift* of sand, sediment, and gravel and assorted materials moving along the beach as **beach drift**—together these materials make up the overall *littoral drift*.

A **tsunami** is a seismic sea wave triggered by an undersea landslide or earthquake. It travels at great speeds in the open sea and gains height as it comes ashore, posing a coastal hazard.

wave (p. 497)
swell (p. 498)
breaker (p. 499)
wave refraction (p. 499)
longshore current (p. 500)
beach drift (p. 500)
tsunami (p. 500)

9. What is a wave? How are waves generated, and how do they travel across the ocean? Does the water travel with

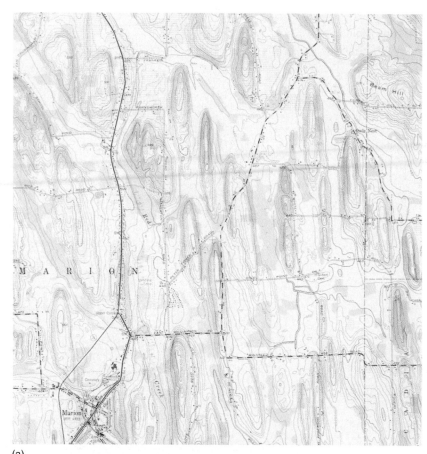

(a)

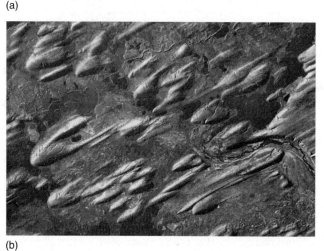

(b)

FIGURE 17.17 Glacially deposited streamlined features.
(a) Topographic map south of Williamson, New York, featuring numerous drumlins. (7.5-minute series quadrangle map, originally produced at a 1:24,000 scale, 10-ft contour interval.) (b) A drumlin field, Snare Lake Canada. [(a) USGS map; (b) photo by National Air Photo Library, NRC, Ottawa, Canada.]

solely on *temperature* and has nothing to do with how much or how little water is present. Two other factors also contribute to permafrost: the presence of fossil permafrost from previous ice-age conditions and the insulating effect of snow cover or vegetation that inhibits heat loss.

Continuous and Discontinuous Zones

Permafrost regions are divided into two general categories: continuous and discontinuous. They merge along a general transition zone. *Continuous permafrost* is the region of severest cold and is perennial, roughly poleward of the −7°C (19°F)

mean annual temperature isotherm (purple area in Figure 17.18). Continuous permafrost affects all surfaces except those beneath deep lakes or rivers. The depth of continuous permafrost may exceed 1000 m (3300 ft), averaging approximately 400 m (1300 ft).

Unconnected patches of *discontinuous permafrost* gradually coalesce poleward toward the continuous zone. Permafrost becomes scattered or sporadic until it gradually disappears equatorward of the −1°C (30.2°F) mean annual temperature isotherm (dark blue area on map). In the discontinuous zone, permafrost is absent on sun-exposed

FIGURE 17.18 Permafrost distribution.

Distribution of permafrost in the Northern Hemisphere. Alpine permafrost is noted except for small occurrences in Hawai'i, Mexico, Europe, and Japan. Sub-sea permafrost occurs in the ground beneath the Arctic Ocean along the margins of the continents as shown. Note the towns of Resolute and Coppermine (now Kugluktuk) in Nunavut (formerly part of NWT), and Hotchkiss in Alberta. A cross section of the permafrost beneath these towns is shown in Figure 17.19. [Adapted from T. L. Péwé, "Alpine permafrost in the contiguous United States: A review," *Arctic and Alpine Research* 15, no. 2 (May 1983): 146. © Regents of the University of Colorado. Used by permission.]

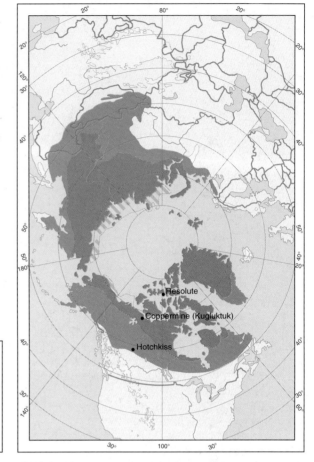

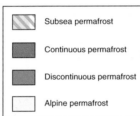

Subsea permafrost

Continuous permafrost

Discontinuous permafrost

Alpine permafrost

south-facing slopes, areas of warm soil, or areas insulated by snow. In the Southern Hemisphere, north-facing slopes experience increased warmth.

As much as 50% of Canada and 80% of Alaska are affected by permafrost of either type. In central Eurasia, the effects of continentality and elevation produce discontinuous permafrost that extends equatorward to the 50th parallel.

Areas of discontinuous permafrost feature a mixture of *cryotic* (frozen) and *noncryotic* ground. In addition to these two types of ground, zones of high-altitude *alpine permafrost* extend to lower latitudes, as shown on the map. Microclimatic factors such as slope orientation and snow cover are important in the alpine environment.

In the Mackenzie Mountains of Canada (62° N), continuous permafrost extends down to an elevation of 1200 m (4000 ft), and discontinuous permafrost occurs throughout. The Colorado Rockies (40° N) experience continuous permafrost down to an elevation of 3400 m (11,150 ft) and discontinuous permafrost to 1700 m (5600 ft).

Behavior of Permafrost We have looked at the spatial distribution of permafrost; let us examine how permafrost behaves. Figure 17.19 is a stylized cross section from approximately 75° N to 55° N, using the three sites located on the map in Figure 17.18. The **active layer** is the zone of seasonally frozen ground that exists between the subsurface permafrost layer and the ground surface.

The active layer is subjected to consistent daily and seasonal freeze–thaw cycles. This cyclic melting of the active layer affects as little as 10 cm (4 in.) depth in the north (Ellesmere Island, 78° N), up to 2 m (6.6 ft) in the southern margins (55° N) of the periglacial region, and 15 m (50 ft) in the alpine permafrost of the Colorado Rockies (40° N).

The depth and thickness of the active layer and permafrost zone change slowly in response to climatic change. Higher temperatures degrade (reduce) permafrost and increase the thickness of the active layer; lower temperatures gradually aggrade (increase) permafrost depth and reduce active layer thickness. Although somewhat sluggish in response, the active layer is a dynamic open system driven by energy gains and losses in the subsurface environment. As you might expect, most permafrost exists in disequilibrium with environmental conditions and therefore actively adjusts to inconstant climatic conditions.

The present trend in concert with global warming is for a deepening of the active layer at high latitudes. Across Canada and Siberia this warming is causing a large release of carbon from peat-rich thawed ground that in turn influences the global greenhouse—a real-time positive feedback mechanism.

A *talik* is unfrozen ground that may occur above, below, or within a body of discontinuous permafrost or beneath a water body in the continuous region. Taliks occur beneath deep lakes and may extend to bedrock and noncryotic soil beneath large, deep lakes (see Figure 17.19). Taliks form

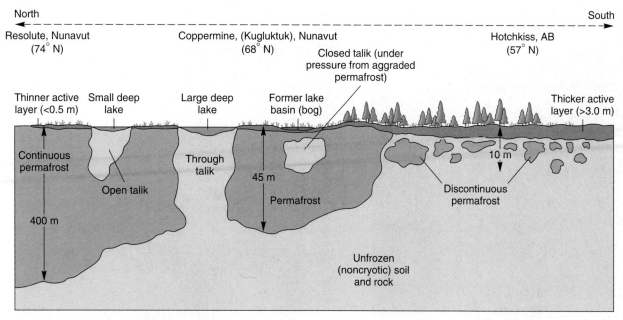

North   South

Resolute, Nunavut (74° N)   Coppermine, (Kugluktuk), Nunavut (68° N)   Hotchkiss, AB (57° N)

Closed talik (under pressure from aggraded permafrost)

Thinner active layer (<0.5 m)   Small deep lake   Large deep lake   Former lake basin (bog)   Thicker active layer (>3.0 m)

Continuous permafrost

Open talik

Through talik

45 m

Permafrost

400 m

Discontinuous permafrost

10 m

Unfrozen (noncryotic) soil and rock

FIGURE 17.19 Periglacial environments.
Cross section of a periglacial region in northern Canada, showing typical forms of permafrost, active layer, talik, and ground ice. The three sites noted are shown on the map in Figure 17.18.

connections between the active layer and groundwater, whereas in continuous permafrost groundwater is essentially cut off from surface water. In this way, permafrost disrupts aquifers and taliks, leading to water supply problems.

Ground Ice and Frozen Ground Phenomena

In regions of permafrost, frozen subsurface water is termed *ground ice*. The moisture content of areas with *ground ice* varies from nearly none in drier regions to almost 100% in saturated soils. From the area of maximum energy loss, freezing progresses through the ground along a *freezing front*, or boundary between frozen and unfrozen soil. The presence of frozen water in the soil initiates geomorphic processes associated with *frost action* and the expansion of water as it freezes (Chapters 7 and 13).

Ground ice may occur as:

- Common *pore ice* (subsurface water frozen in the soil's pore spaces)
- *Lenses* (horizontal bodies) and *veins* (channels extending in any direction)
- *Segregated ice* (layers of buried ice that increase in mass by accreting water as the ground freezes, producing layers of relatively pure ice)
- *Intrusive ice* (the freezing of water injected under pressure, as in a pingo, discussed shortly)
- *Wedge ice* (surface water entering a crack and freezing)

Various aspects of frost action as a geomorphic agent are discussed in Chapter 13 ("Physical Weathering Processes" and the section "Classes of Mass Movements," which describes soil creep). Some forms of ground ice may occur

at the surface, during episodes of *icing*, in which a river or spring forms freezing layers of surface ice. Such icings can occur on slopes as well as level ground. Spring flooding along rivers can result from the presence of icings; it is a particular problem in rivers draining into the Arctic Ocean.

Frost Action Processes The 9% expansion of water as it freezes produces strong mechanical forces. Such frost action shatters rock, producing angular pieces that form a *block field*, or *felsenmeer*. The felsenmeer accumulates as part of the arctic and alpine periglacial landscape, particularly on mountain summits and slopes.

If sufficient water freezes, the saturated soil and rocks are subjected to *frost-heaving* (vertical movement) and *frost-thrusting* (horizontal movement). Boulders and rock slabs may be thrust to the surface. Soil horizons (layers) may be disrupted by frost action and appear to be stirred or churned, a process termed *cryoturbation*. Frost action also can produce contractions in soil and rock, opening up cracks for ice wedges to form. Also, there is a tremendous increase in pressure in the soil as ice expands, particularly if there are multiple freezing fronts trapping unfrozen soil and water between them.

An *ice wedge* develops when water enters a crack in the permafrost and freezes (Figure 17.20). Thermal contraction in ice-rich soil forms a tapered crack—wider at the top, narrowing toward the bottom. Repeated seasonal freezing and melting progressively enlarge the wedge, which may widen from a few millimeters to 5–6 m (16–20 ft) and up to 30 m (100 ft) in depth. Widening may be small each year, but after many years the wedge can become significant, like the sample shown.

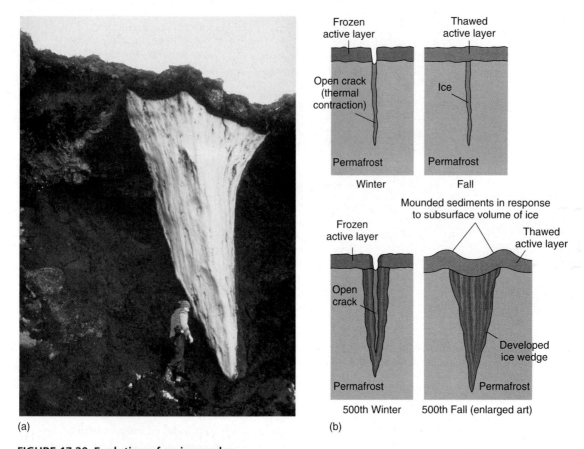

(a)   (b)

FIGURE 17.20 Evolution of an ice wedge.
(a) An ice wedge and ground ice in northern Canada. (b) Sequential illustration of ice-wedge
formation. [(a) Photo by H. M. French. (b) Adapted from A. H. Lachenbruch, "Mechanics of
thermal contraction and ice-wedge polygons in permafrost," *Geological Society of America
Bulletin Special Paper 70* (1962).]

Thin layers of sediment that form foliations in the
wedge mark the annual thickness of ice added. Thus, the
age of a wedge can be determined by counting the foliations, like counting annual accumulations recorded in an ice
core. In summer, when the active layer thaws, the wedge itself may not be visible. However, the presence of a wedge
beneath the surface is noticeable where the ice-expanded
sediments form raised, upturned ridges.

Frost Action Landforms   Large areas of frozen ground
(soil-covered ice) can develop a heaved-up, circular, ice-
cored mound called a *pingo*. It rises above the flat landscape,
occasionally exceeding 60 m (200 ft) height. Pingos rise
when freezing water expands, sometimes as a result of pressure developed by artesian water injected into permafrost
(Figure 17.21).

A *palsa* (from the Swedish word for "elliptical") is a
rounded or elliptical mound of peat that contains thin
perennial ice lenses rather than an ice core, as in a pingo.
Palsas can be 2–30 m (6–100 ft) wide by 1–10 m (3–30 ft)
high and usually are covered by soil or vegetation over a
cracked surface.

The expansion and contraction of frost action results
in the transport of stones and boulders. As the water-ice
volume changes and the ice wedge deepens, coarser parti-

cles are moved toward the surface. An area with a system of
ground ice and frost action develops sorted and unsorted
accumulations of rock at the surface that take the shape of
polygons called *patterned ground*.

Patterned ground may include polygons of sorted
rocks that coalesce into stone polygon nets (Figure 17.22).
Various terms are in use to describe such ice-wedge and
stone polygon forms: nets, circles, hummocks (vegetation-
covered), and stripes (elongated polygons formed on a hill-

FIGURE 17.21 A pingo.
An ice-cored pingo resulting from hydraulic pressures that
pushed the mound upward above the landscape. Coastal
erosion exposed the ice core. This pingo is near Tuktoyaktuk,
Mackenzie Delta, NWT, Canada. [Photo by H. M. French.]

(a)

(b)

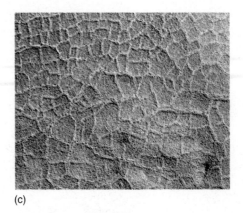

(c)

FIGURE 17.22 Patterned ground phenomena.
(a) Aerial view of polygonal nets formed in Alaska, a fairly widespread frozen-ground phenomenon. (b) Patterned ground in the Dry Valleys of Antarctica. (c) Polygons in the Martian northern plains; each cell averages a little more than 100 m across (300 ft). [(a) Photo from the Marbut Collection, Soil Science Society of America, Inc.; (b) photo by Galen Rowell/Mountain Light Photography, Inc. (c) Image from the *Mars Global Surveyor*, Mars Orbiter Camera, courtesy of NASA/JPL/Malin Space Science Systems, May 1999.]

slope). Ice wedges sometimes are found beneath the perimeter of each cell; however, questions still exist as to how such patterned ground actually develops and the degree to which such ground forms are the vestiges of past eras. Such polygon nets in patterned ground provide vivid evidence of subsurface water ice on Mars (Figure 17.22c). In this 1999 image, the Mars Global Surveyor captured a region of these features on the Martian northern plains.

Hillslope Processes: Solifluction and Gelifluction   Soil drainage is poor in areas of permafrost and ground ice. The active layer of soil and regolith is saturated with soil moisture during the thaw cycle (summer), and the whole layer commences to flow from higher to lower elevation if the landscape is even slightly inclined. Such soil flows are generally called *solifluction*; in the presence of ground ice, the more specific term *gelifluction* is applied. In this ice-bound type of soil flow, movement up to 5 cm (2 in.) per year can occur on slopes as gentle as a degree or two.

The cumulative effect of this landflow can be an overall flattening of a rolling landscape, with identifiable sagging surfaces and scalloped and lobed patterns in the downslope soil movements (Figure 17.23). Other types of periglacial mass movement include failure in the active layer, producing translational and rotational slides and rapid flows associated with melting ground ice. Periglacial mass movement processes are related to slope dynamics and processes discussed in Chapter 13.

Thermokarst Landscapes   As ground ice melts, irregular features develop across the landscape, creating *thermokarst*

FIGURE 17.23 Soil flowage in periglacial environments.
Gelifluction (solifluction) lobes on a hillside near the Yukon–Northwest Territories border. [Photo by Joyce Lundberg.]

topography. These forms result from thermal subsidence and erosion caused by ice-wedge melting and poor drainage. (Note that the term refers only to the topographic style and has nothing to do with the solution processes and chemical weathering that cause limestone karst.)

Thermokarst topography is hummocky, marked by cave-ins, bogs, small depressions, pits, standing water, and small lakes. More thermokarst landforms are found in

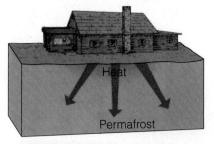
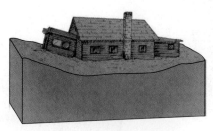

FIGURE 17.24 Permafrost melting and structure collapse.
Building failure due to improper construction and the melting of permafrost, south of Fairbanks, Alaska. [Adapted from U.S. Geological Survey. Photo by Steve McCutcheon, illustration based on U.S. Geological Survey pamphlet "Permafrost" by L. L. Ray.]

Siberia and Scandinavia than in North America. In Canada and Alaska, rounded thaw lakes, or cave-in lakes, are thermokarst features. With the incredibly warm temperatures recorded in the Canadian Arctic in the late 1990s, more thermokarst disruption of surfaces is occurring—leading to highway, railway, and building damage.

Humans and Periglacial Landscapes

In areas of permafrost and frozen ground phenomena, people face several related problems. Because thawed ground above the permafrost zone frequently shifts, highways and

rail lines become warped, twisted, and fail, and utility lines are disrupted. In addition, any building placed directly on frozen ground will "melt" into the defrosting soil, creating subsidence in structures (Figure 17.24).

Construction in periglacial regions dictates placing structures above the ground to allow air circulation beneath. This airflow allows the ground to cycle through its normal annual temperature pattern. Utilities such as water and sewer lines must be built above ground in "utilidors" to protect them from freezing and thawing ground (Figure 17.25). Likewise, the Trans Alaskan oil pipeline was constructed above ground on racks for 675 of its 1285-km (420

Pipeline 1.2 m diameter

Average height 1.5 to 3.0 m

(a) (b)

FIGURE 17.25 Special structures for permafrost.
(a) Proper construction in periglacial environments requires raising of buildings above ground and running water and sewage lines in elevated "utilidors" (Inuvik, Northwest Territories).
(b) Supporting the Trans-Alaska oil pipeline on racks protects the permafrost from heat.
[(a) Photo by Joyce Lundberg; (b) photo by Galen Rowell/Mountain Light Photography, Inc.]

of its 800 mi) length to avoid melting the frozen ground, causing shifting that could rupture the line. The pipeline that is underground uses a cooling system to keep the permafrost around the pipeline stable.

The effects of global warming are already showing up in Siberia, where the active layer is now more than twice the depth it was in the past. Buildings constructed with shallow pilings for support will certainly fail if this trend continues. In an era of global change, the regions of permafrost may provide further climatic indicators.

The Pleistocene Ice Age Epoch

Imagine almost a third of Earth's land surface buried beneath ice sheets and glaciers—most of Canada, the northern Midwest, England, and northern Europe, and many mountain ranges, beneath thousands of meters of ice! This is how it was at the height of the Pleistocene Epoch of the late Cenozoic Era. In addition, periglacial regions along the margins of the ice during the last ice age covered about twice their present areal extent.

The Pleistocene is thought to have begun about 1.65 million years ago and is one of the more prolonged cold periods in Earth's history. It featured not just one glacial advance and retreat, but at least 18 expansions of ice over Europe and North America, each obliterating and confusing the evidence from the one before. Apparently, glaciation can take about 90,000 years, whereas deglaciation is rapid, requiring less than about 10,000 years to melt away the accumulation.

The term ice age is applied to any extended period of cold (not a single brief cold spell). An **ice age** is a period of generally cold climate that includes one or more *glacials*, interrupted by brief warm spells known as *interglacials*. Each glacial and interglacial is given a name that is usually based on the location where evidence of the episode is prominent—for example, "Wisconsinan glacial."

Modern research techniques include the examination of ancient ratios of oxygen isotopes, depths of coral growth in the tropics, analysis of sediments worldwide, and analysis of the latest ice cores from Greenland and Antarctica. These techniques have opened the way for a new chronology and understanding of past climates. Glaciologists currently recognize the Illinoian glacial and Wisconsinan glacial periods, with the Sangamon interglacial between them. These events span the 300,000-year period prior to our present Holocene Epoch (Figure 17.26).

The chart shows that the Illinoian glacial actually consisted of two glacials (occurring during oxygen isotope stages 6 and 8), as did the Wisconsinan (stages 2 and 4), which are dated at 10,000 to 35,000 years ago. The oxygen isotope glacial/interglacial stages on the chart are numbered back to stage 23 at approximately 900,000 years ago. To overcome local bias in core records, investigators correlate oxygen isotope data with other indicators worldwide. (For Web links on glaciers and Pleistocene geology, see **http://research. umbc.edu/~miller/geog111/glacierlinks.htm**.)

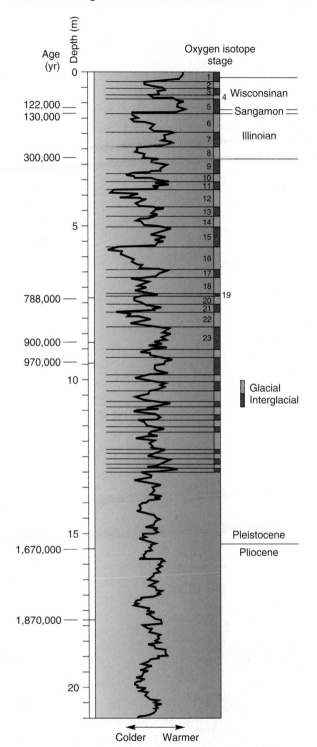

FIGURE 17.26 Temperature record of the past 2 million years.
Pleistocene temperatures, determined by oxygen isotope fluctuations in fossil planktonic foraminifera (tiny marine organisms having a calcareous shell) from deep-sea cores. Twenty-three stages cover 900,000 years, with names assigned for the past 300,000 years. [After N. J. Shackelton and N. D. Opdyke, *Oxygen-Isotope and Paleomagnetic Stratigraphy of Pacific Core V28-239, Late Pliocene to Latest Pleistocene.* Geological Society of America Memoir 145. ©1976 by the GSA. Adapted by permission.]

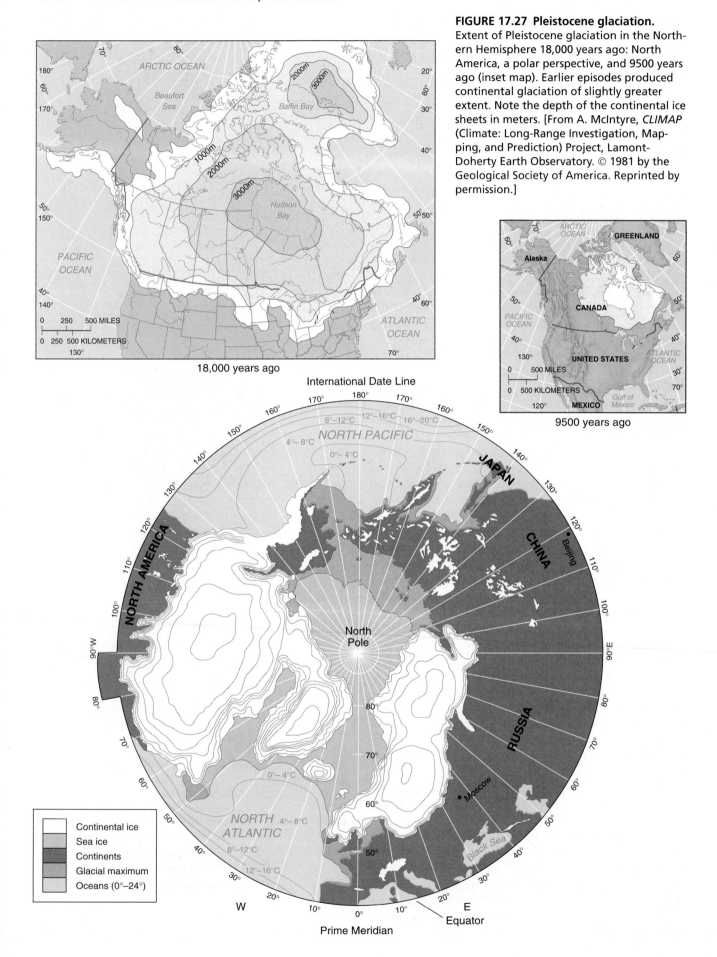

18,000 years ago

FIGURE 17.27 Pleistocene glaciation.
Extent of Pleistocene glaciation in the Northern Hemisphere 18,000 years ago: North America, a polar perspective, and 9500 years ago (inset map). Earlier episodes produced continental glaciation of slightly greater extent. Note the depth of the continental ice sheets in meters. [From A. McIntyre, *CLIMAP* (Climate: Long-Range Investigation, Mapping, and Prediction) Project, Lamont-Doherty Earth Observatory. © 1981 by the Geological Society of America. Reprinted by permission.]

9500 years ago

Continental ice
Sea ice
Continents
Glacial maximum
Oceans (0°–24°)

Changes in the Landscape

The continental ice sheets covered portions of Canada, the United States, Europe, and Asia about 18,000 years ago, as illustrated on the polar map projection in Figure 17.27. The ice sheets ranged in thickness to more than 2 km (1.2 mi). In North America, the Ohio and Missouri River systems mark the southern terminus of continuous ice at its greatest extent during the Pleistocene. The ice sheet disappeared by 7000 years ago.

As both alpine and continental glaciers retreated, they exposed a drastically altered landscape: the rocky soils of New England, the polished and scarred surfaces of Canada's Atlantic Provinces, the sharp crests of the Sawtooth Range and Tetons of Idaho and Wyoming, the scenery of the Canadian Rockies and the Sierra Nevada, the Great Lakes of the United States and Canada, the Matterhorn of Switzerland, and much more. In the Southern Hemisphere there is evidence of this ice age in the form of fjords and sculpted mountains in New Zealand and Chile.

The continental glaciers came and went several times over the region we know as the Great Lakes (Figure 17.28). The ice enlarged and deepened stream valleys to form the basins of the future lakes. This complex history produced five lakes that today cover 244,000 km² (94,000 mi²) and hold some 18% of all the lake water on Earth. Figure 17.28 shows the final formation of the Great Lakes, which involved two advancing and two retreating stages—between 13,200 and 10,000 years before the present. During the final retreat, tremendous quantities of glacial meltwater flowed into the gouged, isostatically (weight of the ice) depressed basins. Drainage at first was to the Mississippi River via the Illinois River, to the St. Lawrence River via the Ottawa River, and to the Hudson River in the east. In recent times, drainage has shifted through the St. Lawrence system.

Study of these glaciated landscapes is important, for we can better understand paleoclimatology (past climates) and discover the mechanisms that produce ice ages and climatic change.

Lowered Sea Levels and Lower Temperatures

Sea levels 18,000 years ago were approximately 100 m (330 ft) lower than they are today because so much of Earth's water was frozen and tied up in the glaciers instead of being in the ocean. Imagine the coastline of New York being 100 km farther east; Alaska and Russia connected by land across the Bering Straits, and England and France joined by a land bridge. In fact, sea ice extended southward into the North Atlantic and Pacific and northward in the Southern Hemisphere about 50% farther than it does today.

Sea-surface temperatures 18,000 years ago averaged between 1.4 C° and 1.7 C° (2.5 F° and 3.1 F°) lower than today. During the coldest portion of the Pleistocene Ice Age, air temperatures were as much as 12 C° (22 F°) colder than today's average, although milder periods ranged to within 5 C° (9 F°) of present air temperatures.

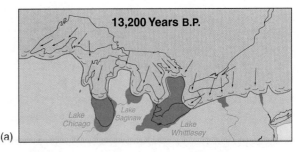

(a)

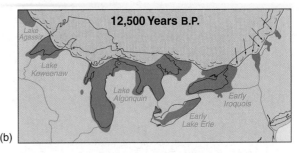

(b)

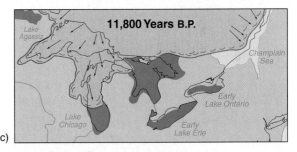

(c)

(d)

FIGURE 17.28 Late stages of Great Lakes formation. Four "snapshots" of the Great Lakes' evolving development during the retreat of the Wisconsinan glaciation. Note the change in stream drainage between (b) and (d). Time is in years before the present. [After *The Great Lakes—An Environmental Atlas and Resource Book*, Environment Canada, U.S. EPA, Brock University, and Northwestern University (Toronto: Environment Canada, 1987), p. 7.]

Paleolakes

Figure 17.29 portrays the American West dotted with large lakes 12,000–30,000 years ago. Except for the Great Salt Lake in Utah (a remnant of the former Lake Bonneville noted on the map) and a few smaller lakes, only dry basins, ancient shorelines, and lake sediments remain today. These ancient lakes are called **paleolakes**. The three photos in Figure 17.29 show what these paleolake sites look like today: the Bonneville Salt Flats in Utah, Mono Lake and its tufa towers in California, and Severe Dry Lake, Utah.

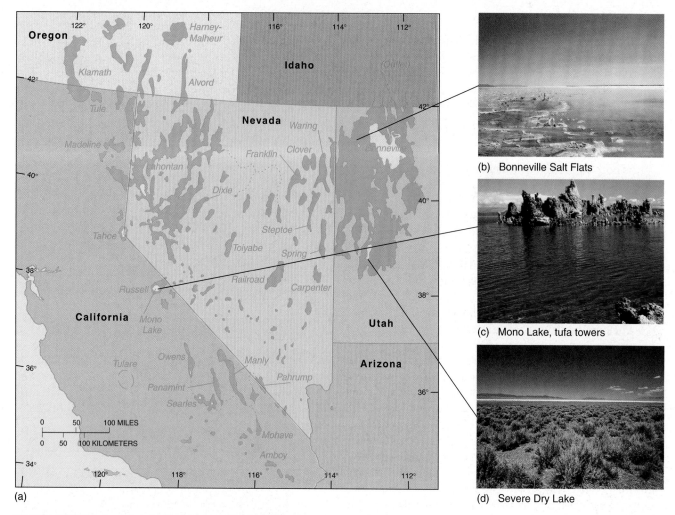

(a)

(b) Bonneville Salt Flats

(c) Mono Lake, tufa towers

(d) Severe Dry Lake

FIGURE 17.29 Paleolakes in the western United States.
Paleolakes of the western United States at their greatest extent 12,000 to 30,000 years ago, a recent pluvial period. Lake Lahontan and Lake Bonneville were the largest. The Great Salt Lake and Severe Dry Lake in Utah are remnants of Lake Bonneville; Mono Lake in California is what remains of pluvial Lake Russell. Modern-day photos of some paleolake locations are in (a) through (c). [After R. F. Flint, *Glacial and Pleistocene Geology*. © 1957 by John Wiley & Sons, Inc. Adapted by permission. Photos by (b) and (c) author; (d) Bobbé Christopherson.]

The term *pluvial* (Latin: "rain") describes any period of wet conditions, such as occurred during the Pleistocene Epoch. During pluvial periods, lake levels increased in arid regions. The drier periods between pluvials are called *interpluvials*. Interpluvials are marked with *lacustrine deposits*, which are lake sediments that form terraces along former shorelines.

Are these lakes of glacial origin like the Great Lakes? Earlier researchers attempted to correlate pluvial and glacial ages, given their coincidence during the Pleistocene. However, few sites actually demonstrate such a simple relation. For example, in the western United States, the estimated volume of melted ice from glaciers is only a small portion of the actual water volume that was in the paleolakes. Also, these lakes tend to predate glacial times and are

correlated instead with periods of wetter climate, or periods thought to have had lower evaporation rates. The term *paleolake* is increasingly used in the scientific literature to describe these lakes and to separate their occurrence from specific glacial stages.

Paleolakes existed in North and South America, Africa, Asia, and Australia. Today, the Caspian Sea in Kazakstan and southern Russia has a level of 30 m (100 ft) below mean world sea level, but ancient shorelines are visible about 80 m (265 ft) above the present lake level.

In North America, the two largest late Pleistocene paleolakes were in the Basin and Range Province of the West—Lake Bonneville and Lake Lahontan. Respectively, these two lakes were eight times and six times the size of their present-day remnants. Figure 17.29 shows these and other paleolakes at their highest level and the few remaining modern lakes in light blue. Much research still is taking place since many of the ancient lakeshores are not yet field mapped.

New evidence reveals that the occurrence of these lakes in North America was related to specific changes in the polar jet stream that steered storm tracks across the region, creating pluvial conditions. The continental ice sheet evidently influenced changes in jet stream position.

The Great Salt Lake, near Salt Lake City, Utah, and the Bonneville Salt Flats (Figure 17.29b) in western Utah are remnants of Lake Bonneville. At its greatest extent this paleolake covered more than 50,000 km^2 (19,500 mi^2) and reached depths of 300 m (1000 ft), spilling over into the Snake River drainage to the north. Today, it is a closed basin with no drainage except an artificial outlet to the west where excess water from the Great Salt Lake is pumped during floods.

Deciphering Past Climates: Paleoclimatology

Glacials and interglacials occur because Earth's climate has fluctuated in and out of warm and cold ages. Evidence for this fluctuation now is being traced in ice cores from Greenland and Antarctica, in layered deposits of silts and clays, in the extensive pollen record from ancient plants, and in the relation of past coral productivity to sea level. This evidence is analyzed with radioactive dating methods and other techniques. One especially interesting fact is emerging from these studies: We humans (*Homo erectus* and *H. sapiens* of the last 1.9 million years) have never experienced Earth's normal (more moderate, less extreme) climate, most characteristic of Earth's entire 4.6 billion year span.

Apparently, Earth's climates slowly fluctuated until the past 1.2 billion years, when temperature patterns with cycles of 200–300 million years became more pronounced. The most recent cold episode was the Pleistocene Epoch, which began in earnest 1.65 million years ago and through which we may still be progressing. The Holocene Epoch began approximately 10,000 years ago, when average temperatures abruptly increased 6 C° (11 F°). The period we live in may represent an end to the Pleistocene, or it may be merely a mild interglacial time. Figure 17.30 details the climatic record of the past 160,000 years.

Medieval Warm Period and Little Ice Age

In A.D. 1001, Leif Eriksson inadvertently ventured onto the North American continent, perhaps the first European to do so. He and his fellow Vikings were favored by a medieval warming episode as they sailed the less-frozen North Atlantic to settle Iceland and Greenland.

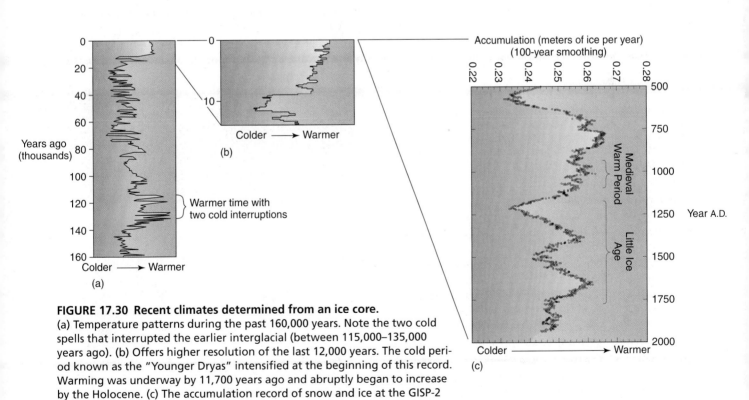

FIGURE 17.30 Recent climates determined from an ice core.
(a) Temperature patterns during the past 160,000 years. Note the two cold spells that interrupted the earlier interglacial (between 115,000–135,000 years ago). (b) Offers higher resolution of the last 12,000 years. The cold period known as the "Younger Dryas" intensified at the beginning of this record. Warming was underway by 11,700 years ago and abruptly began to increase by the Holocene. (c) The accumulation record of snow and ice at the GISP-2 ice-core site is analogous to warmer and colder climatic conditions in Greenland—warmer times yield greater accumulations than colder times because of the moisture capacity of warmer air. Note the Medieval Warm Period and consistent temperatures as compared with the chaotic record of the Little Ice Age. [(a and b) Courtesy of the Greenland Ice Core Project (GRIP); (c) from D. A. Meese and others, "The accumulation record from the GISP-2 core as an indicator of climate change throughout the Holocene," *Science* 266 (December 9, 1994): 1681, ©American Association for the Advancement of Science.]

The mild climatic episode that lasted from about A.D. 800 to 1200 is known as the *Medieval Warm Period*. During the warmth, grape vineyards were planted far into England some 500 km north of present-day commercial plantings. Oats and barley were planted in Iceland, and wheat was planted as far north as Trondheim, Norway. The shift to warmer, wetter weather influenced migration and settlement northward in North America, Europe, and Asia.

However, from approximately 1200–1350 through 1800–1900, a *Little Ice Age* took place. Parts of the North Atlantic froze, and expanding glaciers blocked many key mountain passes in Europe. Snowlines in Europe lowered about 200 m (650 ft) in the coldest years. The Greenland colonies were deserted. Cropping patterns changed, and northern forests declined, along with human population in those regions. In the winter of 1779–1780, New York's Hudson and East Rivers and the entire Upper Bay froze over. People walked and hauled heavy loads across the ice between Staten and Manhattan Islands!

However, the Little Ice Age was not consistently cold throughout its 700-year reign. Reading the record of the Greenland ice cores, scientists have found many mild years among the harsh. More accurately, this was a time of rapid, short-term climate fluctuations that lasted only decades.

Ice cores drilled in Greenland have revealed a record of annual snow and ice accumulation that, when correlated with other aspects of the core sample, is indicative of air temperature. Figure 17.30c presents this record since A.D. 500. The warmth of the Medieval Warm Period is evident, whereas the Little Ice Age appears mixed with colder conditions around 1200, 1500, and after about 1800.

Mechanisms of Climate Fluctuation

What mechanisms cause short-term fluctuations? And why is Earth pulsing through long-term climatic changes that span several hundred million years? The ice age concept is being researched and debated with an unprecedented intensity for three principal reasons: (1) Continuous ice cores from Greenland and Antarctica are providing a new, detailed record of weather and climate patterns, volcanic eruptions, and trends in the biosphere (discussed in News Report 17.2). (2) To understand present and future climate change and to refine general circulation models, we must understand the natural variability of the atmosphere and climate. And (3) global warming, and its relation to ice ages, is a major concern.

Because past occurrences of low temperature appear to have followed a pattern, researchers have looked for causes that also are cyclic in nature. They have identified a complicated mix of interacting variables that appear to influence long-term climatic trends. Let us take a look at several of them.

Climate and Celestial Relations   As our Solar System revolves around the distant center of the Milky Way, it crosses the plane of the galaxy approximately every 32 mil-

lion years. At that time, Earth's plane of the ecliptic aligns parallel to the galaxy's plane, and we pass through regions in space of increased interstellar dust and gas, which may have some climatic effect.

Milutin Milankovitch (1879–1954), a Yugoslavian astronomer who studied Earth–Sun orbital relations, developed other possible astronomical factors. Milankovitch wondered whether the development of an ice age relates to seasonal astronomical factors—Earth's revolution around the Sun, rotation, and tilt—extended over a longer time span (Figure 17.31). In summary:

- Earth's elliptical orbit about the Sun is not constant. The shape of the ellipse varies by more than 17.7 million kilometers (11 million miles) during a 100,000-year cycle, from nearly circular to an extreme ellipse (Figure 17.31a).
- Earth's axis "wobbles" through a 26,000-year cycle, in a movement much like that of a spinning top winding down. Earth's wobble is called precession. As you can see in Figure 17.31b, precession changes the orientation of hemispheres and landmasses to the Sun.
- Earth's present axial tilt of 23.5° varies from 22° to 24° during a 40,000-year period (Figure 17.31c).

Milankovitch calculated, without the aid of today's computers, that the interaction of these Earth–Sun relations creates a 96,000-year climatic cycle. His glaciation model assumes that changes in astronomical relations affect the amounts of insolation received.

Milankovitch died in 1954, his ideas still not accepted by a skeptical scientific community. Now, in the era of com-

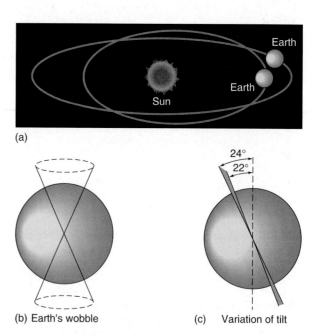

(a)

(b) Earth's wobble   (c)   Variation of tilt

FIGURE 17.31 Astronomical factors that may affect broad climatic cycles.
(a) Earth's elliptical orbit varies widely during a 100,000-year cycle, stretching out to an extreme ellipse. (b) Earth's 26,000-year axial wobble. (c) Variation in Earth's axial tilt every 40,000 years.

GRIP and GISP-2: Boring Ice for Exciting History

The Greenland Ice Core Project (GRIP) was launched in 1989. A site was selected near the summit of the Greenland ice sheet at 3200 m (10,500 ft) so that the maximum thickness of ice history would be drilled through (Figure 1). After 3 years, the drills hit bedrock 3030 m (9940 ft) below the site—or, in terms of time, 250,000 years into the past. The core is 10 cm (4 in.) in diameter.

In 1990, about 32 km (20 mi) west of the summit, the Greenland Ice Sheet Project (GISP-2) began to bore back through time (see Figure 1). GISP-2 reached bedrock in 1993. This core is slightly larger in diameter, at 13.2 cm (5.2 in.), and collects about twice the data as GRIP. The existence of a second core helps scientists compensate for any folds or disturbed sections they encountered below 2700 m, or about 115,000 years, in the first core. Also clarifying the record are correlations with Antarctic ice cores, core samples of marine sediments from the oceans and Lake Baykal in Siberia, and coral studies that indicate past sea levels.

What is being discovered from a 3030-m ice core? Locked into the core are air bubbles of past atmospheres, which indicate ancient gas concentrations. Of special interest are the greenhouse gases, carbon dioxide and methane. Chemical and physical properties of the atmosphere and the snow that accumulated each year are frozen in place.

Pollutants are locked into the core record. For example, during cold periods, high concentrations of dust were

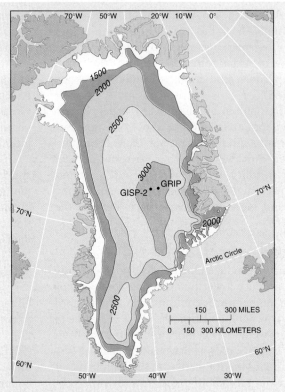

FIGURE 1 Greenland ice core locations.
GRIP is at 72.5° N 37.5° W. GISP-2 is at 72.6° N 38.5° W.
[Map courtesy of GISP-2 Science Management Office.]

present, brought by winds from distant dry lands. An invaluable record of past volcanic eruptions is included in the layers, as if on a calendar. Even the exact beginning of the Bronze Age is recorded in the ice core—about 3000 B.C. When the Greeks and later the Romans began smelting copper, they produced ash and smoke that the winds carried to this distant place. The presence of ammonia indicates ancient forest fires at lower latitudes. And, as an

important analog of past temperatures on the ice-sheet surface when each snowfall occurred, the ratio between stable forms of oxygen is measured.

GRIP and GISP-2 have greatly refined the paleoclimatology of the late Cenozoic Era, helping us to understand Earth's dynamic climate system. Our chances of predicting future patterns have improved, thanks to these efforts.

---

puters, remote-sensing satellites, and worldwide efforts to decipher past climates, Milankovitch's valuable work has stimulated much research to explain climatic cycles and has experienced some confirmation. A roughly 100,000-year climatic cycle is confirmed in such diverse places as ice cores in Greenland and the accumulation of sediment in Lake Baykal, Siberia.

Climate and Solar Variability   If the Sun significantly varies its output over the years, as some other stars do, that variation would seem a convenient and plausible cause of

ice-age timing. However, lack of evidence that the Sun's radiation output varies significantly over long cycles argues against this hypothesis. Nonetheless, inquiry about the Sun's variability actively continues and research using deep-space satellites is underway.

Climate and Tectonics   Major glaciations also can be associated with plate tectonics because some landmasses have migrated to higher, cooler latitudes. Chapters 11 and 12 explain that the shape and orientation of landmasses and ocean basins have changed greatly during Earth's history.

Continental plates have drifted from equatorial locations to polar regions and vice versa, thus exposing the land to a gradual change in climate. Gondwana (the southern half of Pangaea) experienced extensive glaciation that left its mark on the rocks of parts of present-day Africa, South America, India, Antarctica, and Australia. Landforms in the Sahara, for example, bear the markings of even earlier glacial activity. These markings are partly explained by the fact that portions of Africa were centered near the South Pole during the Ordovician Period, 465 million years ago (see Figure 11.16a).

Episodes of mountain building over the past billion years have forced mountain summits above the snowline, where snow remains after the summer melt. Mountain chains influence downwind weather patterns and jet stream circulation, which in turn guides weather systems. More dust was present during glacial periods, suggesting drier weather and more extensive deserts beyond the frozen regions.

Climate and Atmospheric Factors   Some events alter the atmosphere and produce climate change. A volcanic eruption might produce lower temperatures for a year or two. The lower temperatures could initiate a buildup of long-term snow cover at high latitudes. These high-albedo snow surfaces then would reflect more insolation away from Earth to further enhance cooling in a positive feedback system. The eruption of Mount Pinatubo in the Philippines in 1991 caused a temporary cooling and possibly other climatic effects.

The fluctuation of atmospheric greenhouse gases could trigger higher or lower temperatures. An ice core taken at Vostok, the Russian research station near the geographic center of Antarctica, contained trapped air samples from more than 200,000 years in the past. It showed carbon dioxide levels varying from more than 290 ppm to a low of near-ly 180 ppm (Figure 17.32). Higher levels of carbon dioxide generally correlate with each interglacial, or warmer period. In these first few years of a new century, atmospheric carbon dioxide will reach 370 ppm, higher than at any time in the past 240,000 years, principally owing to anthropogenic (human-created) sources.

Climate and Oceanic Circulation   Finally, oceanic circulation patterns have changed. For example, the Isthmus of Panama formed about 3 million years ago and effectively separated the circulation of the Atlantic and Pacific Oceans. Changes in ocean basin configuration, surface temperatures, and salinity and in upwelling and downwelling rates affect air mass formation and air temperature.

Our understanding of Earth's climate—past, present, and future—is unfolding. We are learning that climate is a multicyclic system controlled by an interacting set of cooling and warming processes, all founded on celestial relations, tectonic factors, atmospheric variables, and changes in oceanic circulation.

Arctic and Antarctic Regions

Climatologists use environmental criteria to define the Arctic and the Antarctic regions. The *Arctic region* is defined in Figure 17.33 by the green line. This is the 10°C (50°F) isotherm for July (the Northern Hemisphere summer). This line coincides with the visible treeline—the boundary between the northern forests and tundra. The Arctic Ocean is covered by two kinds of ice: floating sea ice (frozen seawater) and glacier ice (frozen freshwater). This ice pack thins in the summer months and sometimes breaks up. As mentioned in Chapter 10, almost half of the Arctic ice pack has disappeared since 1970 due to regional-scale warming. Scientists are discovering more about the behavior of the Arctic Ocean ice during the last ice age. (See News Report 17.3.)

The *Antarctic region* is defined by the Antarctic convergence, a narrow zone that extends around the continent as a boundary between colder Antarctic water and warmer water at lower latitudes. This boundary follows roughly the 10°C (50°F) isotherm for February (the Southern Hemisphere summer) and is located near 60° S latitude (green line in the figure). The Antarctic region that is covered just with sea ice represents an area greater than North America, Greenland, and Western Europe combined! (For more information on polar region ice, see the National Ice Center at **http://www.natice.noaa.gov/** and the Canadian Ice Service at **http://www.cis.ec.gc.ca/**).

Antarctica is a continent-sized landmass and therefore is much colder overall than the Arctic, which is an ocean. In simplest terms, Antarctica can be thought of as a continent covered by a single enormous glacier, although it contains distinct regions such as the East Antarctic and West Antarctic ice sheets, which respond differently to slight climatic variations. These ice sheets are in constant motion, as indicated in Figure 17.33b.

Ice sheet edges that enter coastal bays form extensive ice shelves, with sharp ice cliffs rising up to 30 m (100 ft)

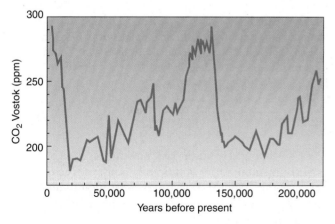

FIGURE 17.32 Vostok ice core and past carbon dioxide concentrations.
Carbon dioxide, trapped in air bubbles in Antarctic ice, was measured from the Vostok ice core. [Adapted by permission from J. Jouzel and others, "Extending the Vostok ice-core record of paleoclimate to the penultimate glacial period," *Nature* 364 (July 29, 1993): 411. © 1993 Macmillan Magazines Ltd.]

FIGURE 17.33 The Arctic and Antarctic regions.
In (a), note the 10°C (50°F) isotherm in midsummer, which designates the Arctic region. In (b), the Antarctic convergence designates the Antarctic region. Arrows on the ice sheet show the general direction of ice movement on Antarctica.

(a)

above the sea. Large tabular islands of ice are formed when sections of the shelves break off and move out to sea (see Figure 10.34). These ice islands can be very large; several in the late 1980s and again in 1995, 1998, 2001, and 2002 exceeded the area of Rhode Island, and one the area of Delaware. The disintegration of some of these ice shelves caused by higher temperatures is discussed in Chapter 10 and illustrated by the Pine Island Glacier in Figure 17.34.

The scientific importance of Antarctica in the words of the people who work there is recorded in the New South Polar Times at **http://www.spotsylvania.k12.va.us/nspt/ home.htm**. A place so remote from civilization is an excellent laboratory for sampling past and present human and natural variables that are transported by atmospheric and oceanic circulation. Far from pollution sources, so cold and dark in winter, and high altitude make it an ideal location for certain astronomical observations. Glacial processes, landforms, and Earth's frozen climatic record in the ice sheet all provide clues for further scientific exploration (Figure 17.35).

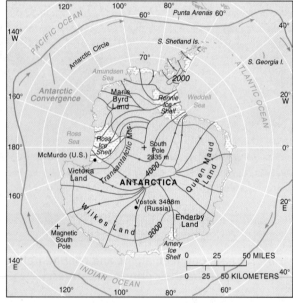

(b)

FIGURE 17.34 The Pine Island Glacier breaks up.
(a) A crack appears in the Pine Island Glacier in December 2000. (b) Between November 4 and 12, 2001, the large crack expanded through the ice shelf at 15 m (50 ft) per day, a surprising rate. The tabular iceberg breaks loose and is 42 km by 17 km (26 mi by 11 mi). Pine Island is Antarctica's largest outlet glacier in terms of discharge and it appears to be the fastest moving. This area is regarded as the weakest part of the West Antarctic Ice sheet, most susceptible to failure. [(a) *Terra* ASTER sensor image courtesy of NASA/GSFC/MITI/ERS-DAC/JAROS, and the U.S./Japan ASTER Science Team. (b) *Terra* MISR sensor image courtesy of NASA/GSFC/LaRC/JPL, and the MISR Team.]

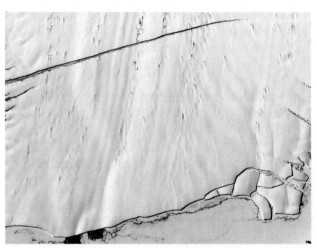

(a)

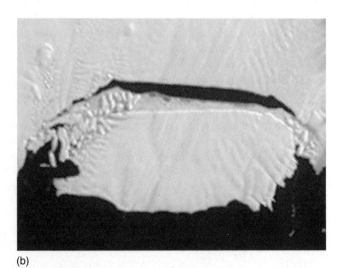

(b)

News Report 17.3

An Arctic Ice Sheet?

Questions remain about the nature of the ice in the Arctic Ocean during the Pleistocene Ice Age. Some scientists believe that the ice was similar to the thin sea ice we see today. Others think that the region was blanketed under deep ice cover. The latter were pleased when sonar images of the ocean floor, published in 1994, showed deep gouges and grooves in bottom sedi-

ments on the floor of the Fram Strait between Greenland and Norway.

Imagine icebergs twice the size of the largest in Antarctic water, extending up to 300 m (1000 ft) high and as much as 700 m (2300 ft) deep! These enormous ice vessels scarred the ocean floor with their deep, icy keels. In another area, the sonar disclosed submerged ridges that were scraped

smooth. These super-icebergs could have originated from a floating ice sheet or from glaciers surrounding the Arctic Basin. Now the question: To what extent were the North American and European ice sheets joined across the Arctic Ocean? Scientists are still investigating this question!

FIGURE 17.35 Scientific base at the South Pole.
Aerial view of the Amundsen-Scott South Pole Station made in 2001. The geodesic dome completed in 1975, 50 m wide and 16 m high (165 ft by 52 ft), shelters buildings from the frigid cold and winds. Twenty-eight scientists and support people work through the winter (February to October) and 130 personnel, or more, research and work there in the brief summer. On the left side you see the first of the new modular buildings, which stand 3 m (10 ft) above the surface to accommodate snow accumulation and protect the ice under the building from melting. Built entirely from materials airlifted to the base, they will be completed in 2006. Above this module is the ceremonial South Pole; slightly to the right is the geographic South Pole. This station is at an elevation of 2835 m (9301 ft). [Photo courtesy of Kristan Hutchison, U.S. Antarctic Program.]

Summary and Review—Glacial and Periglacial Processes and Landforms

● *Differentiate* between alpine and continental glaciers and *describe* their principal features.

More than 77% of Earth's freshwater is frozen. Ice covers about 11% of Earth's surface, and periglacial features occupy another 20% of ice-free but cold-dominated landscapes. A **glacier** is a mass of ice sitting on land or floating as an ice shelf in the ocean next to land. Glaciers form in areas of permanent snow. A **snowline** is the lowest elevation where snow occurs year-round and its altitude varies by latitude—higher near the equator, lower poleward.

A glacier in a mountain range is an **alpine glacier**. If confined within a valley, it is termed a *valley glacier*. The area of origin is a snowfield, usually in a bowl-shaped erosional landform called a **cirque**. Where alpine glaciers flow down to the sea, they calve and form **icebergs**. A **continental glacier** is a continuous mass of ice on land. Its most extensive form is an **ice sheet**; a smaller, roughly circular form is an **ice cap**; and the least extensive form, usually in mountains, is an **ice field**.

glacier (p. 520)
snowline (p. 520)
alpine glacier (p. 520)

cirque (p. 521)
iceberg (p. 521)
continental glacier (p. 522)
ice sheet (p. 522)
ice cap (p. 522)
ice field (p. 523)

1. Describe the location of most freshwater on Earth today.
2. What is a glacier? What is implied about existing climate patterns in a glacial region?
3. Differentiate between an alpine glacier and a continental glacier.
4. Name the three types of continental glaciers. What is the basis for dividing continental glaciers into types? Which type covers Antarctica?

● *Describe* the process of glacial ice formation and *portray* the mechanics of glacial movement.

Snow becomes glacial ice through accumulation, increasing thickness, pressure on underlying layers, and recrystallization. Snow progresses through transitional steps from **firn** (compact, granular) to a denser **glacial ice** after many years.

A glacier is an open system with inputs and outputs that can be analyzed through observation of the growth and wasting of the glacier itself. A **firn line** is the lower extent of a fresh snow-covered area. A glacier is fed by snowfall and is wasted by **ablation** (losses from its upper and lower surfaces and along its margins). Accumulation and ablation achieve a mass balance in each glacier.

As a glacier moves downhill, vertical **crevasses** may develop. Sometimes a glacier will move rapidly in a **glacier surge**. The presence of water along the basal layer appears to be important in glacial movements. As a glacier moves, it plucks rock pieces and debris, incorporating them into the ice, and this debris scours and sandpapers underlying rock through **abrasion**.

firn (p. 525)
glacial ice (p. 525)
firn line (p. 525)
ablation (p. 525)
crevasse (p. 526)
glacier surge (p. 527)
abrasion (p. 528)

5. Trace the evolution of glacial ice from fresh fallen snow.
6. What is meant by glacial mass balance? What are the basic inputs and outputs underlying that balance?
7. What is meant by a glacier surge? What do scientists think produces surging episodes?

● *Describe* **characteristic erosional and depositional landforms created by alpine glaciation and continental glaciation.**

Extensive valley glaciers have profoundly reshaped mountains worldwide, carving **V**-shaped stream valleys into **U**-shaped glaciated valleys, producing many distinctive erosional and depositional landforms. As cirque walls erode away, sharp **arêtes** (sawtooth, serrated ridges) form, dividing adjacent cirque basins. Two eroding cirques may reduce an arête to a saddlelike **col**. A **horn** results when several cirque glaciers gouge an individual mountain summit from all sides, forming a pyramidal peak. An ice-carved rock basin left as a glacier retreats may fill with water to form a **tarn**; tarns in a string separated by moraines are called **paternoster lakes**. Where a glacial valley trough joins the ocean, and the glacier retreats, the sea extends inland to form a **fjord**.

All glacial deposits, whether ice-borne or meltwater-borne, constitute **glacial drift**. Glacial meltwater deposits are sorted and are called **stratified drift**. Direct deposits from ice, called **till**, are unstratified and unsorted. Specific landforms produced by the deposition of drift are **moraines**. A **lateral moraine** forms along each side of a glacier; merging glaciers with lateral moraines form a **medial moraine**; and eroded debris dropped at the glacier's terminous is a **terminal moraine**.

Continental glaciation leaves different features than does alpine glaciation. A **till plain** forms behind end moraines, featuring unstratified coarse till, low and rolling relief, and deranged drainage. Beyond the moraine deposits, **outwash plains** of stratified drift feature stream channels that are meltwater-fed, braided, and overloaded with debris that is sorted and deposited across the landscape. An **esker** is a sinuously curving, narrow ridge of coarse sand and gravel that forms

along the channel of a meltwater stream beneath a glacier. An isolated block of ice left by a retreating glacier becomes surrounded with debris; when the block finally melts, it leaves a steep-sided **kettle**. A **kame** is a small hill, knob, or mound of poorly sorted sand and gravel that is deposited directly by water or by ice in crevasses.

Glacial action forms two types of streamlined hills: the erosional **roche moutonnée** is an asymmetrical hill of exposed bedrock, gently sloping upstream and abruptly sloping downstream; the depositional **drumlin** is deposited till, streamlined in the direction of continental ice movement (blunt end upstream and tapered end downstream).

arête (p. 529)
col (p. 529)
horn (p. 529)
tarn (p. 529)
paternoster lakes (p. 529)
fjord (p. 529)
glacial drift (p. 529)
stratified drift (p. 529)
till (p. 529)
moraine (p. 531)
lateral moraine (p. 531)
medial moraine (p. 531)
terminal moraine (p. 531)
till plain (p. 532)
outwash plain (p. 534)
esker (p. 534)
kettle (p. 534)
kame (p. 534)
roche moutonnée (p. 534)
drumlin (p. 534)

8. How does a glacier accomplish erosion?
9. Describe the evolution of a **V**-shaped stream valley to a **U**-shaped glaciated valley. What features are visible after the glacier retreats?
10. How is an iceberg generated?
11. Differentiate between two forms of glacial drift—till and outwash.
12. What is a morainal deposit? What specific moraines are created by alpine and continental glaciers?
13. What are some common depositional features encountered in a till plain?
14. Contrast a roche moutonnée and a drumlin regarding appearance, orientation, and the way each forms.

● *Analyze* **the spatial distribution of periglacial processes and** *describe* **several unique landforms and topographic features related to permafrost and frozen ground phenomena.**

The term **periglacial** describes cold-climate processes, landforms, and topographic features that exist along the margins of glaciers, past and present. Periglacial regions occupy more than 20% of Earth's land surface. These areas have either near-permanent ice or are at high elevation, and the ground is seasonally snow-free. When soil or rock temperatures remain below 0°C (32°F) for at least 2 years, **permafrost** ("permanent frost") develops. An area of permafrost that is not covered by glaciers is considered periglacial. Note that this criterion is based solely on temperature and has nothing to do with how much or how little water is present. The **active layer** is the zone of seasonally frozen ground that exists between the subsurface permafrost layer and the ground surface.

periglacial (p. 534)
permafrost (p. 534)
active layer (p. 536)

15. In terms of climatic types, describe the areas on Earth where periglacial landscapes occur. Include both higher latitude and higher altitude climate types.

16. Define two types of permafrost, and differentiate their occurrence on Earth. What are the characteristics of each?

17. Describe the active zone in permafrost regions, and relate the degree of development to specific latitudes.

18. What is a talik? Where might you expect to find taliks, and to what depth do they occur?

19. What is the difference between permafrost and ground ice?

20. Describe the role of frost action in the formation of various landform types in the periglacial region.

21. Relate some of the specific problems humans encounter in developing periglacial landscapes.

● *Explain* the Pleistocene ice age epoch and related glacials and interglacials and *describe* some of the methods used to study paleoclimatology.

An **ice age** is any extended period of cold. The late Cenozoic Era has featured pronounced ice-age conditions in an epoch called the Pleistocene. During this time, alpine and continen-tal glaciers covered about 30% of Earth's land area in at least 18 glacials, punctuated by interglacials of milder weather. Beyond the ice, **paleolakes** formed because of wetter conditions. Evidence of ice-age conditions is gathered from ice cores drilled in Greenland and Antarctica, from ocean sediments, from coral growth in relation to past sea levels, and from rock. The study of past climates is *paleoclimatology*.

The apparent pattern followed by these low-temperature episodes indicates cyclic causes. A complicated mix of interacting variables appears to influence long-term climatic trends: celestial relations, solar variability, tectonic factors, atmospheric variables, and oceanic circulation.

ice age (p. 541)
paleolake (p. 543)

22. What is paleoclimatology? Describe Earth's past climatic patterns. Are we experiencing a normal climate pattern in this era, or have scientists noticed any significant trends?

23. Define an ice age. When was the most recent? Explain "glacial" and "interglacial" in your answer.

24. Summarize what science has learned about the causes of ice ages by listing and explaining at least four possible factors in climate change.

25. Describe the role of ice cores in deciphering past climates. What record do they preserve? Where were they drilled?

26. Explain the relationship between the criteria defining the Arctic and Antarctic regions? Is there any coincidence in these criteria and the distribution of Northern Hemisphere forests on the continents?

NetWork

The *Geosystems* Home Page provides on-line resources for this chapter on the World Wide Web. To begin: Once on the Home Page, click on this textbook, scroll the Table of Contents menu, select this chapter, and click "Begin." You will find self-tests that are graded, review exercises, specific updates for items in the chapter, and many links to interesting related pathways on the Internet under "Destinations." *Geosystems* is at **http://www.prenhall.com/christopherson.**

Critical Thinking

A. After checking back issues of the New South Polar Times at **http://www.spotsylvania.k12.va.us/nspt/home.htm,** imagine duty there for yourself. Of the 28 people who winter over at the station, some serve as scientists, technicians, and support staff. The station commander for the 1998 season was Katy McNitt-Jensen, in her third tour at the pole. Remember, the last airplane leaves mid-February and the first airplane lands mid-October—such is the isolation. What do you see as the positives and negatives to such service? How would you combat the elements? The isolation? The cold and dark conditions?

B. Analyze and compare the initial results from ice-core drilling efforts in Greenland and Antarctica as described in this chapter (News Report 17.2 and text). What do they tell us about past climates? How far back do they go in time? Have they found any evidence of humans on Earth in the cores?

C. What is the relationship between knowing about paleoclimates and being able to forecast future climates? Is there a link? Have scientists discovered any cyclic behavior to Earth's past climates that might repeat in the future? Explain.

Career Link 17.1

Karl Birkeland, Avalanche Scientist, Forest Service National Avalanche Center

When snow scientist Karl Birkeland was in second grade, he wrote a little essay called "How Glaciers Form." It reads (spelling corrected): "Glaciers form by large snows, where snow does not melt. The reason glaciers are covered with dirt is because the glaciers shake the canyon and the dirt falls on them. Glaciers must be bigger than the Empire State Building. If they are near the water big icebergs might fall in." Karl added, "So I was thinking about this way back then. I have this framed above my desk, misspellings and all."

Karl completed his undergraduate work at the University of Colorado in environment, population, and biology, and even thought about medical school. His father was a professor of geology at Colorado. A skier since childhood, while finishing his degree he was a member of the ski-patrol. He began to realize that studying the environment, specifically snow science, was what he wanted to do: "Snow science became a focus. Drawing on my ski patrol work where we had to deal with avalanches, I felt I had a good start."

Karl finished his Master's at Montana State University where his thesis topic was "Spatial Variability of Snow Resistance on Potential Avalanche Slopes." He began work as an avalanche specialist in the Gallatin National Forest, Bozeman, Montana, in 1990, where he established the Avalanche Center. He completed his Ph.D. at Arizona State in 1997. In his dissertation, he looked at slopes throughout the Bridger Mountains, Montana, and developed models for assessing variations in snow stability. This was the first time that anyone had tried to map avalanche conditions over a mountain range in one day.

Karl added, "In the Rockies, we have better preserved weak layers within the snowpack. These weak layers can keep the snow packs unstable for long periods of time, even if there hasn't been fresh snowfall." This regional characteristic makes the location of his National Avalanche Center ideal.

FIGURE 1 Karl Birkeland, Avalanche Scientist.
Karl works at the U.S. Forest Service National Avalanche Center.
[Photo by Ginger Birkeland.]

As to his present responsibilities, Karl said, "I'm in charge of transferring new and emerging technologies to the avalanche specialists at the regional avalanche centers around the country. I'm also involved in keeping track of any recent scientific developments in our field that should be conveyed to our avalanche workers." Karl assists the director of the Forest Service National Avalanche Center in coordinating all the avalanche centers.

Karl and his team continue to examine synoptic climatology and analyze air masses to understand resulting snow accumulations. Reconstructing past weather patterns is key. These efforts will produce a classification scheme that categorizes past data into coastal, continental, or intermountain characteristic types. "We can tie this in with weather patterns and predict the kind of avalanches we might expect on a given day." He said the big challenge is to "take scientific work and distill out the day-to-day practical significance for avalanche forecasters." In other words, what is included in this research that can help avalanche forecasters right now?

I kept hearing "spatial" in his work, getting a sense of the geographer in him. "That's me!" he answers, "there is such a need to see things spatially, regionally, with maps as the medium of communication." Geography is in the

family, as Karl's wife is a Ph.D. geographer studying rivers and riparian vegetation across the Colorado Plateau and southwestern deserts.

Karl participates in the biennial International Snow Science Workshop (see http://www.avalanche.org). The key challenge is merging theory and practice. More than 600 scientists attended the 2000 meeting in Montana. The group is working to increase the global database, accident statistics, and forecasting efforts of many scientists in this growing field of interest. The future involves utilization of GIS modeling techniques to better tie together snow pack variables for spatial analysis.

"We want to bring together the operational avalanche workers and the avalanche researchers. The Internet has done a great job at linking this small community of individuals," Karl said. "This communication is needed because the avalanche risk is growing, owing to the increased development and construction in the mountains that are avalanche prone. All in all, the future is quite exciting in snow science!" And, Karl declared, "I continue to be very pro-geography and I like calling myself a geographer! Much of my current research focuses on spatial variations of snow pack properties, a clearly geographic topic that has critical real-world applications for avalanche forecasters."

PART FOUR

Soils, Ecosystems, and Biomes

A misty morning near Heather Meadows and the Nooksack River, below the Mount Baker volcano, Washington. Earth's physical systems interact to produce a unique assemblage of soils and ecosystems in the cool, moist northwest. [Photo by Bobbé Christopherson.]

Earth is the home of the only known biosphere in the Solar System—a unique, complex, and interactive system of abiotic (nonliving) and biotic (living) components working together to sustain a tremendous diversity of life. Energy enters the biosphere through conversion of solar energy by photosynthesis in the leaves of plants. Soil is the essential link among the lithosphere, plants, and the rest of Earth's physical systems. Thus, soil helps sustain life.

Life is organized into a feeding hierarchy from producers to consumers, ending with decomposers. Taken together, the soils, plants, animals, and all abiotic components produce aquatic and terrestrial ecosystems, generally grouped together in various biomes. Today we face crucial issues, principally the preservation of the diversity of life in the biosphere and the survival of the biosphere itself. Patterns of land and ocean temperatures, precipitation, weather phenomena, stratospheric ozone, among many elements, are evolving into different relations. The resilience of the biosphere, as we know it, is being tested in a real-time, one-time experiment. These important issues of biogeography are considered in Part 4.

SOLAR ENERGY

Atmosphere

Hydrosphere

Biosphere

Lithosphere

Vertisols in the Texas coastal plain, northeast of Palacios near the Tres Palacios River, planted with a commercial sorghum crop. Note the dark soil color indicative of Vertisols, wet and shiny from the rains of Tropical Storm Allison, June 2001. [Photo by Bobbé Christopherson.]

18

The Geography of Soils

Key Learning Concepts

After reading the chapter, you should be able to:

- *Define* soil and soil science and *describe* a pedon, polypedon, and typical soil profile.
- *Describe* soil properties of color, texture, structure, consistence, porosity, and soil moisture.
- *Explain* basic soil chemistry, including cation-exchange capacity, and *relate* these concepts to soil fertility.
- *Evaluate* principal soil formation factors, including the human element.
- *Describe* the 12 soil orders of the Soil Taxonomy classification system and *explain* their general occurrence.

Earth's landscape generally is covered with soil. **Soil** is a dynamic natural material composed of fine particles in which plants grow, and it contains both mineral fragments and organic matter. The soil system includes human interactions and supports all human, other animal, and plant life. If you have ever planted a garden, tended a house plant, or been concerned about famine and soil loss, this chapter will interest you. A knowledge of soil is at the heart of agriculture and food production.

You kneel down and scoop a handful of prairie soil, compressing it and breaking it apart with your fingers. You are holding a historical object, one that bears the legacy of the last 15,000 years or more. This lump of soil contains information about the last ice age and intervening warm periods, about distinct and distant source materials, about several physical processes. We are using and abusing this legacy at rates much faster than it formed. Soils do not reproduce, nor can they be re-created.

Soil science is interdisciplinary, involving physics, chemistry, biology, mineralogy, hydrology, taxonomy, climatology, and cartography. Physical geographers are interested in the spatial patterns formed by soil types and the environmental factors that interact to produce them. As an integrative science, physical geography is well suited to the study of soils.

Pedology concerns the origin, classification, distribution, and description of soil (*ped* from the Greek pedon, meaning "soil" or "earth"). Pedology is at the center of learning about soil as a natural body, but it does not dwell on its practical uses. *Edaphology* (from the Greek *edaphos*, meaning "soil" or "ground") focuses on soil as a medium for sustaining higher plants. Edaphology emphasizes plant growth, fertility, and the differences in productivity among soils. Pedology gives us a general understanding of soils and their classification, whereas edaphology reflects society's concern for food and fiber production and the management of soils to increase fertility and reduce soil losses.

In many locales, an *agricultural extension service* can provide specific information and perform a detailed analysis of local soils. Soil surveys and local soil maps are available for most counties in the United States and for the Canadian provinces. Your local phone book may list the U.S. Department of Agriculture, Natural Resources Conservation Service (**http://www.nrcs.usda.gov/**), or Agriculture Canada's Soil Information System (**http://sis.agr.gc.ca/cansis/intro.html**), or you may contact the appropriate department at a local college or university. See the National Soil Survey Center's site at **http://www.statlab.iastate.edu/soils/nssc/**, and for links related to world soils, see **http://www.metla.fi/info/vlib/soils/old.htm**.

In this chapter: The geography of soils deals spatially with a complex substance, the characteristics of which vary from kilometer to kilometer, and even centimeter to centimeter. We begin with soil characteristics and the basic soil sampling and soil mapping units. The soil profile is a dynamic structure, mixing and exchanging materials and moisture across its horizons. Properties of soil include texture, structure, porosity, moisture, and chemistry—all integrating to form soil types. The chapter discusses both natural and human factors that affect soil formation. A global concern exists over the loss of soils to erosion, mistreatment, and conversion to other uses. The chapter concludes with a brief examination of the Soil Taxonomy, the 12 principal soil orders, and their spatial distribution.

Soil Characteristics

Classifying soils is similar to classifying climates, because both involve interacting variables. Before we look at soil classification, let us examine the physical properties that distinguish soils as they develop through time, in response to climate, relief, and topography.

Soil Profiles

Just as a book cannot be judged by its cover, so soils cannot be evaluated at the surface only. Instead, a soil profile should be studied from the surface to the deepest extent of plant roots, or to where regolith or bedrock is encountered. Such a profile, called a **pedon**, is a hexagonal column measuring 1 to 10 m² in top surface area (Figure 18.1). At the sides of the pedon, the various layers of the soil profile are visible in cross section and are labeled with letters. *A pedon is the basic sampling unit used in soil surveys.*

Many pedons together in one area make up a **polypedon**, which has distinctive characteristics differentiating it from surrounding polypedons. A polypedon is an essential soil individual, comprising an identifiable series of soils in an area. It can have a minimum dimension of about 1 m² and no specified maximum size. *The polypedon is the basic mapping unit used in preparing local soil maps.*

Soil Horizons

Each distinct layer exposed in a pedon is a **soil horizon**. A horizon is roughly parallel to the pedon's surface and has characteristics distinctly different from horizons directly above or below. The boundary between horizons usually is distinguishable in the field, on the basis of the properties of color, texture, structure, consistence (meaning soil consistency or cohesiveness), porosity, the presence or absence of certain minerals, moisture, and chemical processes (Figure 18.2). Soil horizons are the building blocks of soil classification.

At the top of the soil profile is the *O* (organic) horizon, named for its organic composition, derived from plant and animal litter that was deposited on the surface and transformed into humus. **Humus** is not just a single material; it is a mixture of decomposed and synthesized organic materials, usually dark in color. Microorganisms work busily on this organic debris, performing a portion of the *humification* (humus-making) process. The *O* horizon is 20%–30% or more organic matter, which is important because of its ability to retain water and nutrients and for the way it acts in a complementary manner to clay minerals.

At the bottom of the soil profile is the *R* (rock) *horizon*, consisting of either unconsolidated (loose) material or consolidated bedrock. When bedrock physically and chemically weathers into regolith, it may or may not contribute to overlying soil horizons. The A, E, B, and C horizons mark differing mineral strata between O and R. These middle layers are composed of sand, silt, clay, and other weathered by-products. (Table 11.3 presents a description of grain sizes for these weathered soil particles.)

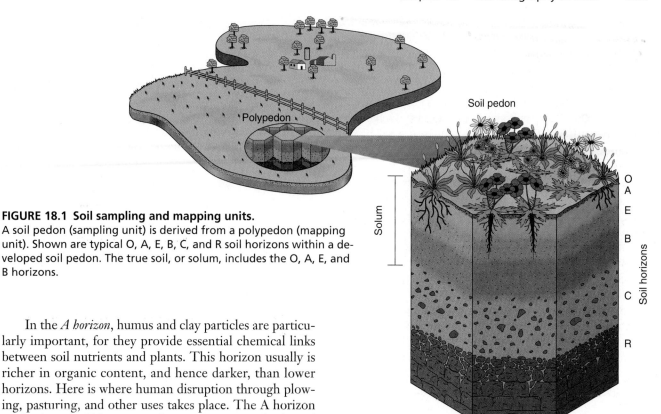

FIGURE 18.1 Soil sampling and mapping units.
A soil pedon (sampling unit) is derived from a polypedon (mapping unit). Shown are typical O, A, E, B, C, and R soil horizons within a developed soil pedon. The true soil, or solum, includes the O, A, E, and B horizons.

In the *A horizon*, humus and clay particles are particularly important, for they provide essential chemical links between soil nutrients and plants. This horizon usually is richer in organic content, and hence darker, than lower horizons. Here is where human disruption through plowing, pasturing, and other uses takes place. The A horizon grades into the *E horizon*, made up of coarse sand, silt, and resistant minerals; generally the E horizon is lighter in color.

From the lighter-colored E horizon, silicate clays and oxides of aluminum and iron are leached (removed by water) and carried to lower horizons with the water as it percolates through the soil. This process of removal of fine particles and minerals by water, leaving behind sand and silt, is termed **eluviation**—thus the E designation for this horizon. The greater the precipitation, the higher the rate of eluviation.

In contrast to the A and E horizons, *B horizons* accumulate clays, aluminum, and iron. B horizons are dominated by **illuviation, a depositional process**. (Eluviation is an erosional removal process; illuviation is depositional.) B horizons may exhibit reddish or yellowish hues because of the illuviated presence of minerals (silicate clays, iron and aluminum, carbonates, gypsum) and organic oxides.

Some materials occurring in the B horizon may have formed in place from weathering processes rather than arriving there by *translocation*, or migration. In the humid tropics, these layers often develop to some depth. Likewise, clay losses in an A horizon may be caused by destructive processes and not eluviation. Research to better understand erosion and deposition of clays between soil horizons is one of the challenges in soil science.

FIGURE 18.2 A typical soil profile.
This is a Mollisol pedon in southeastern South Dakota. The parent material is glacial till, and the soil is well drained. The dark O and A horizons above the #1 transitions into an E horizon. Distinct carbonate nodules are visible in the lower B and upper C horizons. [Photo from the Marbut Collection, Soil Science Society of America.]

The combination of the A and E horizons and the B horizon is designated the **solum,** considered the true definable soil of the pedon. The A, E, and B horizons experience active soil processes (labeled on Figure 18.1).

Below the solum is the *C horizon* of weathered bedrock or weathered parent material. This zone is identified as *regolith* (although the term sometimes is used to include the solum as well). The C horizon is not much affected by soil operations in the solum and lies outside the biological influences experienced in the shallower horizons. Plant roots and soil microorganisms are rare in the C horizon. It lacks clay concentrations and generally is made up of carbonates, gypsum, or soluble salts, or of iron and silica, which form cemented soil structures. In dry climates, calcium carbonate commonly forms the cementing material of these hardened layers.

Soil scientists using the U.S. classification system employ letter suffixes to further designate special conditions within each soil horizon. A few examples include Ap (A horizon has been plowed), Bt (B horizon is of illuviated clay), Bf (permafrost or frozen soil), and Bh (illuviated humus as a dark coating on sand and silt particles). (Soil horizon designations used in the Canadian System of Soil Classification are presented in Appendix B.)

Soil Properties

Soils are complex and varied, as this section reveals. Observing a real soil profile will help you identify color, texture, structure, and other soil properties. A good opportunity to observe soil profiles is at a construction site or excavation, perhaps on your campus, or at a road cut along a highway. The USDA Natural Resources Conservation Service *Soil Survey Manual* (U.S. Department of Agriculture Handbook No. 18, October 1993) presents information on all soil properties. The NRCS publication *Soil Survey Laboratory Methods Manual* (U.S. Department of Agriculture Soil Survey Investigations Report No. 42, v. 3.0, January 1996) details specific methods and practices in conducting soil analysis.

Soil Color

Color is important, for it sometimes suggests composition and chemical makeup. If you look at exposed soil, color may be the most obvious trait. Among the many possible hues are the reds and yellows found in soils of the southeastern United States (high in iron oxides), the blacks of prairie soils in portions of the U.S. grain-growing regions and Ukraine (richly organic), and white-to-pale hues found in soils containing silicates and aluminum oxides. However, color can be deceptive: Soils of high humus content are often dark, yet clays of warm–temperate and tropical regions with less than 3% organic content are some of the world's blackest soils.

To standardize color descriptions, soil scientists describe a soil's color by comparing it with a *Munsell Color Chart* (developed by artist and teacher Albert Munsell in

FIGURE 18.3 A Munsell Soil Color Chart page.
A soil sample is viewed through the hole to match it with a color on the chart. Hue, value, and chroma are the characteristics of color assessed by this system. [Photo courtesy of Gretag Macbeth, Munsell Color.]

1913). These charts display 175 colors arranged by *hue* (the dominant spectral color, such as red), *value* (degree of darkness or lightness), and *chroma* (purity and saturation of the color, which increase with decreasing grayness). A Munsell notation identifies each color by a name, so soil scientists can make worldwide comparisons of soil color. Soil color is checked against the chart at various depths within a pedon (Figure 18.3).

Soil Texture

Soil texture, perhaps a soil's most permanent attribute, refers to the mixture of sizes of its particles and the proportion of different sizes. Individual mineral particles are called *soil separates*. All particles smaller in diameter than 2 mm (0.08 in.), such as very coarse sand, are considered part of the soil. Larger particles such as pebbles, gravel, or cobbles, are not part of the soil. (Sands are graded from coarse, to medium, to fine, down to 0.05 mm, silt to 0.002 mm, and clay to less than 0.002 mm.)

Figure 18.4 is a *soil texture triangle* showing the relation of sand, silt, and clay concentrations in soil. Each corner of the triangle represents a soil consisting solely of the particle size noted (although rarely are true soils composed of a single separate). Every soil on Earth is defined somewhere in this triangle.

Figure 18.4 includes the common designation **loam,** which is a balanced mixture of sand, silt, and clay that is beneficial to plant growth. A sandy loam with clay content below 30% (lower left) usually is considered ideal by farmers because of its water-holding characteristics and ease of cultivation. Soil texture is important in determining water-retention and water-transmission traits.

To see how it works, consider a soil type in Indiana called the *Miami silt loam.* Samples from it are plotted on the soil texture triangle as points 1, 2, and 3. A sample taken near the surface in the A horizon is recorded at point 1, in

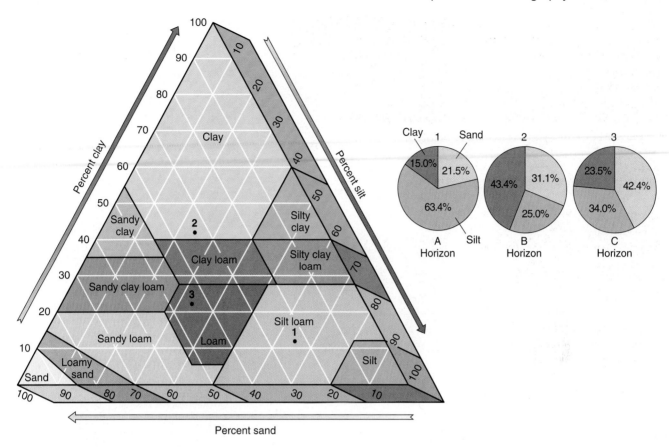

FIGURE 18.4 Soil texture triangle.
Measures of the ratio of clay, silt, and sand determine soil texture. As an example, points 1 (horizon A), 2 (horizon B), and 3 (horizon C) designate samples taken at three different horizons in the Miami silt loam in Indiana. Note the ratio of sand to silt to clay shown in the three pie diagrams. [After U.S. Department of Agriculture, Natural Resources Conservation Service, *Soil Survey Manual*, Agricultural Handbook No. 18, p. 138 (Washington, DC: U.S. Government Printing Office, 1993).]

Table 18.1   Textural Analysis of Miami Silt Loam

Sample Points	% Sand	% Silt	% Clay
1 = A horizon	21.5	63.4	15.0
2 = B horizon	31.1	25.0	43.4
3 = C horizon	42.4	34.0	23.5

Source: J. E. Van Riper, *Man's Physical World*, p. 570. Copyright © 1971 by McGraw-Hill. Adapted by permission.

the B horizon at point 2, and in the C horizon at point 3. Textural analyses of these samples are summarized in Table 18.1. Note that silt dominates the surface, clay the B horizon, and sand the C horizon. The *Soil Survey Manual* presents guidelines for estimating soil texture by feel, a relatively accurate method when used by an experienced person. However, laboratory methods using graduated sieves and separation by mechanical analysis in water allow more-precise measurements.

Soil Structure

Soil *texture* describes the size of soil particles, but soil *structure* refers to the *arrangement* of them. Structure can partially modify the effects of soil texture. The smallest natural lump or cluster of particles is a *ped*. The shape of soil peds determines which of the structural types the soil exhibits: crumb or granular, platy, blocky, prismatic, or columnar (Figure 18.5).

Peds separate from each other along zones of weakness, creating voids (pores) that are important for moisture storage and drainage. Rounded peds have more pore space between them and greater permeability than other shapes. They are therefore better for plant growth than are blocky, prismatic, or platy peds, despite comparable fertility. Terms used to describe soil structure include fine, medium, or coarse. Adhesion among peds ranges from weak to strong.

Soil Consistence

In soil science, the term *consistence* is used to describe the consistency of a soil or cohesion of its particles. Consistence is a product of texture (particle size) and structure

FIGURE 18.5 Types of soil structure.
Structure is important because it controls drainage, rooting of plants, and how well the soil delivers nutrients to plants. The shape of individual peds shown here controls a soil's structure. [Photos by National Soil Survey Center, Natural Resources Conservation Service, USDA, Soil Survey Staff.]

(ped shape). Consistence reflects a soil's resistance to breaking and manipulation under varying moisture conditions:

- A *wet soil* is sticky between the thumb and forefinger, ranging from a little adherence to either finger, to sticking to both fingers, to stretching when the fingers are moved apart. *Plasticity*, the quality of being moldable, is roughly measured by rolling a piece of soil between your fingers and thumb to see whether it rolls into a thin strand.
- A *moist soil* is filled to about half of field capacity (the usable water capacity of soil), and its consistence grades from loose (noncoherent), to *friable* (easily pulverized), to firm (not crushable between thumb and forefinger).
- A *dry soil* is typically brittle and rigid, with consistence ranging from loose, to soft, to hard, to extremely hard.

Soil particles are sometimes cemented together, to some degree. Soils are described as weakly cemented, strongly cemented, or *indurated* (hardened). Calcium carbonate, silica, and oxides or salts of iron and aluminum all can serve as cementing agents. The cementation of soil particles that occurs in various horizons is a function of consistence and may be continuous or discontinuous.

Soil Porosity

Soil porosity, permeability, and moisture storage are discussed in Chapter 9. Pores in the soil horizon control the movement of water—its intake, flow, and drainage—and air ventilation. Important porosity factors are pore *size*, pore *continuity* (whether they are interconnected), pore *shape* (whether they are spherical, irregular, or tubular), pore *orientation* (whether pore spaces are vertical, horizontal, or random), and pore *location* (whether they are within or between soil peds).

Porosity is improved by the biotic actions of plant roots, animal activity such as the tunneling of gophers or worms, and human intervention through soil manipulation (plowing, adding humus or sand, or planting soil-building crops). Much of a farmer's soil preparation work before planting, and for the home gardener as well, is done to improve soil porosity.

Soil Moisture

Reviewing Figures 9.9 and 9.10 (soil moisture types and availability) will help you understand this section. Plants operate most efficiently when the soil is at *field capacity*, which is the maximum water availability for plant use after large pore spaces have drained of gravitational water. Field capacity is determined by soil type. The depth to which a plant sends its roots determines the amount of soil moisture to which the plant has access. If soil moisture is removed below field capacity, plants must exert increased energy to obtain available water. This moisture removal inefficiency worsens until the plant reaches its wilting. Beyond this point, plants are unable to extract the water they need, and they die.

Soil moisture regimes and their associated climate types shape the biotic and abiotic properties of the soil more than any other factor. The U.S. Natural Resources Conservation Service recognizes five soil moisture regimes based on Thornthwaite's water-balance principles (see "The Soil-Water-Budget Concept" in Chapter 9) (Table 18.2).

Soil Chemistry

Recall that soil pores may be filled with air, water, or a mixture of the two. Consequently, soil chemistry involves both air and water. The atmosphere within soil pores is mostly nitrogen, oxygen, and carbon dioxide. Nitrogen concentrations are about the same as in the atmosphere, but oxygen is less and carbon dioxide is greater because of ongoing respiration processes.

Water present in soil pores is called the *soil solution*. It is the medium for chemical reactions in soil. This solution is critical to plants as their source of nutrients, and it is the foundation of *soil fertility*. Carbon dioxide combines with the water to produce carbonic acid, and various organic materials combine with the water to produce organic acids. These acids are then active participants in soil processes, as are dissolved alkalies and salts.

To understand how the soil solution behaves, let us go through a quick chemistry review. An *ion* is an atom, or group of atoms, that carries an electrical charge (examples: Na^+, Cl^-, HCO_3^-). An ion has either a positive charge or a negative charge. For example, when NaCl (sodium chloride) dissolves in solution, it separates into two ions: Na^+ a *cation* (positively charged ion), and Cl^- an *anion* (negatively charged ion). Some ions in soil carry single charges, whereas others carry double or even triple charges (e.g., sulfate, SO_4^{2-}; and aluminum, Al^{3+}).

Ions in soil are retained by **soil colloids**. These tiny particles of clay and organic material (humus) carry a negative electrical charge and consequently attract any positively charged ions in the soil (Figure 18.6). The positive ions, many metallic, are critical to plant growth. If it were not for the negatively charged soil colloids, the positive ions would be leached away by the soil solution and thus would be unavailable to plant roots.

Individual clay colloids are thin and plate-like, with parallel surfaces that are negatively charged (see Figure 18.6). They are more chemically active than silt and sand particles but less active than organic colloids. Metallic cations attach to the surfaces of the colloids by *adsorption* (not *ab*sorption, which means "to enter"). Colloids can exchange cations between their surfaces and the soil solution, an ability called **cation-exchange capacity (CEC)**,

Table 18.2 Principal Soil Moisture Regimes

Regime	Description
Aquic (L. *aqua*, "water")	*The groundwater table lies at or near the surface*, so the soil is almost constantly wet, as in bogs, marshes, and swamps, a reducing environment with virtually no dissolved oxygen present. Commonly, groundwater levels fluctuate seasonally; a small borehole will produce standing, stagnant water.
Aridic (torric) (L. *aridis*, "dry," and L. *torridus*, "hot and dry")	*Soils in this regime are dry more than half the time*, with soil temperatures at a depth of 50 cm (19.7 in.) above 5°C (41°F). In some or all areas, soils are never moist for as long as 90 consecutive days. This regime occurs mainly in arid climates, although where surface structure inhibits infiltration and recharge or where soils are thin and therefore dry, this regime occurs in semiarid regions as well.
Udic (L. *udus*, "humid")	*Soils have little or no moisture deficiency throughout the year*, specifically during the growing season. The water balance exhibits soil moisture surpluses that flush through the soils during one season of the year. If the water balance exhibits a moisture surplus in all months of the year, the regime is called *perudic*, with adequate soil moisture always available to plants.
Ustic (L. *ustus*, "burnt," implying dryness)	*This regime is intermediate between aridic and udic regimes*, it includes the semiarid and tropical wet–dry climates. Moisture is available but is limited, with a prolonged deficit period following the period of soil-moisture utilization during the early portion of the growing season. Temperature is important in determining the moisture efficiency of available water in this borderline moisture regime.
Xeric (L. *xeros*, "dry")	*This regime applies to those few areas that experience a Mediterranean climate*, dry and warm summer, rainy and cool winter. There are at least 45 consecutive dry days during the 4 months following the summer solstice. Winter rains effectively leach the soils.

Source: U.S. Department of Agriculture, *Soil Taxonomy*, Agricultural Handbook No. 436 (Washington, DC: U.S. Government Printing Office, 1975).

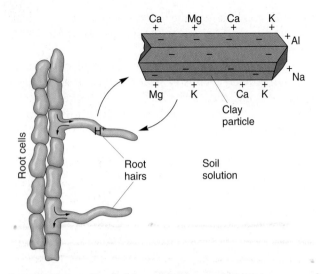

FIGURE 18.6 Soil colloids and cation-exchange capacity (CEC).
This typical soil colloid retains mineral ions by adsorption to its surface (opposite charges attract). This process holds the ions until they are absorbed by root hairs.

which is the measure of soil fertility. A high CEC means that the soil colloids can store or exchange more cations from the soil solution, an indication of good soil fertility (unless there is a complicating factor, such as a soil that is too acid).

Therefore, **soil fertility** is the ability of soil to sustain plants. Soil is fertile when it contains organic substances and clay minerals that absorb water and adsorb certain elements needed by plants. Billions of dollars are expended to create fertile soil conditions, yet the future of Earth's most fertile soils is threatened because soil erosion is on the increase worldwide.

Soil Acidity and Alkalinity

A soil solution may contain significant hydrogen ions (H^+), the cations that stimulate acid formation. The result is a soil rich in hydrogen ions, or an *acid soil*. On the other hand, a soil high in base cations (calcium, magnesium, potassium, sodium) is a *basic* or *alkaline soil*. Such acidity or alkalinity is expressed on the pH scale (Figure 18.7).

Pure water is nearly neutral, with a pH of 7.0. Readings below 7.0 represent increasing acidity. Readings above

7.0 indicate increasing alkalinity. Acidity usually is regarded as strong at 5.0 or lower on the pH scale, whereas 10.0 or above is considered strongly alkaline.

The major contributor to soil acidity in this modern era is acid precipitation (rain, snow, fog, or dry deposition). Acid rain actually has been measured below pH 2.0—an incredibly low value for natural precipitation, as acid as lemon juice. Because most crops are sensitive to specific pH levels, acid soils below pH 6.0 require treatment to raise the pH. This soil treatment is accomplished by the addition of bases in the form of minerals that are rich in base cations, usually lime (calcium carbonate, $CaCO_3$). Increased acidity in the soil solution accelerates the chemical weathering of mineral nutrients and increases their depletion rates.

Soil Formation Factors and Management

Soil is an open system involving physical inputs and outputs. Soil-forming factors are both *passive* (parent material, topography and relief, and time) and *dynamic* (climate, biology, and human activities). These factors work together as a system to form soils. The roles of these factors are considered here and in the soil order discussions that follow.

Natural Factors

Physical and chemical weathering of rocks in the upper lithosphere provides the raw mineral ingredients for soil formation. These are called *parent materials*, and their composition, texture, and chemical nature help determine the type of soil that forms. Clay minerals are the principal weathered by-products in soil. Also important in soil is the organic content present and all that is living in soil—bacteria, algae, fungi, worms, and insects.

Climate types correlate closely with soil types worldwide. The moisture, evaporation, and temperature regimes of climates determine the chemical reactions, organic activity, and eluviation rates of soils. Not only is the present climate important, but many soils exhibit the imprint of past climates, sometimes over thousands of years. Most notable is the effect of glaciations. Among other contributions, glaciation produced the loess soil materials that have been windblown thousands of kilometers to their present locations (Chapter 15).

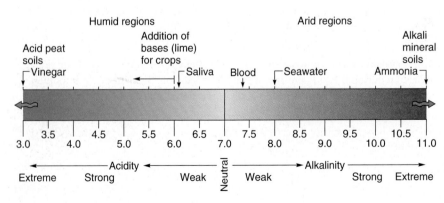

FIGURE 18.7 pH scale.
The pH scale measures acidity (lower pH) and alkalinity (higher pH). (The complete pH scale ranges between 0 and 14.)

News Report 18.1

Soil Is Slipping Through Our Fingers

- The U.S. General Accounting Office estimates that from 3 to 5 million acres of prime farmland are lost each year in the United States through mismanagement or conversion to nonagricultural uses. About half of all cropland in the United States and Canada is experiencing excessive rates of soil erosion—these countries are two of the few that monitor loss of topsoil.

- The Canadian Environmental Advisory Council estimated that the organic content of cultivated prairie soils has declined by as much as 40% compared with noncultivated native soils. In Ontario and Québec, losses of organic content increased to as much as 50%, and losses are even higher in the Atlantic Provinces, which were naturally low in organic content before cultivation.

- A 1995 study completed at Cornell University concluded that soil erosion is a major environmental threat to the sustainability and productive capacity of agriculture worldwide.

- Since 1950, as much as 38% of the world's farmable land has been lost to soil erosion, and the rate continues at 5 to 6 million hectares (about 12 to 15 million acres) per year—560 million hectares, 1380 million acres, to date (World Resources Institute and UNEP, 1997).

- The causes for degraded soils, in order of severity, include: overgrazing, vegetation removal, agricultural activities, overexploitation, and industrial and bioindustrial use (UNEP, 1997).

- The world's human population is growing at the rate of 6.9 million people a month (net increase), increasing the demand for food and agricultural productivity. Especially significant is that proportion of the global population that are adopting a meat-centered diet as American tastes and food outlets spread worldwide—as we see in the next chapter, meat production is an inefficient use of the grain supply.

Vegetation, animal, and bacterial activity determine the organic content of soil. The chemical makeup of the vegetation contributes to the acidity or alkalinity of the soil solution. For example, broadleaf trees tend to increase alkalinity, whereas needleleaf trees tend to produce higher acidity. Thus, when civilization moves into new areas and alters the natural vegetation by logging or plowing, the affected soils are likewise altered, often permanently.

Topography also affects soil formation. Slopes that are too steep cannot have full soil development because gravity and erosional processes remove materials. Lands that are nearly level inhibit soil drainage and can become waterlogged. The compass orientation of slopes is important because it controls exposure to sunlight. In the Northern Hemisphere, a south-facing slope is warmer overall through the year because it receives direct sunlight. Water-balance relations are affected because north-facing slopes are colder, causing slower snowmelt and slower evaporation, providing more moisture for plants than is available on south-facing slopes, which tend to dry faster.

All of the identified factors in soil development (climate, biological activity, parent material, landforms and topography, and human activity) require *time* to operate. Over geologic time we learned that plate tectonics has redistributed landscapes, and thus subjected soil-forming processes to diverse conditions.

The Human Factor

Human intervention has a major impact on soils. Millennia ago, farmers in most cultures learned to plant slopes "on the contour"—to make rows or mounds around a slope at the same elevation, not vertically up and down the slope. Planting on the contour prevents water from flowing straight down the slope and thus reduces soil erosion. It was common to plant and harvest a floodplain but to live on higher ground nearby. Floods were celebrated as blessings that brought water, nutrients, and more soil to the land. Society is drifting away from these commonsense strategies.

A few centimeters' thickness of prime farmland soil may require *500 years* to mature. Yet, this same thickness is being lost annually through soil erosion when the soil-holding vegetation is removed and the land is plowed regardless of topography. Over the same period, exposed soils may be completely leached of needed cations, thereby losing their fertility. Unlike living species, soils do not reproduce nor can they be recreated.

Some 35% of farmlands are losing soil faster than it can form—a loss exceeding 23 billion metric tons (25 billion tons) per year. Soil depletion and loss are at record levels from Iowa to China, Peru to Ethiopia, the Middle East to the Americas. The impact on society is potentially disastrous as population and food demands increase (see News Report 18.1).

Soil erosion can be compensated for in the short run by using more fertilizer, increasing irrigation, and by planting higher-yielding strains. But the potential yield from prime agricultural land will drop by as much as 20% over the next 20 years if only moderate erosion continues. One 1995 study tabulated the market value of lost nutrients and other variables in the most comprehensive soil-erosion study to date. The sum of direct damage (to agricultural land) and indirect damage (to streams, society's infrastructure, and human health) was estimated at more than $25 billion a year in the United States and hundreds of billions of dollars worldwide. (Of course, this is a controversial

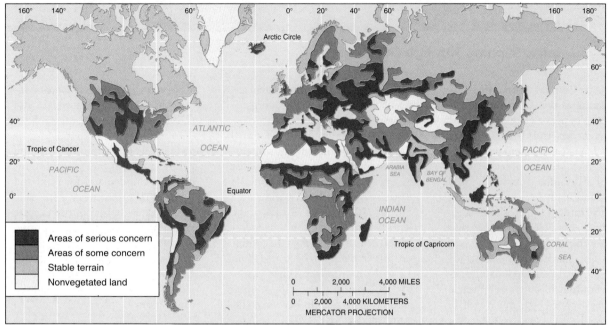

(a)

(b)

FIGURE 18.8 Soil degradation.
(a) Approximately 1.2 billion hectares (3.0 billion acres) of Earth's soils suffer degradation through erosion caused by human misuse and abuse. (b) Typical loss through soil sheet erosion on a northeastern Wisconsin farm. One millimeter of soil lost from an acre weighs about 5 tons. [(a) *A Global Assessment of Soil Degradation*, adapted from United Nations Environment Programme, International Soil Reference and Information Centre, "Map of Status of Human-Induced Soil Degradation," Sheet 2, Nairobi, Kenya, 1990; (b) photo by D. P. Burnside/Photo Researchers, Inc.]

assessment in the agricultural industry.) The cost to bring erosion under control in the United States is estimated at approximately $8.5 billion, or about 30 cents on every dollar of damage and loss. Figure 18.8 maps regions of soil loss.

Our spatial impact on the precious soil resource must be recognized and protective measures must be taken. There is a need for cooperative international action. Critical is a long-term assessment of cost (continued behavior) and benefits (mitigated damage and preservation of the market value of prime soils) of continued practices versus soil conservation actions.

Soil Classification

Classification of soils is complicated by the continuing interaction of physical properties and processes just discussed. This interaction creates thousands of distinct soils, with well over 15,000 soil series identified in the United States and Canada alone. Not surprisingly, different classification systems are in use worldwide. The United States, Canada,

the United Kingdom, Germany, Australia, Russia, and the United Nations Food and Agricultural Organization each have their own soil classification system. Each system reflects the environment of its country. For example, the National Soil Survey Committee of Canada developed a system suited to its great expanses of boreal forest, tundra, and cool climatic regimes (detailed in Appendix B).

Soil Taxonomy

The U.S. soil classification system, *Soil Taxonomy—A Basic System of Soil Classification for Making and Interpreting Soil Surveys*, was published in 1975 and revised in a new edition in 1999. Soil scientists refer to it as **Soil Taxonomy**. Over the years various revisions and clarifications in the system were published in *Keys to the Soil Taxonomy*, now in its 8th edition (1998), which includes all the revisions to the 1975 Soil Taxonomy. Major revisions include the addition of two new soil orders: Andisols (volcanic soils) in 1990 and Gelisols (cold and frozen soils) in 1998. Much of the information in this chapter is derived from these two keystone publications. See http://www.statlab.iastate.edu/soils/nsdaf/ for a copy of this publication and much more regarding soils.

Soil properties and morphology (appearance, form, and structure) actually seen in the field are key to the Soil Taxonomy system. Thus, it is open to addition, change, and modification as the sampling database grows. The system recognizes the importance of interactions between humans and soils and the changes that humans have introduced, both purposely and inadvertently.

The classification system divides soils into six categories, creating a hierarchical sorting system (Table 18.3). The smallest, most-detailed category is the soil series, which ideally includes only one polypedon but may include adjoining polypedons. In sequence from smallest category to the largest, the Soil Taxonomy recognizes *soil series, soil families, soil subgroups, soil great groups, soil suborders, and soil orders.*

Pedogenic Regimes   Prior to the Soil Taxonomy system, **pedogenic regimes** were used to describe soils. These regimes keyed specific soil-forming processes to climatic regions. Although each pedogenic process may be active in several soil orders and in different climates, we discuss them within the soil order where they commonly occur. Such climate-based regimes are convenient for relating climate and soil processes. However, *the Soil Taxonomy system recognizes the great uncertainty and inconsistency in basing soil classification on such climatic variables.* Aspects of several pedogenic processes are discussed with appropriate soil orders:

- **laterization**: a leaching process in humid and warm climates, discussed with Oxisols;
- **salinization**: a process that concentrates salts in soils in climates with excessive POTET (potential evapotranspiration) rates, discussed with Aridisols;
- **calcification**: a process that produces an illuviated accumulation of calcium carbonates in continental climates, discussed with Mollisols and Aridisols;
- **podzolization**: a process of soil acidification associated with forest soils in cool climates, discussed with Spodosols;
- **gleization**: a process that includes an accumulation of humus and a thick, water-saturated gray layer of clay beneath, usually in cold, wet climates and poor drainage conditions.

Table 18.3   U.S. Soil Taxonomy

Soil Category	Number of Soils Included
Orders	12
Suborders	47
Great groups	230
Subgroups	1,200
Families	6,000
Series	15,000

Diagnostic Soil Horizons

To identify a specific soil series within the Soil Taxonomy, the U.S. Natural Resources Conservation Service describes diagnostic horizons in a pedon. A *diagnostic horizon* reflects a distinctive physical property (color, texture, structure, consistence, porosity, moisture) or a dominant soil process (discussed with the soil types).

In the solum (A, E, and B horizons), two diagnostic horizons may be identified: the epipedon and the subsurface. The presence or absence of either of these diagnostic horizons usually distinguishes a soil for classification.

- The **epipedon** (literally, "over the soil") is the diagnostic horizon at the surface where most of the rock structure has been destroyed. It may extend downward through the A horizon, even including all or part of an illuviated B horizon. It is visibly darkened by organic matter and sometimes is leached of minerals. Excluded from the epipedon are alluvial deposits, eolian deposits, and cultivated areas, because soil-forming processes have lacked the time to erase these short-lived characteristics.
- The **diagnostic subsurface horizon** originates below the surface at varying depths. It may include part of the A or B horizon or both. Many diagnostic subsurface horizons have been identified.

The 12 Soil Orders of the Soil Taxonomy

At the heart of the Soil Taxonomy are 12 general soil orders, listed in Table 18.4. Their worldwide distribution is shown in Figure 18.9 and in individual maps with each description. Please consult this table and these maps as you read the following descriptions. Because the Soil Taxonomy evaluates each soil order on its own characteristics, there is no priority to the classification. However, you will find a progression in this discussion, for the 12 orders are arranged loosely by latitude, beginning with Oxisols along the equator as in Chapters 10 (climates) and 20 (terrestrial biomes).

Oxisols   The intense moisture, temperature, and uniform daylength of equatorial latitudes profoundly affect soils. These generally old landscapes, exposed to tropical conditions for millennia or hundreds of millennia, are deeply developed. Soil minerals are highly altered (except in certain newer volcanic soils in Indonesia—the Andisols). Oxisols are among the most mature soils on Earth. Distinct horizons usually are lacking where these soils are well drained (Figure 18.10 on page 571). Related vegetation is the luxuriant and diverse tropical and equatorial rain forest. Oxisols include five suborders.

Oxisols (tropical soils) are so called because they have a distinctive horizon of iron and aluminum oxides. The concentration of oxides results from heavy precipitation, which leaches soluble minerals and soil constituents from the A horizon. Typical Oxisols are reddish (from the iron

FIGURE 18.9 Soil Taxonomy.
Worldwide distribution of the Soil Taxonomy's 12 soil orders. [Adapted from maps prepared by World Soil Resources Staff, Natural Resources Conservation Service, USDA, 1999.]

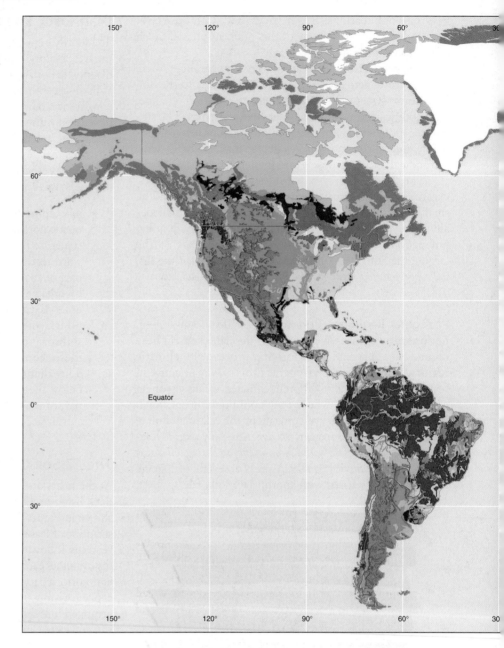

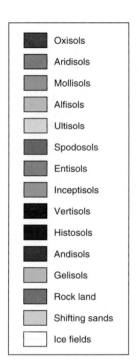

Oxisols
Aridisols
Mollisols
Alfisols
Ultisols
Spodosols
Entisols
Inceptisols
Vertisols
Histosols
Andisols
Gelisols
Rock land
Shifting sands
Ice fields

oxide) or yellowish (from the aluminum oxides), with a weathered clay-like texture, sometimes in a granular soil structure that is easily broken apart. The high degree of eluviation removes basic cations and colloidal material to lower illuviated horizons. Thus, Oxisols are low in CEC (cation-exchange capacity) and fertility, except in regions augmented by alluvial or volcanic materials.

To have the lush rain forests in the same regions as soils poor in inorganic nutrients seems an irony. However, this forest system relies on the recycling of nutrients from soil organic matter to sustain fertility, although this is quickly lost when disturbed.

Consequently, Oxisols have a diagnostic subsurface horizon that is highly weathered, contains iron and aluminum oxides, is at least 30 cm (12 in.) thick, and lies within 2 m (6.5 ft) of the surface (see Figure 18.10). Figure 18.11

illustrates *laterization*, the leaching process that operates in well-drained soils in warm, humid tropical and subtropical climates. If Oxisols are subjected to repeated wetting and drying, an *ironstone hardpan* develops (this is a hardened soil layer in the lower A or in the B horizon—iron-rich and humus-poor, clay with quartz and other minerals). It is called a *plinthite* (from the Greek *plinthos*, meaning "brick"). This form of soil, also called a *laterite*, can be quarried in blocks and used as a building material (Figure 18.12).

Agricultural activities can be conducted with care in these soils. Early cultivation practices, called *slash-and-burn shifting cultivation*, were adapted to these soil conditions and formed a unique style of crop rotation. The scenario went like this: People in the tropics cut down (slashed) and burned the rain forest in small tracts and then cultivated the land with stick and hoe. Mineral nutrients in the or-

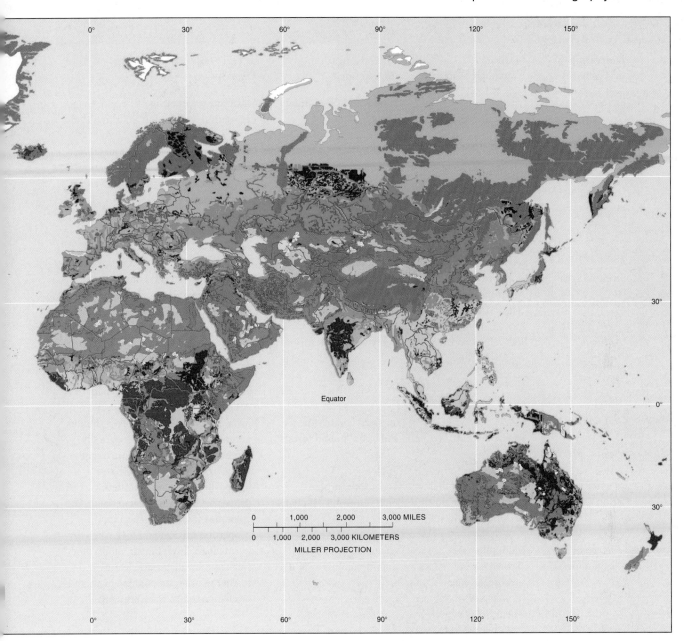

ganic material and short-lived fertilizer input from the fire ash would quickly be exhausted. After several years, the soil lost fertility through leaching by intense rainfall, so the people shifted cultivation to another tract and repeated the process. After many years of moving from tract to tract, the group returned to the first patch to begin the cycle again. This practice protected the limited fertility of the soils somewhat, allowing periods of recovery.

The invasion of foreign plantation interests, development by local governments, vastly increased population pressures, and conversion of vast forest tracts to pasturage, disrupted this orderly land rotation. Permanent tracts of cleared land put tremendous pressure on the remaining forest and brought disastrous consequences. When Oxisols are disturbed, soil loss can exceed a thousand tons per square kilometer per year, not to mention the greatly in-

creased extinction rate of plant and animal species that accompanies such destruction. The regions dominated by the Oxisols and rain forests are rightfully the focus of much worldwide environmental attention.

Aridisols The largest single soil order occurs in the world's dry regions. **Aridisols** (desert soils) occupy approximately 19% of Earth's land surface and some 12% of U.S. land (see Figure 18.9). A pale, light soil color near the surface is diagnostic (Figure 18.13a on page 572).

Not surprisingly, the water balance in Aridisol regions has periods of soil-moisture deficit and generally inadequate soil moisture for plant growth. High potential evapotranspiration and low precipitation produce very shallow soil horizons. Usually there is no period greater than 3 months when the soils have adequate moisture. Lacking water and

Table 18.4   Soil Taxonomy Soil Orders

Order	Derivation of Term	Marbut, 1935 (Canadian System)	General Location and Climate	Description
Oxisols	Fr. *oxide*, "oxide" Gr. *oxide*, "acid or sharp"	Latosols lateritic soils	Tropical soils, hot, humid areas	Maximum weathering of Fe and Al and eluviation, continuous plinthite layer
Aridisols	L. *aridos*, "dry"	Reddish desert, gray desert, sierozems	Desert soils, hot, dry areas	Limited alteration of parent material, low climate activity, light color, low humus content subsurface illuviation of carbonates
Mollisols	L. *mollis*, "soft"	Chestnut, chernozem (Chernozemic)	Grassland soils; subhumid, semiarid lands	Noticeably dark with organic material, humus rich, base saturation high, friable surface with well-structured horizons
Alfisols	Invented syllable	Gray-brown podzolic, degraded chernozem (Luvisol)	Moderately weathered forest soils, humid temperate forests	B horizon high in clays, moderate to high degree of base saturation, illuviated clay accum., no pronounced color change with depth
Ultisols	L. *ultimus*, "last"	Red-yellow podzolic, reddish yellow lateritic	Highly weathered forest soils, subtropical forests	Similar to Alfisols, B horizon high in clays, generally low amount of base saturation, strong weathering in subsurface horizons, redder than Alfisols
Spodosols	Gr. *spodos* or L. *spodus*, "wood ash"	Podzols, brown podzolic (Podzol)	Northern conifer forest soils, cool humid forests	Illuvial B horizon of Fe/Al clays, humus accum.; without structure, partially cemented; highly leached, strongly acid; coarse texture of low bases
Entisols	Invented syllable from *recent*	Azonal soils, tundra	Recent soils, profile undeveloped, all climates	Limited development; inherited properties from parent material; pale color, low humus, few specific properties; hard and massive when dry
Inceptisols	L. *inceptum*, "beginning"	Ando, subarctic brown forest lithosols, some humic gleys (Brunisol, Cryosol with permafrost, Gleysol wet)	Weakly developed soils, humid regions	Intermediate development; embryonic soils, but few diagnostic features; further weathering possible in altered or changed subsurface horizons
Vertisols	L. *verto*, "to turn"	Grumusols (1949) tropical black clays	Expandable clay soils; subtropics, tropics; sufficient dry period	Forms large cracks on drying, self-mixing action, contains >30% in swelling clays, light color, low humus content
Histosols	Gr. *histos*, "tissue"	Peat, muck, bog (Organic)	Organic soils, wet places	Peat or bog, >20% organic matter, much with clay >40 cm thick, surface, organic layers, no diagnostic horizons
Andisols	L. *ando*, "volcanic ash"	—	Areas affected by frequent volcanic activity (formerly within Inceptisols and Entisols)	Volcanic parent materials, particularly ash and volcanic glass; weathering and mineral transformation important; high CEC and organic content, generally fertile
Gelisols	L. *gelatio*, "freezing"	Formerly Inceptisols and Entisols (Cryosols, some Brunisols)	High latitudes in Northern Hemisphere, southern limits near tree line	Permafrost within 100 cm of the soil surface; evidence of cryoturbation (frost churning) and/or an active layer; patterned-ground

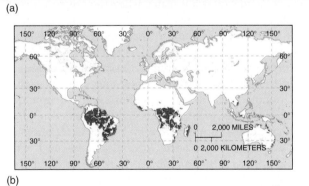

(a)

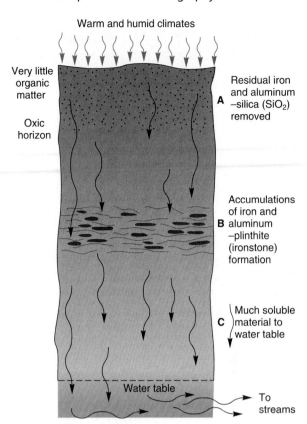

FIGURE 18.11 Laterization.
The laterization process is characteristic of moist tropical and subtropical climate regimes.

FIGURE 18.10 Oxisols.
(a) Deeply weathered Oxisol profile in central Puerto Rico.
(b) General map showing worldwide distribution of these tropical soils. [Photo from the Marbut Collection, Soil Science Society of America.]

FIGURE 18.12 Oxisols are used for building material.
Here plinthite is being quarried in India. [Photo by Henry D. Foth.]

(a)

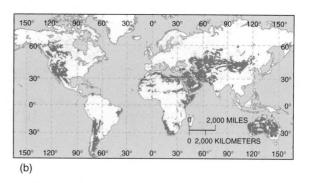

(b)

FIGURE 18.13 Aridisols.
(a) Soil profile from central Arizona. (b) General map showing worldwide distribution of these desert soils. [Photo from the Marbut Collection, Soil Science Society of America.]

FIGURE 18.14 Agriculture in arid lands.
The Coachella Valley of southeastern California is a desert in agricultural bloom; asparagus is growing in the foreground. Providing drainage for excessive water application in such areas often is necessary to prevent salinization of the rooting zone. [Photo by author.]

therefore lacking vegetation, Aridisols also lack organic matter of any consequence. Low precipitation means infrequent leaching, yet Aridisols are leached easily when exposed to excessive water, for they lack a significant colloidal structure.

Salinization is common in Aridisols, resulting from excessive potential evapotranspiration rates in deserts and semiarid regions. Salts dissolved in soil water migrate to surface horizons and are deposited there as the water evaporates. These deposits appear as subsurface salty horizons, which will damage or kill plants when the horizons occur near the root zone. The salt accumulation associated with a desert playa is an example of extreme salinization (see Chapter 15).

Obviously, salinization complicates farming in Aridisols. The introduction of irrigation water may either waterlog poorly drained soils or lead to salinization. Nonetheless, vegetation does grow where soils are well drained and low in salt content. If large capital investments are made in water, drainage, and fertilizers, Aridisols possess much agricultural potential (Figure 18.14). In the Nile and Indus River valleys, for example, Aridisols are intensively farmed with a careful balance of these environmental factors, although thousands of acres of once-productive land, not so carefully treated, now sit idle and salt-encrusted. In California, the Kesterson Wildlife Refuge was reduced to a toxic waste dump in the early 1980s by contaminated agricultural drainage. Focus Study 18.1 elaborates on the Kesterson tragedy. News Report 18.2 describes efforts to deal with salinization.

Selenium Concentration in Western Soils

Irrigated agriculture has increased greatly since 1800, when only 8 million hectares (about 20 million acres) were irrigated worldwide. Today, approximately 255 million hectares (about 630 million acres) are irrigated, and this figure is on the increase (USDA, FAO, 1996). Representing about 16% of Earth's agricultural land, irrigated land accounts for nearly 36% of the harvest.

Two related problems common in irrigated lands are salinization and water logging, especially in arid lands that are poorly drained. In many areas, production has decreased and even ended because of salt buildup in the soils. Examples include areas along the Tigris and Euphrates Rivers, the Indus River valley, sections of South America and Africa, and the western United States.

Irrigation in the West

About 95% of the irrigated acreage in the United States lies west of the 98th meridian. This region is increasingly troubled with salinization and waterlogging problems. In addition, at least nine sites in the West, particularly California's western San Joaquin Valley, are experiencing related contamination of a more serious nature—increasing selenium concentrations. Toxic effects of selenium were reported during the 1980s in some domestic animals grazing on grasses grown in selenium-rich soils in the Great Plains. In California, as parent materials weathered, selenium-rich alluvium washed into the semiarid valley, forming the soils that needed only irrigation water to become productive.

Drainage of agricultural wastewater poses a particular problem in semiarid and arid lands, where river discharge is inadequate to dilute and remove field runoff. One solution to prevent salt accumulations and water-clogged soils is to place field drains beneath the soil to collect gravitational water from fields that have been purposely overwatered to keep salts away from the effective rooting depth of the crops. But agricultural drain water must go somewhere, and for the San Joaquin

Valley of central California this problem triggered a 15-year controversy.

Death of Kesterson

Central California's potential drain outlets are to the ocean, San Francisco Bay, or the central valley. But all these suggested destinations failed to pass environmental impact assessments under Environmental Policy Act requirements. Nonetheless, by the late 1970s, about 80 miles of a drain were finished, even though no formal plan or adequate funding had been completed—a drain was built with no outlet! In the absence of any plan, large-scale irrigation continued, supplying the field drains with salty, selenium-laden runoff that made its way to the Kesterson National Wildlife Refuge in the northern portion of the San Joaquin Valley east of San Francisco. The unfinished drain abruptly stopped at the boundary to the refuge.

The selenium-tainted drainage only took 3 years to destroy the wildlife refuge, which was officially declared a contaminated toxic waste site (Figure 1). Aquatic life forms (e.g., marsh plants, plankton, and insects) had taken in the selenium, which then made its way up the food chain and into the diets of higher life forms in the refuge. According to U.S. Fish and Wildlife Service scientists, the toxicity moved through the food chain and genetically damaged and killed wildlife, including all varieties of birds that nested at Kesterson; approximately 90% of the exposed birds perished or were injured. Because this wildlife refuge was a major migration flyway and stopover point for birds from throughout the West-

ern Hemisphere, this destruction of the refuge also violated several multinational wildlife protection treaties.

Such damage to wildlife presents a real warning to human populations—remember where we are in the food chain. The field drains were sealed and removed in 1986, following a court order that forced the federal government to uphold existing laws. Irrigation water then immediately began backing up in the corporate farmlands, producing both waterlogging and selenium contamination.

Since 1985, more than 0.6 million hectares (1.5 million acres) of irrigated Aridisols and Alfisols have gone out of production in California, marking the end of several decades of irrigated farming in climatically marginal lands. Severe cutbacks in irrigated acreage no doubt will continue, underscoring the need to preserve prime farmlands in wetter regions and to understand essential soil processes.

Frustrated agricultural interests have asked the federal government to finish the drain, either to San Francisco Bay or to the ocean. However, neither option appears capable of passing an environmental impact analysis. Another strategy is to allow irrigated lands to start pumping again into the former wildlife refuge because it is now a declared toxic dump site anyway! There are nine such threatened sites in the West; Kesterson was simply the first to fail. Meanwhile the perimeter fence surrounding Kesterson is posted with signs that read, "Area Beyond This Sign CLOSED, All Public Access Prohibited!"

FIGURE 1 Soil contamination in the wildlife refuge.
Salt-encrusted soil and plants at the contaminated Kesterson National Wildlife Refuge. [Photo by Gary R. Zahm/DRK Photo.]

(a)

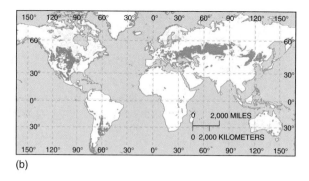

(b)

FIGURE 18.15 Mollisols.
(a) Soil profile from central Iowa. (b) General map showing worldwide distribution of these grassland soils. [Photo from the Marbut Collection, Soil Science Society of America.]

The "fertile triangle" of Ukraine, Russia, and western portions of the Russian Federation is of this soil type. The region stretches from north of the Caspian Sea and Kazakstan westward in a widening triangle toward Central Europe (Figure 18.16). The soils in the fertile triangle are known as the *chernozem* in other classification systems. The remainder of the Russian landscape presents varied soils of lower fertility and productivity.

In North America, the Great Plains straddle the 98th meridian, which is coincident with the 51-cm (20-in.) isohyet of annual precipitation—wetter to the east and drier to the west. The Mollisols here mark the historic division between the short- and tall-grass prairies (Figure 18.17).

Calcification is a soil process characteristic of some Mollisols and adjoining marginal areas of Aridisols. Calcification is the accumulation of calcium carbonate or magnesium carbonate in the B and C horizons. Calcification by calcium carbonate ($CaCO_3$) among others, forms a diagnostic subsurface horizon that is thickest along the boundary between dry and humid climates (Figure 18.18).

When cemented or hardened, these deposits are called *caliche*, or *kunkur*; they occur in widespread soil formations in central and western Australia, the Kalahari region of interior Southern Africa, and the High Plains of the west-central United States, among other places. (Soils dominated by the processes of calcification and salinization were formerly known as pedocals.)

Mollisols Mollisols (grassland soils) are some of Earth's most significant agricultural soils. There are seven recognized suborders, which vary in fertility. The dominant diagnostic horizon is a dark, organic surface layer some 25 cm (10 in.) thick (Figure 18.15). As the Latin name implies (words using the same root, *mollis*, are "mollify" and "emollient"), Mollisols are soft, even when dry. They have granular or crumbly peds, loosely arranged when dry. These humus-rich soils are high in basic cations (calcium, magnesium, and potassium) and have a high cation-exchange capacity and therefore, high fertility. In soil moisture, these soils are intermediate between humid and arid.

Soils of the world's steppes and prairies belong to this soil group: the North American Great Plains, the Pampas of Argentina, and the region from Manchuria in China westward to Europe. Agriculture ranges from large-scale commercial grain farming to grazing along the drier portions of the soil order. With fertilization or soil-building practices, high crop yields are common.

Alfisols Alfisols (moderately weathered forest soils) are spatially the most widespread of the soil orders, extending in five suborders from near the equator to high latitudes. Representative Alfisol areas include Boromo and Burkina Faso (interior western Africa); Fort Nelson, British Columbia; the states near the Great Lakes, and the valleys of central California. Most Alfisols are grayish brown to reddish and are considered moist versions of the Mollisol soil

Drainage Tiles, But Where to Go?

In the California valleys of Imperial, Coachella, and San Joaquin, and in the Wellton–Mohawk district of extreme southern Arizona, the fields have been excavated, drainage tiles (perforated ceramic pipe) laid in place, and the fields replaced. The field drains carry away excess irrigation water, which has been purposely applied to leach salts away from plant roots. The problem has been to find a place to dump the salt- and chemical-laden drainage water.

In the Imperial and Coachella Valleys, the drainage water was sent to the Salton Sea (Figure 1)—a large inland sea formed by an irrigation accident in 1906 (a broken canal levee allowed the Colorado River to flow into the desert of southern California for 18 months). For the Wellton–Mohawk area, the contaminated drainage water was pumped back into the Colorado River just above the Mexican border. By the late 1950s, salty Colorado River water was killing every plant it touched in Mexico.

Today, it passes through a desalinization plant near Yuma, Arizona. The plant went into service in the late 1993, pumping desalted water back to the river and saline sludge directly to the ocean. The irony is that it would be cheaper for the government to buy the subsidized irrigated orchards and fields and allow the land to revert to desert than it is to operate the $484 million desalinization works.

FIGURE 1 Fields drain into canals.
Soil drainage canal collects contaminated water from field drains and directs it into the Salton Sea. [Photo by author.]

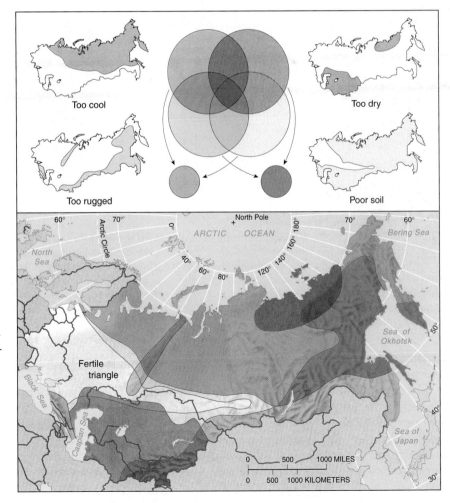

FIGURE 18.16 Soil fertility in Eurasia.
Top: Four main factors influence soil fertility. Bottom: The fertile triangle of central Eurasia. The white area is highest in soil fertility (no negative factors). Areas shown with any color or color combination are limited for the reasons shown. The least-fertile area in Russia is northeastern Siberia. [After P. W. English and J. A. Miller, *World Regional Geography: A Question of Place*, p. 183, ©1989. Adapted by permission of John Wiley & Sons, Inc.]

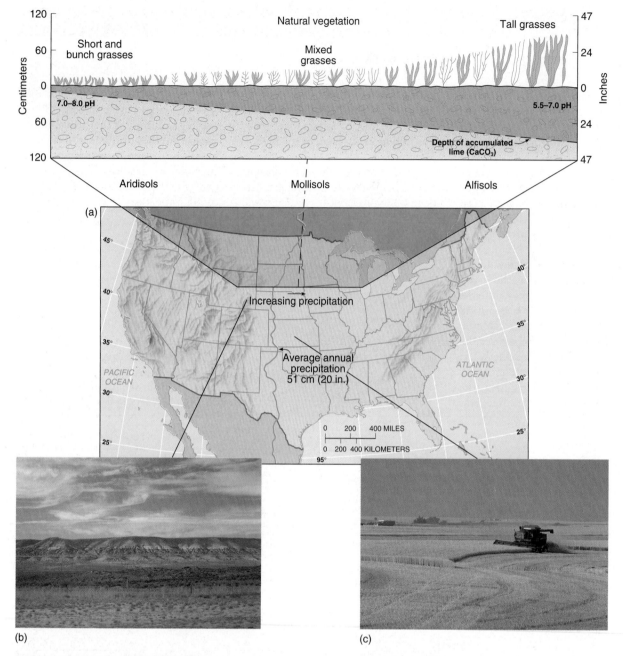

FIGURE 18.17 Soils of the Midwest.
(a) Aridisols (to the west), Mollisols (central), and Alfisols (to the east)—a soil continuum in the north-central United States and southern Canadian prairies. Graduated changes that occur in soil pH and the depth of accumulated lime are shown. (b) Bunch grasses and shallow soils of Wyoming. (c) Wheat harvest from rich Mollisols near Hardtner, Kansas. [(a) Illustration adapted from N. C. Brady, *The Nature and Properties of Soils*, 10th ed., © 1990 by Macmillan Publishing Company, adapted by permission; (b) photo by author; (c) photo by Garry D. McMichael/Photo Researchers, Inc.]

group. Moderate eluviation is present, as well as a subsurface horizon of illuviated clays and clay formation because of a pattern of increased precipitation (Figure 18.19).

Alfisols have moderate to high reserves of basic cations and are fertile. However, productivity depends on moisture and temperature. Alfisols usually are supplemented by a moderate application of lime and fertilizer in areas of active agriculture. Some of the best farmland in the United States

stretches from Illinois, Wisconsin, and Minnesota through Indiana, Michigan, and Ohio to Pennsylvania and New York. This land produces grains, hay, and dairy products. The soil here is an Alfisol subgroup called Udalfs, which are characteristic of humid continental, hot summer climates.

The Xeralfs, another Alfisol subgroup, are associated with the moist winter, dry summer pattern of the Mediterranean climate. These naturally productive soils are farmed

POTET equal to or
greater than PRECIP

Dark
color,
high in
bases

O
A Dense sod
cover of
interlaced
grasses
and roots

E

Calcic
horizon;
possible
formation
of caliche

B Accumulation
of excess
calcium
carbonate

C

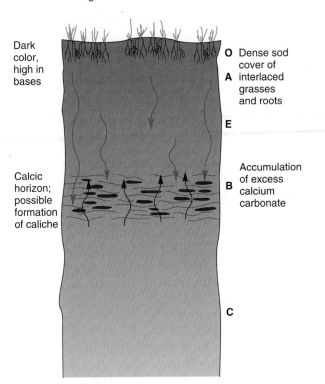

FIGURE 18.18 Calcification in soil.
Calcification process in Aridisol/Mollisol soils occurs in climatic regimes that have potential evapotranspiration equal to or greater than precipitation.

(a)

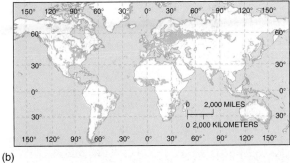

(b)

intensively for subtropical fruits, nuts, and special crops that can grow in only a few locales worldwide (e.g., California grapes, olives, citrus, artichokes, almonds, figs).

Ultisols   Farther south in the United States are the **Ultisols** (highly weathered forest soils) and their five suborders. An Alfisol might degenerate into an Ultisol, given time and exposure to increased weathering under moist conditions. These soils tend to be reddish because of residual iron and aluminum oxides in the A horizon (Figure 18.20).

The relatively high precipitation in Ultisol regions causes greater mineral alteration and more eluvial leaching than in other soils. Therefore, the level of basic cations is lower and the soil fertility is lower. Fertility is further reduced by certain agricultural practices and the effect of soil-damaging crops such as cotton and tobacco, which deplete nitrogen and expose soil to erosion. These soils respond well if subjected to good management—for example, crop

(c)

FIGURE 18.19 Alfisols.
(a) Soil profile from central California. (b) General map showing worldwide distribution of these moderately weathered forest soils. (c) Cultivated Alfisols farmland near Lompoc, California. [(a) Photo from the Marbut Collection, Soil Science Society of America. (c) Photo by Bobbé Christopherson.]

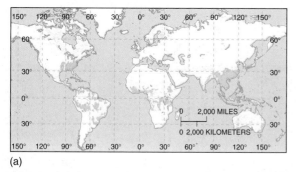

(a)

FIGURE 18.20 Ultisols.
(a) General map showing worldwide distribution of these high-ly weathered forest soils. (b) A type of Ultisol in central Geor-gia bearing its characteristic dark reddish color is planted with pecan trees. (c) Cultivated Ultisols and young peanut plants, near Plains, in west central Georgia. (d) Distinctive reddish soils and the invasive kudzu plant tell us we are in the American southeast. [(b), (c), and (d) photos by Bobbé Christopherson.]

rotation restores nitrogen, and certain cultivation practices prevent sheetwash and soil erosion. Figure 18.20c shows peanut plantings that assist in nitrogen restoration. Much needs to be done to achieve sustainable management of these soils.

(b)

Spodosols The **Spodosols** (northern coniferous forest soils) and their four suborders occur generally to the north and east of the Alfisols. They are in cold and forested moist regimes (*humid continental mild summer Dfb* climates) in northern North America and Eurasia, Denmark, the Netherlands, and southern England. Because there are no comparable climates in the Southern Hemisphere, this soil type is not identified there. Spodosols form from sandy parent materials, shaded under evergreen forests of spruce, fir, and pine. Spodosols with more moderate properties form under mixed or deciduous forests (Figure 18.21).

Spodosols lack humus and clay in the A horizons. An eluviated horizon of sandy and leached clays and iron lies in the A horizon, above a B horizon of illuviated organic matter and iron and aluminum oxides (Figure 18.21c). The surface horizon receives organic litter from base-poor, acid-rich trees (evergreens), which contribute to acid ac-cumulations in the soil. The solution in acidic soils effec-tively leaches clays, iron, and aluminum, which are passed to the upper diagnostic horizon. An ashen-gray color is common in these subarctic forest soils and is characteris-tic of a formation process called *podzolization*. In the Cana-dian system, Spodosols fall within the Podzolic Great Group, as seen in the temperate rain forests of Vancouver Island, British Columbia (Figure 18.21d; and Appendix B, Table 3).

(c)

When agriculture is attempted, the low basic cation content of Spodosols requires the addition of nitrogen, phosphate, and potash (potassium carbonate), and perhaps crop rotation as well. A soil *amendment* such as limestone can significantly increase crop production by raising the pH of these acidic soils. For example, the yields of several crops (corn, oats, wheat, and hay) grown in specific Spo-dosols in New York State were increased up to a third with the application of 1.8 metric tons (2 tons) of limestone per 0.4 hectare (1.0 acre) during each 6-year rotation.

(d)

Entisols The **Entisols** (recent, undeveloped soils) lack vertical development of their horizons. The five subor-ders of this soil group are based on differences in parent materials and climatic conditions, although the presence

(a)

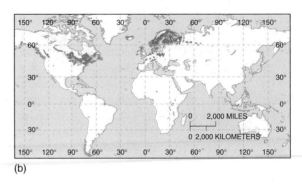

(b)

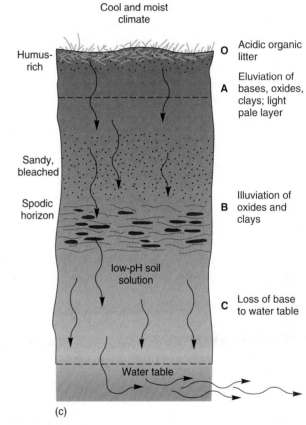

Cool and moist climate

Humus-rich

Sandy, bleached

Spodic horizon

O Acidic organic litter

A Eluviation of bases, oxides, clays; light pale layer

B Illuviation of oxides and clays

low-pH soil solution

C Loss of base to water table

Water table

(c)

(d)

FIGURE 18.21 Spodosols.
(a) Soil profile from northern New York.
(b) General map showing worldwide
distribution of these northern coniferous forest soils. (c) The podzolization
process is typical in cool and moist climatic regimes. (d) Characteristic temperate forest and Spodosols in the cool
moist climate of central Vancouver Island. [(a) Photo from the Marbut Collection, Soil Science Society of America;
(d) photo by Bobbé Christopherson.]

of Entisols is not climate-dependent, for they occur in many climates worldwide. Entisols are true soils, but they have not had sufficient time to generate the usual horizons.

Entisols generally are poor agricultural soils, although those formed from river silt deposits are quite fertile. The same conditions that inhibit complete development also prevent adequate fertility—too much or too little water, poor structure, and insufficient accumulation of weathered nutrients. Active slopes, alluvium-filled floodplains, poorly drained tundra, tidal mud flats, dune sands and erg (sandy) deserts, and plains of glacial outwash all are characteristic regions that have these soils. Figure 18.22 shows an Entisol in a desert climate where shales formed the parent material.

Inceptisols Inceptisols (weakly developed soils) and their six suborders are inherently infertile. They are weakly developed young soils, although they are more developed than the Entisols. Inceptisols include a wide variety of different soils, all holding in common a lack of maturity with evidence of weathering just beginning. Inceptisols are associated with moist soil regimes and are regarded as eluvial because they demonstrate a loss of soil constituents throughout their profile but retain some weatherable minerals. This soil group has no distinct illuvial horizons.

Inceptisols include most of the glacially derived till and outwash materials from New York down through the Appalachians, and alluvium on the Mekong and Ganges floodplains.

Gelisols Gelisols (cold and frozen soils) and their three suborders are new to the Soil Taxonomy and represent inclusion of high-latitude (Canada, Alaska, Russia, and the Antarctic Peninsula) and high-elevation (mountains) soil conditions (Figure 18.23). Temperatures in these regions are at or below 0°C (32°F), making soil development a slow

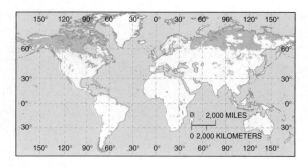

FIGURE 18.23 Gelisols.
General map showing worldwide distribution of these cold and frozen soils.

process and disturbances of the soil long-lasting. Gelisols can develop organic diagnostic horizons because decomposition of materials is slowed by the cold temperatures. Characteristic vegetation is that of tundra, such as lichens, mosses, sedges, and other plants adapted to the harsh cold.

Gelisols are subject to *cryoturbation* (frost churning and mixing) in the freeze–thaw cycle in the active layer (see Chapter 17). This process disrupts soil horizons, pulling organic material to lower layers and rocky C-horizon material to the surface. Patterned-ground phenomena are possible under such conditions.

As we saw in Chapter 17, periglacial processes occupy about 20% of Earth's land surface, including permafrost (frozen ground) that underlies some 13% of these lands. Previously these soils were included in the Inceptisol, Entisol, and Histosol soil orders. In the Canadian System of Soil Classification, these are included in the Crysolic soil order.

Andisols Andisols (volcanic parent materials) and seven suborders occur in areas of volcanic activity. These soils formerly were classified as Inceptisols and Entisols, but in 1990 they were placed in this new order. Andisols are de-

FIGURE 18.22 Entisols.
A characteristic Entisol: young, undeveloped soil forming from a shale parent material in the desert near Zabriskie Point in Death Valley, California. [Photo by author.]

FIGURE 18.24 Andisols in agricultural production.
Fertile Andisols planted with sugar cane on Kaua'i, Hawai'i. The fiftieth state was the second largest U.S. producer of sugar cane after Florida in 1998. [Photo by Wolfgang Kaehler Photography.]

rived from volcanic ash and glass. Previous soil horizons frequently are found buried by ejecta from repeated eruptions. Volcanic soils are unique in their mineral content because they are recharged by eruptions. For an example, see Figure 19.27, recovery and succession of pioneer species in the developing soil north of Mount St. Helens.

Weathering and mineral transformations are important in this soil order. Volcanic glass weathers readily into allophane (a noncrystalline aluminum silicate clay mineral that acts as a colloid) and oxides of aluminum and iron. Andisols feature a high CEC and high water-holding ability and develop moderate fertility, although phosphorus availability is an occasional problem. In Hawai'i, the fertile Andisol fields produce sugar cane and pineapple as important cash crops (Figure 18.24). Andisol distribution is small in areal extent; however, such soils are locally important in the volcanic ring of fire surrounding the Pacific Rim.

Vertisols   Vertisols (expandable clay soils) are heavy clay soils. They contain more than 30% swelling clays (clays that swell significantly when they absorb water), such as *montmorillonite*. They are located in regions experiencing highly variable soil moisture balances through the seasons. These soils occur in areas of subhumid to semiarid moisture and moderate to high temperature. Vertisols frequently form under savanna and grassland vegetation in tropical and subtropical climates and are sometimes associated with a distinct dry season following a wet season. Although widespread, individual Vertisol units are limited in extent.

Vertisol clays are black when wet, but not because of organics: Rather, the blackness is due to specific mineral content. They range from brown to dark gray. These deep clays swell when moistened and shrink when dried. In the drying process, vertical cracks may form, as wide as 2–3 cm (0.8–1.2 in.) and up to 40 cm (16 in.) deep. Loose material falls into these cracks, only to disappear when the soil again expands and the cracks close. After many such cycles, soil contents tend to invert or mix vertically, bringing lower horizons to the surface (Figure 18.25).

Despite the fact that clay soils are plastic and heavy when wet, with little available soil moisture for plants, Vertisols are high in bases and nutrients and thus are some of the better farming soils where they occur. For example, they occur in a narrow zone along the coastal plain of Texas (see chapter-opening photo) and in a section along the Deccan region of India. Vertisols often are planted with grain sorghums, corn, and cotton.

Histosols   Histosols (organic soils), including four suborders, are formed from accumulations of thick organic matter. In the midlatitudes, when conditions are right, beds of

(a)

(b)

FIGURE 18.25 Vertisol.
(a) Soil profile in the Lajas Valley of Puerto Rico. (b) General map showing worldwide distribution of these expandable clay soils. [Photo from the Marbut Collection, Soil Science Society of America.]

former lakes may turn into Histosols, with water gradually replaced by organic material to form a bog and layers of peat (Figure 18.26). (Lake succession and bog or marsh formation are discussed in Chapter 19.) Histosols also form in small, poorly drained depressions, with conditions ideal for significant deposits of sphagnum peat to form. This material can be cut, baled, and sold as a soil amendment. In some locales, dried peat has served for centuries as a low-grade fuel.

The area southwest of Hudson Bay in Canada is typical of Histosol formation. (A complete treatment of the Canadian System of Soil Classification is presented in Appendix B, which includes a map showing the location of these organic soils in Canada.)

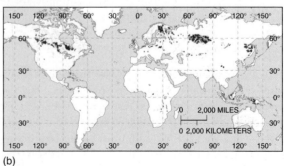

(a)

(b)

FIGURE 18.26 Histosols.
(a) A bog in coastal Maine, near Popham Beach State Park. (b) General map showing worldwide distribution of these organic soils. [Photo by Bobbé Christopherson.]

Summary and Review—The Geography of Soils

● *Define* **soil** and **soil science** and *describe* a pedon, polypedon, and typical soil profile.

Soil is the portion of the land surface in which plants can grow. It is a dynamic natural body composed of fine materials and contains both mineral and organic matter. **Soil science** is the interdisciplinary study of soils involving physics, chemistry, biology, mineralogy, hydrology, taxonomy, climatology, and cartography. *Pedology* deals with the origin, classification, distribution, and description of soil. *Edaphology* specifically focuses on the study of soil as a medium for sustaining the growth of plants.

The basic sampling unit used in soil surveys is the **pedon**. The **polypedon** is the soil unit used to prepare local soil maps and may contain many pedons. Each discernible layer in an exposed pedon is a **soil horizon**. The horizons are designated O (contains **humus**, a complex mixture of decomposed and synthesized organic materials), A (rich in humus and clay, darker), E (zone of **eluviation**, the removal of fine particles and minerals by water), B (zone of **illuviation**, the deposition of clays and minerals translocated from elsewhere), C (*regolith*, weathered bedrock), and R (bedrock). Soil horizons A, E, and B experience the most active soil processes and together are designated the **solum**.

soil (p. 557)
soil science (p. 558)
pedon (p. 558)
polypedon (p. 558)
soil horizon (p. 558)
humus (p. 558)
eluviation (p. 559)
illuviation (p. 559)
solum (p. 560)

1. Soils provide the foundation for animal and plant life and therefore are critical to Earth's ecosystems. Why is this true?
2. What are the differences among soil science, pedology, and edaphology?
3. Define polypedon and pedon, the basic units of soil.
4. Characterize the principal aspects of each soil horizon. Where does the main accumulation of organic material occur? Where does humus form? Explain the difference between the eluviated layer and the illuviated layer. Which horizons constitute the solum?

● *Describe* soil properties of color, texture, structure, consistence, porosity, and soil moisture.

We use several physical properties to classify soils. Color suggests composition and chemical makeup. Soil texture refers to the size of individual mineral particles and the proportion of different sizes. For example, **loam** is a balanced mixture of sand, silt, and clay. Soil structure refers to the arrangement of soil peds, which are the smallest natural cluster of particles in a soil. The cohesion of soil particles to each other is called soil consistence. Soil porosity refers to the size, alignment, shape, and location of spaces in the soil. Soil moisture refers to water in the soil pores and its availability to plants.

loam (p. 560)

5. Soil color is identified and compared using what technique?
6. Define a soil separate. What are the various sizes of particles in soil? What is loam? Why is loam regarded so highly by agriculturists?
7. What is a quick, hands-on method for determining soil consistence?
8. Summarize the five soil-moisture regimes common in mature soils.

● *Explain* basic soil chemistry, including cation-exchange capacity, and *relate* these concepts to soil fertility.

Particles of clay and organic material form negatively charged **soil colloids** that attract and retain positively charged mineral ions in the soil. The capacity to exchange ions between colloids and roots is an ability called the **cation-exchange capacity (CEC)**. CEC is a measure of **soil fertility**, the ability of soil to sustain plants. Fertile soil contains organic substances and clay minerals that absorb water and retain certain elements needed by plants.

soil colloids (p. 563)
cation-exchange capacity (CEC) (p. 563)
soil fertility (p. 564)

9. What are soil colloids? How are they related to cations and anions in the soil? Explain cation-exchange capacity.
10. What is meant by the concept of soil fertility?

● *Evaluate* principal soil formation factors, including the human element.

Environmental factors that affect soil formation include parent materials, climate, vegetation, topography, and time. Human influence is having great impact on Earth's prime soils. Essential soils for agriculture and their fertility are threatened by mismanagement, destruction, and conversion to other uses. Much soil loss is preventable through the application of known technologies, improved agricultural practices, and government policies.

11. Briefly describe the contribution of the following factors and their effect on soil formation: parent material, climate, vegetation, landforms, time, and humans.
12. Explain some of the details that support the concern over loss of our most fertile soils. What cost estimates have been placed on soil erosion?

● *Describe* the 12 soil orders of the Soil Taxonomy classification system and *explain* their general occurrence.

The **Soil Taxonomy** classification system is used in the United States and is built around an analysis of various diagnostic horizons and 12 soil orders, as actually seen in the field. The system divides soils into six hierarchical categories: series, families, subgroups, great groups, suborders, and orders.

Specific soil-forming processes keyed to climatic regions (not a basis for classification) are called **pedogenic regimes: laterization** (leaching in warm and humid climates), **salinization** (collection of salt residues in surface horizons in hot, dry climates), **calcification** (accumulation of carbonates in the B and C horizons in drier continental climates), **podzolization** (soil acidification in forest soils in cool climates), and **gleization** (humus and clay accumulate in cold, wet climates with poor drainage).

The Soil Taxonomy system uses two diagnostic horizons to identify soil: the **epipedon**, or the surface soil, and the **diagnostic subsurface horizon**, or the soil below the surface at various depths. (For an overview and definition of the 12 soil orders, please refer to Table 18.4.) The 12 soil orders are **Oxisols** (tropical soils), **Aridisols** (desert soils), **Mollisols** (grassland soils), **Alfisols** (moderately weathered, temperate forest soils), **Ultisols** (highly weathered, subtropical forest soils), **Spodosols** (northern conifer forest soils), **Entisols** (recent, undeveloped soils), **Inceptisols** (weakly developed, humid region soils), **Gelisols** (cold soils underlain by permafrost), **Andisols** (soils formed from volcanic materials), **Vertisols** (expandable clay soils), and **Histosols** (organic soils).

13. Summarize the basis of soil classification described in this chapter. What have soil scientists revised in the new Soil Taxonomy classification system?

14. What is the basis of the Soil Taxonomy system? How many orders, suborders, great groups, subgroups, families, and soil series are there?

15. Define an epipedon and a diagnostic subsurface horizon. Give a simple example of each.

16. Locate each soil order on the world map and on the U.S. map as you give a general description of it.

17. How was slash-and-burn shifting cultivation, as practiced in the past, a form of crop and soil rotation and conservation of soil properties?

18. Describe the salinization process in arid and semiarid soils. What associated soil horizons develop?

19. Which of the soil orders are associated with Earth's most productive agricultural areas?

20. What is the significance to plants of the 51-cm (20-in.) isohyet in the Midwest relative to soils, pH, and lime content?

21. Describe the podzolization process associated with northern coniferous forest soils. What characteristics are associated with the surface horizons? What strategies might enhance these soils?

22. What former Inceptisols now form a new soil order? Describe these soils as to location, nature, and formation processes. Why do you think they were separated into their own order?

23. Why has a selenium contamination problem arisen in western U.S. soils? Explain the impact of agricultural practices, and tell why you think this is or is not a serious problem.

The *Geosystems* Home Page provides on-line resources for this chapter on the World Wide Web. To begin: Once on the Home Page, click on this textbook, scroll the Table of Contents menu, select this chapter, and click "Begin." You will find self-tests that are graded, review exercises, specific updates for items in the chapter, and many links to interesting related pathways on the Internet under "Destinations." *Geosystems* is at **http://www.prenhall.com/christopherson**.

Critical Thinking

A. Select a small soil sample from your campus or near where you live. Using the sections in this chapter on soil characteristics, properties, and formation, describe as completely as possible this sample, within these limited constraints. Using the general soil map and any other sources available (e.g., local agriculture extension agent, Internet, related department on campus), are you able to roughly place this sample in one of the soil orders?

B. Using the local phone directory or Internet, see if you can locate an agency or extension agent that provides information about local soils and advice for soil management. Is there a place where you can get soil tested?

C. Refer to "Critical Thinking" for Chapter 18 on the *Geosystems* Home Page item 1. Please complete the analysis of the three photographs for environments of deciduous forests of eastern North America, through evergreen forests, and into the arctic tundra. What soil order would you expect in these regions? Why?

Creekfield Lake, really a marsh, is in Brazos Bend State Park, south central Texas. Trees lining the marsh are black willows and the tall cattail-like grasses are southern wild rice. The cloudy sky is from the remnants of Tropical Storm Allison in June 2001. [Photo by Bobbé Christopherson.]

19

Ecosystem Essentials

Key Learning Concepts

After reading the chapter, you should be able to:

- *Define* ecology, biogeography, and the ecosystem concept.
- *Describe* communities, habitats, and niches.
- *Explain* photosynthesis and respiration and *derive* net photosynthesis and the world pattern of net primary productivity.
- *List* abiotic ecosystem components and *relate* those components to ecosystem operations.
- *Explain* trophic relationships in ecosystems.
- *Define* succession and *outline* the stages of general ecological succession in both terrestrial and aquatic ecosystems.

Diversity is an impressive feature of the living Earth. The diversity of organisms, both plant and animal, is a response to the interaction of the atmosphere, hydrosphere, and lithosphere, all powered by solar energy. This interaction produces a variety of conditions within which the biosphere exists. The diversity of life also results from the intricate interplay of living organisms themselves. We, as part of this vast natural complex, seek to find our place and understand the feelings nature stimulates within us.

The physical beauty of nature is certainly among its most powerful appeals to the human animal. The complexity of the aesthetic response is suggested by its wide-ranging expression from the contours of a mountain landscape to the ambient colors of a setting Sun to the fleeting vitality of a breaching whale. Each exerts a powerful aesthetic impact on most people, often accompanied by feelings of awe at the extraordinary physical appeal and beauty of the natural world.*

The biosphere, the sphere of life and organic activity, extends from the ocean floor to about 8 km (5 mi) altitude in the atmosphere. The biosphere includes myriad ecosystems from simple to complex, each operating within general spatial boundaries. An **ecosystem** is a self-sustaining association of living plants and animals and their nonliving physical environment. Earth's biosphere itself is a collection of ecosystems within the natural boundary of the atmosphere and Earth's crust.

Natural ecosystems are open systems for both solar energy and matter, with almost all ecosystem boundaries functioning as transition zones rather than as sharp demarcations. Distinct ecosystems—for example, forests, seas, mountaintops, deserts, beaches, islands, lakes, and ponds—make up the larger whole.

Ecology is the study of the relationships between organisms and their environment and among the various ecosystems in the biosphere. The word *ecology*, developed by German naturalist Ernst Haeckel in 1869, is derived from the Greek *oikos* ("household," or "place to live") and *logos* ("study of"). **Biogeography** is the study of the distribution of plants and animals, the diverse spatial patterns they create, and the physical and biological processes, past and present, that produce Earth's species richness.

The degree to which modern society understands Earth's biogeography and conserves Earth's living legacy will determine the extent of our success as a species and the long-term survival of a habitable Earth:

The time is ripe to step up and expand current efforts to understand the great interlocking systems of air, water, and minerals nourishing the Earth. . . . Moreover, without vigorous action toward that goal, nations will be seriously handicapped in trying to cope with proven and suspected threats to ecosystems and to human health and welfare resulting from alterations in the cycles of carbon, nitrogen, phosphorus, sulfur, and related materials. . . . Society depends upon this life-support system of planet Earth.[†]

*S. R. Kellert, "The biological basis for human values of nature," in S. R. Kellert and E. O. Wilson, eds., *The Biophilia Hypothesis* (Washington, DC: Island Press, 1993), p. 49.
[†]G. F. White and M. K. Tolba, *Global Life Support Systems*, United Nations Environment Programme Information, No. 47 (Nairobi, Kenya: United Nations, 1979), p. 1.

Earth's most influential biotic (living) agents are the humans. This is not arrogance; it is fact, because we powerfully influence every ecosystem on Earth. Since life arose on the planet, there have been six major extinctions. The fifth one was 65 million years ago, whereas the sixth is happening in the present day. Of all these extinction episodes, this is the only one of biotic (living) origin, for the cause is human activity.

In this chapter: We explore ecosystems, and the community, habitat, and niche concepts. Plants are the essential living component in the biosphere, translating solar energy into usable forms to energize life. The role of nonliving systems, including biogeochemical cycles, is examined. We cover the organization of living ecosystems along complex food chains and webs. Ecosystem stability and resilience, and how living landscapes change over space and time through the process of succession, is important. Included is coverage of the effects of global change on ecosystems and rates of succession.

Ecosystem Components and Cycles

An ecosystem is a complex of many variables, all functioning independently yet in concert, with complicated flows of energy and matter (Figure 19.1). An ecosystem includes both *biotic* (living) and *abiotic* (nonliving) components. Nearly all ecosystems depend on a direct input of solar energy; the few limited ones that exist in dark caves, in wells, or on the ocean floor depend on chemical reactions (chemosynthesis).

Ecosystems are divided into subsystems, with the biotic portion composed of producers (plants), consumers (animals), and decomposers (bacteria and fungi). The abiotic flows in an ecosystem include gaseous, hydrologic, and mineral cycles. Figure 19.2 illustrates these essential elements of an ecosystem.

Communities

A convenient biotic subdivision within an ecosystem is a community. A **community** is formed by interactions among populations of living animals and plants at a particular time. An ecosystem is the interaction of many communities with the abiotic physical components of its environment.

Some examples help to clarify these concepts. In a forest ecosystem, a specific community may exist on the forest floor, whereas another community functions in the canopy of leaves high above. Similarly, within a lake ecosystem, the plants and animals that flourish in the bottom sediments form one community, whereas those near the

FIGURE 19.1 The web of life.
"Life devours itself: everything that eats is itself eaten; everything that can be eaten is eaten; every chemical that is made by life can be broken down by life; all the sunlight that can be used is used. . . . The web of life has so many threads that a few can be broken without making it all unravel, and if this were not so, life could not have survived the normal accidents of weather and time, but still the snapping of each thread makes the whole web shudder, and weakens it. . . . You can never do just one thing: the effects of what you do in the world will always spread out like ripples in a pond." [Quotation from Friends of the Earth and Amory Lovins, The United Nations Stockholm Conference, *Only One Earth* (London: Earth Island Limited, 1972), p. 20. Photo by author.]

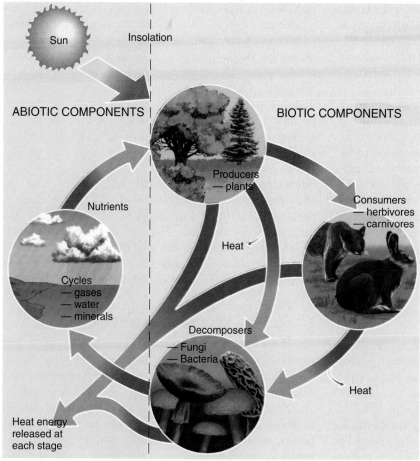

(a)

(b)

FIGURE 19.2 Biotic and abiotic components of ecosystems.
(a) Solar energy is the input that drives the biotic and abiotic components. Heat energy and biomass are the outputs from the biosphere. (b) Biotic and abiotic ingredients operate together to form this delicate forest floor ecosystem with mosses beginning the process of community development. [Photo by Bobbé Christopherson.]

(a)

(b)

FIGURE 19.3 Different forest communities.
(a) Pioneer Mothers' Memorial Forest near Paoli, Indiana, is 36 hectares (88 acres) of old-growth forest, virtually undisturbed since around 1816, featuring black walnut, white oak, yellow poplar, white ash, and beech trees, among others. (b) Turtles bask on a log in a subtropical swamp (low, waterlogged ground) at Juniper Springs Recreation Area, Florida. [Photos by Bobbé Christopherson.]

surface form another (Figure 19.3). A community is identified in several ways—by its physical appearance, the species present and the abundance of each, the complex patterns of their interdependence, and the trophic (feeding) structure of the community.

Within a community, two concepts are important: habitat and niche. **Habitat is the type of environment where an organism resides or is biologically adapted to live.** In terms of physical and natural factors, most species have specific habitat requirements with definite limits and a specific regimen of sustaining nutrients.

Niche (French *nicher,* "to nest") refers to the function, or occupation, of a life form within a given community. It is the way an organism obtains and sustains the physical, chemical, and biological factors it needs to survive. A niche has several facets. Among these are a habitat niche, a trophic (food) niche, and a reproductive niche. For example, the Red-winged Blackbird (*Agelaius phoeniceus*) occurs throughout the United States and most of Canada in habitats of meadow, pastureland, and marsh. This species nests in blackberry tangles and thick vegetation in freshwater marshes, sloughs, and fields. Its trophic niche is weed seeds and cultivated seed crops throughout the year, and during the nesting season it adds insects to its diet—an aspect of its reproductive niche. These birds disperse seeds of many plants during their travels.

Other habitats produce comparable niches. In a stable community, no niche is left unfilled. The *competitive exclusion principle* states that no two species can occupy the same niche (food or space) successfully in a stable community. Thus, closely related species are spatially separated. In other words, each species operates to reduce competition. This strategy in turn leads to greater diversity as species shift and adapt (Figure 19.4).

Some species are *symbiotic,* an arrangement where two or more species exist together in an overlapping relationship. One type of symbiosis, *mutualism* occurs when each organism benefits and is sustained over an extended period by the relation. For example, lichen (pronounced "liken") is made up of algae and fungi living together. The alga is the producer and food source for the fungus, and the fungus provides structure and physical support. Their mutualism allows the two to occupy a niche in which neither could survive alone. Lichen developed from an earlier parasitic relationship in which the fungi broke into the alga cells. Today, the two organisms have evolved into a supportive harmony and symbiotic relationship (Figure 19.5). The partnership of corals and algae discussed in Chapter 16 is another example of mutualism in a symbiotic relationship.

By contrast, another form of symbiosis is a *parasitic* relationship, which may eventually kill the host, thus destroying the parasite's own niche and habitat. An example is parasitic mistletoe (*Phoradendron*), which lives on and may kill various kinds of trees. Some scientists are questioning whether our human society and the physical systems of Earth constitute a global-scale symbiotic relationship of mutualism (sustainable) or a parasitic one (nonsustainable).

Plants: The Essential Biotic Component

Plants are the critical biotic link between solar energy and the biosphere. *Ultimately, the fate of all members of the biosphere, including humans, rests on the success of plants and their ability to capture sunlight.*

Land plants (and animals) became common about 430 million years ago, according to fossilized remains. **Vascular plants** developed conductive tissues and true roots for internal transport of fluid and nutrients. (*Vascular* is from a

FIGURE 19.4 Plants and animals fit specific niches.
(a) Elephant heads (*Pedicularis groenlandica*), a wildflower that grows above 1800 m (6000 ft) in wet mountain meadows. (b) A western yellow-bellied racer (*Coluber constrictor*) lives in prairies and meadows and feeds on insects, lizards, and mice. (c) A dragonfly (*Libellula* sp.) perches atop a reed along a stream, feeding on flying insects. (d) Killdeer chicks (*Charadrius vociferus*) begin life roughing it on an exposed rocky nest, with protective parents creating a noisy diversion nearby. (e) Coral mushroom (*Ramaria* sp.), a fungal decomposer at work on the organic matter accumulated on the forest floor. (f) A large barrel cactus flourishes in the rocky soils and harsh conditions of the desert. [(a, c, e) Photos by Bobbé Christopherson; (b, d, f) photos by author.]

FIGURE 19.5 Symbiosis on the rocks.
Lichen, an example of a symbiotic relationship, specifically mutualism, between fungi and algae. [Photo by Bobbé Christopherson.]

Latin word for "vessel-bearing," referring to the conducting cells.)

In the present day, about 270,000 species of plants are known to exist, and most are vascular. Many more species have yet to be identified. They represent a great untapped resource base. Only about 20 species of plants provide 90% of the world's food—just three, wheat, maize (corn), and rice, comprise half of the food supply. Plants are a major source of new medicines and chemical compounds that benefit humanity. Plants also are the core of healthy, functioning ecosystems that sustain all life.

Leaves are solar-powered chemical factories, wherein photochemical reactions take place. Veins in the leaf bring in water and nutrient supplies and carry off the sugars (food) produced by photosynthesis. The veins in each leaf connect to the stems and branches of the plant and to the main circulation system.

Flows of carbon dioxide, water, light, and oxygen enter and exit the surface of each leaf (see Figure 1.4). Gases flow into and out of a leaf through small pores called **stomata** (singular: *stoma*), which usually are most numerous on the lower side of the leaf. Each stoma is surrounded by guard cells that open and close the pore, depending on the plant's changing needs.

Water that moves through a plant exits the leaves through the stomata and evaporates from leaf surfaces, thereby assisting the plant's temperature regulation. As water evaporates from the leaves, a pressure deficit is created that allows atmospheric pressure to push water up through the plant all the way from the roots, in the same manner that a soda straw works. We can only imagine the complex operation of a 100-m (330-ft) tree!

Photosynthesis and Respiration

Powered by energy from certain wavelengths of visible light, **photosynthesis** unites carbon dioxide and hydrogen (hydrogen is derived from water in the plant). The term is descriptive: *photo-* refers to sunlight, and *-synthesis* describes the "manufacturing" of starches and sugars through reactions within plant leaves. The process releases oxygen and produces energy-rich food for the plant.

The largest concentration of light-responsive, photosynthetic structures (known as *organelles*) in a leaf rests below the leaf's upper layers. These organelle units within cells are called *chloroplasts*, and within each resides a green, light-sensitive pigment called **chlorophyll**. Within this pigment, light stimulates photochemistry. Consequently, competition for light is a dominant factor in the formation of plant communities. This competition is expressed in the height, orientation, distribution, and structure of plants.

Only about one-quarter of the light energy arriving at the surface of a leaf is useful to the light-sensitive chlorophyll. Chlorophyll absorbs only the orange-red and violet-blue wavelengths for photochemical operations, and it reflects predominantly green hues (and some yellow). That is why trees and other vegetation look green.

Photosynthesis essentially follows this equation:

$$6CO_2 + 6H_2O + Light \rightarrow C_6H_{12}O_6 + 6O_2$$

(carbon dioxide)   (water)   (solar energy)   (glucose, carbohydrate)   (oxygen)

From the equation, you can see that photosynthesis removes carbon (in the form of CO_2) from Earth's atmosphere. The quantity is enormous: approximately 91 billion metric tons (100 billion tons) of carbon dioxide per year. Carbohydrates, the organic result of the photosynthetic process, are combinations of carbon, hydrogen, and oxygen. They can form simple sugars, such as *glucose* ($C_6H_{12}O_6$). Plants use glucose to build starches, which are more complex carbohydrates and the principal food stored in plants.

Plants store energy for later use. They consume this energy as needed through respiration by converting the carbohydrates to energy for their other operations. Thus, **respiration** is essentially a reverse of the photosynthetic process:

$$C_6H_{12}O_6 + 6O_2 \rightarrow 6CO_2 + 6H_2O + energy$$

(glucose, carbohydrate)   (oxygen)   (carbon dioxide)   (water)   (heat energy)

In respiration, plants oxidize carbohydrates (stored energy), releasing carbon dioxide, water, and energy as heat. The overall growth of a plant depends on a surplus of carbohydrates beyond what is lost through plant respiration. Figure 19.6 presents a simple schematic of this process, which produces plant growth.

The *compensation point* is the break-even point between the production and consumption of organic material. Each leaf must operate on the production side of the compensation point, or else the plant eliminates it—something each of us has no doubt experienced with a house plant that received inadequate water or light. The difference between photosynthetic production and respiration loss is called *net photosynthesis*. The amount varies, depending on controlling environmental factors such as light, water, temperature, soil fertility, and the plant's site, elevation, and competition from other plants and animals.

Plant productivity increases as light availability increases—up to a point. When the light level is too high, *light saturation* occurs and most plants actually reduce their output in response. Some plants are adapted to shade, whereas others flourish in full sunlight. Crops such as rice, wheat, and sugar cane do well with high light intensity. Figure 19.7 portrays the general energy budget for green plants, showing energy receipt, utilization, and disposition of net primary production.

Net Primary Productivity   The net photosynthesis for an entire plant community is its **net primary productivity**. This is the amount of stored chemical energy (biomass)

FIGURE 19.6 How plants live and grow.
The balance between photosynthesis and respiration determines net photosynthesis and plant growth. [Temperate rain forest photo by Bobbé Christopherson.]

FIGURE 19.7 Energy budget of the biosphere.
Energy receipt, utilization, and disposition of net primary production by green plants.

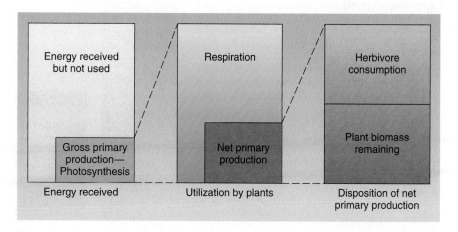

that the community generates for the ecosystem. **Biomass** is the net dry weight of organic material.

Net primary productivity is measured as fixed carbon per square meter per year. ("Fixed" means chemically bound into plant tissues.) Study the map and satellite images in Figure 19.8 and you can see that on land, net primary production tends to be highest between the Tropics of Cancer and Capricorn at sea level and decreases toward higher latitudes and altitudes. Precipitation also affects productivity, as evidenced by the correlations of abundant precipitation with high productivity (adjacent to the equator) and reduced precipitation with low productivity (subtropical deserts). Even though deserts receive high amounts of solar radiation, other controlling factors limit productivity, namely water availability and soil conditions.

In the oceans, differing nutrient levels control and limit productivity. Regions with nutrient-rich upwelling currents generally are the most productive (off western coastlines). The map in Figure 19.8 shows that the tropical ocean and areas of subtropical high pressure are quite low in productivity.

In temperate and high latitudes, the rate at which carbon is fixed by vegetation varies seasonally. It increases in spring and summer as plants flourish with increasing solar input and, in some areas, with more available (nonfrozen) water, and it decreases in late fall and winter. Productivity rates in the tropics are high throughout the year, and turnover in the photosynthesis–respiration cycle is faster, exceeding by many times the rates experienced in a desert environment or in the far northern limits of the tundra. A lush hectare (2.5 acres) of sugar cane in the tropics might fix 45 metric tons (50 tons) of carbon in a year, whereas desert plants in an equivalent area might achieve only 1% of that amount.

Table 19.1 lists various ecosystems, their net primary productivity, and an estimate of net total biomass worldwide—170 billion metric tons of dry organic matter per year. Compare the various ecosystems, especially cultivated

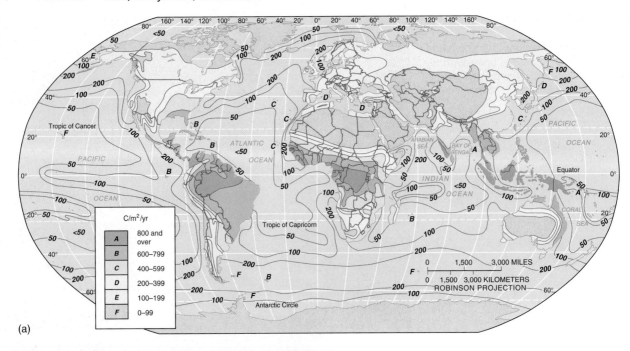

(a)

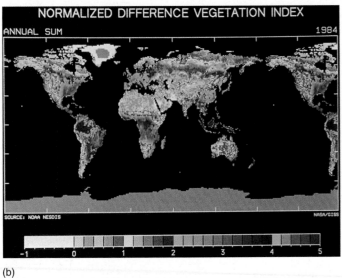

(b)

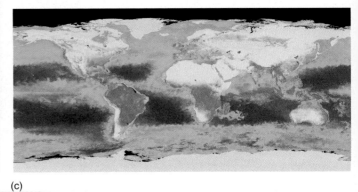

(c)

FIGURE 19.8 Net primary productivity.
(a) Worldwide net primary productivity in grams of carbon per square meter per year (approximate values). (b) Normalized difference vegetation index during 1984. False coloration indicates bare ground in browns, dense vegetation in blues. (c) *SeaWIFS* image of surface chlorophyll concentration for land and sea; intensity of green color represents higher levels of chlorophyll. This is the first continuous record over a three-year period for the oceans.
[(a) After D. E. Reichle, *Analysis of Temperate Forest Ecosystems* (Heidelberg: Springer, 1970). Adapted by permission. (b) NASA/GSFC. (c) *SeaWIFS* image by NASA/GSFC, and ORBIMAGE, Dulles, VA. Used by permission. All rights reserved.]

land (in *italics*) with most of the natural communities. Net productivity is generally regarded as the most important aspect of any type of community, and the distribution of productivity over Earth's surface is an important subject of biogeography.

Abiotic Ecosystem Components

Critical in each ecosystem is the flow of energy and the cycling of nutrients and water in life-supporting systems. These abiotic (nonliving) components set the stage for ecosystem operations.

Table 19.1 Net Primary Productivity and Plant Biomass on Earth				
		Net Primary Productivity per Unit Area (g/m²/yr)[b]		
Ecosystem	**Area (10⁶ km²)[a]**	**Normal Range**	**Mean**	**World Net Biomass (10⁹/tons/yr)[c]**
Tropical rain forest	17.0	1000–3500	2200	37.4
Tropical seasonal forest	7.5	1000–2500	1600	12.0
Temperate evergreen forest	5.0	600–2500	1300	6.5
Temperate deciduous forest	7.0	600–2500	1200	8.4
Boreal forest	12.0	400–2000	800	9.6
Woodland and shrubland	8.5	250–1200	700	6.0
Savanna	15.0	200–2000	900	13.5
Temperate grassland	9.0	200–1500	600	5.4
Tundra and alpine region	8.0	10–400	140	1.1
Desert and semidesert scrub	18.0	10–250	90	1.6
Extreme desert, rock, sand, ice	24.0	0–10	3	0.07
Cultivated land	*14.0*	*100–3500*	*650*	*9.1*
Swamp and marsh	2.0	800–3500	2000	4.0
Lake and stream	2.0	100–1500	250	0.5
Total continental	**149**	—	**773**	**115.17**
Open ocean	332.0	2–400	125	41.5
Upwelling zones	0.4	400–1000	500	0.2
Continental shelf	26.6	200–600	360	9.6
Algal beds and reefs	0.6	500–4000	2500	1.6
Estuaries	1.4	200–3500	1500	2.1
Total marine	**361.0**	—	**152**	**55.0**
Grand total	**510.0**	—	**333**	**170.17**

Source: R. H. Whittaker, *Communities and Ecosystems* (Heidelberg: Springer, 1975), p. 224. Reprinted by permission.

[a]1 km² = 0.39 mi².

[b]1 g per m² = 8.92 lb per acre.

[c]1 metric ton (10⁶g) = 1.1023 tons.

Light, Temperature, Water, and Climate   Solar energy powers ecosystems, so the pattern of solar energy receipt is crucial. Solar energy enters an ecosystem by way of photosynthesis, and heat energy is dissipated from the system at many points. Of the total energy intercepted at Earth's surface and available for work, only about 1.0% is actually fixed by photosynthesis as chemical energy (energy stored as carbohydrates in plants).

The duration of Sun exposure is the *photoperiod*. Along the equator, days are essentially 12 hours long year-round; however, with increasing distance from the equator, seasonal effects become pronounced, as discussed in Chapter 2. Plants have adapted their flowering and seed germination to seasonal changes in insolation. Some seeds germinate only when daylength reaches a certain number of hours. A plant that responds in the opposite manner is the poinsettia (*Euphorbia pulcherrima*), which requires at least 2 months of 14-hour nights to start flowering.

Other components are important to ecosystem processes. Air and soil temperatures determine the rates at which chemical reactions proceed (see Chapter 18). Significant temperature factors are seasonal variation and duration and the pattern of minimum and maximum temperatures (Chapter 5).

Operations of the hydrologic cycle and water availability depend on precipitation and evaporation rates and their seasonal distribution (see Chapters 7, 8, and 9). Water quality—its mineral content, salinity, and levels of pollution and toxicity—is important. Also, regional climates affect the pattern of vegetation and ultimately influence soil development. All of these factors work together to form the limits for ecosystems in a given location.

Figure 19.9 illustrates the general relationship among temperature, precipitation, and vegetation. Can you identify the characteristic vegetation type and related temperature and moisture regime that fits areas you know well, such as a place you have lived or an area around the school you now attend?

Life Zones   Alexander von Humboldt (1769–1859), an explorer, geographer, and scientist, observed that plants and animals recur in related groupings wherever similar

(a)
(b)
(c)
(d)
(e)
(f)
(g)
(h)
(i)

Polar and Alpine
Dry tundra
Moist tundra
Cool
Cool desert
Pine-Spruce-Fir forest
Temperate
Temperate desert
Short grass
Tall grass
Deciduous forest
Hot
Hot desert
Grasslands
Tropical forest
Temperature
Precipitation
Dry
Wet

FIGURE 19.9 How temperature and precipitation affect ecosystems.
(a) Climate controls (wetness, dryness; warmth, cold) and ecosystem types; (b) subtropical desert in Namibia; (c) Sonoran desert of Arizona; (d) cold desert of north-central Nevada; (e) dry tundra of NWT Canada; (f) El Yunque tropical rain forest of Puerto Rico; (g) deciduous trees in fall colors in Ohio; (h) needleleaf forest near Independence Pass, Colorado; and (i) moist tundra of northern Québec. [(b) Photo by NigelDennis/Photo Researchers, Inc.; (c, d) photos by author; (e) photo by John Eastcott/Yva Momatiuk/The Image Works; (f) photo by Tom Bean; (g, h) Bobbé Christopherson; (i) photo by John Eastcott/Yva Momatiuk/Stock Boston.]

conditions occur in the abiotic environment. He described the similarities and dissimilarities of vegetation and the associated uneven distribution of various organisms. After several years of study in the Andes Mountains of Peru, he described a distinct relation between altitude and plant communities as his *life zone concept.* As he climbed the mountains, he noticed that the experience was similar to that of traveling away from the equator toward higher latitudes (Figure 19.10).

This zonation of plants with altitude is noticeable on any trip from lower valleys to higher elevations. Each **life zone** possesses its own temperature, precipitation, and insolation relations and therefore its own biotic communities.

The Grand Canyon in Arizona provides a good example. Plants and animals of the inner gorge at the bottom of the canyon (600 m, or 2000 ft, in elevation) are characteristic of the lower Sonoran Desert of northern Mexico. However, communities similar to those of southern Canadian forests dominate the north rim of the canyon (2100 m, or 7000 ft, in elevation). On the summits of the nearby San Francisco Mountains (3600 m, or 12,000 ft, in elevation), the vegetation is similar to that of the arctic tundra of

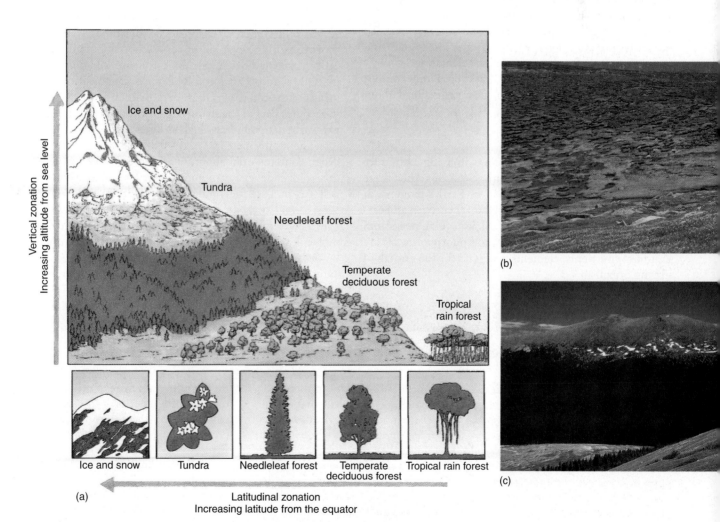

FIGURE 19.10 Vertical and latitudinal zonation of plant communities.
(a) Progression of plant community life zones with changing altitude or latitude. (b) Alpine tundra ecosystem in the Colorado Rockies. (c) The timberline for a needleleaf forest in the Rockies. The line of trees marks the limit of highest continuous forest, whereas the area above it marks the zone above which no trees grow. [Photos by Bobbé Christopherson.]

northern Canada. Needleleaf forest and alpine tundra in the Colorado Rockies at high elevation are pictured in Figure 19.10.

Other Factors   Beyond these general conditions, each ecosystem further produces its own *microclimate*, which is specific to individual sites. For example, in a forest the insolation reaching the ground is reduced and the forest floor is shaded by its own trees. A pine forest cuts light by 20%–40%, whereas a birch–beech forest reduces it by as much as 50%–75%. Forests also are about 5% more humid than nonforested landscapes, have moderated temperatures (warmer winters and cooler summers), and experience reduced winds.

Slope orientation and Sun exposure are important, too, for they translate into differences in temperature and moisture efficiency, especially in middle and higher latitudes. With all other factors equal, slopes facing away from the Sun's rays tend to be moister and more vegetated than slopes facing toward the Sun. (In the Northern Hemisphere, these moister slopes face north, as evidenced by the tendency of moss, a plant that needs high moisture and shade, to grow on the north side of tree trunks.) Such highly localized *microecosystems* are evident where changes in exposure and moisture occur related to geographic orientation.

Even Earth's magnetic field, another abiotic component, plays an interesting role in ecosystems. Read about turtle navigation in News Report 19.1.

Elemental Cycles

The most abundant natural elements in living matter are hydrogen (H), oxygen (O), and carbon (C). Together, these elements make up more than 99% of Earth's biomass; in fact, all life (organic molecules) contains hydrogen and carbon. In addition, nitrogen (N), calcium (Ca), potassium (K), magnesium (Mg), sulfur (S), and phosphorus (P) are significant nutrients, elements necessary for the growth of a living organism.

Several key chemical cycles function in nature. Oxygen, carbon, and nitrogen each have *gaseous cycles*, part of which are in the atmosphere. Other elements have *sedimentary cycles*, which principally involve mineral and solid phases (major ones include phosphorus, calcium, potassium, and sulfur). Some elements combine gaseous and sedimentary cycles. The recycling of gases and sedimentary (nutrient) materials form Earth's **biogeochemical cycles**, because they involve chemical reactions necessary for growth and development of living systems. The chemical elements themselves recycle over and over again in life processes.

Oxygen and Carbon Cycles   We consider these two cycles together because they are so closely intertwined through photosynthesis and respiration (Figure 19.11). The atmosphere is the principal reserve of available oxygen. Larger reserves of oxygen exist in Earth's crust, but they are unavailable, being chemically bound with other elements, especially the silicate (SiO_2) and carbonate (CO_3) mineral families. Unoxidized reserves of fossil fuels and sediments also contain oxygen.

The oceans are enormous pools of carbon—about 42,900 billion metric tons. (Metric tons times 1.10 equals short tons.) However, all of this carbon is bound chemically in carbon dioxide, calcium carbonate, and other compounds. The ocean initially absorbs carbon dioxide by means of the photosynthesis carried on by phytoplankton; it becomes part of the living organisms and is fixed in certain carbonate minerals, such as limestone ($CaCO_3$).

The atmosphere, which is the integrating link in the cycle, contains only about 700 billion tons of carbon (as carbon dioxide) at any moment. This is far less carbon than is stored in fossil fuels and oil shales (13,200 billion metric tons, as hydrocarbon molecules) or in living and dead organic matter (2500 billion metric tons, as carbohydrate molecules). Carbon dioxide is released into the atmosphere by the respiration of plants and animals, volcanic activity, and fossil fuel combustion by industry and transportation.

News Report 19.1

Earth's Magnetic Field—An Abiotic Factor

The fact that birds and bees can detect Earth's magnetic field and use it for finding direction is well established. Small amounts of magnetically sensitive particles in the skull of the bird and the abdomen of the bee provide compass directions for bird migration and for alerting the hive to the location of the latest nectar find.

A team of biologists from the University of North Carolina, Catherine and Kenneth Lohmann, have found that sea turtles can detect magnetic fields of different strengths and

the inclination (angle) of these magnetic fields. This means that the turtles have a built-in navigation system that helps them know where they are on Earth (like our global positioning system, which requires multimillion-dollar satellites). Using magnetic field strengths and inclination, the turtles evidently are aware of their global position (similar to knowing their latitude and longitude). This magnetic map is evidently imprinted in their brain.

Loggerhead sea turtles hatch in Florida, crawl into the water, and spend

the next 70 years traveling thousands of miles between North America and Africa around the subtropical high-pressure gyre in the Atlantic Ocean (see Figure 6.14). The females always return to near where they were hatched to lay their eggs! The researchers think that the hatchlings are imprinted with magnetic data unique to the beach where they hatched and then develop a more global sense as they live a life swimming across the ocean.

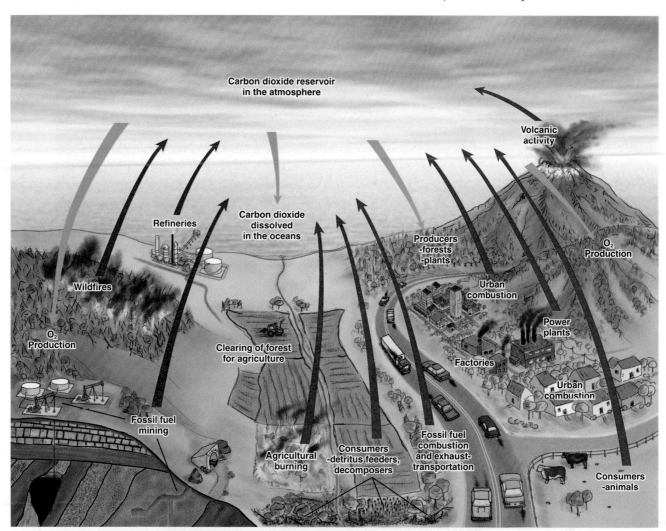

FIGURE 19.11 The carbon and oxygen cycles.
Carbon is fixed through photosynthesis and begins its passage through ecosystem operations. Respiration by living organisms, burning of forests and grasslands, and the combustion of fossil fuels releases carbon to the atmosphere. These cycles are greatly influenced by human activities.

The carbon dumped into the atmosphere by human activity constitutes a vast geochemical experiment, using the real-time atmosphere as a laboratory. Since 1970, we have added to the atmospheric pool an amount of carbon equivalent to more than 25% of the total amount added since 1880; fourfold since 1950.

Global emissions of carbon from fossil fuels declined slightly from a 1997 high of 6.395 down to 6.299 billion metric tons (combined with oxygen in 18.9 billion metric tons of carbon dioxide) in 2000—400% more than the 1.6 billion metric tons of carbon in 1950. About 50% of the carbon dioxide emitted since the beginning of the Industrial Revolution and not absorbed by oceans and organisms remains in the atmosphere, enhancing Earth's natural greenhouse effect. The greenhouse effect was covered more fully in Chapters 3, 5, and 10.

Nitrogen Cycle Nitrogen, which accounts for 78.084% of each breath we take, is the major constituent of the atmosphere. Nitrogen also is important in the makeup of or-

ganic molecules, especially proteins, and therefore is essential to living processes. A simplified view of the nitrogen cycle is portrayed in Figure 19.12.

This vast atmospheric reservoir of nitrogen is inaccessible directly to most organisms. *Nitrogen-fixing bacteria* provide the key link to life, live principally in the soil, and are associated with the roots of certain plants—for example, the *legumes* such as clover, alfalfa, soybeans, peas, beans, and peanuts. Colonies of these bacteria reside in nodules on the legume roots and chemically combine the nitrogen from the air in the form of nitrates (NO_3) and ammonia (NH_3).

Plants use these chemically bound forms of nitrogen to produce their own organic matter. Anyone or anything feeding on the plants thus ingests the nitrogen. Finally, the nitrogen in the organic wastes (sewage, manure) of these consuming organisms is freed by denitrifying bacteria, which recycle nitrogen back to the atmosphere.

To improve agricultural yields, many farmers use synthetic inorganic fertilizers, as opposed to soil-building organic fertilizers (manure and compost). Inorganic fertilizers

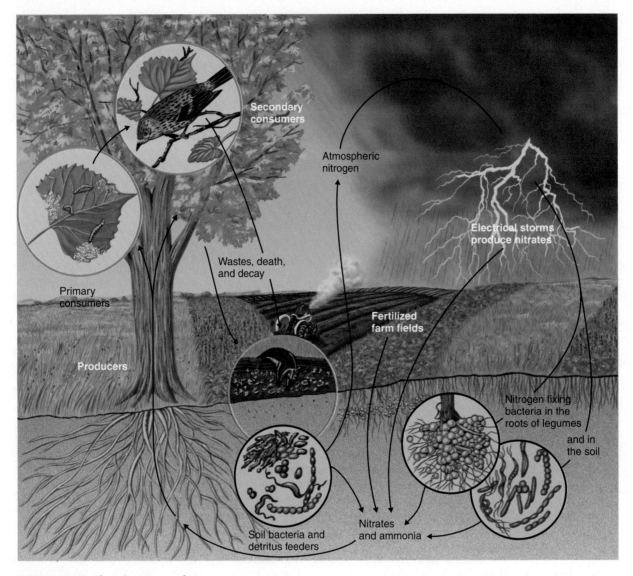

FIGURE 19.12 The nitrogen cycle.
The atmosphere is the essential reservoir of gaseous nitrogen. Atmospheric nitrogen gas is
chemically fixed by bacteria to produce ammonia. Lightning and forest fires produce nitrates,
and fossil fuel combustion forms nitrogen compounds that are washed from the atmosphere
by precipitation. Plants absorb nitrogen compounds and produce organic material. [From
T. Audesirk and G. Audesirk, *Biology: Life on Earth,* 4th ed., Figure 45-10, p. 903. © 1996
Prentice Hall Inc. Used by permission.]

are chemically produced through artificial nitrogen fixation
at factories. The annual production of inorganic fertilizers
far exceeds the ability of natural denitrification systems—
and the present production of synthetic fertilizers now is
doubling every 8 years. Humans presently fix more nitrogen
per year than all terrestrial sources combined.

This surplus of usable nitrogen accumulates in Earth's
ecosystems. Some is present as excess nutrients, washed
from soil into waterways and eventually to the ocean. This
excess nitrogen load begins a water pollution process that
feeds an excessive growth of algae and phytoplankton, increases biochemical oxygen demand, diminishes dissolved
oxygen reserves, and eventually disrupts the aquatic ecosystem. In addition, excess nitrogen compounds in air pollution are a component in acid deposition, further altering

the nitrogen cycle in soils and waterways (see Chapter 3).
For more on this problem, specific to North America, read
News Report 19.2.

Limiting Factors

The term **limiting factor** identifies the one physical or
chemical abiotic component that most inhibits biotic operations, through either its lack or its excess. Here are a
few examples:

- Low temperatures limit plant growth at high eleva-
  tions.
- Lack of water limits growth in a desert.
- Excess water limits growth in a bog.

News Report 19.2

The Dead Zone

The flush of agricultural runoff and fertilizers, animal manure, sewage, and other wastes is carried by the Mississippi River to the Gulf causing huge spring blooms of phytoplankton. By summer the biological oxygen demand of bacteria feeding on the decay exceeds the dissolved oxygen, and hypoxia develops, killing any fish that ventures into the area. These low-oxygen conditions act as a limiting factor on marine life. Figure 1 shows the dead zone, a region of oxygen-depleted (hypoxic) water off the coast of Louisiana in the Gulf of Mexico.

The Mississippi drainage system handles the runoff for 41% of the continental United States. When the 1993 Midwest floods hit, the river's nutrient discharge was such that the size of the dead zone doubled to 17,500 km^2 (6755 mi^2). The worst year was 1999, when it expanded to 20,000 km^2 (7720 mi^2). The agricultural, feedlot, and fertilizer industries dispute the connection between their nutrient input and the dead zone (Figure 2).

In other parts of the world, the connection is established. In Sweden and Denmark, a concerted effort to reduce nutrient flows into their rivers reversed hypoxic conditions in the Kattegat Strait (between the Baltic and North Seas). And, with the fall of the U.S.S.R. and state agriculture in 1990, fertilizer use is down more than 50%, and for the first time in recent decades the Black Sea is getting several months' break from year-around hypoxia off river discharge inlets.

As with most environmental situations, the cost of mitigation is cheaper than the cost of continued damage to marine ecosystems. A cut in nitrogen inflow upstream of 20% to 30% is estimated to raise dissolved oxygen levels by more than 50% in the dead zone region of the Gulf. One government estimate placed the application of excess nitrogen fertilizer at 20% more than needed in Iowa, Illinois, and Indiana. The initial step to resolving this issue might be to mandate applying only the levels of fertilizer needed—a savings to agricultural interests.

(a)

(b)

FIGURE 1 The Gulf Coast dead zone.
(a) The flush of nutrients from the Mississippi River drainage basin enriches the offshore waters in the Gulf of Mexico, causing huge phytoplankton blooms in early spring (red areas). (b) By late summer, oxygen-depleted waters (hypoxic) dominate what is known as the *dead zone*. [*SeaWiFS* image from the NASA/GSFC, courtesy of ORBIMAGE, Dulles, VA.]

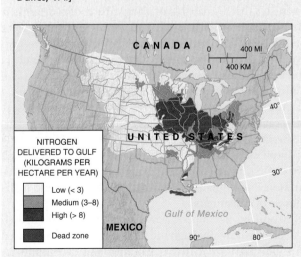

FIGURE 2 Nitrogen sources for the dead zone.
Total nitrogen originating in the upstream watershed that feeds the Mississippi River system. [Adapted from R. B. Alexander, R. A. Smith, and G. E. Schwarz, "Effect of stream channel size on the delivery of nitrogen to the Gulf of Mexico," *Nature* (February 17, 2000); 761; as presented in *Environment,* Jan/Feb 2001, 16.]

- Changes in salinity levels affect aquatic ecosystems.
- Lack of iron in ocean surface environments limits photosynthetic production.
- Low phosphorus content of soils limits plant growth.
- General lack of active chlorophyll above 6100 m (20,000 ft) limits primary productivity.

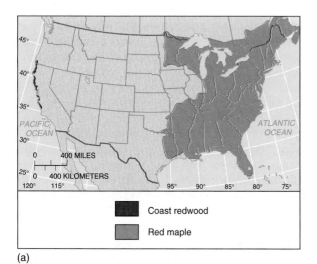

(a)

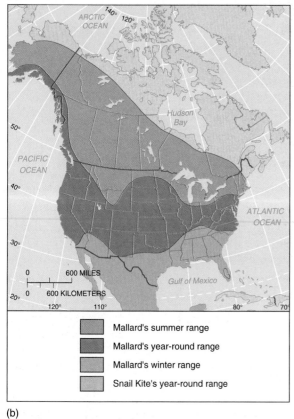

(b)

FIGURE 19.13 Limiting factors affect the distribution of every plant and animal species.
(a) Distributions of coast redwood and red maple demonstrate the effect of limiting factors. (b) So do the distributions of the Mallard duck and the Snail Kite. The Mallard is a generalist and feeds widely. In contrast, the range of its food, a single type of snail, limits the Snail Kite.

In most ecosystems, precipitation is the limiting factor. However, temperature, light levels, and soil nutrients certainly affect vegetation patterns.

Each organism possesses a range of tolerance for each limiting factor in its environment. This fact is illustrated vividly in Figure 19.13, which shows the geographic range for two tree species and two bird species. The coast redwood (*Sequoia sempervirens*) is limited to a narrow section of the Coast Ranges in California, covering barely 9500 km² (less than 4000 mi²), concentrated in areas that receive necessary summer advection fog. On the other hand, the red maple (*Acer rubrum*) thrives over a large area under varying conditions of moisture and temperature, thus demonstrating a broader tolerance to environmental variations.

The Mallard duck (*Anas platyrhynchos*) and the Snail Kite (*Rostrhamus sociabilis*) also demonstrate a variation in tolerance and range (Figure 19.13b). The Mallard, a generalist, feeds from widely diverse sources, is easily domesticated, and is found throughout most of North America in at least one season of the year. By contrast, the Snail Kite is a specialist that feeds only on one specific type of snail. This single food source, then, is its limiting factor. Note its small habitat area near Lake Okeechobee in Florida.

Biotic Ecosystem Operations

The abiotic components of energy, atmosphere, water, weather, climate, and minerals make up the life support for the biotic components of each ecosystem. The flow of energy, cycling of nutrients, and trophic (feeding) relations determine the nature of an ecosystem. As energy cascades through this process-flow system, it is constantly replenished by the Sun. But nutrients and minerals cannot be replenished from an external source, so they constantly cycle within each ecosystem and through the biosphere (Figure 19.14). Let us examine these biotic operations.

Producers, Consumers, and Decomposers

Organisms that are capable of using carbon dioxide as their sole source of carbon are called *autotrophs* (self-feeders), or **producers**. These are the plants. They chemically fix carbon through photosynthesis. Organisms that depend on producers as their carbon source are called *heterotrophs* (feed on others), or **consumers**. Generally, these are animals.

Autotrophs are the essential producers in an ecosystem—capturing light energy and converting it to chemical energy, incorporating carbon, forming new plant tissue and biomass, and freeing oxygen. Solar energy enters each food chain through the producer plant or producer phytoplankton, subsequently flowing through higher and higher levels of consumers.

From the producers, which manufacture their own food, energy flows through the system along a circuit called the **food chain**, reaching consumers and eventually *de-*

FIGURE 19.14 Energy, nutrient, and food pathways in the environment.
The flow of energy, cycling of nutrients, and trophic (feeding) relationships portrayed for a generalized ecosystem. The operation is fueled by radiant energy supplied by sunlight and first captured by the plants. [From T. Audesirk and G. Audesirk, *Biology: Life on Earth,* 4th ed., Figure 45-1, p. 892. © 1996 Prentice Hall Inc. Used by permission.]

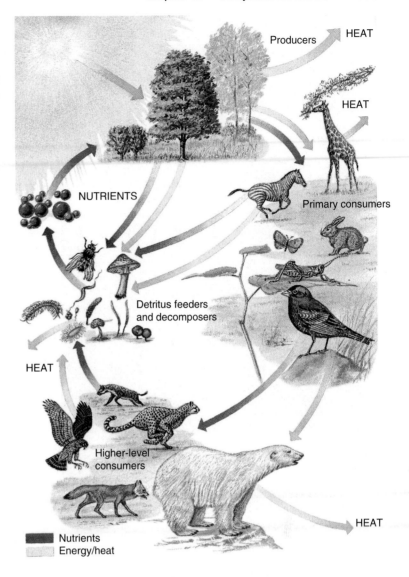

composers. Organisms that share the same basic foods are said to be at the same *trophic level.* Ecosystems generally are structured in a **food web**, a complex network of interconnected food chains. In a food web, consumers participate in several different food chains, comprising both *strong interactions* and *weak interactions* between species in the food web. Figures 19.15 and 19.16 illustrate a simple terrestrial food chain and an oceanic food chain.

Primary consumers feed on producers. Because producers are always plants, the primary consumer is called an **herbivore**, or plant eater. A **carnivore** is a secondary consumer and primarily eats meat. A tertiary consumer eats primary and secondary consumers and is referred to as the "top carnivore" in the food chain. A consumer that feeds on both producers (plants) and consumers (meat) is called an **omnivore**—a role occupied by humans, among others.

Decomposers are the final link in the chain. They renew the entire system by releasing inorganic materials from organic debris. **Decomposers** are bacteria and fungi that digest and recycle the organic debris and waste in the environment. In addition, the *detritus feeders*—worms, mites, termites, centipedes, and others—participate like a small army of workers. Waste products, dead plants and animals, and other organic remains are the principal food source for all these *detritivores.* Inorganic compounds are released in the process and the cycle continues.

Examples of Complex Food Webs

An example of a complex community is the oceanic food web that includes krill, a primary consumer (Figure19.16a). *Krill* is a shrimplike crustacean that is a major food for an interrelated group of organisms, including whales, fish, seabirds, seals, and squid in the Antarctic region. All of these organisms participate in numerous other food chains as well, some consuming and some being consumed.

Phytoplankton begin this chain by harvesting solar energy in photosynthesis. Phytoplankton are eaten by *herbivorous zooplankton* such as krill and other organisms. Krill are eaten by consumers at the next trophic level. Because krill are a protein-rich, plentiful food, increasingly factory ships seek them out, such as those from Japan and

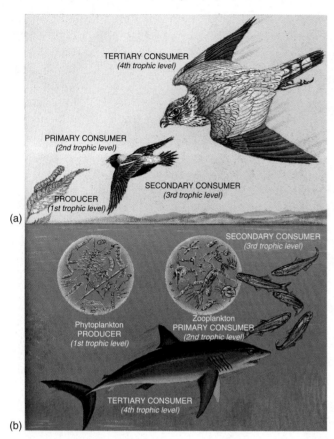

(a)

(b)

FIGURE 19.15 Food chains.
(a) A simplified terrestrial food chain, from leaf to caterpillar to Bobolink to a fast-moving Merlin hawk. (b) A simplified aquatic food chain from phytoplankton to zooplankton to schooling fish and to a mako shark. [(a) and (b) From T. Audesirk and G. Audesirk, *Biology: Life on Earth,* 4th ed., Figure 45-4, p. 895. © 1996 Prentice Hall Inc. Used by permission.]

Russia. The annual krill harvest currently surpasses a million tons, principally as feed for chickens and livestock and as protein for human consumption.

The impact on the food web of further increases in the krill harvest is uncertain, for little is known about the krill reproduction rate. Their long life span (greater than 8 years) and slow growth rate limit the ultimate extent of the resource. In addition, the possible effect on krill of increased ultraviolet radiation resulting from the seasonal "hole" in the ozone layer above Antarctica is under investigation.

Researchers determined the importance of weak (secondary) interactions in a complex food web for island communities in the Gulf of California (Figure 19.16b). The complexity they found produced stability and persistence in the system. Kevin McCann and his research team

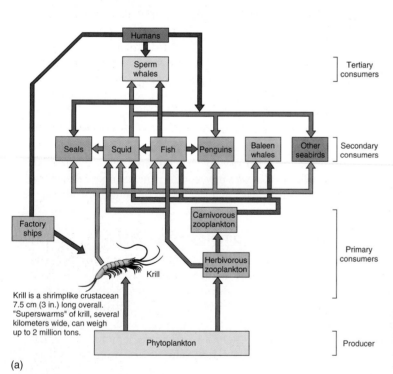

(a)

(b)

FIGURE 19.16 Complex food webs.
(a) The food web in Antarctic waters, from phytoplankton producers (bottom) through various consumers. Phytoplankton begin this chain by storing solar energy using photosynthesis. Krill and other herbivorous zooplankton eat the phytoplankton. Krill in turn are consumed by the next trophic level. (b) A complex food web on islands in the Gulf of California, featuring relationships through both strong and weak interactions. [(b) From Gary A. Polis, "Stability is woven by complex webs," in *Nature* 395 (October 22, 1998): 744–745.]

concluded, "weak interactions may be the glue that binds natural communities together." These findings countered a two-decade-old assumption that complexity is destabilizing and that simple relations are easier for a community to maintain. (See K. McCann, A. Hastings, and G. Huxel, "Weak trophic interactions and the balance of nature," in *Nature* 395 (October 22, 1998): 794–798.)

Efficiency in a Food Web

Any assessment of world food resources depends on the level of consumer being targeted. Let us use humans as an example. Many people can be fed if wheat is eaten directly. However, if the grain is first fed to cattle (herbivores) and then we eat the beef, the yield of available food energy is cut by 90% (810 kg of grain is reduced to 82 kg of meat); far fewer people can be fed from the same land area (Figure 19.17).

In terms of energy, only about 10% of the kilocalories (food Calories, not heat calories) in plant matter survive from the primary to the secondary trophic level. When humans consume meat instead of grain, there is a further loss of biomass and added inefficiency. More energy is lost to

the environment at each progressive step in the food chain. You can see that an omnivorous diet such as ours is quite expensive in terms of biomass and energy.

Food web concepts are becoming politicized as world food issues grow more critical. Today, approximately half of the cultivated acreage in the United States and Canada is planted for animal consumption—beef and dairy cattle, hogs, chickens, and turkeys. Livestock feed includes approximately 80% of the annual corn and nonexported soybean harvest. In addition, some lands cleared of rain forest in Central and South America were converted to pasture to produce beef for export to restaurants, stores, and fast-food outlets in developed countries. Thus, lifestyle decisions and dietary patterns in North America and Europe are perpetuating inefficient food webs, not to mention the destruction of valuable resources, both here and overseas. Scientists are studying the impact on world grain supplies from the spread of American meat-based fast-food restaurants across the globe, and the dietary changes these foster.

Clearly, some food webs are exceptionally simple, such as eating grains directly, whereas others are more complex. The home gardener's tomatoes may be eaten by a tomato hornworm, which is then plucked off by a passing Robin, which is later eaten by a hawk—and so it goes, in endless cycles.

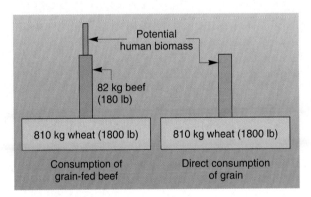

FIGURE 19.17 Efficiency and inefficiency in biomass consumption.
Biomass pyramids illustrate a great difference in efficiency between direct and indirect consumption of grain.

Ecological Relations

A study of food webs is a study of who eats what and where they do the eating. Figure 19.18 charts the summertime distribution of populations in two ecosystems, grassland and temperate forest. The stepped population pyramid is characteristic of summer conditions in such ecosystems. You can see the decreasing number of organisms supported at successively higher trophic (feeding) levels. The base of the temperate forest pyramid is narrow, however, because most of the producers are large, highly productive trees and shrubs, which are outnumbered by the consumers they can support.

FIGURE 19.18 Population distribution in two ecosystems.
Ecological pyramids for 0.1 hectare (0.25 acre) of land show the difference between grassland and forest in the summer. Numbers of consumers are shown. [After E. P. Odum, *Fundamentals of Ecology*, 3rd ed., Figure 3-15a, p. 80. © 1971 Saunders College Publishing. Adapted by permission.]

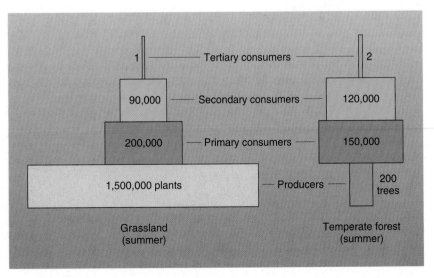

One approach to ecological study is to analyze a community's metabolism. *Metabolism* is the way in which a community uses energy and produces food for continued operation (the sum of all its chemical processes). In 1957,

Howard T. Odum completed one of the best-known studies of this type for Silver Springs, Florida.

Figure 19.19a, an adaptation of his work, shows that the energy source for the community is insolation. The

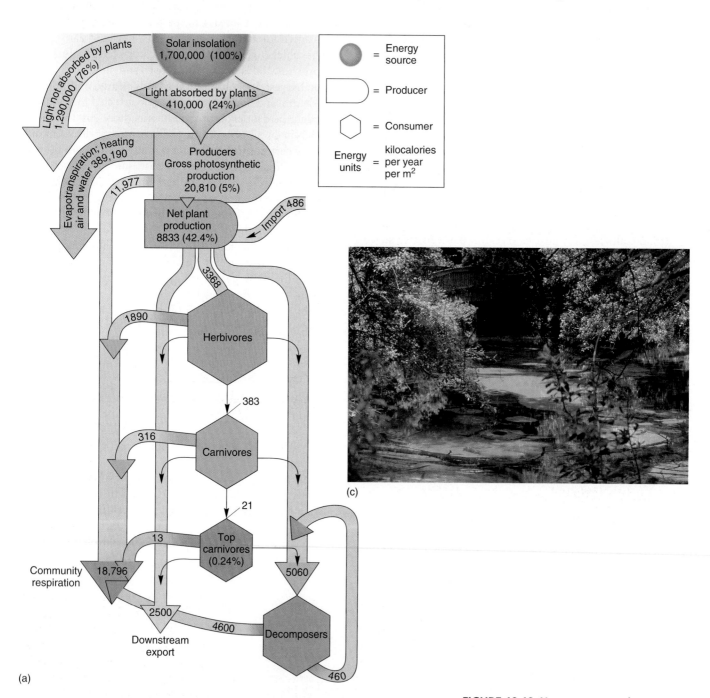

(a)

(b)

(c)

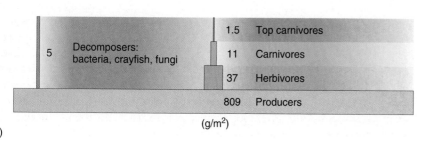

FIGURE 19.19 How energy and nutrients flow through a community.
(a) Analysis of community metabolism. (b) Standing crop distribution in grams per square meter for Silver Springs, Florida. (c) A similar nearby swampy forest ecosystem. [Adapted from E. P. Odum, *Fundamentals of Ecology*, 3rd ed., Figure 6-1a, b, p. 142. © 1971 Saunders College Publishing. Adapted by permission. (c) Photo by Bobbé Christopherson.]

amount of light absorbed by the plants for photosynthesis is 24% of the total amount arriving at the study site. Of the absorbed amount, only 5% is transformed into the community's gross photosynthetic production—and this is only 1.2% of the total insolation input. The net plant production (gross production minus respiration) is 42.4% of the gross production, or 8833 kilocalories per square meter per year. That amount then moves through the food chain of herbivores and carnivores, with the top carnivores receiving only 0.24% of the net plant production of the system.

You can see that, at each trophic level, some of the biomass flows to the decomposers. The largest portion of the biomass is exported from the system in the form of community respiration. In addition, a relatively small portion leaves the system as organic particles by downstream export out of this system. The biomass pyramid shown in Figure 19.19b portrays the same community from the viewpoint of a *standing crop* (all the biomass in a particular environment at one time), which is another way of examining the existing biomass at each trophic level.

Concentration of Pollution in Food Chains

Many farms freely use pesticides on producers to reserve them for eating by selected consumers—such as humans or the animals we feed. When an ecosystem of producers and consumers has chemical pesticides applied, the food web concentrates some of these chemicals. Many chemicals are degraded or diluted in air and water and thus are rendered relatively harmless. Other chemicals, however, are long-lived, stable, and soluble in the fatty tissues of consumers. They become increasingly concentrated at each higher trophic level. Figure 19.20 illustrates this *biological amplification*, or *magnification*.

DDT is one such chemical. As an insecticide, it saved millions of human lives in the tropics by killing the mosquitoes that spread malaria. Crop-destroying insects also were brought under control. However, the persistence of the chemical in the environment and its destruction of established food webs brought many unwanted consequences—death and illness to animals and humans alike. Although banned in some countries (the United States banned it in 1973), it still is widely used in less-developed countries. In addition, some other organic and synthetic chemicals, radioactive debris, and heavy metals such as lead and mercury become concentrated in food webs.

Thus, a food web can efficiently poison the organism at the top. Many species are threatened in this manner, and, of course, humans are at the top of many food chains and can ingest concentrated chemicals in what is consumed.

Ecosystems and Succession

Far from being static, Earth's ecosystems are dynamic (vigorous and energetic) and ever changing. Over time, communities of plants and animals have adapted to great variety,

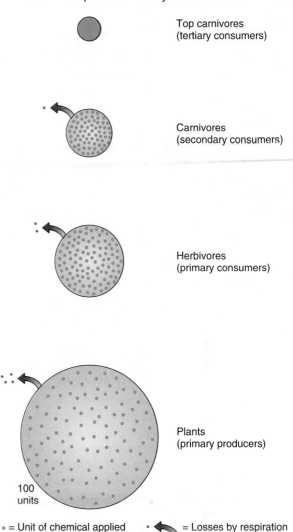

FIGURE 19.20 Concentration of chemical toxins in food chains.
Chemical residues are passed along a simple food chain to the top carnivores, where the accumulation becomes concentrated. This phenomenon is biological magnification.

evolved, and in turn shaped their environments. A constant interplay exists between increasing growth in a community and decreasing growth caused by resistance factors that create limits and disruptions (Figure 19.21). Each ecosystem is constantly adjusting to changing conditions and disturbances in the struggle to survive. The concept of change is key to understanding ecosystem stability.

For most of the last century, scientists thought that an undisturbed ecosystem—whether forest, grassland, aquatic, or other—would progress to a stage of equilibrium, a stable point, with maximum chemical element storage and biomass. Modern research has determined, however, that it is in the intermediate stages of succession the greatest mineral and biomass inventories are in place. Ecosystems do not progress to some static equilibrium conclusion; such is the great diversity of nature. In other words, there is no "balance of nature."

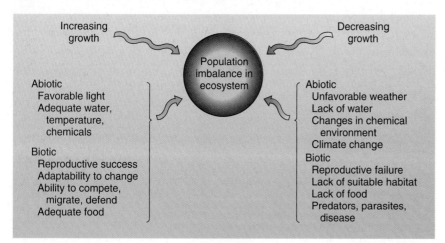

FIGURE 19.21 Factors controlling the population balance of an ecosystem. Potential factors and resistance factors ultimately determine the population of an ecosystem.

Ecosystem Stability and Diversity

In an ecosystem the tendency of species populations to remain stable (*inertial stability*) does not necessarily foster the ability to recover from change (*resilience*). Think of resilience as an ecosystem's ability to recover from disturbance, its ability to absorb disturbance up to some threshold point before it snaps to a different set of relations, thus altering some logical ecological progression.

Examples of communities with high inertial stability include a redwood forest, a pine forest at a high elevation, and a tropical rain forest near the equator. But, cleared forest tracts recover slowly and therefore have poor resilience. Figure 19.22 shows how clear-cut tracts of former forest have drastically altered microclimatic conditions, making

regrowth of the same species difficult. In contrast, a mid-latitude grassland is low in stability; yet, when burned, its resilience is high because the community recovers rapidly.

A critical aspect of ecosystem stability is **biodiversity**, or species richness of life on Earth (a combination of *bio*logical and *diversity*). The more diverse the species population (both in number of different species and quantity of each species), the species genetic diversity (number of genetic characteristics), and ecosystem and habitat diversity, the better risk is spread over the entire community, because several food sources exist at each trophic level. In other words, *greater biodiversity in an ecosystem results in greater stability and greater productivity*. News Report 19.3 presents exciting confirmation of this principle.

(a)

(b)

FIGURE 19.22 Disruption of stable forest communities.
(a) Bowron Clearcut, British Columbia—an example of clear-cut timber harvesting that disrupted a stable community and produced drastic changes in microclimatic conditions. Full exposure to sunlight and wind is different from former shade conditions. (b) A logged pine forest and disruptive logging road. About 10% of the Northwest's old-growth forests remain, as identified through orbital satellite images and GIS analysis. [(a) Photo by Galen Rowell/Mountain Light Photography, Inc. (b) photo by author.]

News Report

News Report 19.3

Experimental Prairies Confirm the Importance of Biodiversity

Field experiments have confirmed an important scientific assumption: that greater biological diversity in an ecosystem leads to greater stability, productivity, and soil nutrient use within that ecosystem. For instance, during a drought, some species of plants will be damaged. In a diverse ecosystem, however, other species with deeper roots and better water-obtaining ability will thrive.

Ecologist David Tilman at the University of Minnesota tracked the operation of 147 grassland plots, each 3 m (9.8 ft) square. Species diversity was carefully controlled on each plot (Figure 1). Plots were sown with different numbers of native North American prairie plant seeds and cared for by a team of 50 people. The results demonstrated that the plots with a more diverse plant community were able to retain and use nutrients more efficiently than the plots with less diversity. This efficiency reduced the loss of soil nitrogen through leaching, thus increasing soil fertility.

Greater plant diversity led to both higher productivity and better resource utilization. Also, total plant cover was found to increase with species richness. Tilman and his colleagues affirm:

> This extends the earlier results to the field, providing direct evidence that the current rapid loss of species on Earth, and management practices that decrease local biodiversity, threaten ecosystem productivity and the sustainability of nutrient cycling. Observational, laboratory, and now field experimental evidence, supports the hypothesis that biodiversity influences ecosystem productivity, sustainability, and stability.[*]

[*]D. Tilman, D. Wedin, and J. Knops, "Productivity and sustainability influenced by biodiversity in grassland ecosystems," *Nature* 379 (February 22, 1996): 720. Also see the summary in *Science* 271 (March 15, 1996): 1497.

FIGURE 1 Biodiversity experiment. Experimental plots are planted with native North American prairie grasses. A total of 147 plots were sown with random seed selections of 1, 2, 4, 6, 8, 12, or 24 species. Experimental results confirm many assumptions regarding the value of rich biodiversity. [Photo by David Tilman/University of Minnesota, Twin Cities Campus.]

The present loss of biodiversity is irreversible, and yet, here we are in the midst of the biosphere's sixth major extinction episode (see Figure 11.1). E. O. Wilson, the famed biologist, stated,

> Biological diversity—the full sweep from ecosystems to species within ecosystems, thence to genes within species—is in trouble. Mass extinctions are commonplace, especially in tropical regions where most of the biodiversity occurs. Among the more recent are more than half the exclusively freshwater fishes of peninsular Asia, half of the fourteen birds of the Philippines island of Cebu, and more than ninety plant species growing on a single mountain ridge in Ecuador. In the United States an estimated 1 percent of all species have been extinguished; another 32 percent are imperiled.[*]

Agricultural Ecosystems   Humans simplify communities by eliminating biodiversity, and in this way we place more ecosystems at risk of harmful change and perhaps failure. An artificially produced monoculture community,

[*]E.O. Wilson, *Consilience, The Unity of Knowledge* (New York: Knopf, 1998), p. 292.

such as a field of wheat, is singularly vulnerable to failure owing to weather or attack from insects or plant disease. In some regions, simply planting multiple crops brings more stability to the ecosystem. This is an important principle of sustainable agriculture, a movement to make farming more ecologically compatible.

A modern agricultural ecosystem not only is vulnerable to failure because of its lack of ecological diversity but also because it creates an enormous demand for energy, chemical pesticides and herbicides, artificial fertilizer, and irrigation water. The practice of harvesting and removing biomass from the land interrupts the cycling of materials into the soil. This net loss of nutrients must be artificially replenished (Figure 19.23). If these subsidies are available, increased yields are possible up to the ecological limits in the system.

Climate Change   Obviously, the distribution of plant species is affected by climate change. Many species have survived wide climate swings in the past. Consider the beginning of the Tertiary Period, 75 million years ago. Warm, humid conditions and tropical forests dominated the land as far north as southern Canada, pines grew in the Arctic, and deserts were few. Then, between 15 and 50 million

(a)

(b)

FIGURE 19.23 Agricultural ecosystems.
(a) The Sacramento Valley of California is one of the most intensively farmed regions of the world. Field crops including irrigated rice, an irrigation canal, and the Sacramento River are in the scene. (b) A pine tree plantation (farm) in Georgia, principally for the production of wood pulp—the forest ecosystem is reduced to efficient rows for clear-cutting when mature. Depleting of soil nutrients over time, this farming practice requires subsidies of chemical fertilizers and herbicides to eliminate competing vegetation. [Photos by Bobbé Christopherson.]

years ago, deserts began developing in the southwestern United States. Mountain-building processes created higher elevations, causing rain-shadow aridity and affecting plant distribution.

Recall, too, that the movement of Earth's tectonic plates created climate changes important in the evolution and distribution of plants and animals (see Figure 11.16a–e). Europe and North America were joined in Pangaea and positioned near the equator, where vast swamps formed (the site of coal deposits today). The southern mass of Gondwana was extensively glaciated as it drifted at high latitudes in the Southern Hemisphere. This glaciation left matching glacial scars and specific distributions of plants and animals across South America, Africa, India, and Australia. The diverse and majestic dinosaurs dispersed with the drifting continents.

The key question for our future is: As temperature patterns change, how fast can plants either adapt to new conditions or migrate through succession (location change) to remain within their shifting specific habitats? Rapid changes in vegetation patterns in northern latitudes since 1980 are indicated from satellite observations. Global warming during spring is reducing snow cover and raising temperatures, allowing spring greening to occur earlier, up to three weeks sooner in some high latitude locations. Likewise the onset of fall is delayed to a later date than previously experienced.

Nina Leopold Bradley is studying the Wisconsin setting her father Aldo Leopold made famous in his book *A Sand County Almanac* (Oxford University Press, 1949). He

kept detailed records of all things dealing with nature and seasonal change on his tract of land. The new findings show that more than one-third of animal and plant species start spring activities 4 weeks earlier than they did 50 years before.

Adaptation to conditions is key to evolution. Through mutation and natural selection, species have either adapted or failed to adapt to changing environmental conditions over millions of years. Current global change is occurring rapidly, at the rate of decades instead of millions of years. Thus, we see die-out and succession of different species along disadvantageous habitat margins. The displaced species may colonize new regions made more hospitable by climate change. Also shifting will be agricultural lands that produced wheat, corn, soybeans, and other commodities. Society will have to adapt to different crops growing in different places.

A study completed by biologist Margaret Davis on North American forests suggests that trees will have to respond quickly if temperatures increase. Changes in the climate inhabited by certain species could shift 100–400 km (60–250 mi) during the next 100 years. Some species, such as the sugar maple, may migrate northward, disappearing from the United States except for Maine, and moving into eastern Ontario and Québec. Davis prepared a map showing the possible impact of increasing temperatures on the distribution of beech and hemlock trees (Figure 19.24).

Changes in the geographical distributions of plant and animal species in response to future greenhouse warming threaten to reduce biotic diversity. . . . The

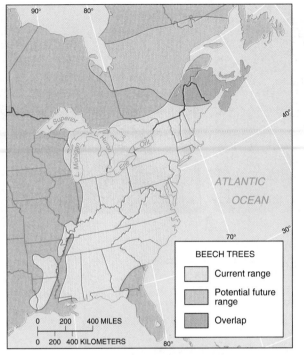

(a)

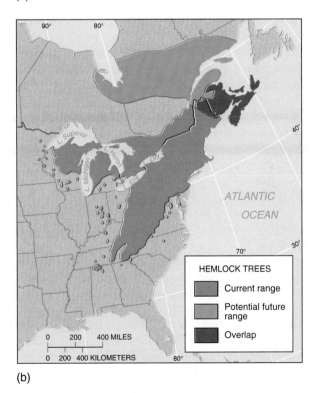

(b)

FIGURE 19.24 Present and predicted distribution of two tree species.
(a) Beech trees and (b) hemlock trees in North America. The potential future range shown for each reflects expected climate change resulting from a doubling of carbon dioxide. These maps are based on forecasts from the Goddard Fluid Dynamics Laboratory (GFDL) general circulation model.
[After M. B. Davis and C. Zabinski, "Changes in the geographical range resulting from greenhouse warming: Effects on biodiversity in forests," in R. L. Peters and T. E. Lovejoy, eds., *Global Warming and Biological Diversity* (New Haven, CT: Yale University Press, 1992), p. 301.]

risk posed by CO_2-induced warming depends on the distances that regions of suitable climate are displaced northward [in the Northern Hemisphere] and on the rate of displacement. . . . If the change occurs too rapidly for colonization of newly available regions, population sizes may fall to critical levels, and extinction will occur.[*]

Ecological Succession

Ecological succession occurs when older communities of plants and animals (usually simpler) are replaced by newer communities (usually more complex)—a change of species composition. Each successive community of species modifies the physical environment in a manner suitable for a later community of species. Changes apparently move toward a more mature condition, although disturbances are common and constantly disrupt the sequence.

Traditionally, it was assumed that plants and animals formed a predictable *climax community*—a stable, self-sustaining, and functioning community with balanced birth, growth, and death—but this notion has been mostly abandoned by scientists. Contemporary conservation biology, biogeography, and ecology assume nature to be in constant adaptation and nonequilibrium. Rather than thinking of an ecosystem as a uniform set of communities, think of ecosystems as a patchwork mosaic of habitats—each striving to achieve an optimal range and low environmental stress.

This is the study of *patch dynamics*, or disturbed portions of habitats. Within an ecosystem, individual patches may arise only to fail later. Succession is the interactions among patches. Most ecosystems are in actuality made up of patches of former landscapes. Biodiversity in some respects is the result of such patch dynamics.

Given the complexity of natural ecosystems, it is obvious that real succession involves much more than a series of predictable stages ending with a specific monoclimax community. Instead, there may be several stages, or a *polyclimax* condition, with adjoining ecosystems, or patches within ecosystems, each at different stages in the same environment. Mature communities are properly thought of as being in dynamic equilibrium. Or at times, these communities may be out of phase and in nonequilibrium with the immediate physical environment because of the usual lag time in their adjustment.

[*]M. B. Davis and C. Zabinski, "Changes in the geographical range resulting from greenhouse warming: Effects on biodiversity in forests," in R. L. Peters and T. E. Lovejoy, eds., *Global Warming and Biological Diversity* (New Haven, CT: Yale University Press, 1992), p. 297.

(a)

(b)

(c)

FIGURE 19.25 Disturbances in ecosystems alter succession patterns.
(a) River flooding along the American River, California. (b) Community development proceeds in a burned forest, progressing perhaps by chance, as species with an adaptive edge, such as fireweed (*Epilobium* sp.), succeed. (c) Unwise practices cause extensive soil erosion on a hog farm in Iowa, 1999; a disrupted prairie landscape. [(a) (b) Photos by Bobbé Christopherson; (c) photo by USDA Natural Resources Conservation Service.]

Succession often requires an initiating disturbance. Examples include wind storms, severe flooding, a volcanic eruption, a devastating wildfire, or an agricultural practice such as prolonged overgrazing (Figure 19.25). When existing organisms are disturbed or removed, new communities can emerge. At such times of nonequilibrium transition, the interrelationships among species produce elements of chance, and species having an adaptive edge will succeed in the competitive struggle for light, water, nutrients, space, time, reproduction, and survival. Thus, the succession of plant and animal communities is an intricate process, both in space and time, and forced by both internal and external variables.

Land and water experience different forms of succession. We first look at terrestrial succession, which is characterized by competition for sunlight, and then at aquatic succession, characterized by progressive changes in nutrient levels.

FIGURE 19.26 Primary succession.
Ferns take hold as primary succession on new lava flows that came from the Kīlauea volcano, Hawai'i. [Photo by Bobbé Christopherson.]

Terrestrial Succession

An area of bare rock or a disturbed site with no vestige of a former community can be a place for **primary succession**, the beginning and development of an ecosystem. Examples include new surfaces created by mass movement of land, areas exposed by a retreating glacier, cooled lava flows and volcanic eruption landscapes, or surface mining and clear-cut logging scars, or an area of sand dunes. Illustrating primary succession are plants taking hold on new lava flows from Kīlauea volcano, Hawai'i, shown in Figure 19.26. In terrestrial ecosystems, succession begins with early species that form a **pioneer community**. You can see these plants taking hold in the photo. Primary succession often begins with lichens, mosses, and ferns growing on bare rock. These early inhabitants prepare the way for further succession to grasses, shrubs, and trees (see Figure 19.5).

More common in nature is **secondary succession**, which begins if some aspects of a previously functioning community are present. An area where the natural community has been destroyed or disturbed, but where the underlying soil remains intact, may experience secondary succession. Examples of secondary succession are most of the areas affected by the Mount St. Helens eruption and blast in 1980 (Figure 19.27). Some soils, young trees, and plants were protected under ash and snow, so community development began almost immediately after the event. About 38,450 hectares (95,000 acres) of trees were blown down and burned, however, and as the photos show, the effects of the eruption were devastating and lingering. Of course, the areas completely destroyed near the Mount St. Helens volcano or those buried beneath the massive landslide north of the mountain became candidates for primary succession.

As succession progresses, soil develops and a different set of plants and animals with different niche requirements may adapt. Further niche expansion follows as the

Landsat-7 8/22/1999

(a) 1979

(b) 1980

(c) 1983

1999

(d) 1983

1999

(e) 1983

1999

(f) 1983

1999

FIGURE 19.27 The pace of change in the region of Mount St. Helens.
(a) The area north of the volcano before the eruption (1979) and (b) after the eruption (1980) experienced profound change in all ecosystems. A series of four photo comparisons made in 1983 and 1999 of matching scenes illustrate the slow recovery of secondary succession: (c) and (d) at Meta Lake; (e) Spirit Lake panorama view (note the mat of floating logs after 19 years); (f) detail of the destroyed forest community. [(a) and all 1983 photos by author; (b) and all 1999 photos by Bobbé Christopherson; *Landsat-7* 8/22/99 satellite image courtesy of NASA/GSFC.]

613

community matures. Succession, which is really the developmental process forming new communities, is a dynamic set of interactions with sometimes unpredictable outcomes, steered by sometimes random, unpredictable events that trigger a threshold leap to a new set of relations—the basis of *dynamic ecology*.

Succession Through the Ice Age Glaciation is a large-scale disturbance that well illustrates how succession operates. As Earth cycled through glacial and interglacial ages (Chapter 17), the ecological succession in all ecosystems was affected repeatedly. Imagine the milder midlatitude climate at the beginning of an interglacial warming, with lush herb and shrub vegetation slowly giving way to pioneer trees of birch, aspen, and pine. Winds and animals dispersed seeds, and changing communities readily spread in response to changing conditions. During the warm interglacial, trees increased shade and the organic content of soil increased. Deciduous trees such as oak, elm, and ash spread freely on well-drained soils; willow, cottonwood, and alder grew on poorly drained land.

As climates cooled with the approach of the next glacial, soils became more acidic, so podzolization (cool, moist, forested) processes dominated (see Chapter 18), helped by the acidity of spruce and fir trees. Ecosystems slowly returned to the vegetation community that existed at the beginning of the interglacial. The increased cold produced open regions of disturbed ecosystems. Freezing and the advance of ice disrupted highly acidic soils.

The Holocene (the past 10,000 years) is apparently atypical, because of the development of human societies and agriculture. Civilization is creating an artificial and accelerated succession, similar in some ways to the closing centuries of an interglacial. The challenge for biogeographers is to understand such long-term developmental change principles, at the accelerated pace that change is occurring in the industrial and post-industrial era.

Wildfire and Fire Ecology Fire and its effects are one of Earth's significant ecosystem and economic processes. In the United States during 2000, about 7.4 million acres, more than double the annual average, burned in wildfires—an incredible 92,000 fires. Over the past 50 years, the role of fire in ecosystems has been the subject of much scientific research and experimentation. Today, fire is recognized as a natural component of most ecosystems and not the enemy of nature that was once thought. In fact, in many forests, undergrowth and surface litter is purposely burned in controlled "cool fires" to remove fuel that could enable a catastrophic and destructive "hot fire." Forestry experts have learned that, when fire-prevention strategies are rigidly followed, they can lead to the abundant undergrowth accumulation that fuels major fires.

Modern society's demand for fire prevention to protect property goes back to European forestry of the 1800s. Fire prevention became an article of faith for forest managers in North America. But, in studies of the longleaf pine forest that stretches in a wide band from the Atlantic coastal plain to Texas, fire was discovered to be an integral part of regrowth following lumbering. In fact, seed dispersal of some pine species, such as the knobcone pine, does not occur unless assisted by a forest fire! Heat from the fire opens the cones, releasing seeds so they can fall to the ground for germination. Also, these fire-disturbed areas quickly recover with protein-rich woody growth, young plants, and a stimulated seed production that provides abundant food for animals.

The science of **fire ecology** imitates nature by recognizing fire as a dynamic ingredient in community succession. The U.S. Department of Agriculture's Forest Service first recognized the principle of fire ecology in the early 1940s and formally implemented the practice in 1972. Controlled ground fires, deliberately set to prevent accumulation of forest undergrowth, now are widely regarded as a wise forest management practice and are used across the country (Figure 19.28).

Nonetheless, after more than a decade of intense forest fires in the western United States, especially those that charred portions of the highly visible Yellowstone National Park, an outcry was heard from forestry and recreational interests (Figure 19.29). The demand was for the Forest Service and ecologists to admit they were wrong and to abandon fire ecology. Critics called fire ecology practice the government's "let it burn" policy.

In its final report on the 1988 Yellowstone fire, a government interagency task force concluded, "an attempt to exclude fire from these lands leads to major unnatural changes in vegetation . . . as well as creating fuel accumulation that can lead to uncontrollable, sometimes very damaging wildfire." Thus, participating federal land managers and others reaffirmed their stand that fire ecology is a fundamentally sound concept.

An increasingly serious problem is emerging as people seek to live in the wildlands near cities. Urban devel-

FIGURE 19.28 Controlled burning.
Fire ecology practices of the U.S. Forest Service. Controlled burning has been used for several decades. [Photo by author.]

(a)

(b)

(c)

FIGURE 19.29 Yellowstone burns.
Yellowstone National Park in northwestern Wyoming: (a) 1988 wildfire progresses up a slope;
(b) trees and shrubs return to Yellowstone with secondary succession. (c) The drama of wild-
fire unfolds in the Bitterroot National Forest in Montana, August 2000, as elk stand in the
Bitterroot River attempting to survive the conflagration. [(a) Photo by Joe Peaco/National
Park Service; (b) photo by Bobbé Christopherson; (c) photo by John McColgan, BLM Alaska
Fire Service.]

opment has encroached on forests, and the added suburban landscaping has created new fire hazards. Uncontrolled wildfires now can destroy homes and threaten public safety. People living in these areas demand fire protection that actually causes undergrowth to accumulate and worsens the risk.

Aquatic Succession

Ecosystems occur in water as well as on land. Lakes, estuaries, ponds, and shorelines are complex ecosystems. The concepts of terrestrial succession that you just read about—stability and resilience, biodiversity, community succession—apply to open water, shoreline, and watershed systems as well.

A lake or pond is really a temporary feature on the landscape, when viewed across geologic time. Lakes and ponds exhibit successional stages as they fill with nutrients and sediment and as aquatic plants take root and grow. This growth captures more sediment and adds organic debris to the system (Figure 19.30). This gradual enrichment in water bodies is known as eutrophication (from the Greek *eutrophos* meaning "well nourished").

In moist climates, a floating mat of vegetation grows outward from the shore to form a bog. Cattails and other marsh plants become established, and partially decomposed organic material accumulates in the basin, with additional vegetation bordering the remaining lake surface. Vegetation and soil and a meadow may fill in as water is displaced; willow trees follow, and perhaps cottonwood trees; and eventually the lake may evolve into a forest community.

The stages in lake succession are named for their nutrient levels: *oligotrophic* (low nutrients), *mesotrophic* (medium nutrients), and *eutrophic* (high nutrients). Greater primary productivity and resultant decreases in water transparency marks each stage, so that photosynthesis becomes

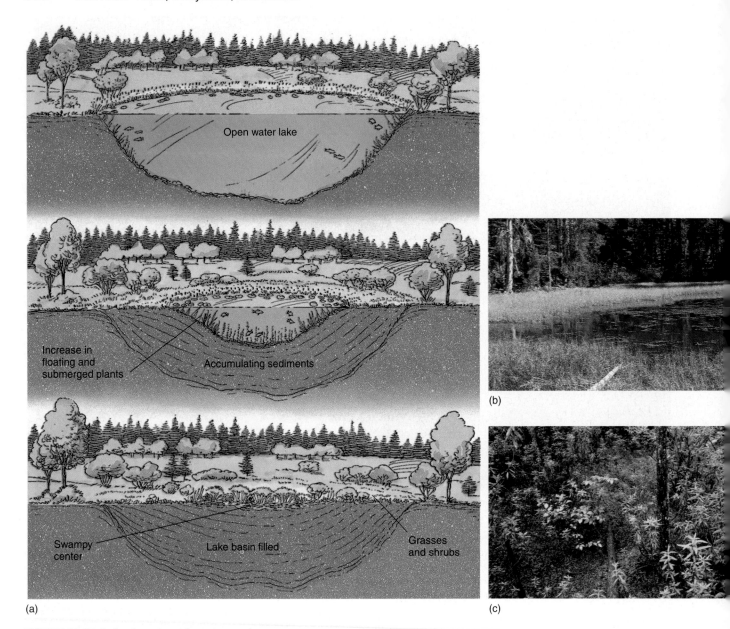

(a)

(b)

(c)

FIGURE 19.30 Lake–bog–meadow succession.
(a) What begins as a lake gradually fills with organic and inorganic sediments, which successively shrink the area of the pond. A bog forms, then a marshy area, and finally a meadow completes the successional stages. (b) Aquatic succession in a mountain lake. (c) The Richmond bog was ocean just 8000 years ago. Frasier River sediment created the mudflats and set the stage for the evolution of this sphagnum bog with acidic soils that "quakes" when you walk on it. Plants in the bog include moss, shore pine, hemlock, blueberry, salal shrubs, and Labrador tea, among others. [Photos by Bobbé Christopherson.]

concentrated near the surface. Energy flow shifts from production to respiration in the eutrophic stage, with oxygen demand exceeding oxygen availability.

Nutrient levels also vary spatially: Oligotrophic conditions occur in deep water, whereas eutrophic conditions occur along the shore, in shallow bays, or where sewage, fertilizer, or other nutrient inputs occur. Even large bodies of water may have eutrophic areas along the shore. As society dumps sewage, agricultural runoff, and pollution in waterways, the nutrient load is enhanced beyond the cleansing ability of natural biological processes. The result is

cultural eutrophication, which hastens succession in aquatic systems. As with all ecosystems, we must be aware of the signals that unwanted change is occurring so that mitigating action can be taken. (Refer to the map of the dead zone in the Gulf of Mexico in News Report 19.2.)

Focus Study 19.1 examines the Great Lakes of North America and their geographically and biologically diverse aquatic ecosystems. These lakes are interrelated with the surrounding terrestrial ecosystems. This is the largest lake system on Earth and is jointly managed by the United States and Canada.

The Great Lakes

The basins of the Great Lakes are gifts of the last ice age to North America (see Figure 17.28). The glaciers advanced and retreated over this region, excavating the basins for five large lakes. This international waterway of lakes and connecting rivers has played an important role in the history of Canada and the United States. Today, about 10% of the U.S. population and 25% of Canada's population live in the Great Lakes drainage basin (Figure 1). Major agricultural regions surround the lakes, as do many centers of industrial activity. Tourism, sport fishing, and maritime commerce are also important.

Society asks much of the Great Lakes: to dilute wastes from cities and industry, to dissipate thermal pollution from power plants, to provide municipal drinking water and irrigation water, and to sustain unique and varied ecosystems—open lake, coastal shore, coastal marsh, lakeplain (former lake bed), inland wetlands, and inland terrestrial (upland areas of forest, prairie, and barrens). Here is a brief profile of this important hydrological and ecological resource. (See Environment Canada's Great Lakes Information Management Resource's site at http://www.cciw.ca/glimr/intro.html and NOAA's Great Lakes Research Lab's site at http://www.glerl. noaa. gov/.)

Geography and Physical Characteristics

The Great Lakes—Superior, Michigan, Huron, Erie, and Ontario—contain 18% of the total volume of all freshwater lakes in the world, some 23,000 km^3 (5500 mi^3) of water. Their combined surface area covers 244,000 km^2 (94,000 mi^2), a little less than the state of Wyoming. The entire drainage basin embraces 528,000 km^2 (204,000 mi^2), or an area about the size of Manitoba.

As we tour the lakes, it is useful to follow along in Figure 1 and to look at their profile in Figure 2a. Lake Superior is the highest in elevation, highest in latitude, deepest, and largest in the

system. It drains through St. Mary's River into Lake Huron. Lake Michigan, the only lake of the five that lies entirely within the United States, is at the same level as Lake Huron because it is joined by the wide connection through the Straits of Mackinac. (This is why, in Figure 2, we combine these two lakes on one hydrograph.)

The St. Clair River, Lake St. Clair, and the Detroit River carry water on to Lake Erie, the shallowest lake in the system. Compared to an average water retention time of 191 years in Lake Superior, Lake Erie has the shortest retention time of 2.6 years. The Niagara River, plunging dramatically over Niagara Falls, transports water into Lake Ontario. Lake Ontario is drained by the St. Lawrence River, which carries the entire discharge of the Great Lakes system to the Gulf of St. Lawrence and eventually into the North Atlantic Ocean. The entire basin extends over 10° of latitude (41° N to 51° N) and 18° of longitude (75° W to 93° W). (For additional Great Lakes information, see http://www.seagrant.wisc. edu/outreach/.)

Because of the large size of the overall Great Lakes system, its associated climate, soils, and topography vary widely. Dominating the north are colder microthermal climates, exposed portions of the Canadian Shield bedrock, and vast stands of conifers and acidic soils. To the south, warmer mesothermal climates and fertile glacially deposited soils provide a vast agricultural base. Farms and urbanization replaced previous stands of mixed forest. Virtually none of the original land cover is undisturbed.

Prevailing winds from the west and seasonal change mix air masses from different source regions, producing variable weather. The presence of the Great Lakes basin strongly influences passing air masses and weather systems producing lake-effect snowfall to the east (see Figure 8.5). Precipitation over the drainage basin feeds the lake storage as part of the renewable hydrologic cycle. Figure 2 illustrates

Great Lakes water levels from 1918 to 2001. Water levels were above average during 1996 and 1997. During 1998 through 2001, water levels lowered to below average levels, reflecting near-drought conditions across the region. (See the Army Corps of Engineers' *Monthly Bulletin of Lake Levels for the Great Lakes, Detroit, MI.*)

Short-term lake levels vary from winter to summer; they are higher in summer after snowmelt and the arrival of maximum-summer precipitation. Over the long term, highest lake levels occur during years of heavier precipitation and during times of cooler temperatures, which reduce evaporation. Wind also affects lake levels. A wind setup (wind-driven water) occurs along the downwind shoreline, as water is pushed higher onshore.

The Great Lakes Ecosystem

The Great Lakes ecosystem should be thought of as young and a fruitful laboratory for ecological studies. These lakes are not simply large bodies of water with a uniform mixture. A stratification occurs in them related to density and temperature differences. In the summer, the surface and shallow waters are warmed and become less dense so that the lakes develop a sharp stratification. The warm surface water and light penetration in the upper layers support most biotic production and adequate dissolved oxygen. This layering affects water quality because it can prevent mixing of pollution and other effluents with cooler bottom water during the summer months.

As the fall season matures, surface water cools and sinks, displacing deeper water and creating a turnover of the lake mass. By midwinter, the temperature from the surface to the bottom is uniformly around 4°C (39°F, the point at which water is densest); temperatures at the surface are near freezing.

The lakes support a food chain of producers and consumers, as does any aquatic ecosystem. Native fish populations have been greatly affected by

(continued)

Focus Study 19.1 *(continued)*

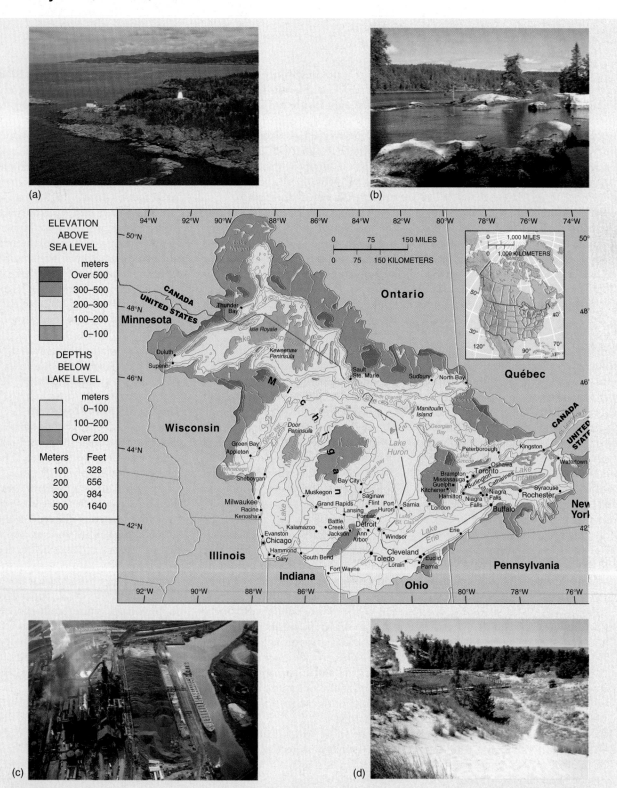

FIGURE 1 The Great Lakes Basin.
Elevations above sea level, depths below lake level, and principal urban areas are shown. The map outline is the limit of the drainage basin and watershed for the Great Lakes system. (a) The rocky shores of Lake Superior, Otter Island lighthouse. (b) Forests along Lake Ontario shores. (c) The Cuyahoga River enters Lake Erie at Cleveland. (d) Sand dunes along the southern shore of Lake Michigan. Note the wooden walkway for pedestrians to prevent erosion of sand by human traffic. [(a) Photo by Carl R. Sams, II/Peter Arnold, Inc.; (b) photo by Wayne Lankinen/DRK Photo; (c) photo by Alex S. MacLean/Peter Arnold, Inc.; (d) photo by Bobbé Christopherson. Map courtesy of Environment Canada, U.S. EPA, and Brock University cartography.]

(continued)

Focus Study 19.1 *(continued)*

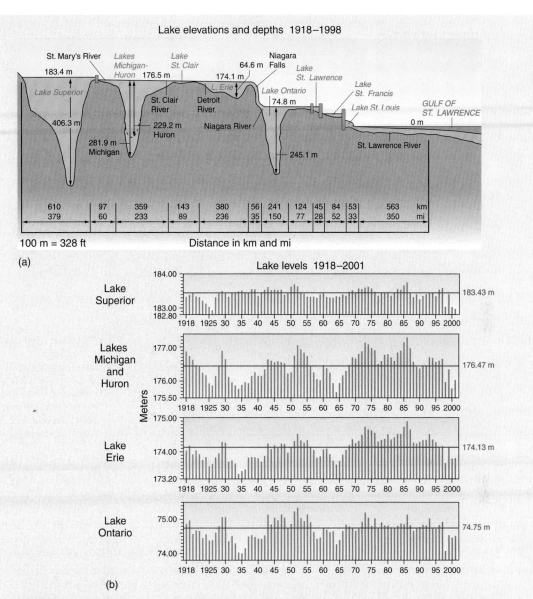

FIGURE 2 Great Lakes elevation profile and hydrographs.
(a) Average depth and lake surface levels, 1918–1998. (b) Annual hydrographs for each of the Great Lakes, 1918–2001. Both figures were prepared using the International Great Lakes Datum of 1985. [Data courtesy of the International Coordinating Committee on the Great Lakes Basic Hydraulic and Hydrographic Data and the Army Corps of Engineers.]

human activities: overfishing, introduced nonnative species, pollution from excess nutrients, toxic contamination of fish, and disruption of spawning habitats.

Peak commercial fishing occurred in the 1880s. The fisheries declined to their lowest levels as water pollution peaked in the 1960s and early 1970s. The U.S. and Canadian governments implemented strong pollution control programs, along with the work of many citizens, industries, and private organizations. These efforts brought the lakes into the present period of recov-

ery. Today, lake trout (in Lake Superior), sturgeon, herring, smelt, alewife, splake, yellow perch, walleye, and white bass are fished. However, health advisories are occasionally issued warning of danger in consuming fish of certain species, sizes, and locales.

Coastline for the five lakes totals 18,000 km (11,000 mi) and is diverse: major dunes and sandy beaches, bedrock shorelines, gravel beaches, and adjoining coastal marsh systems. Marshes can be found at the far western end of Lake Superior in the St. Louis River estuary and at the far east-

ern end of Lake Ontario and its coastal lagoon. Here, as elsewhere, marshes occupy that key interface between the land and the water, storing and cycling organic material and nutrients into aquatic ecosystems.

Human Activities, Land Use, Loss, and Recovery

Figure 3 portrays land use in the Great Lakes Basin. Each activity has its own impact on lake systems. Runoff from agricultural areas contains chemicals, nutrients, and eroded soil. The pulp

(continued)

Focus Study 19.1 (continued)

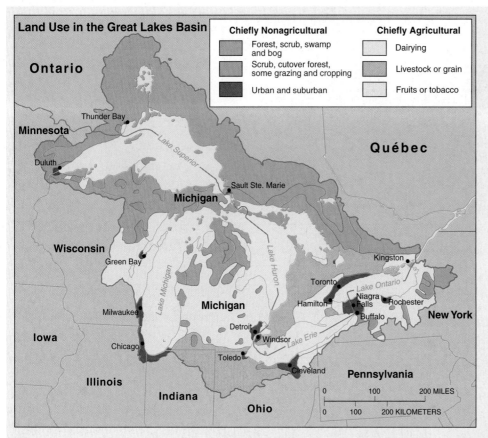

Land Use in the Great Lakes Basin

Chiefly Nonagricultural

- Forest, scrub, swamp and bog
- Scrub, cutover forest, some grazing and cropping
- Urban and suburban

Chiefly Agricultural

- Dairying
- Livestock or grain
- Fruits or tobacco

FIGURE 3 Land use in the Great Lakes Basin.
Generalized distribution of land use, nonagricultural and agricultural, in the Great Lakes Basin. [Map courtesy of Environment Canada, U.S. EPA, and Brock University cartography.]

and paper industry was a major polluter but has improved as more was learned about the dangers of pollution; for example, mercury contamination was halted in the 1970s and better disposal methods for wastes were implemented. The forests were not always treated in a sustainable way in terms of reforestation, and today forests may be regarded as a diminished resource in the basin. Recovery efforts are slow to make up for bad past practices, but progress is underway.

Lake Erie was the first of the Great Lakes to show severe effects of *cultural* eutrophication, for it is the shallowest and warmest of the five lakes. About a third of the overall population in the Great Lakes Basin lives in the drainage of Lake Erie, making it the major recipient of sewage effluent from treatment plants. Over the years, dissolved oxygen levels dropped as biochemical oxygen demand increased, caused by sewage-treatment discharges and algal decay. Algae flourished in these eutrophic conditions and coated the beaches, turning the lake a greenish brown.

The most infamous incident in the 1960s was when the Cuyahoga River (through Cleveland, Ohio) became so contaminated that it caught fire. In 1972, Canada and the United States signed the Great Lakes Water Quality Agreement and the cleanup was under way. The government, spurred by public opinion, was moved to action, knowing that one of these priceless lakes was actually being lost. One example of the success of these efforts is that the input of phosphorus was reduced by 90% (an element with high biological activity that dramatically stimulates unwanted plant growth in aquatic ecosystems).

Concern for the health of the entire Great Lakes ecosystem remains necessary because of the biological amplification of poisons in the food chain. For example, organic chemicals such as PCBs accumulate to dangerous levels. PCBs (polychlorinated biphenyls) are highly viscous, inert fluids used in electrical transformers and hydraulic systems, and they are carcinogenic. Samples reported by Environment Canada show PCBs in phytoplankton

at 0.0025 ppm, in a rainbow smelt at 1.04 ppm, in a lake trout at 4.83 ppm, and higher in the food chain to a herring gull eggshell at 124 ppm! Later additions to the 1972 agreement reduced toxic levels, organic chemicals, and heavy metals entering the lakes. Since humans are at the end of the amplification process, it is in our own best interest to help the continuing recovery of these lakes.

Biological diversity is at the heart of the integrity of the Great Lakes. Surveys are under way to inventory the entire ecosystem, establish a biodiversity baseline, assess current conditions of imperiled species, and strengthen efforts to conserve the Great Lakes system as a productive environment. Both the Canadian and U.S. governments are involved through an international commission, as is the Nature Conservancy through its Great Lakes Program (http://www.tnc.org). The first version (2001) of the program report for Great Lakes biodiversity conservation is complete.

(continued)

Focus Study 19.1 (*continued*)

An ecosystem approach to management of the Great Lakes Basin is the key to recovery. This approach considers the physical, chemical, and biological components that constitute the total system. Thus, a Herring Gull eggshell becomes an indicator of chemical pollution, blue-green algal blooms denote hastening eutrophication, and a loss of species richness and diversity exposes the destruction of habitat. This is where geographic principles of spatial analysis, an integrative tool such as geographic information systems (GIS), and remote-sensing capabilities come into play (Figure 4). A holistic, spatial approach to ecological problems is a marked departure from the past, when pollution problems were considered locally and separately.

FIGURE 4 Orbital view of the Great Lakes.
The Great Lakes region in true color made March 6, 2000, by sensors aboard *Terra*. [*Terra* MODIS sensor image courtesy of NASA/MODIS Land Science Team.]

Summary and Review—Ecosystem Essentials

● *Define* ecology, biogeography, and the ecosystem concept.

Earth's biosphere is unique in the Solar System; its ecosystems are the essence of life. **Ecosystems** are self-sustaining associations of living plants and animals and their nonliving physical environment. **Ecology** is the study of the relationships between organisms and their environment and among the various ecosystems in the biosphere. **Biogeography** is the study of the distribution of plants and animals and the diverse spatial patterns they create.

ecosystem (p. 588)
ecology (p. 588)
biogeography (p. 588)

1. What is the relationship between the biosphere and an ecosystem? Define ecosystem and give some examples.
2. What does biogeography include? Describe its relationship to ecology.
3. Briefly summarize what ecosystem operations imply about the complexity of life.

● *Describe* communities, habitats, and niches.

A **community** is formed by the interactions among populations of living animals and plants at a particular time. Within a community, a **habitat** is the specific physical location of an organism—its address. A **niche** is the function or operation of a life form within a given community—its profession.

community (p. 588)
habitat (p. 590)
niche (p. 590)

4. Define a community within an ecosystem.
5. What do the concepts of habitat and niche involve? Relate them to some specific plant and animal communities.
6. Describe symbiotic and parasitic relationships in nature. Draw an analogy between these relationships and human societies on our planet. Explain.

● *Explain* photosynthesis and respiration and *derive* net photosynthesis and the world pattern of net primary productivity.

As plants evolved, the **vascular plants** developed conductive tissues. **Stomata** on the underside of leaves are the portals through which the plant participates with the atmosphere and hydrosphere. Plants (primary producers) perform **photosynthesis** as sunlight stimulates a light-sensitive pigment called **chlorophyll**. This process produces food sugars and oxygen to drive biological processes. **Respiration** is essentially the reverse of photosynthesis and is the way the plant derives energy by oxidizing carbohydrates. **Net primary productivity** is the net photosynthesis (photosynthesis minus respiration) of an entire community. This stored chemical energy is the **biomass** that the community generates, or the net dry weight of organic material.

> vascular plants (p. 590)
> stomata (p. 592)
> photosynthesis (p. 592)
> chlorophyll (p. 592)
> respiration (p. 592)
> net primary productivity (p. 592)
> biomass (p. 593)

7. Define a vascular plant. How many plant species are there on Earth?
8. How do plants function to link the Sun's energy to living organisms? What is formed within the light-responsive cells of plants?
9. Compare photosynthesis and respiration and the derivation of net photosynthesis. What is the importance of knowing the net primary productivity of an ecosystem and how much biomass an ecosystem has accumulated?
10. Briefly describe the global pattern of net primary productivity.

● *List* abiotic ecosystem components and *relate* those components to ecosystem operations.

Light, temperature, water, and the cycling of gases and nutrients constitute the life-supporting abiotic components of ecosystems. Elevation and latitudinal position on Earth create a variety of physical environments. The zonation of plants with altitude is called **life zones** and is visible as you travel between different elevations.

Life is sustained by **biogeochemical cycles**, through which circulate the gases and sedimentary materials (nutrients) necessary for growth and development of living organisms. The environment may inhibit biotic operations, either through lack or excess. These **limiting factors** may be physical or chemical in nature.

> life zone (p. 597)
> biogeochemical cycles (p. 598)
> limiting factor (p. 600)

11. What are the principal abiotic components in terrestrial ecosystems?
12. Describe what Alexander von Humboldt found that led him to propose the life-zone concept. What are life zones? Explain the interaction among altitude, latitude, and the types of communities that develop.
13. What are biogeochemical cycles? Describe several of the essential cycles.
14. What is a limiting factor? How does it function to control the spatial distribution of plant and animal species?

● *Explain* trophic relationships in ecosystems.

Trophic refers to the feeding and nutrition relations in an ecosystem. It represents the flow of energy and cycling of nutrients. **Producers**, which fix the carbon they need from carbon dioxide, are the plants, including phytoplankton in aquatic ecosystems. **Consumers**, generally animals including zooplankton in aquatic ecosystems, depend on the producers as their carbon source. Energy flows from the producers through the system along a circuit called the **food chain**. Within ecosystems, the feeding interrelationships are complex and arranged in a complex network of interconnected food chains called a **food web**.

The primary consumer is an **herbivore**, or plant eater. A **carnivore** (meat eater) is a secondary consumer. A consumer that eats both producers and consumers is called an **omnivore**—a role occupied by humans. **Decomposers**, the bacteria and fungi, process organic debris and release inorganic materials back to the environment. Detritus feeders, including worms, mites, termites, and centipedes, also process debris. Biomass and population pyramids characterize the flow of energy and the numbers of producers, consumers, and decomposers operating in an ecosystem.

> producers (p. 602)
> consumers (p. 602)
> food chain (p. 602)
> food web (p. 603)
> herbivore (p. 603)
> carnivore (p. 603)
> omnivore (p. 603)
> decomposer (p. 603)

15. What role is played in an ecosystem by producers and consumers?
16. Describe the relationship among producers, consumers, decomposers, and detritus feeders in an ecosystem. What is the trophic nature of an ecosystem? What is the place of humans in a trophic system?
17. What are biomass and population pyramids? Describe how these models help explain the nature of food chains and communities of plants and animals.
18. Follow the flow of energy and biomass through the Silver Springs, Florida, ecosystem. Describe the pathways in Figures 19.14 and 19.19a.

● *Define* succession and *outline* the stages of general ecological succession in both terrestrial and aquatic ecosystems.

A critical aspect of ecosystem stability is **biodiversity**, or species richness of life on Earth (a combination of *bio*logical and *diversity*). The more diverse the species population (both in number of different species and quantity of each species), the species genetic diversity (number of genetic characteristics), and ecosystem and habitat diversity, the better risk is spread over the entire community. The greater the biodiversity within an ecosystem, the more stable and resilient it is, and the more productive it will be. Modern agriculture often creates a nondiverse monoculture that is particularly vulnerable to failure.

Biogeographers trace communities across the ages, considering plate tectonics and ancient dispersal of plants and animals. Past glacial and interglacial climatic episodes have created a long-term succession. Climate change is forcing accelerated succession in ecosystems.

Ecological succession describes the process whereby older communities of plants and animals are replaced by newer communities that are usually more complex. An area of bare rock and soil with no trace of a former community can be a site for **primary succession**. The initial community that occupies an area of early succession is a **pioneer community**. **Secondary succession** begins in an area that has a vestige of a previously functioning community in place. Rather than progressing smoothly to a definable stable point, ecosystems tend to operate in a dynamic condition, with succeeding communities overlapping in time and space. Wildfire is one external factor that disrupts a successional community. Aquatic ecosystems also experience community succession, as exemplified by the eutrophication of a lake ecosystem.

The science of **fire ecology** has emerged in an effort to understand the natural role of fire in ecosystem maintenance and succession.

biodiversity (p. 608)
ecological succession (p. 611)
primary succession (p. 612)
pioneer community (p. 612)
secondary succession (p. 612)
fire ecology (p. 614)

19. What is meant by ecosystem stability?
20. How does ecological succession proceed? What are the relationships between existing communities and new, invading communities?
21. Discuss the concept of fire ecology in the context of the Yellowstone National Park fires of 1988. What were the findings of the government task force?
22. Summarize the process of succession in a body of water. What is meant by cultural eutrophication?

NetWork

The *Geosystems* Home Page provides on-line resources for this chapter on the World Wide Web. To begin: Once on the Home Page, click on this textbook, scroll the Table of Contents menu, select this chapter, and click "Begin." You will find self-tests that are graded, review exercises, specific updates for items in the chapter, and many links to interesting related pathways on the Internet under "Destinations." *Geosystems* is at **http://www.prenhall.com/christopherson.**

Critical Thinking

A. This chapter states, "Some scientists are questioning whether our human society and the physical systems of Earth constitute a global-scale symbiotic relationship of mutualism (sustainable) or a parasitic one (nonsustainable)." Referring to the definition of these terms, what is your response to the statement? How do you equate our planetary economic system with the need to sustain life-supporting natural systems? Mutualism? Parasitism?

B. Over the next several days as you travel between home and campus, a job, or other areas, observe the landscape. What types of ecosystem disturbances do you see? Imagine that several acres in the same area escaped any disruptions for a century or more. Describe what some of these ecosystems and communities might be like.

C. In Chapter 19 of the *Geosystems* Home Page, the first Short Answer exercise asks you to analyze a map of carbon stored in plant material. The map was prepared by determining the dominant vegetation type for "squares" 0.5° in latitude by 0.5° in longitude. The values mapped are kilograms of carbon per square meter of ground area. Where are the highest carbon densities? If you were to harvest 1 m² of land from one of these areas, about how many kilograms of carbon could you extract? How many pounds is this? [1 kg = 2.2 lb.]

An ancient temperate rain forest in the Pacific Northwest in the Gifford Pinchot National Forest features huge old-growth Douglas fir, redwoods, cedars, and a mix of deciduous trees, ferns, and mosses. Only a small percentage of these old-growth forests remain. [Photo by Bobbé Christopherson.]

20

Terrestrial Biomes

Key Learning Concepts

After reading the chapter, you should be able to:

- *Define* the concept of biogeographic realms of plants and animals and *define* ecotone, terrestrial ecosystem, and biome.
- *Define* six formation classes and the life-form designations and *explain* their relationship to plant communities.
- *Describe* 10 major terrestrial biomes and *locate* them on a world map.
- *Relate* human impacts, real and potential, to several of the biomes.

I n 1874, Earth was viewed as offering few limits to human enterprise. But visionary American diplomat and conservationist George Perkins Marsh perceived the need to manage the environment:

> We have now felled forest enough everywhere, in many districts far too much. Let us now restore this one element of material life to its normal proportions, and devise means of maintaining the permanence of its relations to the fields, the meadows, and the pastures, to the rain and the dews of heaven, to the springs and rivulets with which it waters the Earth.*

Unfortunately, his warning was ignored. The clearing of the forests is perhaps the most fundamental change wrought by humans in our transformation of Earth. The development and expansion of civilization was fueled and built on the consumption of forests and other natural resources. This natural "work" of Earth's physical, chemical, and biological systems has an estimated annual worth of some $35 trillion to the global economy.

*G. P. Marsh, *Physical Geography as Modified by Human Action* (New York: Scribner's, 1874), pp. 385–386.

Today, we see that Earth's natural systems do pose limits to human activity. Scientists are measuring ecosystems carefully, and they are building elaborate computer models to simulate our evolving real-time human–environment experiment. In particular, shifting patterns of environmental factors (temperatures and changing frost periods; precipitation timing and amounts; air, water, and soil chemistry; and a redistribution of nutrients). We know from history that the effect of climate change on ecosystems will be significant. Two Harvard researchers assessed the impact of these alterations on ecosystems:

> Based on more than a decade of research, it is obvious that the CO_2-rich atmosphere of our future will have direct and dramatic effects on the composition and operation of ecosystems. According to the best scientific evidence, we see no reason to be sanguine [optimistic] about the response of these habitats to our changing environment.[*]

What will the quotations about Earth's environment be like 125 years from now, about the same time interval since G. P. Marsh made his assessment? Humans now have become the most powerful biotic agent on Earth, influencing all ecosystems on a planetary scale.

In this chapter: We explore Earth's major terrestrial ecosystems, conveniently grouped into 10 biomes. Each biome is an idealized assemblage because much alteration of the natural has occurred. The discussion covers biome appearance, structure, and location, and the present conditions of related plants, animals, and environment, and biodiversity. Table 20.1 brings together most all the information about Earth's physical systems that is presented through the pages of this text. Through biomes we synthesize the physical geography of place and region.

Biogeographic Realms

Figure 20.1 shows Earth's biogeographic realms. The upper map illustrates the botanical (plant) realms worldwide. The lower map shows the zoological (animal) realms. Each realm contains many distinct ecosystems that distinguish it from other realms.

A **biogeographic realm** is a geographic region where a group of plant and animal species evolved. As you can see, these realms correspond generally to continents. The

[*]F. A. Bazzaz and E. D. Fajer, "Plant Life in a CO_2-Rich World," *Scientific American* 256, no. 1 (January 1992): 74.

main topographic barrier that separates these realms is the ocean. Every species attempts to migrate worldwide according to its niche requirements, reproductive success, and competition but is constrained by climatic and topographic barriers. The collective effort of all species to seek optimal habitats and to maximize each personal niche requirement results in the generalized realms shown on the maps. Recognition that such distinct regions of flora (plants) and fauna (animals) exist was an early beginning for *biogeography* as a discipline.

The Australian realm is unique, giving rise to 450 species of *Eucalyptus* among its plants and 125 species of marsupials (animals such as kangaroos, that carry their young in pouches, where gestation is completed). Australia's uniqueness is the result of its early isolation from the other continents (Figure 20.2). During critical evolutionary times, Australia drifted away from Pangaea (see Chapter 11) and never again was reconnected by a land bridge, even when sea level lowered during repeated glacial ages. New Zealand's isolation from Australia also undoubtedly explains why no native marsupials exist in New Zealand.

Transition Zones

Alfred Wallace (1823–1913), the first scholar of *zoogeography*, used the stark contrast in animal species among islands in present-day Indonesia to delimit a boundary between the Oriental and Australian realms (Figure 20.1b). He thought that deep water in the straits had completely prevented species crossover. However, he later recognized that a boundary line actually is a wide transition zone, where one region grades into the other. Boundaries between natural systems are "zones of shared traits." Thus, despite distinctions among individual realms and our discussion of biomes, it is best to think of their borders as transition zones of mixed identity and composition, rather than as rigidly defined boundaries.

A boundary transition zone between adjoining ecosystem regions is an **ecotone**. Because ecotones are defined by different physical factors, they vary in width—think of a transition zone instead of a line. Climatic ecotones usually are more gradual than physical ecotones where differences in soil or topography sometimes form abrupt boundaries. An ecotone between prairies and northern forests may occupy many kilometers of land. The ecotone is an area of tension as similar species of plants and animals compete for the resource base. Scientists have identified these boundary and edge areas as generating biodiversity (a species richness).

Terrestrial Ecosystems

A **terrestrial ecosystem** is a self-sustaining association of land-based plants and animals and their abiotic environment, characterized by specific plant formation classes.

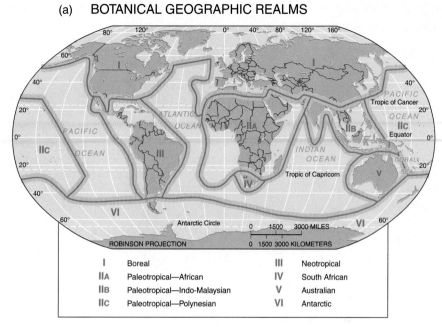

FIGURE 20.1 Biogeographic realms.
(a) Botanical geographic realms (after R. D. Good, 1947). (b) Zoological geographic realms (after L. F. deBeaufort, 1951). [Adapted from E. P. Odum, *Fundamentals of Ecology*, 3rd ed., Figure 14-1. © 1971 Saunders College Publishing. Adapted by permission.]

Plants are the most visible part of the biotic landscape, and they are key members of Earth's terrestrial ecosystems. In their growth, form, and distribution, plants reflect Earth's physical systems: its energy patterns; atmospheric composition; temperature and winds; air masses; water quantity, quality, and seasonal timing; soils; regional climates; geomorphic processes; and ecosystem dynamics. As discussed in the previous chapter, aquatic ecosystems are likewise important and result from the interaction of physical systems (see News Report 20.1).

A brief review of the ways that living organisms relate to the environment and to each other is helpful at this point. Interacting populations of plants and animals in an area form a *community*. An *ecosystem* involves the interplay between a community of plants and animals and their abiotic physical environment. Each plant and animal occupies an area in which it is biologically suited to live—its *habitat*—and within that habitat it performs a basic operational function—its *niche* (see Chapter 19).

A **biome** is a large, stable terrestrial ecosystem characterized by specific plant and animal communities. Each biome usually is named for its *dominant vegetation*, because that is the single most easily identified feature. We can generalize Earth's wide-ranging plant species into six broad biomes: *forest*, *savanna*, *grassland*, *shrubland*, *desert*, and *tundra*. Because plant distributions respond to environmental conditions and reflect variations in climate and soil, the world climate map in Figure 10.5 is a helpful reference

FIGURE 20.2 The unique Australian realm.
The flora and fauna of Australia form a special assemblage of communities. [Photo by Mark Newman/Photo Researchers, Inc.]

for this chapter. (The biome concept can also be applied to aquatic ecosystems: polar seas, temperate seas, tropical seas, seafloor, shoreline, and coral reef.)

More specific characteristics are used for the structural classification of plants themselves. *Life-form designations* are based on the outward physical properties of individual plants or the general form and structure of a vegetation cover. These physical life forms include

- *Trees* (large woody main trunk, perennial; usually exceeds 3 m, or 10 ft, in height)
- *Lianas* (woody climbers and vines)
- *Shrubs* (smaller woody plants; stems branching at the ground)
- *Herbs* (small plants without woody stems; includes grasses and other nonwoody vascular plants)
- *Bryophytes* (nonflowering, spore-producing plants; includes mosses and liverworts)
- *Epiphytes* (plants growing above the ground on other plants, using them for support)
- *Thallophytes* (lack true leaves, stems, or roots; includes bacteria, fungi, algae, lichens)

These plant assemblages in their general biomes are divided into more-specific vegetation units called **formation classes**, which refer to the structure and appearance of dominant plants in a terrestrial ecosystem. Examples are equatorial rain forest, northern needleleaf forest, Mediterranean shrubland, and arctic tundra. Each formation class includes numerous plant communities, and each community includes innumerable plant habitats. Within those habitats, Earth's diversity is expressed in 270,000 plant species. Despite this intricate complexity, we can generalize Earth's numerous formation classes into 10 global terrestrial biome regions, as portrayed in Figure 20.3 and detailed in Table 20.1.

Now let us go on a tour of Earth's biomes, keeping in mind that they synthesize all we have learned about the atmosphere, hydrosphere, lithosphere, and biosphere in the pages of this text. Here we bring it all together.

News Report 20.1

Aquatic Ecosystems and the LME Concept

Oceans, estuaries, and freshwater bodies (streams and lakes) are home to aquatic ecosystems. Poor understanding of aquatic ecosystems has led to the demise of some species, such as herring in the Georges Bank prime fishing area of the Atlantic and an overall decline in fisheries worldwide. One analytical tool is the designation of certain areas as large marine ecosystems (LMEs). The LME was conceived by Kenneth Sherman and Lewis Alexander, scientists with the U.S. Marine Fisheries Service Laboratory and the University of Rhode Island, respectively, to encourage resource planners, managers, and researchers to consider complete ecosystems, not just a targeted species.

An LME is a distinctive oceanic region having unique organisms, floor topography, currents, areas of nutrient-rich upwelling circulation, or areas of significant predation, including human. Examples of identified LMEs include the Gulf of Alaska, California Current, Gulf of Mexico, Northeast Continental Shelf, and the Baltic and Mediterranean Seas. Thirty LMEs, each encompassing more than 200,000 km^2 (77,200 mi^2), are presently defined worldwide.

The largest of 11 government-protected areas in North America is the Monterey Bay National Marine Sanctuary, established in 1992 within the California Current LME. (The Florida Keys Marine Sanctuary is second largest.) Stretching along 645 km (400 mi) of coastline and covering 13,500 km^2 (5300 mi^2), the Monterey Bay sanctuary is home to 27 species of marine mammals, 94 species of shorebirds, 345 species of fish, and the largest sampling of invertebrates in any one place in the Pacific. This represents one of the most species-rich and diverse marine communities on Earth. In the United States these protected areas are administered by NOAA and The National Marine Sanctuaries. Its Web site is at http://www.sanctuaries.nos.noaa.gov/.

Table 20.1   Major Terrestrial Biomes and Their Characteristics

Biome and Ecosystems (map symbol)	Vegetation Characteristics	Soil Orders (Soil Taxonomy and CSSC)	Köppen Climate Designation	Annual Precipitation Range	Temperature Patterns	Water Balance
Equatorial and Tropical Rain Forest (ETR) Evergreen broadleaf forest Selva	Leaf canopy thick and continuous; broadleaf evergreen trees, vines (lianas), epiphytes, tree ferns, palms	Oxisols Ultisols (on well-drained uplands)	Af Am (limited dry season)	180–400 cm (>6 cm/mo)	Always warm (21–30°C; avg. 25°C)	Surpluses all year
Tropical Seasonal Forest and Scrub (TrSF) Tropical monsoon forest Tropical deciduous forest Scrub woodland and thorn forest	Transitional between rain forest and grasslands; broadleaf, some deciduous trees; open parkland to dense undergrowth; acacias and other thorn trees in open growth	Oxisols Ultisols Vertisols (in India) Some Alfisols	Am Aw Borders BS	130–200 cm (<40 rainy days during 4 driest months)	Variable, always warm (>18°C)	Seasonal surpluses and deficits
Tropical Savanna (TrS) Tropical grassland Thorn tree scrub Thorn woodland	Transitional between seasonal forests, rain forests, and semiarid tropical steppes and desert; trees with flattened crowns, clumped grasses, and bush thickets; fire association	Alfisols (dry: Ultalfs) Ultisols Oxisols	Aw BS	9–150 cm, seasonal	No cold-weather limitations	Tends toward deficits, therefore fire- and drought-susceptible
Midlatitude Broadleaf and Mixed Forest (MBME) Temperate broadleaf Midlatitude deciduous Temperate needleaf	Mixed broadleaf and needleleaf trees; deciduous broadleaf, losing leaves in winter; southern and eastern evergreen pines demonstrate fire association	Ultisols Some Alfisols (Podzols, red and yellow)	Cfa Cwa Dfa	75–150 cm	Temperate, with cold season	Seasonal pattern with summer maximum PRECIP and POTET (PET); no irrigation needed
Needleleaf Forest and Montane Forest (NF/MF) Taiga Boreal forest Other montane forests and highlands	Needleleaf conifers, mostly evergreen pine, spruce, fir; Russian larch, a deciduous needleleaf	Spodosols Histosols Inceptisols Alfisols (Boralfs: cold) (Gleysols) (Podzols)	Subarctic Dfb Dfc Dfd	30–100 cm	Short summer, cold winter	Low POTET (PET), moderate PRECIP, moist soils, some waterlogged and frozen in winter; no deficits
Temperate Rain Forest (TeR) West coast forest Coast redwoods (U.S.)	Narrow margin of lush evergreen and deciduous trees on windward slopes; redwoods, tallest trees on Earth	Spodosols Inceptisols (mountainous environs) (Podzols)	Cfb Cfc	150–500 cm	Mild summer and mild winter for latitude	Large surpluses and runoff
Mediterranean Shrubland (MSh) Sclerophyllous shrubs Australian eucalyptus forest	Short shrubs, drought adapted, tending to grassy woodlands and chaparral	Alfisols (Xeralfs) Mollisols (Xerolls) (Luvisols)	Csa Csb	25–65 cm	Hot, dry summers, cool winters	Summer deficits, winter surpluses
Midlatitude Grasslands (MGr) Temperate grassland Sclerophyllous shrub	Tallgrass prairies and shortgrass steppes, highly modified by human activity; major areas of commercial grain farming; plains, pampas, and veld	Mollisols Aridisols (Chernozemic)	Cfa Dfa	25–75 cm	Temperate continental regimes	Soil moisture utilization and recharge balanced; irrigation and dry farming in drier areas
Warm Desert and Semidesert (DBW) Subtropical desert and scrubland	Bare ground graduating into xerophytic plants including succulents, cacti, and dry shrubs	Aridisols Entisols (sand dunes)	BWh BWk	<2 cm	Average annual temperature, around 18°C, highest temperatures on Earth	Chronic deficits, irregular precipitation events, PRECIP $<\frac{1}{2}$ POTET (PET)
Cold Desert and Semidesert (DBC) Midlatitude desert, scrubland, and steppe	Cold desert vegetation includes short grass and dry shrubs	Aridisols Entisols	BSh BSk	2–25 cm	Average annual temperature around 18°C	PRECIP $>\frac{1}{2}$ POTET (PET)
Arctic and Alpine Tundra (AAT)	Treeless; dwarf shrubs, stunted sedges, mosses, lichens, and short grasses; alpine, grass meadows	Gelisols Inceptisols Entisols (permafrost) (Organic) (Cryosols)	ET Dwd	15–180 cm	Warmest months <10°C, only 2 or 3 months above freezing	Not applicable most of the year, poor drainage in summer
Ice			EF			

FIGURE 20.3 The 10 major global terrestrial biomes.

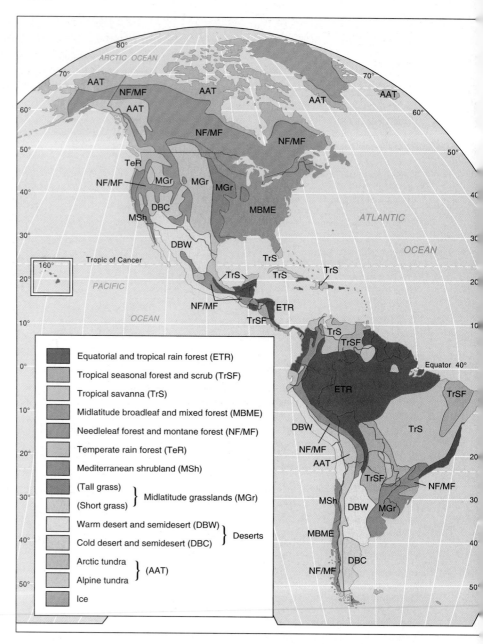

Earth's Major Terrestrial Biomes

Few natural communities of plants and animals remain; most biomes have been greatly altered by human intervention. Thus, the "natural vegetation" identified on many biome maps reflects *ideal potential* mature vegetation, given the environmental characteristics in a region. Even though human practices have greatly altered these ideal forms, it is valuable to study the natural (undisturbed) biomes to better understand the environment and to assess the extent of human-caused alteration.

In addition, knowing the ideal in a region guides us to a closer approximation of natural vegetation in the plants we introduce. In the United States and Canada, we humans are perpetuating a type of *transition community*, somewhere between grasslands and a forest—an artificial successional stage, produced by large-scale interruption

and disturbance. We plant trees and lawns and then must invest energy, water, and capital to sustain such artificial modifications. Or we graze animals and plant crops that perpetuate this disturbed transitional state. When these activities cease, natural succession recovers, and we see the land slowly return to its various vegetation-cover potentials. Some irreversible damage might occur when plants, animals, and organisms are knowingly or accidentally brought into an ecosystem or biome in which they are not native. These exotic species pose an increasing concern to scientists and society—a subject of News Report 20.2.

The global distribution of Earth's major terrestrial biomes, based on vegetation formation classes, is portrayed in Figure 20.3. Table 20.1 describes each biome on the map and summarizes other pertinent information gathered from throughout *Geosystems*.

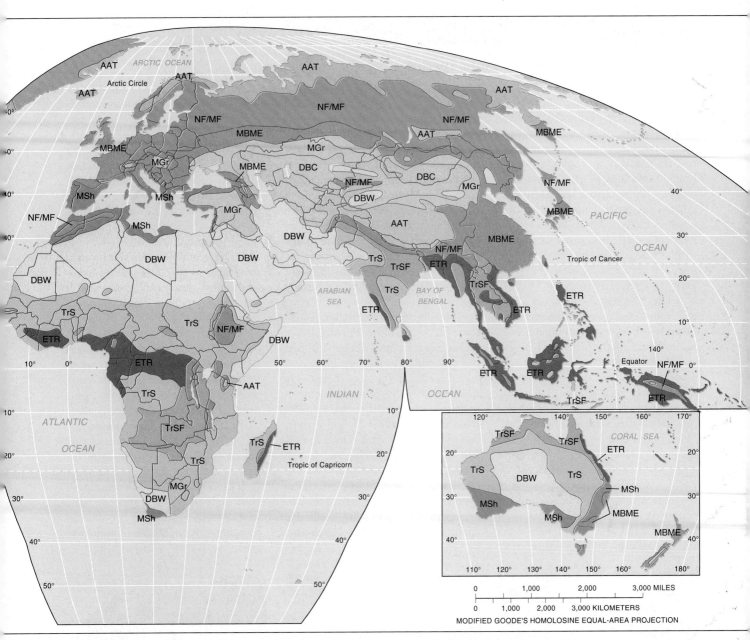

MODIFIED GOODE'S HOMOLOSINE EQUAL-AREA PROJECTION

Now is a good time to refer to "The Living Earth" composite image that appears inside the front cover of this text. A computer artist using hundreds of thousands of satellite images produced this cloudless view of Earth in natural color typical of a local summer day. Compare the map in Figure 20.3 with this remarkable composite image and see what correlation you can make. Then compare the biomes with population distribution as indicated by the nighttime lighting display also on the inside front cover.

Equatorial and Tropical Rain Forest

Earth is girdled with a lush biome—the **equatorial and tropical rain forest**. In a climate of consistent year-round daylength (12 hours), high insolation, average annual temperatures around 25°C (77°F), and plentiful moisture, plant and animal populations have responded with the most diverse expressions of life on the planet.

As Figure 20.3 shows, the Amazon region, also called the *selva*, is the largest tract of equatorial and tropical rain forest. In addition, rain forests cover equatorial regions of Africa, Indonesia, the margins of Madagascar and Southeast Asia, the Pacific coast of Ecuador and Colombia, and the east coast of Central America, with small discontinuous patches elsewhere. The cloud forests of western Venezuela are such tracts of rain forest at high elevation, perpetuated by high humidity and cloud cover. Undisturbed tracts of rain forest are rare.

Rain forests represent approximately one-half of Earth's remaining forests, occupying about 7% of the total land area worldwide. This biome is stable in its natural state, resulting from the long-term residence of these continental plates near equatorial latitudes and their escape from glacial activity.

News Report 20.2

Alien Invaders of Exotic Species

The above title sounds like that of a "B" science fiction movie; instead, it refers to a real problem in the integrity of many ecosystems, both aquatic and terrestrial. Animal and plant species brought or somehow getting into biomes outside their native home is an important subject in biogeography and invasion biology. Niche takeovers by nonnative species can prove damaging to otherwise healthy ecosystems. These intruders are known as *exotic species*, or *alien* or *nonnative species*.

The "Biological Invasions" Web site from Oregon State University, at **http://seagrant.orst.edu/colloquium/index.html**, was the focus of a biology meeting dedicated to these invasive plants, animals, and other organisms. Some states post warning signs at their borders (Figure 1a) while others offer Web sites such as Minnesota's "Prohibited and Noxious Plants by Scientific Name" (**http://www.dnr.state.mn.us/ecological_services/exotics/weedlist00.pdf**).

The African "killer bees" are frequently in the headlines, as are brown tree snakes in Guam, zebra mussels in the Great Lakes, and Eurasian cheat-grass in the Utah desert, to name but a few examples. Brought from Africa to the central coast of Brazil in 1957 to increase honey production, killer bees now have interbred with native bees and range to southern California, Arizona, New Mexico, Texas, and Puerto Rico. More than 1000 people have died from their attacks, although the dangers are in their mass response to disturbances and that some people are allergic to even one bee sting. This is a classic case of geographical diffusion of an exotic species.

Probably 90% of exotic species fail when they try to move into established niches in a community. The 10% that succeed, some 4500 species documented in a 1993 Office of Technology Assessment report, prove damaging to as much as one-fifth of ecosystems they invade. Figures 1b and c show two such alien invaders: purple loosestrife (*Lythrum salicaria*) and kudzu (*Pueraria montana*).

Purple loosestrife was introduced from Europe in the 1800s as a desired ornamental and had some medicinal applications. The plant's seeds also arrived in ships that used soil for ballast.

A hardy perennial, it got loose and invaded wetlands across the eastern portions of the United States and Canada, through the upper Midwest and as far west as Vancouver Island, British Columbia, replacing plants on which native wildlife depend (Figure 1b). The plant is known to infect drier landscapes as well and poses a potential threat to agriculture.

The infamous kudzu was brought from Japan in 1876 for cattle feed, erosion control, and as an ornamental. It has spread to east Texas, southern Pennsylvania and across the South into Florida. The plant can grow 0.3 m (1 ft) on a hot, humid summer day and overtake forests and structures. Serious effort is going into finding some practical use for the prodigious plant, from pulp (treeless paper) to recipes for cooking to the telling of new kudzu jokes.

In learning about biomes, we find out about native plant, animal, and organism locations and the unique place each species occupies. We must avoid knowingly or perhaps unconsciously adding to this escalating problem by bringing exotic species into nonnative sites.

(a)

(b)

(c)

FIGURE 1 Exotic species.
(a) A sign in Wallowa County, Washington state, pleas for awareness and help in this agricultural country, listing noxious (nonnative, often problematic) weeds at the county line. (b) Invasive purple loosestrife near Lake Michigan and the Indiana Dunes National Lakeshore, Indiana (foreground and left center). (c) Kudzu overruns pasture and forest in western Georgia. [Photos by Bobbé Christopherson.]

Rain forests feature ecological niches that are distributed vertically rather than horizontally because of the competition for light. Biomass in a rain forest is concentrated high up in the canopy, that dense mass of overhead leaves. The canopy is filled with a rich variety of plants and animals. Lianas (vines) stretch from tree to tree, entwining them with cords that can reach 20 cm (8 in.) in diameter. Epiphytes flourish there too: Plants such as orchids,

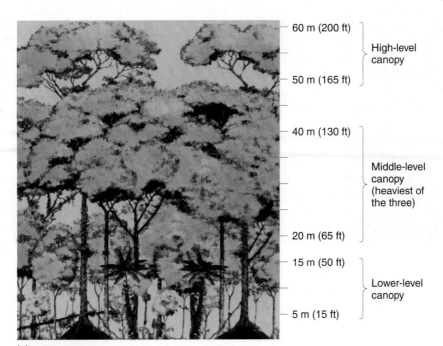

60 m (200 ft)

} High-level
   canopy

50 m (165 ft)

40 m (130 ft)

Middle-level
canopy
(heaviest of
the three)

20 m (65 ft)

15 m (50 ft)

} Lower-level
   canopy

5 m (15 ft)

(a)

FIGURE 20.4 The three levels of a rainforest canopy.
(a) Lower, middle, and high levels of the rain forest. (b) The emergent trees of the high-level canopy are clearly seen above the dense, continuous cover of the middle canopy in a lowland tropical rain forest near the coast of Costa Rica. [Photo by Gregory G. Dimijian/Photo Researchers, Inc.]

(b)

The rain forest canopy forms three levels—see Figure 20.4. The upper level is not continuous but features emergent tall trees whose high crowns rise above the middle canopy. This upper level appears as a broken *overstory* of tall trees breaking through a middle canopy that is nearly continuous (Figure 20.4b). The broad leaves block much of the light and create a darkened *understory* area and forest floor. The lower level is composed of seedlings, ferns, bamboo, and the like, leaving the litter-strewn ground level in deep shade and fairly open.

Aerial photographs of a rain forest, or views along riverbanks covered by dense vegetation, or the false Hollywood-movie imagery of the jungle, makes it difficult to imagine the shadowy environment of the actual rainforest floor, which receives only about 1% of the sunlight arriving at the canopy (Figure 20.5a). The constant moisture, rotting fruit and moldy odors, strings of thin roots and vines dropping down from above, windless air, and echoing sounds of life in the trees together create a unique environment.

The smooth, slender trunks of rainforest trees are covered with thin bark and buttressed by large wall-like flanks that grow out from the trees to brace the trunks (Figure 20.5b). These buttresses form angular, open enclosures, a ready habitat for various animals. There are usually no branches for at least the lower two-thirds of the tree trunks.

The wood of many rainforest trees is extremely hard, heavy, and dense—in fact, some species will not even float in water. (Exceptions are balsa and a few others, which are light.) Varieties of trees include mahogany, ebony, and rosewood. Logging is difficult because individual species are widely scattered; a species may occur only once or twice per square kilometer. Selective cutting is required for species-specific logging, whereas pulpwood production takes everything. Conversion of the forest to pasture usually is done by setting destructive fires.

bromeliads, and ferns live entirely above ground, supported physically but not nutritionally by the structures of other plants. Windless conditions on the forest floor make pollination difficult, so pollination is by insects, other animals, and self-pollination.

(a)

(b)

FIGURE 20.5 The rain forest.
(a) Equatorial rain forest is thick along the Amazon River where light breaks through to the surface, producing a rich gallery of vegetation. (b) The rainforest floor in Corcovado, Costa Rica, with typical buttressed trees and lianas. [(a) Photo by Wolfgang Kaehler Photography; (b) photo by Frank S. Balthis.]

Rainforest soils, principally Oxisols, are essentially infertile, yet they support rich vegetation. The trees have adapted to these soil conditions with root systems able to capture nutrients from litter decay at the soil surface. High precipitation and temperatures work to produce deeply weathered and leached soils, characteristic of the laterization process, with a clay-like texture sometimes breaking up into a granular structure. Oxisols lack nutrients and colloidal material. With much investment in fertilizers, pesticides, and machinery, these soils can be productive.

The animal and insect life of the rain forest is diverse, ranging from small decomposers (bacteria) working the surface to many animals living exclusively in the upper stories of the trees. These tree dwellers are referred to as *arboreal*, from the Latin for "tree," and include sloths, monkeys, lemurs, parrots, and snakes. Beautiful birds of many colors, tree frogs, lizards, bats, and a rich insect community that includes more than 500 species of butterflies are found in rain forests. Surface animals include pigs (bushpigs and the giant forest hog in Africa, wild boar and bearded pig in Asia, and peccary in South America), species of small antelopes (bovids), and mammalian predators (the tiger in Asia, jaguar in South America, and leopard in Africa and Asia).

The present human assault on Earth's rain forests has put this diverse fauna and the varied flora at risk. It also jeopardizes an important recycling system for atmospheric carbon dioxide and a potential source of valuable pharmaceuticals and many types of new foods—and so much is still unknown and undiscovered.

Deforestation of the Tropics

More than half of Earth's original rain forest is gone, cleared for pasture, timber, fuel wood, and farming. Worldwide an area nearly the size of Wisconsin is lost each year (169,000 km², 65,000 mi²) and about a third more is

disrupted by selective cutting of canopy trees, a damage that occurs along the *edges* of deforested areas.

When orbiting astronauts look down on the rain forests at night, they see thousands of human-set fires. During the day, the lower atmosphere in these regions is choked with the smoke. These fires are used to clear land for agriculture, which is intended to feed the domestic population as well as to produce cash exports of beef, rubber, coffee, and other commodities. Edible fruits are not abundant in an undisturbed rain forest, but cultivated clearings produce bananas (plantains), mangos, jack fruit, guava, and starch-rich roots such as manioc and yams.

Because of the poor soil fertility, the cleared lands are quickly exhausted under intensive farming and are then generally abandoned in favor of newly burned and cleared lands, unless fertility is maintained artificially. Unfortunately, the dominant trees require from 100 to 250 years to reestablish themselves after major disturbances. Once cleared, the former forest becomes a mass of low bushes intertwined with vines and ferns, slowing the return of the forests.

Figure 20.6a–c shows satellite false-color images of a portion of western Brazil called Rondônia, recorded in June 1975 (*Landsat 2*), August 1986 (*Landsat 5*), June 1992 (*Landsat 4*), and in (d) a true-color June 2001 (*Terra*, MODIS). These images give a sense of the level of rainforest destruction in progress. You can clearly see encroachment along new roads branching from highway BR364. The *edges* of every road and cleared area represent a significant portion of the species habitat disturbance, population dynamic changes, and carbon losses to the atmosphere—perhaps as much as 1.5 times more impact occurs along these edges than in the tracts of clear-cut logging. The hotter, drier, and windier edge conditions penetrate the forest up to 100 m. Figure 20.6e shows a tract of former rain forest that has just been burned to begin the clearing and road-building process.

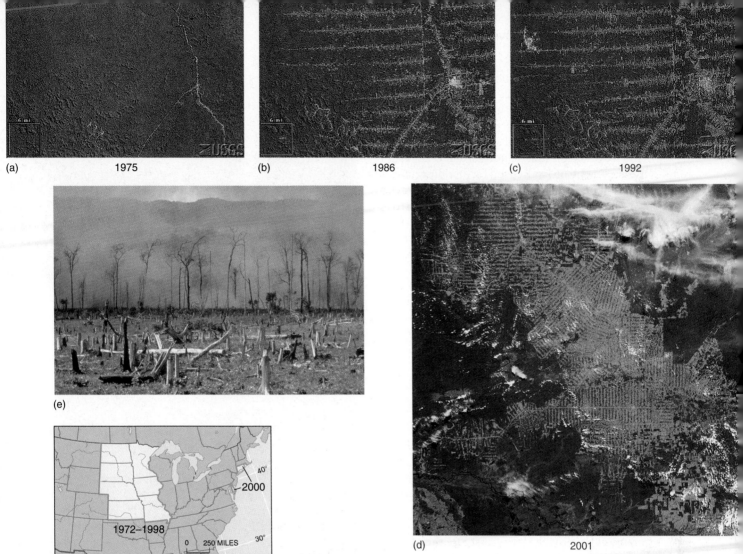

(a)   1975   (b)   1986   (c)   1992

(e)

(f)   Brazil's deforestation areas compared to North America

(d)   2001

FIGURE 20.6 The rain forest diminished over 26 years.
Satellite images of a portion of western Brazil called Rondônia, recorded in false color in (a) June 1975 (*Landsat 2*), (b) August 1986 (*Landsat 5*), (c) June 1992 (*Landsat 4*), and in true color (d) June 1, 2001 (*Terra* MODIS sensor; the area of the other three images is in the upper left). Highway BR364, the main artery in the region, passes through the towns of Jaru and Ji-Paraná near the center of image (d) at about 62° W 11° S. The branching pattern of feeder roads encroach on the rain forest. (e) Surface view of burning rain forest in Guatemala. (f) Deforested areas in Brazil, cleared between 1972 and 2000, and for 2000 only, are compared to North America. [*Landsat* images courtesy of NASA/USGS EROS Data Center; *Terra* MODIS image courtesy of NASA/MODIS Land Rapid Response Team; (e) photo by George Holton/Photo Researchers, Inc.]

Scientists with the Goddard Space Flight Center and the Brazilian government completed a GIS analysis of these rainforest losses to guide policy decisions. In 1998, despite all the efforts to introduce sustainable forestry practices, deforestation increased 27% over 1997—some 16,800 km² (6500 mi²)—and rose again to 19,836 km² lost in 2000, according to the Brazilian Environment Ministry in Brasilia (an area roughly equivalent to Connecticut, Rhode Island, and Delaware combined). Since 1972, this has brought total deforestation to more than 15% of the entire Amazon Basin—more than the area of the combined north-central states of Minnesota, Iowa, Missouri, North and South Dakota, Nebraska, and Kansas (Figure 20.6f), or equivalent to the area of France in 30 years! Remember, in 1970 only about 1% of the Brazilian Amazon had been deforested.

The United Nations Food and Agricultural Organization (FAO) estimates that if this destruction to rain forests continues unabated, these forests will be completely removed by about A.D. 2050! By continent, rain forest losses are estimated at more than 50% in Africa, more than 40% in Asia, and 40% in Central and South America.

To slow this continuing catastrophe of deforestation, in 1985 the FAO, the UN Development Programme, the World Bank, the World Resources Institute, and several nongovernmental organizations initiated the Tropical Forestry Action Plan. This plan, armed with $8 billion in development aid money, instituted an accurate worldwide survey of the rate of deforestation, using satellites to obtain remote measurement of land cover for building more-accurate GIS models. Critics charge that this effort has yet to show progress. (Among many references, see Tropical Rainforest Coalition at http://www.rainforest.org/, World Resources Institute publications on forests at http://www.wri.org/wri/cat-frst.html, the Rainforest Information Centre at http://www.rainforestinfo.org.au/welcome.htm, or the Rainforest Action Network at http://www.ran.org/.)

Focus Study 20.1 looks more closely at efforts to curb increasing rates of species extinction and loss of Earth's biodiversity, much of which is directly attributable to the loss of rain forests.

Tropical Seasonal Forest and Scrub

A varied biome on the margins of the rain forest is the **tropical seasonal forest and scrub**, which occupies regions of low and erratic rainfall. The shifting intertropical convergence zone (ITCZ) brings precipitation with the seasonally shifting high Sun and dryness with the low Sun, producing a seasonal pattern of moisture deficits, some leaf loss, and dry-season flowering. The term semideciduous applies to some of the broadleaf trees that lose their leaves during the dry season.

Areas of this biome have fewer than 40 rainy days during their four consecutive driest months, yet heavy monsoon downpours characterize their summers (see Chapter 6, especially Figure 6.21). The Köppen climates *tropical monsoon Am* and *tropical savanna Aw* apply to these transitional communities between rain forests and tropical grasslands.

Portraying such a varied biome is difficult. In many areas humans disturb the natural biome, so that the savanna grassland adjoins the rain forest directly. The biome does include a gradation from wetter to drier areas: monsoonal forests, to open woodlands and scrub woodland, to thorn forests, to drought-resistant scrub species (Figure 20.7). In South America, an area of transitional tropical deciduous forest surrounds an area of savanna in southeastern Brazil and portions of Paraguay.

The monsoonal forests average 15 m (50 ft) high with no continuous canopy of leaves, graduating into open orchard-like parkland with grassy openings or into areas choked by dense undergrowth. In more open tracts, a common tree is the acacia, with its flat-topped appearance and usually thorny stems. These trees have branches that look like an upside-down umbrella (spreading and open skyward), as do trees in the tropical savanna.

Local names are given to these communities: the *caatinga* of the Bahia State of northeastern Brazil, the *chaco* area of Paraguay and northern Argentina, the *brigalow* scrub

(a)

(b)

FIGURE 20.7 Kenyan landscapes.
Two views of the open thorn forest and savanna near and in the Samburu Reserve, Kenya. [(a) Photo by Gael Summer-Hebden; (b) photo by Stephen J. Krasemann/DRK Photo.]

of Australia, and the *dornveld* of southern Africa. Figure 20.3 shows areas of this biome in Africa, extending from eastern Angola through Zambia to Tanzania; in southeast Asia and portions of India, from interior Myanmar through northeastern Thailand; and in parts of Indonesia.

The trees throughout most of this biome make poor lumber, but some, especially teak, may be valuable for fine cabinetry. In addition, some of the plants with dry-season adaptations produce usable waxes and gums, such as carnauba and palm-hard waxes. Animal life includes the koalas and cockatoos of Australia and the elephants, large cats, rodents, and ground-dwelling birds in other occurrences of this biome.

Tropical Savanna

The **tropical savanna** is large expanses of grassland, interrupted by trees and shrubs. This is a transitional biome between the tropical forests and semiarid tropical steppes

Focus Study 20.1

Biodiversity and Biosphere Reserves

When the first European settlers landed on the Hawaiian Islands in the late 1700s, 43 species of birds were counted. Today, 15 of those species are extinct, and 19 more are threatened or endangered, with only one-fifth of the original species relatively healthy. Humans have great impact on animals and plants and global biodiversity.

Relatively pristine habitats around the world are being lost at unprecedented rates as an expanding human population converts them to agriculture, forestry, and urban centers. As these habitats are altered, untold numbers of species are disappearing before they have been recognized, much less studied, and the functioning of entire ecosystems is threatened. This loss of biodiversity, at the very time when the value of biotic resources is becoming widely recognized, has made it strikingly clear that current strategies for conservation are failing dismally.*

As more is learned about Earth's ecosystems and their related communities, more is known of their value to civilization and our interdependence with them. Natural ecosystems are a major source of new foods, new chemicals, new medicines, and specialty woods, and of course they are indicators of a healthy, functioning biosphere.

International efforts are under way to study and preserve specific segments of the biosphere, among others: the UN Environment Programme, World Resources Institute, The World Conservation Union, Rain Forest Action Network, Natural World Heritage Sites, Wetlands of International Importance, and the Nature Conservancy. An important part of the effort is the setting aside of biosphere reserves, and the focus provided by the World Conservation Monitoring Centre and its IUCN Red List of global endangered species. (See IUCN Red List at **http://www.wcmc.org.uk/data/database/rl_anml_combo.html**, the IUCN at **http://www.iucn.org/**, the endangered species home page of the Fish and Wildlife Service at **http://endangered.fws.gov/**, and the World Wildlife Fund at **http://www.panda.org/**.)

The motivation to set aside natural sanctuaries is directly related to concern over the increase in the rate of species extinctions. We are facing a loss of genetic diversity that may be unparalleled in Earth's history, even compared with the major extinctions over the geologic record.

Species Threatened—An Example

Black rhinos (*Diceros bicornis*) and white rhinos (*Ceratotherium simum*) in Africa exemplify species in jeopardy. Rhinos once grazed over much of the savanna grasslands and woodlands. Today, they survive only in protected districts in heavily guarded sanctuaries and are threatened even there (Figure 1). In 1998, 2599 black rhinos remained, 96% less than the 70,000 in 1960 (source: IUCN/SSC/African Rhino Specialist Group, 1998). Populations have remained statistically stable since 1992, and this count is supported by South Africa's guarding about 50% of the remaining population.

There were 11 white rhinos (northern subspecies) in existence in 1984, increasing to an estimated 29 by 1998 (Congo 25, Côte d'Ivoire 4). Political unrest in the Congo and the area of the Garamba National Park has led to the deaths of an unknown number of these few. The southern white rhino population reached 8431 by 1998.

Rhinoceros horn sells for $29,000 per kilogram! These large land mammals are nearing extinction and will survive only as a dwindling zoo population. The limited genetic pool that remains complicates further reproduction.

Table 1 summarizes the numbers of known and estimated species on Earth (see **http://www.wri.org/biodiv/gba-unpr.html**). Scientists have classified only 1.75 million species of plants and animals of an estimated 13.6

(continued)

*T. D. Sisk, A. E. Launer, K. R. Switky, and P. R. Ehrlich, "Identifying extinction threats," *Bioscience* 44 (October 1994): 592.

FIGURE 1 The rhinoceros in Africa.
Black rhino and young in Tanzania, escorted by Oxpecker birds. These rhinos are quite nearsighted (they can see clearly only up to 10 m); the birds act as the rhino's early-warning system for disturbances in the distance. [Photo by Stephen J. Krasemann/DRK Photo.]

Focus Study 20.1 *(continued)*

Table 1	Known and Estimated Species on Earth				
		Estimated Number of Species		Working	
Species	Number of Known Species	High (000)	Low (000)	Estimate (000)	Accuracy
Viruses	4000	1000	50	400	Very poor
Bacteria	4000	3000	50	1000	Very poor
Fungi	72,000	27,000	200	1500	Moderate
Protozoa	40,000	200	60	200	Very poor
Algae	40,000	1000	150	400	Very poor
Plants	270,000	500	300	320	Good
Nematodes	25,000	1000	100	400	Poor
Arthropods					
Crustaceans	40,000	200	75	150	Moderate
Arachnids	75,000	1000	300	750	Moderate
Insects	950,000	100,000	2000	8000	Moderate
Mollusks	70,000	200	100	200	Moderate
Chordates	45,000	55	50	50	Good
Others	115,000	800	200	250	Moderate
Total	**1,750,000**	**111,655**	**3635**	**13,620**	**Very poor**

Source: United Nations Environment Programme, *Global Biodiversity Assessment* (Cambridge: Cambridge University Press, 1995), Table 3.1–2, p. 118.

million overall; this figure represents an increase in what scientists once thought to be the diversity of life on Earth. And those yet-to-be-discovered species represent a potential future resource for society—the wide range of estimates places the expected species count between a low of 3.6 million and a high of 111.7 million.

Estimates of annual species loss overall range between 1000 and 30,000, although this range might be conservative. The possibility exists that more than half of Earth's present species could be extinct within the next 100 years. The effects of pollution, loss of wild habitat, excessive grazing, poaching, and collecting are at the root of such devastation. Approximately 60% of the extinctions are attributable to the clearing and loss of rain forests alone.

The Question of Medicine and Food

Wheat, maize (corn), and rice—just three grains—fulfill about 50% of human planetary food demands. About 7000 plant species have been gathered for food throughout human history, but more than 30,000 plant species have edible parts. Undiscovered potential food resources are in nature waiting to be found and developed. Biodiversity, if preserved, provides us a potential cushion for all future food needs, but only if species are identified, inventoried, and protected. The same is true for pharmaceuticals.

Nature's biodiversity is like a full medicine cabinet. Since 1959, 25% of all prescription drugs were originally derived from higher plants. In preliminary surveys, 3000 plants have been identified as having anticancer properties. The rosy periwinkle (*Catharanthus roseus*) of Madagascar contains two alkaloids that combat two forms of cancer. (Alkaloids are compounds found in certain plants that help the plant defend against insects and are potentially significant as human medicines; examples are atropine, quinine, and morphine.) Yet, less than 3% of flowering plants have been examined for alkaloid content! It defies common sense

to throw away the medicine cabinet before we even open the door to see what is inside.

Genetic material from the developing world is already making an immense contribution to the food tables and medicine cabinets of the industrialized Northern Hemisphere. Agriculture in developed countries reaps an estimated $5 billion a year from Third World genetic material. . . . One Peruvian variety of tomato has been worth $8 million annually to U.S. tomato processors because it added soluble solid content to tomatoes. . . . And one in four prescription drugs is derived at least in part from plants. . . . Their value to the $60-billion-a-year U.S. pharmaceutical industry is considerable.*

Biosphere Reserves

Formal natural reserves are a possible strategy for slowing the loss of biodiversity and protecting this resource base. Setting up such a biosphere reserve involves principles of *island biogeography*. Island communities are special places for study because of their spatial isolation and the relatively small number of species present. Islands resemble natural experiments, because the impact of individual factors, such as civilization, can be more easily assessed on islands than over larger continental areas.

Studies of islands also can assist in the study of mainland ecosystems, for in many ways a park or biosphere reserve, surrounded by modified areas and artificial boundaries, is like an island. Indeed, a biosphere reserve is conceived as an ecological island in the midst of change. The intent is to establish a core in which genetic material is protected from outside disturbances, surrounded by a buffer zone that is, in turn, surrounded by a transition zone and experimental re-

*J. W. King, "Breeding uniformity: Will global biotechnology threaten global biodiversity?" The *Amicus Journal* 15 (Spring 1993): 26.

(continued)

Focus Study 20.1 *(continued)*

search areas. An important variable to consider in setting aside a biosphere reserve is any change that might occur in temperature and precipitation patterns as a result of global change. A carefully considered reserve could end up outside its natural range as ecotones shift.

Scientists predict that new, undisturbed reserves will not be possible in a decade, because pristine areas will be gone. The biosphere reserve program is coordinated by the Man and the Biosphere (MAB) Programme of UNESCO (http://www.unesco.org/mab/). Nearly 300 such biosphere reserves covering some 12 million hectares (30 million acres) are now operated voluntarily in 76 countries. Not all protected areas are ideal bioregional entities. Some are simply imposed on existing park space and some remain in the planning stage, although they are officially designated. In Canada there are six biosphere reserve sites; see the Canada/MAB Programme site at http://www.cciw.ca/mab/. The United States has 90 biosphere reserves under various jurisdictions. Also,

check another Web site at the University of California at Davis for MAB fauna and MAB flora databases at http://ice.ucdavis.edu/MAB/.

Some of the best biosphere reserve examples have been in operation since the late 1970s and range from the struggling Everglades National Park in Florida, to the Changbai Reserve in China, to the Tai Forest on the Côte d'Ivoire (Ivory Coast), to the Charlevoix Biosphere Reserve in Québec, Canada. Added to these efforts is the work of the Nature Conservancy, which acquires land for preservation as a valuable part of the reserve (http://nature.org/).

The ultimate goal, about half achieved, is to establish at least one reserve in each of the 194 distinctive biogeographic communities presently identified. UNESCO and the World Conservation Monitoring Centre compile the United Nations' list of national parks and protected areas. Presently in these biogeographic communities, 6930 areas covering 657 million hectares (1.62 billion acres) are designated in some protective form, representing about 4.8% of national

land area on the planet. Even though these are not all set aside to the degree of biosphere reserves, they do demonstrate progress toward preservation of our planet's plant and animal heritage.

The preservation of species diversity is a problem that must today be confronted by one species, *Homo sapiens.* . . . The diversity of species is worth preserving because it represents a wealth of knowledge that cannot be replaced. Moreover, today's extinctions are unlike those in previous eras, in which long periods of recovery could follow extinctions. The present situation is an inexorably irreversible one in which human overpopulation will destroy most species unless we plan for protection immediately. Accepting that the goal is worthwhile requires that more energy be devoted to planning and priorities and less to emotionalism and indignation [on all sides].*

*D. E. Koshland, Jr., "Preserving biodiversity," *Science* 253 (August 16, 1991): 717.

---

and deserts. The savanna biome also includes treeless tracts of grasslands, and in very dry savannas, grasses grow discontinuously in clumps, with bare ground between them. The trees of the savanna woodlands are characteristically flat-topped, in response to light and moisture regimes.

Savannas covered more than 40% of Earth's land surface before human intervention but were especially modified by human-caused fire. Fires occur annually throughout the biome. The timing of these fires is important. Early in the dry season they are beneficial and increase tree cover; late in the season they are very hot and kill trees and seeds. Savanna trees are adapted to resist the "cooler" fires.

Forests and elephant grasses averaging 5 m (16 ft) high once penetrated much farther into the dry regions, for they are known to survive there when protected. Savanna grasslands are much richer in humus than the wetter tropics and are better drained, thereby providing a better base for agriculture and grazing. Sorghums, wheat, and groundnuts (peanuts) are common commodities.

Tropical savannas receive their precipitation during less than 6 months of the year, when they are influenced by the shifting ITCZ. The rest of the year they are under the drier influence of shifting subtropical high-pressure cells. Savanna shrubs and trees are frequently *xerophytic*, or

drought-resistant, with various adaptations to protect them from the dryness: small, thick leaves, rough bark, or leaf surfaces that are waxy or hairy.

Africa has the largest area of this biome, including the famous Serengeti Plains of Tanzania and Kenya and the Sahel region, south of the Sahara. Sections of Australia, India, and South America also are part of the savanna biome. Some of the local names for these lands include the *Llanos* in Venezuela, stretching along the coast and inland east of Lake Maricaibo and the Andes; the *Campo Cerrado* of Brazil and Guiana; and the *Pantanal* of southwestern Brazil.

Particularly in Africa, savannas are the home of large land mammals (zebra, giraffe, buffalo, gazelle, wildebeest, antelope, rhinoceros, elephant). They graze on savanna grasses (Figure 20.8) or feed upon the grazers themselves (lion, cheetah). Birds include the Ostrich, Martial Eagle (largest of all eagles), and Secretary Bird. Many species of venomous snakes occur, as does the crocodile.

Unfortunately, in our lifetime we may see the reduction of these animal herds to zoo stock only, because of poaching and habitat losses. The loss of the rhino is discussed in Focus Study 20.1. Establishment of large tracts of savanna as biosphere reserves is critical for the preservation of this biome and its associated fauna.

(a)

(b)

FIGURE 20.8 Animals and plants of the Serengeti Plains.
(a) Savanna landscape of the Serengeti Plains, with wildebeest, zebras, and thorn forest. (b) Near Samburu, Kenya, Grevy's zebra and reticulated giraffe forage. [Photos by (a) Stephen F. Cunha; (b) Galen Rowell/Mountain Light Photography, Inc.]

Midlatitude Broadleaf and Mixed Forest

Moist continental climates support a mixed forest in areas of warm to hot summers and cool to cold winters. This **midlatitude broadleaf and mixed forest** biome includes several distinct communities in North America, Europe, and Asia. Along the Gulf of Mexico, relatively lush evergreen broadleaf forests occur. Northward are mixed deciduous and evergreen needleleaf stands associated with sandy soils and burning. When areas are given fire protection, broadleaf trees quickly take over. Pines (longleaf, shortleaf, pitch, loblolly) predominate in the southeastern and Atlantic coastal plains. Into New England and westward in a narrow belt to the Great Lakes, white and red pines and eastern hemlock are the principal evergreens, mixed with deciduous oak, beech, hickory, maple, elm, chestnut, and many others (Figure 20.9a).

These mixed stands contain valuable timber, but their distribution has been greatly altered by human activity. Native stands of white pine in Michigan and Minnesota were removed before 1910; only later reforestation sustains their presence today. In northern China these forests have almost disappeared as a result of centuries of harvest. The forest species that once flourished in China are similar to species in eastern North America: oak, ash, walnut, elm, maple, and birch.

(a)

(b)

FIGURE 20.9 Animals in the broadleaf mixed forest.
(a) Pioneer Mothers' Memorial Forest near Paoli, Indiana, is 36 hectares (88 acres) of old-growth forest, virtually undisturbed since around 1816, featuring black walnut, white oak, yellow poplar, white ash, and beech trees, among others. (b) A young white-tailed deer grazes in the undergrowth of a mixed forest. [Photos by (a) Bobbé Christopherson; (b) by John Shaw/Tom Stack & Associates.]

This biome is quite consistent in appearance from continent to continent and at one time represented the principal vegetation of the *humid subtropical hot summer Cfa, Cwa* and *marine west coast Cfb, Cwb* (cool summer, winter drought) climatic regions of North America, Europe, and Asia.

A wide assortment of mammals, birds, reptiles, and amphibians is distributed throughout this biome. Representative animals (some migratory) include red fox, white-tail deer, southern flying squirrel, opossum, bear, and a great variety of birds, including tanager and Cardinal (Figure 20.9b). To the north of this biome, poorer soils and colder climates favor stands of coniferous trees and a gradual transition to the northern needleleaf forests.

Needleleaf Forest and Montane Forest

Stretching from the east coast of Canada and the Atlantic provinces westward to Alaska and continuing from Siberia across the entire extent of Russia to the European Plain is the northern **needleleaf forest** biome, also called the **boreal forest** (Figure 20.10). A more open form of boreal forest, transitional to arctic and subarctic regions, is termed the **taiga**. The Southern Hemisphere, lacking *humid micro-*

thermal D climates except in mountainous locales, has no such biome. But **montane forests** of needleleaf trees exist worldwide at high elevation.

Boreal forests of pine, spruce, fir, and larch occupy most of the subarctic climates on Earth that are dominated by trees. Although these forests are similar in formation, individual species vary between North America and Eurasia. The larch (*Larix*) is interesting because it is the rare needleleaf tree that loses its needles in the winter months, perhaps as a defense against the extreme cold of its native Siberia (see the Verkhoyansk climographs and photograph in Figure 10.20). Larches also occur in North America.

The Sierra Nevada, Rocky Mountains, Alps, and Himalayas have similar forest communities occurring at lower latitudes. Douglas fir and white fir grow in the western mountains in the United States and Canada. Economically, these forests are important for lumbering, with trees for lumber occurring in the southern margins of the biome and pulpwood throughout the middle and northern portions. Present logging practices and whether these yields are sustainable are issues of increasing controversy.

In the Sierra Nevada montane forests of California, the giant sequoias naturally occur in seven isolated groves. These trees are Earth's largest (in terms of biomass) living things, although they began as a small seed (Figure 20.11a). Some of these *Sequoia gigantea* exceed 8 m in diameter (28 ft) and 83 m (270 ft) in height. The largest of these is the General Sherman tree in Sequoia National Park (Figure 20.11c); it is estimated to be 3500 years old. The bark is fibrous, a half-meter thick, and lacks resins, so it effectively resists fire. Imagine the lightning strikes and fires that must have moved past the Sherman tree in 35 centuries! Standing among these giant trees is an overwhelming experience and creates a sense of the majesty of the biosphere.

Certain regions of the northern needleleaf biome experience permafrost conditions discussed in Chapter 17. When coupled with rocky and poorly developed soils, these conditions generally limit the existence of trees to those with shallow rooting systems. The summer thaw of surface active layers results in muskeg (moss-covered) bogs and Histolic (organic) soils of poor drainage and stability. Soils of the taiga are typically Spodosols (Podzolic), characteristically acidic and leached of humus and clays. Global warming is altering conditions in the high latitudes, causing increased melting and depth in the active layer. By the late 1990s, some affected forests were beginning to die in response to waterlogging in soils.

Representative fauna in this biome include wolf, moose (the largest deer), bear, lynx, beaver, wolverine, marten, small rodents, and migratory birds during the brief summer season (Figure 20.12). Birds include hawks and eagles, several species of grouse, Pine Grosbeak, Clark's Nutcracker, and several owls. About 50 species of insects particularly adapted to the presence of coniferous trees inhabit the biome.

FIGURE 20.10 Boreal forest of Canada (boreal means "northern"). [Photo by author.]

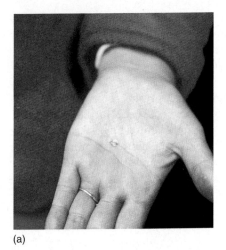

(a)

(b)

(c)

FIGURE 20.11 Sequoia: the seed, the seedling, the giant.
(a) A *Sequoia gigantea* seed. About 300 seeds are in each sequoia cone. (b) Seedling at approximately 50 years of age. (c) The General Sherman tree in Sequoia National Park, California. This tree is probably wider than your classroom. The first branch is 45 m (150 ft) off the ground and is 2 m (6.5 ft) in diameter! [Photos by author.]

FIGURE 20.12 Animals of the northern needleleaf forest.
A bull moose with a full set of antlers grazes along the water's edge. [Photo by Jeremy Woodhouse/PhotoDisc, Inc.]

Temperate Rain Forest

Lush forests at middle and high latitudes are in the **temperate rain forest** biome. In North America, it occurs only along narrow margins of the Pacific Northwest (Figure 20.13a; see also the chapter-opening photo). Some similar types exist in southern China, small portions of southern Japan, New Zealand, and a few areas of southern Chile.

This biome contrasts with the diversity of the equatorial and tropical rain forest in that only a few species make up the bulk of the trees. The rain forest of the Olympic Peninsula in Washington State is a mixture of broadleaf and needleleaf trees, huge ferns, and thick undergrowth. Precipitation approaching 400 cm (160 in.) per year on the western slopes, moderate air temperatures, summer fog, and an overall maritime influence produce this moist, lush

vegetation community. Animals include bear, badger, deer, wild pig, wolf, bobcat, and fox (Figure 20.13b). The trees are home to numerous bird species.

The tallest trees in the world occur in this biome—the coastal redwoods (*Sequoia sempervirens*). Their distribution is shown on the map in Figure 19.13a. These trees can exceed 1500 years of age and typically range in height from 60 to 90 m (200 to 300 ft), with some exceeding 100m (330 ft). Virgin stands of other representative trees, such as Douglas fir, spruce, cedar, and hemlock, have been reduced to a few remaining valleys in Oregon and Washington, less than 10% of the original forest. Replanting and secondary growth forests are predominant. In similar forests in Chile, large-scale timber harvests and new mills began operations in 2000. U.S. corporations are shifting logging operations to these forests in the Chilean Lake District and northern Patagonia.

A study released in 1993 by the U.S. Forest Service noted the failing ecology of these forest ecosystems and suggested that timber management plans should include ecosystem preservation as a priority. The ultimate solution must be one of economic and ecological synthesis rather than continuing conflict and forest losses—the forest can't be all cut down, nor can they be all preserved. The opposing sides in this conflict between logging and society interests should combine their efforts in a sustainable forestry model.

Mediterranean Shrubland

The **Mediterranean shrubland** biome, also referred to as a temperate shrubland, occupies those regions poleward of the shifting subtropical high-pressure cells. As those cells shift poleward with the high Sun, they cut off available storm systems and moisture. Their stable high-pressure

(a)

(b)

FIGURE 20.13 Temperate rain forest.
(a) The temperate rain forest of Macmillan Provincial Park, central Vancouver Island, British Columbia—an old-growth Douglas fir forest, with Western red cedar and hemlock, sword and bracken ferns, and hanging mosses. (b) The Northern Spotted Owl is characteristic of the temperate rain forest. [Photos by (a) Bobbé Christopherson; (b) Greg Vaughn/Tony Stone Images.]

presence produces the characteristic dry summer climate—Köppen's *Mediterranean dry summer Csa, Csb*—and establishes conditions conducive to fire.

Plant ecologists think that this biome is well adapted to frequent fires, for many of its characteristically deep-rooted plants have the ability to resprout from their roots after a fire. Earlier stands of evergreen woodlands (holm or evergreen oak) no longer dominate.

The dominant shrub formations that occupy these regions are stunted and able to withstand hot-summer drought. The vegetation is called *sclerophyllous* (from *sclero,* for "hard," and *phyllos,* for "leaf"). It averages a meter or two in height and has deep, well-developed roots, leathery leaves, and uneven low branches.

Typically, the vegetation varies between woody shrubs covering more than 50% of the ground and grassy woodlands covering 25%–60% of the ground. In California, the Spanish word *chaparro* for "scrubby evergreen" gives us the word **chaparral** for this vegetation type (Figure 20.14). This scrubland includes species such as manzanita, toyon, red bud, ceanothus, mountain mahogany, blue and live oaks, and the dreaded poison oak.

A counterpart to the California chaparral in the Mediterranean region, called *maquis,* includes live and cork oak trees (source of cork), as well as pine and olive trees. The overall similar appearance of the California and Spanish oak savannas is demonstrated in Figure 10.16c and d. In Chile, such a region is called *mattoral*; in southwestern Australia, *mallee scrub.* Of course, in Australia, the bulk of the eucalyptus species is sclerophyllous in form and structure in whichever climate it occurs.

As described in Chapter 10, Mediterranean climates are important in commercial agriculture for subtropical fruits, vegetables, and nuts, with many food types produced

FIGURE 20.14 Mediterranean chaparral.
Chaparral vegetation associated with the Mediterranean dry summer climate. [Photo by Bobbé Christopherson.]

only in this biome (e.g., artichokes, olives, almonds). Larger animals, such as several types of deer, are grazers and browsers, with coyote, wolf, and bobcat as predators. Many rodents, other small animals, and a variety of birds also proliferate.

Midlatitude Grasslands

Of all the natural biomes, the **midlatitude grasslands** are the most modified by human activity. Here are the world's "breadbaskets"—regions that produce bountiful grain

FIGURE 20.15 Grasslands of North America.
Midlatitude grasslands of Alberta, Canada, under cultivation.
[Photo by Fotopic/Omni–Photo Communications, Inc.]

(wheat and corn), soybeans, and livestock (hogs and cattle). Figure 20.15 shows a typical midlatitude grassland in cultivation. In these regions, the only naturally occurring trees were deciduous broadleafs along streams and other limited sites. These regions are called grasslands because of the predominance of grasslike plants before human intervention:

> In this study of vegetation, attention has been devoted to grass because grass is the dominant feature of the Plains and is at the same time an index to their history. Grass is the visible feature which distinguishes the Plains from the desert. Grass grows, has its natural habitat, in the transition area between timber and desert. . . . The history of the Plains is the history of the grasslands.*

In North America, tallgrass prairies once rose to heights of 2 m (6.5 ft) and extended westward to about the 98th meridian, with shortgrass prairies in the drier lands farther west. The 98th meridian is roughly the location of the 51-cm (20-in.) isohyet, with wetter conditions to the east and drier conditions to the west (see Figure 18.17).

The deep, tough sod of these grasslands posed problems for the first European settlers, as did the climate. The self-scouring steel plow, introduced in 1837 by John Deere, allowed the interlaced grass sod to be broken apart, freeing the soils for agriculture. Other inventions were critical to opening this region and solving its unique spatial problems: barbed wire (the fencing material for a treeless prairie); well-drilling techniques developed by Pennsylvania oil drillers but used for water wells; windmills for pumping; and railroads to conquer the distances.

Few patches of the original prairies (tall grasslands) or steppes (short grasslands) remain within this biome. For prairies alone, the reduction of natural vegetation went

*From W. P. Webb, *The Great Plains*, p. 32, ©1931 by Walter Prescott Webb, published by Ginn and Company, Needham Heights, MA.

from 100 million hectares (250 million acres) down to a few areas of several hundred hectares each. The Tallgrass Prairie National Preserve is a 4410-hectare (10,894-acre) section of prairie located in the Flint Hills of Kansas. This is the only national park unit preserving this vestige of the former grasslands that covered much of the North American Great Plains. A state-designated 10-hectare (25-acre) remnant of the "prairie" is located 8 km north of Ames, Iowa. This is a patch of original grassland that never has felt the plow. The map in Figure 20.3 shows the natural location of these former prairie and steppe grasslands.

Outside North America, the Pampas of Argentina and Uruguay and the grasslands of Ukraine are characteristic midlatitude grassland biomes. In each region of the world where these grasslands have occurred, human development of them was critical to territorial expansion.

This biome is the home of large grazing animals, including deer, antelope, pronghorn, and bison (the almost complete annihilation of the latter is part of American history, Figure 20.16). Grasshoppers and other insects feed on the grasses and crops as well, and gophers, prairie dogs, ground squirrels, Turkey Vultures, grouse, and Prairie-Chickens are on the land. Predators include coyote, the nearly extinct black-footed ferret, badger, and birds of prey—hawks, eagles, and owls.

Deserts

Earth's *desert biomes* cover more than one-third of its land area, as is obvious in Figure 20.3. In Chapter 15 we examined desert landscapes and, in Chapter 10, desert climates. On a planet with such a rich biosphere, the deserts stand out as unique regions of fascinating adaptations for survival. Much as a group of humans in the desert might behave with short supplies, plant communities also compete for water and site advantage. Some desert plants, called *ephemerals*, wait years for a rainfall event, at which time their seeds quickly germinate, and the plants develop,

FIGURE 20.16 Bison graze on the grasslands of Montana's National Bison range.
[Photo by Lowell Georgia/CORBIS.]

flower, and produce new seeds, which then rest again until the next rainfall event. The seeds of some xerophytic species open only when fractured by the tumbling, churning action of flash floods cascading down a desert arroyo, and, of course, such an event produces the moisture that a germinating seed needs.

Perennial desert plants employ other adaptive features to cope with the desert. Unique xerophytic features include:

- long, deep taproots (example, the mesquite); succulence (thick, fleshy, water-holding tissue, such as that of cacti);
- spreading root systems to maximize water availability, waxy coatings and fine hairs on leaves to retard water loss, leafless conditions during dry periods (example, palo verde and ocotillo);
- reflective surfaces to reduce leaf temperatures; and
- tissue that tastes bad to discourage herbivores.

The creosote bush (*Lorrea divaricata*) sends out a wide pattern of roots and contaminates the surrounding soil with toxins that prevent the germination of other creosote seeds, possible competitors for water. When a creosote bush dies, surrounding plants or germinating seeds will occupy the abandoned site, but they must wait for infrequent rains to remove the toxins, according to one hypothesis.

The faunas of both warm and cold deserts are limited by the extreme conditions and include few resident large animals. Exceptions are the desert bighorn sheep (in nearby mountains) and the camel, which can lose up to 30% of its body weight in water without suffering (for humans, a 10%–12% loss is dangerous). Some representative desert animals are the ring-tail cat, kangaroo rat, lizards, scorpions, and snakes. Most of these animals are quite secretive and become active only at night, when temperatures are lower. In addition, various birds have adapted to desert conditions and available food sources—for example, Roadrunners, thrashers, Ravens, wrens, hawks, grouse, and nighthawks.

We are witnessing an unwanted expansion of the desert biome discussed in Chapter 15 and Figure 15.24. This process, known as *desertification*, is now a worldwide phenomenon along the margins of semiarid and arid lands. Desertification is due principally to poor agricultural practices (overgrazing and agricultural activities that abuse soil structure and fertility), improper soil-moisture management, erosion and salinization, deforestation, and the ongoing climatic change.

Earth's deserts are subdivided into desert and semidesert associations, to distinguish those with expanses of bare ground from those covered by xerophytic plants of various types. The two broad associations are further separated into warm deserts, principally tropical and subtropical, and cold deserts, principally midlatitude.

Warm Desert and Semidesert
Earth's **warm desert and semidesert** biomes are caused by the presence of dry air and low precipitation from subtropical high-pressure cells. These areas are very dry, as evidenced by the Atacama Desert of northern Chile, where only a minute amount of rain has ever been recorded—a 30-year annual average of only 0.05 cm (0.02 in.)! Like the Atacama, the true deserts of the Köppen *desert BW* classification are under the influence of the descending, drying, and stable air of high-pressure systems from 8 to 12 months of the year.

Remember that these dry regions are defined by low amounts of precipitation that fail to satisfy high amounts of potential evapotranspiration. Deserts receive precipitation that is less than one-half of potential evapotranspiration. Semiarid *steppe BS* climates receive precipitation that is more than one-half of annual potential evapotranspiration.

Vegetation ranges from almost none in the arid deserts to numerous xerophytic shrubs, succulents, and thorn tree forms. The lower Sonoran Desert of southern Arizona is a warm desert (Figure 20.17). This desert landscape features

FIGURE 20.17 Sonoran Desert scene.
Characteristic vegetation in the Lower Sonoran Desert west of Tucson, Arizona (32° N; elevation 900 m, or 2950 ft). [Photo by author.]

the unique saguaro cactus (*Carnegiea gigantea*), which, if undisturbed, grows to many meters in height and up to 200 years in age. First blooms do not appear until it is 50 to 75 years old!

A few of the subtropical deserts—such as those in Chile, Western Sahara, and Namibia—are right on the sea coast and are influenced by cool offshore ocean currents. As a result, these true deserts experience summer fog that mists the plant and animal populations with needed moisture.

The equatorward margin of the subtropical high-pressure cell is a region of transition to savanna, thorn tree, scrub woodland, and tropical seasonal forest. Poleward of the warm deserts, the subtropical cells shift to produce the Mediterranean dry summer regime along west coasts and may grade into cool deserts elsewhere.

Cold Desert and Semidesert The **cold desert and semidesert** biomes tend to occur at higher latitudes. Here, seasonal shifting of subtropical high pressure is of some influence less than 6 months of the year. Interior locations are dry because of their distance from moisture sources or their location in rain-shadow areas on the lee side of mountain ranges; examples are the Sierra Nevada of the western United States, the Himalayas, and the Andes. The combination of interior location and rain-shadow positioning produces the cold deserts of the Great Basin of western North America.

Winter snows occur in the cold deserts, but generally they are light. Summers are hot, with highs from 30° to 40°C (86° to 104°F). Nighttime lows, even in the summer, can cool 10–20 C° (18–36 F°) from the daytime high. The dryness, generally clear skies, and sparse vegetation lead to high radiative heat loss and cool evenings. Many areas of these cold deserts that are covered by sagebrush and scrub vegetation were actually dry shortgrass regions in the past, before extensive grazing altered their ecology. The deserts in the upper Great Basin today are the result of more than a century of such cultural practices.

Arctic and Alpine Tundra

The **arctic tundra** is found in the extreme northern area of North America and Russia, bordering on the Arctic Ocean and generally north of the 10°C (50°F) isotherm for the warmest month. Daylength varies greatly throughout the year, seasonally changing from almost continuous day to continuous night. Winters in this biome, *tundra ET* climate classification, are long and cold; cool summers are brief. The region, except for a few portions of Alaska and Siberia, was covered by ice during all of the Pleistocene glacial episodes.

Intensely cold continental polar air masses and stable high-pressure anticyclones govern tundra winters. A growing season of sorts lasts only 60–80 days, and even then frosts can occur at any time. Vegetation is fragile in this flat, treeless world; soils are poorly developed periglacial surfaces, which are underlain by permafrost. In the summer

FIGURE 20.18 Siberian tundra.
Tundra on the Kamchatka Peninsula, Russia. The Uzon Caldera is in the center-background mountain range. [Photo by Wolfgang Kaehler Photography.]

months, only the surface horizons thaw, thus producing a mucky surface of poor drainage. Roots can penetrate only to the depth of thawed ground, usually about a meter. The surface is shaped by freeze–thaw cycles that create the frozen ground phenomena discussed in Chapter 17.

Tundra vegetation consists of low, ground-level herbaceous plant species, such as sedges, mosses, arctic meadow grass, snow lichen, and some woody dwarf willow (Figure 20.18). Owing to the short growing season, some perennials form flower buds one summer and open them for pollination the next. Animals of the tundra include musk ox, caribou, reindeer, rabbit, ptarmigans, lemmings, and other small rodents (important food for the larger carnivores), wolf, fox, weasel, Snowy Owls, polar bears, and, of course, mosquitoes. The tundra is an important breeding ground for geese, swans, and other waterfowl.

Alpine tundra is similar to arctic tundra, but it can occur at lower latitudes because it is associated with high elevation. This biome usually is described as above the timberline, that elevation above which trees cannot grow. Timberlines increase in elevation equatorward in both hemispheres. Alpine tundra communities occur in the Andes near the equator, the White Mountains of California, the Colorado and Canadian Rockies, the Alps, and Mount Kilimanjaro of equatorial Africa, as well as mountains from the Middle East to Asia.

Alpine meadows feature grasses, herbaceous annuals (small plants), and stunted shrubs, such as willows and heaths. Because alpine locations are frequently windy sites, many plants appear to have been sculpted by the wind. Alpine tundra can experience permafrost conditions. Characteristic fauna include mountain goats, bighorn sheep, elk, and vols (Figure 20.19).

FIGURE 20.19 Alpine tundra conditions.
An alpine tundra and grazing mountain goats near Mount Evans, Colorado, at 3660 m (12,000 ft) elevation. [Photo by Bobbé Christopherson.]

Because the tundra biome is of such low productivity, it is fragile. Disturbances such as tire tracks, hydroelectric projects, and mineral exploitation leave marks that persist for hundreds of years. As development continues, the region will face even greater challenges from oil spills, contamination, and disruption (News Report 20.3).

News Report 20.3

ANWR Faces Threats

Planning continues for oil exploration and development of the Arctic National Wildlife Refuge (ANWR) on Alaska's North Slope. The May 2001 U.S. *National Energy Policy* recommends oil exploration and development in ANWR, using the latest environmentally sensitive techniques. Although the U.S. Senate defeated the plan in 2002—a tentative step.

This pristine wilderness is above the Arctic Circle, bordering on the Beaufort Sea and adjoining the Yukon Territory of Canada (Figure 1). The refuge area sustains almost 200,000 caribou, polar and grizzly bears, musk oxen, and wolves. Some have referred to it as "America's Serengeti," given the annual migration of hundreds of thousands of large animals.

The ANWR remains the only portion of Alaska's Arctic coast that is not open to oil and gas exploration at this time—the other 90% is open. But controversy over this refuge continues, as political and corporate pressures mount to begin oil exploration. In 1995, Congress considered potential ANWR lease revenues in its Budget Resolution, implying a green light for development. A presidential veto stopped this Budget Reconciliation Act. A 1998 USGS assessment disclosed that the oil under the Arctic Refuge coastal plain is likely held in many small reservoirs. Recovering this

oil would require extensive alteration of the Arctic National Wildlife Refuge landscape, a point disputed in the *National Energy Policy.*

Oil extraction could be devastating to the fragile tundra, according to a USGS study released in 2002. The cost of development would price the oil at well over $30 per barrel, far above current market prices. The estimated oil reserve in the ANWR, approximately 3.2 billion barrels (less than a 5-month supply at present U.S. demand levels),

could be offset simply by a small increase in U.S. automobile efficiency. However, 2002 marked another year when mileage efficiency decreased. In 2002, federal funding toward improving automobile mileage efficiency was formally dropped in favor of fuel cell research for the future. Clearly, economic ventures should be weighed against the ecosystem itself, its limitations and its uniqueness, and a total assessment of all costs to the consumer.

FIGURE 1 Arctic National Wildlife Refuge.
In Alaska's Arctic National Wildlife Refuge, Mount Chamberlin, the second highest peak of the Brooks Range, overlooks tundra in the foreground. Is this region destined for petroleum exploration and development or for continued preservation as wilderness? [Photo by Scott T. Smith.]

Summary and Review—Terrestrial Biomes

● *Define* the concept of biogeographic realms of plants and animals and *define* ecotone, terrestrial ecosystem, and biome.

Earth is the only planet in the Solar System with a biosphere. An impressive feature of the living Earth is its diversity, which biogeographers categorize into discrete spatial biomes for analysis and study. The interplay among supporting physical factors within Earth's ecosystem determines the distribution of plant and animal communities. A **biogeographic realm** of plants and animals is a geographic region in which a group of species evolved. This recognition was a start at understanding distinct regions of flora and fauna and the broader pattern of terrestrial ecosystems. A boundary transition zone adjoining ecosystems is an **ecotone**.

A **terrestrial ecosystem** is a self-sustaining association of plants and animals and their abiotic environment that is characterized by specific plant formation classes. A **biome** is a large, stable ecosystem characterized by specific plant and animal communities. Biomes carry the name of the dominant vegetation because it is the most easily identified feature: forest, savanna, grassland, shrubland, desert, tundra.

biogeographic realm (p. 626)
ecotone (p. 626)
terrestrial ecosystem (p. 626)
biome (p. 627)

1. Reread the opening quotation and analysis in this chapter. What clue does it give you to the path ahead for Earth's forests? Is our future direction controllable? Explain.

2. What is a biogeographic realm? How is the world subdivided according to plant and animal types?

3. Describe a transition zone between two ecosystems. How wide is an ecotone? Explain.

4. Define biome. What is the basis of the designation?

● *Define* six formation classes and the life-form designations and *explain* their relationship to plant communities.

Biomes are divided into more specific vegetation units called **formation classes**. The structure and appearance of the vegetation is described: rain forest, needleleaf forest, Mediterranean shrubland, arctic tundra, and so forth. Specific life-form designations include trees, lianas, shrubs, herbs, bryophytes, epiphytes (plants growing above ground on other plants), and thallophytes (lacking true leaves, stems, or roots, including bacteria, fungi, algae, and lichens).

formation classes (p. 628)

5. Distinguish between formation classes and life-form designations as a basis for spatial classification.

● *Describe* 10 major terrestrial biomes and *locate* them on a world map.

Biomes are Earth's major terrestrial ecosystems, each named for its dominant plant community. The 10 major biomes are generalized from numerous formation classes that describe vegetation. Ideally, a biome represents a mature community of natural vegetation. In reality, few undisturbed biomes exist in the world, for most have been modified by human activity. Many of Earth's plant and animal communities are experiencing an accelerated rate of change that could produce dramatic alterations within our lifetime.

For an overview of Earth's 10 major terrestrial biomes and their vegetation characteristics, soil orders, Köppen climate designation, annual precipitation range, temperature patterns, and water balance characteristics, please review Table 20.1.

equatorial and tropical rain forest (p. 631)
tropical seasonal forest and scrub (p. 636)
tropical savanna (p. 636)
midlatitude broadleaf and mixed forest (p. 640)
needleleaf forest (p. 641)
boreal forest (p. 641)
taiga (p. 641)
montane forest (p. 641)
temperate rain forest (p. 642)
Mediterranean shrubland (p. 642)
chaparral (p. 643)
midlatitude grasslands (p. 643)
warm desert and semidesert (p. 645)
cold desert and semidesert (p. 646)
arctic tundra (p. 646)
alpine tundra (p. 646)

6. Using the integrative chart in Table 20.1 and the world map in Figure 20.3, select any two biomes and study the correlation of vegetation characteristics, soil, moisture, and climate with their spatial distribution. Then, contrast the two using each characteristic.

7. Describe the equatorial and tropical rain forests. Why is the rainforest floor somewhat clear of plant growth? Why are logging activities for specific species so difficult there?

8. What issues surround the deforestation of the rain forest? What is the impact of these losses on the rest of the biosphere? What new threat to the rain forest has emerged?

9. What do caatinga, chaco, brigalow, and dornveld refer to? Explain.

10. Describe the role of fire or fire ecology in the tropical savanna biome and the midlatitude broadleaf and mixed forest biome.

11. Why does the northern needleleaf forest biome not exist in the Southern Hemisphere? Where is this biome located in the Northern Hemisphere, and what is its relationship to climate type?

12. In which biome do we find Earth's tallest trees? Which biome is dominated by small, stunted plants, lichens, and mosses?

13. What type of vegetation predominates in the Mediterranean dry summer climates? Describe the adaptation necessary for these plants to survive.

14. What is the significance of the 98th meridian in terms of North American grasslands? What types of inventions enabled humans to cope with the grasslands?

15. Describe some of the unique adaptations found in a desert biome.

16. What is desertification (review from Chapter 15 and this chapter)? Explain its impact.

17. What physical weathering processes are specifically related to the tundra biome? What types of plants and animals are found there?

● *Relate* **human impacts, real and potential, to several of the biomes.**

The equatorial and tropical rain forest biome is undergoing rapid deforestation. Because the rain forest is Earth's most diverse biome and is important to the climate system, such losses are creating great concern among citizens, scientists, and nations. Efforts are under way worldwide to set aside and protect remaining representative sites within most of Earth's principal biomes. These biosphere reserves are coordinated by the Man and the Biosphere (MAB) Programme of UNESCO. Nearly 300 such biosphere reserves, covering some 12 million hectares (30 million acres), are now operated voluntarily in 76 countries.

18. What is the relationship between island biogeography and biosphere reserves? Describe a biosphere reserve. What are the goals?

19. Compare the map in Figure 20.3 with the composite satellite image inside the front cover of this text. What correlations can you make between the local summertime portrait of Earth's biosphere and the biomes identified on the map?

NetWork

The *Geosystems* Home Page provides on-line resources for this chapter on the World Wide Web. To begin: Once on the Home Page, click on this textbook, scroll the Table of Contents menu, select this chapter, and click "Begin." You will find self-tests that are graded, review exercises, specific updates for items in the chapter, and many links to interesting related pathways on the Internet under "Destinations." *Geosystems* is at **http://www.prenhall.com/christopherson.**

Critical Thinking

A. Given the information presented in this chapter about deforestation in the tropics, assess the present situation for yourself. What are the main issues? What natural assets are at stake? Natural resources vs. sovereign state rights? How are global biodiversity and greenhouse warming related to these issues? What is the perspective of the less-developed countries that possess most of the rain forest? What is the perspective of the more-developed countries and their transnational corporations and millions of environmentally active citizens? What kind of action plan would you cast to accommodate all parties? How would you proceed?

B. Using Figure 20.3 (biomes), Figure 10.5 (climates), Figures 10.2 and 9.6 (precipitation), and Figure 8.2 (air masses), and the printed graphic scales on these four maps, consider the following hypothetical. Assume a northward climatic shift in the United States and Canada of 500 km (310 mi) (in other words move North America 500 km north). Describe your analysis of conditions through the Midwest from Texas to the prairies of Canada. Describe your analysis of conditions from New York through New England and into the Maritime provinces. What economic dislocations and relocations do you envision?

Stark desert beauty reminds us of Earth's great diversity, majesty, and the uniqueness of place. Civilizations and technological incarnations may come and go but the biosphere is resilient and persists. In Totem Pole and the Yei-Bi-Chei dancers formation in Monument Valley Tribal Park, we visualize the beauty of the Home Planet. [Photo by author.]

21

Earth and the Human Denominator

Key Learning Concepts

After reading the chapter, you should be able to:

- *Determine* an answer for Carl Sagan's question, "Who speaks for Earth?"
- *Describe* the growth in human population and *speculate* on possible future trends.
- *Analyze* "An Oily Bird" and *relate* your analysis to energy consumption patterns in the United States and Canada.
- *List* the subjects of recent environmental agreements, conventions, and protocols and *relate* them to physical geography and Earth systems science (geosystems).
- *List* 12 paradigms for the 21st century.
- *Appraise* your place in the biosphere and *realize* your physical identity as an Earthling.

During my space flight, I came to appreciate my profound connection to the home planet and the process of life evolving in our special corner of the Universe, and I grasped that I was part of a vast and mysterious dance whose outcome will be determined largely by human values and actions.*

Earth can be observed from profound vantage points, as this astronaut experienced on the 1969 *Apollo IX* mission. Our vantage point in this book is that of physical geography. We examine Earth's many systems: its energy, atmosphere, winds, ocean currents, water, weather, climate, endogenic and exogenic systems, soils, ecosystems, and biomes. This exploration has led us to an examination of the planet's most abundant large animal, *Homo sapiens*.

*Rusty Sweickart, "Our backs against the bomb, our eyes on the stars," *Discovery*, July 1987, p. 62.

We stand in the first few years of the 21st century and a new calendar millennium. The 21st century will be an adventure for the global society, historically unparalleled in experimenting with Earth's life-supporting systems. You will spend the majority of your life in this century. What preparations and "future thinking" are we doing to understand all that is to occur?

In his 1980 book and public television series, *Cosmos*, astronomer Carl Sagan asked:

> What account would we give of our stewardship of the planet Earth? We have heard the rationales offered by the nuclear superpowers. We know who speaks for the nations. But who speaks for the human species? Who speaks for Earth?*

Indeed, who does speak for Earth? We might answer: Perhaps we physical geographers, and other scientists who have studied Earth and know the operations of the global ecosystem, should speak for Earth. However, some might say that questions of technology, environmental politics, and future thinking belong outside of science, and that our job is merely to learn how Earth's processes work and to leave the spokesperson's role to others. Biologist–ecologist Marston Bates addressed this line of thought in 1960:

> Then we came to humans and their place in this system of life. We could have left humans out, playing the ecological game of "let's pretend humans don't exist." But this seems as unfair as the corresponding game of the economists, "let's pretend nature doesn't exist." The economy of nature and the ecology of humans are inseparable and attempts to separate them are more than misleading, they are dangerous. Human destiny is tied to nature's destiny and the arrogance of the engineering mind does not change this. Humans may be a very peculiar animal, but they are still a part of the system of nature.†

A fact of life is that Earth's more-developed countries (MDCs), through their economic dominance, speak for the billions who live in less-developed countries (LDCs). A current reality is that the U.S. defense budget alone is equal to the gross domestic product (GDP) of Russia and its 145 million people. The gross state product of California's 33 million people is greater, by more than $40 billion, than the GDP of China and its 1.27 billion people! The scale of disparity on our home planet is difficult to comprehend.

The fate of traditional modes of life may rest in some distant financial capital. But, economics aside, the reality is that the remote lands of Siberia are linked by Earth systems to the Pampas of Argentina to the Great Plains in North America and to those harvesting grain by hand in the remote Pamirs of Tajikistan (Figure 21.1).

To understand these linkages among Earth's myriad systems is our quest in *Geosystems* (Figure 21.2). We explored the atmosphere, hydrosphere, lithosphere, and biosphere, synthesizing operating systems into the web of life. We now see global linkages among physical and living systems and how actions in one place can affect change elsewhere. In this sense Earth can be compared to a spaceship. In the same way the members of a crew aboard the Space Shuttle are inextricably linked to each other's lives and survival, we too are each connected through the operation of planetary systems. Admittedly, the pace of popular culture makes these connections difficult to perceive.

Growing international environmental awareness in the public sector is gradually prodding governments into action. People, when informed, generally favor environmental protection and public health progress over economic interests. A survey by Louis Harris and Associates confirmed these opinion trends. Majorities in 22 nations are willing to "endorse environmental protection at the risk of

FIGURE 21.1 Lands distant from the global centers of power.
These lands were depopulated under Stalin when they were part of the former USSR. Only in the last decade or so have people returned to this traditional rural landscape in the Pamirs of Tajikistan, beginning life again, so distant from the more-developed countries. [Photo by Stephen F. Cunha.]

*C. Sagan, *Cosmos* (New York: Random House, 1980), p. 329.
†M. Bates, *The Forest and the Sea* (New York: Random House, 1960), p. 247.

Energy-Atmosphere System

Water, Weather and Climate Systems

Geosystems: Our "Sphere of Contents"

Solar energy

Atmosphere

Hydrosphere

Lithosphere

Heat
energy

PART 1: Chapters 2–6

PART 2: Chapters 7–10

Soils, Ecosystems, and Biomes

PART 4: Chapters 18–20

FIGURE 21.2 The spheres within
Geosystems.
Through this text we covered the atmosphere, hydrosphere, lithosphere,
and the synthesizing biosphere—the
culmination of life-sustaining interactions. The systems approach shows us
the flow of energy and matter and the
sequence of events through time, within each sphere and among the spheres
through their complex linkages. Ultimately, *Geosystems* describes the support systems of life. [Atmosphere and
hydrosphere photos by author; lithosphere and biosphere photos by Bobbé
Christopherson.]

Earth-Atmosphere Interface

PART 3: Chapters 11–17

slowing down economic growth." The public seems to
know they are saving money as consumers if things are
done right with the environment.

The Human Count and
the Future

Because human influence is pervasive, we consider the totality of our impact the *human denominator.* Just as the denominator in a fraction tells how many parts a whole is

divided into, so the growing human population and the increasing demand for resources and rising planetary impact
suggest how much the whole Earth system must adjust. Yet,
Earth's resource base remains relatively fixed.

  The human population of Earth passed 6 billion in August 1999; more people are alive today than at any previous point in the planet's long history, unevenly arranged in
192 countries and numerous colonies. In the year leading
up to this milestone, 83,000,000 more people were added—
this is 227,000 a day, or a new U.S. population every 3 years.
Over the span of human history these billion-mark mile-
stones are occurring at closer intervals (Figure 21.3).

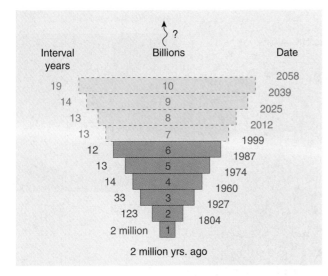

FIGURE 21.3 Human population growth.
The number of years required for the human population to add 1 billion to the count grows shorter and shorter. The new century should mark a slowing of this history of growth if policy actions are taken. Note the population forecasts for the next half century.

Table 21.1	Planetary Population Increase Rates		
	Birth Rate	Death Rate	Natural Increase (%)
World—2001	**22**	**9**	**1.3** (dbl. 47 yrs)
More developed	**11**	**10**	**0.1** (dbl. 809 yrs)
United States	15	9	0.6
Japan	9	8	0.2
United Kingdom	12	11	0.1
Less developed	**25**	**8**	**1.7** (dbl. 45 yrs)
LDC exclu. China	**28**	**9**	**1.9** (dbl. 39 yrs)
Mexico	24	5	2.0
Nigeria	41	14	2.8
India	26	9	1.8

Source: World Population Data Sheet 2001. Population Reference Bureau, Washington, DC, http://www.prb.org.

Thirty-seven percent of Earth's population live in just two countries (21% in China and 17% in India—2.3 billion people combined). Moreover, we are a young planetary population, with some 30% of those now alive under the age of 15 (2001 data from Population Reference Bureau and U.S. Bureau of Census *POPClock Projection:* http://www.census.gov/cgi-bin/popclock).

The present annual global *natural increase* in population is 1.3%, which is the annual difference between the *crude birth rate* (per thousand population) and *crude death rate* (per thousand population). This natural increase, if unchanged, will produce a doubling of world population in 47 years. Growth is not uniformly distributed. *Virtually all new population growth is in the less-developed countries*, which now possess 80.6%, or 4.94 billion, of the total population; the MDCs have the other 19.4%, or 1.2 billion (Figure 21.4). Table 21.1 illustrates this natural increase principle and the differences found in the MDCs and LDCs.

If you consider only the population count, the MDCs do not have a population growth problem. In fact, some European countries are actually declining in growth or are near replacement levels. However, people in the developed world produce a greater impact on the planet per person and therefore constitute a population impact crisis. An equation expresses this impact concept:

$$\text{Planetary impact } (I) = P \cdot A \cdot T$$

where *P* is *population*; *A* is *affluence*, or consumption per person; and *T* is *technology*, or the level of environmental impact per unit of production. The United States and Canada, with about 4.5% of the world's population, produce more than 50% of the world's gross domestic product (almost $10 trillion a year) and use more than double the energy per capita of Europeans, more than 7 times Latin Americans, 10 times Asians, and 20 times Africans. The United States and Canada produce 22 tons of carbon dioxide emissions per person per year, more than 7 times Latin Americans, 18 times Africans, 2.3 times Europeans, and 10 times Asians.

Therefore, consideration by people in the MDCs of the state of Earth systems, natural resources, and sustainability of current practices is critical. Earth systems science is here to provide a high level of planetary monitoring and analysis to assist this process. The MDCs not only should affect change in their own planetary impact but could provide direction for the LDCs, which have a right to move along the road of progress and improve their living conditions. Let us examine this idea of global impact.

FIGURE 21.4 Dhaka, Bangladesh.
The Bangladesh capital typifies the plight of the less-developed world as millions converge on cities to find work and more modern lifestyles. [Photo by Bruce Bander/Photo Researchers, Inc.]

An Oily Bird

At first glance, the chain of events that exposes wildlife to oil contamination seems to stem from a technological problem. An oil tanker splits open at sea and releases its petroleum cargo, which is moved by ocean currents toward shore, where it coats coastal waters, beaches, and animals. In response, concerned citizens mobilize and try to save as much of the spoiled environment as possible (Figure 21.5). But the real problem goes far beyond the physical facts of the spill.

In Prince William Sound off the southern coast of Alaska in clear weather and calm seas, the *Exxon Valdez*, a single-hulled supertanker operated by Exxon Corporation, struck a reef. The tanker spilled 42 million liters (11 million gallons) of oil. It took only 12 hours for the *Exxon Valdez* to empty its contents, yet a complete cleanup is impossible and costs and private claims exceeded $15 billion.

Eventually, more than 2400 km (about 1500 mi) of sensitive coastline was ruined for years to come, affecting three national parks and eight other protected areas. For perspective, had this spill occurred farther south, every beach and bay along the Pacific Coast from southern Oregon to the Mexican border would have been blackened (Figure 21.6).

The death toll of animals was massive: At least 5000 sea otters died, or about 30% of resident otters; about 300,000 birds and uncounted fish, shellfish, plants, and aquatic microorganisms also perished. Sublethal effects,

FIGURE 21.5 An oily bird.
A Western Grebe contaminated with oil from the *Exxon Valdez* tanker accident in Prince William Sound, Alaska. This oily bird is the result of a long chain of events and mistakes. [Photo by Geoffrey Orth/SIPA Press.]

namely mutations, now are appearing in fish. The Pacific herring is still in significant decline, as are the harbor seals; other species are in recovery, such as the Bald Eagle and Common Murre. More than a decade later, oil remains in mudflat and marsh soils and still can be scooped from beneath rocks along shorelines.

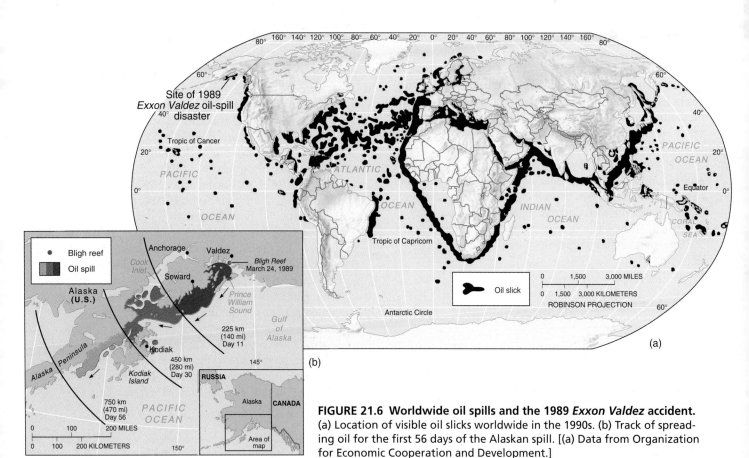

FIGURE 21.6 Worldwide oil spills and the 1989 *Exxon Valdez* accident.
(a) Location of visible oil slicks worldwide in the 1990s. (b) Track of spreading oil for the first 56 days of the Alaskan spill. [(a) Data from Organization for Economic Cooperation and Development.]

The immediate problem of cleaning oil off a bird symbolizes a national and international concern with far-reaching spatial significance. And while we search for answers, oil slicks continue their contamination. In just 1 year following the *Exxon Valdez* disaster, nearly 76 million liters (20 million gallons) of oil were spilled in 10,000 accidents worldwide, as evidenced on the startling map in Figure 21.6a. This is an average of 27 accidents a day, ranging from a few disastrous spills to numerous small ones. There were 50 spills equal to the *Exxon Valdez* or larger since 1970. In addition to oceanic oil spills, people improperly dispose of crankcase oil from their automobiles in a volume that annually exceeds these tanker spills!

The immediate effect of global oil spills on wildlife is contamination and death. But the issues involved are much bigger than dead birds. Let's ask some fundamental questions about the *Exxon Valdez* accident:

- Why was the oil tanker there in the first place?
- Why are only a few of the 28 tankers that traverse the Prince William Sound of the safer double-hulled design?
- Why is petroleum imported into the continental United States from Alaska in such enormous quantity? Is the demand for petroleum products based on real need and the operation of efficient systems?
- The U.S. demand for oil is higher per capita than the demand of any other country. Efficiencies in the U.S. transportation sector again fell in the 2002 fleet of cars and trucks, with SUVs (light trucks) leading in sales. Why?

A combination of economic growth, waste, low prices, and a lack of alternatives has spurred the demand for petroleum. In addition, our land-use policies continue to foster a diffuse sprawl of our population, thereby adding stress to transportation systems. All of these factors increased U.S. demand for oil to more than 6.8 billion barrels a year alone, given a domestic reserve of 110 billion barrels.

Yet, hypocrisy is apparent in our outrage over energy politics and oil-spill accidents. We continue to consume gasoline at record levels in inefficient vehicles, thus creating the demand for oil imports. The task of physical geography is to analyze all the spatial aspects of these events in the environment and the ironies that are symbolized by an oily bird.

The Need for International Cooperation

The idea for a global meeting on the environment was put forward at the 1972 U.N. Conference on the Human Environment held in Stockholm. The U.N. General Assembly in 1987 achieved a landmark in global planning by agreeing to hold the Earth Summit. *Our Common Future*, a book written at the time, set the tone for the 1992 Earth Summit:

The Earth is one but the world is not. We all depend on one biosphere for sustaining our lives. Yet each community, each country, strives for survival and prosperity with little regard for its impact on others. Some consume the Earth's resources at a rate that would leave little for future generations. Others, many more in number, consume far too little and live with the prospect of hunger, squalor, disease, and early death.[*]

The setting in Rio de Janeiro for the 1992 Earth Summit was ironic, for many of the very problems discussed at the conference were evident on the city's streets, with their abundant air pollution, water pollution, toxics, noise, wealth and poverty, and daily struggle for health and education.

Maurice F. Strong, a Canadian and Secretary-General of the UNCED, summarized in his conference address:

The people of our planet, especially our youth and the generations which follow them, will hold us accountable for what we do or fail to do at the Earth Summit in Rio. Earth is the only home we have, its fate is literally in our hands. . . . The most important ground we must arrive at in Rio is the understanding that we are all in this together.

Five agreements were written at the Earth Summit: the *Climate Change Framework*, *Biological Diversity Treaty*, the *Management, Conservation, and Sustainable Development of All Types of Forests*, the *Earth Charter* (a nonbinding statement of 27 environmental and economic principles), and *Agenda 21 for Sustainable Development*. A product of the Earth Summit was the United Nations *Framework Convention on Climate Convention* (FCCC) that led to a series of *Convention of the Parties* (COP) meetings. The *Kyoto Protocol*, agreed to in 1997 to reduce global carbon emissions was finalized in Marrakech, Morocco, in 2001 at *COP-7* (Figure 21.7). As of 2002 the United States was the lone major holdout to the Kyoto Rulebook.

Asking whether the Earth Summit "succeeded" or "failed" is the wrong question. The occurrence of this largest-ever official gathering is a remarkable accomplishment. From the Earth Summit emerged a new organization—the U.N. Commission on Sustainable Development—to oversee the promises made in the five agreements. This momentum lead to Earth Summit 2002 (**http://www.earthsummit2002.org/**) in Johannesburg, South Africa, with an agenda including climate change, freshwater, gender issues, global public goods, HIV/AIDS, sustainable finance, and the five Rio Conventions.

Members of society must work to move the solutions for environmental and developmental problems off the bench and into play. We need international cooperation to consider our symbiotic relations with each other and with Earth's resilient, yet fragile, life-support systems (News Report 21.1).

---

[*]World Commission on Environment and Development, *Our Common Future* (Oxford, UK: Oxford University Press, 1987), p. 27.

A critical corollary to these international efforts is the linkage of academic disciplines. A positive step in that direction is the Earth systems science (geosystems) approach that synthesizes content from across the disciplines to create a holistic perspective. Exciting progress toward an integrated understanding of Earth's physical and biological systems is in progress.

More than a decade ago, *Time* magazine gave Earth an interesting honor and is today a time capsule of activist suggestions (Figure 21.8). As if a victim of short-lived fame, we wonder where Earth awareness fits into the global scheme of things today. Is there a followup in pop culture to making the cover of *Time*?

Twelve Paradigms for the 21st Century

As we conclude, a brief list of the dominant themes and patterns of concern for the 21st century seems appropriate. Hopefully these *paradigms* will provide a useful framework for brainstorming and discussion of the central issues that will affect us in the new century. The paradigms:

1. Population increases in the less-developed countries
2. Planetary impact per person (on the biosphere and resources; $I = P \cdot A \cdot T$)
3. Feeding the world's population

FIGURE 21.7 The Marrakech, Morocco, Climate Summit, November 2001.
The Marrakech Climate Summit (*COP-7*) voted to accept the terms of the Kyoto Protocol. A consensus was reached on the reduction of global carbon dioxide emissions and the Kyoto Rulebook. [Photo courtesy of IISD/ENB-Leila Mead.]

News Report 21.1

Gaia Hypothesis Triggers Debate

Some view Earth as one vast, self-regulating organism. The concept is one of global symbiosis, or mutualism. This controversial concept is called the *Gaia hypothesis* (Gaia was the Earth Mother goddess in ancient mythology). It was proposed in 1979 by James Lovelock, a British astronomer and inventor, and elaborated by American biologist Lynn Margulis.

Gaia is the ultimate synergistic relationship, in which the whole greatly exceeds the sum of the individual interacting components. The hypothesis contends that life processes control and shape Earth's inorganic physical and chemical processes, with the ecosphere so interactive that a very small mass can affect a very large mass. Thus, Lovelock and Margulis think that the material environment and the evolution of species are tightly joined; as species evolve through natural selection, they (in-

cluding us) in turn affect their environment. The present oxygen-rich composition of the atmosphere is given as proof of this coevolution of living and nonliving systems.

From the perspective of physical geography, the Gaia hypothesis permits a view of all Earth and the spatial interrelations among systems. In fact, such a perspective is necessary for analyzing specific environmental issues. Many variables interact synergistically, producing both wanted and unwanted results.

One disturbing aspect of this unity is that any biotic threat to the operation of an ecosystem tends to move toward extinction itself. This trend preserves the system overall. Earth-systems operation and feedback naturally tend to eliminate offensive members. The degree to which humans represent a planetary threat, then, becomes a topic of great concern, for Earth (Gaia) will pre-

vail, regardless of the outcome of the human experiment.

> The maladies of Gaia do not last long in terms of her life span. Anything that makes the world uncomfortable to live in tends to induce the evolution of those species that can achieve a new and more comfortable environment. It follows that, if the world is made unfit by what we do, there is the probability of a change in regime to one that will be better for life but not necessarily better for us.*
>
> The debate is vigorous regarding the true applicability of this hypothesis to nature, or whether it is true science at all. Regardless, it remains philosophically intriguing in its portrayal of the relationship between humans and Earth.

*J. Lovelock, *The Ages of Gaia—A Biography of Our Living Earth* (New York: Norton, 1988), p. 178.

FIGURE 21.8 Earth made the cover of *Time*.
More than a decade ago, global concerns about environmental impacts prompted *Time* magazine to deviate from its 60-year tradition of naming a prominent citizen as its person of the year, instead naming Earth the "Planet of the Year." The magazine devoted 33 pages to Earth's physical and human geography. Importantly, *Time* also offered positive policy strategies for consideration—an interesting time capsule for comparison with real events over the years. [January 2, 1989 issue, Copyright © 1989 The *Time* Inc. Magazine Company. Reprinted by permission.]

4. Global and national disparities of wealth and resource allocation
5. Status of women and children (health, welfare, rights)
6. Global climate change (temperatures, sea level, weather and climate patterns, disease, and diversity)
7. Energy supplies and energy demands; renewables and demand management
8. Loss of biodiversity (habitats, genetic wealth, and species richness)
9. Pollution of air, surface water (quality and quantity), groundwater, oceans, and land
10. The persistence of wilderness (biosphere reserves and biodiversity hot spots)
11. Globalization versus cultural diversity
12. Conflict resolution

Who Speaks for Earth?

Geographic awareness and education is an increasingly positive force on Earth. The National Geographic Society conducts an annual National Geography Bee to promote geography education and global awareness to millions of

FIGURE 21.9 Fifth Annual International Geography Olympiad, 2001.
Students from across the globe met in Vancouver, British Columbia, August 2001, for the fifth Geography Olympiad, sponsored by the National Geographic Society, with Alex Trebek of *Jeopardy!* fame as moderator. Here Alex congratulates the Canadian team on their second-place finish—the U.S placed first, the Hungarian team third in the final standings. The Olympiad is an indicator of growing international geographic awareness. [Photo by O. Louis Mazzatenta, 2001 National Geographic Society.]

6th to 8th graders, their parents, and teachers. The International Geography Olympiad is now an annual event (Figure 21.9). There are presently 60 geographic alliances in 47 states, which coordinate geographic education among teachers and students at all levels: K–12, community college, college, and university. People are learning more about Earth–human relations.

Yet, ideological and ethical differences still remain within society. This dichotomy was addressed by biologist Edward O. Wilson:

> The evidence of swift environmental change calls for an ethic uncoupled from other systems of belief. Those committed by religion to believe that life was put on Earth in one divine stroke will recognize that we are destroying the Creation; and those who perceive biodiversity to be the product of blind evolution will agree. . . . Defenders of both premises seem destined to gravitate toward the same position on conservation. . . . For what, in the final analysis, is morality but the command of conscience seasoned by a rational examination of consequences? . . . An enduring environmental ethic will aim to preserve not only the health and freedom of our species, but access to the world in which the human spirit was born.*

*E. O. Wilson, *The Diversity of Life* (Cambridge, MA: Harvard University Press, 1992), p. 351.

FIGURE 21.10 Secretary-General Kofi Annan addresses the AAG.
U.N. Secretary-General and Nobel Prize winner speaks to more than 3000 geographers at the Association of American Geographers Annual Meeting in New York City, March 1, 2001. [Photo courtesy of the Association of American Geographers, by Kevin J. McCormick.]

United Nations Secretary-General Kofi Annan, recipient of the 2001 Nobel Peace Prize, spoke to the Association of American Geographers annual meeting (Figure 21.10) on March 1, 2001, and offered us this thought:

The idea of interdependence is old hat to geographers, but for most people it is a new garment they are only now trying on for size. Getting it to fit—and getting it imprinted on the mental maps that guide our voices and our choices—is one of the crucial projects of human geography for the 21st century. I look forward to working with you in that all-important journey.

The late Carl Sagan asked, "Who speaks for Earth?" He answered with this perspective:

We have begun to contemplate our origins: starstuff pondering the stars; organized assemblages of ten billion billion billion atoms considering the evolution of atoms; tracing the long journey by which, here at least, consciousness arose. Our loyalties are to the species and the planet. We speak for Earth. Our obligation to survive is owed not just to ourselves but also to that Cosmos, ancient and vast, from which we spring.*

May we all perceive our spatial importance within Earth's ecosystems and do our part to maintain a life-supporting and sustaining Earth for ourselves and countless generations in the future.

*C. Sagan, *Cosmos* (New York: Random House, 1980), p. 345.

Critical Thinking and Learning

1. What part do you think technology, politics, and thinking about the future should play in science courses?

2. Assess population growth issues: the count, the impact per person, and future projection. What strategies do you see as important?

3. According to the discussion in the chapter, what worldwide factors led to the *Exxon Valdez* accident? Describe the complexity of that event from a global perspective. In your analysis, examine both supply-side (corporations and utilities) and demand-side (consumers) issues, as well as environmental and strategic factors. And, what about the oily bird?

4. What is meant by the Gaia hypothesis? Describe several concepts from this text that might pertain to this hypothesis.

5. Relate the content of the various chapters in this text to the integrative Earth systems science concept. Which chapters help you to better understand Earth–human relations and human impacts?

6. After examining the list of 12 paradigm issues for the 21st century, suggest items that need to be added to the list, omitted from the list, or expanded in coverage. Rearrange and organize the list as needed to match your concerns.

7. This chapter states that we already know many of the solutions to the problems we face. Why do you think these solutions are not being implemented at a faster pace?

8. Who speaks for Earth?

 ## NetWork

The *Geosystems* Home Page provides on-line resources for this chapter on the World Wide Web. To begin: Once on the Home Page, click on this textbook, scroll the Table of Contents menu, select this chapter, and click "Begin." You will find self-tests that are graded, review exercises, specific updates for items in the chapter, and many links to interesting related pathways on the Internet under "Destinations." *Geosystems* is at **http://www.prenhall.com/christopherson.**

Robert G. Bailey, Geographer and Ecoregions Author

Robert Bailey is a geographer who studies the spatial patterns of ecosystems. His three books and published maps are titled *Ecosystem Geography* (1996), *Ecoregions: The Ecosystem Geography of the Oceans and Continents* (1998), and *Ecoregion–Based Design for Sustainability* (2002), all published by Springer-Verlag in New York. His *Ecoregion Maps* of the United States and North America was published by the USDA in 1994, 1995, and 1997.

When Robert was a boy, his family made many camping trips around the West, but especially in Colorado. These were long driving trips from California. He remembered, "One time in the 1940s my mother took my sister and me all the way from California to Illinois. That trip exposed me to the land; it gave me a natural curiosity about landscapes."

Robert's undergraduate work was at California State University, Northridge, and he completed his Ph.D. at UCLA, where he examined landslides and related hazards in the Teton National Forest of northwestern Wyoming. "I looked at the factors that led to these landslides and developed a model that might help forecast landslide probabilities," Robert stated. "We tried to define what is commercial forest land. Up to that point everything was thrown in together. We started to recognize that some areas were unstable and prone to landslides or were unproductive and needed protection from disturbance."

"The studies are all based on spatial relationships," Robert explained, "why things are the way they are, and how they relate over space to other things. Landslide study requires a multifactor approach: slopes, geologic

FIGURE 1 Robert G. Bailey, Geographer and *Ecoregions* Author. Inventory and Monitoring Institute, U.S. Forest Service, Fort Collins, Colorado. [Photo from U.S. Forest Service.]

materials, climate, vegetation type, and the manipulation of the system by humans. That's what geography is all about, spatial synthesis."

Robert worked as a hydrologist in Wyoming and in the Lake Tahoe Basin of California. At Lake Tahoe he devised a land capability classification for land-use regulations that became the governing land-use ordinance in the basin. Robert's work helped limit development in the most sensitive areas: "We told them that if they didn't follow these capability classes that they would have trouble with erosion and other factors. The plan was controversial at the time." Here we see the ongoing argument: Are aspects of the environment in the public trust (a common resource pool) or do private rights take precedence (individual appropriation)?

The Tahoe project led to similar work with the U.S. Forest Service in Utah, where Robert expanded his work to a regional scale, synthesizing relationships among landforms, climate, soils, and plant and animal life, and

which resulted in a regional ecosystem map of the United States. In 1978 he moved to the Rocky Mountain Forest and Range Experiment Station in Fort Collins, where he worked on methodology and an integrated approach to classifying land as ecosystems. He has been the geographer and program manager of what is now called the Ecosystem Management Analysis Center, U.S. Forest Service, ever since (see http://www.fs.fed.us/institute). He also taught classes at universities in Utah and Colorado.

As for the next step, he offers, "GIS is an important tool to bring all this together," and, "Perhaps the ecoregions approach can be used to consider the potential effects related to global climate change." His emphatic final assessment is that geography matters. In an article that appeared in *American Forests* (May/June 1994), Robert was referred to as "a maverick geographer with a holistic vision helping the Forest Service and its fellow agencies figure out ecosystems."

Appendix A
Maps in This Text and Topographic Maps

Maps Used in This Text

Geosystems uses several map projections: Goode's homolosine, Robinson, and Miller cylindrical, among others. Each was chosen to best present specific types of data. **Goode's homolosine projection** is an interrupted world map designed in 1923 by Dr. J. Paul Goode of the University of Chicago. Rand McNally *Goode's Atlas* first used it in 1925. Goode's homolosine equal-area projection (Figure 1) is a combination of two oval projections (*homolo*graphic and *sinu*soidal projections).

Two equal-area projections are cut and pasted together to improve the rendering of landmass shapes. A *sinusoidal projection* is used between 40° N and 40° S latitudes. Its central meridian is a straight line; all other meridians are drawn as sinusoidal curves (based on sine-wave curves) and parallels are evenly spaced. A *Mollweide projection*, also called a *homolographic projection*, is used from 40° N to the North Pole and from 40° S to the South Pole. Its central meridian is a straight line; all other meridians are drawn as elliptical arcs and parallels are unequally spaced—farther apart at the equator, closer together poleward. This technique of combining two projections preserves areal size relationships, making the projection excellent for mapping spatial distributions when interruptions of oceans or continents do not pose a problem.

We use Goode's homolosine projection throughout this book. Examples include the world climate map and smaller climate type maps in Chapter 10, topographic regions and continental shields maps (Figures 12.3 and 12.4), world karst map (Figure 13.14), world sand regions and loess deposits (Figures 15.12 and 15.14) and the terrestrial biomes map in Chapter 20.

Another projection we use is the **Robinson projection**, designed by Arthur Robinson in 1963 (Figure 2). This projection is neither equal area nor true shape, but is a compromise between the two. The North and South Poles appear as lines slightly more than half the length of the equator; thus higher latitudes are exaggerated less than on other oval and cylindrical projections. Some of the Robinson maps employed include the latitudinal geographic zones map in Chapter 1 (Figure 1.13), daily net radiation map (Figure 2.11), the world temperature range map in Chapter 5 (Figure 5.17), the maps of lithospheric plates of crust and volcanoes and earthquakes in Chapter 11 (Figures 11.17 and 11.20), and the global oil spills map in Chapter 21 (Figure 21.6).

Another compromise map, the **Miller cylindrical projection**, is used in this text (Figure 3). Examples of this projection include the world time zone map (Figure 1.17), global temperature maps in Chapter 5 (Figures 5.14 and 5.16), two global pressure maps in Figure 6.11, and the global soil maps and soil order maps in Chapter 18 (Figure 18.9). This projection is neither true shape nor true area but is a compromise that avoids the severe scale distortion of the Mercator. The Miller projection frequently appears in world atlases. The American Geographical Society presented Osborn Miller's map projection in 1942.

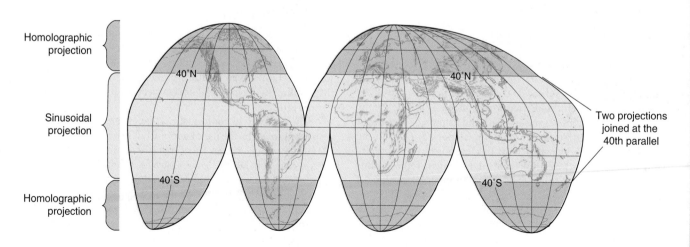

FIGURE 1 Goode's homolosine projection.
An equal-area map. [Copyright by the University of Chicago. Used by permission of the University of Chicago Press.]

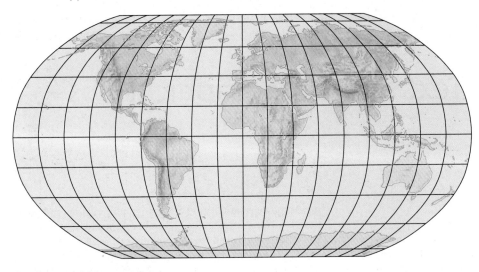

FIGURE 2 Robinson projection.
A compromise between equal area and true shape. [Developed by Arthur H. Robinson, 1963.]

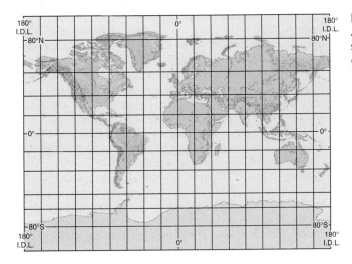

FIGURE 3 Miller cylindrical projection.
A compromise map projection between equal area and true shape. [Developed by Osborn M. Miller, American Geographical Society, 1942.]

Mapping, Quadrangles, and Topographic Maps

The westward expansion across the vast North American continent demanded a land survey for the creation of accurate maps. Maps were needed to subdivide the land and to guide travel, exploration, settlement, and transportation. In 1785 the Public Lands Survey System began surveying and mapping government land in the United States. In 1836 the Clerk of Surveys in the Land Office of the Department of the Interior directed public-land surveys. The Bureau of Land Management replaced this Land Office in 1946. The actual preparation and recording of survey information fell to the U.S. Geological Survey (USGS), also a branch of the Department of the Interior (see **http://mapping.usgs.gov/**).

In Canada, the National Resources Canada conducts the national mapping program. Canadian mapping includes base maps, thematic maps, aeronautical charts, federal topographic maps, and the National Atlas of Canada, now in its fifth edition (see **http://atlas.gc.ca/**).

Quadrangle Maps

The USGS depicts survey information on quadrangle maps, so called because they are rectangular maps with four corner angles. The angles are junctures of parallels of latitude and meridians of longitude rather than political boundaries. These quadrangle maps utilize the Albers equal-area projection, from the conic class of map projections.

The accuracy of conformality (shape) and scale of this base map is improved by the use of not one but two standard parallels. (Remember from Chapter 1 that standard lines are where the projection cone touches the globe's surface, producing greatest accuracy.) For the conterminous United States (the "lower 48"), these parallels are 29.5° N and 45.5° N latitude (noted on the Alber's projection shown in Figure 1.21c). The standard parallels shift for conic projections of Alaska (55° N and 65° N) and for Hawai'i (8° N and 18° N).

Because a single map of the United States at 1:24,000 scale would be more than 200 m wide (more than 600 ft), some system had to be devised for dividing the map into manageable size. Thus, a quadrangle system using latitude and longitude

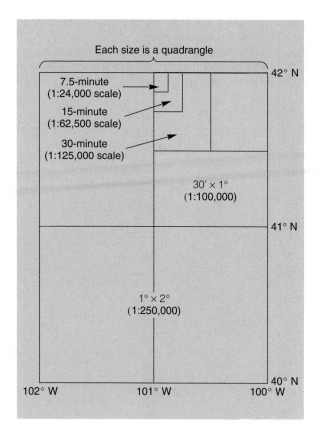

FIGURE 4 Quadrangle system of maps used by the USGS.

coordinates developed. Note that these maps are not perfect rectangles, because meridians converge toward the poles. The width of quadrangles narrows noticeably as you move north (poleward).

Quadrangle maps are published in different series, covering different amounts of Earth's surface at different scales. You see in Figure 4 that each series is referred to by its angular dimensions, which range from 1° to 2° (1:250,000 scale) to 7.5′ × 7.5′ (1:24,000 scale). A map that is one-half a degree (30′) on each side is called a "30-minute quadrangle," and a map one-fourth of a degree (15′) on each side is a "15-minute quadrangle" (this was the USGS standard size from 1910 to 1950). A map that is one-eighth of a degree (7.5′) on each side is a 7.5-minute quadrangle, the most widely produced of all USGS topographic maps, and the standard since 1950. The progression toward more-detailed maps and a larger-scale map standard through the years reflects the continuing refinement of geographic data and new mapping technologies.

The USGS National Mapping Program recently completed coverage of the entire country (except Alaska) on 7.5-minute maps (1 in. to 2000 ft, a large scale). It takes 53,838 separate 7.5-minute quadrangles to cover the lower 48 states, Hawai'i, and the U.S. territories. A series of smaller-scale, more-general 15-minute topographic maps offer Alaskan coverage.

In the United States, most quadrangle maps remain in English units of feet and miles. The eventual changeover to the metric system requires revision of the units used on all maps, with

the 1:24,000 scale eventually changing to a scale of 1:25,000. However, after completing only a few metric quads, the USGS halted the program in 1991. In Canada, the entire country is mapped at a scale of 1:250,000, using metric units (1.0 cm to 2.5 km). About half the country also is mapped at 1:50,000 (1.0 cm to 0.50 km).

Topographic Maps

The most popular and widely used quadrangle maps are **topographic maps** prepared by the USGS. An example of such a map is a portion of the Cumberland, Maryland, quad shown in Figure 5. You will find topographic maps throughout *Geosystems* because they portray landscapes so effectively. As examples, see Figures 13.15 and 13.16, karst landscapes and sinkholes near Orleans, Indiana, and Winter Park, Florida; Figure 14.7, river drainage patterns; Figure 14.21, river meander scars; Figure 15.18, an alluvial fan in Montana; Figure 17.3, glaciers in Alaska; and Figure 17.17, drumlins in New York.

A **planimetric map** shows the horizontal position (latitude/longitude) of boundaries, land-use aspects, bodies of water, and economic and cultural features. A highway map is a common example of a planimetric map.

A topographic map adds a vertical component to show topography (configuration of the land surface), including slope and relief (the vertical difference in local landscape elevation). These fine details are shown through the use of elevation contour lines (Figure 6). A *contour line* connects all points at the same elevation. Elevations are shown above or below a vertical datum, or reference level, which usually is mean sea level. The contour interval is the vertical distance in elevation between two adjacent contour lines (20 ft, or 6.1 m in Figure 6b).

The topographic map in Figure 6b shows a hypothetical landscape, demonstrating how contour lines and intervals depict slope and relief, which are the three-dimensional aspects of terrain. The pattern of lines and the spacing between them indicates slope. The steeper a slope or cliff, the closer together the contour lines appear—in the figure, note the narrowly spaced contours that represent the cliffs to the left of the highway. A wider spacing of these contour lines portrays a more gradual slope, as you can see from the widely spaced lines on the beach and to the right of the river valley.

In Figure 7 are the standard symbols commonly used on these topographic maps. These symbols and the colors used are standard on all USGS topographic maps: black for human constructions, blue for water features, brown for relief features and contours, pink for urbanized areas, and green for woodlands, orchards, brush, and the like.

The margins of a topographic map contain a wealth of information about its concept and content. In the margins of topographic maps, you find the quadrangle name, names of adjoining quads, quad series and type, position in the latitude–longitude and other coordinate systems, title, legend, magnetic declination (alignment of magnetic north) and compass information, datum plane, symbols used for roads and trails, the dates and history of the survey of that particular quad, and more.

Topographic maps may be purchased directly from the USGS or Centre for Topographic Information, NRC (**http://maps.nrcan.gc.ca/main.html**). Many state geological survey offices, national and state park headquarters, outfitters, sports shops, and bookstores also sell topographic maps to assist people in planning their outdoor activities.

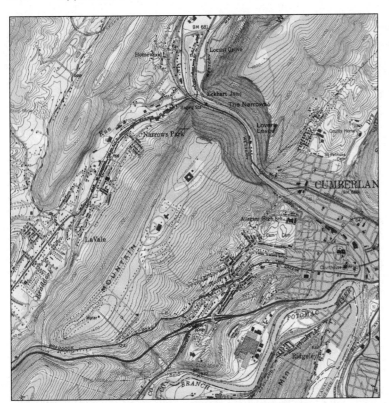

FIGURE 5 An example of a topographic map from the Appalachians.
Cumberland, MD, PA, WV 7.5-minute quadrangle topographic map prepared by the USGS. Note the water gap through Haystack Mountain.

(a)

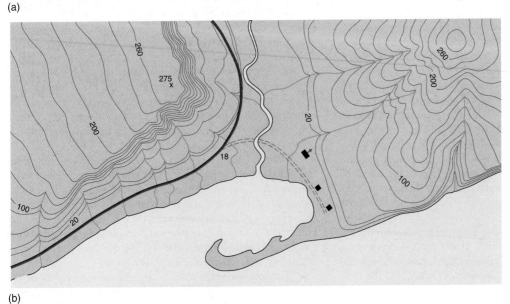

(b)

FIGURE 6 Topographic map of a hypothetical landscape.
(a) Perspective view of a hypothetical landscape. (b) Depiction of that landscape on a topographic map. The contour interval on the map is 20 feet (6.1 m). [After the U.S. Geological Survey.]

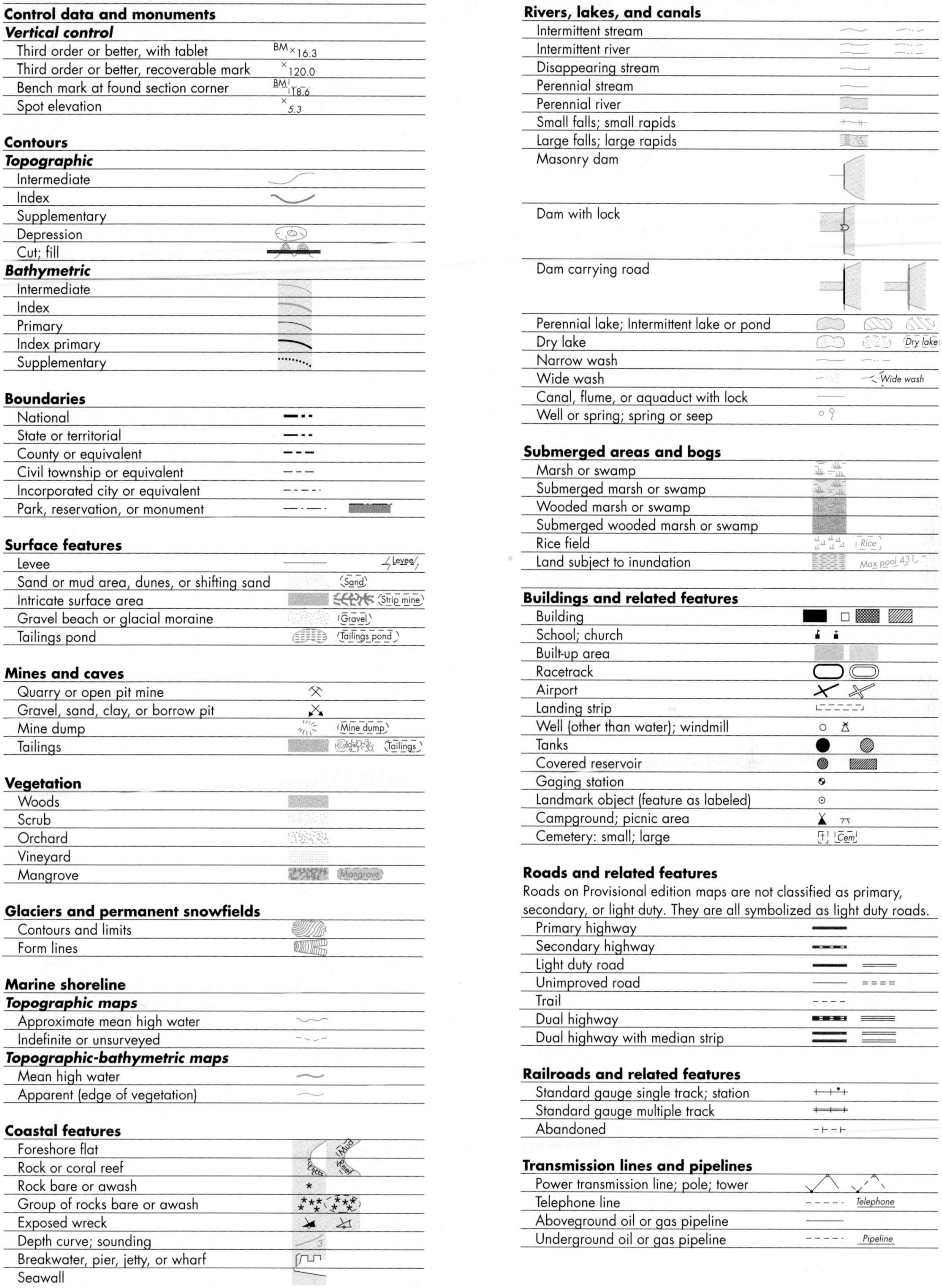

Control data and monuments
Vertical control

Third order or better, with tablet	BM ×16.3
Third order or better, recoverable mark	×120.0
Bench mark at found section corner	BM ×18.6
Spot elevation	×5.3

Contours
Topographic

Intermediate	
Index	
Supplementary	
Depression	
Cut; fill	

Bathymetric

Intermediate	
Index	
Primary	
Index primary	
Supplementary	

Boundaries

National	
State or territorial	
County or equivalent	
Civil township or equivalent	
Incorporated city or equivalent	
Park, reservation, or monument	

Surface features

Levee	Levee
Sand or mud area, dunes, or shifting sand	(Sand)
Intricate surface area	(Strip mine)
Gravel beach or glacial moraine	(Gravel)
Tailings pond	(Tailings pond)

Mines and caves

Quarry or open pit mine	
Gravel, sand, clay, or borrow pit	
Mine dump	(Mine dump)
Tailings	(Tailings)

Vegetation

Woods	
Scrub	
Orchard	
Vineyard	
Mangrove	Mangrove

Glaciers and permanent snowfields

Contours and limits	
Form lines	

Marine shoreline
Topographic maps

Approximate mean high water	
Indefinite or unsurveyed	

Topographic-bathymetric maps

Mean high water	
Apparent (edge of vegetation)	

Coastal features

Foreshore flat	
Rock or coral reef	
Rock bare or awash	
Group of rocks bare or awash	
Exposed wreck	
Depth curve; sounding	
Breakwater, pier, jetty, or wharf	
Seawall	

Rivers, lakes, and canals

Intermittent stream	
Intermittent river	
Disappearing stream	
Perennial stream	
Perennial river	
Small falls; small rapids	
Large falls; large rapids	
Masonry dam	
Dam with lock	
Dam carrying road	
Perennial lake; Intermittent lake or pond	
Dry lake	(Dry lake)
Narrow wash	
Wide wash	(Wide wash)
Canal, flume, or aquaduct with lock	
Well or spring; spring or seep	

Submerged areas and bogs

Marsh or swamp	
Submerged marsh or swamp	
Wooded marsh or swamp	
Submerged wooded marsh or swamp	
Rice field	(Rice)
Land subject to inundation	Max pool 431

Buildings and related features

Building	
School; church	
Built-up area	
Racetrack	
Airport	
Landing strip	
Well (other than water); windmill	
Tanks	
Covered reservoir	
Gaging station	
Landmark object (feature as labeled)	
Campground; picnic area	
Cemetery: small; large	(Cem)

Roads and related features

Roads on Provisional edition maps are not classified as primary, secondary, or light duty. They are all symbolized as light duty roads.

Primary highway	
Secondary highway	
Light duty road	
Unimproved road	
Trail	
Dual highway	
Dual highway with median strip	

Railroads and related features

Standard gauge single track; station	
Standard gauge multiple track	
Abandoned	

Transmission lines and pipelines

Power transmission line; pole; tower	
Telephone line	Telephone
Aboveground oil or gas pipeline	
Underground oil or gas pipeline	Pipeline

FIGURE 7 Standardized topographic map symbols used on USGS maps.
English units still prevail, although a few USGS maps are in metric. [From USGS, Topographic Maps, 1969.]

Appendix B

The Canadian System of Soil Classification (CSSC)

Canadian efforts at soil classification began in 1914 with the partial mapping of Ontario's soils by A. J. Galbraith. Efforts to develop a taxonomic system spread countrywide, anchored by universities in each province. Regional differences in soil classification emerged, further confused by a lack of specific soil details. By 1936, only 1.7% of Canadian soil had been surveyed (15 million hectares).

Canadian scientists needed a taxonomic system based on observable and measurable properties in soils specific to Canada. This meant a departure from Marbut's 1938 U.S. classification. Canada's first taxonomic system was introduced in 1955, splitting away from the soil classification effort in the United States and the Fourth Approximation stage. Classification work progressed through the Canada Soil Survey Committee after 1970 and was replaced by the Expert Committee on Soil Survey in 1978, all under Agriculture Canada.

The **Canadian System of Soil Classification (CSSC)** provides taxa for all soils presently recognized in Canada and is adapted to Canada's expanses of forest, tundra, prairie, frozen ground, and colder climates. As in the U.S. Soil Taxonomy system, the CSSC classifications are based on observable and measurable properties found in real soils rather than idealized soils that may result from the interactions of genetic processes. The system is flexible in that its framework can accept new findings and information in step with progressive developments in the soil sciences.

Categories of Classification in the CSSC

Categorical levels are at the heart of a taxonomic system. These categories are based on soil profile properties organized at five levels, nested in a hierarchical pattern to permit generalization at several levels of detail. Each level is referred to as a category of classification. The levels in the CSSC are briefly described here, as adapted from *The Canadian Soil Classification System*, 2nd ed., Publication 1646 (Ottawa: Supply and Services Canada, 1987), p. 16.

- *Order*: Each of nine soil orders has pedon properties that reflect the soil environment and effects of active soil-forming processes.
- *Great Group*: Subdivisions of each order reflect differences in the dominant processes or other major contributing processes. As an example, in Luvic Gleysols (great group name followed by order) the dominant process is gleying—reduction of iron and other minerals—resulting from poor drainage under either grass or forest cover with Aeg and Btg horizons (see Table 1).

- *Subgroup*: Subgroups are differentiated by the content and arrangement of horizons that indicate the relation of the soil to a great group or order or the subtle transition toward soils of another order.
- *Family*: This is a subdivision of a subgroup. Parent material characteristics such as texture and mineralogy, soil climatic factors, and soil reactions are important.
- *Series*: Detailed features of the pedon differentiate subdivisions of the family—the essential soil-sampling unit. Pedon horizons fall within a narrow range of color, texture, structure, consistence, porosity, moisture, chemical reaction, thickness, and composition.

Soil Horizons in the CSSC

Soil horizons are named and standardized as diagnostic in the classification process. Several mineral and organic horizons and layers are used in the CSSC. Three mineral horizons are recognized by capital letter designation, followed by lowercase suffixes for further description. Principal soil-mineral horizons and suffixes are presented in Table 1.

Four organic horizons are identified in the Canadian classification system. *O* is further defined through subhorizon designations. Note that for organic soils, such layers are identified as *tiers*. These organic horizons are detailed in Table 2.

The Nine Soil Orders of the CSSC

The nine orders of the CSSC, and related great groups, are summarized in Table 3 with a general description of properties, related Great Groups, an estimated percentage of land area for the soil order, a fertility assessment, and any applicable Soil Taxonomy equivalent.

Figure 1 is a generalized map of the distribution of principal soil orders in relation to physiographical regions in Canada. This grouping allows you to easily compare soils across Canada. A summary of the nine soil orders appears in Table 3. Please consult the *National Atlas of Canada*, 5th edition, for a detailed map of Canadian soils (**http://atlas.gc.ca/**).

The *Soil Landscapes of Canada* (SLC) site at **http://sis.agr.gc.ca/cansis/nsdb/slc/intro.html** is most useful! Here you will find a wonderful assortment of landscape and soil profile photographs, arranged geographically across Canada from east to west. The site also has an interactive GIS on-line mapping application. Version 2.2 SLC Component Mapping (December 1996) is operational and involves the CSSC and the Canadian Land Resource

Table 1   Three Mineral Horizons and Mineral Horizon Suffixes Used in the CSSC

Symbol	Mineral Horizon Description
A	Forms at or near the surface; experiences *eluviation*, or leaching, of finer particles or minerals. Several subdivisions are identified, with the surface usually darker and richer in organic content than lower horizons (*Ab*); or a paler, lighter zone below that reflects removal of organic matter with clays and oxides of aluminum and iron leached (removed) to lower horizons (*Ae*).
B	Experiences *illuviation*, a depositional process, as demonstrated by accumulations of clays (*Bt*), sesquioxides of aluminum or iron, and possibly an enrichment of organic debris (*Bh*), and the development of soil structure. Coloration is important in denoting whether hydrolysis, reduction, or oxidation processes are operational for the assignment of a descriptive suffix.
C	Exhibits little effect from pedogenic processes operating in the *A* and *B* horizons, except the process of gleysation associated with poor drainage and the reduction of iron, denoted (*Cg*), and the accumulation of calcium and magnesium carbonates (*Cca*) and more soluble salts (*Cs*) and (*Csd*).

Symbol	Horizon Suffix Description
b	A buried soil horizon.
c	Irreversible cementation of a pedogenic horizon, e.g., cemented by $CaCO_3$.
ca	Lime accumulation of at least 10 cm thickness that exceeds in concentration that of the unenriched parent material by at least 5%.
cc	Irreversible cemented concretions, typically in pellet form.
e	Used with *A* mineral horizons (*Ae*) to denote eluviation of clay, Fe, Al, or organic matter.
f	Enriched principally with illuvial iron and aluminum combined with organic matter, reddish in upper portions and yellowish at depth, determined through specific criteria. Used with *B* horizons alone.
g	Gray to blue colors, prominent mottling, or both, produced by intense chemical reduction. Various applications to *A*, *B*, and *C* horizons.
h	Enriched with organic matter: accumulation in place or biological mixing (*Ab*) or subsurface enrichment through illuviation (*Bh*).
j	A modifier suffix for *e, f, g, n,* and *t* to denote limited change or failure to meet specified criteria denoted by that letter.
k	Presence of carbonates as indicated by visible effervescence with dilute hydrochloric acid (HCl).
m	Used with *B* horizons slightly altered by hydrolysis, oxidation, or solution, or all three to denote a change in color or structure.
n	Accumulation of exchangeable calcium (Ca) in ratio to exchangeable sodium (Na) that is 10 or less, with the following characteristics: prismatic or columnar structure, dark coatings on ped surfaces, and hard consistence when dry. Used with *B* horizons alone.
p	*A* or *O* horizons disturbed by cultivation, logging, and habitation. May be used when plowing intrudes on previous *B* horizons.
s	Presence of salts, including gypsum, visible as crystals or veins or surface crusts of salt crystals, and by lowered crop yields. Usually with *C* but may appear with any horizon and lowercase suffixes.
sa	A secondary enrichment of salts more soluble than Ca or Mg carbonates, exceeding unenriched parent material, in a horizon at least 10 cm thick.
t	Illuvial enrichment of the *B* horizon with silicate clay that must exceed in overlying *Ae* horizon by 3% to 20%, depending on the clay content of the *Ae* horizon.
u	Markedly disrupted by physical or faunal processes other than cryoturbation.
x	Fragipan formation—a loamy subsurface horizon of high bulk density and very low organic content. When dry, it has a hard consistence and seems to be cemented.
y	Affected by cryoturbation (frost action) with disrupted and broken horizons and incorporation of materials from other horizons. Application to *A*, *B*, and *C* horizons and in combination with other suffixes.
z	A frozen layer.

Network (CLRN). The component mapping involves a GIS model consisting of layers that include the major characteristics of soil and land for all of Canada. You can select a spatial area and a variety of attributes to display on the map (drainage class, soil type, rooting depth, local surface form, slope, and vegetation cover, among others).

Table 2   Four Organic Horizons Used in the CSSC

Symbol	Description
O	Organic materials, mainly mosses, rushes, and woody materials
L	Mainly discernible leaves, twigs, and woody materials
F	Partially decomposed, somewhat recognizable L materials
H	Indiscernible organic materials

O is further defined through subhorizon designations:

Of	Readily identifiable fibric materials
Om	Mesic materials of intermediate decomposition
Oh	Humic material at an advanced stage of decomposition—low fiber, high bulk density

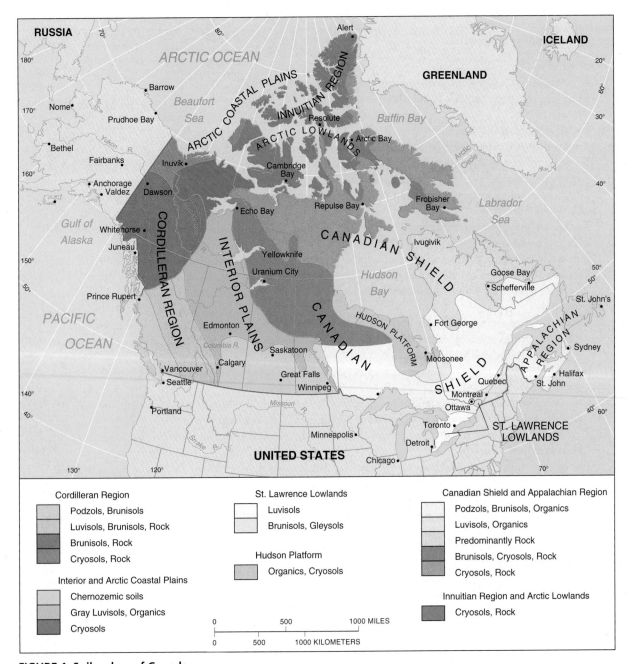

FIGURE 1 Soil orders of Canada.
Principal soil regions of the Canadian System of Soil Classification (CSSC) as related to major physiographic regions. [After maps prepared by the Land Resources Research Institute, Geological Survey of Canada, and the Canadian Soil Survey Committee.]

Table 3 Nine Orders of the Canadian System of Soil Classification

Order Great Group	Characteristics*	Fertility
Chernozemic (Russian, *chernozem*) Brown (more moist) Dark brown Black Dark gray (less moist) (38 subgroups)	Well to imperfectly drained soils of the steppe-grassland-forest transition, Southern Alberta, Saskatchewan, Manitoba, Okanagan Valley, BC, Palouse Prairie, BC. Accumulation of organic matter in surface horizons. Most frozen during some winter months with soil-moisture deficits in the summer. A diagnostic **Ah** is typical (although **Ahe**, **Ap** are present) at least 10 cm thick or 15 cm if disturbed by cultivation. Mean annual temperature $> 0°C$ and usually $< 5.5°C$. (5.1%, 470,000 km²; Soil Taxon. = Mollisols.)	High; wheat-growing
Solonetzic (Russian, *solonetz*) Solonetz Solodized Solonetz Solod (27 subgroups)	Solonetz denotes saline or alkaline soils. Well to imperfectly drained mineral soils developed under grasses in semiarid to subhumid climates. Limited areas of central and north-central Alberta. Noted for a **B** horizon that is very hard when dry but swells to a sticky, low-permeability mass when wet. A saline **C** horizon reflects nature of parent materials. (0.8%, 73,700 km²; Soil Taxon. = Natric horizon of Mollisols and Alfisols.)	Variable (medium) about 50% cultivated, remainder in pasture
Luvisolic Gray brown Luvisol Gray Luvisol (18 subgroups)	Eluviation-illuviation processes produce a light-colored **Ae** horizon and a diagnostic **Bt** horizon. Soils of mixed deciduous–coniferous forests. Major occurrence is the St. Lawrence lowland. Luvisols do not have a solonetzic **B** horizon, evidence of Gleysolic order and gleying, or organics less than in the Organic order. Permafrost within 1 m of surface and 2 m if soils are cryoturbated. (10.3%, 950,000 km²; Soil Taxon. = Boralfs, Udalfs—suborders of Alfisols.)	High
Podzolic (Russian, *podzol*) Humic Ferro-humic Podzol Humo-ferric Podzol (25 subgroups)	Soils of coniferous forests and sometimes heath, leaching of overlying horizons occurs in moist, cool to cold climates. Iron, aluminum and organic matter from **L**, **F**, and **H** horizons are redeposited in podzolic **B** horizon. A diagnostic **Bh**, **Bhf**, or **Bf** is present depending on great group. Dominant in western British Columbia, Ontario, and Québec. (22.6%, 2,083,000 km²; Soil Taxon. = Spodosols, some Inceptisols.)	Low to medium depending on acidity
Brunisolic (French, "brown") Melanic Brunisol Eutric Brunisol Sombric Brunisol Dystric Brunisol (18 subgroups)	Sufficiently developed to distinguish from Regosolic order. Soils under forest cover with brownish **Bm** horizons, although various colors are possible. Also, can be with mixed forest, shrubs and grass. Diagnostic **Bm**, **Bfj**, thin **Bf**, or **Btj** horizons differentiate from soils of other orders. Well to imperfectly drained. Lack the podzolic **B** horizon of podzols although surrounded by them in St. Lawrence lowland. (8.8%, 811,000 km²; Soil Taxon. = Inceptisols, some Psamments [Aquents in Entisols].)	Medium (variable)
Regosolic (Greek, *rhegos*) Regosol Humic Regosol (8 subgroups)	Weakly developed limited soils, the result of any number of factors: young materials; fresh alluvial deposits; material instability; mass-wasted slopes; or dry, cold climatic conditions. Lack solonetzic, illuvial, or podzolic **B** horizons. Lack permafrost within 1 m of surface, or 2 m if cryoturbated. May have **L**, **F**, **H**, or **O** horizons, or an **Ah** horizon if less than 10 cm thick. Buried horizons possible. Dominant in Northwest Territories and northern Yukon, now designated as Cryosols under CSSC. (1.3%, 120,000 km²; Soil Taxon. = Entisols.)	Low (variable)
Gleysolic (Russian, *glei*) Luvic Gleysol Humic Gleysol Gleysol (13 subgroups)	Defined on the basis of color and mottling that results from chronic reducing conditions inherent in poorly drained mineral soils under wet conditions. High water table and long periods of water saturation. Rather than continuous they appear spotty within other soil orders and occasionally may dominate an area. A diagnostic **Bg** horizon is present. (1.9%, 175,000 km²; Soil Taxon. = Various aquic suborders, a reducing moisture regime.)	High to medium
Organic Fibrisol Mesisol Humisol Folisol (31 subgroups)	Peat, bog, and muck soils, largely composed of organic material. Most water-saturated for prolonged periods. Are widespread in association with poorly to very poorly drained depressions, although Folisols are found under upland forest environments. Exceed 17% organic carbon and 30% organic matter overall. (4.2%, 387,000 km²; Soil Taxon. = Histosols.)	High to medium given drainage, available nutrients
Crysolic (Greek, *kyros*) Turbic Cryosol Static Cryosol Organic Cryosol (15 subgroups)	Dominate the northern third of Canada, with permafrost closer to the surface and composed of mineral and organic soil deposits. Generally found north of the treeline, or in fine-textured soils in subarctic forest, or in some organic soils in boreal forests. **Ah** horizon lacking or thin. Cryoturbation (frost action) common, often denoted by patterned ground circles, polygons, and stripes. Subgroups based on degree of cryoturbation and the nature of mineral or organic soil material. (45%, 4,150,000 km²; Soil Taxon. = Cryoquepts, Inceptisols, and pergelic temperature regime in several suborders.)	Not applicable

*Estimated percentage and square kilometers of Canada's land area and Soil Taxonomy equivalent are given in parentheses.

Review Questions

1. Why did Canada adopt its own system of soil classification? Describe a brief history of events that led up to the modern CSSC system.

2. Which soil order is associated with the development of a bog? Explain its use as a low-grade fuel.

3. Describe the podzolization process occurring in northern coniferous forests. What are the surface horizons like? What management strategies might enhance productivity in these soils? Name the soil order for these areas.

4. Compare and contrast Interior Plains soils with those of the southeastern Canadian Shield.

5. What processes inhibit soil development in the extreme north? Explain.

6. Briefly describe what you found on the *Soil Landscapes of Canada* (SLC) Web site. Please try the SLC Component Mapping feature for your specific area or region (http://sis.agr.gc.ca/cansis/nsdb/slc/intro.html).

7. Which of the nine soil orders is characteristic of the area where you live, or where you attend college? How did you determine the answer?

Appendix C

Common Conversions

Metric Measure	Multiply by	English Equivalent
Length		
Centimeters (cm)	0.3937	Inches (in.)
Meters (m)	3.2808	Feet (ft)
Meters (m)	1.0936	Yards (yd)
Kilometers (km)	0.6214	Miles (mi)
Nautical mile	1.15	Statute mile
Area		
Square centimeters (cm^2)	0.155	Square inches (in.^{2})
Square meters (m^2)	10.7639	Square feet (ft^2)
Square meter (m^2)	1.1960	Square yards (yd^2)
Square kilometers (km^2)	0.3831	Square miles (mi^2)
Hectare (ha) (10,000 m^2)	2.4710	Acres (a)
Volume		
Cubic centimeters (cm^3)	0.06	Cubic inches (in.^{3})
Cubic meters (m^3)	35.30	Cubic feet (ft^3)
Cubic meters (m^3)	1.3079	Cubic yards (yd^3)
Cubic kilometers (km^3)	0.24	Cubic miles (mi^3)
Liters (l)	1.0567	Quarts (qt), U.S.
Liters (l)	0.88	Quarts (qt), Imperial
Liters (l)	0.26	Gallons (gal), U.S.
Liters (l)	0.22	Gallons (gal), Imperial
Mass		
Grams (g)	0.03527	Ounces (oz)
Kilograms (kg)	2.2046	Pounds (lb)
Metric ton (tonne) (t)	1.10	Short ton (tn), U.S.
Velocity		
Meters/second (mps)	2.24	Miles/hour (mph)
Kilometers/hour (kmph)	0.62	Miles/hour (mph)
Knots (kn) (nautical mph)	1.15	Miles/hour (mph)
Temperature		
Degrees Celsius (°C)	1.80 (then add 32)	Degrees Fahrenheit (°F)
Celsius degree (C°)	1.80	Fahrenheit degree (F°)
Additional water measurements:		
Gallon (Imperial)	1.201	Gallon (U.S.)
Gallons (gal)	0.000003	Acre-feet

1 cubic foot per second per day = 86,400 cubic feet = 1.98 acre-feet

---

ADDITIONAL ENERGY AND POWER MEASUREMENTS

1 watt (W) = 1 joule/s

1 joule = 0.239 calorie

1 calorie = 4.186 joules

1 W/m^2 = 0.001433 cal/min

697.8 W/m^2 = 1 cal/cm^2 min^{-1}

1 W/m^2 = 2.064 cal/cm^{3}day^{-1}

1 W/m^2 = 61.91 cal/cm^{2}month^{-1}

1 W/m^2 = 753.4 cal/cm^{2}year^{-1}

100 W/m^2 = 75 kcal/cm^{2}year^{-1}

Solar constant:
1372 W/m^2
2 cal/cm^2 min^{-1}

English Measure	Multiply by	Metric Equivalent
Length		
Inches (in.)	2.54	Centimeters (cm)
Feet (ft)	0.3048	Meters (m)
Yards (yd)	0.9144	Meters (m)
Miles (mi)	1.6094	Kilometers (km)
Statute mile	0.8684	Nautical mile
Area		
Square inches (in.^{2})	6.45	Square centimeters (cm^2)
Square feet (ft^2)	0.0929	Square meters (m^2)
Square yards (yd^2)	0.8361	Square meters (m^2)
Square miles (mi^2)	2.5900	Square kilometers (km^2)
Acres (a)	0.4047	Hectare (ha)
Volume		
Cubic inches (in.^{3})	16.39	Cubic centimeters (cm^3)
Cubic feet (ft^3)	0.028	Cubic meters (m^3)
Cubic yards (yd^3)	0.765	Cubic meters (m^3)
Cubic miles (mi^3)	4.17	Cubic kilometers (km^3)
Quarts (qt), U.S.	0.9463	Liters (l)
Quarts (qt), Imperial	1.14	Liters (l)
Gallons (gal), U.S.	3.8	Liters (l)
Gallons (gal), Imperial	4.55	Liters (l)
Mass		
Ounces (oz)	28.3495	Grams (g)
Pounds (lb)	0.4536	Kilograms (kg)
Short ton (tn), U.S.	0.91	Metric ton (tonne) (t)
Velocity		
Miles/hour (mph)	0.448	Meters/second (mps)
Miles/hour (mph)	1.6094	Kilometers/hour (kmph)
Miles/hour (mph)	0.8684	Knots (kn) (nautical mph)
Temperature		
Degrees Fahrenheit (°F)	0.556 (after subtracting 32)	Degrees Celsius (°C)
Fahrenheit degree (F°)	0.556	Celsius degree (C°)
Additional water measurements:		
Gallon (U.S.)	0.833	Gallons (Imperial)
Acre-feet	325,872	Gallons (gal)

Additional Notation

Multiples	Prefixes	
$1{,}000{,}000{,}000 = 10^9$	giga	G
$1{,}000{,}000 = 10^6$	mega	M
$1{,}000 = 10^3$	kilo	k
$100 = 10^2$	hecto	h
$10 = 10^1$	deka	da
$1 = 10^0$		
$0.1 = 10^{-1}$	deci	d
$0.01 = 10^{-2}$	centi	c
$0.001 = 10^{-3}$	milli	m
$0.000001 = 10^{-6}$	micro	μ

Glossary

The chapter in which each term appears **boldfaced** in parentheses is followed by a specific definition relevant to the term's usage in the chapter.

Abiotic (1)   Nonliving; Earth's nonliving systems of energy and materials.

Ablation (17)   Loss of glacial ice through melting, sublimation, wind removal by deflation, or the calving of blocks of ice (see deflation).

Abrasion (14, 15, 17)   Mechanical wearing and erosion of bedrock accomplished by the rolling and grinding of particles and rocks carried in a stream, removed by wind in a "sandblasting" action, or imbedded in glacial ice.

Absorption (4)   Assimilation and conversion of radiation from one form to another in a medium. In the process, the temperature of the absorbing surface is raised, thereby affecting the rate and wavelength of radiation from that surface.

Active layer (17)   A zone of seasonally frozen ground that exists between the subsurface permafrost layer and the ground surface. The active layer is subject to consistent daily and seasonal freeze–thaw cycles (see permafrost, periglacial).

Actual evapotranspiration (9)   ACTET; the actual amount of evaporation and transpiration that occurs; derived in the water balance by subtracting the deficit (DEFIC) from potential evapotranspiration in the surface.

Adiabatic (7)   Pertaining to the cooling of an ascending parcel of air through expansion or the warming of a descending parcel of air through compression, without any exchange of heat between the parcel and the surrounding environment.

Advection (4)   Horizontal movement of air or water from one place to another (compare with convection).

Advection fog (7)   Active condensation formed when warm, moist air moves laterally over cooler water or land surfaces, causing the lower layers of the air to be chilled to the dew-point temperature.

Aggradation (14)   The general building of land surface because of deposition of material; opposite of degradation. When the sediment load of a stream exceeds the stream's capacity to carry it, the stream channel becomes filled through this process.

Air mass (8)   A distinctive, homogeneous body of air that has taken on the moisture and temperature characteristics of its source region.

Air pressure (3)   Pressure produced by the motion, size, and number of gas molecules in the air and exerted on surfaces in contact with the air. Normal sea level pressure, as measured by the height of a column of mercury (Hg), is expressed as 1013.2 millibars, 760 mm of Hg, or 29.92 inches of Hg. Air pressure can be measured with mercury or aneroid barometers (see listing for both).

Albedo (4)   The reflective quality of a surface, expressed as the percentage of reflected insolation to incoming insolation; a function of surface color, angle of incidence, and surface texture.

Aleutian low (6)   See subpolar low-pressure cell.

Alfisols (18)   A soil order in the Soil Taxonomy. Moderately weathered forest soils that are moist versions of Mollisols, with productivity dependent on specific patterns of moisture and temperature; rich in organics. Most wide-ranging of the soil orders.

Alluvial fan (15)   Fan-shaped fluvial landform at the mouth of a canyon; generally occurs in arid landscapes where streams are intermittent.

Alluvial terraces (14)   Level areas that appear as topographic steps above a stream, created by the stream as it scours with renewed downcutting into its floodplain; composed of unconsolidated alluvium (see alluvium).

Alluvium (14)   General descriptive term for clay, silt, and sand, transported by running water and deposited in sorted or semisorted sediment on a floodplain, delta, or stream bed.

Alpine glacier (17)   A glacier confined in a mountain valley or walled basin, consisting of three subtypes: valley glacier (within a valley), piedmont glacier (coalesced at the base of a mountain, spreading freely over nearby lowlands), and outlet glacier (flowing outward from a continental glacier).

Alpine tundra (20)   Tundra conditions at high elevation (see Arctic tundra).

Altitude (2)   The angular distance between the horizon (a horizontal plane) and the Sun (or any point in the sky).

Altocumulus (7)   Middle level, puffy clouds that occur in several forms: patchy rows, wave patterns, a "mackerel sky," or lens-shaped "lenticular" clouds.

Andisols (18)   A soil order in the Soil Taxonomy; derived from volcanic parent materials in areas of volcanic activity. A new order, created in 1990, of soils previously considered under Inceptisols and Entisols.

Anemometer (6)   A device that measures wind velocity.

Aneroid barometer (6)   A device that measures air pressure using a partially evacuated, sealed cell (see air pressure).

Angle of repose (13)   The steepness of a slope that results when loose particles come to rest; an angle of balance between driving and resisting forces, ranging between 33° and 37° from a horizontal plane.

Antarctic high (6)   A consistent high-pressure region centered over Antarctica; source region for an intense polar air mass that is dry and associated with the lowest temperatures on Earth.

Anthropogenic atmosphere (3)   Earth's future atmosphere, so named because humans appear to be the principal causative agent.

Anticline (12)   Upfolded rock strata, in which layers slope downward from the axis of the fold, or central ridge (compare syncline).

Anticyclone (6)   A dynamically or thermally caused area of high atmospheric pressure with descending and diverging air flows that rotate clockwise in the Northern Hemisphere and counterclockwise in the Southern Hemisphere (compare cyclone).

Aphelion (2)   The point of Earth's greatest distance from the Sun in its elliptical orbit; reached on July 4 at a distance of 152,083,000 km (94.5 million mi); variable over a 100,000-year cycle (compare perihelion).

Aquiclude (9)   A body of rock that does not conduct groundwater in usable amounts; an impermeable rock layer, related to an aquitard that slows but does not block water flow (compare aquifer).

Aquifer (9)   A body of rock that conducts groundwater in usable amounts; a permeable layer of rock (compare aquiclude).

Aquifer recharge area (9)   The surface area where water enters an aquifer to recharge the water-bearing strata in a groundwater system.

Arctic and alpine tundra (20)   A biome in the northernmost portions of North America and northern Europe and Russia, featuring low ground-level herbaceous plants as well as some woody plants.

Arête (17)   A sharp ridge that divides two cirque basins. Derived from "knife edge" in French, these form sawtooth and serrated ridges in glaciated mountains.

Aridisols (18)   A soil order in the Soil Taxonomy; largest soil order. Typical of dry climates; low in organic matter and dominated by calcification and salinization.

Artesian water (9)   Pressurized groundwater that rises in a well or a rock structure above the local water table; may flow out onto the ground without pumping (see potentiometric surface).

Asthenosphere (11)   Region of the upper mantle just below the lithosphere; the least rigid portion of Earth's interior and known as the plastic layer, flowing very slowly under extreme heat and pressure.

Atmosphere (1)   The thin veil of gases surrounding Earth, which forms a protective boundary between outer space and the biosphere; generally considered to extend out about 480 km (300 mi) from Earth's surface.

Aurora (2)   A spectacular glowing light display in the ionosphere, stimulated by the interaction of the solar wind with oxygen and nitrogen gases and atoms at high latitudes, called aurora borealis in the Northern Hemisphere and aurora australis in the Southern Hemisphere.

Autumnal (September) equinox (2)   The time around September 22–23 when the Sun's declination crosses the equatorial parallel (0° latitude) and all places on Earth experience days and nights of equal length. The Sun rises at the South Pole and sets at the North Pole (compare vernal equinox).

Available water (9)   The portion of capillary water that is accessible to plant roots; usable water held in soil moisture storage (see capillary water).

Axial parallelism (2)   Earth's axis remains aligned the same throughout the year (it "remains parallel to itself"); thus, the axis extended from the North Pole points into space always near Polaris, the North Star.

Axial tilt (2)   Earth's axis tilts 23.5° from a perpendicular to the plane of the ecliptic (plane of Earth's orbit around the Sun).

Axis (2)   An imaginary line, extending through Earth from the geographic North Pole to the geographic South Pole, around which Earth rotates.

Azores high (6)   A subtropical high-pressure cell that forms in the Northern Hemisphere in the eastern Atlantic (see Bermuda high); associated with warm, clear water and large quantities of sargassum, or gulf weed, characteristic of the Sargasso Sea.

Backswamp (14)   A low-lying, swampy area of a floodplain; adjacent to a river, with the river's natural levee on one side and higher topography on the other (see floodplain, yazoo tributary).

Bajada (15)   A continuous apron of coalesced alluvial fans, formed along the base of mountains in arid climates; presents a gently rolling surface from fan to fan (see alluvial fan).

Barrier beach (16)   Narrow, long depositional feature, generally composed of sand, that forms offshore roughly parallel to the coast; may appear as barrier islands and long chains of barrier beaches (see barrier island).

Barrier island (16)   Generally, a broadened barrier beach (see barrier beach).

Barrier spit (16)   A depositional landform that develops when transported sand or gravel in a barrier beach or island is deposited in long ridges that are attached at one end to the mainland and partially cross the mouth of a bay.

Basalt (11)   A common extrusive igneous rock, fine-grained, comprising the bulk of the ocean floor crust, lava flows, and volcanic forms; gabbro is its intrusive form.

Base level (14)   A hypothetical level below which a stream cannot erode its valley, and thus the lowest operative level for denudation processes; in an absolute sense it is represented by sea level, extending under the landscape.

Basin and Range Province (15)   A region of dry climates, few permanent streams, and interior drainage patterns in the western United States, a faulted landscape composed of a sequence of horsts and grabens.

Batholith (11)   The largest plutonic form exposed at the surface; an irregular intrusive mass (> 100 km^2; > 40 mi^2); it invades crustal rocks, cooling slowly so that large crystals develop (see pluton).

Bay barrier (16)   An extensive barrier spit of sand or gravel that encloses a bay, cutting it off completely from the ocean and forming a lagoon; produced by littoral drift and wave action; sometimes referred to as a baymouth bar (see barrier spit, lagoon).

Beach (16)   The portion of the coastline where an accumulation of sediment is in motion.

Beach drift (16)   Material, such as sand, gravel, and shells, that is moved by the longshore current in the effective direction of the waves.

Beaufort wind scale (6)   A descriptive scale for the visual estimation of wind speeds; originally conceived in 1806 by Admiral Beaufort of the British Navy.

Bed load (14)   Coarse materials that are dragged along the bed of a stream by traction or by the rolling and bouncing motion of saltation; involves particles too large to remain in suspension (see traction, saltation).

Bedrock (13)   The rock of Earth's crust that is below the soil and is basically unweathered; such solid crust sometimes is exposed as an outcrop.

Bermuda high (6)   A subtropical high-pressure cell that forms in the western North Atlantic (see Azores high).

Biodiversity (19)   A principle of ecology and biogeography: the more diverse the species population in an ecosystem (both in number of species, quantity of members in each species, and genetic content), the more risk is spread over the entire community, which results in greater overall sta-

bility, greater productivity, and increased use of nutrients, as compared to a monoculture of little or no diversity.

Biogeochemical cycle (19)   One of several circuits of flowing elements and materials (carbon, oxygen, nitrogen, phosphorus, water) that combine Earth's biotic (living) and abiotic (nonliving) systems; the cycling of materials is continuous and renewed through the biosphere and the life processes.

Biogeographic realm (20)   One of eight regions of the biosphere, each representative of evolutionary core areas of related flora (plants) and fauna (animals); a broad geographical classification scheme.

Biogeography (19)   The study of the distribution of plants and animals and related ecosystems; the geographical relationships with their environments over time.

Biomass (19)   The total mass of living organisms on Earth or per unit area of a landscape; also, the weight of the living organisms in an ecosystem.

Biome (20)   A large terrestrial ecosystem characterized by specific plant communities and formations; usually named after the predominant vegetation in the region (see terrestrial ecosystem).

Biosphere (1)   That area where the atmosphere, lithosphere, and hydrosphere function together to form the context within which life exists; an intricate web that connects all organisms with their physical environment.

Biotic (1)   Living; referring to Earth's living system of organisms.

Blowout depression (15)   Eolian (wind) erosion in which deflation forms a basin in areas of loose sediment. Diameter may range up to hundreds of meters (see deflation).

Bolson (15)   The slope and basin area between the crests of two adjacent ridges in a dry region.

Boreal forest (20)   See needleleaf forest.

Brackish (16)   Descriptive of seawater with a salinity of less than 35‰; for example, the Baltic Sea (contrast to brine).

Braided stream (14)   A stream that becomes a maze of interconnected channels laced with excess sediment. Braiding often occurs with a reduction of discharge that reduces a stream's transporting ability, or with an increase in sediment load.

Breaker (16)   The point where a wave's height exceeds its vertical stability and the wave breaks as it approaches the shore.

Brine (16)   Seawater with a salinity of more than 35‰; for example, the Persian Gulf (contrast to brackish).

Calcification (18)   The illuviated (deposited) accumulation of calcium carbonate or magnesium carbonate in the B and C soil horizons.

Caldera (12)   An interior sunken portion of a composite volcano's crater; usually steep-sided and circular, sometimes containing a lake; also can be found in conjunction with shield volcanoes.

Canadian System of Soil Classification (Appendix B)   A classification system for all soils presently recognized in Canada and adapted to Canada's particular forest, tundra, prairie, frozen ground, and colder climates. The CSSC is organized at five levels of generalization (order, great group, subgroup, family, and series) with each level referred to as a category of classification. The system is described in *The Canadian Soil Classification System*, 2nd edition, published by Agriculture Canada, 1987 (compare Soil Taxonomy; see Gelisols). Also illustrated at the *Soil Landscapes of Canada* (SLC) Web site: http://sis.agr.gc.ca/cansis/nsdb/slc/intro.html.

Capillary water (9)   Soil moisture, most of which is accessible to plant roots; held in the soil by the water's surface tension and cohesive forces between water and soil (see also available water, field capacity, hygroscopic water, and wilting point).

Carbonation (13)   A chemical weathering process in which weak carbonic acid (water and carbon dioxide) reacts with many minerals that contain calcium, magnesium, potassium, and sodium (especially limestone), transforming them into carbonates.

Carbon monoxide (3)   An odorless, colorless, tasteless combination of carbon and oxygen produced by the incomplete combustion of fossil fuels or other carbon-containing substances; toxicity to humans is due to its affinity for hemoglobin, displacing oxygen in the bloodstream; CO.

Carnivore (19)   A secondary consumer that principally eats meat for sustenance. The top carnivore in a food chain is considered a tertiary consumer (compare herbivore).

Cartography (1)   The making of maps and charts; a specialized science and art that blends aspects of geography, engineering, mathematics, graphics, computer science, and artistic specialties.

Catastrophism (11) A philosophy that attempts to fit the vastness of Earth's age and the complexity of the rock record into a very shortened time span through a belief in short-lived and catastrophic worldwide events.

Cation-exchange capacity (CEC) (18) The ability of soil colloids to exchange cations between their surfaces and the soil solution; a measured potential that indicates soil fertility (see soil colloid, soil fertility).

Chaparral (20) Dominant shrub formations of Mediterranean dry summer climates; characterized by sclerophyllous scrub and short, stunted, tough forests; derived from the Spanish *chapparo*; specific to California (see Mediterranean shrubland).

Chemical weathering (13) Decomposition and decay of the constituent minerals in rock through chemical alteration of those minerals. Water is essential, with rates keyed to temperature and precipitation values. Chemical reactions are active at microsites even in dry climates. Processes include hydrolysis, oxidation, carbonation, and solution.

Chinook wind (8) North American term for a warm, dry, downslope air flow; characteristic of the rain shadow region on the leeward side of mountains; known as föhn, or foehn, winds in Europe (see rain shadow).

Chlorofluorocarbons (CFC) (3) A manufactured molecule (polymer) made of chlorine, fluorine, and carbon; inert and possessing remarkable heat properties; also known as one of the halogens. After slow transport to the stratospheric ozone layer, CFCs react with ultraviolet radiation, freeing chlorine atoms that act as a catalyst to produce reactions that destroy ozone; manufacture banned by international treaties.

Chlorophyll (19) A light-sensitive pigment that resides within chloroplasts (organelles) in leaf cells of plants; the basis of photosynthesis.

Cinder cone (12) A volcanic landform of pyroclastics and scoria, usually small and cone-shaped and generally not more than 450 m (1500 ft) in height; with a truncated top.

Circle of illumination (2) The division between light and dark on Earth; a day–night great circle.

Circum-Pacific belt (12) A tectonically and volcanically active region encircling the Pacific Ocean; also known as the "ring of fire."

Cirque (17) A scooped-out, amphitheater-shaped basin at the head of an alpine glacier valley; an erosional landform.

Cirrus (7) Wispy, filamentous ice-crystal clouds that occur above 6000 m (20,000 ft); appear in a variety of forms, from feathery hair-like fibers to veils of fused sheets.

Classification (10) The process of ordering or grouping data or phenomena in related classes; results in a regular distribution of information; a taxonomy.

Climate (10) The consistent, long-term behavior of weather over time, including its variability; in contrast to weather, which is the condition of the atmosphere at any given place and time.

Climatic regions (10) An area of homogenous climate that features characteristic regional weather and air mass patterns.

Climatology (10) The scientific study of climate and climatic patterns and the consistent behavior of weather, including its variability and extremes, over time in one place or region; includes the effects of climate change on human society and culture.

Climograph (10) A graph that plots daily, monthly, or annual temperature and precipitation values for a selected station; may also include additional weather information.

Closed system (1) A system that is shut off from the surrounding environment so that it is entirely self-contained in terms of energy and materials; Earth is a closed material system (compare open system).

Cloud (7) An aggregate of tiny moisture droplets and ice crystals; classified by altitude of occurrence and shape.

Cloud-albedo forcing (4) An increase in albedo (the reflectivity of a surface) caused by clouds due to their reflection of incoming insolation.

Cloud-condensation nuclei (7) Microscopic particles necessary as matter on which water vapor condenses to form moisture droplets; can be sea salts, dust, soot, or ash.

Cloud-greenhouse forcing (4) An increase in greenhouse warming caused by clouds because they can act like insulation, trapping longwave (infrared) radiation.

Col (17) Formed by two headward eroding cirques that reduce an arête (ridge crest) to form a high pass or saddle-like narrow depression.

Cold desert and semidesert (20) A type of desert biome found at higher latitudes than warm deserts. Interior location and rain shadow locations produce these cold deserts in North America.

Cold front (8) The leading edge of an advancing cold air mass; identified on a weather map as a line marked with triangular spikes pointing in the direction of frontal movement (compare warm front).

Community (19, 20) A convenient biotic subdivision within an ecosystem; formed by interacting populations of animals and plants in an area.

Composite volcano (12) A volcano formed by a sequence of explosive volcanic eruptions; steep-sided, conical in shape; sometimes referred to as a stratovolcano, although composite is the preferred term (compare shield volcano).

Conduction (4) The slow molecule-to-molecule transfer of heat through a medium, from warmer to cooler portions.

Cone of depression (9) The depressed shape of the water table around a well after active pumping. The water table adjacent to the well is drawn down by the water removal.

Confined aquifer (9) An aquifer that is bounded above and below by impermeable layers of rock or sediment (see artesian water, unconfined aquifer).

Constant isobaric surface (6) An elevated surface in the atmosphere on which all points have the same pressure, usually 500 mb. Along this constant-pressure surface, isobars mark the paths of upper air winds.

Consumer (19) Organism in an ecosystem that depends on producers (organisms that use carbon dioxide as their sole source of carbon) for its source of nutrients; also called a heterotroph (compare producer).

Consumptive use (9) A use that removes water from a water budget at one point and makes it unavailable further downstream (compare withdrawal).

Continental divide (14) A ridge or elevated area that separates drainage on a continental scale; specifically, that ridge in North America that separates drainage to the Pacific on the west side from drainage to the Atlantic and Gulf on the east side and to Hudson Bay and the Arctic Ocean in the north.

Continental drift (11) A proposal by Alfred Wegener in 1912 stating that Earth's landmasses have migrated over the past 225 million years from a supercontinent he called Pangaea to the present configuration; the widely accepted plate tectonics theory today (see plate tectonics).

Continental effect (5) A quality of regions that lack the temperature-moderating effects of the sea and that exhibit a greater range of minimum and maximum temperatures, both daily and annually (see marine, land–water heating differences).

Continental glacier (17) A continuous mass of unconfined ice, covering at least $50,000 \text{ km}^2$ ($19,500 \text{ mi}^2$); most extensive at present as ice sheets covering Greenland and Antarctica (compare alpine glacier).

Continental platform (12) The broadest category of landform, including those masses of crust that reside above or near sea level and the adjoining undersea continental shelves along the coastline.

Continental shield (12) Generally old, low-elevation heartland regions of continental crust; various cratons (granitic cores) and ancient mountains exposed at the surface.

Contour lines (Appendix A) Isolines on a topographic map that connect all points at the same elevation relative to a reference elevation called the vertical datum.

Convection (4) Transfer of heat from one place to another through the physical movement of air; involves a strong vertical motion (compare to advection).

Convectional lifting (8) Air passing over warm surfaces gains buoyancy and lifts, initiating adiabatic processes.

Convergent lifting (8) Air flows in conflict forces lifting and displacement of air upward, initiating adiabatic processes.

Coordinated Universal Time (UTC) (1) The official reference time in all countries, formerly known as Greenwich Mean Time; now measured by six primary standard atomic clocks, the time calculations are collected in Paris at the International Bureau of Weights and Measures (BIPM), the legal reference for time in all countries, and broadcast worldwide.

Coral (16) A simple, cylindrical marine animal with a saclike body that secretes calcium carbonate to form a hard external skeleton and, cumulatively, landforms called reefs; lives symbiotically with nutrient-producing algae; presently in a worldwide state of decline due to bleaching (loss of algae).

Core (11) The deepest inner portion of Earth, representing one-third of its entire mass; differentiated into two zones—a solid iron inner core surrounded by a dense, molten, fluid metallic-iron outer core.

Coriolis force (6) The apparent deflection of moving objects (wind, ocean currents, missiles) from traveling in a straight path, in proportion to the speed of Earth's rotation at different latitudes. Deflection is to the right in the Northern Hemisphere and to the left in the Southern Hemisphere; maximum at the poles and zero along the equator.

Crater (12) A circular surface depression formed by volcanism; built by accumulation, collapse, or explosion; usually located at a volcanic vent or pipe; can be at the summit or on the flank of a volcano.

Crevasse (17) A vertical crack that develops in a glacier as a result of friction between valley walls, or tension forces of extension on convex slopes, or compression forces on concave slopes.

Crust (11) Earth's outer shell of crystalline surface rock, ranging from 5 to 60 km (3 to 38 mi) in thickness from oceanic crust to mountain ranges. Average density of continental crust is 2.7 g/cm^3, whereas oceanic crust is 3.0 g/cm^3.

Cumulonimbus (7) A towering, precipitation-producing cumulus cloud that is vertically developed across altitudes associated with other clouds; frequently associated with lightning and thunder and thus sometimes called a thunderhead.

Cumulus (7) Bright and puffy cumuliform clouds up to 2000 m in altitude (6500 ft).

Cyclogenesis (8) An atmospheric process that describes the birth of a midlatitude wave cyclone; usually along the polar front. Also refers to strengthening and development of a midlatitude cyclone along the eastern slope of the Rockies, other north–south mountain barriers, and along the North American and Asian east coasts (see midlatitude cyclone, polar front).

Cyclone (6) A dynamically or thermally caused area of low atmospheric pressure with converging and ascending air flows. Rotates counterclockwise in the Northern Hemisphere and clockwise in the Southern Hemisphere (compare anticyclone; see midlatitude cyclone, tropical cyclone).

Daylength (2) Duration of exposure to insolation, varying during the year depending on latitude; an important aspect of seasonality.

Daylight saving time (1) Time is set ahead 1 hour in the spring and set back 1 hour in the fall in the Northern Hemisphere. In the United States and Canada time is set ahead on the first Sunday in April and set back on the last Sunday in October—except in Hawai'i, Arizona, portions of Indiana, and Saskatchewan, which exempt themselves.

Debris avalanche (13) A mass of falling and tumbling rock, debris, and soil; can be dangerous because of the tremendous velocities achieved by the onrushing materials.

Declination (2) The latitude that receives direct overhead (perpendicular) insolation on a particular day; the subsolar point migrates annually through 47° of latitude between the Tropics of Cancer (23.5° N) and Capricorn (23.5° S).

Decomposer (19) Microorganisms that digest and recycle organic debris and waste in the environment, including bacteria, fungi, insects, and worms.

Deficit (9) DEFIC; in a water balance, the amount of unmet (unsatisfied) potential evapotranspiration; a natural water shortage (refer to POTET, or PE).

Deflation (15) A process of wind erosion that removes and lifts individual particles, literally blowing away unconsolidated, dry, or noncohesive sediments (see blowout depression).

Delta (14) A depositional plain formed where a river enters a lake or an ocean; named after the triangular shape of the Greek letter delta, Δ.

Denudation (13) A general term that refers to all processes that cause degradation of the landscape: weathering, mass movement, erosion, and transport.

Deposition (14) The process whereby weathered, wasted, and transported sediments are laid down by air, water, and ice.

Desert biome (20) Arid landscapes of uniquely adapted dry-climate plants and animals.

Desertification (15) The expansion of deserts worldwide, related principally to poor agricultural practices (overgrazing and inappropriate agricultural practices), improper soil-moisture management, erosion and salinization, deforestation, and the ongoing climatic change; an unwanted semipermanent invasion into neighboring biomes.

Desert pavement (15) On arid landscapes, a surface formed when wind deflation and sheetflow remove smaller particles, leaving residual pebbles and gravels to concentrate at the surface; resembles a cobblestone street (see deflation, sheetflow).

Dew-point temperature (7) The temperature at which a given mass of air becomes saturated, holding all the water it can hold. Any further cooling or addition of water vapor results in active condensation.

Diagnostic subsurface horizon (18) A soil horizon that originates below the epipedon at varying depths; may be part of the A and B horizons; important in soil description as part of the Soil Taxonomy.

Differential weathering (13) The effect of different resistances in rock, coupled with variations in the intensity of physical and chemical weathering.

Diffuse radiation (4) The downward component of scattered incoming insolation from clouds and the atmosphere.

Discharge (9) The measured volume of flow in a river that passes by a given cross section of a stream in a given unit of time; expressed in cubic meters per second or cubic feet per second.

Dissolved load (14) Materials carried in chemical solution in a stream, derived from minerals such as limestone, dolomite, or from soluble salts.

Downwelling current (6) An area of the sea where a convergence or accumulation of water thrusts excess water downward; occurs, for example, at the western end of the equatorial current or along the margins of Antarctica (compare upwelling currents).

Drainage basin (14) The basic spatial geomorphic unit of a river system; distinguished from a neighboring basin by ridges and highlands that form divides, marking the limits of the catchment area of the drainage basin, or its watershed.

Drainage density (14) A measure of the overall operational efficiency of a drainage basin; determined by the ratio of combined channel lengths to the unit area.

Drainage pattern (14) A distinctive geometric arrangement of streams in a region; determined by slope, differing rock resistance to weathering and erosion, climatic and hydrologic variability, and structural controls of the landscape.

Drawdown (9) See cone of depression.

Drumlin (17) A depositional landform related to glaciation that is composed of till (unstratified, unsorted) and is streamlined in the direction of continental ice movement; blunt end upstream and tapered end downstream with a rounded summit.

Dry adiabatic rate (DAR) (7) The rate at which an unsaturated parcel of air cools (if ascending) or heats (if descending); a rate of 10 C° per 1000 m (5.5 F° per 1000 ft) (see adiabatic; compare moist adiabatic rate).

Dune (15) A depositional feature of sand grains deposited in transient mounds, ridges, and hills; extensive areas of sand dunes are called sand seas.

Dust dome (4) A dome of airborne pollution associated with every major city; may be blown by winds into elongated plumes downwind from the city.

Dynamic equilibrium (1) Increasing or decreasing operations in a system demonstrate a trend over time, a change in average conditions.

Dynamic equilibrium model (13) The balancing act between tectonic uplift and erosion, between the resistance of crust materials and the work of denudation processes. Landscapes evidence ongoing adaptation to rock structure, climate, local relief, and elevation.

Earthquake (12) A sharp release of energy that sends waves traveling through Earth's crust at the moment of rupture along a fault or in association with volcanic activity. The moment magnitude scale (formerly the Richter scale) estimates Earthquake magnitude; intensity is described by the Mercalli scale.

Earth systems science (1) An emerging science of Earth as a complete, systematic entity. An interacting set of physical, chemical, and biological systems that produce the processes of a whole Earth system. A study of planetary change resulting from system operations; includes a desire for a more quantitive understanding among components, rather than qualitative description.

Ebb tide (16) Falling or lowering tide during the daily tidal cycle (compare flood tide).

Ecological succession (19) The process whereby different and usually more complex assemblages of plants and animals replace older and usually simpler communities; communities are in a constant state of change as each species adapts to conditions; ecosystems do not exhibit a stable point or successional climax condition as previously thought (refer to primary succession, secondary succession).

Ecology (19) The science that studies the relations between organisms and their environment and among various ecosystems.

Ecosphere (1)   Another name for the biosphere.

Ecosystem (19, 20)   A self-regulating association of living plants, animals, and their nonliving physical and chemical environment.

Ecotone (20)   A boundary transition zone between adjoining ecosystems that may vary in width and represent areas of tension as similar species of plants and animals compete for the resources (see ecosystem).

Effusive eruption (12)   A volcanic eruption characterized by low-viscosity basaltic magma and low-gas content, which readily escapes. Lava pours forth onto the surface with relatively small explosions and few pyroclastics; tends to form shield volcanoes (see shield volcano, lava, pyroclastics; compare explosive eruption).

Elastic-rebound theory (12)   A concept describing the faulting process in Earth's crust, in which the two sides of a fault appear locked despite the motion of adjoining pieces of crust, but with accumulating strain they rupture suddenly, snapping to new positions relative to each other, generating an earthquake.

Electromagnetic spectrum (2)   All the radiant energy produced by the Sun placed in an ordered range, divided according to wavelengths.

Eluviation (18)   The removal of finer particles and minerals from the upper horizons of soil; an erosional process within a soil body (compare illuviation).

Empirical classification (10)   A climate classification based on weather statistics or other data; used to determine general climate categories (compare genetic classification).

Endogenic system (11)   The system internal to Earth, driven by radioactive heat derived from sources within the planet. In response, the surface fractures, mountain building occurs, and earthquakes and volcanoes are activated (compare exogenic system).

Entisols (18)   A soil order in the Soil Taxonomy. Specifically lacks vertical development of horizons; usually young or undeveloped. Found in active slopes, alluvial-filled floodplains, poorly drained tundra.

Environmental lapse rate (3)   The actual lapse rate in the lower atmosphere at any particular time under local weather conditions; may deviate above or below the normal lapse rate of 6.4 C° per 1000 m (3.5 F° per 1000 ft). (Compare normal lapse rate.)

Eolian (15)   Caused by wind; refers to the erosion, transportation, and deposition of materials; spelled aeolian in some countries.

Epipedon (18)   The diagnostic soil horizon that forms at the surface; not to be confused with the A horizon; may include all or part of the illuviated B horizon.

Equal area (1)   A trait of a map projection; indicates the equivalence of all areas on the surface of the map, although shape is distorted (see map projections).

Equatorial and tropical rain forest (20)   A lush biome of tall broadleaf evergreen trees and diverse plants and animals, roughly between 23.5° N and 23.5° S. The dense canopy of leaves is usually arranged in three levels.

Equatorial low-pressure trough (6)   A thermally caused low-pressure area that almost girdles Earth, with air converging and ascending all along its extent; also called the intertropical convergence zone (ITCZ).

Erg desert (15)   A sandy desert, or area where sand is so extensive that it constitutes a sand sea.

Erosion (14)   Denudation by wind, water, or ice, which dislodges, dissolves, or removes surface material.

Esker (17)   A sinuously curving, narrow deposit of coarse gravel that forms along a meltwater stream channel, developing in a tunnel beneath a glacier.

Estuary (14)   The point at which the mouth of a river enters the sea, where freshwater and seawater are mixed; a place where tides ebb and flow.

Eustasy (7)   Refers to worldwide changes in sea level that are not related to movements of land but rather to a rise and fall in the volume of water in the oceans.

Evaporation (9)   The movement of free water molecules away from a wet surface into air that is less than saturated; the phase change of water to water vapor.

Evaporation fog (7)   A fog formed when cold air flows over the warm surface of a lake, ocean, or other body of water; forms as the water molecules evaporate from the water surface into the cold, overlying air; also known as steam fog or sea smoke.

Evaporation pan (9)   A weather instrument consisting of a standardized pan from which evaporation occurs, with water automatically replaced and measured; an evaporimeter.

Evapotranspiration (9)   The merging of evaporation and transpiration water loss into one term (see potential and actual evapotranspiration).

Exfoliation dome (13)   A dome-shaped feature of weathering, produced by the response of granite to the overburden removal process, which relieves pressure from the rock. Layers of rock slough (sluff) off in slabs or shells in a sheeting process.

Exogenic system (11)   Earth's external surface system, powered by insolation, which energizes air, water, and ice and sets them in motion, under the influence of gravity. Includes all processes of landmass denudation (compare endogenic system).

Exosphere (3)   An extremely rarefied outer atmospheric halo beyond the thermopause at an altitude of 480 km (300 mi); probably composed of hydrogen and helium atoms, with some oxygen atoms and nitrogen molecules present near the thermopause.

Exotic stream (14)   A river that rises in a humid region and flows through an arid region, with discharge decreasing toward the mouth; for example, the Nile River and the Colorado River.

Explosive eruption (12)   A violent and unpredictable volcanic eruption, the result of magma that is thicker (more viscous), stickier, and higher in gas and silica content than that of an effusive eruption; tends to form blockages within a volcano; produces composite volcanic landforms (see composite volcano; compare effusive eruption).

Faulting (12)   The process whereby displacement and fracturing occurs between two portions of Earth's crust; usually associated with earthquake activity.

Feedback loop (1)   Created when a portion of system output is returned as an information input, causing changes that guide further system operation (see negative feedback, positive feedback).

Field capacity (9)   Water held in the soil by hydrogen bonding against the pull of gravity, remaining after water drains from the larger pore spaces; the available water for plants (see available water, capillary water).

Fire ecology (19)   The study of fire as a natural agent and dynamic factor in community succession.

Firn (17)   Snow of a granular texture that is transitional in the slow transformation from snow to glacial ice; snow that has persisted through a summer season in the zone of accumulation.

Firn line (17)   The snow line that is visible on the surface of a glacier, where winter snows survive the summer ablation season; analogous to a snowline on land (see ablation).

Fjord (17)   A drowned glaciated valley, or glacial trough, along a sea coast.

Flash flood (15)   A sudden and short-lived torrent of water that exceeds the capacity of a stream channel; associated with desert and semiarid washes.

Flood (14)   A high water level that overflows the natural (or artificial) levees along any portion of a stream.

Floodplain (14)   A flat low-lying area along a stream channel, created by and subject to recurrent flooding; alluvial deposits generally mask underlying rock.

Flood tide (16)   Rising tide during the daily tidal cycle (compare ebb tide).

Fluvial (14)   Stream-related processes; from the Latin *fluvius* for "river" or "running water."

Fog (7)   A cloud, generally stratiform, in contact with the ground, with visibility usually reduced to less than 1 km (3300 ft).

Folding (12)   The bending and deformation of beds of rock strata subjected to compressional forces.

Food chain (19)   The circuit along which energy flows from producers (plants), which manufacture their own food, to consumers (animals); a one-directional flow of chemical energy, ending with decomposers.

Food web (19)   A complex network of interconnected food chains (see food chain).

Formation class (20)   That portion of a biome that concerns the plant communities only, categorized by size, shape, and structure of the dominant vegetation.

Friction force (6)   The effect of drag by the wind as it moves across a surface; may be operative through 500 m (1600 ft) of altitude. Surface friction slows the wind and therefore reduces the effectiveness of the Coriolis force.

Frost action (13)   A powerful mechanical force produced as water expands up to 9% of its volume as it freezes. Water freezing in a cavity in a rock can break the rock if it exceeds the rock's tensional strength.

Funnel cloud (8)   The visible swirl extending from the bottom side of a cloud, which may or may not develop into a tornado. A tornado is a funnel cloud that has extended all the way to the ground (see tornado).

Fusion (2)   The process of forcibly joining positively charged hydrogen and helium nuclei under extreme temperature and pressure; occurs naturally in thermonuclear reactions within stars, such as our Sun.

Gelifluction (17)   Refers to soil flow in periglacial environments; a progressive, lateral movement. A type of solifluction formed under periglacial conditions of permafrost and frozen ground (see solifluction).

Gelisols (18)   A new soil order in the Soil Taxonomy, added in 1998, describing cold and frozen soils at high latitudes or high elevations; characteristic tundra vegetation (see Canadian System of Soil Classification for similar types).

General circulation model (GCM) (10)   Complex, computer-based climate model that produces generalizations of reality and forecasts of future weather and climate conditions. Complex GCMs (three-dimensional models) are in use in the United States and in other countries.

Genetic classification (10)   A climate classification that uses causative factors to determine climatic regions; for example, an analysis of the effect of interacting air masses (compare empirical classification).

Geodesy (1)   The science that determines Earth's shape and size through surveys, mathematical means, and remote sensing (see geoid).

Geographic information system (GIS) (1)   A computer-based data processing tool or methodology used for gathering, manipulating, and analyzing geographic information to produce a holistic, interactive analysis.

Geography (1)   The science that studies the interdependence and interaction among geographic areas, natural systems, processes, society, and cultural activities over space—a spatial science. The five themes of geographic education are location, place, movement, regions, and human–Earth relationships.

Geoid (1)   A word that describes Earth's shape, literally, "the shape of Earth is Earth-shaped." A theoretical surface at sea level that extends through the continents; deviates from a perfect sphere.

Geologic cycle (11)   A general term characterizing the vast cycling that proceeds in the lithosphere. It encompasses the hydrologic cycle, tectonic cycle, and rock cycle.

Geologic time scale (11)   A depiction of eras, periods, and epochs that span Earth's history; shows both the sequence of rock strata and their absolute dates, as determined by methods such as radioactive isotopic dating.

Geomagnetic reversal (11)   A polarity change in Earth's magnetic field. With uneven regularity, the magnetic field fades to zero, then returns to full strength but with the magnetic poles reversed. Reversals have been recorded nine times during the past four million years.

Geomorphic threshold (13)   The threshold up to which landforms change before lurching to a new set of relationships, with rapid realignments of landscape materials and slopes.

Geomorphology (13)   The science that analyzes and describes the origin, evolution, form, classification, and spatial distribution of landforms.

Geostrophic wind (6)   A wind moving between areas of different pressure along a path that is parallel to the isobars. It is a product of the pressure gradient force and the Coriolis force (see isobar, pressure gradient force, Coriolis force).

Geothermal energy (11, 12)   The energy in steam and hot water heated by subsurface magma near groundwater. Geothermal energy literally refers to heat from Earth's interior, whereas *geothermal power* relates to specific applied strategies of geothermal electric or geothermal direct applications. This energy is used in Iceland, New Zealand, Italy, and northern California, among other locations.

Glacial drift (17)   The general term for all glacial deposits, both unsorted (till) and sorted (stratified drift).

Glacial ice (17)   A hardened form of ice, very dense in comparison to normal snow or firn.

Glacier (17)   A large mass of perennial ice resting on land or floating shelf-like in the sea adjacent to the land; formed from the accumulation and recrystallization of snow, which then flows slowly under the pressure of its own weight and the pull of gravity.

Glacier surge (17)   The rapid, lurching, unexpected forward movement of a glacier.

Glacio-eustatic (7)   Changes in sea level in response to changes in the amount of water stored on Earth as ice; the more water that is bound up in glaciers and ice sheets, the lower the sea level (compare eustasy).

Gleization (18)   A process of humus and clay accumulation in cold, wet climates with poor drainage.

Global positioning system (GPS) (1)   Latitude, longitude, and elevation are accurately calibrated using this hand-held instrument that calibrates radio signals from satellites.

Goode's homolosine projection (Appendix A)   An equal-area projection formed by splicing together a sinusoidal and a homolographic projection.

Graben (12)   Pairs or groups of faults that produce downward faulted blocks; characteristic of the basins of the interior western United States (compare horst; see Basin and Range Province).

Graded stream (14)   An idealized condition in which a stream's load and the landscape mutually adjust. This forms a dynamic equilibrium among erosion, transported load, deposition, and the stream's capacity.

Gradient (14)   The drop in elevation from a stream's headwaters to its mouth, ideally forming a concave slope.

Granite (11)   A coarse-grained (slow-cooling) intrusive igneous rock of 25% quartz and more than 50% potassium and sodium feldspars; characteristic of the continental crust.

Gravitational water (9)   That portion of surplus water that percolates downward from the capillary zone, pulled by gravity to the groundwater zone.

Gravity (2)   The mutual force exerted by the masses of objects that are attracted one to another; produced in an amount proportional to each object's mass.

Great circle (1)   Any circle drawn on a globe with its center coinciding with the center of the globe. An infinite number of great circles can be drawn, but only one parallel is a great circle—the equator (compare small circle).

Greenhouse effect (4)   The process whereby radiatively active gases (carbon dioxide, water vapor, methane, and CFCs) absorb insolation and reradiate the energy at longer wavelengths, which are retained longer, delaying the loss of infrared to space. Thus, the lower troposphere is warmed through the radiation and re-radiation of infrared wavelengths. The approximate similarity between this process and that of a greenhouse explains the name.

Greenwich Mean Time (GMT) (1)   Former world standard time, now reported as Coordinated Universal Time (UTC) (see Coordinated Universal Time).

Ground ice (17)   Subsurface water that is frozen in regions of permafrost. The moisture content of areas with ground ice may vary from nearly absent in regions of drier permafrost to almost 100% in saturated soils (compare permafrost).

Groundwater (9)   Water beneath the surface that is beyond the soil-root zone; a major source of potable water.

Groundwater mining (9)   Pumping an aquifer beyond its capacity to flow and recharge; an overuse of the groundwater resource.

Gulf Stream (5)   A strong northward-moving warm current off the east coast of North America, which carries its water far into the North Atlantic.

Habitat (19, 20)   A physical location to which an organism is biologically suited. Most species have specific habitat parameters and limits (compare to niche).

Hail (8)   A type of precipitation formed when a raindrop is repeatedly circulated above and below the freezing level in a cloud, with each cycle freezing more moisture onto the hailstone until it becomes too heavy to stay aloft.

Hair hygrometer (7)   An instrument for measuring relative humidity; based on the principle that human hair will change as much as 4% in length between 0% and 100% relative humidity.

Heat (3)   The flow of kinetic energy from one body to another because of a temperature difference between them.

Herbivore (19)   The primary consumer in a food web, which eats plant material formed by a producer (plant) that has photosynthesized organic molecules (compare carnivore).

Heterosphere (3)   A zone of the atmosphere above the mesopause, 80 km (50 mi) in altitude; composed of rarefied layers of oxygen atoms and nitrogen molecules; includes the ionosphere.

Histosols (18)   A soil order in the Soil Taxonomy. Formed from thick accumulations of organic matter, such as beds of former lakes, bogs, and layers of peat.

Homosphere (3)   A zone of the atmosphere from Earth's surface up to 80 km (50 mi), composed of an even mixture of gases, including nitrogen, oxygen, argon, carbon dioxide, and trace gases.

Horn (17)   A pyramidal, sharp-pointed peak that results when several cirque glaciers gouge an individual mountain summit from all sides.

Horst (12)   Upward-faulted blocks produced by pairs or groups of faults; characteristic of the mountain ranges of the interior of the western United States (see graben and Basin and Range Province).

Hot spot (11)   An individual point of upwelling material originating in the asthenosphere, or deeper in the mantle; tends to remain fixed relative to migrating plates; some 100 are identified worldwide; exemplified by Yellowstone National Park, Hawai'i, and Iceland.

Human–Earth relationships (1)   One of the oldest themes of geography (the human–land tradition); includes the spatial analysis of settlement patterns, resource utilization and exploitation, hazard perception and planning, and the impact of environmental modification and artificial landscape creation.

Humidity (7)   Water vapor content of the air. The capacity of the air to hold water vapor is mostly a function of the temperature of the air and the water vapor.

Humus (18)   A mixture of organic debris in the soil, worked by consumers and decomposers in the humification process; characteristically formed from plant and animal litter deposited at the surface.

Hurricane (8)   A tropical cyclone that is fully organized and intensified in inward-spiraling rainbands; ranges from 160 to 960 km (100 to 600 mi) in diameter, with wind speeds in excess of 119 kmph (65 knots, or 74 mph); a name used specifically in the Atlantic and eastern Pacific (compare typhoon).

Hydration (13)   A physical weathering process involving water, although not involving any chemical change; water is added to a mineral, which initiates swelling and stress within the rock, mechanically forcing grains apart as the constituents expand (contrast to hydrolysis).

Hydraulic action (14)   The erosive work accomplished by the turbulence of water; causes a squeezing and releasing action in joints in bedrock; capable of prying and lifting rocks.

Hydrograph (14)   A graph of stream discharge (in cms or cfs) over a period of time (minutes, hours, days, years) at a specific place on a stream. The relationship between stream discharge and precipitation input is illustrated on the graph.

Hydrologic cycle (9)   A simplified model of the flow of water, ice, and water vapor from place to place. Water flows through the atmosphere, across the land, where it is also stored as ice, and within groundwater. Solar energy empowers the cycle.

Hydrolysis (13)   A chemical weathering process in which minerals chemically combine with water; a decomposition process that causes silicate minerals in rocks to break down and become altered (contrast with hydration).

Hydrosphere (1)   An abiotic open system that includes all of Earth's water.

Hygroscopic water (9)   That portion of soil moisture that is so tightly bound to each soil particle that it is unavailable to plant roots; the water, along with some bound capillary water, that is left in the soil after the wilting point is reached (see wilting point).

Ice age (17)   A cold episode, with accompanying alpine and continental ice accumulations, which has repeated roughly every 200 to 300 million years since the late Precambrian era (1.25 billion years ago); includes the most recent episode during the Pleistocene Ice Age, which began 1.65 million years ago.

Iceberg (17)   Floating ice created by calving ice (a large piece breaking off) and floating adrift; a hazard to shipping because about nine-tenths of the ice is submerged and can be irregular in form.

Ice cap (17)   A large dome-shaped glacier, less extensive than an ice sheet ($<$50,000 km^2; $<$19,300 mi^2), although it buries mountain peaks and the local landscape.

Ice field (17)   The least extensive form of a glacier, with mountain ridges and peaks visible above the ice; less than an ice cap or ice sheet.

Icelandic low (6)   See subpolar low-pressure cell.

Ice sheet (17)   An enormous continuous continental glacier. The bulk of glacial ice on Earth covers Antarctica and Greenland in two ice sheets.

Ice wedge (17)   Formed when water enters a thermal contraction crack in permafrost and freezes. Repeated seasonal freezing and melting of the water progressively expands the wedge.

Igneous rock (11)   One of the basic rock types; it has solidified and crystallized from a hot molten state (either magma or lava). (Compare metamorphic rock, sedimentary rock.)

Illuviation (18)   The downward movement and deposition of finer particles and minerals from the upper horizon of the soil; a depositional process. Deposition usually is in the B horizon, where accumulations of clays, aluminum, carbonates, iron, and some humus occur (compare eluviation; see calcification).

Inceptisols (18)   A soil order in the Soil Taxonomy. Weakly developed soils that are inherently infertile. Usually young soils that are weakly developed, although they are more developed than Entisols.

Industrial smog (3)   Air pollution associated with coal-burning industries; it may contain sulfur oxides, particulates, carbon dioxide, and exotics.

Infiltration (9)   Water access to subsurface regions of soil moisture storage through penetration of the soil surface.

Insolation (2)   Solar radiation that is intercepted by Earth.

Interception (9)   Delays the fall of precipitation toward Earth's surface; caused by vegetation or other ground cover.

Internal drainage (14)   In regions where rivers do not flow into the ocean, the outflow is through evaporation or subsurface gravitational flow. Portions of Africa, Asia, Australia, and the western United States have such drainage.

International Date Line (1)   The 180° meridian; an important corollary to the prime meridian on the opposite side of the planet; established by the treaty of 1884 to mark the place where each day officially begins.

Intertropical convergence zone (ITCZ) (6)   See equatorial low-pressure trough.

Ionosphere (3)   A layer in the atmosphere above 80 km (50 mi) where gamma, X-ray, and some ultraviolet radiation is absorbed and converted into infrared, and where the solar wind stimulates the auroras.

Isobar (6)   An isoline connecting all points of equal atmospheric pressure.

Isostasy (11)   A state of equilibrium in Earth's crust formed by the interplay between portions of the lithosphere and the asthenosphere and the principle of buoyancy. The crust depresses under weight and recovers with its removal, for example, the melting of glacial ice. The uplift is known as isostatic rebound.

Isotherm (5)   An isoline connecting all points of equal temperature.

Jet stream (6)   The most prominent movement in upper-level westerly wind flows; irregular, concentrated, sinuous bands of geostrophic wind, traveling at 300 kmph (190 mph). (See polar jet stream, subtropical jet stream.)

Joint (13)   A fracture or separation in rock that occurs without displacement of the sides; increases the surface area of rock exposed to weathering processes.

Kame (17)   A depositional feature of glaciation; a small hill of poorly sorted sand and gravel that accumulates in crevasses or in ice-caused indentations in the surface.

Karst topography (13)   Distinctive topography formed in a region of chemically weathered limestone with poorly developed surface drainage and solution features that appear pitted and bumpy; originally named after the Krš Plateau of Yugoslavia.

Katabatic winds (6)   Air drainage from elevated regions, flowing as gravity winds. Layers of air at the surface cool, become denser, and flow downslope; known worldwide by many local names.

Kettle (17)   Forms when an isolated block of ice persists in a ground moraine, an outwash plain, or valley floor after a glacier retreats; as the block finally melts, it leaves behind a steep-sided hole that frequently fills with water.

Kinetic energy (3)   The energy of motion in a body; derived from the vibration of the body's own movement and stated as temperature.

Köppen–Geiger climate classification (10)   An empirical classification system that uses average monthly temperatures, average monthly precipitation, and total annual precipitation to establish regional climate designations.

Lagoon (16)   An area of coastal seawater that is virtually cut off from the ocean by a bay barrier or barrier beach; also, the water surrounded and enclosed by an atoll.

Landfall (8)   The location along a coast where a storm moves onshore.

Land–sea breeze (6)   Wind along coastlines and adjoining interior areas created by different heating characteristics of land and water surfaces—

onshore (landward) breeze in the afternoon and offshore (seaward) breeze at night.

Landslide (13) A sudden rapid downslope movement of a cohesive mass of regolith and/or bedrock in a variety of mass-movement forms under the influence of gravity; a form of mass movement.

Land–water heating difference (5) Differences in the degree and way that land and water heat, as a result of contrasts in transmission, evaporation, mixing, and specific heat capacities. Land surfaces heat and cool faster than water and have continentality, whereas water provides a marine influence.

Latent heat (7) Heat energy is stored in one of the three states—ice, water, or water vapor. The energy is absorbed or released in each phase change from one state to another. Heat energy is absorbed as the latent heat of melting, vaporization, or evaporation. Heat energy is released as the latent heat of condensation and freezing (or fusion).

Latent heat of condensation (7) The heat energy released to the environment in a phase change from water vapor to liquid; under normal sea-level pressure, 540 calories are released from each gram of water vapor that changes phase to water at boiling; and 585 calories are released from each gram of water vapor that condenses at 20°C (68°F).

Latent heat of sublimation (7) The heat energy absorbed or released in the phase change from ice to water vapor or water vapor to ice—no liquid phase. Water vapor to ice also called deposition.

Latent heat of vaporization (7) The heat energy absorbed from the environment in a phase change from liquid to water vapor at the boiling point; under normal sea-level pressure, 540 calories must be added to each gram of boiling water to achieve a phase change to water vapor.

Lateral moraine (17) Debris transported by a glacier that accumulates along the sides of the glacier and is deposited along these margins.

Laterization (18) A pedogenic process operating in well-drained soils that occur in warm and humid regions; typical of Oxisols. Plentiful precipitation leaches soluble minerals and soil constituents. Resulting soils usually are reddish or yellowish.

Latitude (1) The angular distance measured north or south of the equator from a point at the center of Earth. A line connecting all points of the same latitudinal angle is called a parallel (compare longitude).

Lava (11, 12) Magma that issues from volcanic activity onto the surface; the extrusive rock that results when magma solidifies (see magma).

Life zone (19) A zonation by altitude of plants and animals that form distinctive communities. Each life zone possesses its own temperature and precipitation relations.

Lightning (8) Flashes of light caused by tens of millions of volts of electrical charge heating the air to temperatures of 15,000° to 30,000°C.

Limestone (11) The most common chemical sedimentary rock (non-clastic); it is lithified calcium carbonate ($CaCO_3$); very susceptible to chemical weathering by acids in the environment, including carbonic acid in rainfall.

Limiting factor (19) The physical or chemical factor that most inhibits biotic processes, either through lack or excess.

Lithification (11) The compaction, cementation, and hardening of sediments into sedimentary rock.

Lithosphere (1) Earth's crust and that portion of the upper-most mantle directly below the crust, extending down to about 70 km (45 mi). Some sources use this term to refer to the entire Earth.

Littoral zone (16) A specific coastal environment; that region between the high water line during a storm and a depth at which storm waves are unable to move sea-floor sediments.

Loam (18) A soil that is a mixture of sand, silt, and clay in almost equal proportions, with no one texture dominant; an ideal agricultural soil.

Location (1) A basic theme of geography dealing with the absolute and relative position of people, places, and things on Earth's surface.

Loess (15) Large quantities of fine-grained clays and silts left as glacial outwash deposits; subsequently blown by the wind great distances and redeposited as a generally unstratified, homogeneous blanket of material covering existing landscapes; in China loess originated from desert lands.

Longitude (1) The angular distance measured east or west of a prime meridian from a point at the center of Earth. A line connecting all points of the same longitude is called a meridian.

Longshore current (16) A current that forms parallel to a beach as waves arrive at an angle to the shore; generated in the surf zone by wave action, transporting large amounts of sand and sediment (see littoral current, beach drift).

Lysimeter (9) A weather instrument; device for measuring potential and actual evapotranspiration; isolates a portion of afield so that the moisture moving through the plot is measured.

Magma (11) Molten rock from beneath Earth's surface; fluid, gaseous, under tremendous pressure, and either intruded into existing country rock or extruded onto the surface as lava (see lava).

Magnetosphere (2) Earth's magnetic force field, which is generated by dynamo-like motions within the planet's outer core; deflects the solar wind flow toward the upper atmosphere above each pole.

Mangrove swamp (16) A wetland ecosystem between 30° N or S and the equator; tends to form a distinctive community of mangrove plants (compare salt marsh).

Mantle (11) An area within the planet representing about 80% of Earth's total volume, with densities increasing with depth and averaging 4.5 g/cm^3; occurs between the core and the crust; is rich in iron and magnesium oxides and silicates.

Map (1, Appendix A) A generalized view of an area, usually some portion of Earth's surface, as seen from above at a greatly reduced size (see scale and map projection).

Map projection (1, Appendix A) The reduction of a spherical globe onto a flat surface in some orderly and systematic realignment of the latitude and longitude grid.

Marine effect (5) A quality of regions that are dominated by the moderating effect of the ocean and that exhibit a smaller range of minimum and maximum temperature than continental stations (see continentality, land–water heating differences).

Mass movement (13) All unit movements of materials propelled by gravity; can range from dry to wet, slow to fast, small to large, and free-falling to gradual or intermittent.

Mass wasting (13) Gravitational movement of nonunified material downslope; a specific form of mass movement.

Meandering stream (14) The sinuous, curving pattern common to graded streams, with the energetic outer portion of each curve subjected to the greatest erosive action and the lower-energy inner portion receiving sediment deposits (see graded stream).

Mean sea level (MSL) (16) The average of tidal levels recorded hourly at a given site over a long period, which must be at least a full lunar tidal cycle.

Medial moraine (17) Debris transported by a glacier that accumulates down the middle of the glacier, resulting from two glaciers merging their lateral moraines; forms a depositional feature following glacial retreat.

Mediterranean shrubland (20) A major biome dominated by Mediterranean dry summer climates and characterized by sclerophyllous scrub and short, stunted, tough forests (see chaparral).

Mercator projection (1) A true-shape projection, with meridians appearing as equally spaced straight lines and parallels appearing as straight lines that are spaced closer together near the equator. The poles are infinitely stretched, with the 84th north parallel and 84th south parallel fixed at the same length as that of the equator. It presents false notions of the size (area) of midlatitude and poleward landmasses but presents true compass direction (see rhumb line).

Mercury barometer (6) A device that measures air pressure using a column of mercury in a tube, one end of which is sealed, and the other end inserted in an open vessel of mercury (see air pressure).

Meridian (1) See longitude; a line designating an angle of longitude.

Mesocyclone (8) A large rotating atmospheric circulation, initiated within a parent cumulonimbus cloud at mid-troposphere elevation; generally produces heavy rain, large hail, blustery winds, and lightning; may lead to tornado activity.

Mesosphere (3) The upper region of the homosphere from 50 to 80 km (30 to 50 mi) above the ground; designated by temperature criteria; atmosphere extremely rarified.

Metamorphic rock (11) One of three basic rock types, it is existing igneous and sedimentary rock that has undergone profound physical and chemical changes under increased pressure and temperature. Constituent mineral structures may exhibit foliated or nonfoliated textures (compare igneous rock and sedimentary rock).

Meteorology (8) The scientific study of the atmosphere, including its physical characteristics and motions; related chemical, physical, and geological processes; the complex linkages of atmospheric systems; and weather forecasting.

Microclimatology (4) The study of local climates at or near Earth's surface, or that height above the Earth's surface where the effects of the surface are no longer of affect.

Midlatitude broadleaf and mixed forest (20) A biome in moist continental climates in areas of warm-to-hot summers and cool-to-cold winters; relatively lush stands of broadleaf forests trend northward into needleleaf evergreen stands.

Midlatitude cyclone (8) An organized area of low pressure, with converging and ascending air flow producing an interaction of air masses; migrates along storm tracks. Such lows or depressions form the dominant weather pattern in the middle and higher latitudes of both hemispheres.

Midlatitude grassland (20) The major biome most modified by human activity; so named because of the predominance of grass-like plants, although deciduous broadleafs appear along streams and other limited sites; location of the world's breadbaskets of grain and livestock production.

Mid-ocean ridge (11) A submarine mountain range that extends more than 65,000 km (40,000 mi) worldwide and averages more than 1000 km (600 mi) in width; centered along sea-floor spreading centers (see sea-floor spreading).

Milky Way Galaxy (2) A flattened, disk-shaped mass in space estimated to contain up to 400 billion stars; includes our Solar System.

Miller cylindrical projection (Appendix A) A compromise map projection that avoids the severe distortion of the Mercator projection (see map projection).

Mineral (11) An element or combination of elements that forms an inorganic natural compound, described by a specific formula and crystal structure.

Model (1) A simplified version of a system, representing an idealized part of the real world.

Mohorovičić discontinuity, or Moho (11) The boundary between the crust and the rest of the lithospheric upper mantle; named for the Yugoslavian seismologist Mohorovičić; a zone of sharp material and density contrasts; also known as the Moho.

Moist adiabatic rate (MAR) (7) The rate at which a saturated parcel of air cools in ascent; a rate of 6 C° per 1000 m (3.3 F° per 1000 ft). This rate may vary, with moisture content and temperature, from 4 C° to 10 C° per 1000 m (2 F° to 6 F° per 1000 ft) (see adiabatic; compare dry adiabatic rate).

Moisture droplet (7) A tiny water particle that constitutes the initial composition of clouds. Each droplet measures approximately 0.002 cm (0.0008 in.) in diameter and is invisible to the unaided eye.

Mollisols (18) A soil order in the Soil Taxonomy. These have a mollic epipedon and a humus-rich organic content high in alkalinity. Some of the world's most significant agricultural soils are Mollisols.

Moment magnitude scale (12) An earthquake magnitude scale. Considers the amount of fault slippage, the size of the area that ruptured, and the nature of the materials that faulted in estimating the magnitude of an earthquake—an assessment of the seismic moment. Replaces the Richter scale (amplitude magnitude), especially valuable in assessing larger magnitude events.

Monsoon (6) An annual cycle of dryness and wetness, with seasonally shifting winds produced by changing atmospheric pressure systems; affects India, Southeast Asia, Indonesia, northern Australia, and portions of Africa. From the Arabic word *mausim*, meaning "season."

Montane forest (20) Needleleaf forest associated with mountain elevations (see needleleaf forest).

Moraine (17) Marginal glacial deposits (lateral, medial, terminal, ground) of unsorted and unstratified material.

Mountain–valley breeze (6) A light wind produced as cooler mountain air flows downslope at night and as warmer valley air flows upslope during the day.

Movement (1) A major theme in geography involving migration, communication, and the interaction of people and processes across space.

Mudflow (13) Fluid downslope flows of material containing more water than earthflows.

Natural levee (14) A long, low ridge that forms on both sides of a stream in a developed floodplain; they are depositional products (coarse gravels and sand) of river flooding.

Neap tide (16) Unusually low tidal range produced during the first and third quarters of the Moon, with an offsetting pull from the Sun (compare spring tide).

Needleleaf forest (20) Forests of pine, spruce, fir, and larch, stretching from the east coast of Canada westward to Alaska and continuing from Siberia westward across the entire extent of Russia to the European Plain; called the taiga (a Russian word) or the boreal forest; principally in the D climates. Includes montane forests that may be at lower latitudes at higher elevation.

Negative feedback (1) Feedback that tends to slow or dampen response in a system; promotes self-regulation in a system; far more common than positive feedback in living systems (see feedback loop, compare to positive feedback).

Net primary productivity (19) The net photosynthesis (photosynthesis minus respiration) for a given community; considers all growth and all reduction factors that affect the amount of useful chemical energy (biomass) fixed in an ecosystem.

Net radiation (NET R) (4) The net all-wave radiation available at Earth's surface; the final outcome of the radiation balance process between incoming shortwave insolation and outgoing longwave energy.

Niche (19, 20) The basic function, or occupation, of a life form within a given community; the way an organism obtains its food, air, and water.

Nickpoint (knickpoint) (14) The point at which the longitudinal profile of a stream is abruptly broken by a change in gradient; for example, a waterfall, rapids, or cascade.

Nimbostratus (7) Rain-producing, dark, grayish, stratiform clouds characterized by gentle drizzle.

Nitrogen dioxide (3) A noxious (harmful) reddish-brown gas produced in combustion engines; can be damaging to human respiratory tracts and to plants; participates in photochemical reactions and acid deposition.

Noctilucent cloud (3) A rare shining band of ice crystals that may glow at high latitudes long after Sunset; formed within the mesosphere, where cosmic and meteoric dust act as nuclei for the formation of ice crystals.

Normal fault (12) A type of geologic fault in rocks. Tension produces strain that breaks a rock with one side moving vertically relative to the other side along an inclined fault plane (compare reverse fault).

Normal lapse rate (3) The average rate of temperature decrease with increasing altitude in the lower atmosphere; an average value of 6.4 C° per km, or 1000 m (3.5 F° per 1000 ft). (Compare environmental lapse rate.)

Occluded front (8) In a cyclonic circulation, the overrunning of a surface warm front by a cold front and the subsequent lifting of the warm air wedge off the ground; initial precipitation is moderate to heavy.

Ocean basin (12) The physical container (a depression in the lithosphere) holding an ocean.

Omnivore (19) A consumer that feeds on both producers (plants) and consumers (meat)—a role occupied by humans, among other animals (compare consumer, producer).

Open system (1) A system with inputs and outputs crossing back and forth between the system and the surrounding environment. Earth is an open system in terms of energy (compare closed system).

Orogenesis (12) The process of mountain building that occurs when large-scale compression leads to deformation and uplift of the crust; literally the birth of mountains.

Orographic lifting (8) The uplift of a migrating air mass as it is forced to move upward over a mountain range—a topographic barrier. The lifted air cools adiabatically as it moves upslope; clouds may form and produce increased precipitation.

Outgassing (7) The release of trapped gases from rocks, forced out through cracks, fissures, and volcanoes from within Earth; the terrestrial source of Earth's water.

Outwash plain (17) Glacial stream deposits of stratified drift of meltwater-fed, braided, and overloaded streams; occurs beyond a glacier's morainal deposits.

Overland flow (9) Surplus water that flows across the land surface toward stream channels. Together with precipitation and subsurface flows, it constitutes the total runoff from an area.

Oxbow lake (14) A lake that was formerly part of the channel of a meandering stream; isolated when a stream eroded its outer bank forming a cutoff through the neck of the looping meander (see meandering stream). In Australia known as a billabong (Aboriginal for "dead river").

Oxidation (13) A chemical weathering process in which oxygen dissolved in water oxidizes (combines with) certain metallic elements to form oxides; most familiar is the "rusting" of iron in a rock or soil (Ultisols, Oxisols), which produces a reddish-brown stain of iron oxide (Fe_2O_3).

Oxisols (18) A soil order in the Soil Taxonomy. Tropical soils that are old, deeply developed, and lacking in horizons wherever well drained. Heavily weathered, low in cation exchange capacity, and low in fertility.

Ozone layer (3) See ozonosphere.

Ozonosphere (3) A layer of ozone (O_3) occupying the full extent of the stratosphere (20 to 50 km or 12 to 30 mi) above the surface; the region of the atmosphere where ultraviolet wavelengths of insolation are extensively absorbed and converted into heat.

Pacific high (6) A high-pressure cell that dominates the Pacific in July, retreating southward in the Northern Hemisphere in January; also known as the Hawaiian high.

Paleoclimatology (10, 17) The science that studies the climates, and the causes of variations in climate, of past ages, throughout historic and geologic time.

Paleolake (17) An ancient lake, such as Lake Bonneville or Lake Lahonton, associated with former wet periods when the lake basins were filled to higher levels than today.

Palsa (17) A rounded or elliptical mound of peat that contains thin perennial ice lenses rather than an ice core, as in a pingo. Palsas can be 2 to 30 m wide by 1 to 10 m high and usually are covered by soil or vegetation over a cracked surface.

PAN (3) See peroxyacetyl nitrate.

Pangaea (11) The supercontinent formed by the collision of all continental masses approximately 225 million years ago; named in the continental drift theory by Wegener in 1912 (see plate tectonics).

Parallel (1) See latitude; a line, parallel to the equator, that designates an angle of latitude.

Parent material (13) The unconsolidated material, from both organic and mineral sources, that is the basis of soil development.

Particulate matter (PM) (3) Dust, dirt, soot, salt, sulfate aerosols, fugitive natural particles, or other material particles suspended in air.

Paternoster lake (17) One of a series of small, circular, stair-stepped lakes, formed in individual rock basins aligned down the course of a glaciated valley; named because they look like a string of rosary (religious) beads.

Patterned ground (17) Areas in the periglacial environment where freezing and thawing of the ground create polygonal forms of arranged rocks at the surface.

Pedogenic regime (18) A specific soil-forming process keyed to a specific climatic regime: laterization, calcification, salinization, and podzolization, among others; not the basis for soil classification in the Soil Taxonomy.

Pedon (18) A soil profile extending from the surface to the lowest extent of plant roots or to the depth where regolith or bedrock is encountered; imagined as a hexagonal column; the basic soil sampling unit.

Percolation (9) The process by which water permeates the soil or porous rock into the subsurface environment.

Periglacial (17) Cold-climate processes, landforms, and topographic features along the margins of glaciers, past and present; periglacial characteristics exist on more than 20% of Earth's land surface; includes permafrost, frost action, and ground ice.

Perihelion (2) That point of Earth's closest approach to the Sun in its elliptical orbit; occurs on January 3 at 147,255,000 km (91,500,000 mi); variable over a 100,000-year cycle (compare aphelion).

Permafrost (17) Forms when soil or rock temperatures remain below 0°C (32°F) for at least 2 years in areas considered periglacial; criterion is based on temperature and not on whether water is present (compare frozen ground; see periglacial).

Permeability (9) The ability of water to flow through soil or rock; a function of the texture and structure of the medium.

Peroxyacetyl nitrate (PAN) (3) A pollutant formed from photochemical reactions involving nitric oxide (NO) and volatile organic compounds (VOCs). PAN produces no known human health effect, but it is particularly damaging to plants.

Phase change (7) The change in phase, or state, among ice, water, and water vapor; involves the absorption or release of latent heat (see latent heat).

Photochemical smog (3) Air pollution produced by the interaction of ultraviolet light, nitrogen dioxide, and hydrocarbons; produces ozone and

PAN through a series of complex photochemical reactions. Automobiles are the major source of the contributive gases.

Photogrammetry (1) The science of obtaining accurate measurements from aerial photos and remote sensing; used to create and to improve surface maps.

Photosynthesis (19) The process by which plants produce their own food from carbon dioxide and water, powered by solar energy. The joining of carbon dioxide and hydrogen in plants, under the influence of certain wavelengths of visible light; releases oxygen and produces energy-rich organic material, sugars and starches (compare respiration).

Physical geography (1) The science concerned with the spatial aspects and interactions of the physical elements and processes that make up the environment: energy, air, water, weather, climate, landforms, soils, animals, plants, microorganisms, and Earth.

Physical weathering (13) The breaking and disintegrating of rock without any chemical alteration; sometimes referred to as mechanical or fragmentation weathering.

Pingo (17) Large areas of frozen ground (soil-covered ice) that develop a heaved up, circular, ice-cored mound, rising above a periglacial landscape as water freezes into ice and expands; sometimes results from pressure developed by freezing artesian water that is injected into permafrost; occasionally exceeds 60 m height (200 ft).

Pioneer community (19) The initial plant community in an area; usually found on new surfaces or those that have been stripped of life, as in beginning primary succession; including lichens, mosses, and ferns growing on bare rock.

Place (1) A major theme in geography, focused on the tangible and intangible characteristics that make each location unique; no two places on Earth are alike.

Plane of the ecliptic (2) A plane (flat surface) intersecting all the points of Earth's orbit.

Planetesimal hypothesis (2) Proposes a process by which early protoplanets formed from the condensing masses of a nebular cloud of dust, gas, and icy comets; a formation process now being observed in other parts of the galaxy.

Planimetric map (Appendix A) A basic map showing the horizontal position of boundaries, land-use activities, and political, economic, and social outlines.

Plateau basalt (12) An accumulation of horizontal flows formed when lava spreads out from elongated fissures onto the surface in extensive sheets; associated with effusive eruptions; also known as flood basalts (see basalt).

Plate tectonics (11) The conceptual model and theory that encompasses continental drift, sea-floor spreading, and related aspects of crustal movement; accepted as the foundation of crustal tectonic processes (see continental drift).

Playa (15) An area of salt crust left behind by evaporation on a desert floor usually in the middle of a desert or semiarid bolson or valley; intermittently wet and dry.

Pluton (11) A mass of intrusive igneous rock that has cooled slowly in the crust; forms in any size or shape. The largest partially exposed pluton is a batholith (see batholith).

Podzolization (18) A pedogenic process in cool, moist climates; forms a highly leached soil with strong surface acidity because of humus from acid-rich trees.

Point bar (14) In a stream the inner portion of a meander, where sediment fill is redeposited.

Polar easterlies (6) Variable weak, cold, and dry winds moving away from the polar region; an anticyclonic circulation.

Polar front (6) A significant zone of contrast between cold and warm air masses; roughly situated between 50° and 60° N and S latitude.

Polar high-pressure cells (6) A weak, anticyclonic, thermally produced pressure system positioned roughly over each pole; the region of the lowest temperatures on Earth (see Antarctic high).

Polypedon (18) The identifiable soil in an area, with distinctive characteristics differentiating it from surrounding polypedons that form the basic mapping unit; composed of many pedons (see pedon).

Porosity (9) The total volume of available pore space in soil; a result of the texture and structure of the soil.

Positive feedback (1) Feedback that amplifies or encourages responses in a system (compare to negative feedback, feedback loop).

Potential evapotranspiration (9) POTET, or PE; the amount of moisture that would evaporate and transpire if adequate moisture were available; it is the amount lost under optimum moisture conditions, the moisture demand.

Potentiometric surface (9) A pressure level in a confined aquifer, defined by the level to which water rises in wells; caused by the fact that the water in a confined aquifer is under the pressure of its own weight; also known as a piezometric surface. This surface can extend above the surface of the land causing water to rise above the water table in wells in confined aquifers (see artesian water).

Precipitation (9) Rain, snow, sleet, and hail—the moisture supply; called PRECIP, or P, in the water balance.

Pressure gradient force (6) Causes air to move from an area of higher barometric pressure to an area of lower barometric pressure due to the pressure difference.

Primary succession (19) Succession that occurs among plant species in an area of new surfaces created by mass movement of land, areas exposed by a retreating glacier, cooled lava flows and volcanic eruption landscapes, or surface mining and clear-cut logging scars, or an area of sand dunes, with no trace of a former community.

Prime meridian (1) An arbitrary meridian designated as 0° longitude; the point from which longitudes are measured east or west; Greenwich, England was selected by international agreement in an 1884 treaty.

Process (1) A set of actions and changes that occur in some special order; analysis of processes is central to modern geographic synthesis.

Producer (19) Organism (plant) in an ecosystem uses carbon dioxide as its sole source of carbon, which it chemically fixes through photosynthesis to provide its own nourishment; also called an autotroph (compare consumer).

Pyroclastic (12) An explosively ejected rock fragment launched by a volcanic eruption; sometimes described by the more general term tephra.

Radiation fog (7) Formed by radiative cooling of a land surface, especially on clear nights in areas of moist ground; occurs when the air layer directly above the surface is chilled to the dew-point temperature, thereby producing saturated conditions.

Rain gauge (9) A weather instrument; a standardized device that captures and measures rainfall.

Rain shadow (8) The area on the leeward slope of a mountain range, where precipitation receipt is greatly reduced compared to the windward slope on the other side (see orographic lifting).

Reflection (4) The portion of arriving insolation that is returned directly to space without being absorbed and converted into heat and without performing any work (see albedo).

Refraction (4) The bending effect on electromagnetic waves that occurs when insolation enters the atmosphere or another medium; the same process by which a crystal, or prism, disperses the component colors of the light passing through it.

Region (1) A geographic theme that focuses on areas that display unity and internal homogeneity of traits; includes the study of how a region forms, evolves, and interrelates with other regions.

Regolith (13) Partially weathered rock overlying bedrock, whether residual or transported.

Relative humidity (7) The ratio of water vapor actually in the air (content) compared to the maximum water vapor the air could hold (capacity) at that temperature; expressed as a percentage (compare vapor pressure, specific humidity).

Relief (12) Elevation differences in a local landscape; an expression of local height difference of landforms.

Remote sensing (1) Information acquired from a distance, without physical contact with the subject; for example, photography, orbital imagery, or radar.

Respiration (19) The process by which plants use food to derive energy for their operations; essentially, the reverse of the photosynthetic process; releases carbon dioxide, water, and heat energy into the environment (compare photosynthesis).

Reverse fault (12) Compressional forces produce strain that breaks a rock so that one side moves upward relative to the other side; also called a thrust fault (compare normal fault).

Revolution (2) The annual orbital movement of Earth about the Sun; determines the length of the year and the seasons.

Rhumb line (1) A line of constant compass direction, or constant bearing, which crosses all meridians at the same angle; a portion of a great circle.

Richter scale (12) An open-ended, logarithmic scale that estimates earthquake magnitude; designed by Charles Richter in 1935; now replaced by the moment magnitude scale (see moment magnitude scale).

Ring of fire (12) See circum-Pacific belt.

Robinson projection (Appendix A) A compromise (neither equal area nor true shape) oval projection developed in 1963 by Arthur Robinson.

Roche moutonnée (17) A glacial erosion feature; an asymmetrical hill of exposed bedrock; displays a gently sloping upstream side that has been smoothed and polished by a glacier and an abrupt, steep downstream side.

Rock (11) An assemblage of minerals bound together, or sometimes a mass of a single mineral.

Rock cycle (11) A model representing the interrelationships among the three rock-forming processes: igneous, sedimentary, and metamorphic; shows how each can be transformed into another rock type.

Rockfall (13) Free-falling movement of debris from a cliff or steep slope, generally falling straight or bounding downslope.

Rossby wave (6) An undulating horizontal motion in the upper-air westerly circulation at middle and high latitudes.

Rotation (2) The turning of Earth on its axis; averages about 24 hours in duration; determines day–night relation; counterclockwise when viewed from above the North Pole; from above the equator, west to east or eastward.

Salinity (16) The concentration of natural elements and compounds dissolved in solution, as solutes; measured by weight in parts per thousand (‰) in seawater.

Salinization (18) A pedogenic process that results from high potential evapotranspiration rates in deserts and semiarid regions. Soil water is drawn to surface horizons, and dissolved salts are deposited as the water evaporates.

Saltation (14, 15) The transport of sand grains (usually larger than 0.2 mm, or 0.008 in.) by stream or wind, bouncing the grains along the ground in asymmetrical paths.

Salt marsh (16) A wetland ecosystem characteristic of latitudes poleward of the 30th parallel (compare to mangrove swamp).

Sand sea (15) An extensive area of sand and dunes; characteristic of Earth's erg deserts (contrast to reg deserts).

Saturated (7) Air that is holding all the water vapor that it can hold at a given temperature, known as the dew-point temperature.

Scale (1) The ratio of the distance on a map to that in the real world; expressed as a representative fraction, graphic scale, or written scale.

Scarification (13) Human-induced mass movements of Earth materials, such as large-scale open-pit mining and strip mining.

Scattering (4) Deflection and redirection of insolation by atmospheric gases, dust, ice, and water vapor; the shorter the wavelength, the greater the scattering, thus skies in the lower atmosphere are blue.

Scientific method (1) An approach that uses applied common sense in an organized and objective manner; based on observation, generalization, formulation, and testing of a hypothesis, and ultimately the development of a theory.

Sea-floor spreading (11) As proposed by Hess and Dietz, the mechanism driving the movement of the continents; associated with upwelling flows of magma along the worldwide system of mid-ocean ridges (see mid-ocean ridge).

Secondary succession (19) Succession that occurs among plant species in an area where vestiges of a previously functioning community are present; an area where the natural community has been destroyed or disturbed, but where the underlying soil remains intact.

Sediment (13) Fine-grained mineral matter that is transported and deposited by air, water, or ice.

Sedimentary rock (11) One of three basic rock types; formed from the compaction, cementation, and hardening of sediments derived from other rocks (compare igneous rock, metamorphic rock).

Seismic wave (11) The shock wave sent through the planet by an earthquake or underground nuclear test. Transmission varies according to temperature and the density of various layers within the planet; provides indirect diagnostic evidence of Earth's internal structure.

Seismograph (12) A device that measures seismic waves of energy transmitted throughout Earth's interior or along the coast.

Sensible heat (3) Heat that can be measured with a thermometer; a measure of the concentration of kinetic energy from molecular motion.

Sheet flow (14) Surface water that moves downslope in a thin film as overland flow, not concentrated in channels larger than rills.

Sheeting (13) A form of weathering associated with fracturing or fragmentation of rock by pressure release; often related to exfoliation processes (see exfoliation dome).

Shield volcano (12) A symmetrical mountain landform built from effusive eruptions (low viscosity magma); gently sloped, gradually rising from the surrounding landscape to a summit crater; typical of the Hawaiian Islands (compare to effusive eruption; composite volcano).

Sinkhole (13) Nearly circular depression created by the weathering of karst landscapes with subterranean drainage; also known as a doline in traditional studies; may collapse through the roof of an underground space (see karst topography).

Sleet (8) Freezing rain, ice glaze, or ice pellets.

Sling psychrometer (7) A weather instrument that measures relative humidity using two thermometers—a dry bulb and a wet bulb—mounted side-by-side.

Slipface (15) On a sand dune, formed as dune height increases above 30 cm (12 in.) on the leeward side at an angle at which loose material is stable—its angle of repose (30 to 34°).

Slope (13) A curved, inclined surface that bounds a landform.

Small circle (1) A circle on a globe's surface that does not share Earth's center; for example, all parallels of latitude other than the equator (compare to great circle).

Snowline (17) A temporary line marking the elevation where winter snowfall persists throughout the summer; seasonally, the lowest elevation covered by snow during the summer.

Soil (18) A dynamic natural body made up of fine materials covering Earth's surface in which plants grow, composed of both mineral and organic matter.

Soil colloid (18) A tiny clay and organic particle in soil; provides chemically active sites for mineral ion adsorption (see cation-exchange capacity).

Soil creep (13) A persistent mass movement of surface soil where individual soil particles are lifted and disturbed by the expansion of soil moisture as it freezes, or by grazing livestock or digging animals.

Soil fertility (18) The ability of soil to support plant productivity when it contains organic substances and clay minerals that absorb water and certain elemental ions needed by plants through adsorption (see cation-exchange capacity, CEC).

Soil horizon (18) The various layers exposed in a pedon; roughly parallel to the surface and identified as O, A, E, B, C, and R (bedrock).

Soil-moisture recharge (9) Water entering available soil storage spaces.

Soil-moisture storage (9) Δ STRGE, the retention of moisture within soil; it is a savings account that can accept deposits (soil moisture recharge) or experiences withdrawals (soil moisture utilization) as conditions change.

Soil-moisture utilization (9) The extraction of soil moisture by plants for their needs; efficiency of withdrawal decreases as the soil storage is reduced.

Soil science (18) Interdisciplinary science of soils. Pedology concerns the origin, classification, distribution, and description of soil. Edaphology focuses on soil as a medium for sustaining higher plants.

Soil Taxonomy (18) A soil classification system based on observable soil properties actually seen in the field; published in 1975 by the U.S. Soil Conservation Service, revised in 1990 and 1998 by the Natural Resources Conservation Service to include 12 soil orders (compare Canadian System of Soil Classification, Appendix B).

Soil-water budget (9) An accounting system for soil moisture using inputs of precipitation and outputs of evapotranspiration and gravitational water.

Solar constant (2) The amount of insolation intercepted by Earth on a surface perpendicular to the Sun's rays when Earth is at its average distance from the Sun; a value of 1370 W/m^2, or 1.968 calories /cm^2 per minute; averaged over the entire globe at the thermopause.

Solar wind (2) Clouds of ionized (charged) gases emitted by the Sun and traveling in all directions from the Sun's surface. Effects on Earth include auroras, disturbance of radio signals, and possible influences on weather.

Solifluction (17) Gentle downslope movement of a saturated surface material (soil and regolith), in various climatic regimes, where temperatures are above freezing (compare gelifluction).

Solum (18) A true soil profile in the pedon; ideally, a combination of O, A, E, and B horizons (see pedon).

Spatial (1) The nature or character of physical space, as in an area; occupying or operating within a space. Geography is a spatial science; spatial analysis its essential approach.

Spatial analysis (1) The examination of spatial interactions, patterns, and variations over area and/or space; a key integrative approach of geography.

Specific heat (5) The increase of temperature in a material when energy is absorbed; water has a higher specific heat (can store more heat) than a comparable volume of soil or rock.

Specific humidity (7) The mass of water vapor (in grams) per unit mass of air (in kilograms) at any specified temperature. The maximum mass of water vapor that a kilogram of air can hold at any specified temperature is termed its maximum specific humidity (compare vapor pressure, relative humidity).

Speed of light (2) Specifically, 299,792 kilometers per second (186,282 miles per second), or more than 9.4 trillion kilometers per year (5.9 trillion miles per year)—a distance known as a light year; at light speed Earth is 8 minutes and 20 seconds from the Sun.

Spheroidal weathering (13) A chemical weathering process in which the sharp edges and corners of boulders and rocks are weathered in thin plates that create a rounded, spheroidal form.

Spodosols (18) A soil order in the Soil Taxonomy classification that occurs in northern coniferous forests; best developed in cold, moist, forested climates of *Dfb*, *Dfc*, *Dwc*, and *Dwd*. Lacks humus and clay in the A horizons, with high acidity associated with podzolization processes.

Spring tide (16) The highest tidal range, which occurs when the Moon and the Sun are in conjunction (at new Moon) or in opposition (at full Moon) stages (compare neap tide).

Squall line (8) A zone slightly ahead of a fast-advancing cold front, where wind patterns are rapidly changing and blustery and precipitation is strong.

Stability (7) The condition of a parcel, whether it remains where it is or changes its initial position. The parcel is stable if it resists displacement upward, unstable if it continues to rise.

Stationary front (8) A frontal area of contact between contrasting air masses that shows little horizontal movement; winds in opposite direction on either side of the front flow parallel along the front.

Steady-state equilibrium (1) The condition that occurs in a system when rates of input and output are equal and the amounts of energy and stored matter are nearly constant around a stable average.

Stomata (19) A small opening on the undersides of leaves, through which water and gases pass.

Storm surge (8) A large quantity of seawater pushed inland by the strong winds associated with a tropical cyclone.

Storm tracks (8) Seasonally shifting paths followed by migrating low-pressure systems.

Stratified drift (17) Sediments deposited by glacial meltwater that appear sorted; a specific form of glacial drift (compare till).

Stratigraphy (11) A science that analyzes the sequence, spacing, geophysical and geochemical properties, and spatial distribution of rock strata.

Stratocumulus (7) A lumpy, grayish, low-level cloud, patchy with sky visible, sometimes present at the end of the day.

Stratosphere (3) That portion of the homosphere that ranges from 20 to 50 km (12.5 to 30 mi) above Earth's surface, with temperatures ranging from −57°C (−70°F) at the tropopause to 0°C (32°F) at the stratopause. The functional ozonosphere is within the stratosphere.

Stratus (7) A stratiform (flat, horizontal) cloud generally below 2000 m (6500 ft).

Strike-slip fault (12) Horizontal movement along a faultline, that is, movement in the same direction as the fault; also known as a transcurrent fault. Such movement is described as right lateral or left lateral, depending on the relative motion observed across the fault (see transform fault).

Subduction zone (11) An area where two plates of crust collide and the denser oceanic crust dives beneath the less dense continental plate, forming deep oceanic trenches and seismically active regions.

Sublimation (7) A process in which ice evaporates directly to water vapor or water vapor freezes to ice (deposition).

Subpolar low-pressure cell (6) A region of low pressure centered approximately at 60° latitude in the North Atlantic near Iceland and in the North Pacific near the Aleutians, as well as in the Southern Hemisphere.

Air flow is cyclonic; it weakens in summer and strengthens in winter (refer to cyclone).

Subsolar point (2) The only point receiving perpendicular insolation at a given moment—the Sun directly overhead (see declination).

Subtropical high-pressure cell (6) One of several dynamic high-pressure areas covering roughly the region from 20° to 35° N and S latitudes; responsible for the hot, dry areas of Earth's arid and semiarid deserts (refer to anticyclone).

Sulfate aerosols (3) Sulfur compounds in the atmosphere, principally sulfuric acid; principal sources relate to fossil fuel combustion; scatters and reflects insolation.

Sulfur dioxide (3) A colorless gas detected by its pungent odor; produced by the combustion of fossil fuels, especially coal, that contain sulfur as an impurity; can react in the atmosphere to form sulfuric acid, a component of acid deposition.

Summer (June) solstice (2) The time when the Sun's declination is at the Tropic of Cancer, at 23.5° N latitude; June 20–21 each year (compare to winter solstice).

Sunrise (2) That moment when the disk of the Sun first appears above the horizon.

Sunset (2) That moment when the disk of the Sun totally disappears.

Sunspot (2) Magnetic disturbances on the surface of the Sun; occurring in an average 11-year cycle; related flares, prominences, and outbreaks produce surges in solar wind.

Surface creep (15) A form of eolian transport that involves particles too large for saltation; a process whereby individual grains are impacted by moving grains and slide and roll.

Surplus (9) (SURPL) The amount of moisture that exceeds potential evapotranspiration; moisture oversupply when soil moisture storage is at field capacity; extra or surplus water.

Suspended load (14) Fine particles held in suspension in a stream. The finest particles are not deposited until the stream velocity nears zero.

Swell (16) Regular patterns of smooth, rounded waves in open water; can range from small ripples to very large waves.

Syncline (12) A trough in folded strata, with beds that slope toward the axis of the downfold (compare anticline).

System (1) Any ordered, interrelated set of materials or items existing, separate from the environment, or within a boundary; energy transformations and energy and matter storage and retrieval occur within a system.

Taiga (20) See needleleaf forest.

Talik (17) An unfrozen portion of the ground that may occur above, below, or within a body of discontinuous permafrost or beneath a body of water in the continuous region, such as a deep lake; may extend to bedrock and noncryotic soil under large deep lakes.

Talus slope (13) Formed by angular rock fragments that cascade down a slope along the base of a mountain; poorly sorted, cone-shaped deposits.

Tarn (17) A small mountain lake, especially one that collects in a cirque basin behind risers of rock material, or in an ice-gouged depression.

Temperate rain forest (20) A major biome of lush forests at middle and high latitudes; occurs along narrow margins of the Pacific Northwest in North America, among other locations; includes the tallest trees in the world.

Temperature (5) A measure of sensible heat energy present in the atmosphere and other media; indicates the average kinetic energy of individual molecules within a substance.

Temperature inversion (3) A reversal of the normal decrease of temperature with increasing altitude; can occur anywhere from ground level up to several thousand meters; functions to block atmospheric convection and thereby trap pollutants.

Terminal moraine (17) Eroded debris that is dropped at a glacier's farthest extent.

Terrane (12) A migrating piece of Earth's crust, dragged about by processes of mantle convection and plate tectonics. Displaced terranes are distinct in their history, composition, and structure from the continents that accept them.

Terrestrial ecosystem (20) A self-regulating association characterized by specific plant formations; usually named for the predominant vegetation and known as a biome when large and stable (see biome).

Thermal equator (5) The isoline on an isothermal map that connects all points of highest mean temperature.

Thermokarst (17) Topography of hummocky, irregular relief marked by cave-ins, bogs, small depressions, and pits formed as ground ice melts; an erosion processes caused by ground ice melting. Not related to solution processes and chemical weathering associated with limestone (karst).

Thermopause (2, 3) A zone approximately 480 km (300 mi) in altitude that serves conceptually as the top of the atmosphere; an altitude used for the determination of the solar constant.

Thermosphere (3) A region of the heterosphere extending from 80 to 480 km (50 to 300 mi) in altitude; contains the functional ionosphere layer.

Thrust fault (12) A reverse fault where the fault plane forms a low angle relative to the horizontal; an overlying block moves over an underlying block.

Thunder (8) The violent expansion of suddenly heated air, created by lightning discharges, sending out shock waves as an audible sonic bang.

Tide (16) A pattern of daily oscillations in sea level produced by astronomical relations among the Sun, the Moon, and Earth; experienced in varying degrees around the world (see neap and spring tides).

Till (17) Direct ice deposits that appear unstratified and unsorted; a specific form of glacial drift (compare stratified drift).

Till plain (17) A large, relatively flat plain composed of unsorted glacial deposits behind a terminal or end moraine. Low-rolling relief and unclear drainage patterns are characteristic.

Tombolo (16) A landform created when coastal sand deposits connect the shoreline with an offshore island outcrop or sea stack.

Topographic map (Appendix A) A map that portrays physical relief through the use of elevation contour lines that connect all points at the same elevation above or below a vertical datum, such as mean sea level.

Topography (12) The undulations and configurations, including its relief, that give Earth's surface its texture, portrayed on topographic maps.

Tornado (8) An intense, destructive cyclonic rotation, developed in response to extremely low pressure; generally associated with mesocyclone formation.

Total runoff (9) Surplus water that flows across a surface toward stream channels; formed by sheet flow, combined with precipitation and subsurface flows into those channels.

Traction (14) A type of sediment transport that drags coarser materials along the bed of a stream (see bed load).

Trade wind (6) Wind from the northeast and southeast that converges in the equatorial low-pressure trough, forming the intertropical convergence zone.

Transform fault (11) A type of geologic fault in rocks. An elongated zone along which faulting occurs between mid-ocean ridges; produces a relative horizontal motion with no new crust formed or consumed; strike-slip motion is either left or right lateral (see strike-slip fault).

Transmission (4) The passage of shortwave and longwave energy through space, the atmosphere, or water.

Transparency (5) The quality of a medium (air, water) that allows light to easily pass through it.

Transpiration (9) The movement of water vapor out through the pores in leaves; the water is drawn by the plant roots from soil moisture storage.

Transport (14) The actual movement of weathered and eroded materials by air, water, and ice.

Tropical cyclone (8) A cyclonic circulation originating in the tropics, with winds between 30 and 64 knots (39 to 73 mph); characterized by closed isobars, circular organization, and heavy rains (see hurricane and typhoon).

Tropical savanna (20) A major biome containing large expanses of grassland interrupted by trees and shrubs; a transitional area between the humid rain forests and tropical seasonal forests and the drier, semiarid tropical steppes and deserts.

Tropical seasonal forest and scrub (20) A variable biome on the margins of the rain forests, occupying regions of lesser and more erratic rainfall; the site of transitional communities between the rain forests and tropical grasslands.

Tropic of Cancer (2) The northernmost point of the Sun's declination during the year; 23.5° N latitude.

Tropic of Capricorn (2) The southernmost point of the Sun's declination during the year; 23.5° S latitude.

Tropopause (3) The top zone of the troposphere defined by temperature; wherever −57°C (−70°F) occurs.

Troposphere (3)   The home of the biosphere; the lowest layer of the homosphere, containing approximately 90% of the total mass of the atmosphere; extends up to the tropopause; occurring at an altitude of 18 km (11 mi) at the equator, 13 km (8 mi) in the middle latitudes, and at lower altitudes near the poles.

True shape (1)   A map property showing the correct configuration of coastlines; a useful trait of conformality for navigational and aeronautical maps, although areal relationships are distorted (see map projection; compare equal area).

Tsunami (16)   A seismic sea wave, traveling at high speeds across the ocean, formed by sudden motion in the sea floor, such as a sea-floor earthquake, submarine landslide, or eruption of an undersea volcano.

Typhoon (8)   A tropical cyclone in excess of 65 knots (74 mph) that occurs in the western Pacific; same as a hurricane except for location.

Ultisols (18)   A soil order in the Soil Taxonomy. Features highly weathered forest soils, principally in the humid sub-tropical *Cfa* climatic classification. Increased weathering and exposure can degenerate an Alfisol into the reddish color and texture of these more humid to tropical Ultisols. Fertility is quickly exhausted when Ultisols are cultivated.

Unconfined aquifer (9)   An aquifer that is not bounded by impermeable strata. It is simply the zone of saturation in water-bearing rock strata, with no impermeable overburden and recharge generally accomplished by water percolating down from above.

Undercut bank (14)   In streams a steep bank formed along the outer portion of a meandering stream; produced by lateral erosive action of a stream; sometimes called a cutbank (compare to point bar).

Uniformitarianism (11)   An assumption that physical processes active in the environment today are operating at the same pace and intensity that has characterized them throughout geologic time; proposed by Hutton and Lyell (compare to catastrophism).

Upslope fog (7)   Forms when moist air is forced to higher elevations along a hill or mountain and is thus cooled (compare valley fog).

Upwelling current (6)   An area of the sea where cool, deep waters, which are generally nutrient rich, rise to replace the vacating water; as occurs along the west coasts of North and South America (compare to downwelling current).

Urban heat island (4)   An urban microclimate that is warmer on the average than areas in the surrounding countryside because of the interaction of solar radiation and various surface characteristics.

Valley fog (7)   The settling of cooler, more dense air in low-lying areas; produces saturated conditions and fog.

Vapor pressure (7)   That portion of total air pressure that results from water vapor molecules; expressed in millibars (mb). At a given dew-point temperature the maximum capacity of the air is termed its saturation vapor pressure.

Vascular plant (19)   A plant having internal fluid and material flows through its tissues; almost 250,000 species exist on Earth.

Ventifact (15)   A piece of rock etched and smoothed by eolian erosion—abrasion by windblown particles.

Vernal (March) equinox (2)   The time around March 20–21 each year when the Sun's declination crosses the equatorial parallel and all places on Earth experience days and nights of equal length. The Sun rises at the North Pole and sets at the South Pole (compare to autumnal equinox).

Vertisols (18)   A soil order in the Soil Taxonomy. Features expandable clay soils; composed of more than 30% swelling clays. Occurs in regions that experience highly variable soil moisture balances through the seasons.

Volatile organic compounds (3)   Compounds, including hydrocarbons, produced by the combustion of gasoline, from surface coatings, and from electric utility combustion; participates in the production of PAN through reactions with nitric oxides.

Volcano (12)   A mountainous landform at the end of a magma conduit, which rises from below the crust and vents to the surface. Magma rises and collects in a magma chamber deep below, erupting effusively or explosively and forming composite, shield, or cinder cone volcanoes.

Warm desert and semidesert (20)   A desert biome caused by the presence of subtropical high-pressure cells; dry air and low precipitation.

Warm front (8)   The leading edge of an advancing warm air mass, which is unable to push cooler, passive air out of the way; tends to push the cooler, underlying air into a wedge shape; identified on a weather map as a line with semicircles pointing in the direction of frontal movement (compare to cold front).

A.26

Wash (15)   An intermittently dry streambed that fills with torrents of water after rare precipitation events in arid lands.

Watershed (14)   The catchment area of a drainage basin; delimited by divides (see drainage basin).

Waterspout (8)   An elongated, funnel-shaped circulation formed when a tornado exists over water.

Water table (9)   The upper surface of groundwater; that contact point between the zone of saturation and aeration in an unconfined aquifer (see zone of aeration, zone of saturation).

Wave (16)   An undulation of ocean water produced by the conversion of solar energy to wind energy and then to wave energy; energy produced in a generating region or a stormy area of the sea.

Wave-cut platform (16)   A flat or gently sloping table-like bedrock surface that develops in the tidal zone, where wave action cuts a bench that extends from the cliff base out into the sea.

Wave cyclone (8)   See midlatitude cyclone.

Wavelength (2)   A measurement of a wave; the distance between the crests of successive waves. The number of waves passing a fixed point in one second is called the frequency of the wavelength.

Wave refraction (16)   A bending process that concentrates wave energy on headlands and disperses it in coves and bays; the long-term result is coastal straightening.

Weather (8)   The short-term condition of the atmosphere, as compared to climate, which reflects long-term atmospheric conditions and extremes. Temperature, air pressure, relative humidity, wind speed and direction, daylength, and Sun angle are important measurable elements that contribute to the weather.

Weathering (13)   The processes by which surface and subsurface rocks disintegrate, or dissolve, or are broken down. Rocks at or near Earth's surface are exposed to physical and chemical weathering processes.

West Antarctic ice sheet (10)   A vast grounded ice mass held back by the Ross, Ronne, and Filcher ice shelves in Antarctica, drained by several active ice streams.

Westerlies (6)   The predominant surface and aloft wind flow pattern from the subtropics to high latitudes in both hemispheres.

Western intensification (6)   The piling up of ocean water along the western margin of each ocean basin, to a height of about 15 cm (6 in.); produced by the trade winds that drive the oceans westward in a concentrated channel.

Wetland (16)   A narrow, vegetated strip occupying many coastal areas and estuaries worldwide; highly productive ecosystems with an ability to trap organic matter, nutrients, and sediment.

Wilting point (9)   That point in the soil moisture balance when only hygroscopic water and some bound capillary water remains. Plants wilt and eventually die after prolonged stress from a lack of available water.

Wind (6)   The horizontal movement of air relative to Earth's surface; produced essentially by air pressure differences from place to place; its direction is influenced by the Coriolis force and surface friction.

Wind vane (6)   A weather instrument used to determine wind direction; winds are named for the direction from which they originate.

Winter (December) solstice (2)   That time when the Sun's declination is at the Tropic of Capricorn, at 23.5° S latitude, December 21–22 each year. The day is 24 hours long south of the Antarctic Circle. The night is 24 hours long north of the Arctic Circle (compare to Summer [June] solstice).

Withdrawal (9)   The removal of water from the natural supply, after which it is used for various purposes and then is returned to the water supply.

Wrangellia terrane (12)   One of many terranes that became cemented together to form present-day North America and the Wrangell Mountains; arriving from approximately 10,000 km (6200 mi) away; a former volcanic island arc and associated marine sediments.

Yardang (15)   A streamlined rock structure formed by deflation and abrasion; appears elongated and aligned with the most effective wind direction.

Yazoo tributary (14)   A small tributary stream draining alongside a floodplain; blocked from joining the main river by its natural levees and elevated stream channel (see backswamp).

Zone of aeration (9)   A zone above the water table, which has air in its pore spaces and may or may not have water.

Zone of saturation (9)   A groundwater zone below the water table in which all pore spaces are filled with water.

Index

orographic lifting and, 215, 216–20, 227, 303
systems of, 374–75
temperatures in, 122
topographic definition of, 361
Mountain–valley breezes, 164
Mount Everest, 21, 360, title page
Mount Pinatubo, 11–12, 13, 71, 76, 391, 393–94
atmospheric albedo and, 98, 394
climatic effects, 548
Mount St. Helens, 391–93, 612
Movement, theme in geography, 3, 4
MSL (mean sea level), 493–95
Mudflows, 411, 415, 421
Mudstone, 336, 416
Munsell Color Chart, 560
Mutualism, 590

Nagasaki, Japan, 290
Nappes, 374
National Center for Atmospheric Research, 311
National Center for Geographic Information and Analysis (NCGIA), 34
National Climate Data Center (NCOC), 311
National Council for Geographic Eduation, 3
National Environmental Satellite, Data, and Information Service (NESDIS), 311
National Geographic Society, 3, 658
National Hurricane Center, 234, 236
National Oceanic and Atmospheric Administration (NOAA), 31
National Renewable Energy Laboratory (NREL), 111
National Severe Storms Forecast Center, 227
National Weather Service, 75, 135, 148, 212
Nautical charts, 28
Neap tide, 496
Nebula, 44
Needleleaf (boreal) forest, 297, 534, 641
Negative feedback, 9–10
Net primary productivity, 592–93
Net radiation (NET R), 52, 104–8
Newton, Isaac, 6, 15, 154
NEXRAD (Next Generation Weather Radar) Program, 212
Niagara Falls, 445
Niche, 590, 612, 626, 627
Nickpoints, 444
Nile River, 438, 450, 451
Nimbostratus clouds, 198, 200, 221
NIST-F1, 24
Nitric acid, 83, 84, 86
Nitrogen, 70
Nitrogen cycle, 599–600
Nitrogen oxides, 79, 80, 81, 82, 83
acid deposition and, 83, 84
global warming and, 309, 312
ozone depletion and, 76
trends in, 87
Noctilucent clouds, 71
Nonnative species, 632
Normal lapse rate, 71, 193
North America, air masses and, 213–14
Northeast trade winds, 157
North Pole, 302

Northridge earthquake, 377, 380
North Star (Polaris), 17
Nuée ardente, 392
Numerical weather prediction, 224
Nunatak, 323

Oblate spheroid, 15
Obsidian, 334
Occluded front, 223
Ocean basins, 359
Ocean currents, 170–72
climate change and, 548
coastal environment and, 493
deep, 171–72, 248
forces acting on, 151
longshore, 500, 506
temperature variations and, 126–27
Oceanic crust, 328–29, 335, 340, 342, 347
Oceans, 490. See also Sea level; Seawater
ecosystems of, 490, 603–4, 628
global warming of, 316
in hydrologic cycle, 246
physical structure of, 492
primary productivity in, 593
temperature variations and, 124–28
Ocean trenches, 342, 343, 351
Odum, Howard T., 606
Oil drilling, in Arctic, 647
Oil spills, 655–56
Omnivores, 603, 605
Open systems, 8
Orbit, of Earth, 44, 546–47
Orders, soil, 567, 570
Orders of relief, 359
Organic (O) horizon, 558
Organic soils, 581–82
Orogenesis, 371–75
Orogens, 372
Orographic lifting, 215, 216–20
dry climates and, 303
thunderstorms and, 227
Ortelius, Abraham, 340
Outgassing, 180
Outlet glacier, 523
Outputs, of system, 13
Outwash plains, 534
Oval projection, 27
Overland flow, 252, 432
Oxbow lake, 442, 443, 446
Oxidation, 409
Oxisols, 567–69, 634
Oxygen, 70
in Earth's crust, 332
isotope ratios, 541, 547
from ozone, 74
from photosynthesis, 592
in seawater, 491, 492
Oxygen cycle, 598–99
Ozone, 72
in lower atmosphere, 80, 83
Ozone layer, 70, 71–76
depletion of, 6, 73–76, 312, 604
energy budget and, 102, 103

Pacific high, 159
Pacific Ocean, 170, 182
Pacific plate, 345, 346, 369–70
Pacific rim, 347
Paleoclimatology, 309, 545–48
Paleolakes, 543–45
Paleoseismology, 383

Palsa, 538
Pampas, 644
PAN (peroxyacetyl nitrates), 80, 83
Pangaea, 340, 342–43, 344, 375, 548, 610
Pantanal, 639
Paradigms, 21st century, 657–58
Parallel, 17, 19, 22
Parallel drainage, 436
Parasitic relationship, 590
Parcels, of air, 192–96
Parent rock, 403, 564
Particulate matter (PM), 79, 80, 81, 86, 87
Passive remote sensing, 31–32
Passive solar energy system, 110
Patch dynamics, 611
Paternoster lakes, 529
Patterned ground, 538–39, 580
PCBs (polychlorinated biphenyls), 620
PE. See POTET (potential evapotranspiration)
Peat, 582
Pediment, 483
Pedocals, 574
Pedogenic regimes, 567
Pedology, 558
Pedon, 558, 567
Peds, 567
Percolation, 247, 253, 260
Periglacial landscapes, 534–41
definition of, 534
frost action in, 534, 537–39
ground ice in, 301, 534, 537–40
human development in, 540–41
permafrost in, 534–37
Perihelion, 44
Periods, geologic, 325
Permafrost, 297, 301, 509, 520, 534–37
alpine, 536, 646
definition of, 534–35
human development and, 540–41
in needleleaf biome, 641
soil flows and, 539
Permeability
of rock, 260
of soil, 253–54
Peroxyacetyl nitrates (PAN), 80, 83
Pesticides, in food chains, 607
Petroleum geology, 365
Phanerozoic Eon, 324, 490–91
Phase changes, 184, 185, 186
Photic layer, 125
Photochemical smog, 80, 82–83
Photogrammetry, 30
Photoperiod, 595
Photosynthesis, 8, 9, 592–94, 595, 598
community metabolism and, 607
by phytoplankton, 603
Photovoltaic cells, 111
Phreatophytes, 303, 481
pH scale, 84, 564
Physical geography, 5
Physical weathering, 403, 404–7
Phytoplankton, 601, 602, 603, 620
Piedmont glacier, 521
Pingo, 538
Pinatubo. See Mount Pinatubo
Pioneer community, 612
Place, theme in geography, 3, 4
Plains, 361, 644
Planar projection, 27, 28
Plane of the ecliptic, 44, 55, 56, 546